BUSINESS/SCIENCE/TECHNOLOGY DIVISION
CHICAGO PUBLIC LIBRARY
400 SOUTH STATE STREET
CHICAGO, IL 60605

Atlantic Automobilism

Explorations in Mobility
Series Editors:
Gijs Mom, Eindhoven University of Technology
Mimi Sheller, Drexel University
Georgine Clarsen, University of Wollongong

The study of mobility opens up new transnational and interdisciplinary approaches to fields including transport, tourism, migration, communication, media, technology, and environmental studies. The works in this series rethink our common assumptions and ideas about the mobility of people, things, ideas, and cultures from a broadly understood humanities perspective. The series welcomes projects of a historical or contemporary nature and encourages postcolonial, non-Western, and critical perspectives.

Volume 1
Atlantic Automobilism
Emergence and Persistence of the Car, 1895–1940
Gijs Mom

Atlantic Automobilism
Emergence and Persistence of the Car, 1895–1940

Gijs Mom

berghahn
NEW YORK • OXFORD
www.berghahnbooks.com

Published in 2015 by
Berghahn Books
www.berghahnbooks.com

© 2015 Gijs Mom

All rights reserved. Except for the quotation of short passages for the purposes of criticism and review, no part of this book may be reproduced in any form or by any means, electronic or mechanical, including photocopying, recording, or any information storage and retrieval system now known or to be invented, without written permission of the publisher.

Environment and Society

Library of Congress Cataloging-in-Publication Data

Mom, Gijs, 1949–
Atlantic automobilism : emergence and persistence of the car, 1895–1940 / Gijs Mom.
    pages cm. — (Explorations in mobility ; 1)
    ISBN 978-1-78238-377-2 (hardback : alk. paper) —
    ISBN 978-1-78238-378-9 (ebook)
    1. Automobiles—North America—History—20th century. 2. Automobile travel—North America—History—History—20th century. 3. Automobiles—Social aspects—North America. I. Title.
  TL22.M66 2014
  306.4'6—dc23

2014019632

British Library Cataloguing in Publication Data

A catalogue record for this book is available from the British Library

Printed on acid-free paper.

ISBN: 978-1-78238-377-2 hardback
ISBN: 978-1-78238-378-9 ebook

To Charley, to Clay

Contents

List of Figures and Tables   x

Preface   xiii

Introduction. Explaining the Car: Prolegomena for a History of North-Atlantic Automobilism   1
    Do Narratives Explain?   3
    Constructing a Master Narrative   7
    Developing an Explanatory Toolbox   27
    Conclusions   40

Part I. Emergence (1895–1918)

Chapter 1. Racing, Touring, Tinkering: Constructing the Adventure Machine (1895–1914/1917)   59
    First Phase: Emergence and Roots of the Petrol Car (until 1902)   61
    Second Phase: Resistance Against Elite Touring in Heavy Family Cars (1902–1908)   68
    A First Analysis of Automotive Adventure: The Masculine 'Conquest of Nature'   84
    Third Phase: The 'Small Capitalist' and the 'Average Man' (1908 until the War)   100
    Conclusions   111

Chapter 2. How it Feels to be Run Over: The Grammar of Early Automobile Adventure   133
    Driving and Writing: Analyzing Affinities of Touristic and Artistic Experiences   135
    Autopoetics: Mainstream Authors   141
    The 'Real Unreality of Things': Critical voices from the United Kingdom   148
    Colonialism by Car: Gendered Travel Writing   154
    Male Violence and Aggression: A French-Belgian Group of Writer-Motorists   161

Subliterary Novels: The Williamsons and Youth Novels   170
Flight Forward: The Avant-Garde, Silent Movies, and the
   Celebration of Automotive Violence   176
Tarkington, Cather, and Dreiser: Autopoetics Before America's
   Entry into the War   186
Conclusions   189

Chapter 3. Driving on Aggression: The First World War and the Systems Approach to the Car   227

Preparing for War: Clubs, the Military, and Aggression   228
Preparing for War: Organizing Mobility   236
Mobilization, Immobility, Remobilization: Aggression, Violence, and Atrocities   238
War Trophies: The Truck, Logistics, and Maintenance   242
War Trophies: Thanatourism and Other Adventures   250
Conclusions   267

Part II. Persistence (1918–1940)

Chapter 4. "Why Apologize for Pleasure?" Consuming the Car in Boom and Bust   285

The Car as Commodity: Its Spread Among the Atlantic Middle Class   287
European Car Consumption and 'Americanization': Eagerness Compared   306
The Car as Necessity: A Profile of Car Use in the Interbellum   313
Testing a Different Approach: Migration, Mass Tourism, and the Family Car   328
Conclusions   344

Chapter 5. Translation and Transition: Readjusting the Technology and Culture of Middle-Class Family Adventures   373

Orchestrating Car Technology: Constructing the Closed Automobile   374
The Process of Prosthetization: Mutually Adjusting Skills and Technology   386
Multiple Adventures: Thrills, Skills, and Risks   391
Conclusions   410

Chapter 6. Redefining Adventure: Domesticated Violence and the Coldness of Distance   425

An Avant-Garde in Autopoetic Travel Experience: The Conquest of the 'Periphery'   426

Domesticating Adventure: The Family as Collective Subject, and Speed   439
Flows and Violence: Urban Culture and the Middle-Class Family   460
With or Without a Car: A Women's Adventure?   478
The Ubiquitous Car: A Spectrum of Adventures, Adjusted to Middle-Class Taste   490
The Cult of Cool: Becoming Cyborg   497
Symbolisms and Affinities: Avant-Garde and Popular Culture   508
Conclusions   528

Chapter 7. Swarms Into Flows: The Contested Emergence of the Automobile System — 565

Coping With the Car's Unreliability: Maintenance, Repair, and the Functional Adventure   567
Transnationalizing the Local: Planning and Building National Road Networks   572
Contested Order: Spatial Planners Versus Engineers   590
Rescuing Automotive Adventure: The Construction of Road Safety   594
The Battle of the Systems: Road Versus Rail and the Coordination Crisis   606
Conclusions   617

Conclusion. Transcendence and the Automotive Production of Mobility — 635

Crossing Borders: Half a Century of North-Atlantic Automobilism   636
Crossing Boundaries: Adventure, Fiction, and the Explanation of the Car's Persistence   643
Some Closing Remarks on Methodology and Future Research   650

Bibliography — 660

Index — 737

Figures and Tables

Figures

Figure 0.1. The dual nature of technology and its related practices.	35
Figure 0.2. A taxonomy of car-related practices.	36
Figure 1.1. Modal split in France.	65
Figure 1.2. Car densities (per 1000 inhabitants) in selected countries.	69
Figure 1.3. Members of the *Allgemeiner Deutsche Automobil-Club* (ADAC).	81
Figure 1.4. Car and motorcycle park in France.	82
Figure 1.5. Share of 'rural' automobiles in registrations in the United States.	104
Figure 1.6. Car owners and chauffeurs in New York State.	110
Figure 4.1. Car densities (per one thousand inhabitants) in selected countries until 1940.	289
Figure 4.2. Members of ADAC until 1932.	290
Figure 4.3. Motorcycle densities in the United States and some European countries 1920–1939.	293
Figure 4.4. Motorization (car and motorcycle densities totaled) of Germany, the United Kingdom, and France during the interwar years.	293
Figure 4.5. German car and motorcycle performance (in million passenger-kilometers).	295
Figure 4.6. Average household expenditures in the United States per overlapping decades.	297
Figure 4.7. Car diffusion in the United States and in the country's four regions for five turning-point years.	298

Figures   xi

Figure 4.8. Share of new car owners in domestic sales in the United States, 1921–1941. 300

Figure 4.9a. Car density and urbanization levels. 302

Figure 4.9b. Car densities in states with low, middle, and high income levels. 302

Figure 4.10. Net earnings on gasoline taxes in the United States, 1919–1937. 319

Figure 4.11. Drivers license exams in France. 319

Figure 4.12. Visitors and cars visiting the national parks in the United States. 331

Figure 4.13. Trip information requested by ANWB members, the Netherlands, 1922–1942. 335

Figure 4.14. Persons traveling into Switzerland, 1925–1938. 343

Figure 5.1. Production of open and closed bodies in the United States and Canada (1919–1920: US only), 1919–1931. 384

Figure 7.1. Improving and building the Dutch road system during the interwar years: expenditures by the national government on maintenance and improvement versus new roads and bridges. 583

Figure 7.2. Traffic deaths during the interwar years in two European countries and two European capitals. 596

Figure 7.3. Long-term statistics (1835–1973) of French accidental deaths, including gendering. 597

Figure 7.4. Traffic deaths in France, 1865–1939. 598

Figure 7.5. Mystification of road-death statistics by constructing death rates; example: United States, 1913–1989. 600

Figure 7.6. Car-related fatality rates in the United Kingdom, 1909–1938. 601

Figure 7.7. Gross capital creation in Dutch mobility infrastructure, 1921–1940 in constant guilders (×1000) of 1913. 607

Figure 7.8. Performance (in passenger-kilometers) of the Dutch modal split. 609

Figure 7.9. Buses and trucks in the United Kingdom, 1904–1939. 612

Figure 7.10. Modal split in (semi-)public transport in Germany. 613

Figure 7.11. Modal split in France in billion passenger-kilometers. 614

Tables

Table 0.1. Periodization of Automobile Use in the North-Atlantic World 38

Table 2.1. Taxonomy of Automotive Adventures 190

Table 7.1. PIARC Conference Characteristics, 1908–1938 574

Preface

This is the work of a lifetime but the actual gestation of it took a decade, so where should I start with my thanks and acknowledgments? In the preface to my first monograph on mobility history (*The Electric Vehicle*, 2004) I talked about the friends I left behind (mainly from the field of literary history, my first field of study back in the late 1960s and early 1970s) and the new friends I met (mostly from the field of automotive engineering, my second study, and the history of technology at Eindhoven University of Technology). Since then I had the privilege to found an association (The International Association for the History of Transport, Traffic and Mobility, T²M) and to set up a journal (*Transfers: Interdisciplinary Journal of Mobility Studies*) that helped shape a community of scholars and friends who decisively influenced my thinking about mobility history and its major questions. I am indebted to this network of historians, social scientists, artists, and many others for providing a soundboard for my ideas, just like I (and the writing project I was embarking upon) have hopefully done the same for them. It is in the context of these two scholarly groups (the transnational association and the local university) that the idea of the need for a new synthesis emerged.

This book, thus, represents the next step in my personal life and career: literary analysis. My first book was in fact a (socio-)technical history that ended in a plea for a cultural approach to automobilism in general. While I left the literary field because of what I then felt to be an utterly sterile and far too intellectualistic approach, the transnational history of automobilism provides a unique base for an empirically rich analysis of the 'automotive mood.' Although I do not heed the ambition to contribute to the discourse in literary history, let alone literary theory, I *do* wish to contribute to a new form of mobility history. Not only does literary analysis decisively benefit from extra-literary empirics, it is also true that only through this type of analysis may mobility history ever enter into general history, an ambition common to every historical specialism.

During the construction of this book four places were of major importance to me: my university (Technische Universiteit Eindhoven, TU/e), the Rachel Carson Center (RCC; co-organized by the Deutsches

Museum and Ludwig-Maximilians Universität, LMU), a group of friends from T²M and *Transfers,* and my home.

At TU/e it was especially Johan Schot who recognized the relevance of my project. He not only let me spend an entire year with the RCC, he also made sure that the enormous flood of books, articles, and other sources (more than three thousand, as the bibliography attests) kept flowing in a situation of increasing budget cuts. I thank Peter Smits and his colleagues of the university library for channeling the flow of the Inter Library Loan system, and Mieke Rompen and her colleagues (Iris Custers, Sonja Beekers, Marianne Jonkers) for their support in organizing the many trips abroad to conferences, workshops, and study weeks. Student assistants Jorrit Bakker and Valerian Meijering were also instrumental in keeping the infrastructure functioning, followed, at the final stages of manuscript production, by Texas van Leeuwenstein and Wilco Pesselse. Of my direct colleagues I thank especially Mila Davids, Erik van der Vleuten, Saurabh Arora, Geert Verbong, Rob Raven, Stefan Krebs, Koen Frenken, Anthonie Meijers, and Harry Lintsen for their critical questions whenever I used a Thursday seminar to report on my progress. I also thank the Board of Directors of TU/e for providing me with a very generous grant that helped fund most of my extracurricular activities during three critical years, as well as Bert Toussaint, corporate historian at the – then – Dutch Ministry of Traffic and Water Management, for providing funding of a similar magnitude enabling me to co-organize three important international workshops, the results of which flowed into the following pages.

At RCC I wish to thank first of all my fellow fellows of the first hour: Martin Knoll, Reinhold Reith, and Diana Mincyte, and the fellows and research fellows who joined us later: Heike Egner, Andrew Isenberg, Patrick Kupper, Uwe Lübken, Alexa Weik von Mossner, Lawrence Culver, Gordon Winder, Sherry Johnson, and Cheryl Lousely, as well as PhD students and student assistants Marcus Andreas, Julia Blanc, Andrew Denning, Agnes Kneitz, Marc Landry, Annka Liepold, Pierre Lipperheide, Angelika Möller, Stephanie Rinck, Lisa Spindler, Phillip Stephan, Martin Spenger, Office Manager Korbinian Seitz, and Managing Director Claudia Reusch. Led by two very inspiring directors, Christof Mauch and Helmuth Trischler, my year in Munich was one of the best I had in my career in terms of scholarly support, freedom of research, Alpine hiking, and a regular flow of gentle reminders not to forget the environment. I also thank Frank Uekötter for his interventions during the Thursday seminars at LMU. Editor Katie Ritson deserves a special mention, as she helped me anglicize my English until well after my departure simply because she was interested in my topic. I especially appreciate her successful efforts to help translate the literary quotations.

Of the wider international scholarly community of mobility students I must first and foremost mention my dear friends and colleagues Clay McShane, Georgine Clarsen, Craig Horner, Peter Norton, Steven Spalding, and Rudi Volti, who read, officially reviewed, edited, and commented upon my chapters in a multitude of versions. Although I remain, of course, fully responsible for the entire text and its errors, without their help it would have been much less coherent. Catherine Bertho let me use her Versailles apartment to go through her set of the *Revue du Touring Club de France*, and Nicola Dropmann gave me access to a very rare photocopy of an early motoring book by Mrs. Kennard. Discussions with Mathieu Flonneau, Bruce Pietrykowsky, John Walton, Luísa Sousa, Charissa Terranova, Kurt Möser, Rodrigo Booth, Colin Divall, Mimi Sheller, Peter Merriman, John Urry, Clapperton Mavhunga, Heike Weber, Nanny Kim, Tracy Busch, Jennifer Bonham, Ann Johnson, Nathalie Roseau, Cotten Seiler, and many, many others sharpened my views on the topic at hand. A special mention goes to Karin Bijsterveld, with whom (and her colleagues Eefje Cleophas and Stefan Krebs) I acquired a research project, "Selling Sound," from the Dutch science council NWO, with a separate book on mobile acoustics as a result. Marion Berghahn of Berghahn Books, publisher of *Transfers*, was instrumental in helping to make this project into a real book, for which I owe her very much. It came really as a pleasant surprise that Christof Mauch and Helmuth Trischler spontaneously offered to support the book project financially. At Berghahn, I thank copy editor Mike Dempsey for taming the "beast" (or is it 'beast'?) of my manuscript, and Adam Capitanio and Elizabeth Berg for guiding me through the latest stages of book production.

Speaking of beasts, just like my first book project (which cost me the death of our semi-wild pet-bird Youkeman), I must thank Lapje, an equally semi-wild street cat who kept me company during the long hours at home after my return from Munich when the last two chapters had to be finished in a constant struggle with my teaching obligations at the university. Lapje (who had nothing to do with the death of Youkeman) still lives, although she grew quite old, but I managed to tame her a bit: she scratches my hand (when I try to stroke her) much less than at the start of our acquaintance. Speaking of cats, I cannot finish without thanking my soul mate and partner, Charley, who supported my endeavors through long Skype sessions whenever I was abroad in about ten different countries. I dedicate this book (again) to her, and to my big old friend Clay who was such a crucial intermediary at the start of my academic career when it came to finding a publisher.

—Shanghai, July 2014

INTRODUCTION

Explaining the Car
Prolegomena for a History of North-Atlantic Automobilism

"The novel is history."—Henry James[1]

Linguistic research from 2002 among Dutch parents unearthed that "auto" was among the first fifty words their children produced, with the additional reassuring note that more parents reported "mama" and "papa" (98 percent) than "auto" (76 percent).[2] This book tries to answer the most direct question one can ask about car culture: Why? Why the car (and not, say, the bicycle)? Why in the North-Atlantic realm, and not elsewhere initially?

Answering these 'why' questions was the most complex project I have ever undertaken.[3] Following common practice in general historiography, I first searched for possible answers in the existing car canon (indeed, I needed to construct that canon in the first place, as it has not yet been articulated), evaluated these answers, and, finding that they were not very satisfactory, formulated my own.

By synthesizing the mainly national (hi)stories into a transnational one which I call North-Atlantic, I am fully aware that I have excluded several cultures (like the Brazilian, or the Australian[4]) and that I will only, because of space constraints, focus on the first two phases of such a history (the phases of emergence and persistence, until 1940), leaving the two following phases (of extravagance and doom) for a later study.

Merging national histories (written because of a linguistic disposition of the authors rather than being determined by the topic itself) into a transnational one is a tricky undertaking. Most automobile histories have not been written with the fundamental non-nationality of car culture in mind. And as the transnational approach advocated and practiced in this study is not about substituting the national by a new, higher level of internationality (the North-Atlantic term may suggest this[5]), but about finding common characteristics at the local, *sub-national* level, my story necessarily takes on a kind of abstract, archetypical character.

At times, therefore, it may seem somewhat eclectic. I took from the many national stories those examples that fit my project, emphasizing the commonality of national stories rather than the many differences between these stories.

On the other hand, before I can (re)tell these national stories, it is important to know why they are special and worth telling, and that is why a synthesis is long overdue. That is, if I want to avoid falling in the same double trap over and over again (of which historiography gives ample evidence), namely starting at the nation without further reflection and then comparing the national story with one (American) model, thus equating the West, or the car, with the United States. Both are fateful misrepresentations, as we will see.

'North-Atlantic,' then, is not a programmatic term (like the military saw it when they called their project 'the North Atlantic Treaty Organization' or as some historians termed it when referring to a *longue-durée* process of slave-based economic exchange between four continents on both sides of an ocean[6]), but a loosely used epitheton indicating certain types of Western industrializing and motorizing societies that soon diverged into two separate cultures (a 'Western-European' and a 'North-American') while at the same time actively engaged in a constant transfer of ideas, people, and objects, not the least of them car-related, in the full process of convergence without ever reaching a unified state. Walter Lippmann's 1917 phrase "the Atlantic highway" seems adequate, especially in this mobility context.[7] 'Atlantic,' indeed, connotes a departure from the national and emphasizes "extraordinary sameness" rather than difference, although it follows the *Realpolitik* of the "Atlantic states" rather than the idealized One World fantasies that the term sometimes evokes as well. If there existed "*one* Atlantic civilization" in the past, its very recent emanations most certainly encompassed the car. Following Clifford Geertz, such a culture can only be made meaningful at the local level, so it is there our analysis intends to focus upon.[8]

This book is based on a thorough study of the secondary literature within the history of mobility of seven countries (the Netherlands, Belgium, Germany, France, Switzerland, the United Kingdom, and the United States) and primary research in five of them (excluding Belgium and Switzerland). As such, it is based on my two consecutive research projects of the last ten years, both partially funded by the Dutch Ministry of Traffic and Water Management (*Ministerie van Verkeer en Waterstaat*), in particular its 'agency' *Rijkswaterstaat*, responsible for planning and building the transport infrastructure in the Netherlands.[9] Contrary to my earlier analysis of the electric automobile, however, I decided to focus on as *few* countries as possible, those I know the best.[10] Australia,

New Zealand, and Latin-American countries, and Mexico and Canada were excluded, as well as the European 'periphery' to the North, East, and South.[11] Not only are others more qualified to cover those countries, but including them with some quotations here and there would suggest an encyclopedic ambition, which isn't my intent. This does not mean, of course, that the 'periphery' is not present; it exists in the remote regions within the 'Atlantic' countries themselves (sometimes called 'semiperipheries'[12]), and in the fantasies of those living in the 'core' about 'exotic' destinations abroad.

This study, then, while intending to be synthetic, should be an analysis for others to (partially) confirm or reject. It is a vista on a new type of mobility studies, not an exhaustive overview. It *is* synthetic, but only to open debate, not to close it. The emphasis, therefore, is on Germany, the Netherlands, and the United States, simply because these, with Switzerland and France, have the most developed mobility history research.[13] Instead of expanding the geographic area, I focused on more depth in the 'core Atlantic' by conducting new research in primary sources, mostly trade journals and more unorthodox sources such as novels, poems, films, and songs.[14] Were I lucky enough to spend a year in Paris instead of Munich to write the bulk of the chapters that follow,[15] my narrative would have been different, but I am convinced that the basic storyline would not.

Do Narratives Explain?

When I set about explaining the car, the first set of questions I had to address was: What should be explained? For what period? At what level? To what purpose? Should I explain cars, or *the* car, as a type? These preliminary questions already predetermine the final answers to the 'why' question. A second set of questions are those of method. Both sets will be addressed in this section.

As I have already implicitly answered the question about the level (my answers will necessarily, because of the peculiar transnational approach of this book, be 'typical,' finding the transnational in the local and regional, aggregating to the national when it is possible to do so), I can concentrate on the period and the purpose, as well as on the nature of the artifact to be explained.

As to the latter: my analysis will deal with the 'typical' car of a certain period and a certain place, often simply assuming that the reader knows what type of car I mean.[16] The cultural history of technology I intend to write in the chapters based upon traditional sources (chapters 1, 3 (partly), 4, and 7) assumes a relationship between technology and culture which I will explain later in this introduction.

As to the former, due to my decision to limit my analysis to the first four decades of the previous century, and due to my periodization, which delineates a pre-WWI phase of the car's emergence, followed by a interwar phase of the car's persistence, interrupted by a very crucial First World War, the explanation should cover both phenomena. Most histories that have proposed an explanation, as I will show, have focused on the emergence phase and treated the second one as a phase where the 'toy' of the first naturally evolved into the 'tool' of the second. That in itself apparently figured as a satisfactory explanation. That the car lost its playful aspects and became a utilitarian instrument of transport for physicians, shop owners, and commuters, is often related to the dying culture of an aristocratic (Europe) or monied (US) elite. Once the world around the car was shaped by its use (through the construction of highways and suburbs), it seemed that the historian had not much further explaining to do, at this general level. From there, the 'ambiguity' of mass car ownership with its detrimental effects on urban space and environment was simply left to be described, thus providing a base for the formation of two antagonistic camps of proponents and adversaries. A large part of this book is dedicated to criticizing this 'easy' evolution. Hence the emphasis of my analysis will be on the Interbellum period when the crucial conversion to a ubiquitous vehicle took place. This was all the more possible because the emergence phase was already covered in my *Electric Vehicle* book, albeit from the perspective (to be reformulated in the first chapter) of a non-mainstream vehicle.

Regarding the *purpose* of my explanation, I also have already given part of the answer. I wish to contribute to the existing historiography enabling a new approach to the history of mobility by questioning one of its basic assumptions: the toy-to-tool thesis. At the same time, and not far in the background, resonates a knowledge interest, which is directed to the future rather than the past. Although this book does not make the usefulness of historical research for current policy and planning problems an explicit question, recent discourse about the necessity of reviving what formerly has been called Applied History plays a continuous backstage role as I deal with the reasons and motives for persistent automobile use.[17] From this perspective, the book can be read as a stepping stone to the field of the so-called New Mobility Studies.[18]

The second set of questions that surround the issue of 'why the car?' are theoretical in nature and need some more introduction. "While scientists try to limit themselves to the how of phenomena," Stephen Kern asserts in a recent study about causality in history (which is not the same as causality in the writing of history), "an ultimate why lies behind all their observations and experiments."[19] When the sociologist Richard Miller set out nearly thirty years ago to answer the 'why' question about the experience of mountaineering ("Why do people climb mountains?"), he distinguished between "accounts" as "linguistic devices called into play when taken-for-granted expectations about the way people will act in a given situation are violated," and "explanations" as inquiries that do not require such actions. According to him, the question "Why did you get here at 7:30?" calls for an account, and "What time is it, 7:30?" invites an explanation.[20] This book, in Miller's terms, aims to provide an explanation for the non-controversial parts of car culture, and an account for those parts that may be expected to be controversial, as they advance an argument that deviates from the canon as reconstructed in this introduction. Nowadays, a rigorous distinction between account and explanation cannot be upheld with good reason, but the important point for my argument is that both have a narrative structure: they have an introductory beginning, a middle part, and a concluding end.

"Do narratives explain?" theoretically inclined historians and related philosophers have asked, "and if so how?"[21] According to philosopher of history David Carr, the answer to the first part of this question relates to the "rather obvious fact ... that the narrative mode is very close in form to the structure of action itself, from the agent's point of view." This is exactly what I am after: to find the motives (motivations and motifs) of historical actors that prompted them to engage in automotive culture, or oppose it, motives that go beyond the individual. "Narrative explanation is satisfying," Carr suggests, "precisely because it never strays too far from ordinary discourse." It is based on the existence of a "temporal continuum," whereas causality "is totally out of place here." Other historians, too, argue for "the 'self-explanatory character' of a narrative solidly supported by documentary evidence."[22]

Such opinions have been put forth in the context of an ongoing debate about "the tension between an 'explanatory' and an 'understanding' approach," a tension coined in the German historical *Methodenstreit* as one between *verstehen* and *erklären*. The allocation of both to 'history' and the 'social sciences' became problematic, when 'historical sociologists' argued that "no ... agreement exists" among the social sciences "that causality is 'the first requirement of an adequate explanation.'"[23] Other social scientists (including criminologists, family sociologists, health sociologists, sociologists of education, sociologists that deal with 'life histories,' and all those scholars with "an interest in people's lived experiences and an appreciation of the temporal nature of that experience," including those with "an awareness that the researcher him- or herself is also a narrator") would probably agree on L. and S. Hinchmans' 1997 definition of *the narrative*, as a "discourse... with a clear sequential order that connects events in a meaningful way for a definite audience and thus offer(s) insights about the world and/or people's experiences of it."[24]

Within the debate about 'understanding' versus 'explaining,' David Carr has been criticized by postmodern theorists as someone who adorned 'reality' with a narrative structure that subsequently is only to be 'found' by historians. Such theorists cite Hayden White's famous dictum that "we do not *live* stories" and that projecting narrativity into the past boils down to a reluctance among historians "to accept... responsibility for their narrative constructions."[25] From this perspective, our interest in literature as a historical source is more than a desire for new content; there are also methodological parallels. George Steinmetz observed that "the analysis of narrative has moved gradually from formal literary criticism into fields that emphasize the social role of storytelling." He quoted cultural historian Lloyd Kramer, who claimed that "literary criticism has 'taught historians to recognize the active role of... narrative structures in the creation of reality.'" Quoting Fredric James, he affirms that "narrative is an 'epistemological category traditionally mistaken for a literary form,'" narrative is "one of the abstract or 'empty' coordinates within which we come to know the world, a... form that our perception imposes on the raw flux of reality, giving it... the comprehensive order we call experience."[26] Although such theories still depart from a dichotomy between a 'reality' and 'perception,' which is 'imposed' upon the former, Bruno Latour, too, finds that a description, especially in the case of "socio-technical networks," is valid as an explanation, which "emerges once the description is saturated," thus answering the question of 'why' through a thick description of the 'how.'[27]

On the other hand, telling (hi)stories should not *substitute* for explicit explanations: such stories should allow for explanatory intermezzos, in-between summations of actors and factors that apparently drive the narrative. Paul Roth's 2004 analysis of the Holocaust (for which he claims there is no 'why') may supply a model here. Roth first outlines types of possible explanations for the Holocaust, then puts forward his own explanation on the basis of social-psychological experiments emphasizing situational (behavioral) motives influenced by exogenous factors (dissonance theory), and then criticizes two other explanations, each one situated at the opposite extreme ends of the spectrum of explanations that he presented initially.[28] Although I follow Roth's strategy as much as possible, the problem of using this 'model' for this book is that he analyzes a well-defined sequence of events with clear results that is well placed in time and space (the "completed destruction process" of the Jews, in the words of Hanna Arendt). In contrast, this book tries to explain half a century of mobility in several Atlantic countries. Where the Holocaust was centrally planned (although it could not have been carried out without the consent of many) and ended at a certain point in time (providing historians with a beneficial distance), car culture was certainly not, in both respects. It has been a bottom-up process (pushed onto the table of the authorities, as it were) and continues into the present, becoming ever more controversial as time goes by.

With this in mind, my explanation should address four phenomena: automobilism's emergence (somewhere), its diffusion (or simultaneous emergence elsewhere), its persistence (everywhere), and (largely beyond the scale of this book but clearly prepared during the Interbellum) its growth into an all-pervasive phenomenon. In short, this book traces the domestication of the car, its ascendency to normalcy, and the particular shape its technology and its culture of use took. My question remains, then, how (if ever) have these phenomena so far been explained in the literature?

All historians have theories embedded in their narratives, "even when they choose to express them implicitly rather than explicitly."[29] Because such 'theories' come very close to 'explanation,' in what follows I will seek to make them explicit in order to build upon the historiographical explanatory state of the art in a more transparent way, thus constructing (a proposal to) a canon. My preliminary survey of the historiography revealed that explanations tend to have a disciplinary flavor: technical, economic, systemic, political, social, cultural. Such flavors at the same time mark stages in automobilism's historiography, as the following pages will show.

Constructing a Master Narrative

In the historiography of automobilism, technical and economic approaches were among the earliest and were often linked, mostly through studies of car production. Every Atlantic nation has produced such scholarship, which is not easily separated from an avalanche of amateur histories. The seminal works of James Rood Doolittle (1916) or Ralph Epstein (1928) on the American car industry are examples of such hybrid histories.[30] The first post-WWII academic histories also held to this tradition. The overviews of John Rae provide a well-known example. Considered the founder of American academic scholarship on the history of automobilism, Rae's publications accompanied the emergence of the Society for the History of Technology (SHOT) and are now considered, if at all, as "deep-background reference," a euphemism for being rather dated.[31] Interested in the 'impact' of the car 'on' society (the latter conceived not as something creating the car in the first place, but as an entity to which the car comes as an exogenous 'factor'), and written in an "overwhelmingly positive" style, "openly celebratory in its nostalgia for the freewheeling, largely unregulated, pre-New Deal era of automobility," Rae's work can be read as an almost aggressive defense of American exceptionalism. *The Road and the Car in American Life* (1971), considered to be one of his best studies, contains only four chapters of history (in which he emphasized the systemic character of the automobile complex), the rest being an extensive pamphlet against the anti-car sentiments of the 1960s.[32]

In the historiography of automobilism, James Laux, the 'European Rae,' is often forgotten, although he has had a comparable role for European scholarship.[33] It was however James Flink, seen at the time by mainstream car historians as "the leading iconoclast of automotive studies," whose *America Adopts the Automobile* (1970) was presented as the first 'social history' of the emergence of the automobile in the United States.[34] The differences between Rae and Flink are less spectacular than a later interpretation, perhaps inspired by Flink's own critique of Rae, seemed to suggest.[35] While Rae strengthened his defense of the car industry over his career, Flink shifted his position to a more critical one, arguing that by the 1960s the car had lost its 'democratic' advantages.[36] Flink's *The Car Culture* (1975) integrated this critical stance without bowing too much to those scholars who focused on the pleasurable sides of car culture.[37] This was a milestone in American automotive historiography because of the breadth of his analysis, which deals with buying motives, auto and touring clubs, the industry and the car's technology, the response of local and national governments, and the fledgling institutionalization of an automobile culture that more or less existed before 1908, when production of Ford's Model T began.[38]

Methodologically, Flink pioneered the extensive use of trade and popular journals as historical sources, which he defended by referring to a lack of primary sources and because their reliability seemed plausible to him. His reliance on these journals, however, not only made him develop a predominantly national perspective (not international, nor regional or local), but it also prevented him from including public transportation, resulting in a remarkably biased narrative largely inspired by early American car enthusiasm, which was prolifically celebrated in his sources. Several local and regional American studies were undertaken following Flink's model, such as Norman Moline's study of Illinois and

Robert Ireland's study of North Carolina.[39] Later scholarship in Flink's tradition have corrected those biases to a certain extent, such as through the addition of an international perspective in Rudi Volti's recent overview of *Cars and Culture*.[40]

To summarize Flink's explanation of the car's early success: he 're-constructs' the high expectations among Americans about the imminent coming of the "car for the masses," expectations he took to be barely hampered by "popular prejudices" or "reactionary opposition" against the car.[41] Against the background of such a 'determinism' performed by the historical actors themselves, Flink dedicated an entire chapter (chapter 3) to the "Motives for Adoption."[42] This central chapter reads as a critical comment on the historical actors' discourse on the advantages of the early automobile, resulting in three clusters of factors: economic efficiency, speed, and reliability, as well as additional "nonutilitarian attractions." Close reading of Flink's arguments suggests that economic factors played perhaps a decisive role only in the case of trucks, buses, and car fleets (which were of little significance during this period). He identifies the medical doctor as the quintessential first adopter, soon to be followed by large parts of the urban and rural middle classes. Flink calls their individual (and individualized) solution for the horse substitution problem "characteristically American" and explains its popularity mainly by referring to the (utilitarian) factors of speed, traveling radius, and endurance (in comparison to the horse). Speed, however, was partly offset by the problem of road safety. Flink's critical treatment of the sources clearly relativizes the importance of these utilitarian factors (he suggests on several occasions that it was the *appearance* rather than the reality of these advantages which dominated the discourse), so his secondary, non-utilitarian arguments merit a respectful treatment in any overview of automobilism's historiography. Here, such factors as the love of power, the danger of the automobile ride, private bodily driving sensations, the paradox of the car as a status symbol and a hoped-for object for mass consumption (which was solved by differentiating between types of cars), and the fact that the car was seen as a "panacea for social ills," both urban and rural, stress the importance of expectations in the pioneering phase of automobilism. Buying a car seemed to offer a private and immediate solution to the social ills of the time. While Flink explained the eager reception of the car's promise as a form of American exceptionalism (America's penchant toward individualism), he connected his non-utilitarian factors to a "hedonistic appeal rooted in basic human drives."[43]

Thus, Flink's remarkably versatile early analysis presents the 'impact' of the car upon society as a 'social history,' which explains the American middle class's eager reception of something which seemed to be destined to conquer the continent, by a typical American penchant for individualism on the one hand and an anthropological tendency to hedonism on the other. From such a perspective, 'utilitarian' factors indeed are primary, although Flink acknowledged that they were weak and were supported by non-utilitarian factors.

Rae, Flink, and Volti all are part of the History of Technology, a field that for a long time had a virtual monopoly on academic car studies. Social constructionism brought an end to the focus on artifactual innovation and took automobile scholarship out of the exclusive domain of the history of technology. It allowed for a 'contextualization' that sometimes tended to forget the original object of its study, but in the end enabled a maturing of the field.[44]

The most important correction, however, came from a multitude of studies on the urban car culture. Since 1970, a host of urban-planning historians and, quite recently, Peter Norton's dissertation have shown that American 'enthusiasm' for the automobile was much more fragmented, contested, and ambivalent than Rae and Flink and their followers concluded.[45] Where Flink set out to criticize the car and its proponents himself, later scholars found contemporaries to do the criticizing for them, thus creating a paradigm of 'controversy' and 'ambiguity' around the car. A further correction of Rae and Flink was provided by the emerging scholarship on the electric vehicle.[46]

In this tradition, and clearly intent on separating himself from the "technological determinism" of both Rae and Flink, stands Clay McShane and his seminal study of American car culture during the first half of the twentieth century. If Rae and Flink are the fathers of modern mobility studies, McShane (one of the scholars Rae fulminated against as infected by the 1960s syndrome) is their godfather (all men, so far), and where Flink transcends Rae by introducing social factors, McShane transcends Flink by a socio-cultural approach, which sets itself in the tradition of a remarkably theory-poor American approach to writing history. McShane's popularity among young mobility historians and beyond is not only based on his wit and unorthodoxy, but also on his professionalism as a classic, erudite historian. But his hesitancy regarding theory (instead indulging in Latour's previously quoted 'thick description' of the 'how') makes it hard to formulate his explanation in a couple of sentences. What we can say is that he develops his arguments in an urban setting, bringing urban history and mobility history together in a convincing way, for the first time.

Four new elements figure prominently in McShane's implicit explanatory framework. In the first place, he continues and strengthens Rae's argument about the importance of car culture's infrastructure. Second, about half of his book is dedicated to a cultural analysis of the early car culture in New York, based on a detailed scrutiny of the *New York Times*. Another innovation is his time frame, as he starts his story well before the car, in the 1840s, and only after a hundred pages the car appears on the stage.[47] It allows him to emphasize continuity instead of celebrating 'the invention of the automobile,' continuity mostly with the horse and carriage, but also with tramways and buses. Last, McShane tries, whenever his sources allow, to include references to Europe and break open the deplorable chauvinism of American automotive historiography. In his effort to construct a mostly cultural explanation for the emergence of the car, McShane inserts a unique comparison with the steam car (which he sees disappear because of "public suppression" rather than technical defects), but his explanation of the car's success is to a much larger extent built upon the supporting role of street paving. This approach differs from what this author found when explaining the 'failure' of the electric vehicle and the success of the gasoline car. Also, in the end McShane does not answer so much the 'why' question, but only the 'why America' question, referring to "an interlocking set of economic and regulatory reasons," such as economies of scale and taxes. In particular, the latter factor receives more emphasis in explaining America's exceptionalism than the present study will argue. Also, he emphasizes the ownership of the car (in relation to its use) to a greater extent, not surprisingly because of his focus on urban display. Far ahead of its time, McShane's study is a hallmark of comprehensive history writing, including an analysis of traffic safety, environmental and gender aspects, literature and film, and a conclusion John Rae must have despised: "The modern American metropolis is a socially and politically fragmented, gas-guzzling environmental nightmare."[48]

In Europe, most national academic discussions of the automobile can boast at least one comparably ambitious overview. For Germany there are several candidates, such as Gerhard Horras, with a largely economic approach, or Angela Zatsch, with an institutional analysis, both for the period of the car's emergence. For my purpose, however, Heidrun Edelmann's *Vom Luxusgut zum Gebrauchsgegenstand* (1989) is more appropriate as it offers an analysis of the transition from the car as luxury good into a utilitarian object.[49] Like Horras and Zatsch, it is used as a basic secondary source for any serious history of German automobilism.

Methodologically, Edelman's analysis seems to be a return to a pre-Flinkian phase of the international historiography. But the first paragraph of her conclusions reveals that this emphatically economic analysis presupposes the social actions of many relevant actors. "The history of about one century of the spread of automobiles is characterized, for the initial phase, in Germany as elsewhere, by solving technical-constructional basic problems aimed at improving the vehicles' reliability in order to encourage or invite new customers to buy them. Thereafter the spread of the car and the growth of the industry no longer depended on technological innovations. Socio-economic conditions now became decisive to expand the market."[50] Despite this recognition of the crucial role of socio-economic factors, Edelmann's study uses a simple (if not simplistic) economic model of a balance between supply and demand, which nonetheless seems adequate to explain the lag of the German automobile industry vis-à-vis the two other large European automobile-producing nations (France and the United Kingdom), and especially the United States. More than a decade later, the economist Reiner Flik repeated Edelmann's analysis in a more sophisticated and quantitative way, adding an analysis of the crucial difference between American and German farmers.[51]

Edelmann distinguishes between four phases of diffusion, the latter two occurring after the Second World War: the car for luxury and sporting purposes, the car as a 'time-saving machine' for commerce and professionals, the breakthrough of mass motorization, and the phase of saturation, in which replacement sales dominate. Explaining the transition from 'toy' to 'tool' in the second half of the 1920s by focusing on an inadequate purchasing power among the commercial and professional middle classes does not suffice, Edelmann argues. It was the automobile industry that did not manage to rationalize production and offer a low-priced car, and it was the central government, which in the 1920s adopted a taxation policy that made the *use* of the car too expensive for a middle class that was already impoverished by the war and subsequent inflation. High taxation and high fuel prices, the latter the result of economic support for the powerful agrarian sector, were constitutive of this. Taxation could not be lowered because of the states' (*Länder*) interests in taxation income for their road-improvement programs, hence the dominance of the motorcycle in German motorization during the early 1920s and for the better part of the 1930s. Despite two clear waves of car diffusion in these phases, each resulting in a doubling of the German car fleet (the latter the result of a taxation reform by the National-Socialists), its size still lagged behind the United Kingdom and France by a decade. Edelmann has no doubts about the utilitarian character of the second phase (she quotes statistics to support this), although she admits that it is "out of the question to distinguish between a commercially used automobile and a luxury car. A medical doctor, for instance, used his car for his profession during the week, but on Sunday he drove away with his family, and then it was a luxury vehicle."[52]

Flik and Edelmann, explicitly or implicitly, conceived their analysis with America as a 'model.' They can thus be considered to stand in what one could call a 'diffusionist tradition,' in a double sense. On the one hand, this tradition sees the United States as the basic (and only) model for automobilism, followed after a 'time lag' (which was also a cultural lag) by European countries (and later by non-Atlantic countries as well). If the toy-to-tool myth is the main focus of my study, the America-as-model myth is the second. On the other hand, this diffusionist tradition sees diffusion in general as driven by an economic impetus. As incomes increase, the car 'descends' into ever more layers of an eager population: the 'democratization' of the car is then nothing more than the ongoing spread of the car, first among the middle classes and then the working classes, first in the 'developed' countries, and then in the 'developing' countries.[53] The most extreme example of a diffusionist rationale is Nebojša Nakicenovic's analysis of the post-WWII spread of the car in Italy as an almost perfect process of "homeostasis," like the growth of bacteria in a petri dish.[54] Related arguments have been advanced recently for India and China, and they stand in a long tradition.[55] It is this tradition that also coined the image of the car as a "herald of modernity."[56] As I show in this book, as long as we limit our analysis to the Atlantic realm, the car indeed seems to have played this role, even if its use did not necessarily correlate with individuality as an essential aspect of modernity.

Since Rae, Flink, Edelmann and others in several other car societies, we have witnessed the emergence of a 'cultural turn' in the historiography of automobilism that has challenged the basic socio-economic approach. Even if this turn recognizes, to some extent, the importance of income as a threshold for unfolding the 'car society,' its main corrective argument rests on the insight that people have to develop a desire, an expectation, before they decide to buy a car, and that desire is not the result of a free and autonomous choice (in this respect, the car as a historical object and the ideology of the autonomous consuming subject have a lot in common, as Cotten Seiler recently argued[57]). Even so, the historian still has to answer the 'why' question.

In the effort to find satisfactory answers to why people develop a desire for automobiles, two schools have developed: those who emphasize psychological factors versus those who propose more systemic factors. In the German historiography of the car, for instance, Wolfgang Sachs is an early (and farsighted) example of the former, even if he did not found his analysis upon a sound empirical basis, as has been argued recently. Klaus Kuhm is an internationally less well-known example of the former.[58] Recently, Per Lundin integrated this distinction in his plea for a third, "historical scholarship by moral concerns," which he, rather defensively (but understandably, considering the partisanship reigning in the historiography of mobility) calls "not necessarily a bad thing." He does so by equating the systemic approach with a "critical" tradition of scholarship, while he calls the psychological approach "empathic," resulting in largely negative or positive assessments of the car's role respectively.[59] Although there are also 'critical' studies that emphasize the psychological approach (this study is an example), it cannot be denied that studies emphasizing the "love affair" of Americans with the car tend to accept the basic systemic presuppositions of the car culture.[60] Considering that the accusations against the car become ever more explicitly critical, this tradition has now reached the stage of a disappointed love, as testified by titles such as *Hell on Wheels* (David Blanke) and *Automania* (Tom McCarthy). Whereas the former bases its explanatory framework upon the concept of a "dysfunctional love affair," the latter, as basically an American version of Sachs, also emphasizes that Americans are "in love of cars." Both are driven by the ghosts of American exceptionalism even though they contribute new elements to the explanatory canon. David Blanke distinguishes between four elements of the 'love affair': the excitement of driving, the promise of freedom, the elation about the mastering of technology, and the experience of civic equality. The latter element ("Fordism was one of the most stabilizing social events in modern American history") brings class analysis and arguments of democracy and equality back into the historiography of the car in a novel way, and has been exploited to the full by Cotten Seiler (see below).[61] Tom McCarthy's 2007 study is also innovative because it merges emotions ("self-respect" as motive for American farmers to buy a car) as an explanatory factor with an explicit environmental history (a first in automotive historiography). He also argues that the car's continual obsolescence is not so much a product of the car industry, but is produced by consumers. McCarthy's study is the closest a historian can get to the car buyers' motives without using belletristic literature.[62]

Most studies in the tradition of the 'cultural turn' now recognize the importance of integrating both perspectives, the systemic and the psychological (as does this book), advocating a postmodern multi-perspectivity and, depending on the personal preferences of the author, leaning more toward one or the other approach. Cultural historian Kurt Möser's *Geschichte des Autos* is an eloquent example of a behavioral analysis in a systemic context. His recent *Habilitationsschrift* expands this approach toward other mobility modes (especially aviation), realizing a much-needed de-ghettoization of car historiography from the prison of a single mode (motorized road mobility), while retaining, as a former curator of one of the major German technology museums, his very erudite object-oriented approach.[63]

Möser's ambitions are large: he seeks to write a "comprehensive history (*Totalgeschichte*) of the automobile society." Not only does he wish to cover all relevant aspects of society, but he also suggests that the German case can be a springboard for a European history: "not a global history but with Germany as the point of gravity." He also goes beyond most recent histories through the extension of his analysis into the present time. Möser asserts that "many perspectives" are needed for this, but places emphasis on "our emotional, highly complex, not only practice oriented relationship with the car." "Automobile fascination" in his view is about the "non-rational component of the automobile," because this "was responsible for its breakthrough and world-wide expansion." A focus on "feelings" brings "the handling, driving and living in the mobile private space" into view, and consequently Möser calls the car "a machine to realise desires and passions," placing himself squarely in the tradition of Wolfgang Sachs. But more than Sachs, Möser intends to deal with three "parts" of the automobile: the technology, the interior, and the system.[64]

Putting his analysis into a historiographical context, Möser first criticizes two "myths" around the car: the myth of the superior technology or system, and the myth of a car lobby's conspiracy. Instead, he points to long-term trends, such as the violence and fascination for risk in the early car culture, automobilists' struggle with the state and other road users, the interests of the military in the new vehicle, the importance of what he calls "non-functional use," the United States as an example of car culture for other countries to follow after the First World War and the car as an object of consumption and an aesthetic object.[65] Consequently, Möser does not find the explanation for the car's initial success in the technology, nor in the system, but in the investment with new cultural values by users and non-users alike. According to Möser, that investment started not in Germany, but in France. His emphasis on the immediate driving experience, however, leads him to a shaky analysis: he bases his arguments mainly, like Wolfgang Sachs before him, on secondary sources, often written by amateur historians, or on contemporary popular publications, and largely supported by his own (or at least current-day) experiences. No wonder then that Möser's analysis of the gasoline vehicle is much more convincing than his analysis of the steam car, about which much less has been published. When he is convincing, the abundance of original citations is really impressive, for instance in the case of the "rush" (*Rausch*) of the early automobile driving experience. For the same reason, his analysis of the 1920s is rather thin (no doubt out of a lack of secondary sources), whereas his account of the development of automotive technology is bordering on the journalistic, informed by a vague belief in technological 'progress' and not by recent theorizing about technological innovation.[66]

Thus, Möser has written a generic format for a future study on "collective desires" projected on the car and most of his claims need to be substantiated (or rejected!) by historical research.[67] *Geschichte des Autos* is written from the perspective of cultural sociology rather than as a true historical study. Its strength lies in the emphasis on expectations, almost entirely neglected by others.

As a parallel to Möser's work, Catherine Bertho Lavenir's *La roue et le stylo* (1999; The Wheel and the Pen) on early French bicycle and car culture (with some excursions into Belgium and Italy) is an equally seminal publication, although language prevented both analyses from exerting their deserved influence on the international historiography of automobilism. What Bertho Lavenir contributed mostly to the historiography is a multimedia perspective, emphasizing the parallels between driving and writing, based on an analysis of the journal of the French touring club.[68]

A second expansion after Rae, Flink, and Edelmann is provided by the Swiss economic historian Christoph Merki, who undertook the first systematic cross-national study of automobilism in Germany, Switzerland, and France.[69] Merki's richly documented and topically broad treatment of his main research goal, namely to find "the causes, conditions and developments of motorisation in three European countries," is innovative and traditional at the same time. It is traditional, in that he, like Sean O'Connell for the United Kingdom (as we will see below), takes as his starting point the 'contested modernization' of the consumer society, in which the car can be seen as a "key product" or the "most sacred of modernity" (as articulated by the Swiss historian Christian Pfister and the German philosopher Peter Sloterdijk, respectively). Merki is innovative in his proposal to introduce four new approaches for his explanation of the 'shaky triumph' of the car. First, such a history should not only be a social history, but as the relevant groups within this history are 'represented' by intermediary institutions (*Vermittlungsinstanzen*) such as automobile clubs the latter should be central to the analysis. Second, echoing Rae and McShane, car history cannot be separated from road history and the artifact should be analyzed as part of a socio-technical system, where fuel depots, road signs, maps, garages, legislation, education, tax policy, and insurance are as much a part of the system as the car. Third, road-traffic motorization should be analyzed as a part of the general mechanization of society, as proposed by Siegfried Gideon. And fourth, such a history should be transnational (Merki uses the word *beziehungsgeschichlich* or relational history, quoting Heinz-Gerhard

Haupt and Jürgen Kocka, to distinguish his analysis from comparative history), because the nation state is an unsatisfactory unit of analysis for the explanation of the car's emergence. The interesting promise of such a research program is that, not only is the car contextualized among other transport modes (Merki mentions even roller skates and skateboards), but its history also forms a part of a *longue durée*, because both road network and mechanization predate the automobile by many decades. For such an ambitious program and because history alone does not provide the necessary tools, the historian needs an interdisciplinary methodology, drawing additionally on sociology (such as lifestyle research), institutional economics, and cultural history. While Merki warns against histories that limit themselves "in a culturalistic manner to the analysis of some discourses" (for which he cites Paul Virilio's "empirically poor speculations" as an example), his analysis adds cultural history to traditional socio-economic history.[70] What Merki's excellent historiographical overview also makes clear is that such a historiography, if based only on Anglo-Saxon studies, would be severely limited. Paradoxically, while English is the *lingua franca* in academic history, including mobility, non-Anglo-Saxons have a much broader access to secondary literature than their Anglo-Saxon colleagues, who too often cannot read the products of their colleagues. The result is two separate academic cultures.[71]

Merki's subsequent chapters are less systematic and analytical than his introduction promises, nor does he cover all elements he announces (for instance, the roller skates or skateboards do not return). There is a good reason for that. As Merki himself asserts, his study is, as every 'good history' should be, a "combination of historical understanding [Ranke's *Verstehen*] and causal-analytical methods," a combination which, in the end, the author does not seem to have mastered. Much of the remainder of the book is as *holprig* (shaky) in its structure as Merki's triumph of automobilism.[72] It is therefore difficult to get a grip on Merki's explanation of the car's emergence in the three countries under study, despite the plethora of interesting comparisons and the excellent case studies. From that perspective, Merki's study certainly adds a mature piece of work to the extant historiography. In the end, however, his explanation seems to boil down to the already quite familiar statement that the car emerged as a sporting vehicle used by an urban aristocratic and upper-middle-class elite around 1895, and then became a 'professional machine' for the middle classes, while the same business character was present from the start in the use of the motorcycle, especially in the countryside. Merki's implicit rejection of O'Connell's speculation (see below) about the pleasurable side of car use as also being dominant in the second period of 'persistence' (the interwar phase) seems to be related to his fierce criticism of 'culturalist' speculation on the basis of 'poor empirics.' Like O'Connell, Merki adds culture to socio-economics, but he falls back to a position before O'Connell and Möser in his refusal to take the symbolic aspects of car use seriously. However this may be, it seems that two 'schools' of automotive-historical scholarship are slowly emerging. Both see the car as central to the formation of consumption society (as opposed to seeing the car first and foremost as a transportation device), but differ in the attention they pay to the car as symbol.

In the end, Merki's functionalist approach (stemming from his system perspective) prevents him from seeing the car and the motorcycle as anything other than a substitute for the horse-drawn vehicle and the bicycle, respectively, resulting in the downplaying of the cultural continuities between the bicycle and the automobile movement. The result is nonetheless a valuable addition to a transnational historiography, which allows him to explain the Swiss case much better than any national history could ever achieve. Also his inclusion of road (construction) history has provided a model for a number of European scholars, including the present author.[73] Very recently, Christopher Wells in his dissertation built upon this concept of merging the history of the car with that of the road, enabling him to much better pinpoint the moment (in the 1920s) that the systemic properties of the car culture started to dominate.[74]

O'Connell's study, based on his dissertation from 1996, rates among the best recent historiographies on automobilism.[75] Although generally very positive, reviewers have criticized his study of the emergence of automobility in Britain on several accounts, such as the omission of urban car use (John Griffiths, in the *Journal of Transport History*) and his all-too-easy accusation of his predecessors of falling into the trap of technological determinism (Colin Chant in an H-Net book review).[76] The latter point, which results from O'Connell's rather uncritical embrace of Mary Douglas's and Pierre Bourdieu's consumption studies, is especially serious as it not only unjustifiably depreciates a whole pioneering generation of historians (some of whom, like James Flink, saw themselves as social historians of the car), but also because it strengthens O'Connell's questionable self-definition as a 'social historian.' After all, the emphasis on the 'symbolic values' of the early car as a crucial explanatory factor rather makes him an excellent candidate for a new generation of cultural historians of the car, despite his own labeling of Douglas and Bourdieu as "sociologists of consumption," and despite his embrace of John Urry's *Tourist Gaze*.[77] This is not a semantic quibble, as it throws O'Connell's predecessors into one formless group of determinists, whereas at least two generations of social historians of the car can be distinguished (as O'Connell himself acknowledges in his introduction). After Rae and Flink (and their British counterparts, Harold Perkin's *The Age of the Automobile*, 1976, and Kenneth Richardson's *The British Motor Industry 1896–1939: An Economic and Social History*, 1977), he cites mostly Americans (Howard Preston on Atlanta, Virginia Scharff, Clay McShane, and the only non-Anglo-Saxon, but translated, Wolfgang Sachs).[78]

O'Connell's introduction drafts the main contours of his explanation. Borrowing from social constructivist technology studies and from consumption studies, he is not so much interested in the 'impact' of the car on society (although he mentions this concept on the first page of his book, and the term also figures in the title of his dissertation), but in the "dialectical relationship with social relations such as those of class and gender," while at the same time acknowledging the importance of "contingency."[79] His analysis claims that the middle class had a "polymorphous" shape, and that the car played a crucial role as a "very visible artefact (which) became integral to the arena of middle-class distinction." This was distinct from the "collective gaze" (Urry's term) of working-class people in their holiday trains to the seaside, resulting in the "transfer (of) social space into geographical space" and supporting class segregation. In other words, O'Connell's explanation rests to a large degree on "the symbolic value of car ownership." That is also the reason why he criticized the 1993 econometric study of Sue Bowden and Paul Turner on the 1930s, who concluded on the basis of a comparison between average income and an estimate of the car's purchase and running costs that half a million more than the actual two million Britons should have owned a car by 1939.[80]

O'Connell contributes to the British historiography in two important respects. First, he suggests that the hire-purchase schemes that could have expanded the car market were not successful because of a class bias among manufacturers and retailers, and a middle-class reluctance to use this scheme for leisure car purchase based on 'shame' (so, interestingly, the hire-purchase scheme supported the 'serious side' of mobility). Second, by drawing upon some seventy-five interviews, he concludes that lower-middle-class and skilled workers also were among (sometimes intermittent and certainly second-hand) car owners, developing a different user culture that was characterized by car sharing between friends, colleagues, and family members, facilitated by using inside information on cheap car offers or repairs in the neighborhood of the motor industry, and often used for generating extra income among couples. Both his inclusion of recent sociological theorizing represented by John Urry and oral history form two important methodological innovations to the car's historiography. Remarkably, O'Connell does not question the basic assumption under the econometric analysis, namely that the United States should provide the 'model' on which to judge the UK case. A look over the North Sea would have pointed him to comparable developments on the European continent to those he identified in Britain. Nor does O'Connell draw a simple diffusion curve, which could have suggested a possible periodization of British automobilism. If he, further, had compared that to the American curve, it would have pointed him to salient questions, as we will find out in the next chapter. Nevertheless, because social and cultural factors (such as shame and collective car use) have been added to the well-known economic ones, the diffusion picture becomes more complex, although O'Connell does not deny that car ownership during the 1930s was dominated by the professional and commercial middle classes.[81]

In his central explanatory chapter ("Leisure and the Car"), O'Connell argues (without giving quantitative evidence) that the car in Britain was mainly used for leisure, such as day trips, camping, caravanning, and a remarkably large flow of international tourism. He derives from one of the interviews a tripartite set of 'values' (utility, symbolic, and status values) without being able to quantify them.[82] His main contribution to the historiography is his analysis of what one could call a *use profile* of middle-class leisure, in which 'motoring rituals' dominate. Those rituals come in the shape of motor sport, speed (but like in Flink, heavily tempered by safety concerns), tinkering, embellishments (mascots, striping), illicit sexual activity, and joyriding—the latter, like car theft, claimed to be the ultimate proof of the car having become a consumption object and mainly undertaken by working-class members.

The undeniably innovative role of O'Connell's work lies in the insight that the car has to be analyzed as a commodity: "The leisure uses of the car are central to any explanation of the development of the British car culture." Even more central to O'Connell's explanation is the highly gendered nature of that use. Automobilism, in O'Connell's analysis, should be explained as a masculine middle-class pastime. Regretfully, he does not draw the logical consequence from this very attractive explanation by adding that this also forms the basis for the downplaying, by the motorists themselves, of the pleasurable sides of the car and for the utility myth in general, as women were seen as representing the 'frivolous side of the car,' while men were viewed as 'serious motorists.' The car, then, became the iconic expression of middle-class (and to a lesser extent, working-class) masculinity, and the 'symbolic value' clearly dominated among the three values mentioned.[83] The reformulation of the safety issue (to which O'Connell also dedicates a much more speculative, and less satisfactory chapter) along largely car-friendly lines and the subsequent acceptance of the fatality records as a systemic property, can then be seen as the finishing touch on the 'automobile system' (he enters, contra Griffith, the urban realm in this chapter, emphasizing the diverging interests between local and national governments). Thus, O'Connell provides, in his own words, a picture of the 'contested nature' of several issues already well established within automotive historiography. His main explanatory factor, therefore, remains the basic economic drive of the diffusion process, and he adds a cultural explanation based on the 'symbolic value' of the car, largely for the (professional and commercial) middle classes, which enables him to speculate that leisure use dominated in the use profile of Interbellum automobilism.

Only quite recently, the 'cultural turn' of automobile historiography has moved an extra step toward a 'mediological' analysis of the car experience, an analysis in which the car is seen as a medium rather than a vehicle, which makes it comparable to other media such as film, the telegraph, and books. As already stated, Bertho Lavenir's 1999 analysis of cycle and car tourism can be considered a predecessor in this respect, as she showed how the written accounts of the motorists in touring clubs' journals were an inseparable part of the movement. Writing on the driving experiences had to be learned as much as driving the car.[84] Later, convergent tendencies between the history of mobility and media studies resulted in assuming a more explicit analogy between literary utterances and the car as medium. Providing a possible approach to a new, truly interdisciplinary phase in the car's historiography, this promises to overcome a fatal split between 'transport' and 'communication' that took place around the turn of the nineteenth into the twentieth century. The impulse to restore this split did not come from mobility history but from the history of communication. There, James Carey saw this split as a conceptual as well as historical phenomenon: the telegraph freed communication from its dependence upon transport. "Messages could be transferred without the aid of transportation technologies such as trains, coaches, ships, and so on." Since then, two separate disciplines dealt with material and virtual mobility respectively, until the 'cultural turn' triggered convergence as communication history started to see its subject shift toward rituals rather than transmission and transport, a development comparable to what, somewhat later, started among mobility historians. This convergence may lead, perhaps, to a reversal of the direction of dependence, as "transportation has become increasingly dependent upon communication technologies." It may also lead to a "theory of mobile relations" (*Theorie beweglicher*

Relationen), where virtual and 'real' 'vehicles,' wired and wireless apparatuses, and the movement of bodies, goods, and messages are treated on an equal, symmetrical basis. Building such a theory also requires that we address "the problems of communicating mobility," as Nigel Thrift posits, claiming that we need "a change in style of 'writing,'" including "other forms of expression than the written word, for example all kinds of visual and oral cues." In the period covered by this book we observe only some hesitant recognitions of this process of convergence, most eloquently expressed by an avant-garde of motorists-writers (see chapter 2).[85]

One of the best examples of this approach is a recent dissertation by German *Volkskunde* student Andrea Wetterauer.[86] Her dissertation, which offers a full-blown analysis of a well-defined corpus of primary sources (the 'feuilleton' and the travel annex of the *Frankfurter Zeitung* between 1923 and 1929) in the historiography of the automobile, is embedded in a different scholarly tradition: the German *Volkskunde* (ethnology, material culture), particularly the school of the university of Tübingen. We are far from Rae and Flink here. Apart from criticizing the tradition of technological explanation (the attraction of the car as a self-propelled, cocooning vehicle), Wetterauer places her study in opposition to the German tradition of the systems approach of automobilism (the so-called *Technikgenese*) as developed by the sociologist Werner Rammert in general, and applied to the automobile by Andreas Knie and Weert Canzler of the Berlin Science Centre (*Wissenschaftszentrum Berlin*). In fact, this sociological approach (with some interest in long-term aspects of the analysis) has a much longer tradition, starting with Hans-Magnus Enzensberger in the 1950s and heavily resting upon Marxist concepts of the circulation of commodities and the accompanying tendencies of social differentiation and individualization. While the Berlin group of 'historical sociologists' is especially strong in explaining the emergence of the car, Wetterauer's study focuses upon its 'cultural appropriation' by a new group of the post-WWI, non-technically educated middle class, thus addressing the 'persistence' period rather than the 'emergence' phase, although her analysis results, as we will see in chapter 6, in the conclusion that German travel writing in the 1920s seemed a literal repetition of the emergence experiences of an earlier period. Wetterauer's approach uses concepts from recent consumption studies, such as the "return of the actor" and especially the role of intermediary actors (*Vermittlungsinstanzen*, i.e., the press). It reveals a sensibility for the distinguishing aspects of human practices as developed by Pierre Bourdieu as well as the 'microstudies' of power relations as proposed by Michel Foucault, transferred into German cultural theory by Stefan Beck and Rudy Koshar, as Wetterauer explains in her excellent overview of the rich sociological thinking in Germany about the car since the 1970s.[87] Beck, it should be said, translated Bourdieu's mostly discursive approach into an approach governed by discourse and activities (*Handlungen*). Wetterauer is interested in the creation of the car as a "medium for experience and observation" (*Erfahrungs- und Wahrnehmungsmittel*). She is also interested in the production of "distance" (Bourdieu) and "discretion" (Georg Simmel) during the car's appropriation by the new user groups. As we will see in chapter 2 (where we will use Wetterauer's analysis to get a grip on our set of literary utterances about the car), as early as the 1920s Simmel had analyzed the car as an "extension of the body (and of) the private space" and also introduced concepts such as "stylization" (*Stilisierung*) and "aesthetisation" to characterize the middle-class relationship with the emerging technological environment, especially in the cities.[88] With Wetterauer's cultural-anthropological approach (and her use of Simmel), we are back at seeing automobilism as a fundamentally urban phenomenon.

In her overview of the history of German automobilism, Wetterauer leans heavily on Heidrun Edelmann and the sociologist Thomas Krämer-Bodoni (1971), while she also re-tells the canonized history of German tourism, in order to embed her analysis into a history of travel culture.[89] The results of her analysis convincingly show that technological and systemic approaches should be complemented with an analysis of the practices and sensory experiences afforded by the automobile and created during the cultural appropriation process, especially from the middle of the 1920s onward. In other words, she replaces the psychological factors in the dichotomy between 'psychology' and 'system' by factors related to social practices, thus opening the study of car history to approaches that try to overcome the mind-body dualism. Like Merki, she also concludes that the new experience of car travel is reduced to a "road experience" through which new spaces are apprehended. Her analysis further confirms the results already proposed earlier by others (like Kurt Möser who, remarkably, is not cited), such as the 'adventurous' character of early automobilism, as well as the importance of tinkering, maintenance and technical expertise in general (elements of automobilism also investigated by Kathleen Franz and Kevin Borg for the United States).[90]

Her study, however, also leaves open the question about the precise relationship between the feuilleton and travelogue authors and their readers or car users. Wetterauer seems to imply that these authors should more or less be considered as ideal typical road users, but she does not elaborate further on this. What she in the end studies is the overt, not the covert culture in the sense of Leo Marx.[91] The study is also a remarkable proof of my thesis, expressed earlier, about a ghettoization of modal and national approaches to the history of mobility, not only of a European history vis à vis an American history, but also of a German historiography vis à vis the other important European historiographies. Wetterauer only uses German secondary sources, and German translations of foreign sources.

Wetterauer is therefore not aware of a contemporaneous 'boom' of American car-related scholarship from the last half decade or so, from historians but also from STS (Science, Technology and Society) scholars, American and cultural studies colleagues.[92] No doubt, automobile scholarship in the United States is in the full swing of a cultural and political turn, not only undertaken by historians of technology or mobility, but also by scholars less versed in the intricacies of transport and mobility matters.[93] If there is one dominant characteristic of the historiography of the last decade, it is the spectacular spread of the automobile and automobilism as a topic among scholars who have no institutional or bibliographical ties to mobility history. Despite the inaccuracies and empirical errors that sometimes slip in, those new perspectives are enriching recent historiography that hitherto suffered from an increasing ghettoization of competing modalities. They are opening up new trajectories of research and questions, such as trucking, road building, safety, racing, the environment, landscapes—insights and topics I will gratefully integrate into my subsequent analysis.[94]

One of the books within this US wave received a critical reception organized by the journal *Technology and History*. The comments on Cotten Seiler's *Republic of Drivers* by two Americans and one European is the first systematic effort to evaluate the usefulness of different arguments to explain the emergence and especially the persistence of the automobile.[95] Seiler weighs the "utility thesis" against "automobility's thrilling nature" and settles on the fundamental consumptive character of automobilism, calling the car "a depoliticized analogue of republican selfhood." This came about because the car became "the culture's most powerful signifier of identity and status." Consumption here connotes hedonism and utilitarianism, pleasure and seriousness, and often pleasure hidden as seriousness. Although his explanation is still caught in the national canon, he is thus able to explain the persistence, but not the emergence of automobilism, or, differently put, the role of the car once purchased. This implies that people are tricked into buying the car. They purchased it as an identity and status enhancer, while in the final analysis it strengthened (and perhaps even rescued) capitalism.[96]

In the *Technology and History* debate, Catherine Gudis, who herself took an active part in the recent American car-book boom with her study of car consumption, advertising, and the landscape, stresses the importance of agency (which she finds lacking in Seiler's analysis), while Jeremy Packer, also a participant in the boom with an analysis of the medial character of the car, emphasizes the missing technology in Seiler's book. "Did automobility arise," he asks, "'in the last instance' because of changes in the nature and structure of capitalism [as Seiler argues] or did the automobility apparatus come into being as the 'multifaceted, coordinated network of power' that 'bridges the ground between the more textual "discourse" and the materiality of practice.'" Whatever the answer to this truly enigmatic question, Seiler's seminal analysis has certainly put the political role of automobilism (including its relationship with democracy) on the scholarly agenda, bringing at the same time a welcome theoretical reflection inspired by Bourdieu and Foucault in the debate. These developments differentiate the new scholarship from earlier analyses of politics and policies of automobilism. Explaining the car by emphasizing its "meliorative" role in the "crisis of legitimacy in turn-of-the-century capitalism brought about by a Taylorist transformation of production" and emphasizing the consolatory function of automobilism "as a means to governing populations" is certainly an attractive proposition as it combines the systemic and the addiction approaches of earlier historiography.[97]

From here, it is only a small step to an approach that further widens the perspective from the car to mobility, and even "mobilities" as forms of movement and non-movement in a society that increasingly is characterized by fluidity and the destabilization of "sessility."[98] Historians start to respond to this scholarship too, for instance Mathieu Flonneau who, as an outsider in the *Technology and History* forum, criticizes the hidden exceptionalism in this 'American' debate by pointing to the different role of individualism on both sides of the Atlantic: "In Europe, a driver's individualism has been seen as a threat to the community, whereas in the United States, it constituted a building block of the community, and is thus viewed favorably."[99]

Flonneau's scholarship stands in the tradition of McShane's urban history, but where McShane uses New York as a *pars pro toto* for the United States (at least the East Coast), Flonneau's analysis of Paris (containing about a fifth of French car registrations in 1913) does not make a connection to the national level.[100] As the birthplace of automotive culture, his analysis of Paris thus becomes as iconic as the city itself, as a representation (if not a celebration) of what he calls "car civilization," which he seems to defend against "car adversaries" he does not clearly identify. Flonneau's contribution to the explanatory canon is clear-cut: the entire (urban) society wanted the car as an icon of (American-style) modernism and there was hardly any resistance. It was the car's speed and its control that formed the basis for the "passions" built around the car and its "triumph." Thus, his scholarship does not fit Per Lundin's scheme (although it *does* adhere to Lundin's appeal for a "moral history"), as it is about the system, which he defends as an expression of democracy and republicanism. This is especially clear in his more pamphlettist publications, such as *L'Autorefoulement et ses limites* (2010; The limits of Auto(mobile)-Repression) dealing with the "self-exorcism" of those suffering from "autophobia," provocatively embracing both the Parisian collective bicycle program and the "pleasure of regularly leaving Paris ... in a car!"[101] *Les cultures du Volant* (2008; The Cultures of the Steering Wheel) is meant to be the French equivalent of Wolfgang Schivelbusch's analysis of railway culture and refers to many of the same French literary sources analyzed in the present book as well. Flonneau's books are written to demonstrate the "historical thickness" of the "automotive desire." As a historian, he wants to demonstrate a thesis, which he is eager to defend on national radio and television, and in French-style polemics with those who tend to attack the car as a basic modernist project. He also experiments with new typographical forms to enliven his public role, like structuring the content overview such that the consecutive chapter titles form the word *automobile*.[102]

Seiler's reply to his opponents in the *Technology and History* forum is remarkable, and may have been inspired by Flonneau's defense of the liberal project of the modernist car. Seiler declares to have a problematic relationship with what he calls "the more straightforward manner of historians of technology (e.g. James Flink, John Rae, and Clay McShane) on whom we nonetheless rely," and he refers to Gudis's "mercenary attitude" which he supports: "beyond what automobility can tell us about national (and increasingly global) values, priorities, structures, institutions, and configurations of power, we really do not give a damn about it." Even in the full-blown 'car-society' United States it is apparently still not *bon-ton* to study automobilism without at least a disclaimer that one really does not love the beast.[103] It seems that the historiography of the car, at last, enters a phase in which an open and fierce debate is possible about the very roots of both "car passions" and "car society."

As I try, after this very partial overview of half a century of car-related scholarship, to summarize its main explanatory thrust, it would not be too far-fetched to speak of a 'master narrative' that pervades most of the literature. Most *national* histories of the car describe how a wealthy elite developed a taste for travel, by train or carriage at first, and then switched to the car as soon as this became available around the turn of the twentieth century, after they had practiced the new 'sport' of racing and touring on the bicycle from the 1880s. When word of the 'movement' reached the United States, according to this historiography a process of 'democratization' started, based on a relatively cheap, reliable, mass-produced vehicle that precipitated and facilitated changes in urban structure (suburbanization) as well as rural landscape (road building). Once firmly established as a societal 'necessity,' that 'car society' began its triumphant diffusion over the industrialized and industrializing world, to Canada, Australia, and New Zealand, and back to Europe. Lagging behind by a quarter century, Europe then caught up after the Second World War and joined the global mass motorization that characterized the second half of the twentieth century.

The 'master narrative' fails to explain the undeniable persistence of the motorization of individual mobility despite a spate of seemingly rational arguments in favor of its abolishment or at least its limitation. These arguments were not unique to the post-1970s period when environmentalism and concerns about global energy procurement became vocal, but can be traced back to the very beginnings of automobility. However, it is also possible to read the 'master narrative' as an abstraction of a multitude of national subplots, illustrating a variety of 'multiple modernities' within an overall process of Western modernization.[104] The connection between these modernities, their interdependence, and mutual shaping are virtually absent from the historiography of mobility, thus strengthening the main characteristics of the master narrative.

In this study, I will not emphasize the multiplicity of modernity. I will reserve a critique of the 'car as herald of modernity' paradigm for a later study. Instead, I will construct a typical counter-narrative that, however, allows for important differences between local cultures. Such a counter-narrative should overcome two shortcomings of the current master narrative. The first is the assumption, based on the linearity principle of the master narrative, that Europe 'lagged behind' the American 'model' by a quarter century, a diffusionist assumption which fails to explain why all major events, conflicts, and controversies happened at more or less the same time on both sides of the ocean.[105] In addition to this 'America-as-model' myth, as we have called it, there is the 'toy-to-tool' myth: the assumption that automobilism owes its success mainly to the substitution of a 'luxury' vehicle by a 'utilitarian' vehicle, which resulted from a functional shift from 'pleasure' to 'necessity.'

How, then, do we intend to explain 'the car' through the mist of these two myths?

Developing an Explanatory Toolbox

Any explanation of the emergence and especially the persistence of automobilism should take into account those weaknesses in the master narrative. In the following chapters I intend to overcome these weaknesses first and foremost on the basis of a rereading of the existing, 'national' utterances of mobility-history scholarship alone, from a transnational and cultural perspective. In chapters 1 and 4 I will offer a synthesis of the state of the art, enriched with new research (mostly including trade journals and contemporary publications such as reports and monographs). These largely descriptive chapters will then be followed by analyses of the motivations of the historical actors in chapters 2, 3, 5, and 6, based on new sources (literature, songs, film).

If explaining is not (only) about causation, then turning to belletristic literature as a historical source (and other artistic utterances, such as film and songs) seems to offer a promising extension to our methodological toolbox. Some scholars have already shown that such an approach can be fruitful.[106] "A history, unlike a novel in the realist genre," literary theorist Perez Zagorin asserts, "consists not only of specimens of narrative but of arguments, footnotes of documentation and justification, acknowledgements of what is known, discussions and evaluations of sources and evidence, and critiques of the view of other scholars."[107] Literature tells stories by claiming and exposing the typicality of a specific event or person. According to German author Robert Musil, himself an active theoretician of literature, literature does not show the general, "but single cases in the complicated sounds of which the generalities resonate."[108] From this perspective, literature's contribution to the narrative explanation of automotive culture promises to effectively support the 'typical' transnational history of the car culture this book intends to pursue.

Within literary theory since the 1960s, literary utterances (especially belletristic literature) have been 'dethroned' as entrances into a humanist universality by putting them in a context of other forms of communications.[109] Since then, text-immanent criticism has been replaced, in reaction to nineteenth-century positivism and historicism, by sociological approaches to literary production (e.g., Lucien Goldmann) and reception (e.g., Hans Robert Jauss, Umberto Eco). At the same time, in adjacent disciplines such as cultural studies and media studies, the deconstruction of these utterances resulted in literature's new status of being just a marginal part of a culture of mass consumption. This development, however prolific it was and is in opening up exciting new vistas of analysis, also led to a loss of the specific insights that belletristic literature (and related artistic utterances in music and art) could provide. This claim is based on a methodological shift from a mentality-based analytical stance toward a social-practice oriented approach,[110] in which literary utterances are considered to provide an entrance into the motivational backgrounds of the attitudes and ideologies of the specific high-middle-class group that started to use automobiles as new tools of perception at the same time as they developed new literary conventions. In other words, the claim of this book is that the production of early automobility and the revolutionary shifts in literary conventions not only took place in parallel, they also were mutually constitutive, not only as 'texts,' but also, and especially, as 'social practices.' I do not believe, as Hayden White argues, that speaking about history is only possible in the ways literature and art speak about reality; some objects (among them experiences) do exist beyond texts.[111] Nor do I believe that the development of the car as a commodity, which is one of the crucial functional complexes the car acquired and became adorned with during the first forty years of its existence, can be reduced to a "consumption of places" as the "spatial turn" in the social sciences would have it. In the same vein, the 'car as perception device' cannot be reduced to a "tourist gaze" (derived from Foucault's "medical gaze"), although the acknowledgement that in certain circumstances the plurisensorial perception is reduced to a visual gaze can be used with great benefit, including in this study.[112] Instead, the inclusion of the artifact in a culture characterized by experiences and practices, prompted me to call the car an "adventure machine."[113] The present study can be read as an effort to further qualify and detail that "automotive adventure": to add new aspects to its tripartite structure of an adventure in time, in space and in functionality (as will be explained in the first chapter), to investigate its development over time and, most of all, to acknowledge the existence of "multiple adventures," including a female one next to a masculine one. This study will show how important it is to 'historicize' concepts like "consumption of spaces" or "tourist gaze." This is all the more important as the spatial adventure started to dominate only during the Interbellum, while the reduction of the multisensorial perception to a "gaze" really was enabled by the closing of the car body in the second half of the 1920s.

Apart from my theoretical justification for the use of literary and other artistic sources, there is also a pragmatic argument. Whatever one's position in the 'understanding' versus 'explanation' controversy, fiction has been (and still is) used as historical source by historians. Dickens's *A Tale of Two Cities* (1859), for instance, did not so much function as a source to better understand the tale's topic (the Revolution) but rather as an insight into "mid-nineteenth-century British attitudes towards it." Fitzgerald's *The Great Gatsby* (1925) has been used as "a historical cameo not of American thinking but of American feeling." But the theorist who put forward these examples at the same time hints at a possible "circularity of argument (for which is the chicken here, and which the egg? Do we not get our impressions of a period's 'atmosphere' and 'feeling' from precisely those artifacts—books, paintings, music, and so on—which we then retrospectively claim as supporting evidence?)." Should we therefore restrain from using literature as historical source? I think not at all. "The essential distinction between history and fiction, as long maintained, has been rendered untenable," Beverley Southgate concludes. "It is now clear that all history is fictional, in the sense that it is a literary (rhetorical, aesthetic) construction based on evidence that is itself of *inevitably* questionable reliability. If, as a literary discourse, it bears any relationship to the 'facts' of what actually happened, we can never know what that relationship is; for we can never *know*—have certain knowledge of—the past by which to test the validity of any descriptions of, or assertions about, it."[114]

Even if narrative structure is not to be found in historical 'reality' ('the past'), it can put forward explanatory claims in which literature (and other artistic 'texts') plays a role. According to Jon Adams, "narrative is not a fictional technique per se. It is an organizing device: the ordering of events into a structure that readers can 'follow' by presenting a series of events and imputing a motive (in the case of agency) or mechanism by which two or more chronologically sequential events stand in causal relation (causal ligatures need not be explicit; narrative is the sequential ordering of events, and thus causally may remain implicit). Thus, narrative links events causally in a chronological structure."[115] For Michel de Certeau, "fiction is the repressed other of historical discourse." No wonder, then, that scholars of development studies recently published a plea to include literary sources in their scholarship.[116]

There are at least three (and perhaps four) arguments why literary and other artistic or sub-artistic sources can and should be used in historical research, especially on the automobile. The first argument is that a content analysis of such utterances can provide two types of information: "situational details" necessary to make the fictional world imaginable, and information on "attitudes, emotions, ways of thinking and social relations."[117] Recent theorizing in the realm of media and art studies reveals that a third type of information may be even more important: 'affinities' between literary and societal structures (Lucien Goldmann's 'structural homologies') such as those found by Wolfgang Schivelbusch and Marc Desportes in their erudite analysis of mobility in history.[118] The constant change in point of view in Valéry Larbaud's novel *Fermina Márquez* (1910), for instance, can be related to the emergence of the cinema and of the new mobility modes of the bicycle, the motorcycle, and the car.[119] This kind of information often cannot readily be taken from the novel's content, and the novel itself does not have to deal at all with cars or mobility in general, but the 'grammars' of both the novel and the motorists' experience resemble each other, no doubt because they have been constructed at the same time, within the same culture, and often by similar actors. One further claim of this book is that this last level of analysis, which promises to generate new insights, cannot be (or at least so far has not been) found through the study of regular, mainstream historical sources. I will therefore dedicate much space to the 'affinity' between writing and driving (see chapter 2). Although "fiction is not a substitute for systematically accumulated, certified knowledge," an introduction in *Sociology through Literature* remarks, "it provides the social scientist with a wealth of sociologically relevant material.... The creative imagination of the literary artist has achieved insights into social processes which have remained unexplored in social science."[120] The same applies to history. These insights bounce back to the regular sources, however. Once literary texts have sharpened our mind we tend to find similar experiences in sub-literary and non-literary texts. In short, the first argument in favor of literature as historical source is that it provides information (in both content and symbol) that is not easily found in more traditional sources, such as archival documents and trade journals. As far as mobility historians have used literary sources, they emphasized the first and second type of information mentioned earlier, giving literature an embellishing role of the historical narrative. Conversely, literary or cultural historians who have ventured into the merging of literary and mobility history often lack the knowledge or skill to engage in a dialogue with the existing body of mobility-historical scholarship, resulting in surprising generalizations, such as that the literature on the car has been one large 'Quest of the Self.'[121]

A second argument in favor of combining literary and mobility history is that this type of sources is welcome for (re)constructing historical situations for which, for a variety of reasons, there are few other available written sources. Attention to literary texts as additional historical sources, for example, has spread among historians of the Middle Ages because of the scarcity of regular sources. Despite the problematic referentiality of imaginative texts (especially the ambiguous use and lack of historical specificity of metaphors) premodern poems provide a tool for an archeology of experiences in order to "reenact the past," seen by some historians as the key to historical understanding.[122] The scarcity of documents on the motivations of early car use seems to invite a similar archaeology of experiences. Peter Gay, in his seminal multivolume Freudian analysis of the bourgeois experience in the nineteenth century, benefits (as I do in this book) from the fact that the nineteenth-century novel was written by and for, as well as read by, the same class he sets himself to analyze. But whereas Gay had the luxury of abundance for his theme of 'love,' I have to search for utterances that deal with car use beyond the literary versions of travel accounts. Nonetheless, the situation appears to be much less gloomy than earlier literary or cultural historians dealing with the car culture thought. There is an abundance of highbrow literature that dealt with mobility themes, as we will see in chapters 2, 3, and 6.

Gay defends his preference for "masterpieces" against "an article of faith among sociologists of literature," that non-literary texts "are closer to the naked wishes [of the bourgeoisie] than the searching, discriminating, imaginative exploration that the serious writer tends to undertake." Opting instead for "first-rate fictions," he declares, "can be immensely rewarding (even to the historian) since they dig deep, see a great deal, and do not evade complexities."[123] A similar strategy is deployed by Leo Marx, who reconstructs the pastoral disposition of eighteenth- and nineteenth-century American society largely from literary sources.[124] Gay does not solve the problem of how he knows that the information in the novels 'mirrors' elements of bourgeois reality (Marx does not have to solve this as he is interested in the opposite problem, the 'impact' of society on literary imagery, although the implication, of course, is the same, that this imagery influences the actual shaping of society itself). A more sophisticated justification for the use of literature as historical source is given by the German literary historian Michael Titzmann, who offers the example of the literary discourse on incest in Goethe's times, where literature (identified by Titzmann as those texts that have been seen as such by contemporaries, plus those texts that have the same structure) acquired the function of "prepar[ing] new types of knowledge" that could not have been achieved in other types of discourse (such as the law).[125]

Much has been written about artists as practitioners of "higher forms of perception." A "sharp gift of observation" would enable them "despite their subjective point of view, to reach representations of reality which are often nearer to historical reality than the results of the social sciences. The reason for this may well be the graphical and enlivened character of their literary texts. They also put people at the center of their considerations."[126] According to Gilles Deleuze and Félix Guattari, artists "think with affects and percepts rather than concepts. And novelists try to think what philosophers and scientists exclude from their speculations and inquiries as the unthinkable: the particular, the singular, the non-identical, the contingent, that which cannot be fitted into the systems of society and the patterns of orderly conceptual thinking."[127] Thus, literature functions as an intermediary between structural and experiential history.[128]

"The power of novels is the result of an illusion of omniscience," a more recent "sensual" assessment of the working of literature concludes. "Novels do things that life cannot: they relate the actions of others directly to their thoughts and consciousness—in other words, to ours, since a human being has no consciousness but her or his own." Quoting Roland Barthes, the representative of "structuralist narratology," this assessment compares "narration" with "a huge traffic-control center" as narrative practice is not descriptive but predictive.[129] As historical discourse cannot reach the "real," according to Barthes, "it cultivates 'the reality effect.'" It does so, among others, through "plot," a specific temporal and often causal ordering of events and experiences, resulting in an "explanation by emplotment." In the case of our history of automobilism the emplotment often takes the form of romance (first phase: emergence) turned into tragedy (second phase: persistence), but elements of comedy and satire are always present.[130] And as imagination and expectations are drivers of human experience, the "affinities" between writing and driving experiences are striking indeed, and can be mobilized to gain insight on the motives and emotions involved in both.[131]

A third reason why belletristic literature is a promising source to investigate car driving experience is that the 'aesthetic experience' forms a crucial part of the consumption history of the car, and it is belletristic literature that seems to be able, as no other type of 'text,' to express this experience. At first sight, this looks like the kind of circular argument (the chicken-and-egg problem) identified previously, until one realizes that being aesthetic, in this view, is defined as "pleasurable to the senses." In other words, the reason why literary sources can function as historical sources is the fundamental 'intermediality' of the production of texts and films and the production of driving experiences, an intermediality based on the body and its senses. There are many more senses than the traditional five (or six, if one includes the "sixth sense," the mind). Remarkably, a surprisingly large number of them are related to movement, such as equilibrioception (balance), proprioception (limb movement), and kinaesthesia (acceleration) and perhaps even magnetoreception (direction).[132] In the following chapters I will often observe the intimate merging of body and machine, giving the car the role of both a "prosthesis" and a "channel" of communication.[133] From this perspective, based on the media theory of Marshall McLuhan, media are "externalized apparatus(es) of sensorial organs, whose evolution cannot remain without consequences for perception and appropriation of the world." Based on the theory of Friedrich Kittler, literary history can be approached as media history (or at least medialized history).[134]

A fourth argument in favor of using artistic utterances as privileged sources might be the "ontological" (in philosopher Peter Sloterdijk's words) "purely being-in-movement" as characteristic of modernity, a characteristic also pervading modern literature with "movement in multiple forms as a sense-giving and structure-providing property."[135] However, from Stephen Kern to Hartmut Rosa, the speed and "acceleration" of modernity is so ubiquitous that they threaten to lose their analytical power, so much so that Rosa, in his seminal study of modernity, can simply refrain from empirically grounding the thesis of speed increase.[136] And although there is no doubt that social scientists have for all too long neglected the increasing 'fluidity' of modern societies and the recent recognition, among geographers and sociologists alike, of the crucial importance of 'mobilities' can only be welcomed,[137] it seems that 'speed' should be treated as an *explanandum* rather than an *explanans*, especially in a historical treatment of utterances in which 'speed' is so much tested, rejected, enjoyed—and modified. Speed, in our analysis, will be reexamined and reconnected to bodily experience related to being in the car.

Aesthetic experience—indeed, experience in general—is a corporeal practice encompassing "the dialectic relationship between actor and society" and includes layers, as Reinhart Koselleck posits, "that are beyond linguistic scrutiny."[138] This relationship is multisensorial and can be approached in a way that allows the classical dichotomy between mind and body to be overcome.

Constitutive of the relationship between mind and body is the role of the artifact, or more precisely, the interaction between user (and nonuser and, to a certain extent, the producer) and the artifact. For the last quarter century, several traditions of research have studied the characteristics of artifact use. The artifact mediates between the "activities" or "practices" of the human and some non-human actors (such as horses), or between these actors and their environment. Apart from the fact that this basic circumstance potentially turns the history of automobilism into a truly 'environmental history' (with 'nature' just as mobile as the car), experience characterizes a dynamic process of appropriation in which language plays a constitutive role, because the "seeming authenticity of emotions and experiences (*Erlebnisse*) and events" comes to us mediated through the textual structure.[139] Especially those theories that Nigel Thrift gathers under the term "non-representational" (Vygotsky, Leonti'ev, Luria, Gibson, Bakhtin) seem to promise interesting results, including the anthropological theory of mobility developed by Tim Ingold on the basis of James Gibson's work; Thrift's "modest theory" (as he calls is, as opposed to "a fully finished theoretical programme") is concerned with "a rather different notion of 'explanation,' which is probably best likened to understanding a person, a phenomenalism of character that involves, more than other approaches, empathies and ethical components: "One reads the story of the life of a person. One follows the story,

Explaining the Car 35

one travels for awhile together with that individual and eventually gains understanding of him or her. When understanding has been achieved one discovers that one can tell a story."[140] A "history of experiences" of automobilism would allow the systematic inclusion of technology into a socio-cultural history, such as recently has been attempted by Christian Kehrt for airplane pilots in Germany during the First World War and the subsequent Interbellum phase. Routines in dealing with the vehicle (for users and non-users alike) are especially ingrained in bodily behavior when technology resists the flow of gestures, and "a discrepancy between expectation and experience" occurs.[141] There are many parallels with related approaches such as design theory, the performative variants of cultural studies (represented by Pierre Bourdieu and James Clifford) and the anthropology of artifact "appropriation."[142] Design theorists distinguish between three "levels" of "product experience": emotional, aesthetic, and meaning, and some of them even speak of an "experience of love" when analyzing user-product interaction.[143] Literary history and theory itself also increasingly deal with the "cultural appropriation" of things, including cars.[144]

Experiences can be analyzed as "social practices." The act of 'experiencing' can be defined as "a permanent process of appropriation" in which perception, interpretation and acting are mutually coordinated."[145] According to Theodore Schatzki's "action theory," the unit of analysis is not the "mind," nor any "objective structure," but practice as "a nexus of doings and sayings."[146] In a later publication, Schatzki distinguished between practices as "open, spatial-temporal manifolds of activity that are carried on by multiple individuals," and "material arrangements . . . of humans, artifacts, organisms, and things," and it is history of "the nexuses of practices and arrangements" that constitute the history of mobility and automobilism.[147] From this perspective, early automobile culture can be approached as a "collective activity system," as a "community of actors who have a common object of activity," a system that "connects the psychological, cultural, and institutional perspectives to analysis."[148] Practices are mostly mobile practices.[149]

To analyze the activity system of the early car user culture (including the non-users), we can now formulate the schematic shown in figure 0.1.

Intermediary actors

Producer → Practices → Properties — Functions ← Practices ← User

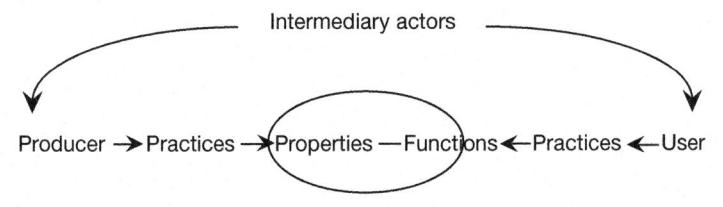

Figure 0.1. The dual nature of technology and its related practices.

According to this schematic, the artifact (the oval shape) is characterized by (technical) 'properties' and (relational) 'functions.' The latter term relates to the artifact's potentiality ("affordance," in the words of the psychologist Gibson) to enable a socially meaningful use.[150] In other words, properties enable ('afford') functions, but they do not prescribe them. Nor do functions prescribe properties: the same (or nearly the same) functions can be made possible by different sets of properties. Conversely, not all functions are enabled by properties. The relation between properties and functions is 'fuzzy,' and is not to be found within the artifact. The relationship has to be construed, by designers and by users, both in the act of designing the artifact and in the act of its use.[151] Functions are made visible (and analyzable) by users' activities. A possible matrix of such activities (derived from Schatzki) is given in figure 0.2.

	direct	indirect
dispersive	1 steering	2 singing
integrative	4 driving	3 touring

Figure 0.2. A taxonomy of car-related practices.

Some activities, called 'dispersive,' relate to a single and linear chain of activities, such as 'steering' or 'singing (in the car).' While the former is performed in direct relation to the handling of the car, the latter is not specific to car use; one can sing everywhere, but 'singing in the car' is an important historical activity (for instance when cruising with the family through an American landscape) that gives this singing a special cultural significance. Other practices, called 'integrative,' relate to complex activities, such as 'driving' and 'touring.' Here again, the former relates to the handling of the car itself, whereas the latter can be performed on a bicycle or in a train, although of course, touring by car is not the same as touring on a bicycle. They can be called "nested affordances."[152] An important element of this activity system is 'delegation' to non-humans, a preferential topic in Actor-Network Theory and related theories. According to Reijo Miettinen, Bruno Latour's cases (hotel key weight, traffic light, road bump, the latter two a replacement of a policeman) show only a partial delegation of the direct activities (the third and fourth quadrant in the matrix of figure 0.2); after all, the policeman is not replaced, he is still around, as he does much more than forcing motorists to drive slowly.[153]

The content of the matrix is time and place dependent, and could be used as a heuristic device for changes in automobile-user culture. These changes, however, are not totally fluid; it is to be expected that changes in user culture stabilize into 'normal use' characteristic for certain periods, the very basis of the development of 'skills' and (paradigmatic) 'attitudes.'[154]

Another type of practices relates to the production of emotions, and feelings and motivations (in the literal sense as 'drivers' of actions) in general. They are, indeed, as much bodily activities as 'gestures' are. Emotions are produced, and (according to the classic 'action theories' of Leont'ev, Vygotsky, and others[155]) their production, and the experiences of the environment in general, depend on the sensory system of the human body. Whereas the practices in the matrix above can readily be analyzed on the basis of the usual empirical material, emotions and motives are the most difficult to grasp empirically. It is here that literary texts seem exceptionally suitable for research by the historian. Emotional experiences are difficult to express in words, either because the senses are differently structured to language, or because their expression is discouraged in a 'civilized' society.[156] Literature, then, is a promising source because its primary practices are, firstly, to connect the different senses, mainly through the use of metaphors, and, secondly, to express this in a culturally acceptable (even if controversial) form. It does so by making particular experiences 'universal,' a highly ideological term in classic literary theory and meanwhile interpreted as being an expression of a middle-class 'universality.' However, it is exactly the parallel between this literary, middle-class universality and the universality of the 'universal car' for the middle-class family that will form the backbone of the analysis in this book.

I call the resulting car culture 'automobilism' and not 'automobility,' as has recently been suggested by sociologists making the 'mobility turn.' Automobility has been coined as the core concept in the new sociological analysis of the car society. There, it has been defined as a not very historically founded "set of political institutions and practices" as well as "an ideological ... or discursive formation" and "a set of ways of experiencing the world." John Urry, one of the driving scholars behind this new approach, distinguished between six "components" of automobility, varying from the car as product, as consumption good, as system, to the car as quasi-private mobility, as culture, and as a means to use environmental resources.[157] Geographers have meanwhile joined this endeavor and have argued, for instance, that "automobility" and "national identity" are closely related.[158] But so are consumption and national identity, a second connection exposed in this study through the production of stereotypes around car technology and car driving.[159] Remarkably, when Urry or his collaborators in several basic documents propose a taxonomy of theories and methods, history, or any acknowledgement of a long-term development, is absent.[160] In a more recent publication, *After the Car*, Urry's six characteristics of automobility appear again in a not-so-successful effort to historicize them, at the end of a very brief (and very unconvincing) historical overview of the car, complete with a revival of General Motors' "conspiracy" to promote the bus against the streetcar, long since challenged by American historical scholarship.[161]

Thus, 'automobilism' better expresses the dynamism of the culture over time, while the '-ism' suffix also suggests that we are dealing with a 'movement,' as a 'community' that moves and at the same time is moved in the emotional sense by its relation to the car ('transport' has the same double connotation); it suggests the collectiveness of something which has been systematically reduced, in the historiography of mobility and most particular in literary histories, as a fundamentally individual (if not individualistic) practice.

The new sociology and geography of 'mobilities' provided a welcome antidote to conventional transport theories that still deal with transport as a "derived demand." Although here, too, the paradigm starts to shift toward the recognition that movement causes pleasure and trip motivation thus should deal with emotions as well, this book can be read as one long argument against this conventional transport theory.[162]

Table 0.1. Periodization of Automobile Use in the North-Atlantic World

Period	User Group	Sensorium	Technology	User Function / Practice
Until WWI	aristocracy and upper middle class	sound	open car: (clam) shell	tinkering, speeding, 'speedy' touring around town in 'nature'
Interbellum	(new) middle class: 'white collar' / angestellte	vision	closed car: capsule	family outings (distancing) and some business
Post-WWII until 1975	'mass motorization'	smell, touch	'affordable family car': cocoon/ corridor	holidays and commuting (quest for self)

Table 0.1 provides an overview of a periodization of this culture, distinguishing between the dominant users, the specific phase in car technology, the most common user function, and the senses deployed to experience this function. According to this table we have a first pe-

riod until the end of the First World War, when the upper middle class used mainly open cars to indulge in tinkering and 'speedy' touring in the peri-urban countryside. Although it is difficult to single out one sense as characteristic for this phase, as automobilism remained a multisensorial experience, making sound the dominant sense reflects in this period the difficulty car technology posed for indulging in 'nature,' despite the fact that only during the Interbellum did noise abatement become a major engineering activity among automotive engineers, as we will see in chapter 5. Several early motorists noticed how the attention paid to the sounds of the car often prevented the travelers from enjoying the landscape.

During the second phase, the Interbellum, car use was expanded toward the middle levels of the middle class (most notably the 'new middle classes' such as a large part of the 'white collars' in the United States and the higher echelons of the *angestellte* in Germany). At the same time the user experience shifted toward the nuclear family that indulged in holiday trips. A somewhat differentiated user culture emerged, because the average, white middle-class 'father' sometimes also used the car during the year for commuting, business trips, deliveries, or visits of patients. Vision as dominant sense now reflected the shift from a temporal and functional to a spatial adventure.

The third period, the phase of mass motorization after the Second World War until the breaking of the spell of the car in the 1970s' ecological and energy crises, is the most difficult to characterize in a few words. 'Affordable family cars,' also driven now by lower middle classes and workers, were used about one-third of the time for commuting and two-thirds for recreational and 'social' purposes, as we will argue in chapter 4. Besides vision and sound, smell and touch in addition now formed a part of the sensorial tools used to produce the 'car experience,' the former registering pollution, the latter as a token of the explosion of 'comfort' as a constitutive part of the car's marketing. Car use also got further differentiated into subcultures, as did their literary representation. Highbrow literature, loyal to its mission, now followed the single (if not lonely) motorists and their 'Quest of the Self' as well as their search for a national (in the United States) or Western (in Europe) identity. Literature thus became less 'representative' of the auto-mobile crowd, indicating that our method will be more fruitful for the first two phases, while the insights of media studies and film studies will be of more use to study the third period, especially its more orally grounded popular culture.[163] Literature also became less representative because of the further spread of the car beyond the borders of the North-Atlantic realm, necessitating and enabling at the same time a new, non-Western vision on postwar automobilism. This will be the subject of a later study, but will be hinted at in the concluding chapter.

Conclusions

This introduction argues that it makes sense to explain automobilism in an explicit, although not necessarily causal way. It also suggests that the chances of success of such an explanation will dramatically increase when unorthodox sources are taken into account, most particularly those fictional sources that provide insights into motives and emotions as drivers of behavior and attitudes. Including those sources within historical analyses is possible through the methodological bridge of 'social practice theory,' which enables the holistic analysis of corporeal, sensorial experiences.

These sources have been selected by taking the (never explicitly defined) 'canon' as a basis, which I constructed by collecting the ubiquitous mentions of such sources (often introduced as embellishment of the main story) in the mobility history canon, as well as the dozen or so literary approaches to mobility history in Germany, France, the United Kingdom, and the United States. In the course of my research I extended the resulting list through studying general national literary histories as well as the reviews, comments, and analyses by literary and media scholars of the books in the original 'canon,' thus identifying related sources. All in all, about 150 novels, short stories, and poems have been listed for the first period until WWI, and about 250 for the second, while in addition 50 youth novels, 50 movies, and 25 songs were identified. Chapter 2 contains an introduction into the analytical methodology I applied to a selection of these sources. Whereas for the first period I was mostly interested in the parallels between writing and driving (and their common background: modernity), for the second period I focused upon the experiences of the middle-class nuclear family.

In this book I limit the application of such an analysis to the two first periods of twentieth-century automobilism in the North-Atlantic realm, containing its emergence (until the First World War, chapters 1 and 2) and its subsequent persistence (during the Interbellum, chapters 4, 5, 6, and 7), interrupted by a war (chapter 3). The coverage of both periods is symmetrical: chapters 1 and 4 describe the development of automobilism on both sides of the Atlantic, whereas chapter 2 (and for the persistence period, chapters 5 and 6) analyzes the constantly evolving 'automotive adventure' on the basis of belletristic and other sources. Because the First World War was so important as a catalyst to harden tendencies already present during the first phase, a separate chapter has been dedicated to this inter-phase (chapter 3). Also, the emergence of a 'car system,' the first contours of a future 'car society' during the second phase (as a crucial anticipation of the third period), necessitates a separate chapter on the inclusion of the car in a 'traffic flow,' in which other 'vehicles' (bus, truck, bicycle, moped, pedestrians) participate and from which some of them will be excluded during the 1930s through the construction of a high-speed road network (chapter 7). The conclusions allow us to come back to our initial questions formulated in this introduction and to make up the balance of nearly half a century of automobilism in the North-Atlantic realm.

Notes

1. Henry James, "The Art of Fiction" (1884), quoted in Beverley Southgate, *History Meets Fiction* (London: Pearson Education, 2009), 27.
2. Michael Persson, "Techniek raakt basale babysnaar," *Technisch Weekblad* (2 February 2008), 13; personal communication Liesbeth Schlichting (who wrote the dissertation containing the data), 13 June 2011.
3. In this book I use single quotation marks to emphasize the typical character of a term or phrase not directly derived from a single source. Referenced quotations carry double quotation marks.
4. There exists some rationale in excluding Australia from my synthesis, as its history may well have followed another 'model' than the Atlantic one. See Georgine Clarsen, "Automobility 'South of the West': Toward a Global Conversation," in Gijs Mom et al. (eds.), *Mobility in History: Themes in Transport (T²M Yearbook 2011)* (Neuchâtel: Alphil, 2010), 25–41. A one-week stay in Wollongong, where I co-organized on 15 and 16 December 2010, with my friend Georgine Clarsen, a workshop on 'gendered mobility' which was instrumental in getting my ideas about 'female adventurousness' in shape, also convinced me of leaving out Australia altogether, in the reassuring hope that she would in due course publish her new book on the specificities of Australian frontier and settler car cultures. For a recent Brazilian automotive history see Guillermo Giucci, *The Cultural Life of the Automobile: Roads to Modernity*, Llilas Translations from Latin America Series (Austin: University of Texas Press / Teresa Lozano Long Institute of Latin American Studies, 2012; transl. from *A vida cultural del automóvil: Rutas de la modernidad cinética* (2007) by Anne Mayagoitia and Debra Nagao).
5. I make the same "abbreviation of convenience" from "North-Atlantic" to "Atlantic" as Daniel Rodgers does in his seminal analysis of the "emerging transnational social-political networks" of urban planning. See Daniel T. Rodgers, *Atlantic Crossings: Social Politics in a Progressive Age* (Cambridge, MA, and London: The Belknap Press of Harvard University Press, 1998), 33.
6. For the latter historians, the "Atlantic World is a relational concept. It refers not to a place . . . , but to a sum of relations across the ocean." José C. Moya, "Modernization, Modernity, and the Trans/formation of the Atlantic World in the Nineteenth Century," in Jorge Cañizares-Esguerra and Erik R. Seeman (eds.), *The Atlantic in Global History 1500–2000* (Upper Saddle River, NJ: Pearson Ed-

ucation/Prentice Hall, 2007), 179–197, here 179. Such scholarship tends to decenter the 'West' rather than put it in the center, as will be done in this book. The transport revolutions these scholars refer to are in shipping and railroads, not the car (ibid., 182).

7. Lippmann quoted in Bernard Bailyn, *Atlantic History: Concept and Contours* (Cambridge, MA, and London: Harvard University Press, 2005), 7.
8. Marco Mariano, "Introduction," in Mariano (ed.), *Defining the Atlantic Community: Culture, Intellectuals, and Policies in the Mid-Twentieth Century* (New York and London: Routledge, 2010), 1–10, here 5; on sameness, see ibid., 7 (quoting Daniel Rodgers); Bailyn, *Atlantic History*, 8–9 (*Realpolitik* and One World), 25 (civilization; italics in original); Clifford Geertz, *The Interpretation of Cultures* (1973), quoted in Mehdi Parvizi Amineh, "Er bestaat helemaal geen islamitische beschaving," *Volkskrant* (11 October 2001).
9. The first research project resulted in the publication, in Dutch, of a two-volume handbook co-authored with Ruud Filarski, former director of the Department on Knowledge Development at the Advisory Service Traffic and Transport (*Adviesdienst Verkeer en Vervoer*, AVV): Filarski and Mom, *Van transport naar mobiliteit: De transportrevolutie (1800–1900)* (Zutphen: Walburg Pers, 2008), and Mom and Filarski, *Van transport naar mobiliteit: De mobiliteitsexplosie (1895–2005)* (Zutphen: Walburg Pers, 2008). The second research project was the expansion of this handbook into a comparison with six other countries (five European countries along with the United States). The six reports, three each written by Filarski and this author, dealt with six topics considered crucial in the history of mobility: the emergence of the railways in the nineteenth century; the emergence of the urban tramway; the emergence of the automobile; the coordination crisis (the struggle between road and rail); freight transport since the 1920s; and mass motorization. The three reports on the automobile and freight, written by this author, form one of the bases of the present book.
10. Gijs Mom, *The Electric Vehicle: Technology and Expectations in the Automobile Age* (Baltimore, MD: Johns Hopkins University Press, 2004).
11. For a definition of the European *periphery*, including those countries in the South that "seem to have missed out on the first great globalization boom," which also enabled automobilism, see Kevin H. O'Rourke and Jeffrey G. Williamson, *Globalization and History: The Evolution of a Nineteenth-Century Atlantic Economy* (Cambridge, MA, and London: The MIT Press, 2000), 18, 21 (quote). The term is controversial: Kostas Gavroglu et al., "Science and Technology in the European Periphery: Some Historiographical Reflections," *History of Science* 46 (2008), 153–175.
12. Manuela Boatcă, "Semiperipheries in the World-System: Reflecting Eastern European and Latin American Experiences," *Journal of World-Systems Research* 12, no. 11 (December 2006), 321–346.
13. For an overview of the state of the art of mobility history and the history of the automobile in many countries, see: Mom et al. (eds.), *Mobility in History*, and Mom, Gordon Pirie, and Laurent Tissot (eds.), *Mobility in History: The State of the Art in the History of Transport, Traffic and Mobility* (Neuchâtel: Alphil, 2009), as well as Peter Norton et al. (eds.), *Mobility in History: Reviews and Reflections (T²M Yearbook 2012)* (Neuchâtel: Alphil, 2011). See especially the contributions on the United Kingdom, France, and Germany in these yearbooks.
14. This research was done in two other projects: a Fulbright exchange professorship grant spent at the University of Michigan-Dearborn in the fall of 2006

(upon invitation by Bruce Pietrykowski) and a grant from the Dutch national science council NWO with Karin Bijsterveld, University of Maastricht, called "Selling Sound: The Standardization of Sound in the European Car Industry and the Hidden Integration of Europe," undertaken from 2007 to 2011 together with Stephan Krebs and Eefje Cleophas.

15. I cannot imagine, however, that I had encountered elsewhere such a extremely stimulating environment as at the Rachel Carson Center at Ludwig-Maximilians-Universität in Munich, led by Christoph Mauch and Helmuth Trischler during my one-year stay from October 2009.

16. On the ways to arrive at a car 'type' on the basis of a car population, seen from an evolutionary point of view, see Gijs Mom, "Constructing the State of the Art: Innovation and the Evolution of Automotive Technology (1898–1940)," in Rolf-Jürgen Gleitsmann and Jürgen E. Wittmann (eds.), *Innovationskulturen um das Automobil: Von gestern bis morgen; Stuttgarter Tage zur Automobil- und Unternehmensgeschichte 2011* (Stuttgart: Mercedes-Benz Classic Archive, 2012), 51–75.

17. Gijs Mom, "'Historians Bleed Too Much': Recent Trends in the State of the Art in Mobility History," in Norton et al. (eds.), *Mobility in History*, 15–30. Also see Colin Divall, "Mobilizing the History of Technology," *Technology and Culture* 51, no. 4 (October 2010), 938–960.

18. Gijs Mom (together with Georgine Clarsen et al.), "Editorial," *Transfers* 1, no. 1 (Spring 2011), 1–13.

19. Stephen Kern, *A Cultural History of Causality: Science, Murder Novels, and Systems of Thought* (Princeton, NJ, and Oxford: Princeton University Press, 2004), 1.

20. Richard G. Mitchell, Jr., *Mountain Experience: The Psychology and Sociology of Adventure* (Chicago and London: The University of Chicago Press, 1983), 137–138.

21. David Carr, "Narrative Explanation and its Malcontents," *History and Theory* 47 (February 2008), 19–30, here 19.

22. Carr, "Narrative Explanation and its Malcontents," 20 (obvious), 21 (discourse), 22 (continuum, causality); self-explanatory: A. A. van den Braembussche, "Historical Explanation and Comparative Method: Towards a Theory of the History of Society," *History and Theory* 28, no. 1 (February 1989), 1–24, here 3. The quoted texts in this book may deviate slightly from the original in the use of capital instead of lowercase letters (or vice versa). These deviations have not been marked to avoid the repeated use of square brackets.

23. Van den Braembussche, "Historical Explanation and Comparative Method," 8; Craig Calhoun, "Explanation in Historical Sociology: Narrative, General Theory, and Historically Specific Theory," *American Journal of Sociology* 3 (November 1998), 846–871, here 852 (*Methodenstreit*), 863 (no agreement).

24. Jane Elliott, *Using Narrative in Social Research: Qualitative and Quantitative Approaches* (Thousand Oaks, CA: Sage, 2009; reprint of 2005 edition), 6; L. P. Hinchman and S. K. Hinchman (eds.), *Memory, Identity, Community: The Idea of Narrative in the Human Sciences* (New York, 1997), quoted on 3.

25. Southgate, *History Meets Fiction*, 14 (White; italics in original), 15 (critique on Carr).

26. George Steinmetz, "Reflections on the Role of Social Narratives in Working-Class Formation: Narrative Theory in the Social Sciences," *Social Science History* 16, no. 3 (Fall 1992), 489–516, here 495 (Kramer), 496 (James).

27. Bruno Latour, "Technology is Society Made Durable" (1991), quoted in Mika Pantzar, "Do Commodities Reproduce Themselves Through Human Beings? Toward an Ecology of Goods," *World Futures* 38 (1993), 201–224, here 221.
28. Paul A. Roth, "Hearts of darkness: 'perpetrator history' and why there is no why," *History of the Human Sciences* 17, no. 2/3 (2004), 211–251. Also see his "How Narratives Explain," *Social Research* 56, no. 2 (Summer 1989), 449–478.
29. Calhoun, "Explanation in Historical Sociology," 855.
30. James Rood Doolittle, *The Romance of the Automobile Industry: Being the Story of its Development—its Contribution to Health and Prosperity—its Influence on Eugenics—its Effect on Personal Efficiency—and its Service and Mission to Humanity as the Latest and Greatest Phase of Transportation* (New York, 1916); Ralph C. Epstein, *The Automobile Industry: Its Economic and Commercial Development* (Chicago and New York, 1928).
31. Which means that they are not cited anymore. David N. Lucsko, "John Bell Rae and the Automobile: 1959, 1965, 1971, 1984 (Classics Revisited)," *Technology and Culture* 50, no. 4 (October 2009), 894–914, here 912.
32. John B. Rae, *American Automobile Manufacturers: The First Forty Years* (Philadelphia and New York, 1959). For a critique of the "impact-imprint model" (coined by Claude Fischer), see Ronald Kline, *Consumers in the Country: Technology and Social Change in Rural America* (Baltimore, MD, and London: Johns Hopkins University Press, 2000) 127.
33. James M. Laux, *In First Gear: The French Automobile Industry to 1914* (Liverpool, 1976); James M. Laux, *The European Automobile Industry* (New York, Toronto, Oxford, Singapore, and Sydney, 1992).
34. James J. Flink, *America Adopts the Automobile, 1895–1910* (Cambridge, MA, and London: MIT Press, 1970). Iconoclast: Donald Finley Davis, *Conspicuous Production: Automobiles and Elites in Detroit, 1899–1933* (Philadelphia: Temple University Press, 1988) 17.
35. Lucsko, "John Bell Rae and the Automobile," 907. Lucsko also cites Clay McShane (ibid., 912), who accused both of "technological determinism." Flink himself states that Rae and he "about 90 percent agreed." Greg Thompson, "'My Sewer'; James J. Flink on His Career Interpreting the Role of the Automobile in Twentieth-century Culture," in Peter Norton et al. (eds.), *Mobility in History: The Yearbook of the International Association for the History of Transport, Traffic and Mobility*, vol. 4 (New York and Oxford: Berghahn Journals, 2013), 3–17, here 16.
36. James J. Flink, "Three Stages of American Automobile Consciousness," *American Quarterly* 24 (October 1972), 451–473; James J. Flink, "The Car Culture Revisited: Some Comments on the Recent Historiography of Automotive History," in David L. Lewis and Laurence Goldstein (eds.), *The Automobile and American Culture* (Ann Arbor: The University of Michigan Press, 1983), 89–104. For a critique of Rae's industry—and automobile-friendly account—see David J. St. Clair, *The Motorization of American Cities* (New York, Westport, CT, and London: Praeger, 1986), 84; for a critique of Rae's and Flink's approach of the car as a "democratic technology," see Davis, *Conspicuous Production*, 17.
37. James J. Flink, *The Car Culture* (Cambridge, MA, and London: MIT Press, 1975).
38. In his Introduction, Flink also announces his next book, dealing with the 1920s. Perhaps because he saw this book as the first in a series of two, he ends his first in a remarkably abrupt way, without proper conclusions. See also Greg Thompson, "'My Sewer.'"

39. Norman T. Moline, *Mobility and the Small Town, 1900–1930: Transportation Change in Oregon, Illinois* (Chicago: The University of Chicago, 1971); Robert E. Ireland, *Entering the Auto Age: The Early Automobile in North Carolina, 1900–1930* (Raleigh: Division of Archives and History, North Carolina Department of Cultural Resources, 1990).
40. Rudi Volti, *Cars and Culture: The Life Story of a Technology* (Westport, CT, and London: Greenwood Press, 2004).
41. Flink, *America Adopts the Automobile*, 34 (masses and prejudices), 68 (opposition).
42. Ibid., 87–112.
43. Ibid., 91 (American), 100 (hedonistic).
44. See, for instance, Margaret C. Jacob, "Science Studies after Social Construction: The Turn toward the Comparative and the Global," in Victoria E. Bonnell and Lynn Hunt (eds.), *Beyond the Cultural Turn: New Directions in the Study of Society and Culture* (Berkeley, Los Angeles, and London: University of California Press, 1999), 95–120, and Trevor Pinch, "The Social Construction of Technology: A Review," in Robert Fox (ed.), *Technological Change: Methods and Themes in the History of Technology* (Amsterdam: Harwood Academic Publishers, 1996); for a nice 'car study' from the history of technology perspective, see Ronald Kline and Trevor Pinch, "Users as Agents of Technological Change: The Social Construction of the Automobile in the Rural United States," *Technology and Culture* 37, no. 4 (October 1996), 763–795.
45. For a treatment of the car as an 'agent' in urban historiography, see, typically, Mark S. Foster, *From Streetcar to Superhighway: American City Planners and Urban Transportation, 1900–1940* (Philadelphia: Temple University Press, 1981), later corrected by the same author in "The Role of the Automobile in Shaping a Unique City: Another Look," in Martin Wachs and Margaret Crawford (eds.), *The Car and the City; The Automobile, The Built Environment, and Daily Urban Life* (Ann Arbor: The University of Michigan Press, 1992), 186–193. Peter David Norton, *Fighting Traffic: The Dawn of the Motor Age in the American City* (Cambridge, MA, and London: The MIT Press, 2008). For a summation of Norton's main and highly original argument, which has so far not received the attention it deserves, see his "Street Rivals: Jaywalking and the Invention of the Motor Age Street," *Technology and Culture* 48, no. 2 (April 2007), 331–359. Partisanship is one of the evils of the historiography of the automobile, more so than in other domains of history. See for an example Bruce Epperson's review of Norton's book in *Technology and Culture* 50, no. 1 (January 2009), 235–237. For Norton's answer, see ibid., 50, no. 4 (October 2009), 982–984, and Epperson's response on 984–986. For a critique of partisanship in the history of mobility, see Gijs Mom, Colin Divall, and Peter Lyth, "Towards a Paradigm shift? A Decade of Transport and Mobility History," in Mom, Pirie, and Tissot (eds.), *Mobility in History*, 13–40.
46. Rae saw the electric vehicle as a dead-end technology: John B. Rae, "The Electric Vehicle Company: A Monopoly that Missed," *Business History Review* 29, no. 4 (December 1955), 298–311; this was corrected by David A. Kirsch and Gijs Mom, "From Service to Product Based Mobility Concepts: Technical Choice and the History of the Electric Vehicle Company," *Business History Review* 76 (Spring 2002), 75–110.
47. This later led to a separate publication on the demise of the horse as an urban 'tool.' See Clay McShane and Joel A. Tarr, *The Horse in the City: Living Machines*

in the Nineteenth Century (Baltimore, MD: The Johns Hopkins University Press, 2007).
48. Clay McShane, *Down the Asphalt Path: The Automobile and the American City* (New York: Columbia University Press, 1994), 98 (suppression), 113 (interlocking set), 148 (ownership), 228 (nightmare). For a comparable shift from the 'why' to the 'why America' question, see Christopher Wells, "Car Country: Automobiles, Roads, and the Shaping of the Modern American Landscape, 1890-1929" (dissertation, University of Wisconsin-Madison, 2004), 2.
49. Gerhard Horras, *Die Entwicklung des deutschen Automobilmarktes bis 1914* (Munich, 1982); Angela Zatsch, *Staatsmacht und Motorisierung am Morgen des Automobilzeitalters* (Konstanz, 1993); Heidrun Edelmann, *Vom Luxusgut zum Gebrauchsgegenstand: Die Geschichte der Verbreitung von Personenkraftwagen in Deutschland* (Frankfurt: Verband der Automobilindustrie, 1989).
50. Edelmann, *Vom Luxusgut zum Gebrauchsgegenstand*, 231.
51. Reiner Flik, *Von Ford lernen? Automobilbau und Motorisierung in Deutschland bis 1933* (Cologne, Weimar, and Vienna: Böhlau Verlag, 2001).
52. Edelmann, *Vom Luxusgut zum Gebrauchsgegenstand*, 12-13 (four phases), 17 (purchase power), 234 (inflation); last quote, 20.
53. For a critique of this approach, see Gijs Mom, "Frozen History: Limitations and Possibilities of Quantitative Diffusion Studies," in Ruth Oldenziel and Adri de la Bruhèze (eds.), *Manufacturing Technology: Manufacturing Consumers; The Making of Dutch Consumer Society* (Amsterdam: Aksant, 2008), 73-94.
54. Nebojša Nakicenovic, "The Automobile Road to Technological Change: Diffusion of the Automobile as a Process of Technological Substitution," *Technological Forecasting and Social Change* 29 (1986), 309-340.
55. For a classic identification of a "mobility gap" of developing countries, see Wilfred Owen, *Transportation and World Development* (Baltimore, MD: The Johns Hopkins University Press, 1987), 8.
56. For instance, ushered in by Richard Overy, "Heralds of Modernity: Cars and Planes from Invention to Necessity," in Mikuláš Teich and Roy Porter (eds.), *Fin de siècle and its legacy* (Cambridge and New York: Cambridge University Press, 1990), 54-79, here 73.
57. Cotten Seiler, *Republic of Drivers: A Cultural History of Automobility in America* (Chicago and London: The University of Chicago Press, 2008).
58. Wolfgang Sachs, *For the Love of the Automobile: Looking back into the History of Our Desires*, transl. Don Reneau (Berkeley, Los Angeles, and London: University of California Press, 1992); German edition: *Die Liebe zum Automobil: Ein Rückblick in die Geschichte unserer Wünsche* (Reinbeck bei Hamburg: Rowohlt Verlag, 1984); Klaus Kuhm, *Das eilige Jahrhundert: Einblicke in die automobile Gesellschaft* (Hamburg: Junius Verlag, 1995); Klaus Kuhm, *Moderne und Asphalt: Die Automobilisierung als Prozess technologischer Integration und sozialer Vernetzung* (Pfaffenweiler: Centaurus-Verlagsgesellschaft, 1997). Empirical: Sasha Disko, "Men, Motorcycles and Modernity: Motorization in the Weimar Republic" (unpublished dissertation, New York University, May 2008), 15. Disko gives a good overview of the recent, mostly German historiography on the car (ibid., 15-22), but she overlooks Möser.
59. Per Lundin, "Driven by Morality: Systems and Users in the Historiography of the Car in Sweden," in Norton et al. (eds.), *Mobility in History*, 119-131, here 129.
60. See also Peter Norton, "Americans' Affair of Hate with the Automobile: What the 'Love Affair' Fiction Concealed," in Mathieu Flonneau (ed.), *Automobile: Les*

Explaining the Car

cartes des désamour; Généalogies de l'anti-automobilisme (Paris: Descartes & Cie, 2009), 93–104.
61. David Blanke, *Hell on Wheels: The Promise and Peril of America's Car Culture, 1900–1940* (Lawrence: University Press of Kansas, 2007), 77.
62. Tom McCarthy, *Auto Mania: Cars, Consumers, and the Environment* (New Haven, CT, and London: Yale University Press, 2007), 40 (self-respect), 79 (obsolescence).
63. Kurt Möser, *Geschichte des Autos* (Frankfurt and New York: Campus Verlag, 2002); Kurt Möser, *Fahren und Fliegen in Frieden und Krieg: Kulturen individueller Mobilitätsmaschinen 1880–1930* (Ubstadt-Weiher: Verlag Regionalkultur, 2009).
64. Möser, *Geschichte des Autos*, 12 (*Totalgeschichte*), 11 (many perspectives; fascination), 12 (desires), 15 (gravity).
65. Möser, *Geschichte des Autos*, 16 (myths), 18 (trends). Möser also coined the idea of a "dark side of the automobile," emphasizing the importance of the military (and a militaristic attitude) in the emergence and persistence of automobilism. See his seminal article, "The Dark Side of 'Automobilism,' 1900–30: Violence, War and the Motor Car," *The Journal of Transport History*, Third Series, 24, no. 2 (September 2003), 238–258, which functioned as an inspiration for chapter 2.
66. Möser, *Geschichte des Autos*, 33 (France), 41 (contemporary), 52–55 (steam car), 72–77 (*Rausch*), 275 (progress). There are some exceptions to this general conclusion, however: his chapter 7 on the connection between automotive and military culture is the first such study based on original sources. Many of the problems signaled here have been addressed in the *Habilitationsschrift* (especially through the excellent analysis of the bodily unity of the pilot and its machine, an inspiration for parts of chapter 5), but this does not result in a new explanation, compared to *Geschichte des Autos*. There is a remarkable parallel with the analysis in Christian Kehrt's recent dissertation, *Moderne Krieger: Die Technikerfahrungen deutscher Militärpiloten 1910–1945* (Paderborn, Munich, Vienna, and Zürich: Ferdinand Schöningh, 2010).
67. Möser, *Geschichte des Autos*, 210.
68. Catherine Bertho Lavenir, *La roue et le stylo: Comment nous sommes devenus touristes* (Paris: Editions Odile Jacob, April 1999).
69. Christoph Maria Merki, *Der holprige Siegeszug des Automobils 1895–1930: Zur Motorisierung des Strassenverkehrs in Frankreich, Deutschland und der Schweiz* (Vienna, Cologne, and Weimar: Böhlau, 2002). For my earlier review of Möser's and Merki's books, see *Technology and Culture* 45 (2004), 195–197.
70. Merki, *Der holprige Siegeszug des Automobils*, 15 (goal), 16 (Pfister and Sloterdijk), 20 (*Vermittlungsinstanzen*), 22 (Gideon, Haupt, and Kocka; skates), 23 (cultural history; causal-analytical), 25 (Virilio).
71. This is, for instance visible in the International Association for the History of Transport, Traffic and Mobility T^{2}M, founded in 2003, where European scholarship dominates. See Mom, "'Historians Bleed Too Much.'"
72. Merki, *Der holprige Siegeszug des Automobils*, 23.
73. Gijs Mom and Laurent Tissot (eds.), *Road History: Planning, Building and Use* (Lausanne: Alphil, 2007).
74. Wells, "Car Country"; also see his "The Road to the Model T: Culture, Road Conditions, and Innovation at the Dawn of the American Motor Age," *Technology and Culture* 48, no. 3 (July 2007), 497–523.
75. Sean O'Connell, *The Car and British Society: Class, Gender and Motoring, 1896–1939* (Manchester and New York: Manchester University Press, 1998).

76. See John Griffiths's review in the *Journal of Transport History* 21, no. 1 (March 2000), 112–114; Colin Chant's H-Net book review is from November 2000 (http://www.h-net.org/reviews/showrev.php?id=4689, last accessed on 8 July 2011). O'Connell's accusation is all the more unjustified, as it was James Flink who provided the first template, as it were, for a social history of the car which runs remarkably parallel to O'Connell's story, much more than to comparable national stories of the European continent. O'Connell does not justify his nearly exclusive choice for the American case as a basis for comparison, but I suspect it is language.
77. John Urry, *The Tourist Gaze: Leisure and Travel in Contemporary Societies* (Thousand Oaks, CA: Sage, 1990).
78. O'Connell, *The Car and British Society*, 5.
79. Ibid., 6. O'Connell cites both McShane and Sachs to suggest that the car could also have been abandoned (ibid.). O'Connell's dissertation title is "The Social and Cultural Impact of the Car in Interwar Britain" (ibid., 235).
80. Ibid., 8 (middle class), 79 (collective gaze), 93 (geographic space), 78 (symbolic value). Bowden and Turner's analyses belong to the very best the 'hardcore' diffusionist school has ever produced. See S. M. Bowden, "Demand and Supply Constraints in the Inter-War UK Car Industry: Did the Manufacturers Get it Right?" *Business History* 33, no. 2 (April 1991), 241–267; Sue Bowden and Paul Turner, "Some Cross-Section Evidence on the Determinants of the Diffusion of Car Ownership in the Inter-War UK Economy," *Business History* 35, no. 1 (1993), 55–69. For a comparably detailed analysis of US car diffusion, see the (up to now nearly completely neglected) dissertation of George Kirkham Jarvis, "The Diffusion of the Automobile in the United States: 1895–1969" (unpublished dissertation, University of Michigan, Ann Arbor, 1972). See Mom, "Frozen History," for more examples from other countries.
81. O'Connell, *The Car and British Society*, 35.
82. Ibid., 94. As a matter of fact, O'Connell remains a bit vague toward the importance of 'utility' in the use profile. He quotes McShane to show that the American car's popularity lay in "its symbolic qualities as much as its utilitarian ones" (ibid., 6). And in his conclusions he distinguishes (without any comment) between "usefulness" and "utility": "The car's increasing usefulness for middle-class Britons, both as a utilitarian tool (for work or travel) and as a symbolic one (to express social status and taste, or to travel to places which expressed both of these factors) saw influential sections of opinion swing against significant restrictions on motoring" (ibid., 218).
83. Ibid., 78 (leisure use), 220 (frivolous), 221 (masculinity).
84. Bertho Lavenir, *La roue et le stylo*. Also see Catherine Bertho Lavenir, "Manières de circuler en France depuis 1880," *Le Mouvement Social* no. 192 (July–September 2000), 3–8.
85. Jeremy Packer and Craig Robertson, "Introduction," in Jeremy Packer and Craig Robertson (eds.), *Thinking with James Carey: Essays on Communications, Transportation, History* (New York: Peter Lang, 2006), 1–9, here 5 (conceptual and historical split); James Hay, "Between Cultural Materialism and Spatial Materialism: James Carey's Writing about Communication," in Packer and Robertson, *Thinking with James Carey*, 29–55, here 29 (ritual); Jeremy Packer, "Rethinking Dependency: New Relations of Transportation and Communication," in Packer and Robertson, *Thinking with James Carey*, 79–99, here 80 (telegraph), 81 (increasingly dependent); theory of mobile relations: Regine Buschauer, *Mobile*

Räume: Medien- und diskursgeschichtliche Studien zur Tele-Kommunikation (Bielefeld: transcript Verlag, 2010), 18; Nigel Thrift, *Spatial Formations* (Thousand Oaks, CA: Sage, 1996), 302.
86. Andrea Wetterauer, *Lust an der Distanz: Die Kunst der Autoreise in der "Frankfurter Zeitung"* (Tübingen: Tübinger Verein für Volkskunde, 2007).
87. Stefan Beck, *Umgang mit Technik: Kulturelle Praxen und kulturwissenschaftliche Forschungskonzepte* (Berlin: Akademie Verlag, 1997). For Koshar's analysis of German travel culture, see Rudy Koshar, "Cars and Nations: Anglo-German Perspectives on Automobility between the World Wars," *Theory, Culture & Society* 21, no. 4/5 (2004), 12–14; Rudy Koshar, *German Travel Cultures* (Oxford and New York: Berg, 2000); Rudi Koshar, "Germans at the Wheel: Cars and Leisure Travel in Interwar Germany," in Rudy Koshar (ed.), *Histories of Leisure* (Oxford and New York: Berg, 2002), 215–230; Rudy Koshar, "On the History of the Automobile in Everyday Life," *Contemporary European History* 10, no. 1 (2001), 143–154; Rudy Koshar, "Organic Machines: Cars, Drivers, and Nature from Imperial to Nazi Germany," in Thomas Lekan and Thomas Zeller (eds.), *Germany's Nature: Cultural Landscapes and Environmental History* (New Brunswick, NJ, and London: Rutgers University Press, 2005), 111–139.
88. Wetterauer, *Lust an der Distanz*, 10 (*Vermittlungsinstanzen*), 41 (Enzensberger; *Wahrnehmungsmittel*), 44 (return of the actor), 87 (Simmel), 90 (stylization).
89. Thomas Krämer-Bodoni, Herbert Grymer, and Marianne Rodenstein, *Zur sozioökonomischen Bedeutung des Automobils* (Frankfurt: Suhrkamp Verlag, 1971).
90. Wetterauer, *Lust an der Distanz*, 153 (road experience); Kathleen Franz, *Tinkering: Consumers Reinvent the Early Automobile* (Philadelphia: University of Pennsylvania Press, 2005); Kevin L. Borg, *Auto Mechanics: Technology and Expertise in Twentieth-Century America* (Baltimore, MD: The Johns Hopkins University Press, 2007).
91. Leo Marx, *The Pilot and the Passenger: Essays on Literature, Technology, and the Culture in the United States* (New York and Oxford: Oxford University Press, 1988). I thank Kurt Möser for bringing this study to my attention.
92. Examples of this boom only since 2007 (so after the publication of Wetterauer's dissertation; but see, for instance, Gudis (2004), mentioned in note 95) are Blanke, *Hell on Wheels*; Michael R. Fein, *Paving the Way: New York Road Building and the American State, 1880–1956* (Lawrence: University Press of Kansas, 2008); Shane Hamilton, *Trucking Country: The Road to America's Wal-Mart Economy* (Princeton, NJ, and Oxford: Princeton University Press, 2008); John A. Jakle and Keith A. Sculle, *Motoring: The Highway Experience in America* (Athens and London: University of Georgia Press, 2008); Brian Ladd, *Autophobia: Love and Hate in the Automotive Age* (Chicago and London: University of Chicago Press, 2008); David N. Lucsko, *The Business of Speed: The Hot Rod Industry in America, 1915–1990* (Baltimore, MD: The Johns Hopkins University Press, 2008); Christof Mauch and Thomas Zeller (eds.), *The World Beyond the Windshield: Roads and Landscapes in the United States and Europe* (Athens, OH, and Stuttgart: Ohio University Press / Franz Steiner Verlag, 2008); Norton, *Fighting Traffic*; Susan Sessions Rugh, *Are We There Yet? The Golden Age of American Family Vacations* (Lawrence: University Press of Kansas, 2008); Jeremy Packer, *Mobility Without Mayhem: Safety, Cars, and Citizenship* (Durham, NC, and London: Duke University Press, 2008); Tom Vanderbilt, *Traffic: Why We Drive the Way We Do (and What It Says About Us)* (London: Allen Lane / Pen-

guin Books, 2008); Heitmann, *The Automobile and American Life*; McCarthy, *Auto Mania;* William J. Mitchell, Christopher E. Borroni-Bird, and Lawrence D. Burns, *Reinventing the Automobile: Personal Urban Mobility for the 21st Century* (Cambridge, MA, and London: The MIT Press, 2010).

93. A case in point is a recent overview of car automation in Jeremy Packer, "Becoming Bombs: Mobilizing Mobility in the War of Terror," *Cultural Studies* 20, no. 4/5 (July–September 2006) 378–399.

94. On the threatening ghettoization of transport and mobility history, see Gijs Mom, "What kind of transport history did we get? Half a century of *JTH* and the future of the field," *Journal of Transport History* 24, no. 2 (September 2003), 121–138.

95. Seiler, *Republic of Drivers;* also see the *History and Technology* Forum: Martin Collins, "Introduction," *History and Technology* 26, no. 4 (December 2010), 359–360; Jeremy Packer, "Automobility and apparatuses: Commentary on Cotten Seiler's *Republic of Drivers*," 361–368; Catherine Gudis, *Buyways: Billboards, Automobiles, and the American Landscape* (New York and London: Routledge, 2004), 369–378; and Mathieu Flonneau, "Read Tocqueville, or drive? A European perspective on US 'automobilization,'" 379–388.

96. Seiler, *Republic of Drivers,* 38 (utility thesis), 39 (thrilling nature), 41 (signifier).

97. Packer, "Automobility and apparatuses," 363; Seiler, *Republic of Drivers,* 41 (meliorative), 65 (governing populations), 151 (consolation). For a pathbreaking and rare earlier analysis of automobile politics, see William Plowden, *The Motor Car and Politics in Britain* (Harmondsworth and Ringwood, Australia: Penguin, 1971). For a German study in this tradition, see Michael Hascher, *Politikberatung durch Experten: Das Beispiel der deutschen Verkehrspolitik im 19. und 20. Jahrhundert* (Frankfurt and New York: Campus Verlag, 2006).

98. John Urry, *Consuming places* (London and New York: Routledge, 1995). Also see Zygmunt Bauman, *Liquid Life* (Cambridge and Malden, MA: Polity Press, 2006; first ed. 2005). For an interesting introduction into the (history of the) "metaphor of mobility," including its subversive roots (the mobility of the crowd, or the "mob"), see Nancy Koppelman, "One for the Road: Mobility in American Life, 1787–1905" (unpublished dissertation, Emory University, 1999), chapter 1, 22–66; "mob" on 34.

99. Flonneau, "Read Tocqueville, or drive?" 383.

100. Mathieu Flonneau, *Paris et l'automobile: Un siècle de passions* (Paris: Hachette Littératures, 2005). This book was based on his three-volume dissertation, from which some later books were also partially derived.

101. Mathieu Flonneau, *L'Autorefoulement et ses limites: Raisonner l'impensable mort de l'automobile* (Paris: Descartes & Cie, 2010), 25 (autophobia), 82 n. 1 (pleasure), 85 (triumph).

102. Mathieu Flonneau, *Les cultures du volant: Essai sur les mondes de l'automobilism XXe–XXIe siècles* (Paris: Éditions Autrement, 2008), 13 (Schivelbusch), 183 (thickness). One of his adversaries is Luc Boltanski, who analyzed the French car culture in the 1970s from a class perspective. See Luc Boltanski, "Les usages sociaux de l'automobile: concurrence pour l'espace et accidents," *Actes de la recherche en sciences sociales* 1, no. 2 (1975), 25–29. Experimental content overview in Flonneau, *L'Autorefoulement.*

103. Cotten Seiler, "Author response: The ends of automobility," *History and Technology* 26, no. 4 (December 2010), 389–397, here 392.

104. On the concept of 'multiple modernities,' see Mike Featherstone (ed.), *Global Culture: Nationalism, Globalization and Modernity* (Thousand Oaks, CA: Sage, 1990).
105. The diffusionist argument is still very much alive among social scientists, recently under the guise of a "viral" phenomenon. John Urry, "The 'System' of Automobility," *Theory, Culture and Society* 21, no. 4/5 (2004), 25–39, here 27. This 'viral' is an allusion to the 'contact diffusion' paradigm within diffusion studies. See Mom, "Frozen History."
106. Möser, *Fahren und Fliegen in Frieden und Krieg*, for instance, uses belletristic literature extensively as historical source. For a reversed plea to analyze the influence of the car on literature, see Jennifer Shepherd, "The British Press and Turn-of-the-Century Developments in the Motoring Movement," *Victorian Periodicals Review* 38, no. 4 (Winter 2005), 379–391, here 388.
107. Quoted in: Lubomír Doležel, *Possible Worlds of Fiction and History* (Baltimore, MD: The Johns Hopkins University Press, 2010), 27.
108. Claudia Lieb, *Crash: Der Unfall der Moderne* (Bielefeld: Aisthesis Verlag, 2009), 235 (my translation).
109. For a brief overview of this development among many others, see for instance William H. Sewell, Jr., "The Concept(s) of Culture," in Victoria E. Bonnell and Lynn Hunt (eds.), *Beyond the Cultural Turn: New Directions in the Study of Society and Culture* (Berkeley, Los Angeles, and London: University of California Press, 1999), 35–61.
110. Andreas Reckwitz, "Toward a Theory of Social Practices: A Development in Culturalist Theorizing," *European Journal of Social Theory* 5, no. 2 (2002), 243–263.
111. Hayden White, *Tropics of Discourse: Essays in Cultural Criticism* (Baltimore, MD, and London: The Johns Hopkins University Press, 1986; first ed. 1978). As Wolfgang Iser (referring to Wittgenstein) asserts, "Fictional discourse is self-referential and shares with everyday utterances the aspect of symbol use, but not the aspect of empirical recourse to objects." Wolfgang Iser, "Texts and Readers," *Discourse Processes* 3, no. 4 (1980), 327–343, here 328. On experience as a pre-discursive concept, see Joan Scott, "The evidence of experience," in Gabrielle M. Spiegel (ed.), *Practicing History: New Directions in Historical Writing after the Linguistic Turn* (New York and London: Routledge, 2005), 199–216, here 208.
112. Medical gaze: Urry, *The Tourist Gaze*, 1. For a more detailed discussion of the visual bias, see chapter 2.
113. Gijs Mom, "Civilized adventure as a remedy for nervous times: Early automobilism and fin the siecle culture," *History of Technology* 23 (2001), 157–190.
114. Southgate, *History Meets Fiction*, 8 (Dickens), 9 (Fitzgerald and chicken), 195 (concludes; italics in original).
115. Jon Adams, "Real Problems with Fictional Cases," in Peter Howlett and Mary S. Morgan (eds.), *How Well Do Facts Travel? The Dissemination of Reliable Knowledge* (Cambridge and New York: Cambridge University Press, 2011), 167–191, here 170.
116. David Lewis, Dennis Rodgers, and Michael Woolcock, "The Fiction of Development: Literary Representation as a Source of Authoritative Knowledge," *Journal of Development Studies* 44, no. 2 (February 2008), 198–216.
117. W. P. Blockmans, C. A. Davids, and E. K. Grootes, "Inleiding," *Tijdschrift voor Sociale Geschiedenis* 10, no. 35 (August 1984), 223–227 (introduction to a special issue on literature as historical source).

118. Wolfgang Schivelbusch, *Geschichte der Eisenbahnreise: Zur Industrialisierung von Raum und Zeit im 19. Jahrhundert* (Munich and Vienna, 1977); Marc Desportes, *Paysages en mouvement: Transports et perception de l'espace XVIIIe–XXe siècle* (n.p. [Paris]: Gallimard, 2005).
119. Douwe Fokkema and Elrud Ibsch, *Modernist Conjectures: A Mainstream in European Literature 1910–1940* (London: C. Hurst & Company, 1987), 127.
120. Lewis A. Coser (ed.), *Sociology Through Literature: An Introductory Reader* (Englewood Cliffs, NJ: Prentice-Hall, 1963), 3.
121. For instance in Ronald Primeau, *Romance of the Road: The Literature of the American Highway* (Bowling Green, OH: Bowling Green State University Popular Press, 1996).
122. Ann Rigney, "De lokroep van het verleden; Literatuur als historische bron" (The Seductive Call of the Past; Literature as Historical Source), *Feit & fictie* 4, no. 3 (Summer 1999), 82–98; on problematic referentiality, see Philip Stewart, "This Is Not a Book Review: On Historical Uses of Literature," *Journal of Modern History* 66 (September 1994), 521–538.
123. Peter Gay, *The Bourgeois Experience, Victoria to Freud: Volume II: The Tender Passion* (New York and Oxford: Oxford University Press, 1986), 433–434. Gay's analysis of amorous and erotic bourgeois experience benefits tremendously from an enormous amount of work done by psychoanalysts beginning with Freud himself. For a comparable claim by literary historian Cecilia Tichi, see her *Shifting Gears: Technology, Literature, Culture in Modernist America* (Chapel Hill and London: The University of North Carolina Press, 1987), 27, where she states that novelists have the "task" to "represent the vanguard of contemporary consciousness" and to "validate" the world.
124. Leo Marx, *The Machine in the Garden: Technology and the Pastoral Idea in America* (Oxford and New York: Oxford University Press, 2000; first ed. 1964).
125. Michael Titzmann, "Kulturelles Wissen—Diskurs—Denksystem: Zu einigen Grundbegrifffen der Literaturgeschichtsschreibung," *Zeitschrift für französische Sprache und Literatur* 99 (1989), 47–61.
126. David Bakhurst, "Ilyenkov on Aesthetics: Realism, Imagination, and the End of Art," *Mind, Culture, and Activity* 8, no. 2 (2001), 187–199, here 193 (perception); Elisabeth Tworek-Müller, *Kleinbürgertum und Literatur: Zum Bild des Kleinbürgers im bayerischen Roman der Weimarer Republik* (Munich: tuduv-Verlagsgesellschaft, 1985), 2 (my translation).
127. The quote is a paraphrase by André Bleikasten, "Faulkner and the New Ideologues," in Donald M. Kartiganer and Ann J. Abadie (eds.), *Faulkner and Ideology: Faulkner and Yoknapatawpha, 1992* (Jackson: University Press of Mississippi, 1995), 3–21, here 16.
128. Rüdiger Hachtmann, "Tourismusgeschichte—ein Mauerblümchen mit Zukunft! Ein Forschungsüberblick," in H-Soz-u-Kult 06.10.2011, http://hsozkult.geschichte.hu-berlin.de/forum/2011-10-001 (last accessed 6 October 2011), 15.
129. Ala Alryyes, "Description, the Novel and the Senses," *Senses & Society* 1, no. 1 (March 2006), 53–70, here 54 (omniscience), 62 (predictive). Structuralist: Doležel, *Possible Worlds of Fiction and History*, 15.
130. Doležel, *Possible Worlds of Fiction and History*, 20.
131. Gijs Mom, "'The future is a shifting panorama': The role of expectations in the history of mobility," in Weert Canzler and Gert Schmidt (eds.), *Zukünfte des Automobils: Aussichten und Grenzen der autotechnischen Globalisierung* (Berlin: edition sigma, 2008), 31–58. On the role of expectations in the "dynamics

of technology" in general, see Harro van Lente, *Promising Technology: The Dynamics of Expectations in Technological Developments* (Delft: Eburon, 1993). For the concept of "affinity," see chapter 2.
132. "Sense" (http://en.wikipedia.org/wiki/Sense, last accessed on 21 June 2010).
133. Fabio Crivellari et al., "Einleitung: Die Medialität der Geschichte und die Historizität der Medien," in Crivellari et al. (eds.), *Die Medien der Geschichte: Historizität und Medialität in interdisziplinärer Perspektive* (Konstanz: UVK Verlagsgesellschaft, 2004), 9–45, here 23. For the car as prosthesis, see, for instance, Charles Grivel, "D'un écran automobile," in Jochen Mecke and Volker Roloff (eds.), *Kino-/(Ro)Mania: Intermedialität zwischen Film und Literatur* (Tübingen: Stauffenburg Verlag, 1999), 47–77, here 69; and on the gun, the camera, and the airplane as prosthetic, see Robert Dixon, *Prosthetic Gods: Travel, Representation and Colonial Governance* (St Lucia QLD: University of Queensland Press in association with the API Network, 2001).
134. Fabio Crivellari and Marcus Sandl, "Die Medialität der Geschichte: Forschungsstand und Perspektiven einer interdisziplinären Zusammenarbeit von Geschichts- und Medienwissenschaften," *Historische Zeitschrift* 277 (2003), 619–654, here 623 (MacLuhan), 625 (Kittler).
135. Hans Ulrich Seeber, *Mobilität und Moderne: Studien zur englischen Literatur des 19. und 20. Jahrhunderts* (Heidelberg: Universitätsverlag Winter, 2007), 11; Sloterdijk quoted on 17.
136. Stephen Kern, *The Culture of Time and Space 1880–1918* (London, 1983); Hartmut Rosa, *Beschleunigung: Die Veränderung der Zeitstrukturen in der Moderne* (Frankfurt: Suhrkamp Verlag, 2005), 161.
137. One of the best overviews of the 'new mobilities' paradigm is given by Peter Adey, *Mobility* (London and New York: Routledge, 2010).
138. Dialectic: Nikolaus Buschmann and Horst Carl, "Zugänge zur Erfahrungsgeschichte des Krieges: Forschung, Theorie, Fragestellung," in Nikolaus Buschmann and Horst Carl (eds.), *Die Erfahrung des Krieges: Erfahrungsgeschichtliche Perspektiven von der Französischen Revolution bis zum Zweiten Weltkrieg* (Paderborn, Munich, Vienna, and Zürich: Ferdinand Schönigh, 2001), 11–26, here 17; Reinhart Koselleck, "'Neuzeit': Zur Semantik moderner Bewegungsbegriffe," in Koselleck, *Vergangene Zukunft: Zur Semantik geschichtlicher Zeiten* (Frankfurt: Suhrkamp Verlag, 1979), 300–348, here 300.
139. Charlotte Heymel, *Touristen an der Front: Das Kriegserlebnis 1914–1918 als Reiseerfahrung in zeitgenössischen Reiseberichten* (Berlin: Lit Verlag Dr. W. Hopf, 2007), 21–22. On "mobile nature," see Mark Fiege, "The Weedy West: Mobile Nature, Boundaries, and Common Space in the Montana Landscape," *Western Historical Quarterly* 36, no. 1 (Spring 2005), 22–47, here 42.
140. Thrift, *Spatial Formations*, 6–7 (non-representational), 30 (modest); Tim Ingold, *The Appropriation of Nature: Essays on Human Ecology and Social Relations* (Manchester: Manchester University Press, 1986), 2.
141. Christian Kehrt, *Moderne Krieger: Die Technikerfahrungen deutscher Militärpiloten 1910–1945* (Paderborn, Munich, Vienna, and Zürich: Ferdinand Schöningh, 2010), 19 (socio-cultural), 35 (discrepancy).
142. William H. Sewell, Jr., "The Concept(s) of Culture," in Gabrielle M. Spiegel (ed.), *Practicing History: New Directions in Historical Writing after the Linguistic Turn* (New York and London: Routledge, 2005), 76–95, here 83.
143. Paul Hekkert, "Design aesthetics: Principles of pleasure in design," *Psychology Science* 48, no. 2 (2006), 157–162; Beatriz Russo and Paul Hekkert, "On the

Conceptualization of the Experience of Love: The Underlying Principles," paper submitted to Designing Pleasurable Products Conference—DPPI'07 (http://studiolab.io.tudelft.nl/russo/stories/storyReader$20, last accessed 18 March 2011); see also Beatriz Russo Rodriguez, *Shoes, Cars and Other Love Stories: Investigating the Experience of Love for Products* (Delft: VSSD, 2010).

144. See, for instance, the German research project "Travelling Goods, Travelling Moods" on the 'cultural appropriation' of several commodities, including the car, led by Christian Huck at the University of Kiel. Christian Huck and Stefan Bauernschmidt (eds.), *Travelling Goods, Travelling Moods: Varieties of Cultural Appropriation (1850–1950)* (Frankfurt and New York: Campus Verlag, 2012).

145. Barbara Korte, "Erfahrungsgeschichte und die 'Quelle' Literatur: Zur Relevanz genretheoretischer Reflexion am Beispiel der britischen Literatur des Ersten Weltkriegs," in Jan Kusber et al. (eds.), *Historische Kulturwissenschaften: Positionen, Praktiken und Perspektiven* (Bielefeld: transcript Verlag, 2010), 143–159, here 146.

146. Theodore R. Schatzki, *Social Practices: A Wittgensteinian Approach to Human Activity and the Social* (Cambridge: Cambridge University Press, 1996).

147. Theodore R. Schatzki, "Nature and Technology in History," *History and Theory*, Theme Issue 42 (December 2003), 82–93, here 84.

148. Reijo Miettinen, "The Riddle of Things: Activity Theory and Actor-Network Theory as Approaches to Studying Innovations," *Mind, Culture, and Activity* 6, no. 3 (1999), 170–195, here 174.

149. Tim Cresswell and Peter Merriman, "Introduction: Geographies of Mobilities—Practices, Spaces, Subjects," in Cresswell and Merriman (eds.), *Geographies of Mobilities: Practices, Spaces, Subjects* (Farnham and Burlington: Ashgate, 2011), 1–15, here 5.

150. Edward Reed and Rebecca Jones (eds.), *Reasons for Realism: Selected Essays of James J. Gibson* (Hillsdale, NJ, and London: Lawrence Erlbaum Ass., 1982).

151. Gijs Mom, "Translating Properties into Functions (and Vice Versa): Design, User Culture and the Creation of an American and a European Car (1930–1970)," *Journal of Design History* 20, no. 2 (2007), 171–181. The following paragraphs are based on this article.

152. Auke Jan Koop Pols, *Acting with Artefacts* (n.p. [Eindhoven]: Eindhoven University of Technology, Simon Stevin Series in the Philosophy of Technology), chapter 4, 59–76. The proposed matrix can even be given a third, normative dimension, by distinguishing between 'good' and 'bad' use, as proposed by philosophers of technology (see Peter Kroes and Anthonie Meijers, "Introduction: The dual nature of technical artefacts," *Studies in History and Philosophy of Science* 37 [2006], 1–4).

153. Miettinen, "The Riddle of Things," 180.

154. This 'normal use' is an application of Kuhn's concept of 'normal science,' applied by Peter Hugill to analyze technological change in airplanes and cars. See Peter J. Hugill, "Technology and Geography in the Emergence of the American Automobile Industry, 1895–1915," in Jan Jennings (ed.), *Roadside America: The Automobile in Design and Culture* (Ames: Iowa State University Press, for the Society for Commercial Archaeology, 1990), 29–39.

155. A. N. Leont'ev, *Activity, Consciousness, and Personality*, transl. Marie J. Hall (Englewood Cliff, NJ: Prentice Hall, 1978). On these early Soviet theorists of action theory from 1920s, see especially Michael Cole, "Cross-Cultural Research in the Sociohistorical Tradition," *Human Development* 31 (1988), 137–157; James V.

Wertsch (ed.), *Culture, Communication, and Cognition: Vygotskian Perspectives* (Cambridge and New York: Cambridge University Press, 1985).
156. For an application of the 'civilization theory' on road behavior, see Norbert Elias, "Technization and Civilization," *Theory, Culture & Society* 12 (1995), 7–42.
157. Steffen Böhm et al., "Introduction: Impossibilities of automobility," *The Sociological Review* 54, no. s1 (October 2006), 3–16, here 3. This introduction was part of a special issue of *The Sociological Review* dedicated to "Automobility," containing also John Urry, "Inhabiting the car," 17–31.
158. Tim Edensor, "Automobility and National Identity: Representation, Geography and Driving Practice," *Theory, Culture & Society* 21, no. 4/5 (2004), 101–120. The French geographer Jean-Luc Piveteau already in 1990 distinguished between three geographical dimensions in analyzing the car: "corps, habilitation, espace" (body, dwelling, space). Jean-Luc Piveteau, "La voiture, signe et agent d'une nouvelle relation de l'homme à l'espace," *Cahiers de l'Institut de Géographie* 7 (1990), 45–55.
159. Oliver Kühlschelm, "Konsumgüter und Nation: Theoretische und methodische Überlegungen," *Österreichische Zeitschrift für Geschichtswissenschaften* 21, no. 2 (2010), 19–49.
160. For instance, in Mimi Sheller and John Urry, "The new mobilities paradigm," *Environment and Planning A* 38 (2006), 207–226, where "six bodies of theory" are listed (Simmel, STS, spatial turn, recentering the body, social networks, complex systems) and seven methods (observation of people's movement, mobile ethnography, time-space diaries, cyber-research [imaginative and virtual mobilities], multimedia methods [including literature, for the "affective dimension"], memory research, examination of transfer points such as airports). On history added as an afterthought, without much specificity, see the programmatic first editorial of the journal *Mobilities*: Kevin Hannam, Mimi Sheller, and John Urry, "Editorial: Mobilities, Immobilities and Moorings," *Mobilities* 1, no. 1 (March 2006), 1–22, here 15: "We hope to have made a strong case for . . . an approach that offers both theoretical and methodological purchase on a wide range of urgent contemporary issues, as well a new perspectives on certain historical questions." These "certain historical questions" are not further specified. Geographers seem to be less hesitant to use history, although they may run the risk of treading on paths already trodden, such as is the case in the historical parts of Peter Merriman, *Mobility, Space and Culture* (London and New York: Routledge, 2012). See, for instance, the excellent analyses by Tim Cresswell and Peter Merriman: Cresswell, *On the Move: Mobility in the Modern Western World* (New York and London: Routledge, 2006); Merriman, *Driving Spaces: A Cultural-Historical Geography of England's M1 Motorway* (Malden, MA, and Oxford: Blackwell, 2007); Cresswell and Merriman (eds.), *Geographies of Mobilities: Practices, Spaces, Subjects* (Farnham and Burlington: Ashgate, 2011). For an effort to historicize Urry's "tourist gaze" concept, see Cord Pagenstecher, *Der bundesdeutsche Tourismus: Ansätze zu einer Visual History: Urlaubsprospekte, Reiseführer, Fotoalben 1950–1990* (Hamburg: Verlag Dr. Kovač, 2003), especially 470.
161. Kingsley Dennis and John Urry, *After the Car* (Cambridge and Malden, MA: Polity Press, 2009), 28–39 ("conspiracy" on 35). For the "Snell myth" about General Motors being responsible for the demise of the streetcar, see Robert C. Post, *Urban Mass Transit: The Life Story of a Technology* (Westport, CT, and London: Greenwood Press, 2007), 153–156.

162. Robin Law, "Beyond 'women and transport': Towards new geographies of gender and daily mobility," *Progress in Human Geography* 23, no. 4 (1999), 567–588, here 580 (derived demand).
163. For popular culture being largely oral (referring to Carlo Ginzberg), see Davide Panagia, *The Political Life of Sensation* (Durham, NC, and London: Duke University Press, 2009), 51.

PART I

Emergence (1895–1918)

CHAPTER 1

Racing, Touring, Tinkering
Constructing the Adventure Machine (1895–1914/1917)

> "Oh... if he were rich... an automobile! Nothing would do but an automobile! ... It would make a poet of him."[1]

In 1902 Otto Julius Bierbaum, an established German writer known to belong to the *Münchener Jugendstil* movement, undertook a journey to Italy in a car borrowed from the Adler company. Popular among contemporaries and historians of mobility alike, and followed, during the remainder of his literary career, by several shorter pieces celebrating the delights of automobile travel, his *empfindsame Reise* (sentimental journey) was conceived in the 'old-fashioned' form of letters to friends, a 250-page-long attack on the train, and a eulogy to the 'freedom of movement' the automobile afforded instead. The book offered an overt morality: the joy of the car trip, Bierbaum concluded, resided in movement itself, and in order to enjoy this as well as the landscape they were traversing, the "nearly addictive" attraction of speed had to be avoided. "*Lerne zu reisen, ohne zu rasen!*" (learn to travel without racing) he advised, and he likened the caresses of touring to the sensations of drinking tea, whereas racing reminded him of the whiplashes (*Peitschenhiebe*) and the rush (*Rausch*) of alcohol consumption.[2]

Half a decade later Filippo Tommaso Marinetti used the car precisely *for* the whiplashes and the rush it provided. Trying since 1902 "to free the Italian lyrical genius from its traditional and commercial shackles" in his international journal *Poesia*, Marinetti sent a pamphlet containing a "Manifeste du Futurisme" to a host of international newspapers. Several published it or commented on it, before it was printed on the first page of the Parisian newspaper *Le Figaro* on 20 February 1909. Later Marinetti would remember that he had been in doubt whether to use the word "dynamisme" instead of "futurisme." Adorned with a new prologue, Marinetti's text became even more iconic in the historiography of the early automobile than Bierbaum's account. Referring to an exciting car drive (no chauffeur, of course) from half a year back through city streets, ending in a crash in the outskirts of Milan, the manifesto proclaimed "to sing the love of danger," of "courage, audacity and revolt" and "the beauty of speed" that makes the racing car with its engine running on "machine-gun fire," "more beautiful than the Victory of Samothrace."[3]

Both Bierbaum and Marinetti, icons if not clichés of a new automotive age, saw themselves as modernists. While this is clear for the declared anti-traditionalist Marinetti, it is perhaps less well known how Bierbaum expressed a "virile contempt for the spineless bourgeois" as "the embodiment of all that is mediocre" and unable to see that "danger is the spice of life."[4] Nevertheless, within the space of a single decade, *biedermeier* travel delicacies had apparently been replaced (or accompanied) by raw-vitalistic death wishes. How can we explain this rapid change of artistic attitude, which was at least inspired, if not invited, by the use of automobiles? Are we witnessing the resonance of class differences here, or of nationalities, generations, or political stance? Did automotive technology develop so fast as to afford self-steered urban aggression and violence where, only a decade before, it only permitted the romance of the country road under the guidance of a chauffeur? Or are these musings mere coincidences of a much larger variety of utterances of early automotive culture?

To complicate matters even more: well after Marinetti had cried out his automotive death wish and Europe had slipped into a war, a much more tranquil Theodore Dreiser, on the other side of the Atlantic, convinced an artist friend to make a car trip into the country of their youth. The trip from New York to Indiana was undertaken with a chauffeur called Speed, who was just as much a part of the conversation as the others. Like Bierbaum, Dreiser called himself a "sentimentalist" and like Bierbaum and many other European colleagues he opposed racing on the road.[5]

We will analyze in the next chapter the differences between Bierbaum, Marinetti, and Dreiser and their respective literary colleagues in more detail. In this chapter we will approach their experiences as a part of a much wider 'movement' of emergence of the earliest automobile culture in the North-Atlantic realm, based on the traditional sources such as trade journals and reminiscences, supported by the existing historiography. This emergence occurred in three consecutive phases leading up to the outbreak of the First World War, 1914 for the European car culture and 1917 for the United States. The first phase, ending shortly after the turn of the century, was a gestation period in which all things still seemed possible, both technically and culturally. The second was an intermediate phase, lasting until nearly the end of the first decade, in which a certain type of use, of a certain type of car, was celebrated by a particular kind of user (very wealthy elites touring in very heavy family cars), while the third was the brief period, abruptly interrupted in Europe by the war but lived to its climax in the United States, of tourism in lighter, cheaper cars undertaken by some upper echelons of the middle class ("small capitalists" or "the average man," as they were called at the time in Europe and the United States respectively) ranked immediately "below" the elite of the previous phase. We will argue that we need the accounts of both the Bierbaums and the Marinettis to explain the motives behind the shaping of prewar European mobility culture, which would have such a lasting influence on Atlantic automobilism of the following century, even if the Dreisers seemed inclined to follow a different path.

First Phase: Emergence and Roots of the Petrol Car (until 1902)

It would be futile to attempt to pinpoint the precise beginning of the earliest car culture, both in time and geographic place: its fuzzy gestation rather characterizes the pervasive urge to motorize, a sentiment that was felt all over the Western world. After the surge of national chauvinism around the car's centenary celebrations in 1986, academic historiography seems to have deliberately avoided the issue of the exact start of the 'movement,' thus leaving it to the popular publicists to construct a deterministic, linear genealogy based on the eventual winner, the gasoline car. The historiographical shift from invention and production to use and non-use[6] has meanwhile led to the consensus that automobile culture started in the second half of the 1890s as a sporting culture while its cradle was France, mostly Paris and the wealthy *Côte d'Azur*, whereas a large part of the car's technical properties, especially its engine, were provided by German engineers and companies. By positioning this story so close to the turn of the century we tend to forget, as a recent French dissertation warns, that the car, however defined, was already at least a generation old before it really entered into history's limelight.[7] Indeed, automotive historiography would be well served to take this 'prehistory' into account and make the transition toward automotive mobility less abrupt than it is often presented. "The motor car has not imposed new values so much as it has reinforced old," a historian of tourism concluded.[8]

Indeed, the early automobile emerged as an amalgam, a bricolage, a true hybrid, both technically and culturally. Technically, the tubing and wheels of the light voiturette version were taken from the bicycle, while the heavier versions inherited the body and wheel suspension from carriage technology, and the tiller steering, on some types, was to be found on some ships, such as barges. Culturally, some lineages were problematic, such as the (undelivered) promise of a horseless variety for business purposes, as advertised by Carl Benz. The bicycle culture, then a decade old as a fledgling mass movement, provided the dominant influence, no doubt also because many early car pioneers indulged in the emerging bicycle culture in their youth. The bicycle incorporated horse culture (some of the earliest bicycles had been called hobby or dandy horses) and some of the horse sport's cultural paraphernalia adorned early car culture as well, from the cavalier attitude, to the view from on high, looking down on the commoners. Perhaps due to this hybridity, the car's early role was very flexible, and as a "collective symbol" its functions and its properties were quite diverse.[9]

The resulting, 'typical' transnational Panhard-Daimler hybrid (or Mercedes model, as it is also called: engine in front, rear wheels driven through a longitudinal propeller shaft) emerged from a further, 'internal' struggle between three propulsion alternatives, where steam seemed to be best suited for heavy loads, and electric propulsion lent itself especially for fleet applications such as taxicabs and municipal vehicles.[10] Electric vehicle proponents tend to place the start of automobilism in the early 1880s in Paris, when electrical engineer Nicolas-Jules Raffard (on a horse streetcar) and luxury-carriage builder Charles Jeantaud (on a converted coach) began experimenting. Steam car enthusiasts can even boast of a much earlier start: in the 1820s in the United Kingdom, large steam buses were developed, shortly before most engineers concluded that such heavy vehicles functioned better on rails. A second wave of lighter steam cars during the 1880s made steam the leading candidate promising individual motorization at the end of the nineteenth century, especially in the United States where it dominated the earliest private user culture of the passenger car, rather than electric propulsion (as often has been stated). One can even go further back in mobility history to identify isolated (mostly steam-propelled) examples of the wish, the dream, to motorize, a lineage that inspired at least one historian to see mobility as an 'anthropological constant.'[11]

Quite surprisingly, perhaps, around the beginning of the twentieth century, and at least by 1902, it was clear that the gasoline car had won, despite its unreliability and its confusing array of shapes and technical layouts. As I have argued elsewhere, the unreliability was so pervasive in the early automotive movement that one has to conclude that the car's frequent defects functioned as a catalyst rather than a hindrance for the emergence of a mostly masculine user culture that largely developed around the car as an "adventure machine."[12] In this culture, then, the adventure in functionality (tinkering) was as important as the two other aspects: the adventure in time (racing) and in space (touring), if not more so: early motorists observed that they were listening to the engine and trying to pick up the car's mechanical signals rather than enjoying the countryside. Or they were negotiating poor roads. Caring for the car was not only "convenient," but also "a pleasure and a source of relaxation."[13]

Contrary to many historians,[14] the emerging car culture was urban in ownership, but extra-urban (inter-urban, rural) in use and in quantitative terms it was very successful. While the North Atlantic world had 15,000 cars in 1900 (one-third in the United States, with the rest in Europe, mostly France, and a few elsewhere), a decade later there were 800,000 motorized vehicles, while at the outbreak of the First World War two million were counted, 65 percent in the United States, mostly built by Henry Ford.[15]

In order to prevent the result of the struggle between steam, petrol, and electricity becoming the teleological endpoint of a linear 'history' projected back in time, it should be contextualized within several layers of technical and socio-cultural developments, such as the state of the art of general mobility and, in a still wider circle, the fin-de-siècle culture within industrializing Western societies in a general context of globalization.[16] To start with the most central layer of the mobility context, it is remarkable that from midcentury every generation had been witness to a fundamental innovation: the surge in horse and carriage use in the 1850s, the massive expansion of railway networks in the 1860s at the level of the national state, the local and regional tramway networks in the 1880s, and the 'bicycle craze' in the 1890s, which were followed, in an ever-increasing tempo, by the automobile, the metro, the submarine, the steamship, the balloon, and the airplane, each new vehicle adding a new dimension to the mobility spectrum rather than substituting for an older technology. Indeed, there can be no doubt that the (at the moment of the introduction of the car still raging) 'bicycle craze' functioned in creating a set of people, a group culture, and individual experience as a basis for the fledgling automobile culture. Substitution is also a problematic concept, because the urban "horse economy" was already stagnating (probably because of a lack of space) well before motorized vehicles could function as an alternative.[17]

This is not to say, of course, that the bicycle should be analyzed as a step-up to the car. On the contrary, in accordance to David Edgerton's *Shock of the Old,* the bicycle followed its own trajectory, in its own subculture, just as the car, a decade later, would become 'old-fashioned' (but remained very much alive) when the airplane emerged as the newest 'adventure machine.'[18]

With a 'prehistory' of nearly a century and accelerated through an avant-garde of a mass movement in the 1880s by middle-class boys, and some girls, of school age, the bicycle culture came to rest on the same tripartite adventure (racing, touring, tinkering) as the automobile later would.[19] The bicycle's institutional movement soon split into two opposing but fundamentally related adventurous cultures of racing and touring. The latter was institutionalized in touring clubs (many of them founded in the early 1880s) that emphasized the non-commercial and amateur character of their activities. These grew at an astonishing rate into true mass associations.[20] In Germany in 1905, bicycle clubs (with forty thousand members) were by far the largest sport organizations, before soccer and light athletics.[21]

Despite the emphasis, by contemporaries and historians alike, on the quintessential 'modern' character of the bicycle's individualized use, many early trips were group experiences, reminiscent of military practices with uniforms, a group leader, and the use of military terminology and of maps of military origin, at least in Europe. In their mobile pattern with their "free associationism," they resembled more a "swarm" (in Henry Adams's terms, in his description of groups of tourists in which women seem to feel particularly at ease: "ephemeral like clouds of butterflies in season") than a collection of independent "monads," and more a mobile group without a center than a movement in the classical sense, such as a workers' union.[22] In other words: there was no intrinsic 'liberal' property hidden somewhere in the bicycle's technical structure: both its collective and individualistic deployment required active development, and both trajectories were initially still uncharted even if the combination of the promise of individual independence and an embeddedness in republican associationism seemed to be the bicycle's mainstream characteristic.

In a wider context, the bicycle culture should be placed against a background of team sports, which makes the heavy cultural influence of the British bicycle movement on both the early European and the American mobility cultures understandable. Sport and tourism for the middle classes would have been unthinkable without the processes of nineteenth-century industrialization and urbanization, both underway in Great Britain long before the Continent, or before the United States for that matter. Tourism as 'circular mobility' (traveling for its own sake from A to A, returning at the spot where one left) has preindustrial (pre-railway) roots, and can be traced back to pedestrianism, while "the railway excursionist went from A to B and back again on his return ticket."[23]

One neglected cultural root of automobilism is horse racing and carriage racing.[24] But in a deeper seam, walking can justifiably be seen as the basic root of automobilism, as the late-eighteenth-century practice initiated in the United Kingdom as an alternative to the Grand Tour by "educated gentlemen" with the intent to "level with the poor" was a political act: the middle class first had to free itself from societal constraints through the practice of pedestrianism before it could start the long sequence of accelerated mechanization.[25]

From the turn of the century onward, the bicycle started to diffuse among the lower echelons of the middle classes and some upper layers of the workers. Women (and the family culture they brought with them) influenced the originally male high-wheel culture decisively once the 'safety bike' got generally accepted in the second half of the 1880s with its wire-spoked wheels of equal size, 'diamond' tubular frame, and pneumatic tires meant for 'speedy touring.' This, according to a recent analysis, "prompted a 'masculinity crisis,'" resulting, among others, in fraternal bonding among men as a reaction against "feminized version(s) of masculinity" young boys learned from women. This is all the more likely as other research has meanwhile unearthed that the "spheres" of men and women were "far from separate."[26] The 'democratization' toward workers and women prompted *la Petite Reine* to show the highest performance (both in vehicle kilometers and in passenger kilometers) of the entire mobility spectrum (next to horse traction, of course) until well into the 1920s (figure 1.1).[27]

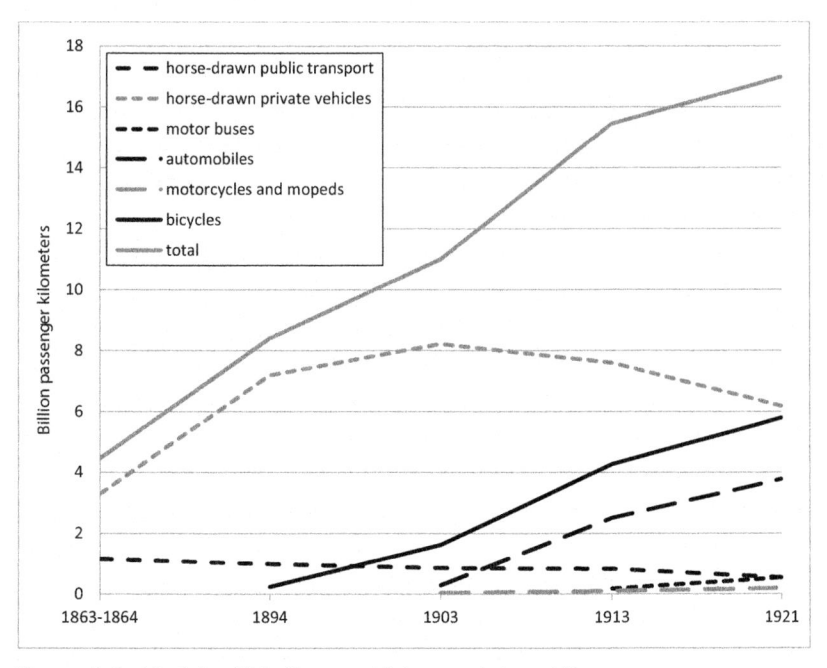

Figure 1.1. Modal split in France. Note: x axis is not linear.
Source: Orselli, 76, table 16.

Although more research is needed, it seems likely that the high price of the earliest automobiles made the car culture more aristocratic than the bicycle culture and more geared toward racing rather than to the touring aspect of the emergent culture. This made it easier to borrow from an aristocratic horse-racing culture including its chivalrous manners of fair play, honor, and male competitiveness. European and American elites did not differ very much in their attitude toward the car: their aversion to 'commerce' and their chivalrous 'amateurism' permeated the early sporting culture.[28] The European aristocracy, from whom these values had been taken, learned to stay in power by adapting to (and absorbing) the upcoming *grande bourgeoisie*, as much as the latter championed aristocratic culture. This convergence, seen by some as a betrayal of the bourgeoisie's cultural ambitions,[29] was especially apparent in England (where Joseph Schumpeter noticed their "active symbiosis"). In many countries (except in England and France) the pre-industrial ruling classes also "maintained their primacy in political society." In France (where noble titles were abandoned during the first years of the Revolution) the aristocracy, "as if to compensate for its absolute political fall and relative economic decline, ... became more self-consciously mannered and proud-minded than any other European nobility ... Rather than standing out as decadent, corrupt, idle, and vain, the French nobility dazzled Paris and foreign notables with its charm, elegance, and finesse."[30] One study claims that the French aristocracy, alienated ever more "from the direction of affairs," excelled in "the cult of prowess, where there is an ostentatious disregard by the actor of the limits to action placed by problems of personal safety and the scarcity of means: the aristocrat does not take cover or collect his change." French aristocrats, this study concludes, "transform all of consumption into prowess.... Motility became graceful gesture, eating became an occasion for display and fine taste, clothing became an occasion for esthetic arrangements of cloth of varying texture and color, aggression became an occasion for the deadly ceremony known as the duel, sex became a fine art or the occasion for a conquest often fraught with great risks to one's life or pocketbook." But in the end, the family's prestige made the "joie de vivre" into a joy of temperance (*mesure*). In other words, the French aristocrat was interested in transforming as "much of life as possible into esthetic experiences."[31] This attitude, aimed at "the immediate satisfaction of the actor" is basically narcissistic: it "puts the self in the center of the world." Of the sports, aristocrats preferred those that allowed "disinterested activities" such as hunting, horse riding, and fencing, sports that also allowed to savor "the taste of risk, of the challenge, the cult of prowess and the tour de force, sometimes also of originality, and a very vivid feeling of superiority." Charles-Louis Baudry de Saunier, founder of a literary review and author of several handbooks on cycling and automobilism, is the iconic example of an aristocrat venturing into more risky and ostentatious sports. In 1909, more than a quarter of the French Automobile Club were nobles.[32] This "post-feudal" or "bourgeoisified" aristocracy played a decisive role in the fledgling French automobile movement, including its highly artisanal production sector.

It is in this context important to stress that until the outbreak of the First World War, despite the noise made by several avant-gardes (and the subsequent attention they had from historians) and despite the observation by historians that Europe "had never before be more magnificent and powerful," the continent was still largely agricultural—even Germany with its extremely rapid industrialization and urbanization. This explains why the post-feudal nobilities and landed elites (and some relevant high-ranking civil and military officials) still ruled this world not because of a conspiracy, but "because of their still massive, if slowly decreasing, economic weight." Moreover, during the first decade of the new century, shifting economic power was not translated into equivalent shifts in political power. Until far into the century the countryside was overrepresented in the national parliaments.[33] In other words, there are arguments which support the observation (formulated in the next chapter) that the automobile as we know it today was in the end a trick played upon an unconfident bourgeoisie by a declining, but still powerful aristocracy: it is the powerful mixture of modernism and nostalgia, of progressivism and conservatism, of new values molded after the old of which the fledgling automobile culture became a part. The bourgeoisie, indeed, was a modernizing class: after the train, photography, the bicycle, and now the car, "the presumption of the superiority of the modern was strong among the middle classes.... Free-market economics blended with the gospel of work was as close as the Victorian middle class came to an ideology of its own." The car fitted nicely into a new culture of adolescence as "a time of self-discovery and rebellion [within the bourgeoisie], a stage of life in which the son (but usually not the daughter) could cultivate a sense of idealism and revel in a feeling of freedom to choose a life different from his father's, though he seldom ultimately did so."[34] For Germany, it is perhaps best to call the car a "plutocratic" (rather than an aristocratic) status symbol: the "*Herrenfahrer*" (just as the "*Herrenreiter*" on the horse) was a "liberal bourgeois," belonging to the "upper 10,000," exactly the number of Germans who in 1910 owned a car for pleasure and sport, according to the statistics.[35]

Despite their fundamental parallels, the European fins-de-siècle diverged in some subtle but important respects into distinct cultures, and they, in their turn, differed from an American culture. In all these countries, however, we find a "composite elite" in all kinds of class and group mixtures.[36] Perhaps the emergence of the automobile can best be called a liberal project, with all the contradictions associated with turn of the century liberalism. This liberalism, in Europe, was strong in France, but weak in Germany, with a not very self-conscious bourgeoisie (and not very prone to conspicuous consumption). Whereas France had a relatively big (and liberal) white-collar class, in Germany the working class was larger, with a cleavage between entrepreneurs and civil servants.[37] It was this 'liberal' class that, more than the aristocracy with its 'blood ties' and more than the workers with their common wage dependency, had to depend on a common moral or lifestyle, or culture, consisting of a sensitivity for competition, a need for individual achievement, professional ethics as well as ideologies and symbols of 'community' (*Gemeinschaft*).[38]

Second Phase: Resistance Against Elite Touring in Heavy Family Cars (1902–1908)

It is, at the current state of the art in mobility history, not possible to distinguish, at a transnational level, between first- and second-phase automotive cultures, although this author's impression is that the pioneers before 1902 were single, adventurous men, whereas after 1902 the couple and the extended family dominated the 'movement.' However this may be, it was the long gestation period of car technology together with the catalyst of a highly publicized emergent bicycle culture that prompted, shortly after 1900, a sudden burst of enthusiasm for the new 'sport' of automobilism. From the onset, close cooperation between national clubs resulted in a well-established exchange of club journals, the organization of international events such as races and tours, the founding of international associations of touring and automobile clubs, such as LIAT (Ligue Internationale des Associations Touriste, 1899) and AIACR (Association Internationale des Automobile-Clubs Reconnus, 1904), respectively, as well as the establishment of transnational associations for road building (Permanent International Association of Road Congresses PIARC, 1908) and cooperation between municipal authorities (International Union of Local Authorities IULA, 1913). Also, multinational car companies played a role in this, as well as a transatlantic flow of engineers and engineering knowledge.[39]

Despite the remarkable uniformity of the early car culture local experiences and traditions aggregated into national styles. If car density

Racing, Touring, Tinkering

(cars per capita) is an indication of the existence of these national styles (see figure 1.2), then German, Italian, and Dutch urban elites were certainly slow followers of the Parisian 'model.' In the Netherlands, this elite lacked the strong aristocratic element so influential in France, the United Kingdom, and the United States for the initial startup of the movement, and it could not fall back on a strong domestic industrial tradition, hence it could not enjoy the benefits of a fledgling national car industry. Also, the Dutch national mobility culture was dominated by a travel culture based on the very dense railway network and rapidly 'democratizing' bicycle culture, the latter under the strong leadership of a touring club (ANWB, founded in 1883) that gradually managed to take on an intermediate role between the central government and the highly

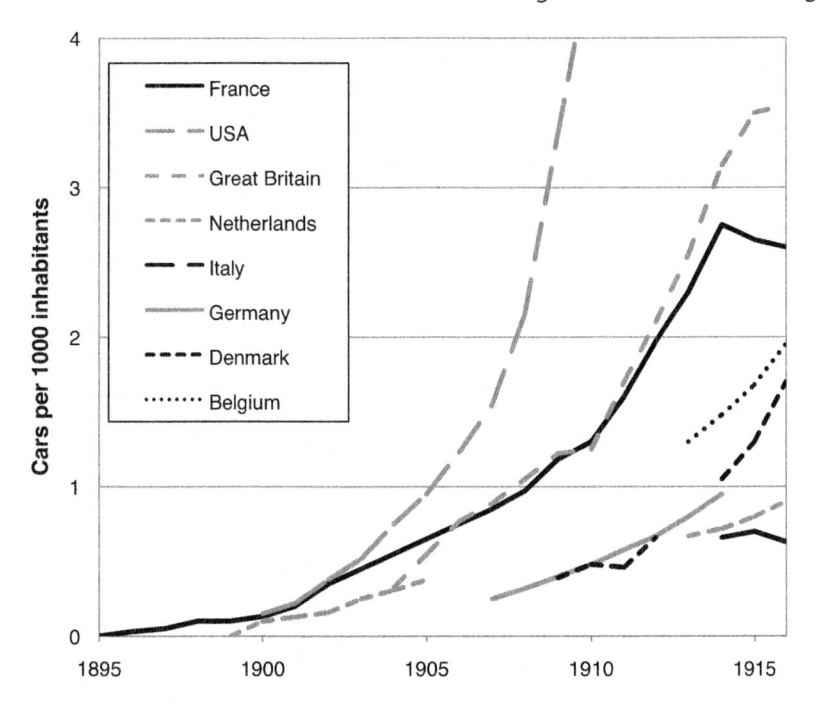

Figure 1.2. Car densities (per 1000 inhabitants) in selected countries.
Source: Number of vehicles (European countries): B. R. Mitchell, *International Historical Statistics Europe 1750–1993*, vol. 4 (London and New York: MacMillan Reference LTD / Stockton Press 1998). Number of vehicles (United States): US Department of Transportation, Federal Highway Administration, Office of Highway Policy Information, Highway Statistics Series; Population: Mitchell (1998) and US Census Bureau, International Data Base. I would like to thank Luísa Sousa (New University of Lisbon) and Hanna Wolf (formerly Eindhoven University of Technology) for collecting and ordering the data and producing this graph.

urbanized countryside through its structure of 'consuls' recruited from the local elite. In this respect, Dutch mobility culture resembled the one in Germany, where the aristocracy, mostly living outside the big cities, does not seem to have played an initiatory role. The German urban middle class also seems to have been less innovative than their British and French counterparts.[40]

All in all, 'European' car diffusion shows an extremely variegated picture per country (resulting in 1916 in a bandwidth between Italy and Sweden, with 0.52 and 0.59 cars per thousand inhabitants respectively, and the United Kingdom, with seven times as many), but their common difference vis-à-vis the United States, with a car density ten times as high as that of the United Kingdom, seems to justify the oft-repeated thesis of the 'American model' and its European bandwagons. This study, however, challenges this iconic dichotomy in automotive historiography on the basis of several arguments. First, a recent quantitative analysis identifies at least three different models of car diffusion: an Anglo-Saxon one (including Australia, Canada, Great Britain, and Argentina), a (northern) European one (the Scandinavian countries and Ireland), and a heterogeneous cluster of all other motorizing countries, a mix of diffusion patterns that, although they are similar, take place in cultures so different that one is tempted to refrain from using the concept of a 'model' on the basis of car-density figures alone.[41] Second, at a disaggregate level of provinces, *länder* and states, the picture appears even more complex than hitherto acknowledged. The differences between northeastern and western American states on the one side and southern states on the other were also very large, while in several highly urbanized European regions car densities soon became comparable to many American regions.[42] And third, earlier work from this author suggests that an alternative distinction explaining motorization better is based on the rate of urbanization, resulting in calling some countries (or parts thereof) 'empty' (characterized by low population densities) and others 'full.'[43]

Clearly, a national frame of reference cannot cope with such an explanation, although this perspective has been very persistent if only because of the pivotal role of the wealthy midwestern farmers in the United States who started to motorize during the second half of the 1900s. Since they practiced a highly mechanized form of agriculture with much tinkering, they knew how to adapt the unreliable petrol car to their needs and were eager to overcome their geographical isolation in a basically 'empty country,' in just the same way as they embraced the telephone, modern plumbing, and electricity during the same period. "Americans beyond New England," a recent social history explains, "lived on scattered homesteads rather than in compact villages," and only later "congealed" into local communities, hamlets, and towns.[44] Thorsten Veblen, whom one could characterize in this context as a philosopher of the Midwest, found the country town to be typical for the American culture of his time. It was there, in the small towns and on farms, where car densities started to grow precipitously.[45]

In Europe as a whole, such a wealthy non-metropolitan class did not exist. While car densities in northern agrarian states in the United States rapidly became the highest in the nation, in Germany—where this has been researched as nowhere else in Europe—agricultural *länder* were the poorest and the least industrialized, and certainly far from motorized, and car ownership remained an urban phenomenon until far in the Interbellum.[46]

In France, a very detailed early (1905) study of the spread of the car suggested a more complex picture, probably because it focused on the local and regional level. Paris was the undeniable center of gravity, showing the highest density of car ownership in its urbanized surroundings (in that respect France resembled the United States, where until 1905 almost half of the cars were owned by residents of greater New York). At the same time, car diffusion in the area around Lyon bore more resemblance to the Dutch situation, while the rural West was more akin to the German picture.[47] In the department of Ille-et-Vilaine in Brittany, early diffusion was most intense in midsized towns along the coast, both by English tourists and local residents of, for example, St. Malo.[48] Just as in the Rhône valley around Lyon, the car was most popular in an urban environment, where the aristocracy, new professionals (medical doctors, veterinarians, lawyers), and shop owners were among the pioneers. The membership of the Automobile Club Dauphinois in 1900 consisted primarily of large and small manufacturers and entrepreneurs, followed by owners of shops, hotels, and cafés. Only then came medical doctors and other "free professions" such as journalists, lawyers, and artists. In 1903 and 1904 these groups still formed two-thirds of the membership (which meanwhile had tripled to about two hundred), but the group of physicians had increased in size.[49] In the department of Cher, too, professionals formed the majority among the pioneers of motorcycle drivers.[50]

The role of medical doctors in fueling early motorization has meanwhile acquired a nearly iconic status in automotive historiography, in the United States especially emphasized by John Rae and James Flink (who perhaps echo the car-friendly trade journals too closely), but for Europe this factor is still not very well researched.[51] French medical doctors in and around Lyon did not motorize in villages, but did so in larger towns. This is consistent with the Dutch picture, where there are signs that countryside doctors could simply not afford the expensive contraptions. In Paris in 1905, only 1.2 percent of car registrations were in the hands of medical doctors, while they owned one-third of all cars in France registered for business purposes. In the British county of Essex, on the other hand, physicians were by far the most frequent owners (more than one-tenth) of the 536 registrants of 1904.[52]

However this may be, in France the countryside started to motorize in the period 1901–1905. After that period, there was an inverse relation between motorization and urbanization, just as in the United States, which reinforces the impression that, overall, France belonged to the group of 'empty' countries, despite the myopia of contemporary observers and historians alike who focused and still focus on the exceptionalism of the Parisian region. Although the country's urban growth was much less than in the United Kingdom and in Germany, at the same time France formed a part of a European model in the making, as motorization in agrarian departments clearly had an urban flavor, as it was concentrated in villages and small towns. In 1905, 37 percent of the cars were registered in "*communes rurales*" (with less than five thousand inhabitants) where two-thirds of the French population still lived; ownership (in cars per capita) was very low, while it was highest in Paris, followed by the larger towns with ten to twenty thousand inhabitants.[53] For Germany, Reiner Flik advances an economic argument to explain this: German farmers simply could not afford to buy cars, even if they had wanted to. An excellent 1973 French analysis by Nicolas Spinga argues that, when urban and rural elite markets approached saturation in 1907, rural motorization stagnated because the French car industry could not or would not develop a "French Ford" and the countryside "refused to Americanize." French rural motorization only started in earnest after WWI, just as in most other European countries, and probably like the United States where the share of the farmer in prewar car diffusion has been typically overstated.[54]

Who were these early users, and what cars did they use? The excellent French statistics show a remarkable high share of 'professional use': 20 percent of national registrations in 1900, 30 percent in 1905, and more than 50 percent in 1912. Likewise, German national statistics between 1907 and 1914 show a large minority of about 40 percent of pure "sport and pleasure" users (they were the largest category in the registration statistics), while commercial deployment (of which three-quarters were motorcycles) took the same share and "other professions" such as medical doctors took about 12 percent (governments at all levels used the rest). Half of the vehicles of these "professions" were motorcycles and 80 percent of the cars were light cars. About half of the commercially used vehicles were in Prussia, mostly in the industrialized Ruhr area, while Berlin counted surprisingly few "commercial" vehicles. The registration directory of British Essex also had half of its cars having "a professional or commercial status."[55] But such statistics should be treated with care, as owners tended to overemphasize car purchase as an act of necessity to secure favorable taxation, although for Germany it seems fair to conclude, because of its high share of motorcycles in commercial applications, that the number of "pleasure cars" was much higher than the number of commercially deployed cars, by about six to four.[56] And, as we will see later in a better-documented period, utilitarianism often functioned as an alibi for largely recreational use.

Aggregate statistics tend to hide such issues, and the answer should come from local, detailed studies of registration documents such as has been done for Orléans, where 70 percent of all cars were designated as pure "pleasure vehicles."[57] In three counties in West Wales "professional" registrations never exceeded 10 to 20 percent of "private" ones before 1914, but if the former were combined with "commercial" registrations (garages, hotels, grocers) they would represent about 40 percent of the total. Medical doctors (were they better paid in Anglophone countries?) and shop owners were the largest groups of owners, but farmers were hardly represented, not surprising for the United Kingdom, where 80 percent of the population lived in towns by 1914.[58]

Petrol car technology supported this culture, not only because it was so much less reliable than the electric vehicle, and hence formed an ideal basis for technology-inspired male bonding, but also because early engine technology favored a constant, high-speed driving style rather than a defensive, stop-and-start practice as would have been necessary in town streets where horse traction, trams, pedestrians, and bicyclists followed their own 'rules' of navigation. Car speeds, the Belgian minister of public work argued during a Senate debate, could not simply be brought down to the low values local authorities often prescribed. The German car club journal concurred: a speed of 4 km/h as enforced in certain municipalities was impossible to maintain, and also made the engine overheat because of a lack of cooling air. Another German motorist who conducted some personal tests stated that a new car in its "most careful and tame pace" still showed a value of 20 km/h on its speedometer.[59] From this perspective the accelerator pedal fooled the early user (and still does so): while it gives the impression that by pushing the pedal down one 'produces' more speed (by opening the throttle valve, as the automobile pioneers knew), this valve in the intake system of the engine was originally put in there to *close* the channel compared to an application in an industrial stationary setting where such a throttle was not necessary, because such an engine ran at a constant speed regulated by a 'governor' working on the basis of centrifugal weights.[60] From this perspective it may not come as a surprise that early motorists and their clubs were as much fascinated (if not obsessed) by the braking power of their cars than by their acceleration, a fact that also may be interpreted as a co-construction by critical non-users.

The difficulty of gearing down (and the reluctance of the industry to implement a solution for this) may have strengthened the largely negative response by non-users in the countryside. In the United States this resistance soon disappeared largely due to a mass movement of domestic touring soon to be dominated by farmers, although this issue, belonging to the accepted canon, would deserve some further scholarly probing: in New York a "Committee of Fifty" used stopwatches to show that cars were speeding in town. The Long Island Highway Protective Society was one of the more aggressive anti-speed "vigilante groups" that resorted to "illegal tactics" such as puncturing tires of speeding cars. Similar opinions could be found among farmers, whose "hostility toward cars reached a peak during the period 1904–1906, when the Farmers Institute of Indiana asked for a ban on cars from using roads" and some wanted to boycott political candidates owning cars. Farmers in Rochester, Minnesota, plowed roads to inhibit motorists to use them. In Evanston, Illinois, they formed an Anti-Automobile League. It was, in the words of one historian of the modernization of the American countryside, a real "crusade against cars," more so than against the telephone, for instance. As we will see in chapter 4, even in the early 1920s American cities were the stage of well-spread skepticism against the car. Nevertheless, Europe seems to have known a more persistent resistance movement, especially in Germany, where the resistance in the countryside has been studied in detail. There, a couple was beheaded in the neighborhood of Berlin in 1913 by a rope stretched across the road. A member of the Bavarian parliament reported about "great agitation" among the rural population which feared the motorists "much more ... than the previous street robbers, who only took money and things, but the motorists also endanger life." An automotive journal observed in 1905 a "car hostile attitude among large masses of the population." One German car owner commented on the conflict between drivers and bystanders: during a Prinz-Heinrich contest all kind of objects were thrown at his head, including flowers made heavier with pieces of wood. This motorist sometimes thought he confronted "lynch justice." In France, too, a motorist was beheaded by a rope across the road, near Profondeville in 1907, but the size of the protest is controversial: recently, a historian found "only" ten messages of popular antagonism against motoring in the automotive press for 1904. In the same year as the Berlin episode, a cable mounted across the main road from Brussels to Namen, near the village of Rhisnes, injured a Belgian couple. There are also reports from the Canadian countryside of similar events. The rural resistance in Ontario seems to have been especially fierce, while the mayor of Winnetka, Illinois, had a rope stretched across the road to stop speed offenders. In another state, a motorist was shot at. Anecdotal evidence in all national car histories support the impression of a large and fierce resistance, which in Switzerland, with its strong regional autonomy, resulted in a ban on automobiles altogether in the eastern canton of Graubünden (Grison) until the mid-1920s, without any discernible negative effects on the tourist trade. Calling the population of this largely agrarian canton backward does not help, as by 1912, 40 percent of Switzerland's entire surface was forbidden for the "folk enemy automobile."[61] In the United Kingdom the counties of Surrey and East and West Sussex were notoriously anti-car.[62]

Resistance was not limited to the countryside: in early summer 1902, "mobs of boys in New York sabotaged autos by stoning them; news of these incidents appeared almost daily." Motorists driving through urban working-class neighborhoods "were so harassed by stone throwing that drivers demanded police protection."[63] The anti-car resistance was not anti-technological, but inspired by anger about the intrusion of arrogant, aggressive upper-class urbanites into recreational and social street space, and by fear for limb and life.[64] Here, too, the anti-bicycle movement had started this tradition, with stone throwing, "disturbances," and "acts of sheer banditism." Former United Kingdom Prime Minister Arthur Balfour, to name only one famous example, was hit on the cheek by stones when biking back from a tennis match near Cannes.[65]

That a skeptical French historian referred to this as "sheer banditism" suggests an answer to the vexing question of how to judge this resistance. Also, the enigmatic "organizational weakness of the protest," observed by other historians, fits this answer, which (despite the much broader criticism from parts of the local and national bureaucracies, and the police and justice) should be seen as a form of class struggle with a typical anarchist, socially marginal aspect. Not only the German motorist who wrote a pamphlet *Der Krieg gegen das Auto* (The War Against the Car) but also the later American President (and then—1906—Governor of Princeton University) Woodrow Wilson saw class antagonisms shimmer through the protests. Wilson not only warned against the "socialistic feeling" that ostentatious car culture could stir up, he also—less well known—gave the remedy: "I am a Southerner and know how to shoot." An American motorist saw stoning of a motor car as "the fruits of the Anarchistical utterances," a clear reference to "anxieties about lower-class families, youth and juvenile crime (that) appeared throughout Western Europe and the United States during this period."[66] Also the use of words like *apaches* indicates the type of protesters motorists were describing: "knife-wielding muggers" in a period in which "hooliganism" seemed to take on alarming proportions. "Fin-de-siècle culture was remarkable for its questioning of the rationality and superiority of mainstream European culture," a study of urban and rural "hooliganism" concludes, and the same applies to the United States. "What is less well known is that a similar challenge was being posed from below. At the same time that fin-de-siècle intellectuals were exploring the newly discovered realms of irrationality, violence, and sexuality, hooliganlike waves of crime were presenting the same issues on the public streets." Hooligans were marginal to the urban and rural lower classes. Their behavior was no ordinary crime: "Without an audience, hooligans would not have existed.... Their acts took on special meaning and power because they performed them in a way that provoked a potent response." For the period under consideration there is a special truth to this, as not only hooligans, but also other groups displayed provocative behavior, including avant-garde artists: "As the public sphere widened, hooligans and avant-garde artists were not the only ones confronting old standards with new kinds of public behavior. All sorts of people entered the public sphere for the first time or on a new scale: women, elected politicians, workers, professionals, and social reformers of many kinds"—and motorists, one is inclined to add: ostentatious display and raw protest were two sides of the same coin. "The abhorrence of cars," a recent study of "Autophobia" concludes, "is inseparable from their appeal."[67] For historians of technology, the conclusion to be gained from this episode is even more revealing: it shows how the so-called non-users co-constructed car technology and culture, for instance through the introduction of speed limits, which were first imposed by municipalities (if they had the power) and then by the central state.

Gradually, this anarchistic element of car resistance subsided and a political struggle occurred, especially between 1905 and 1910 when most national parliaments started to debate the necessity of automobile taxation and traffic regulation. This heavily under-researched aspect of early automobilism worked out differently in each country, but in most the controversy shifted from a class antagonism to a tension between an urban and a rural culture, and even to a tension between 'situational groups' of motorists against non-motorists. Often, the worker parties and the unions stood on the side of 'progress,' but in any case were against 'plebeian' Luddism. In the Bavarian parliament, for instance, a coalition of urban left and right opposed rural representatives assembled in "*das Zentrum*" (the Catholic Party) who, according to one member, still followed the "war path against the automobile," because it interfered with rural road practices during harvest time: "if we don't get a law against them, we will simply gun those guys down."[68]

No wonder, then, that automobile and touring clubs all over the Western world tried to present the car as a "necessity," or a "productive luxury" (as German motorists argued).[69] In all these countries the central states appeared remarkably receptive to accommodating the car. Whatever the specific outcome, most national governments de facto recognized at least the car's future potential by restricting its use, regulating its speed, and obliging drivers to register. In most Atlantic countries this recognition-cum-regulation was endorsed by the local clubs, if only to keep the regulation out of the hands of the lower governmental echelons.[70]

And yet, it is remarkable that in this early phase, the car enjoyed greater importance in societal discourse than its quantitative or economic presence would suggest, even in France, perhaps the only country outside the United States where a consensus existed about the economic importance of the fledgling automobile and related industries. But there, too, just as in other countries, expectations played a crucial role in the history of individual motorized mobility. They help explain why elite car cultures in European countries had so much in common. The extensive reports on foreign developments in the national public sphere (especially through the club journals) functioned as a sketch of future domestic developments.[71] Emulation was rampant.

To cite the Dutch example: in 1899, when not more than a hundred cars operated in the country, the Minister of *Waterstaat* (Water Management) Cornelis Lely described to a skeptical parliament a future in which automobiles would run over well-paved roads at high speeds, uniting national spaces in an unprecedented way.[72] The role of Dutch motorization within the European context did not stop at the country's borders. The small size of the country soon directed the automobile and touring clubs' attention across its national borders, and gave it an international flavor. Not only was Amsterdam chosen as the halfway meeting point for one of the first international automobile races organized by the French automobile club in 1898 (not by coincidence, the accompanying media event resulted in the founding of the Dutch automobile club NAC). The touring club ANWB was also instrumental in initiating the first meetings which resulted in the founding of the *Ligue Internationale des Associations Touriste* (LIAT) for which it would act as secretariat. While in other European countries the national automobile club developed into one of the major players (if not *the* major player), Dutch mobility culture was characterized by a relatively small, but loud and mainly upper-middle-class (rather than aristocratic) urban automobile culture (concentrated in and around the NAC) and a much broader, deliberately anti-commercial (and, hence, anti-racing) tourism movement, espousing travel by train, on foot, and by bicycle, all closely managed by a touring club with solid roots in the provincial urban elites. Much less than in neighboring countries, the touring club stood for a largely non-motorized tourism based on hiking, camping and bicycling until well after the First World War.[73]

The readers' questions sent to ANWB's *De Kampioen* (The Champion, founded in 1885) and its *Technisch Bijblad* (Technical Appendix, introduced in 1907) rarely addressed the car itself. While the ANWB's officials were focusing on motorization in the journal's editorials, the club's membership seemed first and foremost interested in shaping its increasing leisure time in which the car only played a marginal role. By the end of the war, when a ban on petrol made driving virtually impossible, a poem appeared in *De Kampioen* that opined that the disappearance of the car and the motorcycle from the Dutch streets made hardly any difference in how people lived their daily life.

Car-related practices, then, were mostly mediated through the automobile clubs and its members, rather than the touring clubs. The NAC in 1910 represented about one-quarter to one-eighth (depending on the method of estimation[74]) of all Dutch automobile owners. The car adventure, as represented in club journals, now has been analyzed for several countries, notably the United States, France, and the Netherlands.[75] Even in the Dutch case, which seems to be most comparable to the "modest motoring"[76] of the British, fascination for the masculine, violent aspects of car driving trickled down in the contributions sent in by readers of the Dutch automobile club's magazine *De Auto*. By and large, the editorials and other contributions in *De Auto* confirm the international picture of the pioneer car movement: elitist (defining themselves as 'travelers' rather than 'tourists') and interested in the car as a tripartite 'adventure machine' (racing, speedy touring, and tinkering, the latter mostly done by the chauffeur/mechanic). On the one hand, 'Grand Tourism' (across the Dutch borders) was reported as a discovery of the picturesque, always mixed with practical tips on special spots to visit and advices for behavior in strange environments. Travelogues on tours of several days within the country largely followed the same schedule (and photographs of 'picturesque' scenes often accompanied them): point of departure, route followed, distance and speed traveled, defects encountered, the condition of the road, anecdotes about people met along the voyage, vistas along the route, practical suggestions how to deal with hostility in certain towns, return to the point of departure. In general, the tours had no clear destination, although the aim of the tour was often expressed in terms of a "flight from the busy city" or "vagabonding through the prettiest parts of our country." Remarkably, the chauffeur was given much less attention than might have been expected, perhaps also because the journal was shared by motorcyclists who clearly had a greater inclination to tinkering, and knowledge of their vehicle in general.

The elitist character of the early car and motorcycle culture became especially clear during 1906 and shortly thereafter, when the Dutch national government responded to protests in the countryside about the "racing maniacs" by proposing legislation that curtailed road speed considerably, especially in towns. The outcry among motorists reveals that they considered speed to belong to the very essence of what automobilism was all about. The resulting aggression was equally revealing. Now, trip reports started to appear in which the countryside was depicted as full of "barbarians," "cannibals," "apaches," and "*kaffers*" (a racist name for an African tribe evolving into a general epithet for blacks, and then for everyone allegedly not endowed with bourgeois civilization). Tellingly, as early as 1910, 27,000 of the national total of 160,000 lower-court decisions were for misbehavior in road traffic (including offenses by bicyclists).

Anecdotal evidence from NAC's *De Auto* and from other sources makes it plausible that owner-drivers were a minority and that the 'chauffeur/mechanic' was the norm, driving ever-larger automobiles (up to two tons and above) for the entire family.[77] The problem with these chauffeurs was, as the German car club's journal explained, not so much that they needed two types of technical skills, for repairing and maintenance and for driving, but that their social skill was more akin to that of a valet than a mechanic.[78] As in France, and especially in Germany, Dutch motorists were also confronted with aggression and the deployment of power, by country people and their stone-throwing children, by local police officers who organized 'speed traps,' and by municipal councils who declared their streets unfit for the new vehicles. The speed traps seem a UK invention, since local authorities (including judges) had a direct interest in the considerable income to be gained from fines, to be reinvested in repair of the roads destroyed by speeding motorists.[79] And just as in those countries, the Netherlands too saw an event that offered a glimpse of the future 'utility' of the new vehicle: the railway strike in 1903 that NAC members defused by driving policemen and mail bags through the country at speeds normally never allowed.[80] As in other countries, the self-confidence of the early motorists was revealing: they (at least those who were organized, and their circles of friends) saw themselves as the harbingers of 'modernity,' bringing speed, efficiency, and technology to backward countryside areas. It becomes apparent especially in *De Auto* that automobile pioneers did not doubt for a moment that the future was theirs. If the role of the state in this first phase was important at all, its importance mainly is to be found in this early legitimation of motorists' claim upon the use of the road, brought about by the ideological and personal overlap between the higher level of the state and the car movement.

In Germany, this intermediary phase was clearly delineated right from the beginning by the kaiser's decision to start backing the fledgling movement and invest in a 'stable' of cars.[81] In Germany, too, the founders of the *Deutsche Automobil-Club* (DAC, 1899) were members of the nobility, high-ranking military, factory owners and landowners, bankers and diplomats. They soon got imperial backing (and renamed themselves *Kaiserliche Automobil-Club*) and did not shy away from an openly commercial coalition with the associations of automobile and aviation manufacturers.[82] The competing mass organization, the *Allgemeiner Deutsche Automobil-Club* (ADAC) was founded (in 1903, initially under a different name) by motorcycle owners (with a lot of help from the NSU bicycle and motorcycle company). It soon opened its ranks for low-priced "*Volksautomobile*" (people's cars) owners as well. The latter represented the majority among the members, indicating that the bulk of the motorcycle owners were non-organized. Members were recruited mainly from the "*Mittelstand*" (there were hardly any members of the nobility), among whom those with commercial interests ("*Kaufleute*"), as well as "independent professionals" (lawyers, notaries, architects, medical doctors) and civil servants were very well represented. The ADAC, like comparable middle-class clubs in other European countries, grew precipitously: within ten years it boasted a membership of 28,000, representing 37 percent of all German motorized individuals.[83] The acceleration in membership growth after 1909, until interrupted by the outbreak of the war, is comparable to that in the Netherlands. Figure 1.3 is a nice illustration of the three-pronged structure of the pre-WWI period: a first burst until 1905, interrupted by stabilization with slow growth into the promise of a real breakthrough.

In other European countries a different mix of car and touring clubs led to a great diversity of local and regional car cultures, the most extreme of which is perhaps the one in Norway where the car, initially very sparsely diffused, appeared as a utilitarian contraption rather than a 'pleasure vehicle,' which one recent analysis explains as the result of the largely agrarian context.[84] At the other extreme, the national government

Racing, Touring, Tinkering

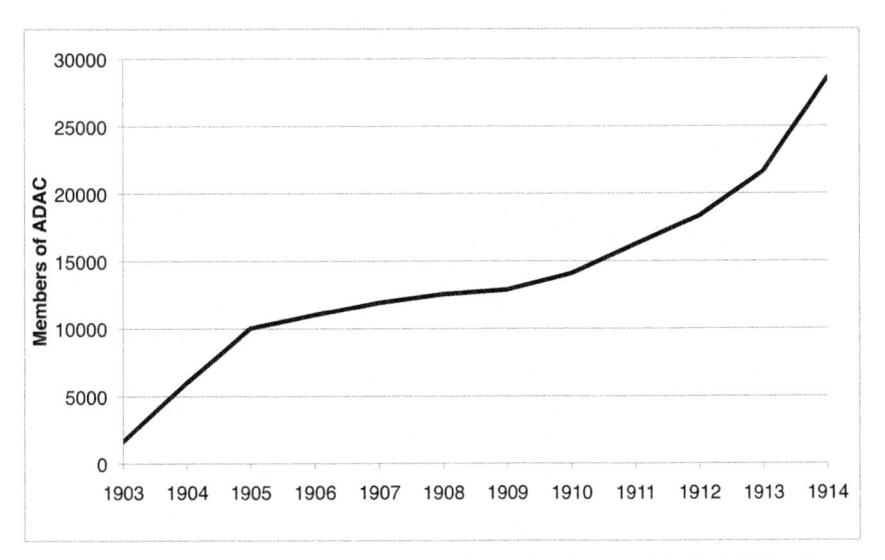

Figure 1.3. Members of the *Allgemeiner Deutsche Automobil-Club* (ADAC). *Source:* Seherr-Thoss, 27.

of heavily urbanized Belgium (where as early as 1909 three-quarters of the senate owned an automobile and the national automobile club's leaders were dominated by the nobility) did not introduce restrictive legislation before the First World War.[85] In all European countries similar dual institutional structures of elitist, racing sport–oriented automobile clubs versus a (tendentially lower-) middle-class mass organization emerged, often accompanied by a third independent center around a motoring personality with a journalistic bent such as Gustav Braunbeck in Germany (*Das Schnauferl*), Louis Baudry de Saunier in France (*Omnia*), Lord Montagu in Great Britain (*The Car, Illustrated*), or Tom Schilperoort in the Netherlands (*Auto-Leven*). If a generalization of a highly variegated institutional landscape in Europe is allowed, one could say that automobile clubs were elitist, bad organizers, arrogant, and prone to racing, whereas touring clubs were excellent organizers, skeptical toward commerce, and more inclined to promote 'tourism' in which the car only played a marginal role. Several of the latter had a hard time convincing their members that automobilism was nothing more than an extension of the bicycle sport and some of them lost interest during the middle phase dealt with here, because of the dominance of ever heavier, expensive vehicles. This is nicely illustrated by the tapering off of the ADAC membership curve between 1905 and 1909 (see figure 1.3). But never was the affinity between sport and media broken, as this was based upon an "intrinsic kinship" between these media and speed. While the institutional press fit a tradition of sports publicity, emphasizing the club spirit that should be used to gain social and political clout, the 'independents' (regrettably under-researched) more sought what we now would call 'lifestyle' and 'human interest,' each 'flavored' after the local press culture. Montagu used his extensive network to fill his heavily illustrated and 'glossy' magazine with elite gossip, while Baudry de Saunier shaped his journal into a technical course. Schilperoort played the bohemian, paying special attention to the subversive role of women (for instance as early motorcycle drivers) whereas Braunbeck filled his *"fliegende Blätter für Sport-Humor"* with caricatures and comics. Like Schilperoort, he later steered his journal toward depicting the 'elegant world,' devoting a remarkable space to the 'new woman.'[86]

A comparable outside (and equally under-researched) role was played in this early culture by motorcyclists, often much larger as a group than motorists. The motorcycle dominated motorization in Germany, the United Kingdom, France, Switzerland, and the Netherlands as the Great War approached, but not in the United States. For instance, in Washington, DC, in 1908, 1000 automobiles were registered, but only 300 motorcycles. In 1909, 123,000 cars were produced in the United States, and only 19,000 (15 percent) motorcycles. The Federation of American Motorcyclists desperately proposed to "burn leather coats and britches of members" to improve the "careless appearance" of motorcyclists.[87] In France, registered cars overtook motorcycles in numbers as early as 1907 (see figure 1.4), but in several European countries the dominance of the

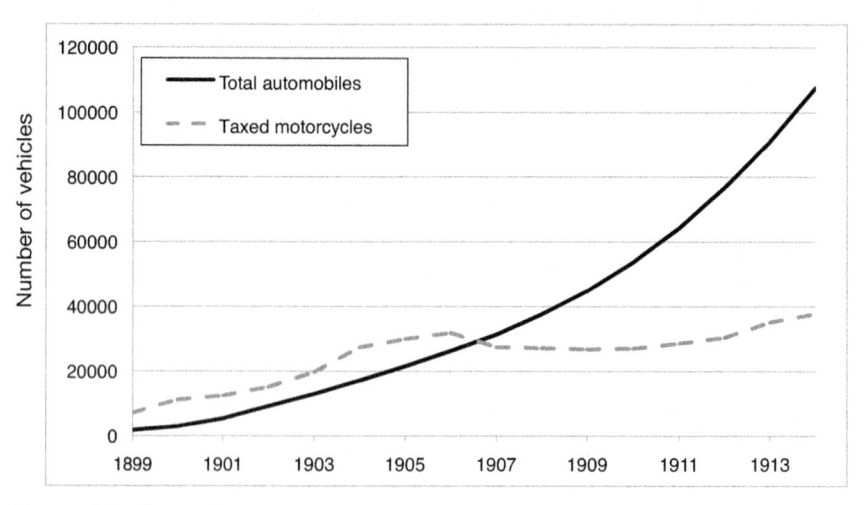

Figure 1.4. Car and motorcycle park in France.
Source: Orselli, 212, table 33.

motorcycle lasted until the 1920s, with Germany, an extreme case, even experiencing a revival during the 1930s, especially in the countryside.[88]

The little-studied motorcycle culture was important from an anthropological point of view. Not only culturally but also technically a successor of the bicycle rather than the carriage, it was less elitist, more prone to the mixed use of pleasure and utility. In the Dutch case it was more frequently the object of tinkering, as the act of purchase implied that one had to cope without a chauffeur. As such, the motorcycle was the carrier and the expression of a form of adventurous practice, which gained all the more importance once the car started to acquire utilitarian characteristics around 1910. Even the Dutch version of the classical first adopter, the countryside medical doctor, preferred the motorized two-wheeler (in this case perhaps also because of a lack of income), and around this vehicle a culture of tinkering developed comparable to what American historians have described for their country's early car culture.[89] It was motorcycle tinkering culture that, for the first time on a massive scale, brought repair and maintenance out of the factory when local blacksmiths had not yet acquired the relevant skills nor had the auto mechanic shop appeared.[90] In the Netherlands, it was the motorcycle (produced locally more than cars, but many also imported from the United Kingdom and Germany) that prompted most of the technical questions sent to the journals of ANWB. Many of these questions were geared toward the problem of how to construct a safe connection to a sidecar, allowing a couple to indulge in touring. If the question section of *De Kampioen* and its *Technisch Bijblad* can stand for the Dutch motorcycle culture in general, it shows also that the preferred use of the motorcycle was for recreational purposes, a practice best called 'transcending the local.' The desire to leave the locality only to return there after a while (performing trips from A to A rather than from A to B) revealed itself most prominently in the explosion of questions in 1917–1918 (at the moment the Dutch government prohibited car and motorcycle use) on how to build and use radios; even the questions about the battery were often not aimed at a deployment within the car or the motorcycle, but at powering a radio.[91] Apparently, the 'sport' of tinkering on and actually building radios and using them to 'tour' the airwaves was closely intertwined with the desire to physically tour the countryside, and briefly gained prominence when the latter was prohibited. The history of transport, which separated itself from the history of communication since the late nineteenth century, tends to overlook this 'intermodality' between virtual and physical mobility.[92]

In the United States the outsider role was played by the magazine *Outing*, among others. From the columns of this "Illustrated Monthly

Magazine of Recreation," which started as an outlet for bicycle culture, it is possible to reconstruct the changing ideology of the American automobile movement on the East Coast: the shift from elite racing with large cars toward long-range touring in ever-lighter and less costly cars accompanied by a gradual shortening of the touring trips performed by "the man of moderate means."[93] American owners imported the large, heavy cars from Europe, especially France (despite the protectionist tariff of 45 percent) where they supported a speed craze on the smooth roads.[94]

Going through largely the same phases as in Europe (although touring in the immediate vicinity of towns started later, in about 1902), American automobilism spread much faster among the urban middle class. Envious of the hedonism of the upper classes as well as the working classes, the new and old middle classes developed "new standards of masculine and feminine beauty," got caught up in a "cult of fitness" but still felt "clearly uneasy about the place of pleasure in life." In their "crusades against drink, prostitution and divorce," the Progressivist reformers welcomed the car as a means of "family restoration," but they could not prevent that Victorian thrift yielded to consumerism. Right at the period 1904–1906, when rural resistance reached its height, long-range touring started off. At the same time the urban luxury car market reached saturation and farmers started to motorize.[95] It led to a craze of continental crossings, some by women (and some of these women on motorcycle), the equivalent of European "transcontinental" tours to Africa, or the highly publicized race between Beijing and Paris.[96] As a confirmation of the two turning points that mark the beginning and end of this intermediary phase, the first continental crossing occurred in 1903, whereas in 1908 the first family crossed the continent in a car, starting off a craze of coast-to-coast pleasure trips. The first fully female cross-country trip was undertaken in 1909 by Alice Ramsey ("born mechanical," as she wrote in her reminiscences half a century later) with three friends, in a Maxwell, a trip that took forty-one days. Sponsored by the *New York Globe* and sent off from New York by female car racer Joan Cuneo, among others, Ramsay had to ask permission from her husband. By 1911 such trips were "undertaken so frequently ... that they no longer attract attention."[97]

A First Analysis of Automotive Adventure: The Masculine 'Conquest of Nature'

What functionalities did the early car's properties allow? What practices did its technical structure afford? If we want to understand early automobilism, we will have to identify the main practices drivers developed in this period.[98]

Academic scholarship on early car mobility has dedicated surprisingly little research to racing, or, in a wider orbit, to competitive motorized sports and contests, including reliability runs, speed-record contests, 'corsos,' and gymkhanas, all practices on which the first phase of automobilism was predominantly built.[99]

Much more work has been done on the second main practice in this earliest period: touring in the countryside, although the underdeveloped state of racing research has prevented most analysts from comparing those two closely intertwined practices, and hence overlooked how borrowing from racing decisively shaped touring, especially because both practices shared a high esteem for speed and the thrill of acceleration. Tinkering, the third adventurous aspect of early automobilism, was a matter of course for a touring party, even (or perhaps especially) if one delegated this part of the adventure to a driver-mechanic. This seems to have been much more prevalent in Europe than in the United States with its high labor costs and its egalitarian self-help attitude toward technology, especially among farmers. When by the end of the first decade the French Touring Club (counting 130,000 members in 1912) started to regain interest in car touring again (after it had, just as its Dutch counterpart, lost interest when automobiles became heavier and heavier after the turn of the century to transport aristocratic and bourgeois families), its members started to demand tinker-friendly products, pointing at components that were not accessible for the mechanically inclined motorist.[100]

Tourism by bicycle and car has at least four roots: the discovery of 'nature,' the emergence of organized leisure (recreation), consumption and its therapeutic functions, and imperialism.

The rediscovery of 'nature,' with or without a chauffeur, showed a strained relationship with the other two adventurous functions: tinkering (and in general the fascination for the noises and odors of the mechanism) stood in the way of enjoying the environment, and speeding had a similar effect. When the German emperor decided to make a tour in the countryside, his marine cabinet chef remembered, he drove so fast "that his ability to perceive the lovely details in the image of the landscape was made impossible, and besides one was in the cars behind constantly enveloped in a thick cloud of dust."[101] Also, touring was mostly a collective experience for a small group (extended family, a party of several couples). Hence, it is important to stress the Foucaultian function of the clubs. According to the *Ossavatore Romana* of the Vatican, the bicycle sport was "a veritable anarchy in the world . . . ; it is a hermaphrodite, undefinable, which escapes from all laws of movement, of traction, of transport."[102] Associationism thus had a function to counteract the seemingly intrinsic individualism of the bicycle. As the French media historian Catherine Bertho Lavenir argues, the French Touring Club, founded in 1890 and modeled after the British example (Cyclist Touring Club, founded 1878), adapted a military model of organization and touristic practices (she speaks of a "collective pleasure"). The clubs functioned as a pseudo-parliament in their effort to reconcile the different interest groups, such as tourists, local inhabitants, and others. When it included car touring again in its activities, the French sporting *petit-bourgeois* embarked on a political project that not only included the invention of a regional French cuisine, but also the reforestation of the French Alps and the reorganization of the regional *hotellerie*.[103] Similarly, Claude Fischer has recently argued that (lower-) middle-class voluntarism tempered American 'individualism' in a period of an explosion of clubs, fraternities, and lodges and an urban "bachelor subculture" served by "saloons, billiard halls, and brothels." Only later was this "networked individualism" of the clubs partly taken over by the "companionate model" of the nuclear family, living more and more in suburbs. Fischer concludes that American iconic individualism "did not turn (Americans) into free lovers, free thinkers, ramblers, rebels, or anarchists; they remained by Western standards remarkably committed to family, church, community, job, and nation—quite bourgeois."[104]

The new, 'bourgeois' tourism started on foot: it was organized pedestrianism in nature (made possible on a massive scale by the railways, such as the Sunday Tramps in the United Kingdom) that defined the 'touring adventure' (*Wandern* in Germany, rambling in the United Kingdom) as a safe practice, as a combination of "discovery" and "reassurance," the latter performed by the guides (the Baedekers in Germany, the Murrays and the translated Baedekers, that eventually pushed Murray from the market in the United Kingdom, the Michelin guides later in France) and the infrastructural support from the clubs. Tourism in the United Kingdom was performed mostly by the "young thinking men and women" of the "urban clerical and artisanal classes." A contemporary found them "decently clad and fed," enjoying "a home and a friendly circle, who knows how to respect himself and be respected by others." They also liked walking and had a surprisingly "large number of women" among their members. This was part of a "'middle classification' of travel."[105] Alpinism was among the first of these practices, where the journey became a goal in and of itself. The departure, the leaving behind of the daily life, became at least as important as the destination, where the tourist could indulge in his or her temporary adventure with the return in mind, a crucial difference from other forms of travel such as migration. This "pseudo-departure" (leaving home in the secure knowledge of returning soon) and the subsequent fundamental "liminality" of the travel experience (as a "social and spatial separation from the normal"), form the basis of the fundamental "deviant" character of the tourist practice, and, according to John Urry, constitutes the "tourist gaze."[106] As we will see in the next chapter, the tourist adventure had a high level of self-centeredness, and functioned on the basis of a 'distancing' of 'nature' from the gazing subject, including the exotic 'Other.' In this process the bicycle and especially the car fitted perfectly, as these vehicles enabled what the wandering group desired: indulging in the unexpected, but at a safe distance.

To put bicycle and car tourism in perspective it is useful to emphasize that both cultures were built upon an already well-established massive phenomenon, mostly carried by the railways. Shortly before the First World War, Blackpool boasted of receiving nearly four million visitors annually.[107] For a century the European continental *hotellerie* mirrored itself on British middle-class standards of comfort and cleanliness, a cultural transfer of enormous proportions (in the process the British tourists brought their sports to the continent; British subjects founded AC Milan, hence its anglicized name). This changed with the advent of the bicycle, when the thousands of British bicyclists, roaming the countryside of Brittany, the Netherlands, and as far as the Rhine in Germany (brought there by train), were joined by tens of thousands of continental Europeans during the bicycle craze of the 1890s, shortly before tourism really exploded. In 1911 more than half of all inhabitants in England and Wales went to the seaside at least once, while the number of cross-Channel passengers increased by more than 50 percent during the first decade of the century, although a significant part of this flow consisted of British residents of France and their visitors. An inner struggle accompanied the decision to indulge in pleasure: tourists needed an 'alibi,' which many found in either their health or their education, although it is not clear whether this also applied to the "Cook's Vandals," the people traveling with Thomas Cook, "ridiculed for their ignorance and vulgarity." Thus, "wildness" (as the American naturalist John Muir said) was increasingly seen as a "necessity," an inescapable locus for regeneration of society, part of an "elementary human need" (*Lebensbedürfnis*). Many tourists gave in to "an urge towards the periphery" where they could gaze at 'the Other.'[108]

There is a second root of tourism (next to wandering pedestrianism), which has to be sought in the emergence of leisure time (during the week, rather than during the year as annual holiday) and its gradual separation from work, as an increasingly necessary locus for regeneration of the self after the alienation during the hours of work. Recreation started as a "large city movement" and was thus immediately linked with urbanization and industrialization.[109] For the United States, the emergence of the "Sunday drive" has been located in the period immediately following the Civil War, a habit that seems to have been introduced by the German American community. The idea of a necessity is crucial here: "People injured on Sunday in accidents caused by poor roads, defective transportation machinery, or negligent workers could recover damages only if they could prove that they were traveling for charity or necessity.... Others injured on streetcars, in private carriages, or while walking faced questions from the court about why they were traveling. If their answers even hinted at pleasure, then it was assumed they were violating the religious and legal order to rest on Sunday." One historical study of American leisure claims that initially middle-class Americans used their cars on Sunday for the spin into nature, a typical example of "family-centered activities," "preferring to use horse-drawn carriages on weekdays."[110] In this context the bicycle, and especially the car as it enabled to cover large distances, was literally a liaison between 'civilization' and 'nature,' between the present and the past (because the trip into nature was invariably experienced as a journey into civilization's memory) in the process creating a dichotomy in the minds of the motorists that since then has been extensively criticized by environmental historians. 'Nature,' according to these historians, is not only an umbrella concept that hides a multiplicity of 'natures' (in different geographies, time periods, and for different people), but is also a construct, an idealized state put in opposition to 'culture,' thus enabling it to be seen as something to be conquered, domesticated, consumed, and plundered.[111]

Few motorists would have realized that by "enjoying nature" or by "reveling in nature's beauties," they were enhancing this fateful construction full of dominant male sexual and therapeutic connotations equating women, nature, and landscapes. Indeed, travel accounts testified of male dominance, which is, according to current sociology of mobility, closely related to the "priority apparently given to the malevolent power of the visual sense," a power of colonialism and racial oppression that goes hand in hand with a commodification of 'nature,' and the emergence of a dominant "pleasure principle" in tourism. The 'landscape' (and the scape in general, including the cityscape) becomes a "way of seeing the world," a synonym for "nature as spectacle" and the result of an appropriation of spaces through "consumption." Hence, the landscape has a dominantly pictorial, if not "painterly" connotation.[112]

This phase was characterized by the founding of conservative-modernist institutions such as *Bund Heimatschutz* (1904) in Germany and the Prussian department for *Naturdenkmalpflege* (1906), the National Trust (1895) in the United Kingdom, the *Société pour la Protection des Paysages* (1901) in France and the National Game, Bird, and Fish Protection Association in the United States. The first international conferences on *Heimatschutz* were in 1909 (Paris) and 1912 (Stuttgart). Associations to indulge in 'nature' were founded at the same time, such as the German *Wandervogel* (1901) and the Boy Scouts (1909) in the United Kingdom (and later in the United States). The movement for 'natural monuments' would grow into a typical European, transnational campaign, with a slightly anti-American flavor, as it rejected the national park approach.[113]

In an influential analysis of the Anglo-Saxon pastoral ideal the literary historian Leo Marx has tried to make plausible how the gaze upon 'nature' was shaped by the reading of its belletristic descriptions. Intent on analyzing "literary responses to the onset of industrialism in America," and at the same time "an enthusiastic wilderness camper and amateur ornithologist," Marx's hunt for metaphors resulted in the finding of a very potent symbol governing his entire analysis: the machine intruding in the garden, "its meaning ... carried not so much by express ideas as by the evocative quality of the language, by attitude and tone."[114]

What Marx reconstructed was the decline of an old, harmonious pastoral motif and its replacement by a much more complex, dualistic ideal in which the 'machine' (standing for "crude, masculine aggressiveness in contrast with the tender, feminine, and submissive attitudes traditionally attached to the landscape") and 'nature' are somehow, and somewhat uneasily, combined. With industrialization, a new ecological consciousness emerged, which substituted the metaphor of the wilderness for that of the garden, and which shifted from the shepherd to the farmer as the central actor. The emergence of an urban, industrialized culture in the midst of the agricultural taming of the American wilderness resulted in a "symbolic middle landscape created by mediation between art and nature," between the artificial and the unspoiled, leading to a new, "technological sublime." But the paradox of civilization and nature, of progressives and neo-romantics, cannot be resolved. The German historian Rolf Peter Sieferle sees both groups involved in the construction of "mirrored Utopias (*spiegelbildlich verkehrte Utopien*), sometimes projected into the future, and sometimes in the past."[115]

Marx does not connect the concept of this "middle landscape" to the obvious emergence of the middle class, an addition we will use to link up with the automotive symbolic 'conquest of nature' after the turn of the century. The middle landscape was in fact the result of a merger of "the arts of travel, poetry, painting, architecture, and gardening," a landscape that gave as much pleasure as it provided utility, that is, profit to the owner. Students of Marx have meanwhile extended his analysis by providing a typology of 'middle landscapes' after 1900, including a hyper-pastoral type, where the pastoral is fully simulated, in a Baudrillardian, hyper-real way.[116]

Connected to this middle landscape was the theory of the "middle state," defined by Richard Price, friend of presidents Jackson and Jefferson, as "the happiest state of man ... between the *savage* and the *refined*, or between the wild and the luxurious state." This middle state had a definitely anti-European flavor, which sets the American "common man" as an independent, anti-intellectual farmer against the misery of the European workman. Within this middle landscape the train appeared, "flying from place to place with the speed of thought." One wonders, then, what the bicycle at the end of the century, and the car immediately following, contributed to this "rhetoric of the technological sublime," after the train already had harvested the "incredulity" of speed. As a matter of fact, the joy of speed has a much longer tradition, going back to the carriage days of Samuel Johnson, who called going at full speed in a carriage "one of the greatest pleasures in life." In the following chapter we will go into detail about the bicycle's contribution, which consisted in the individuality of spatial roaming in a collective setting, whereas the car then would combine the two: the temporal and spatial aspects of "adventure."[117]

Much more soon appeared in this middle landscape than only the train. Preceded by "'literary commuters,' those idealists whose essays laid the spiritual foundation for an age of suburbs," middle-class America discovered a way to reconcile 'civilization' and 'nature' in a seemingly permanent way by creating 'suburbs.' One of these literary commuters was the naturalist John Burroughs, who, with his forty volumes as guidance, "brought us home in time for tea," thus emphasizing the limited adventurousness of the undertaking. Another middle landscape was the urban park, originated in the United Kingdom but from the middle of the nineteenth century imported into the United States as "landscape gardening" under the aegis of Frederick Law Olmsted. Cemeteries were also "naturalized" and made into tourist attractions, "for a quiet stroll or a family picnic," or for a presidential Memorial Day address, as Theodore Roosevelt did in 1904. While previously a rural elite, especially in Europe, had indulged in a specifically "urbanized" experience of the spa, by the end of the century nature experience had been shaped into a desire of all urbanites.[118]

While in America many within the middle classes embraced this middle landscape, in Europe other concepts of nature prevailed. In the

United Kingdom, for example, the urbanized countryside had been 'ordered' into many individually owned enclosures, and a pronounced anti-urban attitude mixed with agrarianism and romanticism prevailed while national parks were regionally governed. Similarly, in France, 'nature' was seen as the "regional picturesque," the "local charm" with its "natural monuments." Just as astronomers meet the past when piercing far into space, tourists could encounter the "old France" in their visits to the remote region. Likewise, for British tourists after the middle of the nineteenth century, the entire holiday on the continent was a journey in the past because of Britain's advanced industrialization and urbanization. Conversely, early European travelers visited the United States as a journey into the future.[119]

In the United States it became clear that having the machine in the garden could quickly deteriorate 'nature' in the national parks. One of the park movement initiators, the naturalist John Muir, opined in 1900 that the "scenery habit, even in its most artificial forms, mixed with spectacles, silliness and kodaks" was part of the American way of life. From the middle of the first decade, the national parks' popularity received a definitive boost toward true "entertainment parks" as a result of the 'See America First' movement, which, not by chance, coincided with the promotion of the Lincoln Highway. This road, "a people's highway" running from east to west over the entire continent, was not only meant "for the motorist who wants to see his own country first," but also to connect the farmer to "a great transcontinental trunk line." The western boosting campaign, not a "unified movement with clearly defined and articulated aims" but meant to show that American "scenery" was at least as spectacular as Europe, became nationally important when the outbreak of the First World blocked tourist access to Europe, a golden opportunity to keep home a large part ($150 million) of the annual expenditure abroad of nearly half a billion dollars.[120] The first parks opened to the car were Mount Rainier (1908), Yosemite (1913), and Yellowstone (1915); in 1920 a park-to-park highway was "dedicated" to connect most of the famous parks in the West. The roads to Mount Rainier, constructed with horse traction in mind, had a double function: access to the scenery and channeling of the flow of tourists, necessary because the park was the holiday backyard of cities like Seattle and Portland. When one automobile club member described "America's Newest Playground Open to Motorists," he implicitly referred to motoring tourists' habit of simply leaving the road and literally drive 'into nature': Glacier National Park had been opened to the car because "so rugged is the surface, cars will probably never be able to travel in all parts of the scenic preserve."[121]

Despite the emphasis placed among tourism historians on the anti-European connotation of the 'See America First' campaign, it does not seem that it decreased the enthusiasm and the eagerness of the elite to visit Europe by car, even if they tended to lose their immigrant American chauffeurs who often left the touring party to stay back home in France. When they decided to hire French chauffeurs, they risked experiencing a similar fate in the French provinces.[122] Paris especially was an irresistible destination: the touristic discourse on Europe in the United States abounds with anecdotes about Americans who "unknowingly" drank and danced with prostitutes in Paris only to be "saved" at the last minute by fellow Americans. For American men and women of the wealthy classes Paris became a "temple [to] the twin goddesses of Gayety [sexuality] and Beauty." It also saw an increasing number of "new American girls" indulging in their version of the European adventure, although they were warned not to behave like Daisy Miller, who gave Henry James's 1878 novel its title and who died after a flirtatious holiday in France. During the early 1900s, the *Quartier Latin* counted more than three thousand American students alone; the Boulevard Montparnasse was a center of "American Bohemianism," including female homosexuals such as Gertrude Stein, and some blacks, such as Josephine Baker. For the motorists among them, the French straight roads were a paradise of touring and especially racing, and Americans "did little to dispel (the) illusion of a nation of millionaires."[123]

For some historians, the middle class "seemed happy," a "collective happiness" expressed in leisure activities such as sports, tourism, and the taking of photographs as an "addiction to amassing memories of pleasure." An English clerk expressed this double-bind of happiness in a poem in which "he blessed the groaning, stinking cars / That made it doubly sweet to win / The respite of the hours apart / From all the broil and sin and din / Of London's damned money mart." Integrating 'stinking cars' into one's leisurely hours as a compensation for work led to a revision of the speed and nervousness of the era, and to "middleclass parishioners ... being lured away from the kirk by motoring and golf," as an English minister lamented in 1911.[124] In the period "between Bismarck and Hitler," as Joachim Radkau calls it in his exemplary analysis of the "experiences" of the German middle classes, the bicycle and the car played a central role in the neurasthenia discourse: they evolved into remedies for, rather than a cause of, nervousness and hysteria, at least as long as nothing interfered, such as an accident. According to some psychiatrists, the motorized versions in particular (motorcycle and automobile) functioned as male vibrators, as tools of self-therapy for a narcissistic type of human being who used his sport and its accompanying noises, smells, dust, and bodily dangers to offload his frustrations onto other people, mostly non-users. The bicycle and especially the car turned the "nervous lifestyle" from a disease into a *"Leid-Lust-Phänomen,"* a phenomenon of lustful suffering, of angst-lust in a period that had been diagnosed by the American Henry Adams as ruled by a "law of acceleration." Not coincidentally, Adams was a "convinced motorist" (he bought a Mercedes when in France in 1904) who formulated his "idea of Paradise" as "a perfect automobile going 30 miles an hour on a smooth road to a 12th-century cathedral." According to French medical research the car could even be seen as a tool for physical health as the number of red blood cells appeared to increase considerably during a car trip.[125] From this perspective, the car emerged as a primary example of the therapeutic characteristics of consumption, the third root (after love of nature and leisurely recreation) of bicycle and car tourism.[126]

A fourth root of bicycle and car tourism were the multiple national imperialist 'projects.' The car emerged at a moment that the middle class became increasingly uneasy about the upcoming working class, and felt haunted by a fear for revolution. "Whoever wants to avoid civil war, must become an imperialist," Cecil Rhodes opined in 1894. In an intricate way the imperialist mood got intertwined with fascination for 'nature,' the urge (for men) to follow a 'strenuous life' built around sports (which was given a comparable 'channeling' function of aggression[127]), the display of power and violence, the bicycle, and especially the car. The 'imperialist adventure' under the guise of the 'civilizing mission' should certainly be added to the shortlist of early automotive adventures, as we will see in the next chapter. The explorations undertaken by car tourists filled the gaps of the railway network. It has to be seen in the context of several national imperialist projects, of which, in the case of the United States, President Theodore Roosevelt was the iconic representative. Awarded the Nobel Prize for Peace in 1906, his actions in Cuba, Mexico, and the Philippines were part of the same Western pattern of aggression against the 'other' and against 'nature' as Robert Baden-Powell, whose initiative of the Boy Scouts had to result (as he declared in 1911) in the "hardening of the nation or the building up of a self-reliant, energetic manhood." The German scouts, according to their national journal in 1914, entertained "colourful thoughts of distant coasts and lands; we have a dark and secret inkling that we are approaching new fields of war and work."[128] Indeed, Europe had its 'frontier' too, at a continental scale between 'civilization' and its southern 'periphery,' and nationally between a city culture and the countryside.

Was the automotive adventure a purely masculine movement? At least until 1916, the car movement was not only an elitist affair looking for a way to get embedded in society's infrastructure and systemic ordering, but also one practiced by white, male-dominated, heterosexual couples and 'families.' The exclusion of non-whites was a matter of course in this racialized movement (just as neurasthenia was considered to be an illness of the white man), while the role of women still is a matter of controversy for this early period. As Henry Adams posited: "the typical American man had his hand on a lever and his eye on a curve in his road," while "his living depended on keeping up an average speed of forty miles an hour, tending always to become sixty, eighty, or a hundred." But "he could not run his machine and a woman too; he must leave her, even though his wife, to find her own way, and all the world saw her trying to find her way by imitating him."[129] This stereotypical image of the lonely motoring man, resonating among several historians of mobility for this early period, forgets that women were often occupying the passenger seat. Thus, Michael Berger argues that "some women were kept off the road for a generation by folklore."[130] Likewise, some European tourism historians see women dominant in the "stationary holiday," while others (including this author) argue that they soon formed a majority in several collective forms of travel such as trains and buses to holiday destinations. Historians with a more multimodal vision point at the liberating role of the bicycle in this phase: the Dutch feminist Aletta Jacobs (who planned to sign up as a boy to travel to America where she wanted to become a carriage driver) made holiday trips with her husband to be, as far as Switzerland. After she had visited Colorado in 1914, where women had gained voting rights for the first time, she complained, back in the Netherlands, that "her carriage driver, who earns the salary that makes him a voter in *her* service, is given that right while *she* is allowed to watch on."[131]

Most gendered histories of mobility start in the early years of the Interbellum, and those who, despite the dearth of sources, venture in the earliest period follow either one of two possible chains of argumentation. In her analysis of gendered travel writing, Susan Bassnett argues that as "the essence of adventure lies in taking risks and exploring the unknown, ... it is hardly surprising to find that early travel accounts tended for the most part to be written by men, who moved more freely in the public sphere." Women, therefore, may be the bearers of a subversive form of mobility because of their exceptional attributes of a nonviolent, or at least a more defensive and protective, form of driving, if they managed to get access to the steering wheel. According to some contemporaries, woman was a better cyborg than man: "She acquires the 'feel' of the mechanism more readily, she detects more quickly the evidence of something out of adjustment, and altogether she drives more gently and with more delicate technic—all of which adds peculiarly to her pleasure and satisfaction in motoring." And this commentator added (before the electric starter, which supposedly made driving by women possible): "Evidently unusual physique is not necessary for the woman motorist."[132] Such assessments were the predecessor of postwar arguments that women were safer motorists, and hence, perhaps, also the basis for stereotyping them as poor drivers that, according to Michael Berger, was not yet the accepted image before the war.[133] For the American Kate Masterson, for example, who wrote in 1910 a pamphlet against "The Monster in the Car," the "new delirium, the mysterious intoxication of the devil wagon" had also enthralled some women, who "have made guys of themselves, donning dresses of leather, hideous goggles, shapeless veils and hoods. . . . Everything that tends to feminine enjoyment must have the scent of gasoline in it." Now, "even the middle-class girl with a beau" had become an adventurer, but the "speed madness" had made them indifferent: "women have never looked so unmoved on the torture of the wounded and bruised flesh—horses, cows, watch-dogs, flung dying to the side of the road, while the motor-man with a howl of laughter put [sic] on more power to escape infuriated pedestrians and onlookers at such wholesale murder of dumb animals."[134]

On the other hand, and clearly as an expression of the second feminist wave, some analysts argue that women were as good as men, and they focus on the factors that prevented them from playing this role to the full. Thus, while some historians observe an underrepresentation of women among electric vehicle drivers and owners, others argue that their numbers were surprisingly high given their societal submissive position.[135] By the end of the period dealt with in this chapter, a surprisingly high share (25 percent) in the membership of the Automobile Club of America's "Bureau of Tours" were women. New York also knew a Women's Automobile Club (its president, Alice Ramsey, was the first woman to cross the American continent).[136] One analyst, Cotten Seiler, even goes as far as to claim that the car started off not as a masculine, but as a feminine vehicle that in the course of the first decade became de-feminized, so to speak. Likewise, Sean O'Connell posits for the United Kingdom that "it was commonly felt that men used cars for utilitarian and business reasons," the dominant pleasure character of early automobilism must then have been a result of "feminine mercurality and frivolity."[137] In the same vein, while Genevieve Wren demonstrates that more than five thousand American women took out auto patents in the decade before 1922, Georgine Clarsen dedicates an entire book arguing that women "loved cars" as much as men, and that mobility historiography have neglected their roles as active agents (as mechanics in garages run by women in the United Kingdom and Australia, as engineers, supervisors, and workers in a car factory run by women in Scotland during the war, and as drivers of suffragettes across America). Criticizing feminism's second wave, she takes "bodily differences between men and women" as a basis for describing a corporeal struggle for automotive skills that took more effort for women than was the case for men. The "suffrage auto tour" by Florence Luscomb and Margaret Foley between 1911 and 1915 was, indeed, full of adventures, including a car race with an opponent in the elections, a car chase, tales of maintenance and repair, and poetry, for instance when Foley in the *Boston American* wrote about "the sunset sky stretching on beyond the treetops [which] gave a splendid reflection of the vivid colouring. As the sunset faded, the full moon rose bright in the eastern sky, and all the evening we rode through a dazzling white country that seemed part of a fairyland."[138]

Ironically, the electric vehicle has been the victim of this controversy about women's 'exceptionalism,' because, in order to claim that women just wanted the same as men, the electric vehicle has either to be ignored again (as mainstream mobility history has done so for the bulk of the twentieth century) or to be depicted as an 'inferior technology,' a projection of a later argument by gasoline car interests. Elsewhere I have shown that this was certainly not the case for this early period (on the contrary: electric vehicles performed better than their gasoline counterparts) and that it is very likely that women initially *did* use electric vehicles more than men, for instance in a flat city with lots of charging stations like Chicago.[139] When the Dutch feminist Aletta Jacobs visited Paris in 1890, she marveled at a steam car on the Champs Élysées, but she added (fully neglecting the gasoline option): "Who knows what miracles electricity will bring us in this area?"[140] However this may be, the role of the car in the lives of white elite women was as important as it was for men: both used the car as a temporary escape from the home, while for women the liberating affordances of the car and its relation to the struggle for voting rights cannot be overestimated. But there is still a consensus among historians that women (and African Americans) used "the driver's seat as a sort of podium from which they staked their citizenship claims."[141] President Theodore Roosevelt's daughter Alice was certainly not the only one "who stirred up considerable controversy when she drove from Newport to Washington alone, an act which she admitted was a mild protest against her father's notion of femininity." Recent research in film studies (applied to an early movie, *Mabel at the Wheel* [1914], by the actress and director Mabel Normand) claims that "auto-erotics" were as important in the earliest car-driving practices for women as they were for men. In other words, the automobile was, in the fantasies of men and women alike, made into "an apt vehicle for modernist individualism," despite its often collective context of use, but this happened with "a gendered twist," enough to see much of women's use of the car as subversive. In short, the car's gendered symbolism was utterly confusing: "Machinery has always been the province of men," Deborah Clarke concludes, "but here was a machine that was both phallic and female. (T)o make matters more confusing, the car, as a powerful machine, also remains to a degree masculine. It is masculine in power yet feminine in being a body that is ridden and mastered."[142] To add to this confusion even more: nobody has researched the homoerotic relationship between men and their cars. One needs not be a Freudian to see this relationship emerge in men's handling of the gear lever. Men cannot have it both ways: making the car feminine the moment they get into the car is an analytical weakness that contradicts with the phallic projection upon the same vehicle, unless one includes homoeroticism in the equation. No wonder, then, that men were especially receptive to a violent externalization of such confusing experiences (as we will see in the next chapter), and that women, as "the family's consumer," could easily take on the role of a more benign, more protective way of dealing with the new toy.[143] The presidential Taft family was certainly not representative of automobilism in the United States around 1910, but the role division between husband and wife seemed typical for the elite's relation with automobiles: While the president insisted on buying a White steamer with the $12,000 Congress had granted him, he left the selection of the rest of the 'stable' to Mrs. Taft. She in turn chose a representative, expensive Pierce Arrow limousine for her husband, and a landaulet of the same brand for herself (the company provided a chauffeur with the cars), while she bought a Baker electric for those times she wanted to drive on her own. To complete the collection, the White House also purchased an electric van, plus a charging station.[144]

The Automobile Club of America boasted in 1910 that more than two thousand of its members were involved in "extensive touring."[145] The touring adventure as created by the auto club's members was generally modest, quiet, and slowly evolving toward a family event, but the trip accounts, in all their bread-and-butter style printed in *The Club Journal* (later renamed *Motor Travel*), were nonetheless full of tongue-in-cheek disdain of the 'Other' and ventilated a colonial attitude toward the 'conquest' of foreign territory. The 'Other' not only included the non-user who, if he would ask you to stop, "nine times out of ten ... is either drunk or dishonest," but also included animals, as was revealed when *The Club Journal* reproduced a letter sent by George Bernard Shaw to a British car journal on killing animals by car, followed by a wave of reactions from American motorists. Some of them stated proudly that their radiator in summer resembled "scrambled eggs seasoned with feathers," illustrating the old saying that "one cannot make an omelet without breaking eggs." Many of them also liked to drive around with open exhaust, in the belief that the decrease of back-pressure on the pistons would enhance hill-climbing, but the Club, through its Technical Department, tried to convince its members that the gain was only 5 percent at most, so the noisy habit could (and should) stop, in cities at least. Legislation would soon ban this practice, although the first traffic regulation in Glacier National Park required motorists to honk at every corner.[146]

The "Joy of Touring," the journal explained in an editorial, could be organized: "With a good car, a competent driver, congenial company, an itinerary that has been planned to give the maximum of good roads, fine scenery and interesting country, touring is clearly unrivaled as a sport, a pastime and a means of furnishing an antidote to the cares and worries, business and otherwise, of our life to-day." On the other hand, the adventure should not be circumscribed too much: "We do not wish to suggest," the journal quoted *Country Life,* "that a fixed route is essential. On the contrary, there may be a certain fascination in wandering haphazard through the length and breadth of the land with never a thought for the morrow and scarcely more than a vague idea of the town or village in which the coming night will be spent." But the American version of the automobile adventure was stuffed with 'comfort,' especially now that "the days of perpetual breakdowns and involuntary stoppages are past," an overly optimistic assessment of the car's technical reliability, as we will see in the following chapters. Proper planning, diligent maintenance, and an adequate set of accessories minimized discomfort, so much so that by 1910 a "party of six, including my wife and two daughters" could undertake a transcontinental trip with "comfort and quick progress" in a car adjusted by the owner for this particular purpose.[147] With the support of the automobile club, car-touring adventure could now be "roughed up" a bit: "America's available touring ground has become so extensive it is almost impossible for a tourist to accomplish anything satisfactory without expert direction and advice. It is easy enough to follow the trail of motor dust which rises along the beaten paths in May, but this is not touring. It is merely a daily dash for a hotel; a game of chase or 'follow the leader.'" With the help of the Bureau of Tours such a tourist now "deserts the caravan of 'dust chasers' for the cool and shady 'unfamiliar roads,'" so that "an element of exploration is added to a journey and when the traveler has returned to his home after a motoring vacation, he finds he has seen something and been somewhere out of the ordinary," even if he has followed the increasing number of "Yellow Arrow A.C.A. road signs."[148]

The "*Mittelstand*" car culture of the ADAC represents the other extreme of the Atlantic adventurous spectrum in this period: the club's journal, *Der Motorfahrer*, continuously emphasized the collective aspect of this culture, the touring experience of the team as an expression of a "common will" behind which the individual interests (and the "unrelenting competition of all against all") should withdraw. By performing automobile sport, "the individual harvests the honor, but the common good, the people, enjoys the benefits."[149] Technical unreliability potentially harmed the excursion of the "*Ortsgruppe*" (local group), hence German car culture, more than any other national culture, put a large trust in experts. The "*Autlergemeinschaft*" (motorist community) developed a militarized group culture in which the individual member could experience "the most perfect happiness" (*vollkommenster Glück*).[150] "Discipline" was necessary to tame the younger generation of motorists who liked to race. Instead, the collective adventure should be experienced as a "struggle" against "adverse road problems" (*widrige Strassenschwierigkeiten*) and against the resistance of the 'Others' who were mostly to be found in the countryside.[151]

An important element of these 'Others' was "the child in the countryside," who threw stones, strung ropes across the road, or simply played in the street, because he or she was either "malignant by nature" or "whipped up by short-sighted parents," or because he or she acted out of "a lack of intelligence." Yet more 'Others' were seen, from a distance, in foreign regions, such as the inhabitants of the Pyrennees with their "shabby look."[152] In addition, the road hogs within the movement itself had to be disciplined. Well-behaving motorists found it almost unsupportable for the state to treat them as potential criminals, whose personal judgment was not trusted anymore, but whom the law punished as soon as some impersonal threshold (such as a speed limit) was crossed. This was all the more painful as they feared special "stop-watch societies" (*Stoppuhrvereinen*) of civilians enabling small municipalities to earn thousands of marks.[153] Nonetheless, they experienced this highly restricted collective sport as adventurous, opposite to the "decent ... bourgeois" (*brave Bürger*), a feeling mostly derived from a mixture of manliness and homoeroticism. This was expressed, for instance, by a semi-anonymous "H.H." in 1914, who praised the "collective experience ... granted by the car trip in and of itself, [which] makes us happy, merry and free, only through the movement of driving.... One feels what may come, but one does not know what it will be.... At the steering wheel of the car the strong man becomes even stronger, the sensitive, enjoying man sniffs the air as if on a hunt, his nerves taut and fine-tuned vibrating like fine membranes."[154] This adventure of the "sportman and nature lover," especially on long-range tours, was held in check because it provided only a pseudo-departure: "if we motorists do not feel fully at ease in 'foreign' countries, this feeling of being without a home (*Heimatlosigkeit*) disappears immediately when we in the morning ... climb into our car. The engine is started, the car moves and then one hears in four voices (if it is a four-seater): 'Oh, in the car it is so beautiful, because it is as if we are at home.'"[155]

Mischievous behavior such as smuggling of small objects over European borders augmented the adventure. It gives "us" (as part of a movement, no doubt) the feeling, when keeping a "hellish speed," that "although one is a fully civilized, tamed European, there still is a shot of healthy, overflowing wildness in us." Behind the wheel, "we hunt for what we lack.... We feel good and free, healthy and sublime, as a completion of our selves.... In aeroplanes, air ships, motor yachts, automobiles and motorcycles (we) speed away on a quest for adventure, greatness, the unexpected and the extra-ordinary." Part of this sublime feeling was the enjoyment of skill in diagnostics, repair, and maintenance, and in the "art" of driving.[156]

Third Phase: The 'Small Capitalist' and the 'Average Man' (1908 until the War)

From the end of the first decade of the new century the fledgling automobile culture started to shift toward more 'modest' forms, the result of an expansion of the user group toward the middle strata of the middle class. This shift was also visible in the technology as lighter, cheaper cars came on the market following the Panic of 1907. This hit the French car industry especially hard, as it was mostly geared toward an elite clientele, producing very heavy family touring cars for six or more passengers. This market now seemed to be saturated. As I have argued elsewhere, the recession not only led to a production crisis and an effort to diversify toward utilitarian vehicles (trucks, buses, and taxicabs, which had the additional advantage of counteracting the seasonality of the pleasure-car market) and less costly passenger cars. It was also accompanied by a safety crisis that was related to changes in the discourse about speed and in its turn translated into the technical crisis of the pneumatic tire. The latter contributed considerably to the costs of the car's upkeep and these costs increased exponentially with the weight and the speed of the car. The tire problem, for which specialists at first discussed an alternative in the form of the 'elastic wheel' (its spokes replaced by springs), was solved through a 'technical fix' by reinforcing the tire, first with soot and then by integrating textile cords in the tire body. The issue of speed, however, was much more difficult to resolve, as it was always heavily loaded with social tensions. Speed seemed related to danger, especially for non-users: in Berlin in 1907 a quarter of the registered cars had been in accidents (two years later this percentage was 12.1 for the entire German *Reich*), while in 1911 the Automobile Club of the French department Sarthe fabricated figures meant to prove that the horse caused many more accidents than the car. The figures the club produced, however, reveal when recalculated that the 'lethality' of the car was twice as high as that of the horse.[157] Despite these alarming statistics, motorists were not prepared to give in over speed because they considered risk to be the quintessential characteristic of their new sport. In the French department of Le Loiret the number of municipalities limiting car speeds peaked in 1907. Motorists, a French historian of speed posits, believed that they had a "droit à la vitesse" (a right to speed).[158] The publicity around car safety gave the movement a bad reputation also in the United States, where the *New York Times* dedicated ever more space to accidents and safety regulation (increasing from 28 percent of all car-related articles in 1900 to 37 percent in 1915). In the latter year more than half of the articles on cars were negative in tone or content.[159]

An early German analysis, undertaken by a specialist after the first alarming figures became known, revealed that more than one-third of the 145 deaths were children, and that only 22 of the deaths resulted from vehicle-to-vehicle collisions. In 110 cases the courts did not see a reason to prosecute the drivers. Remarkably, the vehicles involved in accidents differed from the distribution of car ownership: "pleasure cars" (which formed 40 percent of the registered fleet) caused one-third of the fatalities, while commercial vehicles (also 40 percent of the fleet) only caused 15 percent. Public road transport, which only represented 5 percent of the German cars, caused the bulk of the lethal accidents. Apparently, not only did the context of the accident play a role (urban, rural), but also the frequency of the vehicle's use. While nearly half of the accidents occurred in Berlin, where 39 percent of the German cars operated, these accidents resulted in only 17 percent of the deaths. The accidents were caused by only 7 percent of the cars. Accidents that occurred on rural roads and on village streets accounted for 20 percent. The complaints about the 'dangerous automobile' in the countryside thus had a solid basis in accident statistics: the higher speed, the intrusion of an urban machine into a quite different culture, may well have resulted in more deaths per car than in large towns. In a situation still largely hostile toward motorists, police reports on the accidents revealed that "driving too fast and not using the horn" caused most accidents (both lethal and non-lethal). Later, for a period on which more data will be available, we will investigate the safety situation of road traffic in more detail (see chapter 7), but it is worthwhile to see how at this very early moment already the first efforts were undertaken to suppress the seriousness of the situation by rhetoric that any neutral onlooker immediately recognized as absurd. When the car club's journal at the end of the year made up the balance of the safety turmoil it observed with relief that the "accident frequency" had decreased compared to the year before. Implicitly this suggested that purchasing an automobile diminished danger, as every new car helped decrease the accident ratio.[160]

There are many more indications in the mobility history literature that the first really deep-going turning point in automobilism's history should be located in or around 1908. In the United States, somewhat exaggerating, as we will see in chapter 4, Peter Hugill declared the car to be regarded as "an absolute necessity of American countrylife" by the end of the first decade. In Massachusetts cars surpassed horse traffic on main roads in numbers shortly after 1909. According to the recollection of General Motors' President Alfred Sloan Jr., in 1908 the American class market turned into a mass market, which was symbolized by the Model T. A second-hand car market started to flourish around 1910. There is no 'gap' to be seen here with Europe: there too the 'democratization' of the car started around the year 1908. The turning point can there even be seen in the motorcycle statistics. For instance, in the French department Cher registrations for this vehicle type jumped between 1908 and 1909 and provoked a growth curve much steeper than before.

In the Netherlands some evidence exists that after 1910 even the most aggressive car promoter, the automobile club, started to change the tone of its journal: new, younger, less elitist authors managed to express some empathy toward less privileged compatriots who mostly experienced motordom as a source of dust and accidents. The car club's opinion thus converged a bit toward the petty bourgeois ideology of the touring club, which generally supported (or secretly endorsed) the press campaign against 'road hogs.' The Dutch touring club had lost interest in automobilism after very heavy, expensive automobiles dominated the market, but expressed new hope that the car might soon become available in a more compact, cheaper form for the 'small capitalist.' Our analysis of Dutch automobilism suggests that, had the Great War not intervened, the Netherlands would have witnessed a first massive wave of motorization, which was not only postponed by about a decade, but was also decisively altered by what would happen during the war. The same can be said about German car culture: after 1909 ADAC membership rose quickly from 13,000 to 28,000 in 1914 (see figure 1.3). Expert observers saw the market for "the small, light car" emerge when motorcycle owners stepped over to the car.[161] Automobile club membership rose to 8000. This coincided with the emergence of smaller, cheaper cars that were widely discussed as the harbinger of a "people's car."[162]

Shortly before the war, even the Pope gave his blessing, in his *Benedicio vehiculi seu currus*.[163]

While most of the European countries plunged into a war, the United States definitively established its own car culture in an unprecedented wave of diffusion among the urban middle class and farmers. In 1910 farmers owned only 23,000 of the half million cars registered; ten years later one-third of American farms were equipped with an automobile, with a much higher percentage (above 50 percent) in the Midwest. Indeed, 1910 marked the first outburst of car diffusion in America, ironically identified by *Outing* when it observed that in that year "approximately 150,000 American citizens will abandon their antipathy to automobiles and become the worst, because the newest, of motor maniacs." Twenty thousand of these drove light, "low-priced cars," costing less than $2,000; the average selling price that year was $1,200, the result of a number of cheap 'runabouts' that sold for $500 or less. The outburst also created a market for second-hand cars, "snapped up eagerly by the farmers of the Middle West."[164] For 1911 the *American Agriculturist* even claimed that farmers formed the majority of car buyers.[165] While Americans had used the years between 1903 and 1908 to start substituting horse-drawn by motorized vehicles, now a "boom in all transportation vehicles began."[166]

Despite initial resistance against the "citified automobile," farmers were considered to be excellent motor pioneers because they were used to tinkering with mechanical equipment, including the gasoline engine.[167] They thus could apply the unreliable car to a host of alternative uses, from farm-to-market transport, to visiting friends and relatives in town, to assisting in field and household work.[168] Three factors supported rural motorization during the 1910s: an explosion of rural income (17 percent between 1908 and 1918), a lowering of car prices by a factor of three to four (including Ford's Model T; in contrast, between 1900 and 1908, car prices had doubled on average), and a quadrupling of the car's longevity to acceptable levels.[169]

In relative terms, there was nothing special about this: as we have seen in Germany and the Netherlands the years immediately before the outbreak of the war promised to be the starting years of a first massive diffusion wave as well (in Germany operating costs had halved in the last eighteen months before the war), and in France, too, despite a continuously diminishing relative growth rate of the car park, the absolute growth figures between 1908 and 1914 had never been as high, with a peak of 10,000 new cars in 1913.[170]

On the other hand, several crucial differences with a 'European car culture' became visible, such as the pivotal role of the midwestern farmer and the important fact that the United States benefitted from 3.5 extra years of peace, in which the European need for (military) vehicles kept increasing. Until America's entry into the war, however, the absolute number of farmers motorizing remained modest, and the car remained for the most part an urban (and to a certain degree suburban) phenomenon (see figure 1.5).[171]

While Europe kept fascinating the American elite, the new middle-class motorists chose to 'see America first,' and the elite joined them when war broke out in Europe. In 1910, for the first time, Americans drove more cars to visit the national parks than carriages; in 1916 visitors in cars outnumbered train travelers in Yellowstone.[172] These dates, at the end of the first decade and halfway through the second, are no coincidences. It is remarkable that the United States gave itself any time to come to grips with 'nature' before motorization. Theodore Roosevelt, president from 1901 to 1909, is a pivotal figure here: the "gentleman sportman," active birder, and dedicated hunter strenuously opposed the automobile and oil interests (he rejected a check for $100,000 from

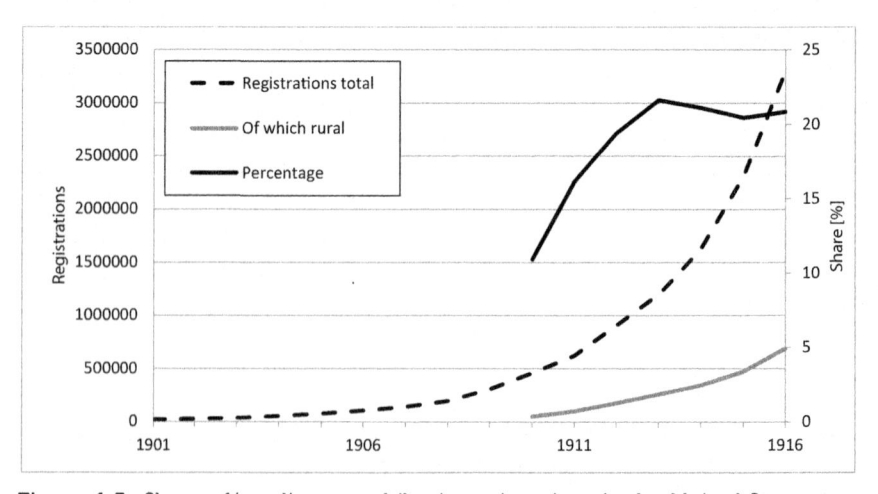

Figure 1.5. Share of 'rural' automobiles in registrations in the United States.
Source: Bromley, 387, appendix D.

Standard Oil before his reelection in 1904). He also opposed giving cars access to the parks, just as he disdained the telephone. In 1908 he inaugurated a Country Life Commission only to find farmers who had conquered 'nature' fully equipped with harvesters, bailing machines, stationary internal combustion engines, and soon cars, too.

Roosevelt's handpicked successor, William Howard Taft, was the first to bring automobiles into the White House. He declared: "I am sure the automobile coming in as a toy of the wealthier class is going to prove the most useful... to all classes, rich and poor." His example was immediately followed by congressmen who started filling up that institution's parking spaces but who up to then had not dared to discuss the automobile publicly. When Taft's son nearly killed an Italian immigrant road worker in 1910, no "political fallout" occurred, and Taft gave the victim $500 in cash and a trip to his family in Italy. In 1911 Taft even accepted an invitation to the elite automobile club's annual banquet. He planned to make this into "an occasion of international importance," where the ACA's officers and nearly 1,000 members (as well as 300 female guests sitting in separate boxes), heard the president remind his hosts that he himself had long held "a sort of spirit of intolerance... toward the horrible looking machine that the automobile then was to the ordinary eye." His presence now at the banquet was a clear sign that he endorsed the ACA president's opinion about the car as by now "almost one of the necessities of life and commerce." But Taft did not answer the question of whether the nation or states should pay for an improved road network. Clearly, America had not yet made up its mind how to tackle the "automobile problem," although the poem that closed the banquet ("The Automobile Speaks") expressed a clear vision on the direction of the solution:

> They say I have no style. They may!
> What's style to me! I don't eat hay,
> Nor prance. Lugs have for me no lure,
> No powdered wig on my chauffeur!
> Plain goods, I glide where pride is rife,
> The herald of the simpler life.

While Taft still had to refer to class antagonism, his successor in 1912, Woodrow Wilson, shifted the attention of the national government from the car, as vehicle, to the system. Both Taft and Wilson planned to appear at the "First Annual Session" of the American Road Congress to be held on 30 September—5 October 1912 in Atlantic City, where the trump card of the "great waste" and "the great losses that are being sustained by farmers and consumers in the hauling of crops" from the farm to the city was played out. The plea had the full support of the president of the Southern Railway, who hoped to fill up his empty freight cars if the roads to the shipping points were improved. Therefore, for the United States the year 1916 (with two million cars on the American roads, and shortly before America entered the European war) was a turning point toward a phase in which the car would directly confront the train, threatening to take over the latter's long-range transport function. In 1916 Congress approved the National Park Service and the first Federal Aid Highway Act. Both would help American car tourism to explode in the 1920s. These milestones were the endpoint of several developments underway at the local and regional level: New York State had started to invest heavily in its road system by 1914, the year when local streetcar companies began to feel the emergence of massive competition. As a result, like in other Western countries, the elite began to see aviation as the ultimate adventure, but the *New York Times* also observed a return to a "vogue of the horse." When, in 1910, former President Theodore Roosevelt was invited to take part in a four-minute flight, the man who had barred cars from the parks stammered: "War, army, aeroplane, bomb!"[173]

While diehards among the old elites fled into aviation as a new playing field for masculine adventure, the 'See America First' campaign resulted in an increase in attendance of the national parks, where the 'new motorist' could have his weekend spin in the 'wilderness.' Denver established the first autocamp in 1915. Between 1916 and 1923 the number of visitors to the national parks tripled to 1.3 million, while visitors to the national forests increased nearly fivefold to 10.6 million. Nearly 90 percent of them came by car, most of them farmers and other Americans from rural towns.[174] "Motor gypsying ... de luxe" required proper preparation and a new attitude. "Nine tourists out of ten travel too fast," a representative of the American Automobile Club opined in 1911, probably realizing that his plea was largely in vain, "(they) make daily runs of too great length and fail to enjoy the local color and historical values of the places they pass." Now, ACA members also wished to receive a quick answer to the question: "Where can I go for an afternoon's spin?" By that time, the automobile club faced a broad-based competitor, the American Automobile Association (AAA, founded in 1904) that soon counted about twenty thousand members (against ACA in the neighborhood of three thousand) and started to organize races and rallies and to publish touring books and maps. By 1910, 225 local clubs were affiliated to AAA. At the same time *Outing* advised its readers to "confine your touring to trips for which the most complete route books and road maps have been prepared," a form of modest adventure recommendable "unless you wish to benefit your fellow motorists by becoming an automobile explorer."[175]

By 1910 pleasure trips in general tended to be shorter, an indication of the shift from an elite to a middle-class culture (and of the preponderance of an urban user pattern), more often than not performed in runabouts, sought by "the early buyers of automobiles ... for pleasure, for recreation. But within the last year or so a class has been coming up which buys the automobile for utility's sake with recreation thrown in as a pleasing and possible side issue." This "new motorist" indulged more and more in weekend spins into the American 'wilderness' and in the suburban countryside, including the ersatz natural environment of new parks, forests, and parkways.[176]

It was Henry Ford who benefitted the most from this trend toward lighter and cheaper cars. Preceded by attentive manufacturers who up to then had produced horse-drawn high-wheeled 'buggies' and by others who had experimented with small and simple 'runabouts,' Ford's Model T, as a "Universal Car," and as an "'Americanized' French technology," adjusted for the poor American roads, remained until 1914 (when the automated production on the basis of the assembly line started in earnest) a part of a mostly exclusive and elitist culture. The Model T had "a reputation as a man's car."[177] This culture developed on the basis of an American car technology that had more or less caught up with the European example by 1908 against a background of "a revulsion against European cultural leadership" among parts of the elite.[178] At the same time, the failure of the cyclecar (a very light alternative to the Model T) between 1913 and 1915 to conquer the American market showed how the definition of what an automobile was all about, was already pretty much solidified at that moment: a cheap car was not a stripped car, but a vehicle that possessed all 'normal' components (just like in the heavy cars of the previous phase) but less luxuriously designed.[179]

Despite the efforts to construct a utilitarian discourse around the gasoline car, one should not exaggerate its reliability, however, and certainly not on American roads: of the eighty-one cars taking part in the 1908 Glidden Tour reliability run from Cleveland and Chicago to New York, thirty-six got stuck in the "desert of dust" and the "oceans of mud" with broken springs, bent axles, and burned-up brakes on roads that would "by any European government, be closed to travel." One of the participants died after skidding into a ditch and overturning.[180] The poet and physician William Carlos Williams, who bought his first Ford in 1911 ("A beauty with brass rods in front holding up the windshield, acetylene lamps, but no starter!") in winter time "crank(ed) the car for twenty minutes, until I got it going, then, in a dripping sweat, leave the engine running, go in, take a quick bath, change my clothes then sally forth on my calls."[181] Still, by the end of the first decade the functional adventure of tinkering seems to have become manageable to the extent that thorough periodical maintenance could reduce it, although *Outing* warned readers that, if they wanted to drive their cars themselves ("careful and speedy"), they "must study to acquire the degree of Doctor of Automobiles," a pun referring to the medical doctor's need to develop "trouble diagnosis." Modern motorists "must know [their] machine."[182] As late as the beginning of 1914, William Vanderbilt experienced a puncture every 750 km during his winter trip through France and Italy.[183] And as long as the functional adventure was enjoyed as much as the other two adventurous aspects (speeding and touring), it could compensate for many weaknesses of the car through the acquisition of maintenance skills. Hence, the journalist Luigi Barzini could optimistically conclude in his report on the Beijing to Paris contest that "frequent breakdowns ... are due to neglect and lack of skill on the drivers' part rather than to any innate weakness of the vehicle."[184] For those American motorists who did not want to depend on chauffeurs, tinkering and a visit to the local mechanic's shop was the only solution. The "most wonderful thing about an automobile," *Outing* concluded in 1911 (but it meant the *gasoline* automobile), "is its almost infinite capacity to endure cruel and inhuman treatment."[185] In 1909, at the height in the number of reliability contests (which, according to a recent quantitative analysis, lasted from 1908 until 1911), car manufacturer Charles Duryea opined that the "novelty of the automobile has largely worn off." In 1912 the Glidden Tour was discontinued and the number of speed contests started to soar.[186]

But despite the fledgling robustness of the car's technology, a protective "system" slowly began to be built around the exploring tourist. Whoever wished to pioneer a more utilitarian deployment (such as commuting, which according to some historians already started in 1905[187]) still had to engage in the functional adventure: the "salaried man" who in 1910 bought "a sturdy, heavily built, reliable car of moderate power" (a runabout of $1,450), "knew nothing of machinery or mechanical principles" and spent two days "studying the machinery." And although after six thousand miles of daily use only one breakdown (a broken wire in the ignition) occurred, he emphasized the importance of thorough maintenance: "To be a successful driver requires some intelligence, some enthusiasm, not for going at great speed, but for knowing the engine is working at its greatest efficiency, some love for wheels and gears and valves, and some degree of willingness to get into the grease and grime to right some part if it needs it. The man who merely knows how to wiggle the levers and turn the steering wheel is missing much of the joy of motoring." A clearer picture of the emerging mixed use of the car, in which the utilitarian side gave as much 'joy' as the pleasurable side, could hardly have been formulated. For carmaker Hiram Maxim, a ride "taken just before business ... serves as an admirable awakening for a busy day and repeated after business gives an exciting finish." For one American novice in motoring, the skills of both handling the car ("The throttle or the switching off of the ignition altogether are the only legitimate means of cutting down your engine's speed" and "The general principle is to change as little as possible on an upgrade") and driving it ("learn to ride backward") generated pleasure, in each application. But the ultimate pleasure was the car's leisure deployment: "Also I am getting much more of the out-of-doors," he added. "I frequently leave home a little early in the morning that I may have a longer drive. Every Sunday and holiday, when the weather conditions are favorable, my wife and I spend on the road, exploring new parts of the adjacent country, drinking in the fresh air and the fine scenery of the mountains, and returning at the end of the day better in body and mind for our outing, and blessing the men who have invented and perfected the automobile."[188]

In the tricky department of technical reliability, the threat of the electric vehicle, and its ease of handling, was constantly around the corner, even in a gasoline-engine paradise like the ACA. A remarkably open attitude toward the electric car in the club's journal led to confessions such as this: "To one accustomed (to) driving a gas car, the entire absence of the subconscious strain of bearing in mind the carburetor, the oil gauges, the exhaust, the temperature of the radiator, the gear shifts, etc., is, indeed a revelation, for, with the electric car there are no complications of this sort. Just turn on the current and the car glides along, leaving the operator with an absolutely free and tranquil mind to admire the scenery and enjoy the tour." In the United States a somewhat different culture compared to the one in Europe was in the making. This can also be seen in the way the "chauffeur question" was energetically dealt with, namely by the setting up of a separate chauffeur department at the club headquarters in New York City. Complete with a fully equipped repair division with even a facility to manufacture the club's own tires, this was a measure against the widespread "corruption" among chauffeurs and tire dealers as well as the joyriding by chauffeurs.[189] At the height of the first elitist wave of motorization, about two-thirds of the car owners in New York State also registered a chauffeur, but after 1908 this share quickly decreased to one-third, a clear sign that another type of owner was joining the group of car users (figure 1.6). Nevertheless, the journal *Horseless Age* counted 100,000 chauffeurs nationally, as late as 1911 (at a moment that there were hardly 450,000 cars in the country).[190]

Owners, through their club, now started to formulate wishes as to the car's physical as well as social construction, revealing a trajectory

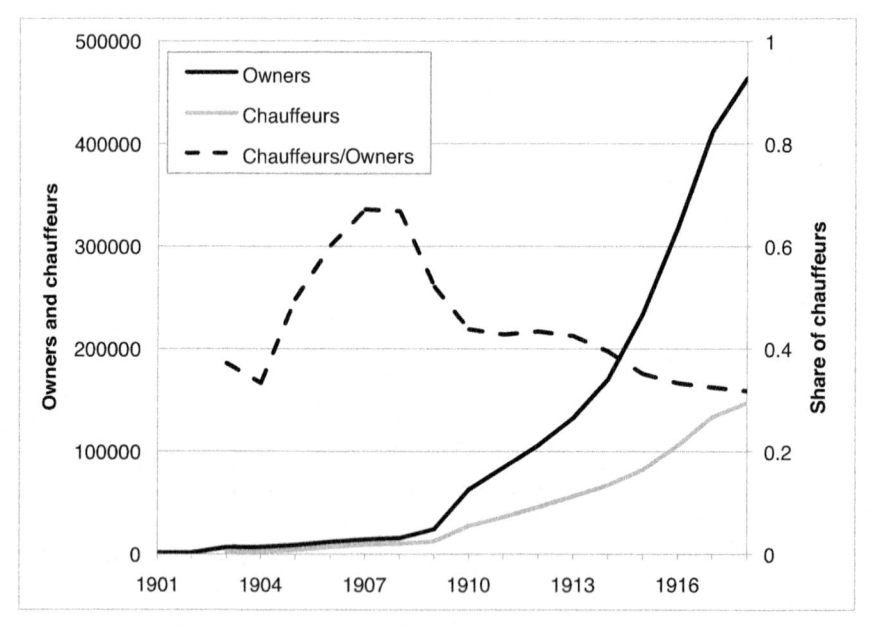

Figure 1.6. Car owners and chauffeurs in New York State.
Source: *Motor Travel* 10, no. 8 (January 1919), 34.

slowly deviating from the European example. Despite the emphasis on 'efficiency,' the 'co-construction' of American automobile technology was at least as much aimed at 'comfort,' which at this stage meant a preference for a large, relatively low-revving engine that could operate over a wide range of car speeds without having to change gears, the lowest speed to be driven in highest gear being, as one overview of the "ideal car" revealed, as low as 4 mph (6.4 km/h). Such an 'elastic engine,' as automotive engineering calls it nowadays, was quite different from the European practice, where it was recommended to use the first and second gears extensively in order to reach a high average touring speed. The difference can be traced back to the emerging tax laws in Europe, which levied a tax on cylinder size as a measure of the car's power, resulting in high-revving engines with a relatively narrow elasticity, necessitating, as during a car race, a prolific use of the gearbox. When German Opel placed a 1908 advertisement in the *Allgemeine Automobil-Zeitung* quoting the French journal *L'Auto* that had set up a survey among its readers about their wishes as to the car's layout, a "European car" seemed to be in the making as well. The German journal showed how 2567 French motorists (half of them coming from commerce or the medical trade) opted for a 12- to 16-hp four-cylinder engine with magnet ignition and sparkplugs, a pump for water cooling, a conical clutch, a three-speed transmission with a cardan axle to the rear wheels, equipped with wheels of the same size and a double-phaeton body.[191] The latter enabled the social construction of a "speed community" (*Geschwindigkeitsgemeinschaft*) in the car as all passengers now were looking in the same forward direction, contrary to the conversational layout of the vis-à-vis structure.[192]

But despite the differences with Europe, a study of the ACA's journal shortly before the war reveals that even in the United States an elitist car culture existed, although it was rather small (several thousands of motorists), probably mostly concentrated in the New York City metropolitan area. The 'See America First' campaign was largely ignored by this group and the tongue-in-cheek attitude toward the problems the motoring movement provoked among non-users also reveals the similarities with Europe, although the American version only arose in full around 1910, a decade behind Europe. What is remarkably different, however, is the surprisingly meager information in the journal on technical matters and the almost exclusive attention for touring, roads (for touring), and legislation (for touring). This difference from most of the automobile club journals in Europe can be explained by pointing at the high number of garages (23,868, where owners stored cars when not in use) and of "automobile machine shops" (12,171, on average 1 repair facility for every 300 cars) in the United States in 1917, exclusive of nearly 26,000 car and truck dealers.[193]

Conclusions

North Atlantic automobilism was a bundle of several, remarkably similar "national" movements (in their turn consisting of a bundle of local subcultures) undertaken by internationally oriented elites and borrowing from a culture developed during the previous decade of the 1890s by the bicycle movement. The continuity in road mobility is much more important than hitherto recognized in the historiography: the car, when it came, did not "irrupt" onto empty roads.[194] The previous analysis allows us to be much more precise than in earlier publications[195] as to the periodization of and geographic specialization in the emergence of the 'adventure machine.'

Aristocrats and members of the high bourgeoisie shaped the car into an adventure machine as an intermediary between their urban culture and 'nature,' in three phases. During the first phase they developed the petrol version into a racing machine within a highly associationist setting led by automobile clubs. A second phase, starting shortly after the turn of the century, saw these cars develop into heavy, but speedy touring machines, often driven by a chauffeur, who spent a large part of his daily routine maintaining an unreliable technology. Although reliability runs tried to show that the technology was improving year by year, tinkering was undoubtedly part of automotive adventure in this phase.

The Panic of 1907 triggered a turning point in several respects. Signs of market saturations led manufacturers to develop lighter, cheaper, and more reliable cars that were eagerly bought by a new middle layer of the middle classes (and including in the United States some parts of wealthy midwestern farmers). These same buyers started to use them to commute or to bring produce to nearby town markets, but generally took over most of the adventurous aspects of automobilism of the previous phase, led now by touring clubs as mass associations that filled the role once held by elite car clubs. In most countries, this third phase also started a process of recognition of the movement by the national states, through legislation which also tried to alleviate all too obvious class aspects, such as narcissistic and dangerous speeding in the countryside. More than class antagonism, however, pre-WWI automobilism was an element in the societal division between urban and rural culture.

How was this automotive adventure shaped? It was not only highly political, but also heavily gendered. It had to cope with fierce resistance from rural and some urban antagonists (mostly of an anarchistic flavor) because of its elitism, but also with a skeptical, if not subversive attitude from women because of its masculinity. The adventure was one of speedy touring accompanied by a willingness to tinker. White middle-class males formed by far the majority of the movement that in many countries, had the war not intervened, would have witnessed a wave of massive diffusion into the lower echelons of the middle class. But while in Europe this process was suddenly interrupted by the war, in the United States a domestic long-range tourism movement enabled the automobile to become a 'necessity' to visit the national parks, go on a Sunday or afternoon spin, and here and there combine that pleasurable deployment with commuting from the suburbs.

In short, explaining the emergence of automobilism rests on the concept of the tripartite adventure machine (racing, touring, tinkering), enriched with additional adventures such as courting, conspicuous consumption, and touristic 'imperialism', as we will see in the next chapter. The car seemed to fit in a fin-de-siècle and a Belle Époque culture of increasing middle-class leisure and consumption, but this was not an automatic process. Users and non-users alike had to act, had to go out motoring and resisting, in order for automobilism to take the shape it took, which differed per country, but which had so much in common that one can speak of an Atlantic culture with a beginning cleavage between a European and an American subculture based on a first divergence of automotive technology. The practices of users and non-users were based on (purchasing and rejecting) motivations worthy of further analysis, as this should answer the question why the Dutch impressionist novelist Louis Couperus, quoted at the beginning of this chapter, saw one of his protagonists turn into a poet once he started to desire possessing and driving in an automobile. In other words: we have now explained the emergence of the car and of automobilism only halfway. We have made plausible *that* the adventurous car appeared in this culture, and also to a certain extent why it was so utterly attractive to many who had the money. In the next chapter we will continue our analysis of early automotive adventure based on another type of source—fiction—in order to give more depth to the answers to the 'why' question through an investigation of the motivations (the motives and motifs) of the main actors, especially the car drivers and their companions.

Notes

1. Louis Couperus, *De boeken der kleine zielen: het late leven, eerste deel* [The books of the small souls; first part] (Amsterdam, n.d. [1901]), 7–8, 31 (my translation; all non-English quotes have been translated by me, with major support from Katie Ritson, editor at the Rachel Carson Center in Munich).
2. Otto Julius Bierbaum, *Eine empfindsame Reise im Automobil von Berlin nach Sorrent und zurück an den Rhein: In Briefen an Freunde beschrieben* (Munich: Albert Langen—Georg Müller Verlag, 1979; reprint of the 1903 edition), 21 (Adler), 7 (old-fashioned), 20 (freedom), 21 (addictive), 24 (whip). His publisher, August Scherl Verlag, borrowed the Adler in his name. A second edition was published in 1906 as part of a collection called *Mit der Kraft*. Dushan Stankovich, *Otto Julius Bierbaum—eine Werkmonographie* (Bern and Frankfurt: Verlag Herbert Lang, 1971), 40, 42. Remarkably, Stankovich dedicates hardly any space to Bierbaum's travel account, which in form as well as content was nothing more than "a mimicry of old models. The car only functioned as a funny reinforcement" (ibid., 51).
3. F. T. Marinetti, "The Founding and Manifesto of Futurism 1909," in Umbro Apollonio (ed.), *Futurist Manifestos* (New York: The Viking Press, 1973). Manifesto translated by R. W. Flint. "La nouvelle formule de l'Art-Action": Giovanni Lista, *F. T. Marinetti: L'Anarchiste du futurisme*, Les Biographies de Séguier (Paris: Séguier, 1995), 77. For a detailed analysis of the manifesto, see, for instance, Marjorie Perloff, *The Futurist Moment: Avant-Garde, Avant Guerre, and the Language of Rupture* (Chicago and London: The University of Chicago Press, 1986), chapter 3, "Violence and Precision: The Manifesto as Art Form" (p. 80–115).
4. Quoted in Peter Gay, *The Bourgeois Experience, Victoria to Freud: Volume III: The Cultivation of Hatred* (New York and London: W. W. Norton Company, 1993) 97.

5. Theodore Dreiser, *A Traveler at Forty* (Urbana and Chicago: University of Illinois Press, 2004; first ed. 1913); Theodore Dreiser, *A Hoosier Holiday*, with illustrations by Franklin Booth and an introduction by Douglas Brinkley (Bloomington and Indianapolis: Indiana University Press, 1997; first ed. 1916), 29 (conversation), 53 (chauffeur), 61, 204 (slum).
6. The term *use* has been deliberately chosen, as the 'practices' invented and developed around the automobile should not be reduced without investigation into 'driving.' Motorists and their opponents did much more than that: they made love, slept in, showed off, and phantasized about the car; they used it as weapon and as metaphor.
7. Jean Orselli, "Usages et usagers de la route: Pour une histoire de moyenne durée, 1860–2008" (unpublished dissertation, Université de Paris I Panthéon Sorbonne, 2009). On early Parisian car culture, see Mathieu Flonneau, *Paris et l'automobile: Un siècle de passions* (Paris: Hachette Littératures, 2005), 23–45.
8. John A. Jakle, "Landscapes Redesigned for the Automobile," in Michael P. Conzen (ed.), *The Making of the American Landscape* (Boston, London, Sydney, and Wdellington: Unwin Hyman, 1990), 293–310, here 293.
9. Siegfried Reinecke, *Autosymbolik in Journalismus, Literatur und Film: Struktural-funktionale Analysen vom Beginn der Motorisierung bis zur Gegenwart* (Bochum : Universitätsverlag Dr. N. Brockmeyer, 1992), 15.
10. For the following, see Gijs Mom, *The Electric Vehicle: Technology and Expectations in the Automobile Age* (Baltimore, MD: Johns Hopkins University Press, 2004), 54–55 and passim. On the early American steam car, see Andrew Jamison, *The Steam-Powered Automobile: An Answer to Air Pollution* (Bloomington and London: Indiana University Press, 1970). See also Rudi Volti, *Cars and Culture: The Life Story of a Technology* (Westport, CT, and London: Greenwood Press, 2004).
11. Wolfgang König, *Geschichte der Konsumgesellschaft* (Stuttgart: Franz Steiner Verlag, 2000), 330. The best source on early British steam propulsion in road vehicles is still K. Kühner, *Geschichtliches zum Fahrzeugantrieb* (Friedrichshafen, 1965). "Up to 1901 or 1902, in fact, steam cars were in the majority" (David T. Wells, "The Growth of the Automobile Industry in America," *Outing* 51, no. 2 [November 1907], 207–219, here 214), a conclusion also reached in Mom, *The Electric Vehicle*, 31–32.
12. Gijs Mom, "Civilized adventure as a remedy for nervous times: Early automobilism and fin the siecle culture," *History of Technology* 23 (2001) 157–190.
13. Stephen L. McIntyre, "'The Repair Man Will Gyp You': Mechanics, Managers, and Customers in the Automobile Repair Industry, 1896–1940" (unpublished dissertation, University of Missouri-Columbia, May 1995), 40.
14. Greg Thompson, "'My Sewer': James J. Flink on His Career Interpreting the Role of the Automobile in Twentieth-century Culture," in Peter Norton et al. (eds.), *Mobility in History: The Yearbook of the International Association for the History of Transport, Traffic and Mobility*, vol. 4 (New York and Oxford: Berghahn Journals, 2013), 3–17, here 5: "The automobile was most widely adopted first in rural areas—widely adopted because it served a real function there."
15. Jean-Philippe Matthey, "Un aspect de la motorisation dans le canton de Vaud: la question de l'interdiction dominicale de la circulation automobile (1906–1923)" (unpublished MA thesis, Université de Lausanne, Faculté des Lettres, 1992), 13.
16. Valeska Huber, "Multiple Mobilities: Über den Umgang mit verschiedenen Mobilitätsformen um 1900," *Geschichte und Gesellschaft* 36 (2010), 317–341, here 319.

17. Gijs Mom, "Compétition et coexistence: La motorisation des transports terrestres et le lent processus de substitution de la traction équine," *Le Mouvement Social* no. 229 (October/December 2009), 13–39.
18. Manuel Stoffers, "Cycling Cultures: Review Essay," *Transfers* 1, no. 1 (Spring 2011), 155–162.
19. Anne-Katrin Ebert, "Ein Ding der Nation? Das Fahrrad in Deutschland und den Niederlanden, 1880–1940. Eine vergleichende Konsumgeschichte" (doctoral dissertation, Universität Bielefeld, Fakultät für Geschichtswissenschaft, Philosophie und Theologie, January 2009), 38; for an English abstract, see Anne-Katrin Ebert, "Cycling towards the Nation: The Use of the Bicycle in Germany and the Netherlands, 1880–1940," *European Review of History* 11, no. 3 (2004), 347–364; for very young early French wanderers (twelve to fifteen years of age), see Catherine Bertho Lavenir, *La roue et le stylo: Comment nous sommes devenus touristes* (Paris: Editions Odile Jacob, April 1999), 288.
20. Orselli, "Usages et usagers de la route," 110.
21. Christiane Eisenberg, *"English sports" und Deutsche Bürger: Ein Gesellschaftsgeschichte 1800–1939* (Paderborn, Munich, Vienna, and Zürich: Ferdinand Schöningh, 1999), 236. See also Robert A. Smith, *A Social History of the Bicycle: Its Early Life and Times in America* (New York: American Heritage Press, A Division of McGraw-Hill Book Company, 1972); and Rüdiger Rabenstein, *Radsport und Gesellschaft: ihre sozialgeschichtlichen Zusammenhänge in der Zeit von 1867 bis 1914* (Hildesheim, Munich, and Zurich, 1991).
22. Nancy Koppelman, "One for the Road: Mobility in American Life, 1787–1905" (unpublished doctoral dissertation, Emory University, 1999), 244, figures 4.3 and 4.4; Ebert, "Ein Ding der Nation?" 36 (uniforms), 171 (horse sport); Henry Adams, *The Education of Henry Adams: Edited with an Introduction by Jean Gooder* (New York and London: Penguin Books, 1995; first ed. 1907), 421.
23. Ian Ousby, *The Englishman's England: Taste, Travel and the Rise of Tourism* (Cambridge and New York: Cambridge University Press, 1990), 18.
24. Clay McShane and Joel A. Tarr, *The Horse in the City: Living Machines in the Nineteenth Century* (Baltimore, MD: The Johns Hopkins University Press, 2007), 92–95.
25. Robin Jarvis, *Romantic Writing and Pedestrian Travel* (Houndsmill, London, and New York: MacMillan Press / St. Martin's Press, 1997), 27–28, 34. See also Joseph A. Amato, *On Foot: A History of Walking* (New York and London: New York University Press, 2004); and Dirk Schümer, *Zu Fuss: Eine kurze Geschichte des Wanderns* (Munich: Malik/Piper Verlag, 2010).
26. Philip Gordon Mackintosh and Glen Norcliffe, "Men, Women and the Bicycle: Gender and the Social Geography of Cycling in the Late Nineteenth-Century," in Dave Horton, Paul Rosen, and Peter Cox (eds.), *Cycling and Society* (Aldershot and Burlington: Ashgate, 2011), 153–177, here 155–156; separate: Betsy Klimasmith, *At Home in the City: Urban Domesticity in American Literature and Culture, 1850–1930* (Durham: University of New Hampshire Press; published by the University Press of New England, 2005), 7.
27. Orselli, "Usages et usagers de la route," 76 (table 16) and 214 (table 36). The term *Petite Reine* (little queen) stems from the journalist Pierre Giffard, although there is also an etymology that goes back to the Dutch Queen Wilhelmina, an enthusiastic bicyclist. Jean-Paul Laplagne, "La femme et la bicyclette à l'affiche," in Pierre Arnaud and Thierry Terret (eds.), *Histoire du sport féminin: Tome I: Le sport au féminin: histoire et identité* (Paris and Montreal: L'Harmattan, 1996),

83–94, here 93. On women's influence on early bicycle culture, see Sammy Kent Brooks, "The Motorcycle in American Culture: From Conception to 1935" (unpublished dissertation, George Washington University, Washington, DC, 1975), 29.
28. For an investigation into American urban elites, see, for instance, Frederic Cople Jaher, *The Urban Establishment: Upper Strata in Boston, New York, Charleston, Chicago, and Los Angeles* (Urbana, Chicago, and London: University of Illinois Press, 1982).
29. Such as H. G. Wells as quoted in Andrew Gavin Altman, "Motoring for the Million? Cars and Class in Pre-1952 Britain" (unpublished PhD thesis, Boston College, Graduate School of Arts and Sciences, 1997), 181.
30. Reinhard Bendix, "Tradition and Modernity Reconsidered," *Comparative Studies in Society and History* 9, no. 3 (1967), 292–346, here 326; Jürgen Kocka, "The Middle Classes in Europe," *The Journal of Modern History* 67, no. 4 (December 1995), 783–806, here 802; Arno J. Mayer, *The Persistence of the Old Regime: Europe to the Great War* (London: Croom Helm, 1981), 12–13 (Schumpeter and convergence), 105 (mannered), 127 (primacy); noble titles: Heinz-Gerhard Haupt, "Der Adel in einer entadelten Gesellschaft: Frankreich seit 1830," in Hans-Ulrich Wehler (ed.), *Europäischer Adel 1750–1950* (Göttingen: Vandenhoeck & Ruprecht, 1990), 286–305, here 289; aversion: ibid., 297, and Ileen Montijn, *Leven op stand 1890–1940* (Amsterdam: Thomas Rap, 1998), 21; chivalrous: Philipp Blom, *The Vertigo Years: Europe, 1900–1914* (New York: Basic Books, 2008), 246.
31. T.R. Pitts, "The Bourgeois Familiy and French Economic Retardation" (unpublished PhD dissertation, Harvard University, 1957), 205–210.
32. Monique de Saint Martin, *L'Espace de la noblesse* (Paris: Éditions Métaillié, 1993), 133, 146.
33. Mayer, *The Persistence of the Old Regime*, 9 (post-feudal), 23 (Germany), 56 (car industry), 82 (quotation), 168–169 (overrepresentation); magnificent: Volker Berghahn, *Europa im Zeitalter der Weltkriege: Die Entfesselung und Entgrenzung der Gewalt* (Frankfurt: Fischer Taschenbuch Verlag, 2002), 9.
34. Geoffrey Crossick, "The Emergence of the Lower Middle Class in Britain: A Discussion," in Crossick (ed.), *The Lower Middle Class in Britain 1870–1914* (London: Croom Helm, 1977), 11–60, here 33; Lawrence James, *The Middle Class: A History* (London: Little and Brown, 2006), 235 (superiority), 240 (ideology).
35. Eisenberg, *"English sports" und Deutsche Bürger*, 238–240.
36. Kocka, "The Middle Classes in Europe," 802. American historiography is inclined to deny the existence of an aristocracy, but see Edward Pessen, *Riches, Class and Power Before the Civil War* (Lexington, MA: Heath, 1973), referred to in Laurence Veysey, "The Autonomy of American History Reconsidered," *American Quarterly* 31, no. 4 (Autumn 1979), 455–477, here 469, who for an earlier phase documents "the existence of a continuous elite of established wealth, surprisingly 'European' in character and socially isolated from everyone else."
37. Hartmut Kaelble, *Nachbarn am Rhein: Entfremdung und Annäherung der französischen und deutschen Gesellschaft seit 1880* (Munich: Verlag C.H. Beck, 1991), 62, 66, 72, 81–83; Franz Urban Pappi, "The Petite Bourgeoisie and the New Middle Class: Differentiation or Homogenisation of the Middle Strata in Germany," in Frank Bechhofer and Brian Elliott (eds.), *The Petite Bourgeoisie: Comparative Studies of the Uneasy Stratum* (New York: St. Martin's Press, 1981), 105–120. For an exemplary overview of ruling class formation in Germany, see

Hans-Ulrich Wehler, "Klasenbildung und Klassenverhältnisse: Bürger und Arbeiter 1800–1914," in Jürgen Kocka with Elisabeth Müller-Luckner, *Arbeiter und Bürger im 19. Jahrhundert: Varianten ihres Verhältnisses im europäischen Vergleich* (Munich: R. Oldenbourg Verlag, 1986; Schriften des Historischen Kollegs, Kolloquien 7), 1–27.
38. Eisenberg, *"English sports" und Deutsche Bürger*, 17.
39. For an overview of this movement in the Netherlands, see Gijs Mom and Ruud Filarski, *Van transport naar mobiliteit: De mobiliteitsexplosie (1895–2005)* (Zutphen: Walburg Pers, 2008). The following paragraphs are based on this publication. On PIARC, see Gijs Mom, "Building an Automobile System: PIARC and Road Safety (1908–1938)," proceedings of the PIARC conference, Paris, 17–23 September 2007 (on CD-ROM). On IULA, see Oscar Gaspari, "Cities against States? Hopes, Dreams and Shortcomings of the European Municipal Movement, 1900–1960," *Contemporary European History* 11, no. 4 (2002), 597–621.
40. Montijn, *Leven op stand*, 279; Blom, *The Vertigo Years*, 29; on the nobility of the north German plains, see Richard Hamilton, *Who Voted for Hitler?* (Princeton, NJ: Princeton University Press, 1982), 22. For early German motorization, see Angela Zatsch, *Staatsmacht und Motorisierung am Morgen des Automobilzeitalters* (Konstanz, 1993); Gerhard Horras, *Die Entwicklung des deutschen Automobilmarktes bis 1914* (Munich, 1982). In larger countries the decentralized 'grip' of automobile clubs on the entire nation took place through an association with local and regional automobile clubs, especially in France, Germany, and the United States. Less innovative: Anette Gudjons, "Die Entwicklung des 'Volksautomobils' von 1904 bis 1945 unter besonderer Berücksichtigung des 'Volkswagens': Ein Beitrag zu Problemen der Sozial-, Wirtschafts- und Technikgeschichte des Automobils" (unpublished dissertation, Universität Hannover, 1988), 33.
41. Hanna Wolf, *Following America? Dutch Geographical Car Diffusion, 1900 to 1980* (Eindhoven: ECIS, 2010), 38.
42. Jarvis, "The Diffusion of the Automobile in the United States."
43. Gijs Mom, "Mobility for Pleasure: A Look at the Underside of Dutch Diffusion Curves (1920–1940)," *TST Revista de Historia: Transportes, Servicios y Telecomunicaciones* no. 12 (June 2007), 30–68.
44. Claude S. Fischer, *Made in America; A Social History of American Culture and Character* (Chicago/London: The University of Chicago Press, 2010), 107. On the motorization of the American farmer, see Ronald Kline, *Consumers in the Country: Technology and Social Change in Rural America* (Baltimore, MD, and London: Johns Hopkins University Press, 2000); Michael L. Berger, *The Devil Wagon in God's Country: The Automobile and Social Change in Rural America, 1893–1929* (Hamden, CT: Archon Books, 1979); Reynold M. Wik, *Henry Ford and Grass-roots America* (Ann Arbor: The University of Michigan Press, 1972).
45. Christopher Wells, "Car Country: Automobiles, Roads, and the Shaping of the Modern American Landscape, 1890–1929" (dissertation, University of Wisconsin-Madison, 2004), 188; Thorstein Veblen, *Absentee Ownership: Business Enterprise in Recent Times: The Case of America*, with a new introduction by Marion J. Levy (New Brunswick, CT, and London: Transaction Publishers, 2009; first ed. 1923), 140. On Veblen as philosopher of the Midwest, see the review of Stephan Truninger, *Die Amerikanisierung Amerikas: Thorstein Veblens amerikanische Weltgeschichte* (Münster, 2010), by Sebastian Voigt on http://hsozkult.geschichte.hu-berlin.de/rezensionen/2010-4-142 (last accessed 24 November 2010).

46. Reiner Flik, *Von Ford lernen? Automobilbau und Motorisierung in Deutschland bis 1933* (Cologne, Weimar, and Vienna: Böhlau Verlag, 2001); Wolf, *Following America?* 11.
47. [Félicien] Hennequin, *L'évolution automobiliste en France de 1899 à 1905* (Rapport de la Commission extraparlementaire de la circulation des automobiles) (Paris: Imprimerie Nationale, 1905); Clay McShane, *Down the Asphalt Path: The Automobile and the American City* (New York, 1994), xiii. For the Lyon region, see Étienne Faugier, "L'introduction du système automobile et ses impacts sur les campagnes du département du Rhône entre 1900 et 1939" (unpublished MA thesis in Histoire Contemporaine [Études Rurales], Université Lumière Lyon 2, Département d'histoire, 2007).
48. Bruno Gauthier, "Conducteurs et marché de l'automobile en Ille-et-Vilaine (1908–1918)" (Mémoire de maîtrise d'Histoire contemporaine, Université de Haute-Bretagne Rennes II, Département Histoire, October 2000), 141.
49. Elodie Arriola, "Des automobiles et des Hommes: Les débuts de l'Automobile Club Dauphinois (1899–1904), Volume I" (MA thesis 1, "Sciences humaines et sociales"; Grenoble, Université Mendès-France, 2007–2008), 30–47 (this author's calculation). In Ille-et-Vilaine, too, it was shop owners and not the medical doctors that functioned as pioneers. Gauthier, "Conducteurs et marché de l'automobile en Ille-et-Vilaine (1908–1918)," 68.
50. Antoine Dubois, "La motocyclette, modernité et véhicule social: émergence et resistance, l'exemple du département du Cher, 1899–1914" (Thèse de Maîtrise, Université François Rabelais U.E.R., Tours 1989), 61.
51. American medical doctors, in large numbers, seem to have been among the pioneers, if one can believe an advertisement from Maxwell-Briscoe Motor Company that boasted that 15,550 of its 47,000 "users" were physicians. "Notice to automobile buyers: Maxwell leadership proven" (advertisement), *The Club Journal* 3, no. 18 (9 December 1911), 570–571. On medical doctors as the iconic first users, see, for instance, James J. Flink, *The Car Culture* (Cambridge, MA, and London: MIT Press, 1975), 26.
52. Hennequin, *L'évolution automobiliste*, 11; Thomas Peter Maloney, "Essex motor car registrations, 1904," (unpublished MSc thesis, University of London, 1986), 26 (table 5).
53. Orselli, "Usages et usagers de la route," 204; urban growth: Mayer, *The Persistence of the Old Regime,* 71. Austria can also be grouped among the "empty countries" (ibid., 72).
54. Nicolas Spinga, "L'introduction de l'automobile dans la société française entre 1900 et 1914: Étude de presse" (unpublised MA thesis, Histoire contemporaine, Université de Paris X—Nanterre, Année 1972–1973), 37; Spinga's conclusion is typical for car historiography in the 1970s when all motorization processes were measured against the American 'model.' Spinga's findings are largely confirmed by Orselli, "Usages et usagers de la route," 207. German rural motorization only overtook urban motorization in the mid-1930s; see Hermann Röttger, "Die Struktur der deutschen Automobil-Einzelhandels mit Personenkraftwagen" (dissertation, Universität Köln, 1940), 4.
55. Gudjons, "Die Entwicklung des 'Volksautomobils,'" 38 (table 1); "Das Bestand an Kraftfahrzeugen in Deutschland," *Allgemeine Automobil-Zeitung* 9, no. 13 (27 March 1908), 36–42; Maloney, "Essex motor car registrations, 1904," 4. For a recent confirmation of my thesis "that the car was predominantly used for leisure and pleasure and this as a commodity and luxury and not as a utility,"

see Christiane Katz, "Städtische Pfade—Die Automobilisierung der Stadtverwaltung und Berufsfeuerwehr in Aachen vor dem Ersten Weltkrieg," *Zeitschrift des Aachener Geschichtsvereins* 113/114 (2012), 207–230, here 215.
56. Spinga, "L'introduction de l'automobile dans la société française." For an example of fraud in licensing for the 'series W' type (cars to be sold by manufacturers, dealers, and body makers) in France 1913, see Gauthier, "Conducteurs et marché de l'automobile en Ille-et-Vilaine (1908–1918)," 27. "Zeitgemässe Glossen zur Automobilsteuer," *Allgemeine Automobil-Zeitung* 9, no. 36 (4 September 1908), 31–33, here 31.
57. Hervé Debacker, "L'automobile à Orléans de 1897 à 1913: les débuts d'un nouvel espace identitaire" (Mémoire de maîtrise dirigé par Marie-Claude Blanc-Chaleard, Université d'Orléans, Département d'histoire, Année universitaire 1998–1999), 75. I thank Étienne Faugier (University of Lyon and Laval University, Canada) for providing me with a copy of this source.
58. Baron F. Duckham, "Early Motor Vehicle Licence Records and the Local Historian," *The Local Historian* 17 (1987), 351–357. The figures should be interpreted with care as they document transfers of registrations rather than actual car ownership. Eighty percent: Friedrich Lenger, "Der Stadt-Land-Gegensatz in der europäischen Geschichte des 19. und 20. Jahrhunderts—ein Abriss," *comparativ: Zeitschrift für Globalgeschichte und vergleichende Gesellschaftsforschung* 18, no. 2 (2008), 57–70, here 60, 67.
59. Donald Weber, "Automobilisering en de overheid in België vóór 1940: Besluitvormingsprocessen bij de ontwikkeling van een conflictbeheersingssysteem" (dissertation, University of Ghent, 2008), 271; L. S.-G., "Das Automobil in der bayerischen Abgeordnetenkammer," *Allgemeine Automobil-Zeitung* 9, no. 5 (31 January 1908), 35–39, here 38; Walter Oertel, "Militärautomobilistische Neuigkeiten," ibid. 9, no. 7 (14 February1908), 37–38, here 37; H. Conrads, "Zur Frage der Höchstgeschwindigkeit der Kraftfahrzeuge in geschlossenen Ortschaften," *Der Motorfahrer* (27 August 1909), 741–743, here 742–743.
60. Gijs Mom, *The Evolution of Automotive Technology: A Handbook* (forthcoming).
61. Uwe Fraunholz, *Motorphobia: Anti-automobiler Protest in Kaiserreich und Weimarer Republik* (Göttingen: Vandenhoeck & Ruprecht, 2002); Hayagreeva Rao and Jitendra V. Singh, "The Construction of New Paths: Institution-Building Activity in the Early Automobile and Biotechnology Industries," in Raghu Garud and Peter Karnoe (eds.), *Path Dependence and Creation* (Mahwah, NJ: Lawrence Erlbaum Associates, 2001), 243–267, here 252–253; Anti-Automobile League: Hal S. Barron, *Mixed Harvest: The Second Great Transformation in the Rural South, 1870–1930* (Chapel Hill and London: The University of North Carolina Press, 1997), 194; crusade: Kline, *Consumers in the Country*, 57; Weber, "Automobilisering en de overheid in België vóór 1940," 235; Lothar Diehl, "Tyrannen der Landstrassen: Die Automobilkritik um 1900," *Kultur & Technik* no. 3 (1998), 51–57, here 51; Frank Uekötter, "Stark im Ton, schwach in der Organisation: Der Protest gegen den frühen Automobilismus," *Geschichte in Wissenschaft und Unterricht: Zeitschrift des Verbandes der Geschichtslehrer Deutschlands* 54, no. 11 (2003), 658–670, here 661; "Das Bewerfen von Automobilisten mit Blumensträussen," *Allgemeine Automobil-Zeitung* 9, no. 18 (1 May 1908), 56–57; Spinga, "L'introduction de l'automobile dans la société française," 114; Donald F. Davis, "Dependent Motorization: Canada and the Automobile to the 1930s," *Revue d'études canadiennes* 21, no. 3 (Fall 1986), 106–132, here 124; Stephen James Davies, "Ontario and the Automobile, 1900–1930: Aspects of

Technological Integration" (unpublished doctoral dissertation, McMaster University, October 1987), 302–303; Michael L. Bromley, "Scorching Through 1902: 'The Automobile Terror'; The Year in Automobiles and Deaths in *The New York Times*," *Automotive History Review* (Summer 2003), 20–25, here 22; "Motorist Shot At," *The Club Journal* 2, no. 1 (9 July 1910), 253; Christoph Maria Merki, *Der holprige Siegeszug des Automobils 1895–1930: Zur Motorisierung des Strassenverkehrs in Frankreich, Deutschland und der Schweiz* (Vienna, Cologne, and Weimar: Böhlau, 2002), 147–166; Matthey, "Un aspect de la motorisation dans le canton de Vaud," 22; Orselli, "Usages et usagers de la route," 335.

62. William Plowden, *The Motor Car and Politics in Britain* (Harmondsworth and Ringwood, Australia: Penguin, 1971), 91.
63. Koppelman, "One for the Road," 326; Enda Duffy, *The Speed Handbook: Velocity, Pleasure, Modernism* (Durham, NC, and London: Duke University Press, 2009), 128.
64. This has been analyzed in an exemplary manner for New York City in McShane, *Down the Asphalt Path*, chapter 7 (125–148).
65. Orselli, "Usages et usagers de la route," 120; Richard Mullen and James Munson, *"The Smell of the Continent": The British Discover Europe 1814–1914* (London, Basingstoke, and Oxford: Macmillan, 2009), 200–202.
66. Uekötter, "Stark im Ton," 659; S. Daule, *Der Krieg gegen das Auto* (Leipzig, n.d.); Duffy, *The Speed Handbook*, 128; Bromley, "Scorching Through 1902," 23; Wilson: Michael L. Bromley, *William Howard Taft and the First Motoring Presidency, 1909–1913* (Jefferson, NC, and London: McFarland, 2003), 95.
67. Apaches: Orselli, "Usages et usagers de la route," 253; Joan Neuberger, *Hooliganism: Crime, Culture, and Power in St. Petersburg, 1900–1914* (Berkeley, Los Angeles, and London: University of California Press, 1993), 69 (from below), 159 (crime), 226 (n. 23; muggers), 275 (audience); Jon Savage, *Teenage: The Creation of Youth Culture* (London: Pimlico, 2007), 41–48; Brian Ladd, *Autophobia: Love and Hate in the Automotive Age* (Chicago and London: University of Chicago Press, 2008), 177. 'Apache' was a media creation, probably inspired by French fascination with American Indians and applied to a Parisian youth gang in 1900 (Savage, *Teenage*, 46).
68. L. S.-G., "Das Automobil in der bayerischen Abgeordnetenkammer," *Allgemeine Automobil-Zeitung* 9, no. 5 (31 January 1908), 35–39, here 35–36; unions and progress: Rolf Peter Sieferle, *Fortschrittsfeinde? Opposition gegen Technik und Industrie von der Romantik bis zur Gegenwart* (Munich: Verlag C.H. Beck, 1984), 78. According to Wells, "Car Country," 252, the resistance against the car cannot be explained as a class issue alone. It was chauffeurs as much as owner-drivers who caused the problems, and their behavior affected everyone, of all classes.
69. Koppelman, "One for the Road," 290; "Zeitgemässe Glossen zur Automobilsteuer," *Allgemeine Automobil-Zeitung* 9, no. 36 (4 September 1908), 31–33, here 31; on luxuries becoming a "necessity," see Fischer, *Made in America*, 74.
70. For instance, "W.W. Miller tells why Reasonable Speed Laws are Preferable to Arbitrary Laws," *The Club Journal* 4, no. 12 (23 November 1912), 454.
71. On the crucial role of expectations in the history of technology, see Gijs Mom, "'The Future is a Shifting Panorama': The Role of Expectations in the History of Mobility," in Weert Canzler and Gert Schmidt (eds.), *Zukünfte des Automobils: Aussichten und Grenzen der autotechnischen Globalisierung* (Berlin: edition sigma, 2007), 31–58. On the economic role of the French car industry in 1908, much more than that of the German, see "Frankreichs Aussenhandel mit Kraft-

fahrzeugen 1908," *Allgemeine Automobil-Zeitung* 10, no. 9 (26 February 1909), 30–33, here 31.
72. A year later, Britain's Prime Minister Balfour ushered the same vision: "great highways [will be] constructed for rapid motor traffic and confined to motor traffic." Quoted in Peter Scott, "Public-Sector Investment and Britain's Post-War Economic Performance: A Case Study of Roads Policy," *Journal of European Economic History* 34, no. 2 (2005), 391–418, here 394.
73. This is a correction to Dutch historiography of the car published so far, and is based on a content analysis of ANWB's journal *De Kampioen*. For a part of this analysis, see Herman Vink et al., 'History of innovation' (final student report, Eindhoven University of Technology, Department *Technische Innovatiewetenschappen*, December 2007; in author's possession). For the uncorrected interpretation, which accords the touring club much more defining power over the car movement before the First World War, see Gijs Mom, Peter Staal, and Johan Schot, "Civilizing Motorized Adventure: Automotive Technology, User Culture and the Dutch Touring Club as Mediator," in Ruth Oldenziel and Adri de la Bruhèze (eds.), *Manufacturing Technology: Manufacturing Consumers: The Making of Dutch Consumer Society* (Amsterdam: aksant, 2008), 141–160.
74. Mom and Filarski, *De mobiliteitsexplosie*. The following is based on this book.
75. The best sources for this are Bertho Lavenir, *La roue et le stylo*, and, for the United States, Warren James Belasco, *Americans on the Road: From Autocamp to Motel, 1910–1945* (Cambridge, MA, and London: The MIT Press, 1979).
76. Craig Horner, "'Modest Motoring' and the Emergence of Automobility in the United Kingdom," *Transfers* 2, no. 3 (Winter 2012), 56–75.
77. The only other European country where indirect evidence exists about the predominance of the chauffeur/mechanic is the United Kingdom, where in 1910 "it was noted that there were far more owner-drivers than before—perhaps half of the total." Plowden, *The Motor Car and Politics in Britain*, 94.
78. Rudolf Kemp, "Das deutsche Chauffeurwesen," *Allgemeine Automobil-Zeitung* 9, no. 17 (24 April 1908), 51–55, here 55.
79. Plowden, *The Motor Car and Politics in Britain*, 88.
80. In the United States the 1906 San Francisco earthquake played a comparable role. James J. Flink, *America Adopts the Automobile, 1895–1910* (Cambridge, MA, and London: MIT Press, 1970), 98. In Australia it was a rail strike, also in 1903, with a similar effect. Susan Priestley, *The Crown of the Road: The Story of the RACV* (Melbourne: The MacMillan Company of Australia, 1983), 6. In France it was the Marne taxis of 1914 (see chapter 3).
81. Wolfgang König, *Wilhelm II. und die Moderne: Der Kaiser und die technisch-industrielle Welt* (Paderborn, Munich, Vienna, and Zürich: Ferdinand Schöningh, 2007), 205–209, 219.
82. *100 Jahre AvD: 100 Jahre Mobilität* (Königswinter: Heel Verlag, 1999).
83. Barbara Haubner, *Nervenkitzel und Freizeitvergnügen: Automobilismus in Deutschland 1886–1914* (Göttingen: Vandenhoeck & Ruprecht, 1998), 68–94. *"Mittelstand": 50 Jahre ADAC im Dienste der Kraftfahrt* (Munich: ADAC, n.d. [1953]), 65; NSU: Hans-Christoph Graf von Seherr-Thoss, *75 Jahre ADAC 1903–1978: Tagebuch eines Automobilclubs* (Munich: ADAC Verlag, 1978), 10; membership in 1914: twenty-one; member types taken from a list of new members in 1913: "Neuanmeldungen als Club-Mitglieder des A.D.A.C.," *Der Motorfahrer* (19 December 1913), 1692. In the largest German bicycle league, the *Deutscher*

Radfahrerbund, also more than half of the members were businessmen ("Kaufleute"). Ebert, "Ein Ding der Nation?" 39.
84. Bård Toldnes, "Indtil Automobilerne har gaat sin rolige men sikre seiersgang"- Integrering av ny teknologi i perioden 1895–1926 ["Until the automobiles goes from strength to final victory—integration of new technology in the period 1895–1926"] (Trondheim: NTNU, 2007), 318 (English version of chapter 14, conclusions, in possession of this author; I thank Bård Toldnes for providing me with this document).
85. Weber, "Automobilisering en de overheid in België vóór 1940," 137, 293.
86. Haubner, Nervenkitzel und Freizeitvergnügen, 95–103; kinship ("dem Wesen... verwandt"): Eisenberg, "English sports" und Deutsche Bürger, 229.
87. Bromley, William Howard Taft, 389; Brooks, "The Motorcycle in American Culture," 55 (coats), 107 (15 percent).
88. Flik, Von Ford lernen? 80–85. In fifty-nine French departments in 1905, the motorcycle dominated. Hennequin, L'évolution automobiliste, 89. Orselli, "Usages et usagers de la route," 212 (table 33), 214 (table 34). Frank Steinbeck's Das Motorrad: Ein deutscher Sonderweg in die automobile Gesellschaft (Stuttgart: Franz Steiner Verlag, 2012; Vierteljahresschrift für Sozial- und Wirtschaftsgeschichte—Beihefte, eds. Günther Schulz et al., Band 216) came too late to be included in my analysis, but he does not seem to provide a radically different perspective.
89. Kathleen Franz, Tinkering: Consumers Reinvent the Early Automobile (Philadelphia: University of Pennsylvania Press, 2005); Kevin L. Borg, Auto Mechanics: Technology and Expertise in Twentieth-Century America (Baltimore, MD: The Johns Hopkins University Press, 2007).
90. Gudjons, "Die Entwicklung des 'Volksautomobils,'" 87.
91. Vink et al., "History of innovation."
92. For a comparable argument, see Claude S. Fischer and Glenn R. Carroll, "Telephone and Automobile Diffusion in the United States, 1902–1937," American Journal of Sociology 93, no. 5 (March 1988), 1153–1178.
93. Outing started as The Wheelman, a bicycle-touring magazine subsidized by the Pope company. Ellen Gruber Garvey, "Reframing the Bicycle: Advertising-Supported Magazines and Scorching Women," American Quarterly 47, no. 1 (March 1995), 66–101, here 82, 91. By outsider it is meant that Outing did not support car fanatics: the car was seen as just one of the possible tools for a healthy outdoors experience, for instance when a trip to Florida was recommended to be done either in a car or a boat. E. P. Powell, "Truck Farming in Florida," Outing 53, no. 4 (February 1909), 622–627, here 623; moderate: G. F. Carter, "Automobiles for Average Incomes," Outing 55, no. 4 (January 1910), 410–419, here 412.
94. Frank Presbey, "Automobiling Abroad: Practical Suggestions to those contemplating a Tour," Outing 52, no. 2 (May 1908), 245–250, here 245.
95. Michael McGerr, A Fierce Discontent: The Rise and Fall of the Progressive Movement in America, 1870–1920 (Oxford and New York: Oxford University Press, 2003), 62 (envious), 63 (fitness and standards), 71 (uneasy), 87 (crusades); Joseph Anthony Interrante, "A Movable Feast: The Automobile and the Spatial Transformation of American Culture 1890–1940" (PhD thesis, Department of History, Harvard University, Cambridge, MA, June 1983), esp. chapter 2, 58–118. Peri-urban touring: Robert Bruce, "The Range of Automobile Touring," Outing 49, no. 6 (March 1907), 810–813, here 810. For a comparable phasing as we follow in this chapter, see Peter J. Hugill, "The Rediscovery of America: Elite

Automobile Touring," *Annals of Tourism Research* 12, no. 3 (1985), 435–447. Hugill bases his periodization on an analysis of another "independent" outsider: *Country Life in America*, as well as a local newspaper, *The Cazenovia Republican*.

96. Curt McConnell, *Coast to Coast by Automobile: The Pioneering Trips, 1899–1908* (Stanford, CA: Stanford University Press, 2000); Curt McConnell, *"A Reliable Car and a Woman Who Knows It": The First Coast-to-Coast Auto Trips by Women, 1899–1916* (Jefferson, NC, and London: McFarland, 2000); Adeline and Augusta Van Buren crossed the continent on a motorcycle in 1916; Netzley, *The Encyclopedia of Women's Travel and Exploration*, 211. For examples of European long-range touring, see Pierre de Crawhez, *Les Grands Itinéraires en Automobile à travers l'Europe, l'Algérie et la Tunisie: Annuaire de l'A.C.N.L.* (Namen: Bertrand, 1906); Arnold Holtz, *Im Auto zu Kaiser Menelik* (Berlin-Charlottenburg: VITA, Deutsches Verlagshaus, n.d. [1908]); Beijing-Paris: Mathieu Flonneau, *Les cultures du volant: Essai sur les mondes de l'automobilism XXe-XXIe siècles* (Paris: Éditions Autrement, 2008), 68–73.
97. McConnell, *Coast to Coast by Automobile*, 5; attention: C. F. Carter, "What an Automobile Can Do," *Outing* 57, no. 4 (January 1911), 412–421, here 420; Alice Huyler Ramsey, *Veil, Duster, and Tire Iron* (self-published, Covina, CA, 1962; first ed. 1961), 11.
98. For a similar plea, that "travel practices might be conveniently grouped according to style," see Judith Adler, "Origins of Sightseeing," *Annals of Tourism Research* 16 (1989), 7–29, here 7.
99. The best overview of the very early 'chivalrous' phase is Christoph Maria Merki, "The Birth of Motoring out of Sport: Car Racing as a Public Relations Strategy, 1894–1905," in Laurent Tissot and Béatrice Veyrassat (eds., with the collaboration of Michèle Merger and Antoine Glaenzer), *Technological Trajectories, Markets, Institutions. Industrialized Countries, 19th–20th Centuries: From Context Dependency to Path Dependency* (New York and Oxford: Peter Lang, 2002), 227–249. Also see his *Der holprige Siegeszug des Automobils 1895–1930*. For the United States an extensive treatment of the earliest *Times Herald* race is given by Richard P. Scharchburg, *Carriages Without Horses* (Warrendale, 1993).
100. Bertho Lavenir, *La roue et le stylo*, 177–178; 1912: Patrick Young, "*La Vieille France* as Object of Bourgeois Desire: The Touring Club de France and the French Regions, 1890–1918," in Rudy Koshar (ed.), *Histories of Leisure* (Oxford and New York: Berg, 2002), 169–189, here 185 (n. 1). Co-construction by the users has, among historians of technology, increased in popularity because of a redefinition of technology, encompassing parts of the system and culture in which the artefact is embedded. Limited to the artefact itself, such co-construction did often not go beyond the location and shape of the technology, such as in the case of the French Touring Club. On the co-construction of automotive artefacts proper, see also Gijs Mom, "Translating Properties into Functions (and Vice Versa): Design, User Culture and the Creation of an American and a European Car (1930–1970)," *Journal of Design History* 20, no. 2 (2007), 171–181.
101. Wolfgang König, "Wilhelm II. und das Automobil: Eine Technik zwischen Transport, Freizeitvergnügen und Risiko," in Gunter Gebauer et al. (eds.), *Kalkuliertes Risiko: Technik, Spiel und Sport an der Grenze* (Frankfurt and New York: Campus Verlag, 2006), 179–198, here 189.
102. Bertho Lavenir, *La roue et le stylo*, 92.

103. Ibid., 70, 96, 106, 113 (collective pleasure), 126, 213, 218 (political project), 234 (cuisine), 253 (Alps).
104. Fischer, *Made in America*, 122–123, 130–133, 146, 158–159 (quotation).
105. Harvey Taylor, *A Claim on the Countryside: A History of the British Outdoor Movement* (Edinburgh: Keele University Press, 1997), 161–162, 165. Anecdotal evidence from the United Kingdom in 1897 suggests a female membership of 20 percent (ibid., 183). Push from market, middle classification: Jan Palmowski, "Travels with Baedeker—The Guidebook and the Middle Classes in Victorian and Edwardian Britain," in Rudy Koshar (ed.), *Histories of Leisure*, Leisure, Consumption and Culture Series (Oxford and New York: Berg, 2002), 105–130, here 119, 105.
106. Urry, *The Tourist Gaze*, 1–3; Nancy Tillman Romalov, "Mobile Heroines: Early Twentieth-Century Girls' Automobile Series," *Journal of Popular Culture* 28, no. 4 (Spring 2005), 231–243, here 240.
107. John Kimmons Walton, "The Social Development of Blackpool 1788–1914" (unpublished doctoral dissertation, University of Lancaster, July 1974), 263 (table 5.2).
108. Bertho Lavenir, *La roue et le stylo*, 61 (reassurance); Mullen and Munson, *"The Smell of the Continent,"* 197, 254, 313; Sunday Tramps: Michael Bunce, *The Countryside Ideal: Anglo-American Images of Landscape* (London and New York: Routledge, 1994), 114; more than half: David W. Lloyd, *Battlefield Tourism: Pilgrimage and the Commemoration of the Great War in Britain, Australia and Canada, 1919–1939* (Oxford and New York: Berg, 1998), 15, 18–19 (Vandals); Lutz, "Der subjektive Faktor," 229, 241 (necessity), 240 (periphery), 242 (departure, self-centered), 249 (Other); Karl Ditt, "Naturschutz zwischen Zivilisationskritik, Tourismusförderung und Umweltschutz: USA, England und Deutchland 1860–1970," in Mathias Frese and Michael Prinz (eds.), *Politische Zäsuren und gesellschaftlicher Wandel im 20. Jahrhundert: Regionale und vergleichende Perspektiven* (Paderborn: Ferdinand Schöningh, 1966), 499–533, here 503 (Muir); residents of France: Palmowski, "Travels with Baedeker," 107.
109. Jesse Frederick Steiner, *Americans at Play: Recent Trends in Recreation and Leisure Time Activities* (New York and London: McGraw-Hill, 1933), 9.
110. Alexis McCrossen, *Holy Day, Holiday: The American Sunday* (Ithaca, NY, and London: Cornell University Press, 2000), 79 (Germans), 86 (quotation), 84 (family-centered), 90 (carriages); Fischer, *Made in America*, 129, refers to a "continental Sunday" (see also 132).
111. David Louter, *Windshield Wilderness: Cars, Roads, and Nature in Washington's National Parks* (Seattle and London: University of Washington Press, 2006), 61 (past); Dietmar Sauermann, "Das Bürgertum im Spiegel von Gästebüchern des Sauerlandes," in Dieter Kramer and Ronald Lutz (eds.), *Reisen und Alltag: Beiträge zur kulturwissenschaftlichen Tourismusforschung* (Frankfurt: Institut für Kulturanthropologie und Europäische Ethnologie der Universität Frankfurt am Main, 1992), 81–99. For a typology of human–nature relations, see Ernst Oldemeyer, "Entwurf einer Typologie des menschlichen Verhältnisses zur Natur," in Götz Grossklaus and Ernst Oldemeyer, *Natur als Gegenwelt: Beiträge zur Kulturgeschichte der Natur* (Karlsruhe: von Loeper Verlag, 1983), 15–42.
112. Phil Macnaghten and John Urry, *Contested Natures* (London: Sage, 1999), 14–15, 24 (pleasure principle), 105 (appropriation), 113 (spectacle), 174 (painterly), 202 (walking); seeing the world: Michael P. Conzen (ed.), *The Making of the American Landscape* (Boston, London, Sydney, and Wellington: Unwin Hyman,

1990), 1. See also Adler, "Origins of Sightseeing." The Australian mobility historian Georgine Clarsen warns that motorists from other continents than North America and Europe developed different adventures: "the greater emotional proximity of colonization in Australia means that non-Indigenous narratives of travel often express unease with the landscape—an unnerving sense of uncertainty, foreboding or imminent crash." Georgine Clarsen, "Mobility in Australia: Unsettling the settled," in Gijs Mom, Gordon Pirie, and Laurent Tissot (eds.), *Mobility in History: The State of the Art in the History of Transport, Traffic and Mobility* (Neuchâtel: Alphil, 2009), 123–128, here 123–124.

113. Ditt, "Naturschutz zwischen Zivilisationskritik, Tourismusförderung und Umweltschutz," 509, 519; Bertho Lavenir, *La roue et le stylo*, 242–243, 319; NGBFPA: Brinkley, *The Wilderness Warrior*, 259. Anti-American: Patrick Kupper, "Translating Yellowstone: Early European National Parks, Weltnaturschutz and the Swiss Model," in Bernhard Gissibl, Sabine Höhler, and Patrick Kupper (eds.), *Civilizing Nature: National Parks in a Global Historical Perspective* (New York and Oxford: Berghahn Books, 2012), 123–139, here 126; Paris and Stuttgart: Sieferle, *Fortschrittsfeinde?* 168.

114. Leo Marx, *The Machine in the Garden: Technology and the Pastoral Idea in America* (Oxford and New York: Oxford University Press, 2000; first ed. 1964), 367 (industrialism), 369 (ornithologist), 193 (meaning).

115. Marx, *The Machine in the Garden*, 29, 42, 71 (quotation); Sieferle, *Fortschrittsfeinde?* 159.

116. Torben Huus Larsen, *Enduring Pastoral: Recycling the Middle Landscape Ideal in the Tennessee Valley* (Amsterdam and New York: Rodopi, 2010), 16, 176.

117. Marx, *The Machine in the Garden*, 79 (Arcadia), 83 (painting), 89 (travel), 94 (pleasure), 105 (middle state; italics in original), 128–131 (anti-European), 230 (sublime), 196 (incredulity), 200 (poetry); Johnson quoted by Valerie Larbaud, "La lenteur," in Larbaud, *Aux couleurs de Rome* (Paris: Gallimard, 1938), 141–152, here 141 (my translation). On the mythical roots of speed as related to the *"thumos"* of the Greek, the *"furor"* of the Romans, the *"Wut"* of the Germans, and the *"ferg"* of the Scandinavian peoples, see Frédéric Monneyron and Joël Thomas, *L'Automobile: Un imaginaire contemporain* (Paris: Éditions Imago, 2006), 115.

118. Bunce, *The Countryside Ideal*, 141, 145 (Olmsted); Lloyd, *Battlefield Tourism*, 22 (Roosevelt); Peter J. Schmitt, *Back to Nature: The Arcadian Myth in Urban America* (New York: Oxford University Press, 1969), 21 (commuters), 24 (Burroughs); rural elite: Urry, *The Tourist Gaze*, 5. On train commuting in the United Kingdom before WWI, see Carol Dyhouse, "Mothers and Daughters in the Middle-Class Home, c. 1870–1914," in Jane Lewis (ed.), *Labour and Love: Women's Experience of Home and Family, 1850–1940* (Oxford: Basil Blackwell, 1986), 31–47, here 30–31.

119. Bunce, *The Countryside Ideal*, 4, 14; Ditt, "Naturschutz zwischen Zivilisationskritik, Tourismusförderung und Umweltschutz," 509; Young, "*La Vieille France* as Object of Bourgeois Desire," 175; Mullen and Munson, "The Smell of the Continent," 3; future: Adler, "Travel as Performed Art," 1375.

120. Marguerite S. Shaffer, "Seeing America First: The Search for Identity in the Tourist Landscape," in David M. Wrobel and Patrick T. Long (eds.), *Seeing and Being Seen: Tourism in the American West* (Lawrence, KS: University Press of Kansas, 2001), 165–193, here 165–166; Marguerite S. Shaffer, *See America First: Tourism and National Identity, 1880–1940* (Washington, DC, and London: Smithsonian Institution Press, 2001), 27; half a billion: Ditt, "Naturschutz zwischen Zivilisa-

tionskritik, Tourismusförderung und Umweltschutz," 505. The Lincoln Highway originated by a private association in 1913, was fully operational by 1919 and ran from New York City to San Francisco, 3389 miles. Carey S. Bliss, *Autos Across America: A Bibliography of Transcontinental Travel: 1903–1940* (Los Angeles: Dawson's Book Shop, 1972), xvii; McConnell, *Coast to Coast by Automobile*, 305; Henry B. Joy, "The Lincoln Highway: What it Means to American Civilization and American Wealth," *The Club Journal* 5, no. 17 (22 November 1913), 407–409.

121. Bunce, *The Countryside Ideal*, 28; Louter, *Windshield Wilderness*, 13–27, 42; Ditt, "Naturschutz zwischen Zivilisationskritik, Tourismusförderung und Umweltschutz," 504 (n. 14); Schmitt, *Back to Nature*, 154–155; last quotation: "Glacier National Park," *The Club Journal* 4, no. 4 (25 May 1912), 95–97, here 95.

122. See for instance "How To See Europe: Members Who Have Toured the Continent Give Those Going Abroad the Benefit of Their Experiences," *The Club Journal* 3, no. 2 (29 April 1911), 108–109; R. Burnham Moffat, "Costs of Motoring Abroad" (letter to the editor), *The Club Journal* 3, no. 15 (28 October 1911), 481–482.

123. Harvey Levenstein, *Seductive Journey: American Tourists in France from Jefferson to the Jazz Age* (Chicago and London: The University of Chicago Press, 1998), 191–213 (Gayety: 209; Bohemianism: 207; millionaires: 213).

124. James, *The Middle Class*, 393–395 (happiness and poem), 319 (minister).

125. Joachim Radkau, *Das Zeitalter der Nervosität: Deutschland zwischen Bismarck und Hitler* (Munich and Vienna, 1998), 27, 190–191. Adams: Bromley, *William Howard Taft*, 97–98; "Das Automobil als Heilmittel," *Allgemeine Automobil-Zeitung* 9, no. 39 (25 September 1908), 50–51, here 50; "Zeitgemässe Glossen zur Automobilsteuer," *Allgemeine Automobil-Zeitung* 9, no. 36 (4 September 1908), 31–33.

126. T. J. Jackson Lears, "From Salvation to Self-Realization: Advertising and the Therapeutic Roots of Consumer Culture, 1880–1930," in Richard Wightman Fox and T. J. Jackson Lears (eds.), *The Culture of Consumption: Critical Essays in American History, 1880–1980* (New York: Pantheon Books, 1983), 1–38, here 11, 15.

127. James, *The Middle Class*, 299.

128. Jackson Lears, *Rebirth of a Nation: The Making of Modern America, 1877–1920* (New York: HarperCollins, 2009), 203, 279–286; James, *The Middle Class*, 308 (Baden-Powell), 362–363 (frontier); German scout journal (*Der Feldmeister*) quoted in Alan Kramer, *Dynamic of Destruction: Culture and Mass Killing in the First World War* (Oxford and New York: Oxford University Press, 2007), 161.

129. Henry Adams, *The Education of Henry Adams: Edited with an Introduction by Jean Gooder* (New York and London: Penguin Books, 1995; first ed. 1906), 422 (reference to Adams found in Laura L. Behling, "'The Woman at the Wheel': Marketing Ideal Womanhood, 1915–1934," *Journal of American Culture* 20, no. 3 [Fall 1997], 13–30, here 13); neurasthenia: Gail Bederman, *Manliness & Civilization: A Cultural History of Gender and Race in the United States, 1880–1917* (Chicago and London: The University of Chicago Press, 1995), 87.

130. Michael L. Berger, "Women drivers! the emergence of folklore and stereotypic opinions concerning feminine automotive behavior," *Women's studies international forum* vol. 9, no. 3 (1986), 257–263, here 261.

131. Hasso Spode, "Der Aufstieg des Massentourismus im 20. Jahrhundert," in Heinz-Gerhard Haupt and Claudius Torp (eds.), *Die Konsumgesellschaft in*

Deutschland 1890–1990: Ein Handbuch (Frankfurt and New York: Campus Verlag, 2009), 114–128, here 119; majority: Palmowski, "Travels with Baedeker," 115; Afke van der Toolen, "The Making of Aletta Jacobs," Historisch Nieuwsblad (September 2009), 34–39, here 35 (italics in original; article is in Dutch); Mineke Bosch, Een onwrikbaar geloof in rechtvaardigheid: Aletta Jacobs 1854–1929 (n.p.: Balans, 2005), 279–283

132. Susan Bassnett, "Travel Writing and Gender," in Peter Hulme and Tim Youngs (eds.), The Cambridge Companion to Travel Writing (Cambridge: Cambridge University Press, 2002), 225–241, here 225; Robert Sloss, "What a Woman can do with an Auto," Outing 56, no. 1 (April 1910), 62–69, here 68–69, 67.
133. Berger, "Women drivers!" 258.
134. Kate Masterson, "The Monster in the Car: A Study of the Twentieth-Century Woman's Passion for the Motor Speed-Mania and Its Attendant Evils and Vagaries," Lippincott's Monthly Magazine 86 (10 August 1910), 204–211, here 204 (goggles), 207 (bruised flesh), 209 (beau and gasoline).
135. As does Virginia Scharff, Taking the Wheel: Women and the Coming of the Motor Age (Albuquerque, NM: UNM Press, 1992), 49.
136. "Bureau of Tours Notices," Motor Travel 8, no. 11 (April 1917), 30: percentage calculated on the basis of a list of seventy-three new subscribers, eighteen of which were women, most of them married; Ramsay: Patricia D. Netzley, The Encyclopedia of Women's Travel and Exploration (Westport, CT: Oryx Press, 2001), 210.
137. Sean O'Connell, "Gender and the Car in Inter-war Britain," in Moira Donald and Linda Hurcombe (eds.), Gender and Material Culture in Historical Perspective (London and New York: MacMillan Press / St. Martin's Press, 2000), 175–191, here 184; Cotten Seiler, Republic of Drivers: A Cultural History of Automobility in America (Chicago and London: The University of Chicago Press, 2008), 50, 54 (frivolity).
138. Genevieve Wren, "Women in the World of Autos: 5000+ Women Granted Auto Patents between 1912–22," Wheels (June 1994), 7–8; Clarsen, Eat My Dust, 7–9, 158; Foley quoted in Tim Cresswell, On the Move: Mobility in the Modern Western World (New York and London: Routledge, 2006), 214.
139. Mom, The Electric Vehicle.
140. Aletta Jacobs, Herinneringen: Autobiografie (n.p. [Amsterdam]: Contact, 1995; first ed. 1924), 227.
141. Seiler, Republic of Drivers, 67.
142. Romalov, "Mobile Heroines," 234; Jennifer Parchesky, "Women in the Driver's Seat: The Auto-erotics of Early Women's Films," Film History 18, no. 2 (2006), 174–184, here 175; Deborah Clarke, Driving Women: Fiction and Automobile Culture in Twentieth-Century America (Baltimore, MD: The Johns Hopkins University Press, 2007), 12 (twist), 23 (mastered).
143. Michele Ramsey, "Selling Social Status: Woman and Automobile Advertisements from 1910–1920," Women and Language 28, no. 1 (Spring 2005), 26–38, here 30.
144. Bromley, William Howard Taft, 378.
145. 2000: "Activities of the Club," The Club Journal 2, no. 1 (16 April 1910), 5–10, here 9. At that moment the club counted only 1700 members, but people could subscribe to the Bureau of Tours separately (The Club Journal 3, no. 1 [15 April 1911], 12). A year later the membership had doubled, to 3073 members (The Club Journal 4, no. 8 [20 July 1912], 206); it had dropped by 100 two years later (The Club Journal 6, no. 1 [April 1914], 48). For an example of such a touring ac-

count, see William Jarvie, "From Richmond Over Proposed Highway to Florida, Part I," *The Club Journal* 2, no. 4 (28 May 1910), 138–140.
146. Drunk: "What Would You Do If You Were Held Up?" *The Club Journal* 3, no. 11 (2 September 1911), 399; "How Many Dogs Have You Kiled?" *The Club Journal* 2, no. 25 (18 March 1911), 995–999, and no. 26 (1 April 1911), 1069–1072, here 1069; "Technical Department," *The Club Journal* 3, no. 17 (25 November 1911), 538–539, here 539; "Glacier National Park," 97.
147. "The Joy of Touring," *The Club Journal* 2, no. 5 (11 June 1910), 171; *Country Life*: "To Enhance the Pleasure of a Tour," *The Club Journal* 2, no. 8 (23 July 1910), 303, and no. 9 (6 August 1910), 349. Richard M. Hurd, "A Family Trans-Continental Trip," *The Club Journal* 2, no. 11 (3 September 1910), 421–425, here 421.
148. "In the Bureau of Tours," *The Club Journal* 2, no. 20 (7 January 1911), 762.
149. Oscar Schubert, "Wie demoliere ich meinen Motor am schnellsten?" *Der Motorfahrer* (29 January 1909), 70–71; quotation: Wa. Ostwald, "Zwecksport: Einige Geleitworte zum A.D.A.C.-Protesttage in Eisenach," *Der Motorfahrer* 12, no. 29 (18 July 1914), 21–24.
150. Hermann Hasenauer, "Das Lächeln des Autlers," *Der Motorfahrer* (19 December 1913), 1698–1699, here 1698.
151. F., "Paedagogicum," *Der Motorfahrer* 11, no. 2 (10 January 1914), 13–14; struggle: "Die Winterprüfungsfahrt des A.D.A.C. im Oberharz 1. und 2. Februar 1914," *Der Motorfahrer* 11, no. 7 (14 February 1914), 15–16.
152. Paul Schneider, "Kind und Automobil," *Der Motorfahrer* (23 April 1909), 306–307, here 306; Richter, "In den Pyrenäen," *Der Motorfahrer* 11, no. 15 (11 April 1914), 29–30, here 30.
153. H. Conrads, "Der Krebsschaden der Autofallen," *Der Motorfahrer* 12, no. 29 (18 July 1914), 25–28, here 27.
154. Bourgeois: T., "Heimfahrt," *Der Motorfahrer* 11, no. 7 (14 February 1914), 18–19; H. H., "Fahrten zu Zweiten," *Der Motorfahrer* 11, no. 15 (11 April 1914), 21–22.
155. Sportsman: Hanns Withalm, "Auto-Kulturelles," *Der Motorfahrer* 11, no. 23 (6 June 1914), 19–20, here 19; quotations: "Reisekultur," *Der Motorfahrer* 11, no. 22 (30 May 1914), 27–28, here 27
156. Skills: Withalm, "Auto-Kulturelles," 20; quotations: H. H., "Sportliche Abenteuer," *Der Motorfahrer* 12, no. 32 (8 August 1914), 13–14; smuggle: "Reisekultur," 28.
157. Haubner, *Nervenkitzel und Freizeitvergnügen*, 130; Claudia Lieb, *Crash: Der Unfall der Moderne* (Bielefeld: Aisthesis Verlag, 2009), 63 (n. 21); "Accident Statistics in France," *The Club Journal* 3, no. 2 (29 April 1911), 92: since 1907, 68 car accidents had resulted in 14 dead and 84 injured, while 857 horse-related accidents had caused 96 deaths and 859 injured. Both the relative numbers of injuries (per accident) and of deaths are higher for the car, the latter as much as twice as high (0.2 versus 0.1 deaths per accident). The lethality of accidents in New Jersey in 1910 and 1911 was much lower: 0.07. See "News of Good Roads Movements," *The Club Journal* 4, no. 11 (31 August 1912), 285. At that moment, Paris traffic was much denser than the traffic in any other city, including New York: in the Rue the Rivoli in 1913, more than 33,000 vehicles passed, more than four times as many as on Fifth Avenue: "Vehicular Congestion in Paris," *The Club Journal* 5, no. 2 (26 April 1913), 37. On the Panic of 1907, see for instance Robert H. Wiebe, *The Search for Order 1877–1920* (New York: Hill and Wang, 1967), 201; and, in general, Charles P. Kindleberger, *Manias, Panics, and Crashes: A History of Financial Crises* (London and Basingstoke, 1978).

158. In another department, Indre-et-Loire, the highest number of municipalities limiting car speeds appeared in 1903 and 1905. Orselli, "Usages et usagers de la route," 241–242; right: Christophe Studeny, L'invention de la vitesse: France, XVIIIe–XXe siècle (n.p. [Paris]: Gallimard, 1995), 315.
159. I thank Clay McShane for conducting this survey and sharing his information with me. [Clay McShane], "New York Times Index: Content analysis" (n.d.).
160. "Die Automobilunfälle des Jahres 1907," *Allgemeine Automobil-Zeitung* 9, no. 14 (3 April 1908), 36–37; "Die 'Ursachen der Autounfälle' nach amtlicher Darstellung," *Allgemeine Automobil-Zeitung* 9, no. 25 (19 June 1908), 66; "Die Lehren der Automobil-Unfallstatistik," *Allgemeine Automobil-Zeitung* 9, no. 12 (20 March 1908), 57–58; "Die Automobilunfälle," *Allgemeine Automobil-Zeitung* 10, no. 13 (26 March 1909), 29–33, here 29.
161. B. Martini, *Leichte Wagen bis inkl. 10 Steuer-PS* (Berlin: Richard Carl Schmidt & Co., 1914; third edition rewritten by C. O. Ostwald), 11, 15.
162. Sloan and second-hand: Lendol Calder, *Financing the American Dream: A Cultural History of Consumer Credit* (Princeton, NJ: Princeton University Press, 1999), 185–186; Massachusetts: Frederic W. Cron, "Highway Design for Motor Vehicles—A Historical Review," *Public Roads* 38, no. 3 (December 1974), 85–95, here 92 (table 1); Hugill, "The Rediscovery of America," 443; Dubois, "La motocyclette, modernité et véhicule," 72; 13,000: Conrads, "Zur Frage der Höchstgeschwindigkeit," 741; 28,000: Wa. Ostwald, "Zwecksport," *Der Motorfahrer* 12, no. 29 (18 July 1914), 21–24, here 24; people's car: Wolfgang Vogel, "Lilliputaner (Wirkliche Volksautomobile)," *Der Motorfahrer* (4 December 1908), 1112–1114; R. K.-L., "Der Siegeszug des kleinen Wagens," *Der Motorfahrer* 11, no. 1 (3 January 1914), 15–19.
163. Vittorio Marchis, "Not cars alone: An Italian story," in *Motu proprio: ACI's first hundred years, heading onward* (Rome: Automobile Club d'Italia, October 2005), 87–94, here 89. The Benedicio was published in the Roman Ritual in 1913, and again a decade later.
164. Carter, "Automobiles for Average Incomes," 416; one-third: Edmund de S. Brunner and J. H. Kolb, *Rural Social Trends* (New York and London: McGraw-Hill Book Company, 1933), 63. In 1904 the "cheap runabout" sold for $650 to $750 and "accounted for the success of Michigan's automobile industry." Donald Finley Davis, *Conspicuous Production: Automobiles and Elites in Detroit, 1899–1933* (Philadelphia, PA: Temple University Press, 1988), 43.
165. Quoted in Reynold M. Wik, *Henry Ford and Grass-roots America* (Ann Arbor: The University of Michigan Press, 1972), 21.
166. Martha L. Olney, *Buy Now Pay Later: Advertising, Credit, and Consumer Durables in the 1920s* (Chapel Hill and London: The University of North Carolina Press, 1991), 47.
167. Howard Lawrence Preston, *Dirt Roads to Dixie: Accessibility and Modernization in the South, 1885–1935* (Knoxville: The University of Tennessee Press, 1991), 48.
168. When Flink, *The Car Culture*, 51, talks about the existence of a "fairly reliable family car" as early as 1908, he obviously refers to a definition of reliability guaranteeing the basic use of the car as a car, but the subsequent emergence of an extensive service network, the existence of a broad tinkering culture in the Interbellum and the myriad complaints among motorists during the interwar years (not least those owning a second-hand car) drove maintenance costs up to a level that determined largely the adoption or non-adoption of the car by low-income users. A history of automobile reliability is long overdue, but see

Kevin L. Borg, *Auto Mechanics: Technology and Expertise in Twentieth-Century America* (Baltimore, MD: The Johns Hopkins University Press, 2007), and Kathleen Franz, *Tinkering: Consumers Reinvent the Early Automobile* (Philadelphia: University of Pennsylvania Press, 2005); see also Joseph J. Corn, "Work and Vehicles: A Comment and Note," in Martin Wachs and Margaret Crawford (eds., with the assistance of Susan Marie Wirka and Taina Marjatta Rikala), *The Car and the City: The Automobile, The Built Environment, and Daily Urban Life* (Ann Arbor: The University of Michigan Press, 1992), 25–38.

169. Kline, *Consumers in the Country*, 66–79; Robert Paul Thomas, *An Analysis of the Pattern of Growth of the Automobile Industry 1895–1929* (New York: Arno Press, 1977), 91–93; doubled: Walton Hamilton et al., *Price and Price Policies* (New York and London: McGraw-Hill, 1938), 29.

170. Orselli, "Usages et usagers de la route," 212 (table 33; this author's calculations); W. O. (= Ostwald), "Goldene Zeiten," *Der Motorfahrer* 12, no. 28 (11 July 1914), 27–28, here 27.

171. Bromley, *William Howard Taft*, 387; density: George Kirkham Jarvis, "The diffusion of the automobile in the United States: 1895–1969" (unpublished dissertation, University of Michigan, Ann Arbor, 1972), 182–183 (table 30).

172. Louter, *Windshield Wilderness*, 19, 27; Katharine M. Christopher, "Motor Fashions," *The Club Journal* 6, no. 8 (November 1914), 559–562, here 559: "the automobilist who usually made an autumn tour abroad has become, through force of circumstances, a follower of the 'See America First' movement."

173. Stephen Fox, *John Muir and his Legacy: The American Conservation Movement* (Boston and Toronto: Little, Brown and Company, 1981), 121–124; Douglas Brinkley, *The Wilderness Warrior: Theodore Roosevelt and the Crusade for America* (New York: Harper, 2009), 384, 396 (telephone), 576 (check), 784 (commission); Schmitt, *Back to Nature*, xxi; Fischer, *Made in America*, 147; two million: Jakle, "Landscapes redesigned for the automobile," 294; Taft: Bromley, "Scorching Through 1902," 24, and Bromley, *William Howard Taft*, 5 (toy), 91, 276–277 (son's accident; Roosevelt), 359 (vogue); for Taft's speech, see ibid., 371–376; "The Twelfth Annual Banquet," *The Club Journal* 3, no. 20 (6 January 1912), 599–610, here 599 (guests), 601 (necessities), 604 (intolerance), 603 (state), 607 (poem); 1916: Paul S. Sutter, *Driven Wild: How the Fight against Automobiles Launched the Modern Wilderness Movement* (Seattle and London: University of Washington Press, 2002), 25; "Preparations Completed for Opening of First Road Congress," *The Club Journal* 4, no. 13 (28 September 1912), 341–342, here 341; John N. Carlisle, "Market Roads for Farmers," *The Club Journal* 6, no. 5 (August 1914), 360–361; Olaf F. Larson and Thomas B. Jones, "The Unpublished Data from Roosevelt's Commission on Country Life," *Agricultural History* 50 (October 1976), 583–599.

174. Borscheid, *Das Tempo-Virus*, 311; Robert Sloss, "Camping in an Automobile," *Outing* 56, no. 2 (May 1910), 236–240; Interrante, "A Movable Feast," 92, 96–97.

175. Gypsying: "Nomadic Touring: Or How to Avoid the 'Comforts' of the Country Tavern," *The Club Journal* 3, no. 1 (15 April 1911), 41–42, here 42; too fast: "Racing Tourists," *The Club Journal* 3; "Der Streit zwischen dem American Automobile Club und der American Automobile Association," *Allgemeine Automobil-Zeitung* 9, no. 19 (8 May 1908), 60–61. The term *modest motoring* is borrowed from Craig Horner, who is researching the British car diffusion toward the middle classes. See Horner, "'Modest Motoring.'" Spin: "A Tour on the Tory Isle," *The Club Journal* 3, no. 13 (30 September 1911), 445–448, here 445.

225 clubs: Hayagreeva Rao, "Institutional activism in the early American automobile industry," *Journal of Business Venturing* 19 (2004), 359–384, here 365.
176. Caspar Whitney, "The View-Point," *Outing* 51, no. 3 (December 1907), 373–376, here 373–374; shorter trips: Hugill, "The Rediscovery of America," 437; new motorist: Robert Sloss, "Camping in an Automobile," *Outing* 56, no. 2 (May 1910), 236–240, here 236.
177. Christopher W. Wells, "The Road to the Model T: Culture, Road Conditions, and Innovation at the Dawn of the American Motor Age," *Technology and Culture* 48, no. 3 (July 2007), 497–523, here 515, 519–520 (n. 46); Cornelius W. Hauck, "1907–1911: The Highwheeler Era," *Antique Automobile* 27, no. 2 (March/April 1963), 72–79; man's car: Kline, *Consumers in the Country*, 69.
178. John Higham, "The Reorientation of American Culture in the 1890's," in John Weiss (ed.), *The Origins of Modern Consciousness* (Detroit, MI: Wayne State University Press, 1965), 25–48, here 45.
179. On the cyclecar, see Wells, "Car Country," 201–202.
180. M. Worth Colwell, "The Worst Roads in American: Discovered by the Motorists who were Contestants in the Recent Fourth Annual Tour of the American Automobile Association," *Outing* 51, no. 2 (November 1907), 246–249. Catch-up in quality: Caspar Whitney, "The View-Point," *Outing* 51, no. 4 (January 1908), 495–498, 516, here 495–496; start of long-range touring: Bruce, "The Range of Automobile Touring," 810. On the Glidden Tour, as a "race" for "tourists," who had to keep a certain preset speed, see James F. Moore, "Predators and Prey: A New Ecology of Competition," *Harvard Business Review* (May–June 1993), 75–86. On the 1911 edition the tour manager and president of AAA was killed (ibid., 41). The race was increasingly dominated by manufacturer's teams, and was meant to "[convince] the average American that automobile travel could be undertaken without undue hazard." The average preset speed led to speeding "to make up for lost time," leading to a "marvelously narrow escape from death." Alvin W. Waters, "The Last of the Glidden Tours: Minneapolis to Glacier Park, 1913." *Minnesota History* (March 1963), 205–215, here 209, 213, 215.
181. [William Carlos Williams], *The Autobiography of William Carlos Williams* (New York: New Directions Books, 1967), 127.
182. Robert Sloss, "How to Find the Motor Trouble," *Outing* 56, no. 4 (July 1910), 476–482, here 476: Robert Sloss, "Taking Care of Your Own Auto," *Outing* 55, no. 2 (November 1909), 240–246, here 240–241; periodic maintenance: Augustus Post, "Automobile Opportunities," *Outing* 53, no. 4 (February 1909), 651–652.
183. William K. Vanderbilt, Jr., "Ideal Winter Touring: Part XI: Bougie to Algiers, Marseilles and Paris," *The Club Journal* 5, no. 26 (28 March 1914), 850–855, here 855.
184. Marchis, "Not cars alone: An Italian story," 89.
185. Carter, "What an Automobile Can Do," 420.
186. Rao, "Institutional activism in the early American automobile industry," 372–373 (figure 1).
187. Hugill, "The Rediscovery of America," 443.
188. Maxim quoted in: Seiler, *Republic of Drivers*, 46; James S. Madison, "How I Made My Car Pay For Itself," *Outing* 55, no. 4 (January 1910), 439–442; throttle: Robert Sloss, "Driving an Automobile with Brains," *Outing* 56, no. 3 (June 1910), 336–341, here 340, 341.
189. Quotation: William C. Trewin, "The Electric Car for Touring," *The Club Journal* 6, no. 3 (June 1914), 184–186, here 186; chauffeur question: "Protecting the

Owner Benefits the Maker," *The Club Journal* 2, no. 24 (4 March 1911), 935–936, here 936; on the American chauffeur problem, see, in general, Borg, *Auto Mechanics*. Own tyres: special issue on "Facts about the Club," *The Club Journal* 2, no. 26 (1 April 1911), section "Supply Department," 1032–1033, here 1033. "Corrupt Dealers and Chauffeurs Rob Motor Car Owners," *Motor Travel* 8, no. 6 (November 1916), 29; "First Anti Joy Riding Conviction," *The Club Journal* 3, no. 4 (27 May 1911), 207.
190. McIntyre, "'The Repair Man Will Gyp You,'" 42, 318.
191. "Frankreichs Urteil!" *Allgemeine Automobil-Zeitung* 9, no. 12 (20 March 1908), 20.
192. Kurt Möser, *Fahren und Fliegen in Frieden und Krieg: Kulturen individueller Mobilitätsmaschinen 1880–1930* (Ubstadt: Verlag Regionalkultur, 2009), 217.
193. John R. Eustis, "Motorists Mobilized," *The Club Journal* 9, no. 1 (June 1917), 10–12, here 11. A rare analysis of new car owners in Los Angeles reveals that, as early as 1914, one-fifth of them were workers (and 15 percent women), while the largest group (28 percent) came from the lower middle class, consisting mostly of businessmen and salesmen. Interestingly, although "almost all cars in the [Los Angeles] survey were reputedly purchased for business, at least 40 percent of all mileage driven was for pleasure." Marlou Belyea, "The Joy Ride and the Silver Screen: Commercial Leisure, Delinquency and Play Reform in Los Angeles, 1900–1980" (unpublished dissertation, Boston University Graduate School, 1983), 171–173 (table I).
194. Orselli, "Usages et usagers de la route," 400.
195. Mom, *The Electric Vehicle;* Mom, "Civilized adventure as a remedy for nervous times."

CHAPTER 2

How it Feels to be Run Over
The Grammar of Early Automobile Adventure

> "They call, the road consenting, 'Haste!'—
> Such as delight in dust collected—
> Until arrives (I too have raced!)
> The unexpected."—Rudyard Kipling[1]

In May 1912 the professional car thief Jules Bonnot and an accomplice, hiding in a house in the Parisian suburb Choisy le Roi, were dynamited by the police while thirty thousand Parisians looked on, together with an army of reporters and several "moving-picture machines." Bonnot, a mechanic who had worked in several car factories (and had been a chauffeur for Arthur Conan Doyle), was the first in France to use a car for a robbery. He preferred luxurious Delaunay-Bellevilles and De Dion-Boutons. Upon learning about the shootout the *New York Times* had cabled its Parisian correspondent to hire no less than Maurice Leblanc himself, "creator of the world-famous character, 'Arsène Lupin,'" the gentleman-thief and detective, to give a hands-on account of the last days of two members of the Bonnot gang, part of the so-called illegal anarchist movement.[2] Who could better be trusted to write a *true* story about two *real* bandits, the responsible journalists at the newspaper must have thought, than the man who so successfully had created a *fictional* hero?

Bonnot's deployment of the automobile was—as a niche practice of what one might call criminal racing—one of the extreme examples of automotive adventure, something made popular by the car chases in early cinema: as such it provides a powerful example of fiction supporting societal practices and vice versa. Others like Otto Julius Bierbaum preferred a rather more modest adventure, one that was spatial rather than temporal, even if his self-declared Biedermeier touring sometimes slid into speeding, as we will see later.[3] Between these two extremes a spectrum of automotive adventures was created, commented upon, and resisted, of which the traditional sources, used in the previous chapter, provide only a superficial impression. Relying on such sources, in an earlier publication I found automotive adventure was especially attractive to men (and some equally adventurous women) because of its tripartite challenge to masculine control of pioneer technology and its utter lack of predictability as to safety (when racing), destination (when touring), and reliability (when maintaining the car's technology and negotiating with poor road surfaces, leading to frequent forced stops due to defects or impassable terrain).[4] This fundamental uncertainty (as an antidote to the 'boredom' and 'decadent luxury' of the age) formed the real thrill of the early movement, as excellently expressed by Kipling's quote at the start of this chapter. In other words: together with the artifact in steel, rubber, wood, and glass, another immaterial artifact was produced by the 'movement.'

Qualifying the automotive adventure in this earliest phase beyond the bare acknowledgment of the masculinity of its tripartite structure is hampered by a lack of sources, and yet only this information allows us to satisfactorily explain the emergence of the automobile culture. In a period before the First World War in which the 'automobile system' was hardly mature, automobile use was directly embedded in 'nature,' not yet mediated by the infrastructures and social phenomena that we now refer to under the label 'traffic,' and that later would provide the historian with alternative access routes into the world of the motorist, of a more sociological nature. The emphasis, in this first period, on the direct experience of the car trip released experiences and emotions that the average motorist, regardless of his education, could hardly express on paper. No wonder then, that this reflective aspect of the automobile adventure, the 'account,' was often left to the specialist, the literary professional who supposedly was trained in putting sensory experiences in writing. Several such professionals were among the earliest participants in the automobile culture, in a period "of public respect for literature unique in modern times.... Except for sex and drinking, amusement was largely found in language formally arranged, either in books and periodicals or at the theater and music hall, or in one's own or one's friend's anecdotes, rumors, or clever structuring of words," literary historian Paul Fussell concluded.[5]

Qualifying early automotive adventure, then, necessitates the analysis of belletristic literature, even with this literature's bias toward the individuality of emotion. After all, automobile sport was often a *collective* practice in car and touring clubs and even in the case of individual car trips it was the small group—of middle-class friends, of the aristocratic extended family, of the courting couple, the coincidental 'party'—that shaped the adventurous experience.

In this chapter we will focus on this type of source, but we will also look into a fledgling pop culture of mass-produced narratives—such as youth literature and the well-known flow of fictional books by the Williamson couple—as well as early motion pictures to get a feel also for how nonusers experienced automobile adventure. In other words, how was the car able to function for many as a carrier of fantasies about future possession and use, yet also serve as a source of fear of being a victim of the car's seemingly inherent violence—or: how it felt to be run over.[6]

Driving and Writing: Analyzing Affinities of Touristic and Artistic Experiences

The body of analysis, then, covers the canon[7] as well as some significant utterances little known in the English-language historiography of the car, especially those made by a group of French and Belgian motoring literati active during the first decade of the new century.[8] This was well before the celebrated car enthusiasts of the Futurists and the Expressionists came to the fore. This French-Belgian group seems to have been more in touch with the ambiguous driving forces of the fledgling car culture than the Futurists. Given the fact that large-scale automobilism started in France, it is remarkable that mobility history has not taken into account what these men (nearly all men, indeed) had to say. Likewise, there has been little investigation of the German Expressionist poetics of the car, especially outside of German-language scholarship.

Taken together, this body of literature allows the construction of a 'grammar of the car adventure,' through three levels of analysis: the level of the narrative content, the level of the symbolic world, and the level of 'rhyme,' 'structural analogy,' 'analogy relation' or 'affinity' between the production of the texts and the production of automotive experiences.[9] These are levels that—put in this order—draw ever-wider contextualizing circles around these utterances of car adventure. The reader can access the first level through a form of "close but 'suspicious' reading."[10] Analysis of the second level benefits from semiological insights and discourse analysis.[11] The third level has up to now hardly been investigated, but it is the most promising and the most interesting (and the most challenging) since the first two levels do not necessitate belletristic utterances while the third does. This level is proposed here on the basis of some exemplary analyses—rather than any well-developed theory, although lately some related work has appeared in the field of media studies.[12] Especially the work by Charles Grivel is of importance here, as he characterizes the role of the car in film as "not primarily thematic," but as a "vision machine," while he also equates writing with driving, calling the car a "writing machine" (*Schreibmaschine*) as well.[13] There is, however, also a *historical* reason for this affinity: both the motoring movement and the press had much in common. They "catered to the same 'modern tastes' for novelty, flexibility, and technological advancement." In some cases they even went one small step further, such as at British car proponent Alfred Harmsworth's *Daily Mail,* that was very reluctant to cover car accidents: in such a case the arguments in favor of traditional sources over belletristic sources become futile, as deliberately selective media compete with fictional accounts.[14]

The most famous of these analyses is Wolfgang Schivelbusch's account of early train travel and his emphasis upon the 'panoramatic' experience that 'rhymes' with the emergence of the panorama as a new visual, urban form of entertainment prior to the appearance of cinema. According to Schivelbusch, citing contemporary witnesses, train travel experience was panoramatic because the train speed blurred the foreground (which was in fact a 'sideground,' the train driver's foreground being unaccessible to the passenger's gaze) into an impressionist landscape of streaks and points, and privileges a wider, sideways view on the landscape for which, his analysis suggests, the traveler had been trained during his or her visits to the panorama, which found its artistic expression in impressionist painting.[15]

The train window shows objects and scenes that are occurring simultaneously, as a sequence of events prefiguring the movie. Travelers perceived the movement of the train as a movement of the landscape, suggesting a static observer, a phenomenon we will later recognize as inversion. What is crucial here is that the technology affords the staging of a new landscape, an "artificial environment."[16] Thus, the train is an experience machine.

Although Schivelbusch's approach has since been criticized because he neglected similar experiences reported by earlier carriage passengers, and also because his train window view resembles the diorama as a nineteenth-century "protocinematic" entertainment technique more than the panorama, his analysis has influenced a whole generation of mobility historians and many others.[17] Among them is Marc Desportes, who, as a French Schivelbusch, extended the latter's analysis to cover the bicycle and the automobile in both an urban and an extra-urban, highway setting, drawing upon spatial and art theories as well as phenomenology (Husserl). Educated as a *Ponts et Chaussée* engineer in urban planning, Desportes used the concept of 'affinity' for the parallelisms between the techniques of the artistic and the car-trip production, although not in a systematic way and certainly not as part of a theory.[18] His analysis relates to concepts of intermediality, as the car appears as a medium of perception and communication just as novels, pieces of music, paintings, and movies. Like Schivelbusch, Desportes claims that the pre-industrial traveler sees a landscape not as a site but as a "genre de tableau," for which the onlooker developed a sensibility through poetry, the art of gardening, and painting.[19]

Desportes's 'template model' assumes a one-directional flow from the artist to the traveler, in a master-pupil relationship. Indeed, he cites travelers who described their experience by explicitly referring to painters' tableaux of landscapes. Art critics actually recommended using the painters' vision as guide for looking at the environment, but it is not clear exactly how this worked. One has to assume, for instance, that travelers had actually viewed the works of these painters, which is, of course, possible because both groups belonged to the same, small societal elite, and it is well known that in these early periods travelers were familiar with the main artistic works. However, I prefer a more balanced view, assuming that both travelers and painters used the same *cultural* resources, and that travelers were as active in constructing their landscape experience as artists. While artists produced their paintings and novels (which we can 'read'), travelers produced their experiences (volatile as memories and most often not recorded), drawing upon their common eagerness to indulge in mechanized motion. This second assumption takes for granted the perhaps somewhat romantic idea that artists were better observers, more trained in working with visual perception and more aware of the cultural and artistic conventions of the time, although the perspective they developed was a specific one tainted by class, ethnicity, and gender. After all, observing and developing techniques to transfer their perception into words, or sound, or onto canvas, was the core of their profession. This is also why artistic work, and literary work in particular, is an interesting historical source, especially when the historian wants to reconstruct experiences that were so dependent on a verbalization of sensory experiences. This was especially true for the period under consideration because the professional observers we will analyze did not differ that much in social and cultural background, ethnicity and gender from the larger group of motorists we aim to study.

Desportes' concept of affinity, therefore, attracts as an analytical tool because it rests upon the sensory parallelism between artistic and traveling practices. Although others had noted the parallelism between sightseeing and reading before,[20] Desportes points to senses other than the visual as well. For instance, quoting Goethe, who likens to a dance the way humans explore a given space, he points at the isomorphism of architecture and the locomotive characteristics of the human body. The automobile, and the cinema that emerged with it, broke with this rhythm of the pedestrian in that it made the view on the environment abstract, without depth, as an uninterrupted sequence of images. The landscapes, also the urban ones, lost their aura. Using an infrastructure designed in an earlier period, for another type of traveler, the automobile introduced the "radical subjectivism" of an onlooker for whom the motion picture, through a fast, traveling vision, informed and educated the experience of the environment. The early introduction of the "traveling camera," the camera that moves along with the main protagonists, illustrates this intermedial affinity in a very material way, even in its name.[21] This loss of aura has incited one scholar of tourism, Jonas Larsen, to propose replacing John Urry's "tourist gaze" by a more adequate "travel glance," because the former would give "preference to immobility" while the latter produces "visions that disappear as soon as they are seen." Larsen introduces the concept of "aesthetic affinity" for this, which is based upon "a way of seeing that, unlike the gaze, does not try to fix the fleeting, but finds aesthetic pleasure in the velocity." Although Larsen, like Urry, seems to underestimate the multisensorial character of the landscape experience (leading to a kind of visionary reductionism), but most of all neglects (unlike Urry) the fact that motorists are static relative to the vehicle while driving through the landscape (thus experiencing a curious hybrid sensation of mobility and immobility at the same time), his proposal leads to a very welcome variation in the concept of affinity, for different vehicles in different periods.[22]

The movement of the train, Schivelbusch states, "straight as a die and monotonous, is experienced as *pure* movement, cut free from the space through which it goes."[23] Interpreted like this, Schivelbusch presents the railroad as a materialization of Newton's theory of movement (as long as one does not disturb the movement of a mass, it goes in a straight line, ever faster), while I observe that Desportes focuses on vehicles such as the bicycle and the automobile whose movement suggests a Brownian motion: particles at a subatomic level moving through space, like a random walk, within a swarm. We will later, in part II, elaborate on the 'swarming' movement of cars and motorcycles in traffic on well-populated roads. For this phase, the swarm as style of movement only seems to apply to pedestrians in big towns. There, social life, in Hobbes's definition (who thus appropriated the swarm metaphor as a concept of liberalism), looked like a "homogenous swarm of incoherent, aimless perpetuations of momentum that had no capacity for growth, for fulfillment, or for rest," or to humankind in general, which in the words of entomologist William Wheeler showed "a kind of interstitial swarming, which resembles that of the honey bee."[24]

Like Schivelbusch, Desportes draws upon the seminal analysis by Leo Marx that has functioned among many historians as an exemplary way to make sense of a pastoral, Anglo-Saxon ideology in the eighteenth and nineteenth century, through an erudite grasp of literary and other forms of art, in particular painting.[25] Indeed, the erudite, transdisciplinary stance seems crucial for this type of analysis, as it has to reconstruct rather elusive experiences from sources that are often hardly thematically connected. Others who have emulated this type of analysis, such as Wolfgang Sachs and Kurt Möser in Germany or Catherine Bertho Lavenir in France and Cecilia Tichi in the United States, are all well versed in a form of cultural analysis that manages to contextualize automobile culture in at least a national substrate of general history, allowing mobility history to escape from a potential ghetto of fixation on either the vehicle or its culture.[26] Recently, from the side of environmental history, Marx has been criticized for neglecting the technical character of nature, thus creating a false dichotomy between technology and nature. This critique is justified, but did not result so far in convincing counter-narratives, so Marx's analysis must and can still function as a starting point for an investigation of the mediating role of the car, even in a situation in which the garden has entered the machine as well.[27]

In his effort to "describe and evaluate the uses of the pastoral ideal in the interpretation of American experiences," Marx distinguishes between two kinds of pastoralism emanating from two types of sources, "the richest, most coherent literary materials" and "the mass media [and its] advertising copywriters." Loyal to his profession as literary scholar intent on analyzing "literary responses to the onset of industrialism in America," but at the same time "an enthusiastic wilderness camper and amateur ornithologist," Marx's dichotomy results in a "popular and sentimental" view on the one hand and an "imaginative and complex" assortment of "collective symbols" on the other. Instead of satisfying himself, however, with the assertion "that 'literature' embodies a more sensitive and precise, a 'higher,' mode of perception," Marx set himself the task of "defining the complex relation between serious literature and the larger body of meanings and values, the general culture, which envelops it." He does so by a hunt for metaphors (our second level of analysis, firmly based on the first), after he, quite elegantly, zooms in for a brief moment on the nature experience of the American novelist Nathaniel Hawthorne. In *Sleepy Hollow* (1844), through his narrator's harmonious unity with nature, Hawthorne recounts how the lyrical 'I' suddenly hears the whistle of a distant locomotive. Marx's hunt results, as we now know, in the finding of a very potent symbol governing his entire analysis: the machine intruding in the garden, "its meaning ... carried not so much by express ideas as by the evocative quality of the language, by attitude and tone." Thus, in the words of Desportes, by intermedial affinity. According to Marx, then, highbrow literature, and art in general, "manage to qualify, or call into question, or bring irony to bear against the illusion of peace and harmony in green pasture. And it is this fact that will enable us, finally, to get at the difference between the complex and sentimental kinds of pastoralism."[28]

Not everybody saw the industrial revolution as "a railway journey in the direction of nature": Marx points out that his analysis is clearly biased toward the negative reactions to industrialization (in a hardly noticed section at the end of his book he explicitly deals with the car, for instance in F. Scott Fitzgerald's *The Great Gatsby* of 1925), while crediting other types of sources such as newspapers, especially those for "the governmental, business, and professional elites" (which he promises to analyze as well, but in the end does not) with providing a much more positive picture. The "wider, more popular sphere" was even "celebratory." For those groups of nature lovers, Marx concludes, from an analysis of Herman Melville's *Moby-Dick*, "the romantic attitude toward external nature is finally narcissistic," an assessment we will use as well to characterize automotive experience of the environment.[29]

In the end, Marx's analysis, despite its inspirational brilliance, is a work of literary history rather than social history, let alone of history of technology or mobility history. Initiated before the founding of the Society for the History of Technology, his analysis repeatedly warns against the mixing of the pastoral ideal with reality (warnings that subside after the first hundred pages), but he confines himself to literary and other texts, well contextualized of course, assuming a general knowledge among his readers of pastoral praxis. He therefore cannot deal with the most challenging question of all, which he nonetheless set out to answer: "My chief interest had been [at the point when he started to write his dissertation in the 1930s on which his later book is based] the interplay between imaginative literature and social change. ... Though poetry and fiction are not very helpful in establishing the historical record as such, they are singularly useful, I learned, in getting at the more elusive, intangible effects of change—its impact on the moral and aesthetic, emotional and sensory, aspects of experience."[30] It is this emotional and sensory experience, of a small group of early artistic motorists during the first decade of the twentieth century, that we now intend to chart, using the same type of sources drawn on by Leo Marx, and using a similar technique of thematic analysis as an entrance into the symbolic world of automotive adventure. If it is true that "fiction (articulates) social relationships," as it allows the reader to experiment with alternative scenarios of the 'I,' then this analysis is all the more promising, as the borders between fictional and 'real' car use are being blurred in the social practices of the writer-motorists under consideration.[31]

Autopoetics: Mainstream Authors

The body of literature I wish to analyze, I call autopoetics, a subset of mobility-poetics in which cars are the 'vehicle' of the narrative.[32] This is a concept borrowed from literary analyst Lawrence Buell, who uses the term "environment-poetics" in his effort to establish the subfield of "eco-criticism," literary and cultural analysis of environmental history. I use the adjective 'autopoetic' to refer to passages in works that, viewed as a whole, cannot be said to belong to the set of autopoetics, such as Marcel Proust's passage on the steeples of Martinville or the use of the car as a status symbol in Thomas Mann's *Königliche Hoheit*.[33]

My analysis covers German, French, Italian, Belgian (Flemish and Walloon), Dutch, English, and American literature, fitting in a tradition of travel writing, but certainly not all: the short stories by Maurice Leblanc, sketches by Rudyard Kipling, some poems by Apollinaire and the German Early Expressionists and not primarily peripatetic adventures of boys and girls in youth novels clearly belong to the genre of autopoetics, but they are written, it seems, with a much different intent than a 'Quest of the Self' or a metaphor for life. Nevertheless, it cannot be denied that the modern travelogue in the period under study here underwent a shift from the Other to the Self, "from a depiction of the picturesque to a questioning of the traveler's experience of the world."[34] Travel writing per se, then, is not the best entrance into autopoetics, even though travelogues functioned as "inquiries into how to write literature in the twentieth century." They are not the best entrance also because of their neglect of the technological and infrastructural aspects of travel, which blocks our view of the crucial bodily experience.[35] That does not mean, of course, that this analysis does not benefit from recent insights from travel-writing studies, such as the connection with colonial history and folklore studies. It benefits too from the debate about authenticity and fictionality, as well as attempts at periodization that distinguish between a realistic, didactic phase (1880–1900), a phase in which authors add subjectivity and literariness (up to the First World War), and literary accounts that dominated an Interbellum period.[36] But while literary travel writing primarily focused on the relation of its object of study with the literary canon (and claims, questionably, that leisure travel had to cease in times of mobilization or war), the present analysis aims at connecting textual and adventurous car practices. We will therefore, rather eclectically and iconoclastically, take what we need from these sources with the sole purpose of coming to grips with the motivations of the men and women who were shaping and being shaped by these early experiences.[37]

Another tradition that influenced autopoetics is the adventure novel, which went through a brief phase of revival before it evaporated into myriad forms of popular culture such as films and dime novels, a process in which autopoetics participated. The adventure novel reduces societal complexity to a binary opposition of life and death and good and bad that is of broad appeal.[38] In them, the plot dominates and the characters are "warlike and ... formidable."[39] Such novels were attractive because they focused upon "the aesthetic and ludic aspects of adventure." They were anti-environmentalist *avant la date* ("when an adventure is underway, forests are burned, habitats demolished, and picturesque settings polluted without a moment's hesitation"), while travel seems a "precondition" of adventure, the length of the road an indication of the adventurousness of the journey. "Adventure is the irruption of hazard, or fate, in daily life, where it introduces an overthrow (*bouleversement*) that make death possible, probable, present, until the dénouement that triumphs over it." Adventure is related to the future.[40] It is (*ad venio* or *at venture*, literally: "whatever chance brings") about physical challenge and bodily risks, but there is also a psychological dimension, including the "magic world" of "dreamlike" states, which have a "reality of their own." Such a condition, one literary scholar concludes, is easier to be found in "our 'bad taste'... in pulp magazine stories, second-rate movies, sports events; or else in another register, fast cars, gambling casinos, hikes in the wilderness: small vertical escapes from the chain gang of our days; parentheses of unreal intensity." In the words of another analyst, this "anarchic dream of heroic energies and escape," forms the basis of transcendence and the sublime, if not sublimation. After all, it was Sigmund Freud who distinguished between two types of adventure, the death drive of the "daredevils" and the "sensation seekers who search for thrills through dangerous activities which marginally put their lives at risk," which he lumped together into a "category of sublimation since such participants might be taking these risks to substitute for either of the two other opposing instincts [eros and thanatos]." Adventure, then, is especially a "*rite de passage* from white boyhood into white manhood," although, as we will see later, girls and women managed to define their own adventures. Adventure stories are "overwhelmingly conservative," in the words of George Orwell, "censored in the interests of the ruling class."[41] This was an interpretation à la Werner Sombart, who in his seminal *Der Moderne Kapitalismus* had analyzed the capitalist class as driven by adventure and preparedness to take risk, in contrast to Max Weber, who emphasized its "spirit of calculability."[42]

The adventure and its literary representation experienced a revival in the same phase that saw the emergence of the car. Its 'classic' period was the second half of the nineteenth century and has been explained with reference to the "development of empires and of science" in France and Great Britain, and the West in the United States.[43] For Georg Simmel, who dedicated an essay to the phenomenon, there existed a fundamental "connection" (*Beziehung*) between the adventurer and the artist: in both their products full life can be experienced. There is also a fundamental relation between love and adventure (for the man at least, women, according to Simmel, experience love along different categories), while adventure at the same time is concerned with form rather than content. Adventure, Simmel concluded, is near universal: we are all in a way adventurers.[44] In other words, the affinity between life and art, driving and writing, is based on this universality. Friedrich Nietzsche preferred adventure over domesticity, the latter suppressing the former (with the bicycle and the car, we tend to add, as a mediator between them). Nietzsche is also the thinker comparing the adventurer to the rebel and the criminal who disturbs "domestic activity," which he hated. The adventurer, for Nietzsche, liked "repetition, not progression" and "the 'moment,' not time, is the medium of his nature."[45]

It was in the cradle of the modern adventure novel, France (which shared this epithet with the United Kingdom, where the link between adventurousness and early capitalism was established, the latter celebrating him "who risks a great loss in the hope of a great gain") that a new literary theory of the *"roman d'aventure"* was advanced. André Gide, who struggled in this phase (1909–1914) with his desire to write such a novel (in vain), and his group of literati around the journal *Nouvelle Revue Française* redefined the protagonist of such novels as a "whole man, in his distinct individuality; 'living,' not just 'feeling'" and adorned with a *"génie naïf et ignorant,"* for whom the future of life is as much a continuous surprise as the writing of the novel is for the author. This "emotion of 'unrest'" (*inquiétude*) led the novelist to avoid fixing on one particular form. "I love the play, the unknown, the adventure," Gide declared. Although Gide was not particularly a car adept (as we will see), his theory of the adventure novel provided the basis for the affinity between the motorist and the novelist: neither of them knows what comes next.[46] The affinity between motorist and novelist was enhanced by literary strategies such as "mimetic writing, whereby words translate senses and emotions through their sound and rhythm. The quickened beatings of the heart are, for example, conveyed by the repetition of the verb which expresses them."[47] Such adventurous experiences, later literary historians conclude, were subsequently internalized in highbrow literature as a Quest of the Self, while they "reached an extraordinary peak during the 1920s and 1930s, in pulp magazines like the Doc Savage series, in Westerns, in the soaring cult of the movie stars, in the creation of instant legends around figures like Rupert Brook and Lawrence in England, and Lindbergh in the United States." Adventure, this commentator concludes, "is clearly a form of 'experiential transcendence.'"[48]

Instead of an adventure novel, France got an automotive adventure, which was preceded by a bicycle adventure. We will not analyze in depth the bicycle literature here, although we will later dedicate space to Maurice Leblanc, because he fits so well in the early automotive literary discourse (see below). Neglecting this culture in an analysis of car culture fully, however, would bias our story toward the masculine side of mobility history, as it seems that what the car initially was for men, the bicycle meant for women.[49] In Émile Zola's novel *Paris* (1897; part of the *Les trois villes* trilogy), several elements we will later identify as part of autopoetics are already present, such as the flight metaphor (the cyclists "fly like birds," and they drive "on velvet" on paved roads) and hunger for speed and the inversion it causes: "that vivid thrill of speed . . . while the gray road fled under our feet and the trees, on both sides, turn like the spoke of a fan that one opens."[50]

At the same time the cycle culture was also a male effort to recapture lost terrain, and to regain the role of a mobile avant-garde. The American writer Willa Cather's short story "Tommy, the Unsentimental" (1896, published first in the *Home Monthly*) is a case in point, as it not only played with gender roles (she "looked aggressively masculine and professional when she bent her shoulders and pumped like that" on her bike), but it also was, at this moment in her career, still filled with Bergsonian vitalism and optimism about the future role of technology. Cather had her androgynous protagonists race against electric streetcars and compete with and save weak men.[51] Sherwood Anderson, too, dedicates one short story in his retrospective collection on *Winesburg, Ohio* (1919) to a female adventurer who "startled the town by putting on men's clothes and riding bicycle down Main Street."[52]

By the turn of the century bicycle stories seem to have forgotten about the craze around the novelty of a decade ago. Jerome K. Jerome's famous *Three Men on the Bummel* (1900) shares with many automotive travelogues the nineteenth-century convention of not referring to means of travel.[53] The novel is special because of its reflective character:

Jerome interrupts the narrative to warn the reader that there "will be no useful information in this book," and "no scenery."[54] Jerome's book appeared five years after Frances Willard's account of her struggle to learn cycling, and four years after H. G. Well's sympathetic account of a young white-collar cyclist who discovers a new world during his first cycling holiday.[55] Also written after the bicycle craze, but in a much more harsh and pitiless style, is Alfred Jarry's *Le Surmâle* (1902; The Supermale).[56] Himself a bicycle rider (and "devout alcohol user"), Jarry describes a contest between five cyclists and a train, the cyclists being paced first by a car (now depicted as an explosive shell, now as a carapace), then by a flying machine to a speed of 300 km/h, meanwhile consuming artificial food. No wonder that the story ends with the death of the protagonist, and that it has been read by several French avant-garde writers as surrealism *avant la date*.[57]

It was perhaps no coincidence that it was a woman, using her husband's name, who wrote the most convincing and remarkably early transitional novel between the cycling and the car culture. Mrs. Edward Kennard (Mary, wife of a country gentleman and a justice of the peace) wrote "absolutely conventional" novels on horse culture before, but she slowly started to address the delicate topic of the New Woman who loved to deal with nuts and bolts in cycling (two novels) and in cars. *The Motor Maniac*, starring the "exhilaratingly anarchic and curiously infantile" Mrs. Jenks, as one literary analyst opines, who indulges, right at the start of the autopoetic tradition, in the functional adventure of tinkering and maintenance, and the amused golf-addict husband who observes her.[58] "She had read that it required a trained electrician to repair an induction coil if wrong," but she and her driver managed to repair it themselves: "Altogether they spent a very happy day." Mrs. Jenks's "poetry of motion" is appreciated at high speed (about 45 km/h), or as fast as "an express train," which combines the experience of flying and gliding at the same time. She buys a motorcycle to experience speed independently of men's gaze, and movement itself, not the landscape, commands her attention: "her eyes were glued to the road ahead." Hardly distinguishable as a woman from a distance, Mrs. Jenks has to compete with a handsome, much younger woman, but in the end succeeds in interesting her husband in the motoring movement, through racing and her own intermediary role.[59] Mary Kennard seems to suggest that flirtation has no home in motoring: the 'solid society' of married couples, once converted, will carry the movement headed by women. No flappers needed!

Also very early was Mary Kennard's compatriot Rudyard Kipling. From 1900 on, Britain's "poet motorate" wrote a motoring column, "Musings on Motors," in the *Daily Mail*, a newspaper known to be outright pro-car.[60] For Kipling, Nobel Prize winner for literature in 1907, who himself toured in an American Locomobile steamer (and after 1902 in a British Lanchester, after 1911 in a Rolls Royce, always with a chauffeur, and often adventuring onto the Continent), the car had a clear societal benefit. He evidently wished to convey this to his readers when telling the story of the rescue of a sick child in one of his many short stories related to the car, a rescue that allows the narrator to go racing round the countryside, even if it is only to fetch a doctor. But the child dies. Several other stories deal with the questionable reliability of the car, which incites Kipling to compose a paean to the beauty of technological skill through an almost erotic unity with the "she" of technology. This alludes to a cyborg motif we will encounter often in the course of this period, while the motif of the driver as musician will return in Proust:

> I had seen Kysh drive before, and I thought I knew the Octopod [the Lanchester], but that afternoon he and she were exalted beyond my knowledge. He improvised on the keys—the snapping levers and quivering accelerators—marvellous variations, so that our progress was sometimes a fugue and sometimes a barn-dance, varied on open greens by the weaving of fairy rings. When I protested, all that he would say was: 'I'll hypnotise the fowl! I'll dazzle the rooster!' or other words equally futile. And she—oh! that I could do her justice!—she turned her broad black bows to the westering light, and lifted us high upon hills that we might see and rejoice with her. She whooped into veiled hollows of elm and Sussex oak; she devoured infinite perspectives of park palings; she surged through forgotten hamlets, whose single streets gave back, reduplicated, the clatter of her exhaust, and, tireless, she repeated the motions.[61]

Kipling's car poems as collected in *The Muse Among The Motors 1900–1930*, are all written in the style of great poets from the past (including the couplet with which we started this chapter, after Q. H. Flaccus) and nicely illustrate the paradoxical but widely acknowledged ability of the car to evoke times past.[62]

That link to the past, as we saw in the previous chapter, was also what made the car so attractive to Otto Julius Bierbaum. The account of his *empfindsame Reise* (sentimental journey) to Italy begins with a small, automotive manual-style introduction to car technology, as if to show readers that he, despite the presence of the chauffeur, is well prepared for his trip in what contemporaries no doubt saw as a largely *technological* vehicle. It was Bierbaum who made the connection, anticipating several historians, between 'passive movement' in the car and the vibration machine of the Swedish 'therapist,' Zander.[63] This vibration caused an "inner massage," which, combined with the "air wave bath" (*Luftwel-*

lenbad) and the relaxing effect of the changing landscape, especially benefitted the health of the older man.

But despite his good intentions to travel rather than race, even Bierbaum could not resist the 'invitation,' the 'script' of the car to speed up, although he laid the blame on the technology ("it seemed as if the engine enjoyed to be able to work again"). As seen in the previous chapter, where we dealt with the problem of maintaining a very low speed, this was not such a strange idea. Bierbaum also recorded the distance traveled and calculated the average speed. Satisfied, he concluded that the car (traveling at an average of less than 25 km/h) was "*herrlich bei Rhythmus*" (at a lovely pace), which was, he conceded, "a totally unique feeling, nearly something addictive."[64] Bierbaum's title also reveals his lust for adventure, because Laurence Sterne's 1768 *Sentimental Journey Through France and Italy* was a story of erotic and amorous experiences, "adventures" as Sterne called them. But while Sterne told his readers that in such adventures "the devil was in me," Bierbaum saw the devil creep in the car that started speeding up against his will—so it seemed. And while Sterne decided to tell a history of the starling during a "blank" in his travelogue (created because "there was nothing in this road, or rather nothing which I look for in travelling"), Bierbaum's sentimental journey is all about the trip itself.[65]

In another short story Bierbaum explained how he nonetheless, like his fellow motorists, became the object of aggression by "philistines," despite his declared modesty in terms of speed: "I have never in my life been cursed at so often as during my car trip of 1902. All German dialects from Berlin to Dresden, Vienna and Munich to Bozen were represented and all Italian tongues between Trient and Sorrento—not to count the muttered curses, such as: shaken fists, tongues stuck out, backsides displayed in a lewd gesture, and more."[66] This is perhaps why he goes on to imagine a "fantasy car" driven by the devil, painted red,

> not propelled by petrol, but by the eruptive fury of slanderous people, whose souls were locked into the energy tank where they brought each other to an explosion. Thanks to that, the car could move a thousand kilometers an hour, but it also stank a hundred times more than a conventional automobile. It had a big light in front and two smaller ones somewhat to the back on the side. The one in front glared green with such a shocking intensity that any flowers touched by the beam, withered. It was not acetylene that made the lamp glow, but envy. The light on the right was red and flashing, radiating a huge devouring heat. It was hatred which fed it. The left side light gave a sallow, blue and cold beam which made everything seem dead, pathetic, and insignificant. . . . For the brake pad leather, the devil used the many skins, pressed over each other of those

people who, during their lifetime, had succeeded in disrupting the work of bright and good minds, finding justification in the opinion of the majority of tyrannic fools.

A mass consisting of human brains fills the tires on this hellish car, while a paste of spinal marrow covers the tire.[67] Bierbaum, seen in automotive historiography as an obvious, if not boring, source, reveals here surprisingly that even the biedermeier cannot always hide his aggressive feelings, especially not when fantasizing about car driving. He will not be the only one.

Pre-1914 mainstream belletristic literature, so far as it included the car in its universe, emphasized the possession rather than the use of the car, illustrating, as it were, Thorstein Veblen's 'conspicuous consumption.' Thomas Mann's *Königliche Hoheit* (1909), for instance, described the car as an instrument of distinction, status, and power, and as a mediator between two lovers: the daughter of the owner-billionaire and the prince of the fictional kingdom, to which the billionaire travels for health reasons. The car functions as a token of extreme wealth, plain and simple. Remarkably, however, Mann drafted passages of texts in 1906 that are extremely critical of the car: "Do you love noisy adventure? Accidents? Bloody and turbulent events? Rowdy events where popular humor blossoms? Terror scenes reported by the police? Does a child have to be whimpering under car wheels ... does at least a streetcar have to get derailed or a cow break its leg before your pedestrian interests are awoken? Go, go! What a crude and debased taste." It was, of course, no coincidence that Mann changed his mind during the period in which we saw (discussed in the previous chapter) a new type of middle-class user hesitantly join the aristocrats and very wealthy bourgeois, leading to a further marginalization of resistance against the car. Mann's book sold well: 148,000 copies until 1932, excluding several translations.[68]

The 'Real Unreality of Things': Critical Voices from the United Kingdom

While Thomas Mann self-censored his critical stance toward the car, quite a few voices critical of the early car appeared in the United Kingdom, perhaps because the British revered the pastoral idyll highly and they could not accept a machine in the garden without a fight. That does not mean, of course, that there were no fierce car critics elsewhere, the most eloquent of them being Léon Bloy, French "author of the extreme right, vaguely anarchist, with raging anti-semitism," in short a "Christian Céline," with a certain "sympathy" for the Bonnot gang, "who cause panic among the bourgeois." For Bloy in his diaries, "all what is modern is of the devil," and certainly the car, this "hideous machine... that stinks and kills" and that emerged in the same phase that "stupidity" (*crétinisme*) was on the rise. Motorists are "assassins with malice afore thought." Like so many of his colleagues, Bloy also got the chance, in 1906, to experience the car trip and he immediately grasps why it is so attractive: "I understand the sort of physical pleasure generated by the vibrations and the rapid pace; but... the ugliness dominates. One knows the atrocious misuse of this hideous and homicidal machine, ruinous to the intelligence as well as to the body, which makes our delicious roads of France as dangerous as the embankments of hell and which one can never loathe enough."[69]

But the critical literati dominated in Great Britain. H. G. Wells, in his story "The History of Mr. Polly" for instance, lets his protagonist, a clerk, hate the Rooseveltian "strenuous life," but has him leave his wife and become a tramp, thus finding a "nonviolent answer to a problem confronting citizens of developed countries: how to prevent sedentary, predictable lives from losing all luster and adventure. For many men, breaking out of the pattern meant violence—in sports, hunting, war."[70] Mr. Polly did not have access to a car to accomplish such a feat, but E. M. Forster's protagonists in *Howards End* (1910) did. That did not help much, either. Forster was hostile toward urban modernity. In this critical evaluation of the British upper and middle classes he saw the "throbbing, stinking car" as a symbol of a new nomadic culture that would destroy human communication. The narrator's ecological criticism of the car is rather unusual in a largely positive reception among the European literati: "And month by month the roads smelt more strongly of petrol, and were more difficult to cross, and human beings heard each other speak with great difficulty, breathed less of the air, and saw less of the sky. Nature withdrew: the leaves were falling by midsummer; the sun shone through dirt with an admired obscurity."[71] So much for the machine in the British garden.

Forster uses the car to distinguish between the provincial lower middle-class couple Charles and Dolly on the one side, and the bourgeois and wealthy Wilcoxes (who own a car) on the other, sharing with his readers the sadness for the loss of a pre-industrial Britain and at the same time exposing the hypocrisy of car ownership and use:

"Ours is such a careful chauffeur, and my husband feels it particularly hard that they should be treated like road-hogs."

"Why?"

"Well, naturally he—he isn't a road hog."

"He was exceeding the speed-limit, I conclude. He must expect to suffer with the lower animals."

Mrs. Wilcox was silenced.[72]

In Forster's universe, characterized as "typically British" by at least one critic, speed was not only rejected as dangerous to children and chickens, but also because it was an "enemy of contemplation and carefully considered actions." Remarkably, one of the protagonists abhors a car trip exactly because of the inversion effect we will analyze later: she finds that the scenery "heaved and merged like porridge."[73]

Criticism can also be read between the lines of the Scot Kenneth Grahame's famous *The Wind in the Willows* (1908) where the wealthy Mr. Toad, the "Terror of the Highway," and one of the few tormented characters in the novel, stirs up the rather quiet pastoral setting of the wood where he and his fellow animals live. Begun as a clerk at the Bank of England, and a keen sportsman who enjoyed walking and boating, Grahame started writing fragments of his story in 1904 as education for his handicapped son, whose behavior resembled Toad's recklessness and selfishness (and who committed suicide when he was nineteen years old). Grahame literally brings the mobile machine into the woods, depicted no doubt to resemble his own "mediaeval paradise" of his youth in Oxford. Toad, as an "archetypal road hog," is an animal full of contradictions, good tempered but easily roused and naive, loved and despised by his friends, laughed at by his enemies. He is a central figure but one without a real mission and as such is the essential indeterminable middle-class hero. Grahame criticizes the car because it strengthens irrationality and may even cause insanity or at least addiction. That does not prohibit the narrator from extolling the poetry of motion, or stressing the necessity of masculine skill (in a fully male context) in handling the automobile, but in the end Toad's best friends must detoxify him, to counteract the 'conversion effect' of the car: once behind the steering wheel, however fierce one's previous criticism of the "monster," one is hooked.

Rat, another animal who doesn't feel at home and wants to leave the woods, finds a cure by writing poetry. For Toad, though, the only remedy is solitary confinement and withdrawal of his 'dope.' Grahame's message is clear: poetry and car driving both offer escape, but the former is much more socially acceptable, and personally healthier. He ends up in prison, but when he escapes in women's clothes, he manages to get hold of the steering wheel of a car, only to end his escapade in a ditch:

> Toad went a little faster; then faster still, and faster. He heard the gentlemen call out warningly, "Be careful, washerwoman!" And this annoyed

him, and he began to lose his head. The driver tried to interfere, but he pinned him down in his seat with one elbow, and put on full speed. The rush of air in his face, the hum of the engine, and the light jump of the car beneath him intoxicated his weak brain, ... "Ho! ho! I am Toad, the motor-car snatcher, the prison-breaker, the Toad who always escapes! Sit still, and you shall know what driving really is, for you are in the hands of the famous, the skillful, the entirely fearless Toad!"

After this liberating eruption of narcissism, which we will not easily find in the columns of car- and touring-club journals, Toad realizes that his case is hopeless. Grahame plays with what many at the time considered to be the addictive effect of car driving, a process that works through the body and makes one dizzy. Others explored the full spectrum of the relationships between the car passenger, driver, and the environment, and some even tried to find advantages, although they often confirmed Grahame's musings.

Bierbaum's devilish car fantasy resembles that other car-devil story written by Marie Corelli, the British occult new age champion, only one of a large group of esoteric and Theosophic authors of that period, but in her case more popular (in sold copies) than any other of her contemporaries, including H. G. Wells, Rudyard Kipling, and Arthur Conan Doyle. Written in 1901 as part of *A Christmas Greeting,* but republished with color drawings by her friend Arthur Severn, *The Devil's Motor* has been called "the first 'green' motoring novel," while Corelli herself (her name a pseudonym for Mary Mackay, the illegitimate daughter of a Scottish poet and his servant) has been portrayed as an eccentric man-hating lesbian and critic of "he-females" on bicycles. Mainstream critics attacked her "overly melodramatic and emotional writing" as combining "the imagination of a Poe with ... the mentality of a nursemaid." The critics' misogyny against a woman whom at least one biographer calls a feminist (albeit a reactionary one), may have drawn inspiration from Corelli's nickname, the "female Haggard," a nickname that suggested that young women found in Corelli's work "the same sort of thrill boys sought in *King Solomon's Mine*" by the Victorian adventure writer Henry Rider Haggard. Haggard was a worthy opponent, as his best-selling novels of "lost-race romance" (such as *She: A History of Adventure,* 1887) were characterized by a "combative response to an emergent feminist culture" and the threat of the "New Woman." Corelli's "heretically religious" story, which did not sell as well as her other work, takes the motif of flight, ubiquitously present in pre–World War autopoetics, literally: the car, with its "stench and ... muffled roar," driven by the devil himself, races through the night skies. "The forest dropped like broken reeds— the mountains crumbled into pits and quarries, the seas and rivers, the lakes and waterfalls dried up into black and muddy waters, and all the land was bereft of beauty." And somewhat further into the story the flight metaphor reaches its ultimate zenith, illustrated by a nightmarish color plate by Severn and prefiguring, so it seems, the scenarios of one-minute silent movies, as we will analyze later: "Like lightning now the great Car tore through space—its flaring lamps flashing, its wheels grinding with the sullen noise of a bursting volcano—and amidst cries and shrieks indescribable, it leaped, as it were, from peak to peak of toppling clouds that towered above and around it like mighty mountains." Such roaring and thundering was a different sound than Forster's 'hum of the engine.' Before the "Demon of the Car" disappears into "the fathomless abyss of the Unseen and the Unknown," taking the entire world with it, it has "ruin(ed) Mankind": "Now shall you forget that God exists! Now shall you all have your own wild way, for Your way is My way! . . . And, mingling with the grinding roar of its wheels came other sounds— sounds of fierce laughter and loud cursing—yells and shrieks and groans of torture—the screams of the suffering, the sobs of the dying—and as the Fiend drove on with swiftly quickening fury, men and women and little children were trampled down one upon another and killed in their thousands, and the Car was splashed thick with human blood."[74]

The car as a devil's vehicle was a popular motif in this early phase full of resistance. Jules Verne described in 1904, in his *Maître du monde* (Master of the World), a car that ran 240 km/h and which he called "the devil's chariot, fired with the flames of hell and steered by Satan himself."[75] And yet, in the same year of *The Devil's Motor*'s publication, Corelli bought a new Daimler, hired a chauffeur, and had a little seat installed for her Yorkshire terrier. With her companion Bertha van der Vyver, she used the car to drive on holiday.[76]

Between the enthusiasm of a Bierbaum or a Kipling and the rejection of a Corelli or Forster, the only mainstream author that seems to take a middle ground of recognizing the car's sensory and experiential potential but at the same time acknowledging its detrimental effects on nonusers and the wider environment, was Vernon Lee, the pseudonym of Violet Paget, "one of the most popular travel writers at the turn of the century," and author, among many others, of *The Sentimental Traveller* (1908).[77] Paget was born at a chateau of English expatriates in France. Living most of her life near Florence, Italy, she was until recently one of those forgotten "female aesthetes" and Victorian intellectuals, an expert (like her example Walter Pater) of the Renaissance, and more recently celebrated as a lesbian feminist ("dressed à la garçonne") and a "New Woman" novelist of "supernatural fiction." Part of an oeuvre characterized as written in "an age of transition between romanticism and modernism," her short story "The Motor-car and the Genius of Places" not only stands out as an effort in psychological landscape description (based on her theory of psychological aesthetics, in which she introduced the German concept of *Einfühlung,* or empathy), it was also one of the few texts explicitly opposing the idea of the car as an adventure machine, or perhaps, the first effort to define a female adventure (if we do not count Corelli's motor devil as a female adventure fantasy). "They took me yesterday on a long drive in their motor-car along the Hog's Back, through Guildford," her story begins, "and a score of other places, all of which I do not believe in." For her, the "moral mission" of the car in the future is that "it will be a thing of honourable utility, not swagger, and within the reach of many. For instead of travelling, like irresponsible outlaws, imprisoned between fences and embankments, it takes us into the streets and on to the roads where people are moving about naturally." Although in her mind the car still remains the colonial surveying machine to observe the 'natural' Other, her approach to that Other was quite different from many of her male autopoetic colleagues. The car's moral mission "makes us slacken and deflect for waggons and go-carts, nay, stop short, decently, for children and dogs, feeling the claims of other life than ours, and suggesting that remote districts and foreign lands are not our tea-gardens and racecourse; for I fear that railways have merely diminished the sense of enlarged brotherhood which should come from reasonable travel." This "reasonable travel," then, "will reinstate the decorous sense of mystery connected with change of place." To experience that mystery we do not need high speed: "Apart from the rapture of mere swift movement, which I neither feel nor regret not feeling (there are so many possible exhilarations in life without verging on drunkenness), is the sense of triumph over steepness, flying uphill with the ease we are accustomed to only in rushing down." Lee was well aware of the "grievances against the motor-car," to which she responds, "it is not good, I am afraid, dear friends, to scatter people along roads and cover them with the dust of our wheels; there is a corresponding scattering of our soul, and a covering of *it* with dust." Her critical attitude toward an irresponsible use of the car 'rhymes' with her attitude toward aestheticism, which she starts to reject in this phase as she "became suspicious of exquisite pleasures for their own sake." For Lee, moreover, violence is "wastefulness," and waste is against "progress." But as if to support her American fellow author Edith Wharton (who admired her art and whose work we will examine later), she praises the car as a vehicle of fantasy: "No other mode of travel has ever given me so fantastic a sense of the real unreality of things, of their becoming only because we happen to see them."[78] This "real unreality of things" was indeed the contribution of the car to the new, modern way of perceiving the environment; it seems to be the very base of the fascination with the car and partly may explain the successful emergence of automobilism, which was a movement of self-awareness.

If we are looking for ways to nuance the tripartite automotive adventure, it might be found here: the car's thrill lays in its being a motor of fantasy, of imaginary adventure. Whether in favor or against, poetry and car driving seem to draw from the same source.

Colonialism by Car: Gendered Travel Writing

Bierbaum and Kipling formed the tip of an iceberg of lesser-talented reflective travelers, although Eugen Diesel, the son of the engine inventor, achieved some national fame with his account of a car trip to Italy with his father.[79] Less well known but probably quite popular at the time (because the journals of their respective car clubs first published their stories) were travel accounts such as Paul Konody's *Through the Alps to the Apennines* (with a party of four, including a painter, an architect, and a photographer) or Charles L. Freeston's guide to 'conquering' the Alpine roads (which made it into a German translation), or Francis Miltoun's accounts of trips through the Netherlands ("almost an automobilist's paradise") and Italy. While Miltoun drove himself (like most Americans, even when they 'did' Europe), Konody used a chauffeur (in a steam car, aptly named "Cricket" because of the hissing sounds it made). Both were anti-urban (anti-industrialization), anti-train, both used the car as a protective 'shell' against 'the Other' (even if it was only their bad breath), both found the resistance by the local population to be quite modest (thus exposing themselves as experienced, well-behaving motorists), both were hunting for the picturesque and the sublime, both described (borrowing from impressionist literary examples) the car's movement as a flight ("a magic carpet," says Konody) and both cannot resist the invitation to race, to compete with fellow road users. Both endure the functional adventure (defects, tinkering, maintenance), and Konody explains why it is no impediment for a successful trip, as "it is one of the peculiarities of motoring, caused perhaps by the healthy and exhilarating rush of pure air into the lungs, that unpleasant experiences are soon forgotten." Konody did what many Americans planned: to "abandon ourselves to the physical delight of unimpeded speed on an absolutely straight switchback run through rural France." For Miltoun his trip is a pilgrimage rather than just travel, and Italy is "a vast kinetoskope of heterogeneous sight" rather than just a landscape, as the car enables him to see "the real Italy, the old Italy." He warns his fellow countrymen to desist in their habit of throwing pennies to the locals, as a way of wading through the potentially hostile periphery of Europe.[80]

Throwing coins at the Other was exactly what the francophone Belgian short-story writer Eugène Demolder did when he undertook a trip to Spain in a car with a chauffeur called "Marius" in the summer of 1905: "While we put our luggage on the car, I take pleasure in throwing coins at the boys who are watching us.... The lads throw themselves upon them: jostlings, caprioles, shouts, struggles, butts and legs in the air, scrimmages of black necks and naked bronzed feet."[81] Demolder's journey in the footsteps of Théophile Gautier, the great French romantic author of *Voyage en Espagne* (1843), provided a kaleidoscope of the adventures during long-range car travel in the first decade of the last century, complete with a staccato style suggesting rapid movement: "We are touring! ... The car swallows the road. Its speed pushes the trees aside rapidly: by the thousands they pass us on both sides. Some, with pitiful branches, are desperate in their being fixed to the ground. They would like to tear themselves from the terrain, fly away too, like us, like the birds.... Charming car: docility of a beaten dog, supple like a field snake, quickness of lightning—and on top of that, quiet and discrete. Thirty horsepower."

The account of the journey in "our little rolling salon" with a closed body, the "red Baedeker" near at hand, is a mixture of poetic landscape and town descriptions, interlaced with the intricacies of the trip. To these belong its high speed (70 km/h; "*à folle vitesse*"; "we scorch the road"), the narrator explaining the "iron bowels" of the car "beast" that sings when it starts, and the necessity of braking on the engine, "gas closed," on a steep descent. He instructs Marius to slow down but expresses his admiration for Marius's skills, meanwhile enjoying the landscape's varied color palette, the tire defects, complaining that by and by Europe becomes smaller (a gypsy dance he sees in Spain was already performed at the Paris Exhibition of 1900), or about the anthropomorphizing of the car and the many defects of cars passed in France on the way back home.[82]

But most remarkably the trip resembles, at times, a colonial survey into the land of the Other, the staring 'mob' from which a girl breaks away, smiling, before "she insults us in a guttural voice." The foreigners (in their own country) have an "arrogant attitude," but dance as "little fauns." He compares a boy guide's behaviour in Granada to the amusing curiosity of an ape, while the erotic dances of bands of Gypsies are attractive because they are not yet civilized. Some of the Others throw stones, and a man of fifty, enraged because of the disturbance caused to his two mules, is relieved of his stick at gun point, justified by a sarcastic "We did not touch any of those three animals." He experiences this colonial and violent aspect of the trip as an adventure, important and ubiquitous enough in autopoetics, I conclude, to create a subcategory of 'colonial aggression' within the wider category of the spatial adventure. Looking back at the trip the narrator says: "And those hundred villages that the auto entered while slowing down, with a bit of anxiety: what kind of reception would the inhabitants have in store for us?"[83]

More and less talented narrators by the dozens wrote similar travel accounts. For instance Demolder's compatriot baron Pierre de Crawhez made dozens of trips to the European peripheries (including the Northern parts of Africa). His account evolved over the years into a travel guide with hotel tips, distance tables and maps published by the local car club.[84] For motoring art lovers at the other extreme there was the reenactment of a Grand Tour to Italy, camera at the ready, in which the critiques of art and architecture by the American archeologist Dan Fellows Platt fully overshadowed the description of the journey.[85]

Colonialism by car in this prewar period, in which European exoticism went through a last peak, was rare, testifying to the car's limited reliability, although the Paris to Beijing race, and some unique examples of 'autoimperialism' expressed the urge with which the "expansionist dream" was pursued. That was because "on the outskirts of Empire ... the machine is relatively impotent and the individual is strong," an excellent base for adventure through whatever mobile medium. When the German missionary doctor Rudolf Frisch traveled more than 2000 km through the British colony of Gold Coast and the German colony of Togo (current-day Ghana) in 1909 and 1910, he used a bicycle (which gave the local population enough time to empty their villages upon his arrival, in order to avoid being forced to perform carrier duties). France's famous travel writer Victor Segalen (who went to China) and Blaise Cendrars (with his experimental poem *La prose du Transsibérien*) used the train, but they were no less fascinated by speed. Such adventures were important because "imperial relations may have been established initially by guns ... but they were maintained in their interpellative phase largely by textuality."[86]

Female travel writers seemed less inclined to include an open celebration of aggression and violence in their narrative, as two other travel accounts, Edith Wharton's *A Motor-Flight through France* (1908) and Margarete Winter's *5000 Kilometer Autofahrt ohne Chauffeur* (1905) testify. While the former (Edith Wharton was the pseudonym of Edith Newbold Jones, born 1862 in New York City) was a well-known Pulitzer Prize–winning author born to wealthy parents who spent much of her adult life in France, the latter, who left hardly any trace other than that she belonged to a family of textile and leather manufacturers, even left her name off the book's cover.[87] Winter's was a bread-and-butter travel account. What makes her story special is that she drove herself and took her son along as a mechanic. Describing her kin as "free *Herrenmenschen*" (as opposed to the aristocratic motorists who left the work on the car to others) her story contained all the familiar ingredients: advice to other motorists, praise of the car (an Opel) as an adventure machine, and comments on a largely friendly population except for the Belgians and especially the Dutch (most particularly the Frisians), who were utterly hostile to the small European nomadic community of wealthy motorists. Most of all, her story is one of technical unreliability, so much so that she includes a listing of all thirty-five tire punctures at the end of the book. The narrative itself abounds with many other technical mishaps, frequent enough to prevent the travelers from enjoying the landscapes. Winter's adventures are the most spartan account of automobility during the first phase. The narrative's themes of the joys of nature and architecture (the painter's gaze) yield on nearly every page to two other motifs: the skill of driving and repairing, and the practice of negotiating the road and its curves, preventing the protagonists from reaching full speed. The story provides evidence that car driving on its own following extensive itineraries was only attractive to those (whether men or women) who really enjoyed a tripartite adventure, including the functional part: Margarete Winter was technically au courant (she knows, for instance, that the mechanical strain on the tires is largest when accelerating from a standstill) while her son executed the repairs. It was sport rather than tourism, although here, too, all the well-known ingredients were present: the aversion to monotony (the scenery has to change literally every minute in order to avoid the motorists becoming bored), the restlessness forcing the driver to speed up to compensate for lost time, the competition with other tourists leading to a form of covert racing, the stereotyping of the foreigner, the aggressive metaphors in describing the "conquest" of nature, the hostility to the train and the anthropomorphizing of the car, the flight metaphor. There is even a so-called inversion to be found, practiced famously by Marcel Proust, as we will see later: the dome of Ulm "grows" while they drive toward the town.[88]

Quite the opposite was true for Wharton's account: nowhere in her three trips through France between 1904 and 1907 (recorded first in *Atlantic Monthly* and subsequently published as a book in 1908), frequently using the same routes as Winter, does she mention her chauffeur: we know his name from her autobiography, and although one of the recent editors of her book (Julian Barnes) refers to "literary decorum" as preventing attention to such mundane things as tire punctures, oil changes and "half a dozen servants [sent] ahead by train," Wharton's suppression of the automotive experience is so extreme that it allows her to present the trip as an act of "reading," as if we travel through a history book brought about by Wharton's study of church architecture and her pictorial description of nature in the jargon of a painter. A "scholar-adventurer" with an "un-American and unfeminine" erudition, using a bicycle when doing research for one of her books, Wharton "creates two adventures: one of geographical and another of intellectual discovery."[89]

Meandering between impressionist nature description and the prose of the guidebooks consulted ("As a centre for excursions there is no place like [Pau] in France" is one of the few mundane slips of Wharton's tongue), the "descent to Rouen" (How? On foot? In the car?) is characterized as "poetry" (which is not the same as a poetic description of the descent) and the climb of a certain hill is "almost a drop back to prose," while she praises the coastline around Saint Tropez because of its "Virgilian breadth of composition." It is as if we are hardly moving in physical reality, instead making a spiritual journey, a journey in reality-as-poetry, an impression supported by the fact that there are hardly any human actors mentioned.[90] It is the road that "carries one" through the landscape. Not even her fellow travelers are mentioned, one of them being the admired Henry James (who had published *A Little Tour in France* in 1884).

Although commentators always quote her very first sentences with their Bierbaum-like attack on the railway and praise of the car's role in "restoring the romance of travel," the car as actor is fundamentally absent from the narrative: practices are described as functions (climbing) rather than properties (of the car). From one of her biographies we know the make of the car (a Panhard-Levassor) and from this source we also know that after her three trips through France and especially after her divorce in 1913, "15 months of frenetic traveling" with several of her friends followed, through seven countries and on three continents, experiences that flowed into her eight (another biographer counts five) nonfiction travelogues, a series that closed, quite suddenly, with a never-published fragment called "A Motor-Flight Through Spain" (1926). Wharton loved the "wonderful adventure" in her car which produced "*bon movement* of the imagination" and thus created an affinity between her simultaneous production of driving and writing experiences, so much so that often even the 'I' disappears: "One must cross the central mass of the Cévennes." Relics are seen by "the observer," while the traveler is called "the pilgrim" (rather than "the tourist").[91] This has the effect of universalizing the experience of nature's sublime (making it independent from Wharton's gaze), visible only through the cocooning effect of the car as observed by editor Julian Barnes. "Wharton's book," he says, "chronicles peasant faces glimpsed in doorways and the flushed servant at the auberge, but it's significant that the two largest human presences in her text are both long dead: George Sand and Madame de Sévigné." Although her car was of the open type, her husband closed the construction, adding electric light and "every known accessorie and comfort." This provides protection against an "inhospitable village" they traverse and shields the passengers against "some unfavourable comment on the part of the opposition."[92] What makes this a motorcar travelogue, then, is not so much the description of the flight movement of the body, or the often-cited advantages of the car over the train. Except for the very first sentences (and a remark about the unifying effect of the car on French *hôtellerie*, as well as an equally isolated remark on the role of the car in avoiding the railway station) it is the visions—the view of the landscape, the approach of a city, in the sense analyzed by Desportes—that constitute this text, including the evocation of "the incidental graces of the foreground." The writer is literally *in* the landscape, moving fast (supported by Proustian inversion: "the white road ... swept us ... to Beauvais"), unmediated by the car, so it seems, let alone by a chauffeur.[93] This has the effect of changing the relationship between time and space, an effect, it is suggested implicitly, only possible through the automobile. Traveling to remote parts of France initiates a journey through time, resulting in an apparently very attractive and sometimes confusing mix of the present and the past, of reality and fiction, producing a "rush of associations" as she observes herself. This effect is deliberately enhanced as Wharton treats reality as fiction, for instance when she describes a young woman stepping out of a house as a figure stepping out of a novel by George Sand. But it is only because of the absent car that Wharton was able to see what she describes, at the tempo she narrates. Autopoetics, in other words, written by a driver-author, does not have to deal explicitly with the car at all: "Wharton's aesthetic of travel," a recent analysis concludes, "is an effect of a new mode of mobility that inspires a new, modernist aesthetic. The proliferation of points of view, the imprinting of multiple layers of time, the preoccupation with formal qualities abstracted from the passing landscape—these are features of the modernist aesthetic in the early decades of the century. These features of mobility suggest how implicated new modes of technology are in new modes of representation." The paradox of Wharton's experience, then, is that she leaves the "rampant modernization" of the

United States behind to find this modernism in a "premodern France," as so many Americans of her class did in this phase.[94] In other words, the adventure in time (one aspect of the tripartite adventure) is also experienced in a totally different way from the car as speed machine. In Wharton's travelogue, as in many others, the car is implicitly used as a time machine, a machine that allows travel between the now and the past, and back. Apart from the colonial aggression signaled previously, this is another extension of the automotive adventure we owe to the reading of autopoetics.

Wharton's seemingly quiet adventures reveal again a hidden and basically violent aspect of early car culture: not only does the "elitist refusal of the limitations of the railways and cry for freedom" lead to the surreptitious taking of towns (somewhere she uses the word "attack" for this practice), but the entire expedition into the European southern periphery works as "an almost imperialist appropriation on the part of the writer." This includes the distanced description of the Other from the safe interior of the automotive cocoon, which functions as an isolated and distant viewpoint, a shell that moves (for the outsiders) and is static (for the passengers, seen from the interior) at the same time.[95] In this sense, Wharton's narrative fits in the tradition of the imperialist and colonialist "adventurer: that individual who challenges the unknown, the new frontier, the personally dangerous" with the aim of integrating newly acquired territory.[96]

The plea by Vernon Lee and others for modest motoring, reminiscent of the way that several touring clubs tried to resist what they saw as the violent speed orgies of the car clubs, was a far cry from the hundreds and hundreds of travel accounts of shorter trips by lesser artistic souls, sent to the club journals of every motorizing nation, amateur author-travelers who used the stories and poems presented here as a model and who in their "paradoxical pleasures" of minor inconveniences (such as feeling hungry) experienced similar emotions to their highbrow compatriots.[97] Such travelogues easily slid into boring, Baedeker-like, matter-of-fact descriptions of roads, landscapes, people, and the technical mishaps of the car. By the start of the second decade the travelogues sent into the Dutch auto-club journal, for instance, had become so "threadbare" that the editor invited literary figures—presumably better equipped to express the very essence of adventurous traveling—to contribute. This new genre, as yet unnoticed by Dutch literary historiography, started with a novella contest among the journal's readers. Most of the winning novellas play with the erotic risk of car speed, and exhibit (mostly male) fantasies of power in a world in which old values vanish and women emancipate. The cyborg topos (the lust for feeling a bodily "unity" with the car) and the neurasthenia theme (the eulogy of "practice," presenting car driving as a healthy, nerve-strengthening activity providing escape from the daily routine) were present in many of these stories, all embedded in an orgy of Impressionist neologisms and sensory experiences. Car driving, in short, made the poet emerge within the man. One of the winners, the well-known journalist Leo Faust, gave his female protagonist the following text:

> When you sit behind the steering wheel you are forced to focus fully on the car. No time to think of anything else, so you don't, automatically. You become, as it were, one with the motor which works on your nerves as a healthy sleep. A horse thinks for you. I have composed whole fantasies sitting behind a horse. The engine doesn't think, *he* simply obeys, strict and immediate, as a good soldier, even if sent into disaster. The constant consciousness of this responsibility is educating, would you believe?

"Do you venture with me into the most dangerous? ... Do you dare to go to the ultimate with me?" the driver asks. "In God's name, faster," she begs, shortly before "the machine shocks into wreckage." In other stories, when the car's speed was a bit lower, there was the ubiquitous flight metaphor that at the same time presented the driving experience as a cradling in the mother's womb.

> That was joy, to be propelled unnoticed, as on a sea of oil, without any vibrations or jolts, being pulled by something invisible under your own power, of which you are lord and master, that listens to a gentle foot pressure, a finger's touch.

Car-driving, according to this fantasy, was "a kind of light-hearted drunkenness, a certain ozone flush."[98]

Male Violence and Aggression:
A French-Belgian Group of Writer-Motorists

In this surprisingly widespread and uniform chorus of car adventure eulogies a group of French writers stood apart as perhaps the most eloquent and influential. The most celebrated of all was Marcel Proust, whose father was a medical doctor and co-author of the first French publication on neurasthenia, a disorder that afflicted Marcel as well as one of his famous literary creations, Aunt Léonie. Much has been written within the discipline of literary history about Proust's church spires of Martinville. Proust described his automobile trip in 1907 in a short story ("*Impressions de route en automobile*") published in the Parisian newspaper *Le Figaro* in the same year. Tellingly, he changed the car into a horse-drawn carriage when he decided to include the scene in his *À la recherche du temps perdu*, on which he would start working the following year, a change made as if to emphasize the universality of his experience, independent of the vehicle used. At the moment he sees the steeples moving, he borrows a pencil and starts to write, for the first time: writing and the ambiguous experience (or illusion) of standing still while the surrounding world moves, are triggered simultaneously, by the moving car. At the same time it resembles Bierbaum's modernist-romantic efforts to restore the romance of travel through the car; in fact, Proust's entire project, as the title of his opus magnum declares, is an effort to generate remembrance through modern means, as is suggested by the recent debate about his Impressionism being, at closer look, only a part of a meandering movement including anti-Impressionism. The car as time machine, indeed, the fifth aspect of our adventure concept.

After a long period of inactivity, mostly spent in bed, Proust hired a car from a friend, the owner of the Unic taxicab and rental car company in Paris, replete with a chauffeur who drove him "from one medieval church to another" in Normandy. Meanwhile he resided in a new grand-hotel in Cabourg. According to the best analysis of Proust's experiences, it was these car trips that not only made Proust into "*un fervent de l'automobilisme*," but also gave him his first "impressions of speed and parallax vision that would become important features of his presentation of multiple perspectives in the novel." The novel's fame—the first part appeared in 1913 (*Du côté de chez Swann*, privately published, after rejection by all major Parisian publishers), the rest after the war, totaling eleven volumes—ensures that for generations after its author, readers did and probably will continue to remember the quintessentially timeless role of the 'vehicle' in enabling a special awareness of our environment. Not many readers though will know that an extraordinary use of drugs enhanced Proust's experience of speed: the drinking of seventeen cups of coffee daily, which he believed alleviated his asthma attacks. Whether his trip to Caen, interrupted by a mechanical defect, was also facilitated by extensive coffee consumption is not known, but on this trip he witnessed "the steeples of the cathedral shift position as he approached them in the car," something that later found its way into his novel, experienced in this case from a carriage near the fictional town of Martinville.[99] Literary critics have called this literary technique 'inversion.' It belongs to the third level of analysis (on affinity between the car driving and the writing/reading experience) discussed earlier, and although it was not new (Schivelbusch already pointed at the phenomenon experienced from a train, as we saw), it certainly was an attractive and powerful way of appropriating the environment, especially during automobile travel: the narcissistic perspective of the observer who places himself at the static center stage, seeing the environment pass by at high speed, has been evoked time and again by all kinds of observers of the automotive experience.[100] It expresses the remarkable combination of mobility (of the car vis-à-vis the environment) and immobility (of the driver and passengers vis-à-vis the car) that seems so typical of the car, and sets the car in the tradition of the carriage rather than the bicycle and the train. What is more: the esthetization of the experience of inversion by Proust sharpens our analytical eye for similar techniques deployed consciously and unconsciously by subliterary writers when dealing with the car-touring experience. Such an experience is not a peaceful act, incidentally: Proust who, according to some accounts, regularly visited male brothels where he required bloody and brutal scenes enabling him to reach an orgasm, describes his narrator, Marcel, approaching a town by car as the act of rape, where "the cityscape becomes the mobile prey." "In the morning," Proust wrote to a friend, he got "an insane urge to rape the little sleeping cities (read city and not little sleeping girls!)"[101]

As in the case of Wharton, Proust's mobility cannot be considered as that of the pedestrian, a conclusion true for his entire work as much as for his life as a "patrician bourgeois."[102] He saw the car trip as "a better instrument [than the train] to gain true insight," because in the car one can "follow the gradual changes" from afar and at close range.[103] But this comes at a price: in his novel, Proust paints a nostalgic picture of a Paris where horses drew erstwhile sumptuous carriages, while now, *"hélas!"* vulgar automobiles were driven by "moustached mechanics" (instead of proud drivers or *cochers*). And although he gained new visionary potential through the car (one of his critics speaks of "radical impressionistic poetics" enabled by "the automobile as vehicle of perception and imagination"), he also called it "a cage of crystal and steel," as if to emphasize the process that we will later identify as cocooning, interpreted as a process of a loss of sensory input. In the opposite direction to Thomas Mann (who, as one may recall, accommodated the car in his work after initial rejection), Proust's attitude toward the car went through three phases, from euphoria, via the car's smells that trigger involuntary memories and thus as a tool for romantic and nostalgic remembrance, to a vehicle of fleetingness and transience. At first sight, this position seems to converge with the much more critical André Gide, who saw in the car an instrument of superficial perception:

> The young whom I have known to be the most fanatical in automobilism also were the less curious in the journey itself. The pleasure here is not in the traveling, nor even to arrive quickly in those places where nothing else draws us to—but precisely in going fast. And although one experiences here inartistic, anti-artistic sensations as deeply as in alpinism, one has to concede that they are intense and irreducible; the period that saw them arise will feel its consequences; it's the period of impressionism, of the quick and superficial vision.[104]

Gide's position, however, that automobilism affords a superficial, and thus 'inartistic' experience, an interest in speed rather than the journey, is surprisingly one-sided as it reduces the automotive experience to vision. Literary historians suffer from similar restrictions, as much less attention in literary circles has been given to the nonvisual experiences described in the *Figaro* text, strengthening the bias of the visual vis-à-vis the other senses. This text indeed deals for a considerable part of its printed space with the sounds and the bodily experience of vibrations. In this fragment Proust likens his chauffeur, a "nun of speed" in his long coat and goggles, to an organist pulling out the stops and making music as he handles the gearbox, a vision that was very appropriate with the Martinville church steeples in sight, one which resonates with Kipling's similar image we saw before. Later in the text, Proust calls the sound of the horn "nearly human" and he likens it to Wagner's music. Somewhere in the novel, it is the smell of petrol that triggers an entire chain of memories. In other words, Proust described the trip (whether by carriage or by car) as an experience of the entire bodily sensorium, and not as a pair of eyes on wheels, as Gide seems to do. This seems to me to be the crucial difference between the car's *fervents* and their opponents, the full bodily cyborg-like unity with the road vehicle, versus a train-like reduction to (sideways) vision on smooth, hardly noticeable rails. From the beginning Proust had the capacity, as one of his biographers writes, to enjoy without limits, to mobilize all his senses, or at least many of them, at the same time. And as Proust's project was about the power and the interaction of these senses (the synesthesic description of a change in color as a crescendo in music, to give only one example), the role of the car as an instrument in reshaping the mobile sensorium becomes all the more interesting. This is especially true if one realizes that the eventual passage in the novel on the Martinville steeples contains a long, literal borrowing from the 1907 *Figaro* piece, between quotation marks, which makes one ask: what is fact and what is fiction in this context, and who is the narrator of these lines, the fictional "Marcel" or the real Marcel Proust?[105] Again, it transpires that the car trip, when experienced through the entire body, through all senses, enables a poetic attitude.

Proust and Gide were not part of the core group of young Parisian artists who shared a passion for sports, the bicycle and automobile sports in particular.[106] Prominent among them was Maurice Leblanc, dandy, novelist, and writer of short stories, who, before his creation Arsène Lupin, modeled after the popular anarchist folk hero Alexandre Jacob, made him rich and famous, wrote a short novel on the bicycle (which he praised because of its "anarchic" characteristics). *Voici des ailes,* a "snobbish novel" as one critic said, was pre-published in part in *Gil Blas,* where he wrote a column on bicycling. It appeared as a book in 1898. Leblanc (like many literary people in Paris) participated in the bicycle and automobile movements from an early date (he may even have been involved in the first car race from Paris to Bordeaux and back in 1894) and started to write for *L'Auto* in 1903. His writing being one of the early examples of an effective combination of highbrow and lowbrow cultures, Leblanc also represented more perhaps than anyone else the fundamental continuity between the two cultures of bicycling and automobilism. *Voici les ailes* was an immediate success (it went through eight editions during the first year alone) and gave Leblanc the reputation of being the "messiah" of the movement. The bicycle, Leblanc later commented, "was our aeroplane. . . . There has not been enough said about the chief distraction of our generation, when the true chain-driven bicycle appeared; we were drunk on speed!" As the title itself proclaimed, *Voici des ailes* (Look! Wings!) is a book-long eulogy of flight, of the aesthetics of the bicycle trip in the literal sense: an enumeration of the sensations of the cyborg-bicyclist *avant la lettre* where the vehicle takes on a role as active as the one played by the human actors. The bicycle trip was "a pleasure of real art"—"a new emotion" related to "the idea of speed," enabled through "an extreme harmony, a grace made of force and of lightness and, I emphasize, a special beauty."[107] That was quite a different tone than Jerome K. Jerome's ironic novel that would appear two years later and illustrates the divergent pace of national cultures.

Proust's plurisensory, bodily experience starts as a holistic and biomechanic (and erotic, as we will see) one in Leblanc's novel. Illustrated in art nouveau style by Lucien Métivet, the novel tells the story of two couples on a tour through Normandy lasting several weeks. As modern-day swingers attracted by each other's female partner, the male bicyclists become lovers and poets at the same time. Their adventure is erotic in a triple sense, mixing the conquest of the women with that of nature and of the bike's technology. The novel's modernity becomes visible in its emphasis on skill, both in using the bicycle in the right way and in attracting a woman. Leblanc describes the skill of seduction as suggestively as the skill of bicycling: the discovery and the shaping of the bicycle adventure is like the discovery and the shaping of love. One of Leblanc's protagonists declares the bicycle (and the locomotive) more beautiful than the car because the latter "hides its organs shamefully. One doesn't know what it wants. It seems incomplete. One awaits something. It gives the impression of being made to carry, not to advance." The bicycle, however, is "an improvement of [man's] body . . . , a pair of legs faster than he is normally equipped with. He and his machine are only one, and not two different beings such as man and horse, two opposed instincts; no, it's a single being, an automaton in one piece. There is no man and there is no machine, there is a faster man." The "monotonous clicking of the chain, it's the palpitation of its heart, the discrete sound of the tires on the ground, it's the throbbing of the blood in its veins." The bicycle, in other words, "is the great liberator" because it is "stronger than sadness, stronger than boredom, as strong as hope. . . . It helps us escape from the past." The bicycle, thus, is a *moral* vehicle, for the good and generous, the noble soul and the enthusiast. The flight the bicycle allows us enables us to leave the face of the earth, to gain an ironic perspective on the world and its wickedness and stupidities.[108]

Leblanc published his "sporting fairytales" in *L'Auto*, the same journal where one of his best friends, the Belgian writer Henry Kistemaeckers, also contributed. One of these fairytales, the opening story of a later book from 1904 dealing with automobilism, introduces a large, "monstrous" car, taken for a suicide ride with the car owner's wife tied to the grill: "And both they [the car and the owner] went, the two monsters, they went into death, drunk on speed, angered by hatred, triumphant as if they felt protected by the woman's head that swayed in front of them, by that woman's body, fragile and tender, which took the first shock, like a siren attached to the bow of a ship."[109] This remarkable early nearcyborg of man, woman, and machine (the man becoming a monster shaped after that of the car, the woman its bumper) filled with violent energy toward environment, women, and children, would remain a potent symbol of mobile masculinity for the rest of the century, and seems to define the crucial difference from the bicycle-cyborg so eloquently depicted in *Voici les ailes*: motorization introduces violence as it enables the magical multiplication of the touch of a finger, the pressure of a foot, a phenomenon that men in particular seem to have appropriated eagerly. Remarkably, Leblanc's metaphor may have inspired the German painter Hubert von Herkomer (who organized between 1905 and 1907 three reliability runs carrying his name) in designing the cover of a menu for a dinner during one of the contests, on which a blindfolded female nude is tied to the front of a car.[110]

The French-Belgian group of artist-motorists was more than just a situational group: Leblanc's sister, the actress and singer Georgette, married Maurice Maeterlinck, Belgian novelist and playwright and Nobel Prize winner for literature in 1911. Like Leblanc, Maeterlinck was an enthusiastic sportsman, who practiced boxing, motorcycling, and automobilism. In 1903 the revenues from his books and money inherited from his father enabled Maeterlinck to buy a villa in Nice, where several of his friends, among them the Belgian writer Cyriel Buysse, would visit him. Afflicted like Proust by neurasthenia in his early years, Maeterlinck became part of a socialist and anarchist circle of friends, including the anarchist Barrès and the writer Octave Mirbeau, who also suffered from neurasthenia in the 1890s. Maeterlinck developed a tense relationship with his family, with his native country as well as with the Catholic Church (which put his essay "*La Mort*" on the Index, the list of prohibited books). Maeterlinck's ticket for admission to the pool of autopoetics was a short story, "*En automobile*" (By Car), in which all of the well-known stylistic figures surrounding the car appeared and which may have functioned as a model for Proust's publication in *Le Figaro*. Describing a tour without a real destination (and without a chauffeur) in Normandy, the Parisian backcountry for the wheeled machine of literary motorists, he characterizes the car as an animated machine, half-human, a savage animal to be domesticated, allowing an acceleration (intensification) of experience. The trip and its angst-lust stand for life itself. The car, not the "prison" of the train, is a uniter of people. Interrupted by a brief digression on automotive technology (the inner landscape), the relation with the outer landscape is inverted, as in Proust: "On both sides of the road the wheat fields were peacefully flowing like green rivers," while "the road came towards me in a cadenced movement." Meanwhile, the protagonists experience not only images but also sounds, such as the "singing" of the car and the ever-moving environment in which the leaves "mumble . . . the eloquent psalms of Space."[111] It was literary convention that invited writers to make leaves mumble, even if this was impossible to hear while the car was driving. Maeterlinck also wrote an essay on the accident, and the role of the car and the motorcycle as "marvellous instruments of desperation . . . which question in a most effective way Fortune's role in the great game of life and death."[112]

The Flemish Cyriel Buysse fled Catholic Belgium at the same time as Maeterlinck, but only for half the year when he, after 1896, lived in The Hague with his Dutch wife. Attacked by Catholics and Flemish nationalists alike for his "modest naturalist" novels, "our Maupassant" (as Maeterlinck called him) was an active sportsman, practicing skiing, golf, rowing, and skating. Living on a comfortable income from his family's manufacturing activities and through his marriage, he owned two cars (one of them a Belgian Minerva) that he used extensively to entertain his friends and to write several travel accounts.[113] Like Bierbaum in Germany, Buysse represented the quiet, 'conservative-modern' version of car-trip experience, mixing the fascination for the machine with an anti-urban attitude, and like Leblanc for the French magazine *L'Auto*, was regularly asked to contribute to the Dutch automobile club's journal *De Auto*, a fact up to now ignored by Dutch literary scholarship. The beginning of one of his first car stories ("De vrolijke tocht" [The Merry Journey] 1911) describes a trip through France in October 1910 with three women, and no chauffeur, very typical of the earliest forms of literary autopoetics and deserves a lengthy quotation as it employs about every trope we have identified so far in our analysis:

> The automobile flies through the land. Mountains, valleys, towns, villages, fields, forests and rivers, all rustles past in high speed, as the flash-visions of a cinematograph. It wouldn't be worth the effort to absorb and represent them again if the car were like a train, stopping only on fixed spots, or as a cinematograph who develops his images one after the other, super-fast. But the car is a free bird, able to fly and to sit, a fantasist or a visionary, a runner or a dreamer, a poet or a slogger, in faithful correspondence with the will or the fantasy of the driver. All roads, or nearly all, lay invitingly open to him. Never is he bound to keep to the geometrically-straight line of two tracks. The region is beautiful; let us purr softly, admire, enjoy, dream. The road is long, the landscape without color, monotonous, boring: let's devour that part at full speed, and enjoy ourselves nonetheless, enjoy the speed for speed's sake, with the little thrill of fear and danger connected to it.[114]

Written in an impressionistic style (as testified by the frequent use of neologisms, such as "flash-visions") and in a cinematographic tempo reminiscent of the movie camera the story begins by lambasting the train (*and*, remarkably, the movie camera) as enemies of poetic, "free" landscape experience as they force the traveler (*and* the viewer) along a path others designed. The narrator praises the individuality of the car 'flight,' which also helps overcome the boredom of a monotonous stretch of the journey by indulging in the fear-lust of speed. The automotive adventure is erotic and feeds on constant change, which Buysse does not find in the "cosmopolitan ... monotony" of towns, especially not in the luxurious seaside resorts. He concentrated this 'conservative modernism,' mixed with some anti-Americanism and a disgust for 'new women,' in a most powerful way in his best-known short story, *Per auto* (By Car), published in 1913. The story describes, among other things, a trip through Picardy, Normandy and Brittany with his brother-in-law and the Parisian writers Emile Claus and Léon Bazalgette. For an owner-driver versed in car maintenance, Buysse's adventure included a functional element: in his account of a trip to the South of France with Maeterlinck and his wife (also included in *Per auto*), the narrator and his friend revel in their pride at being "tinkerers who haphazardly rummaged a bit" while mending their punctured tire. But there are limits: in the 1920s Buysse will abandon car driving out of disgust for the car's ongoing technical unreliability, making it unfit for long-distance travel. By then, car driving was not fun anymore, because of emergent mass tourism and subsequent traffic congestion and so he relied on the guidance of a chauffeur for his last car adventures.[115] Buysse is a typical example of those pioneers who were motivated by the 'individual freedom' the car afforded the elite and who refused to bow to the car's domestication during the following phase and did or could not escape into aviation to continue the adventure.

It was this elitist attitude (against the mainstream trend that slowly started to abandon the chauffeur) that Henry Kistemaeckers parodied in his frivolous (but no less successful in terms of sales) *Monsieur Dupont chauffeur: Nouveau roman comique de l'automobilisme* (1908). Badly composed (at midway the novel suddenly becomes a collection of short impressions), the book is a lighthearted description of the Parisian *milieu*, with its hierarchical car culture. Kistemaeckers also describes a conversation among car components who make fun of *petit-bourgeois* who have "enlarged themselves" by buying a *voiturette*. An automobile *salon* is presented as an art *vernissage*, while in another story a landscape painter substitutes a car for his painting tools: his "passion for speed" makes his brushes "harden into permanent immobility." And, of course, there is the usual bashing of the "vulgar" train, of materialist Americans and their desire for "comfort," the violent image of the car as a grenade, the accidents that are part of the automotive adventure. There is also the engine as a source of music (the "devilish" car emits "a terrible noise of metal shaken by furious lunatics," but the attentive ear also discerns "the beating of cymbals" somewhere) and the car is even likened to an organ, the pedals being those of a piano. Kistemaeckers's "borrowing" from Proust (or is it the other way around?) is so overwhelming that he even speaks of the car that, driven at a "crazy" speed, "gives the illusion of making up for lost time (*temps perdu*)." Here, too, the automotive experience is much more than a visual adventure: as well as the speed vibrations, the landscape, and the music, there is also the smell of the machine.[116] In all its confused fun and criticism, Kistemaeckers declares the culture of the new users ready to join the aristocratic motorists of the first phase.

In the same vein, Paul Adam, popular author of colonial adventure novels featuring a protagonist who later was the model for the Hollywood hero Indiana Jones, friend of Rudyard Kipling and like him positioned on the political right, praised the car ("beast of sheet metal and steel") through a description of a race. Another member of the French-Belgian group, Tristan Bernard, known for his vaudeville plays, managed a velodrome in Paris and also wrote satirical pieces on the early automotive culture, especially on the car's technical unreliability and the role of the chauffeur.[117] In his *Sur les grands chemins* he included a short story in which a man uses the maintenance of his car as an alibi to stay away from his wife (ending in the joke that he uses his wife as claxon by pinching her hard on the leg), while in another story he satirizes the new practice of keeping statistics of distances and time during the trip. Bernard sings the praise of the car as civilizer (of the hotels in remote parts of France) and as an addictive (showing withdrawal symptoms when having to go on foot). Elsewhere in the collection the narrator declares his disgust with the car "promenade," as he only loves "the journey, the expedition, the adventure," while in yet another story he describes the formation of "two parties" in the car: the speed lovers and the careful drivers—the narrator's wife of course belonging to the latter camp. That was two years after he had argued, in *Les Veillées du chauffeur* (1909), that the car was characterized by its defects, not its running.[118]

Subliterary Novels: The Williamsons and Youth Novels

Even more light-hearted, if subliterary, are the novels and short stories that capitalized on the bodily thrill of the car experience, for adults and children alike. The genre was dominated, during this and a part of the next phase, by Alice Williamson's prolific production. On arriving in London from America on a small inheritance, Alice Livingston managed to convince the newspaper magnate Alfred Harmsworth of the *Daily Mail* (and "some of my young unmarried editors") to engage her as a serial writer, agreeing to produce the remarkable amount of seven thousand words per day. She married Charlie Williamson, himself a magazine editor, but who had "formerly been connected with a firm of autocar manufacturers." In her autobiography, Williamson relates how she wanted to make lots of quick money and how she succeeded by engaging her husband as co-author. Harmsworth gave Livingston a quick lesson in serial writing: no humor, "plenty of action, but more important, plenty of love; a strong curtain for each chapter." No "character-drawing."[119]

The novels blurred the distinction between fiction and travel account: most of them are based on trips undertaken by Charlie and Alice (and in the 1920s and 1930s following Charlie's death by Alice alone). Photographs taken during these trips enhanced their authenticity.[120] This remarkable combination of (pictorial) fact and (literary) fiction gives their stories a peculiar perspective on reality, as if experienced in a dream, which was exactly the intended function of this type of serial literature, and made it the subliterary equivalent of Edith Wharton's artistic trip through France-as-novel. Their first novel, *The Lightning Conductor*, partly written (in the traditional form of letters sent to relatives) during their honeymoon to France and Italy, became "an astounding success." Published in October 1902, a second edition followed in May 1903 and by the end of 1907 seventeen editions had appeared; there was also a Broadway show with the same title in 1905. All in all, twenty editions can be counted, totaling more than a million copies. This does not include the many translations. Most of their books appeared in Dutch translation, for instance, and to enhance their authenticity further, the luxury Dutch make of Spyker replaced the Napier car in the first book.[121]

The Lightning Conductor deals with the controversial theme of the fraudulent chauffeur, the aristocrat surreptitiously taking on the role of a driver for the protagonist and her female chaperone, which gives the adventure described in these types of novels a sixth, erotic element of roleplay (in addition to the spatial, temporal, functional, colonial, and time-traveling adventures) through the attraction between two people seemingly of different classes: seemingly, that is, for the woman invariably is the victim of the deception. Of course, the 'real gentleman' cannot hide behind the chauffeur's role, but the conflict within the woman between her emerging love and the social constraints of class barriers carries the story until the happy ending. In the process the heroine learns to drive, acquiring the skills of handling the car and the man at the same time, thus making the novel into a female counterpart of Leblanc's skillful men in their conquest of women and machines. A clever mix of 'new woman' (as opposed to "an early-Victorian female") and traditional roleplaying (including the chaperone), the Williamson novels clearly represented the first phase of autopoetic writing where the driver, as a chauffeur from a different class, mediates between the traveling party and 'nature' (including the monuments to the past and its ruins and castles), often as an omniscient guide only there to please the object of his conquest. Williamson's delight in this roleplaying also reveals itself in a scene where a snobbish French motorist (clad in "leather cap with ear-flaps, goggles and mask, a ridiculously shaggy coat of fur, and long boots of skin up to his thighs—a suitable costume for an Arctic explorer, but mighty fantastic in a mild French winter") shows "how quickly one can be unclassed," a conclusion she reaches by explicitly using the term "Others" (with capital 'O') of whom she forms a part, while the chauffeurs and mechanics are seen as mere "others" (*nous autres*).

The same narrative structure determined another Williamson product, *The Botor Chaperon*, depicting a trip (in a motor boat, here and there interrupted by a car trip) through the Netherlands where the deceiver makes sure that he determines every minute detail of the itinerary of the trip, thus presenting it as an exercise in life itself. "Never was man in better mood for the rush and thrill of the motor than I, after the conquering of Miss van Buren," the aristocratic *jonkheer* Brederode (who owns more than one car) muses, and his object of conquest, Nell van Buren, half-Dutch, describes to him her readiness to be conquered, by referring to her fear-lust during the car trip: "'I never was in a car until the other day with my cousin,' said she, in the same carefully unconscious tone, 'And I'm afraid in my feet and hands now; but the rest of me is enjoying it awfully.'" Brederode is satisfied: "She trusted my skill, she was not really afraid, but only excited enough to forget her stiffness." Even more: Nell enjoys the cyborg feeling. Talking about the car she muses: "I *know* it's [the car's] alive. Feel its heart beat. What [the dog] Tibe is to [the chaperone] Lady MacNairne, *Lorelei* [the car] is going to be to me."[122]

In *The Lightning Conductor* the heroine writes to her father: "'This is life!' said I to myself. It seemed to me that I'd never known the height of physical pleasure until I'd driven in a motor-car. It was better than dancing on a perfect floor with a perfect partner to *pluperfect music; better than eating when you're awfully hungry; better than holding out your hands to a fire when they're numb with cold; better than a bath after a hot, dusty railway journey. I can't give it higher praise, can I?[123]

Many novels followed, all of them praising the liberating effect of the car, among them titles further exploiting *Lightning Conductor*'s success, such as *The Lightning Conductor Discovers America* and *The Lightning Conductress*, followed, ironically, in the year of her sudden death, by *The Lightning Conductor Comes Back.*[124]

The function of such novels was exactly to enhance the motoring experiences by doubling it: most of the adventures we already know from reading the club journals, but they are intensified. This is the case, for instance, with a less car-centered book from their series, *The Princess Passes* (probably 1903), mostly dedicated to a hike in the Alps, but only after the conversion of the protagonist from a car hater into a car lover, simply by being in the car on the road to the Alps, enjoying "a new and singular sense of well-being." From then on, the protagonist is "sniffing adventure with the sense of petrol." The drive through a European version of Marx's middle landscape: the Alpine valleys resemble "a vast engineering workshop," which "gave a certain majestic romance of its own." When the protagonist then is allowed to drive himself his initiation into a spectrum of exciting skills is described in detail (beginning with steering alone, then extended with the gear-shifting, then the functional adventure of maintenance), leading to his conclusion that "I could imagine that more than half the fun of owning a motor car would lie in understanding the thing inside and out." It is the cyborg-thrill that makes the protagonist feel like a second Toad: "'Full speed ahead, then!' said Jack. I took him at his word. I could have shouted for joy. Mercédès was mine, and I was Mercédès." The trip feels like "gliding, swanlike" but "I did well-nigh run over a chicken."[125]

In one crucial aspect this type of novel did not differ very much from highbrow literature: the close relationship between reading (as a reproduction of experiences contained in the text of the novels) and motoring. This is made explicit in Alice Williamson's autobiography:

> Charlie and I had wonderful journeys, in various makes of motor cars, and we wove them into novels. I did the writing, but he planned our journeys, and took as many notes as I did. Then he would give me his notes to match with mine, and it was strange how two people saw so many different things, formed so many different impressions, in the same automobile and on the same road. I never could have written such books without Charlie's advice, so it was right indeed that our names should be together. We laughed sometimes because, if we tried even to say three words about adventures of ours to friends at home, their eyes would become empty windows; yet the same people would spend money to buy our books!

This is a revealing observation indeed: only the act of reading brings back the experience of driving, makes the adventures into experiences worth being involved in. Could this be explained by the peculiarities of the practice of reading, allowing the reader to stage her own cultural and psychological appropriation, which is perhaps harder to achieve in conversation? The difference between the conversation and the written text may well have been that in the latter two, simultaneous experiences were merged into one narration. This remarkable affinity between reading (and hence writing) and driving made the Williamsons rich, living as they did in a villa on the French Riviera where they met "the great Maeterlinck." Alice was fascinated, like Charlie, by the casino, and they got gambling advice from "J.M. Keynes, the great economist." There, when Charlie was bored by Alice's constant writing, the couple entertained Kipling, Lloyd George, and many members of royalty, most of whom lived near Monte Carlo. Their popularity was enhanced by translations of their works in several languages, provoking visits from "German tourists, large, fat families, [who] used to stream into our garden and picnic on sausages, dropping crumbs on the marble fountain-ledge, or in the pergola."[126]

Many more autopoetic novels à la Williamson can be identified, such as Edward Field's *A Six-Cylinder Courtship,* with comparable erotics spun around a millionaire figuring as chauffeur who is well-versed in automotive technology, problems with the police over speeding ("What's the use of having six cylinders if a fellow isn't allowed to enjoy them?"), all adventures experienced in a very expensive French car that in the end explodes in a fire from which the hero rescues the woman "as if she'd been a child." The novel even contains a Proustian passage about the car as a musical instrument, of which "the coils hummed merrily to a six-cylinder accompaniment.... Even so, the music is monotonous, you say. Why not vary it, then? You wish a change in *tempo*? Certainly. A sextet of cylinders will obey the throttle as readily as your trained musician obeys his conductor's baton; one can manage a beautiful *crescendo* whenever one pleases; an artistic *diminuendo* may be introduced at any moment. If it is your desire to climb yonder hill *pianissimo,* try the third speed; if you prefer to mount it *fortissimo,* engage a second speed and the muffler cut-out."[127]

Another very popular genre, youth novels, capitalized on another aspect of the automotive experience: its violence. If there was one literary genre that celebrated the car's violent relationship with the environment as one high adventure, it was the enormous flood of juvenile novels, especially in the United States. In this respect youth novels (especially for boys) fall into the category of "adventure novels" identified at the beginning of this chapter. Such novels "view(ed) nature as a source of plunder." No wonder, then, that the war in Europe formed an ideal place for high-tech youth adventures. The Motor Boys series, for instance, written by Clarence Young, was one of dozens of new juvenile series that appeared between 1900 and 1917, "sentimental, family-centered novels" (all offshoots of the boys' and girls' serials that started to appear at the end of the nineteenth century and in their turn "longer, more respectable and more expensive versions" of the post–Civil War dime novels) with such heroes and heroines as Frank Merriwell, Dorothy Dale, Billie Bradley, Ruth Fielding, and Polly Pendleton. Several volumes of the Motor Boys series used the European war as a background for adventures, as did some of the girls' series. For instance: *The Khaki Girls behind the Lines* was part of a series specially written to capitalize on the connection between women and the war. This particular volume was all about "the chance to be in the big fight." These girls' books were "feminist but not radical, muting the call to revolution to a subliminal whisper of female competence," with girls "strong enough to take care of themselves, ... often stronger than boys." In series such as the Motor Maids (1911–1917, written by Katherine Stokes), the Motor Girls (1910–1917, Margaret Penrose) and the Automobile Girls (1910–1913, Laura Dent Crane) girls "largely celebrated the advent of the car." The acquisition of technical and driving skills thrilled them as much as the boys. Boys' books, the limited scholarship on this under-researched topic concludes, are more violent than the girls' books, while the latter stress social manners as much as adventure.[128] Some scholars derive from their analyses and their comparisons between boys' and girls' series that the car had a "liberating" function for nonhegemonic groups and classes, such as women, in a time that "male subjectivities were construed as universal or integrated selves whose opposite was not the female subject but rather a simulacrum, that is, an imperfect double of masculinity."[129]

"To read the juvenile literature of the nineteen-teens and twenties," a recent analysis remarks, "one would think that the nation was overrun with touring cars filled with teens on mysterious adventures." And it was not just cars that carried these adventures: ships, submarines, and electric hydroplanes all functioned as vehicles in both senses for tales of expectations written for nonusers, but very active participants indeed, fantasizing about their soon-to-be-realized part in this fascinating technicized world of "formulaic simplicity." This was a motherless world in the case of boys' novels, designed "to keep at bay the middle-class curse of 'overcivilization,' which threatened to turn a whole generation of early twentieth-century young men into sissies."[130]

Tom Swift was such a hero in what has been called an "Edisonade," an inventor adventure modeled on Thomas Edison and Henry Ford published from 1910 (*Tom Swift and His Motor Cycle*) through to the 1970s, in five series totaling more than one hundred volumes, translated into several languages (mostly for readers in Scandinavian countries, and in France) and sold over twenty million copies. Appearing under the collective pseudonym of Victor Appleton, the series was conceived and written by Edward Stratemeyer (of the book-packaging company Stratemeyer Syndicate), Howard Garis, and Stratemeyer's daughter, Harriet Stratemeyer Adams. Criticized as "anti-intellectual" and "blatantly racist" in its portrayal of workers, Jews and African-Americans (not a surprise, given that the hero was modeled on Henry Ford), the books convey a picture of middle-class, moderate adventure, now and then interrupted in a near-Brechtian style by a direct address to the reader that he or she should not forget to buy the other stories. A board game (1966) grew from the series as well as a television show, and in 2008 a feature film was announced.[131]

Despite his name, Tom Swift is not a speed addict, but his opponents often are. In *Tom Swift and his Electric Runabout*, the hero develops a new battery for an electric car literally copied from Edison's nickel-iron battery (complete with its tubular structure and shiny metal container, its quick charging capabilities, and explicitly called "a wonder") but here reformulated into a tool to win races. The car is so silent when racing at a top speed of 128 km/h, that it nearly causes an accident.[132]

In 1914 alone Tom Swift novels sold more than 150,000 copies in the United States. A library survey in 1926 claimed that 98 percent of public school children had read at least one. It seems that the hardly analyzed Motorcycle Chums series (written by Andrew Carey Lincoln) was less politically correct than the car-related novels, and did not conform to one scholar's conclusion that youth novels portrayed "not the eccentric but the type."[133] In *The Motorcycle Chums Stormbound* (1914) the protagonists expose a "somewhat antisocial behavior": they are ticketed for speeding but manage to speed away unharmed. Their analysis is long overdue, because in the United States "in adult fiction the motorcycle seemingly played no part" in this phase. Gang boys declared in the early 1920s that they read Tarzan, Rover Boys, Boy Scouts, and Tom Swift. They were probably not aware that the many Boy Scout novels (as part of "outdoor fiction" such as The Outdoor Chums, The Outdoor Girls, The Ranger Boys, and The Camp Fire Girls) were explicitly introduced in opposition to Tom Swift's urban lifestyle. This was a reflection of a controversy within the Boy Scout and Girls Guides movements between a "pioneering" and "roughing-it" wing on the one side and "a movement dedicated to urban social work" on the other.[134]

Flight Forward: The Avant-Garde, Silent Movies, and the Celebration of Automotive Violence

Literary studies of early automobilism have often neglected the kinds of plebeian narratives such as those expressed in movies, focusing instead on the avant-garde, especially Futurism. Such studies miss the surprisingly widespread character of a basic 'grammar of motoring' on which, in their turn, the avant-garde artists could thrive and develop their peculiar dialect of obsession with speed, violence, and death. There could be no vanguard without the more modest forms of motoring violence and aggression, despite the indignation that avant-garde utterances on car culture caused among the more tranquil members of society.

Any treatise on the autopoetic avant-garde should begin with the captivating poetry of Guillaume Apollinaire (pseudonym for Wilhelm Albert Apolinary de Wąż-Kostrowicki) who in 1913 published *Alcools* (a collection of poetry written between 1898 and 1913), just before volunteering in the French army. In this publication, his poem "Zone" (which he started working on in 1898) begins with the unforgettable image of the Eiffel Tower as a shepherdess looking down on the Seine bridges with their "bleating herds" of cars, in a city where "even the automobiles seem to be antique." In his *Calligrammes, poèmes de la paix et de la guerre 1913–1916* (published in 1918, the year he died of Spanish influenza) appears a poem with lines in the form of a car: "I shall never forget that night ride in silence, O somber departure where 3 of our headlights died, o tender night before the war ... and 3 times we stopped to change flat tires."[135]

The avant-garde character in terms of autopoetics of the writers in this section reveals itself mainly in how they take the side of the automobile club and follow this institution's disdain for the common man. Apparently, literary and motoring avant-gardes (the latter defined as indulging in the lust for speed and violence) went hand in hand. Octave Mirbeau, who started buying cars at an astonishing rate from 1900 on when he was established as an accomplished writer, was a "red millionaire," adopting the ideas of the intellectual anarchist Jean Grave, which made him the subject of a police investigation after a bloody attack on the Spanish King Alfonso in 1905.[136] Mirbeau had published several "scandalous" novels for which his "dialectic philosophy" that only a terrorist campaign could bring about social change was the inspiration: *Le Jardin des supplices* (1899) contains a discussion among male intellectuals about the "universal instinct of murder," while *Le Journal d'une femme de chambre* (1900) "expose[d] the fraud of bourgeois modesty."[137] Such novels made him into an artists' artist, praised by the young generation of French and Belgian artists (some of whom he "launched," such as Maeterlinck, but he was also one of the first to draw attention to Van Gogh and Gauguin).[138]

The loosely, but spiritually firmly interconnected group of experimenters in mobility and aesthetics decisively shaped early automobile culture, and the way one ought to talk (and write) about it especially. Or, more precisely, talk and write through mobility since a number of their writings were not so much characterized by the presence of an automobile as by a literary technique of observing the new mobile reality. Their opinion was respected, as can be seen from *Les annales politiques et littéraires* of November 1907, which printed their short contributions, including such items as a photograph of Maeterlinck and his wife Georgette Leblanc in their car. Leblanc praised the car for its violent properties as "a marvelous tool to solve a situation in a tragic manner," while Mirbeau explained that the car "forms a part of my life as a matter of course. It is my life, my artistic and spiritual life as much as my home." Well before Marinetti's Futurist Manifesto, Mirbeau declared that the automobile to him was "more dear, more useful, more filled with knowledge than my library."[139] With that point of view Mirbeau put himself at one extreme on a scale of largely positive auto-appropriation: the panorama of Belgian and French author-drivers in the *annales politiques et littéraires* made visible what many observers have since repeated, that there were "two series of contradictory representations," or, as Mirbeau himself accurately observed in one of his novels: "not only am I the Element, as confirmed by the Auto Club, that is the beautifully blind and brutal force that ravages and destroys, but I am also Progress, as suggested by the Touring Club, that is the organizing and conquering Force, which, among other civilisatory benefits, ... distributes English-style water closets complete with a manual to the small hotels of the most remote provinces."[140]

Mirbeau's iconoclastic *La 628-E8* (the title being the number on the license plate of his CGV car, an abbreviation of Charron, Girardot et Voigt) was one of his last publications, followed only by *Dingo,* in which not his car but his dog figured as protagonist. *La 628-E8* described (and fantasized about) two trips Mirbeau made through Belgium, the Netherlands, and Germany in 1905 together with friends, including the painter Pierre Bonnard whose drawings illustrated a special edition of the book.[141] It first appeared in 1907 after it had been published partially in *L'Auto* and in *L'Illustration.* Respected colleagues like Valery Larbaud (who characterized it as "pornography of the lowest level") and André Gide ("a child writing overheatedly, without much thought") criticized it, but it sold 43,000 copies during his lifetime. Some modernist contemporaries praised it and later critics have increasingly held it up as the "epitaph of the novel," a violent and aggressive destruction of the linear narrative into "a succession of Impressionist tableaux," a fact totally lost in the selective English translation of 1989. The destruction of the novel was expressed in an "atomisation of the senses" leading to "a new poetics, of speed, movement, a novel way of describing the world *from within.*"[142]

Mixing eroticism and violence as constitutive of the conquest, Mirbeau, known for his very direct reactions to his own feelings (as Gide commented), was an excellent witness of what the automobile could do to the man, and vice versa. His narrative was a "hallucination and a travelogue" at the same time, "a kaleidoscope of memories, and a montage of dreams," which makes it understandable why he disliked the linearity of the train as the car allowed "spontaneous detours" and stops "in order to make associational observations." In other words: "Mirbeau identifies his text with the vehicle it praises": it is a "novel-as-car." "Mirbeau adopts the model of fictional narrative as car travel," or "the narrative is controlled . . . by the car-author." As such, the text lends itself to an analysis at the second, symbolic level and even the third, intermedial level distinguished at the beginning of this chapter: the car became a grenade, a war tool in an environment characterized by Proustian inversion: "The plain seemed moving, tumultuous" while the road behind the car was littered with "broken vehicles and dead beasts." Like Paul Adam, who in his *Le troupeau de Clarisse* (1904), along with its fantasies about killing someone, dedicated an entire chapter to a description of the car as a giant carapace (*scarabée géant*) and a terrible tank (*notre char terrible*), or Valerie Larbaud who in his famous *A.O. Barnabooth* (1913) described a car trip by some super-rich motorists who, just for fun, decide to drive into a populated part of Nice to terrorize the residents, Mirbeau's dehumanized description of "a reality explored by smelling, hunting, killing, and driving" was possible only because of his "automotive disorientation." This made his fiction—like the car—into a direct "vehicle of aggression."[143]

Some biographers have explained this psychologically, referring to the sexual abuse he suffered at the Jesuit school he attended, leading to a "retreat inside the armor of . . . cynicism," claiming that for Mirbeau, "love's perversion entails a loss of affect, a horror of women, a pathologizing of the sexual response, guilt eroticized as masochism." Mirbeau's heroes are "smooth, hard, and cool," and the model for his art is "catastrophe, an apocalyptic machine, savagery run amok . . . , the car as a roaring purveyor of road-kill."[144] Mirbeau's juvenile memories may have inspired his automotive delirium, but the connection between car enthusiasm and anarchism reached beyond his individual biography to include several other members of the French-Belgian group, on the political Left and on the Right. The violent attacks on the train are just another aspect of the way these writer-motorists read the 'script' of the automobile, as a celebration of male aggression, misogyny, and condescension to the petty bourgeois, precisely at the moment that the latter started to discover the car as something for them to appropriate.

If Marinetti knew how to provoke, Mirbeau had done it earlier. Doubtless, Mirbeau's violent rhetoric influenced Marinetti, but calling Mirbeau a Futurist *avant la lettre* would deny the latter's cynical pessimism and his criticism of "the obsessional fetishisation of the future" of Marinetti *cum suis*.[145]

In his pre-Futurist work, Marinetti's symbolist poetry, full of art nouveau embellishments, explicitly included the automobile as a new vehicle for experiencing the environment. Wallowing like Mirbeau in masculine and misogynistic aggression and violence (and like him one of the first car owners in Milan), Marinetti instrumentalized the car as a tool of power and self-liberation.[146] In his *"La Mort tient le volant. ...."* (Death at the steering wheel), published in 1907 in his own journal *Poesia* and included as epilogue in his collection *La Ville charnelle* (The Carnal City) space has to be killed in order for the lyrical 'I' to be liberated from earthly constraints, in an extreme example of what we have been identifying throughout this chapter as flight. Written on the occasion of a race in Brescia, Marinetti depicts the racing cars as "metallic jaguars" whose drivers, with whom the narrator identifies, defy death by escaping from the earth, a feat only possible through sheer will power.[147]

The same Nietzschean vitalism and the same relation between flight and drunkenness, aggressiveness and masculinity, can be found in Marinetti's poem "À Mon Pégase," also to be found in *La Ville charnelle* of 1908:

> Release the brakes!...You cannot?...
> Then break them!... [...]
> Hurrah! Repugnant earth does not constrain me anymore!
> At last I loosen myself and fly already
> into exhilarating abundance[148]

In fact, the car was only a substitute for the real adventure machine, the plane.[149]

Other Futurists (who were called "Automobilists" by their British artistic rivals, the Vorticists[150]) also found the car to be an ideal vehicle for their toying with aggression and death, such as Gabriele D'Annunzio (we will encounter his work again in chapter 4), who in his *Forse che si forse che no* (Perhaps, perhaps not) added sexuality to the mix of blood, drunken destruction, and reality-denying flight: "'Now I have your life in my hands! Like this wheel,'" the protagonist Paolo says to his lover, Isabella.

> "Sure." "I can destroy it!" "Sure." "I can hurl it into the dust, against the rocks! In one second I can transform you and me in a single bloody tangle." "Sure." Stooping forward she repeated the same hissing syllables, with a mix of excitement and voluptuousness. And indeed, the blood of both flowed faster and revolted: one against the other they seemed to flare up and ignite, like the petrol mixture that through the sparks of the magnet in the cylinder under the long-stretched hood exploded. "Die, die!"[151]

When Marinetti, barely a year later, published his Futurist Manifesto, it was only a small step to plead for the destruction of linguistic syntax as well, and eliminate adjectives and adverbs, an experiment fully explored by a later generation of (New Objectivist) writers.

The connection between violence and cars also fascinated Russian Futurists, although they were less belligerent than their Italian colleagues: Vladimir Mayakovsky's poem "We also want meat" ends thus: "When you tear along in a car hundreds of persecuting / enemies, there's no point in sentimentalizing: 'Oh a chicken was / crushed under the wheels.'"[152] His poem "The Huge Hell of the City" (1913) describes cars as "ruddy devils" with "explosive shells," while "In an Auto" offers an analysis of affinities between car driving and poetry as the poem contains broken rhymes suggesting fragments of conversation overheard from a passing car.[153]

Similar feelings were expressed in Early German Expressionism (*Frühexpressionismus*), a movement that got its start in 1910.[154] Expressionism is important for mobility history because, with Futurism, it was the first literary movement that consistently defined the car as an urban vehicle.[155] While Marie Holzer in her small essay "Das Automobil" thrilled to the vitalistic potential of the car ("The car is the anarchist among the vehicles"), Hans Flesch von Bruningen's short story "Der Satan" presented a prince in a scarlet car, driving at 200 km/h, who "ran down three people before he reached the Hotel Adlon," while his poem "Totentanz" (Death Dance) declared that "the automobile trumpeted violence."[156] The radical leftist Oskar Kanehl lambasted the aggressive behavior of the car vis-à-vis its environment:

> Like predators they pounce
> on the innocent landscape.
> We savage the forests
> in the neck
> and maul them in our jaws.
> We toss the towns
> behind us like toys. [...]
> Ahead of us, ahead of us
> ever new land leaps towards us
> uiii, uiii,
> and we devour it.

Others discussed the car's questionable safety, for instance by depicting a near-accident with a motorbus (Gottfried Kölwel's "So stand ich vor dem Sterben...." [I stood face to face with death]). Kanehl also described the car accident from the perspective of a nonuser, a boy, in his poem "Schrei eines Überfahrenen" (The cry of one run over).[157]

In her analysis of German expressionist poetry Dorit Müller concludes that it conveyed a different message from the texts of non-literary publicists, relating fear, uncertainty, and confusion, which prevented the narrator from enjoying the "flow of fresh air" and the "massage vibrations" of the softly oscillating engine, or the thrills of the scenic landscape. The car, according to this body of literature, has two effects: its speed causes confusion and uncertainty because it generates too many impressions, while at the same time it enables, through its multiplication of mechanical energy, an intensification of experience, lust, emancipatory feelings, and personal power mixed with fear and violence.[158] The power fantasy and the feeling of freedom, up to then reserved for the elite, could now be bought in a metal box.[159]

Müller's analysis of Expressionist novels also conveys a shift in symbolism from the car as an aggressive weapon or a grenade, to a cocoon, a place protective from the violence of war, a shift we will explore fully in the next chapter (where I will use the term 'capsule' for the second phase of cocooning). Meanwhile, the automobile started conquering the city, and its speed and violent behavior generated feelings of resistance, of a failed symbiosis, making the car into an enemy, or at least a counter-world ("*Gegenwelt*"[160]). The freedom of the car, these stories revealed, led in the end to lunacy, obscenity, and incest, as in Carl Einstein's novel *Bebuquin oder die Dilettanten des Wunders* (1912; Bebuquin, or the Dilettantes of Wonder), and to expressions of resistance against bourgeois society, as in Flesch von Brunningen's "Der Satan," where Satan and his sister sexually unite during the car trip, making the car into an instrument in the hands of the powerful. In Einstein's novel "the car functions as a kind of extension of those passionately irrational forces that declare themselves enemy of the existing society."[161] The film critic Einstein constructed his protagonist (whose name should be pronounced as a French word) "not as an individual, a personality or a subject, but as a collection of properties and 'elements'" as they were to be found in the modern metropolis. On one of the cars driving through Paris, a screen ("*Kintop*") is mounted showing "how the actress Fredegonde Perlenblick undressed, bathed and went to bed."[162] Through Expressionism's integration of the car into an urban culture, for the first time the car could be experienced as part of a fledgling system, a system that developed to its full extent during the following period.[163]

Many of the expressionist poems, essays, and stories dealt with here appeared first in the anarchist Franz Pfemfert's politico-literary magazine *Die Aktion*, whose contributors often belonged to an urban *bohème* that defined itself as opposed to the prewar bourgeois society and that opposed the war after 1914. The prose of the Parisian correspondent for German newspapers René Schickele in particular extolled the driving experience as a sensual act of fear-lust (and the car interior as an intimate, erotic space), which made the car user prone to deploy violence and spread destruction, as if he (mostly a 'he,' indeed) were at war. His *Benkal der Frauentröster* (Benkal, consoler of women) describes, one year before the real one, a war between the "Kremmen" and the "Langnasen" (the Longnoses) during which the protagonists perform "racing trips through moonlit nights" complete with flight metaphors, accidents, and Proustian inversions.[164]

Thus, this overview of European autopoetics leads to the conclusion that at least three possible attitudes can be distinguished among author-drivers, typically represented by Mirbeau (aristocratic self-mockery), Proust (middle-class synthesis of rationality and sensation, celebrated as 'true modernism') and Marinetti (a plebeian flight forward in violence). Whatever model the literary intellectuals stood for, they found in their cars a road toward what they experienced as full life, overcoming their "social and moral isolation."[165]

Popular culture, especially the silent movies, drew from the same source. The movie audience, consisting of the majority of (urban) nonusers of the car, whose car-related preoccupations were nothing but fearful, likewise was fueled by a fascination for violence, bloody accidents, explosions, kidnappings and, not to be forgotten, car chases. While the arrival of a train in the Lumière brothers' movie of 1895 reputedly scared the audience as it threatened to run off the screen, the car has gotten much less attention as one of the central drivers of early one-minute movies. From the beginning of cinema, film-history scholars have asserted that film was much more emotionally effective in its "appropriation mechanism" than literature. Of the 20,000 pre-WWI American films, 2 percent (about 500) "made significant use of the automobile." After 1914 car-related movies comprised two-thirds of the total for that period.[166]

Cars not only appeared as ideal actors of violent sequences, but they were literally made a part of the production of cinema experience through the technique of phantom rides, where directors placed the camera on automobiles to provide the perspective of the driver while moving through an urban environment. At the Coney Island amusement park Tim Hurst Auto Tours offered a movie theater in the shape of a giant motorbus seating sixty persons. It created, as the advertisement said, a "perfect illusion" because of a "correct combination of the vibratory and gently oscillating motions of a high-grade touring car."[167] Humor (a way to vent anxieties), horror (a democratized form of the sublime), and slapstick effects relived the tensions caused by the appearance of the new technology, as in Cecil Hepworth's *Explosion of a Motor Car* or in *How it Feels to be Run Over* (both 1900) where human limbs fly through the air and trickery simulates the amputation of a character's legs by a car. At the end, the car threatens to "run over" the audience in the same manner as the Lumière brothers' train.[168] One silent film scholar identified more than 110 movies (more than one-fifth of the 500 prewar movies) depicting crashes, almost all of them a "happy accident." For the United Kingdom between 1900 and 1906, a considerable number of comic films with traffic accidents have recently been studied.[169] They make us see early cinema "less as a medium of vision than as a feeling machine," or at least providing a multisensorial experience.[170] The relationship (and affinity) between film viewing and car driving was even more direct if one realizes that early film projection was only possible with the support of trained mechanics, and even then there seem to have been many accidents (in 1897 a film caught fire in Paris in a full theatre, killing 120 [mostly] women and children in the audience), events "repressed ... in most theoretical and historical accounts of the medium." Tinkering characterized automobilism as much as film: Méliès, for instance, built his own camera.[171]

In Germany, Oskar Messter was the prewar champion of auto movies, with titles such as *Das Automobil der Zukunft* (1910; The Car of the Future), *Die verwechselten Wagen* (1911; The Swapped Car), *Der Werdegang eines Daimler-Motors* (1912; The Emergence of a Daimler Car), and *Eine wilde Jagd im Automobil* (1913; A Wild Hunt in a Car, which may have inspired Hermann Hesse's hunt on cars; see chapter 5). There were many more, however, such as *Strassenraub per Automobil* (1905; Street Robbery by Car) and *Das geheimnisvolle Auto* (1914; The Secret Car).[172] Every North-Atlantic country must have experimented with early movies and cars; the ten minutes of *Captain Deasy's daring drive in his tourist automobile on the ballast of the cogwheel railway from Caux to Rochers de Naye above the Lake of Geneva*, shot in 1903 and released by British Mutoscope & Biograph Company, are a case in point. The 3.5-mile-long stunt, undertaken by a former officer of the British army in a Swiss Martini adjusted in its gearing, was filmed by a traveling camera mounted on a small rail wagon.[173] In the United States Thomas Edison's *Automobile Parade* (1900) was a rudimentary documentary on the (automotive) toys of the Newport elite, while the Hollywood one-reeler *A Unique Race between Elephant, Bicycle, Camel, Horse and Automobile* (1902) already showed the direction that the American movie thematic would also follow: a year later the same Biograph label produced *Runaway Match* (1903) while the highway robbery appeared in *The Gentlemen Highwaymen* (1905) and D. W. Griffith produced another car race in *Drive for Life* (1909). Another title was *Dashed to Death* (1909). That same year Edison released *Happy Accidents* (1909) as always depicting victims that did not suffer from the crashes. *Traffic in Souls* (1913) suggests the seductive effect of the car upon women to go into prostitution, whereas *A Race for Life* (1913) showed Indianapolis ace Barney Oldfield racing against a train while rescuing a woman.[174] Slapsticks and comedies were especially popular, including the Keystone Kops (1912–1917) and Charlie Chaplin's 1912 French debut *La course d'auto*. The fascination with the early movies grew from the double effect of their "tempo" (the fast sequencing of events) and their being a "stenogram of life," easily transferred to the automobile: the experience of flight, for instance, so ubiquitous in literary utterances, appeared literally as a trip through the sky in Georges Méliès's *Le voyage à travers l'impossible* (1904; The Impossible Journey), complete with, of course, a crash at the end. Spectacular accidents were, certainly for the majority of nonusers in the audience, representative of early automobile culture. For them, the car chase, soon to be integrated in the emerging detective movies as well as the slapstick movies, represented the ultimate experience of modernity the nonuser could produce while watching. The car, and the user culture built up around it, seemed to lend itself as an ideal object for slapstick films: the extremely short takes of Mack Sennett's "Stepping on the Gas" (1912–1917), for instance, 'rhyme' with the desire for change and stand against the boredom of the travel accounts. The 'traveling camera' was another case in point. From that perspective, film is the cheap version of the car: the automotive adventure of the elite can be reenacted, experienced, in the movies. The Atlantic fascination with the violent death of the Bonnot gang was just the real-world equivalent of that experience.[175] That experience, however, differed fundamentally depending on where the actors stood in this process: as onlookers or as drivers. Whereas the earliest movie theater experience was typically collective, the driving experience seemed to emphasize an individualistic aspect, something scholars of early film have identified in early entertainment. "Dance halls, vaudeville theaters, movie palaces, all ballparks emphasized the importance of personal gratification." And while the earliest movies did not even mention the actors' names, by 1913 the 'star' principle had begun to emerge, supported technically by the "close-up shot."[176]

A second strand of car movies emerged from a totally different background: the lecture circuit, illustrated by lantern slides, immensely popular in the United States after the Civil War. When some of the lecturers around the turn of the century started to substitute movies for the slides, it appeared that many of these movies deal explicitly with movement, as titles in early catalogues suggest. Edward McDowell, one of the lecturers who would later become a movie maker, filmed such early movies as: "Locomotives Laboring Up Steep Mountain Grades," "Ocean Steamer in a Storm, Tossing and Plunging through the Heavy Sea," and "Railway Panorama of the White Pass, Alaska." Later he included the colonial trip in his lectures, adding shots from Samoa (1903) picturing "savage life" and bearing such titles as: "Samoan Dance by Twenty Men and Women," "Canoe Race," "Waterfall," "Food Offering and Processional March," "Diving Scene by Thirty-Five Native Boys," and, as if to remind his audience how he got there, a "Firing Gun on a U.S. Man-of-War."

After the second decade of the new century, films like these became separated from the lecture circuit, especially when they developed into travel and nature films of several reels in length, such as "Paul J. Rainey's African Hunt." They became accepted as a separate genre when the war broke out and American filmmakers hurried to Europe to shoot such titles as "With the German Army" and "Great Italian-Austrian Struggle."[177] By that time movies had gained a crucial role in the American travel and camping movement, especially in their support of the See America First movement.[178]

Tarkington, Cather, and Dreiser: Autopoetics Before America's Entry into the War

While Europe stumbled into a war, the culture around the car in the United States took quite a different trajectory. Before the United States entered the war, not much literary appropriation of the automobile can be observed, apart from a couple of plays on Broadway, over 120 popular songs published and performed between 1905 and 1907 alone, and a host of juvenile literature.[179] Indeed, from a very early date onward the car formed a part of popular song culture: from Rudolph Anderson's "Love in an Automobile" (who invites Daisy to "a honeymoon in a cozy automobile") to Gus Edwards and Vincent Bryan's 1905 hit "In My Merry Automobile." Many of these songs stress the erotic connotation of the intimacy the car promised. Kerry Mills's "Take Me Out for a Joy Ride," in which a woman declares "I'm as reckless as can be," is a case in point.[180]

As far as literary autopoetic fragments were produced, just like in Europe (in Mann's *Königliche Hoheit*, as we saw) the car was depicted as a token of the wealth of the elite Other, such as by Booth Tarkington in his monumental *The Magnificent Ambersons* (1918). Although from the viewpoint of the novel's critical reception it belongs to the second phase, its content has been conceived and written in or from 1916, while the story evolves clearly in the prewar period.[181] It describes the transformation from a horse-drawn society to the first contours of a car society and as such played a role in the post-WWI canonization of the latter: the "coming of the new speed." The novel showed both the fundamental continuity between the two societies ("driving a dog-cart at criminal speed," for instance), and its major breaks, symbolized in the rise of the Morgan family and the demise of the Ambersons (who saw "too much overalls and monkey-wrenches and grease" in the new world), sealed, of course, by a grave car accident, which "rolled over the Ambersons." The car, tellingly, appears on the scene with its "exhaust racketing like a machine gun gone amuck."[182] "I think," Eugene Amberson predicts, "men's minds are going to be changed in subtle ways because of automobiles." Amberson is using the writer's hindsight, which makes the novel, in this respect, clearly belong to the second phase, while "the new spirit of citizenship" he sees emerging is analogous to the ideals of the businessmen with their optimism "to the point of belligerence." In *The Magnifent Ambersons* the Babbitts, who will be criticized and satirized in the next phase, are born.[183]

Willa Cather, whom we met already as a writer of a short story on cycling, also lets her heroine Thea in her novel *The Song of the Lark* (covering the period from the 1880s to 1909, published in 1915) witness the transition from a carriage society in Colorado to New York with its motorized taxicabs and its city noise "strident with horns." In her short story "The Bohemian Girl" (1912, published in *McClure Magazine,* where she worked as an editor from 1906 to 1912), Old Lady Ericson, at nearly seventy, "never lets another soul touch her car. Puts it into commission herself every morning, and keeps it tuned up by the hitch-bar all day." One of Cather's "superwomen," she "left a cloud of dust and a trail of gasoline behind her," and although everyone in her family is afraid "that thing'll blow up and do Ma some injury yet, she's so turrible venturesome."[184]

Because of the scarcity of examples of literary appropriation of early automobilism, the book-length account of a car trip in 1915 by Theodore Dreiser, together with the illustrator Franklin Booth (owner of a Pathfinder, an open tourer) and a chauffeur, stands out as an important object of comparison. Indeed, it is certainly no coincidence that it appeared more than a decade after the European avant-garde discovered the car. True, Henry Adams had in *The Education of Henry Adams* (published on his own account in 1906 and accessible to the public in 1918) quite optimistically given the car, because of its speed, the role of visiting the "virgin" (meaning: all the good aspects of the country as opposed to the technology), even if "this was the form of force which Adams most abominated."[185] But like Adams, Dreiser had made an extensive excursion across the European continent in 1911 and 1912 (of which his *A Traveller at Forty* gives an account) during which he visited the "underside of Europe that Baedeker does not talk about—sleazy cafés where prostitutes hang out, working-class slums, drab bars where poor people drink themselves into numbness," which made him a critical advocate of American democratic culture. His work, too, reflects the transitional phase toward automobilism, as his *Sister Carrie* (1901), which made him famous, uses a buggy to drive along the Chicago lake shore with a neighbor, while in *The Titan* (1914) the "traps, Victorias, carriages and vehicles of the latest designers" were used not only for conspicuous consumption (emphasizing possession rather than use), but also rudimentary cyborg feelings as the reigns transmitted "a mysterious vibrating current" from the protagonist's fingers into the horses.[186] This was the era of the "American Renaissance" (President Taft defeated in the elections, the new President Wilson representing hope for a new age) in which Dreiser played his role, regularly visiting the left-leaning bohemian Village culture of New York, and addicted to sex rather than cars.[187] Indeed, it seems that American writers sought their adventure elsewhere, in sex like Dreiser, or in appropriating the car and developing driving skills rather than its representation in literature.

Perhaps that is why Dreiser's travel account is so much less aggressive and less elitist than those of his European counterparts. Praised by one critic because it did not contain his usual dose of sexuality, *A Hoosier Holiday* (1916) presents a critical eulogy of the American middle class (its prosperity, its democratic instinct) and (to borrow a term introduced fifty years later by Leo Marx) the American middle landscape, a mix of the natural sublime and the machine. The trip from New York to Dreiser's place of birth in Indiana (the Hoosier State) is undertaken with a chauffeur called Speed (a former test driver for a car manufacturer), who is just as much a subject of the conversation as the others. They also stop to gaze at factories and visit slums. On several occasions Dreiser declares his belief in rational planning and technology (but the throwing of their empty beer bottles in the river passes without comment). As a somewhat reluctant member of the New York bohème in Greenwich Village and a leftist "radical without a program" (but enthused by the Bolshevik revolution) he shared with his European colleagues a disdain for the "naïve, so gauche, so early Victorian" life of the middle classes ("Nothing could be duller, safer, more commonplace"). During his car trip he personally witnesses and comments positively on the spread of the car among midwestern farmers, and also he appreciates the "picturesqueness" of "ships and cranes and great moving derricks which formed a kind of filigree of iron in the distance with all the delicacy of an etching." And as if to formulate a counterargument to the European fascination for the changes in perception wrought by the car, he commonsensically declares: "Mechanicalizing the world does not, cannot, it seems to me, add to the individual's capacity for sensory response. Life has always been vastly varied. How, by inventing things, can we make it more so? ... We are capable of feeling so much and no more," only to end his argument by saying: "I am not all-wise and I do not know."

Despite these important differences with a European culture (in which, as we have seen, quite a few Americans participated), the similarities are not to be forgotten. Like Bierbaum, Dreiser calls himself a "sentimentalist" and like Bierbaum and many other European colleagues he opposes racing on the road ("one of those fierce souls who cover a thousand miles a day on a motorcycle. They terrify me"). And like his European colleagues he likens the sound of the motor to music:

> Tr-r-r-r-r—a line of small white cottages facing a stream and a boy scuffing his toe in the warm, golden dust—oh, happy boyland!—and then, T-r-r-r-r-r—but why go on? It was all beautiful. It was all so refreshing. It was all like a song only—T-r-r-r-r-r—and here comes another great wide spreading view, which Franklin wishes to sketch.... Then T-r-r-r-r-r, and we were on again at about thirty-five miles an hour.[188]

Conclusions

Analyzing the previous selection of a sample of around 150 poems, collections of poetry, stories, novels, and travelogues within their multilayered contexts of pre-WWI autopoetics, published by around fifty authors from seven countries results in the identification of a French-Belgian group of *autoliterati*, a mixture of anarchists and radicalists on the one hand, and vitalists and activists on the other. The anarchist background of several of these avant-garde figures (including Early German Expressionists, as we saw) is remarkable: anarchism's "analysis and condemnation of society in its totalizing purity resembled the comprehensive moral visions of the imaginative artist," so much so that it even attracted conservative writers (such as Léon Bloy, as we saw). These artists' and anarchists' "belief in the possibility of ultimate individual liberty" inspired the fledgling movement of automobilism, in a very direct, personal way. Easy violence was part and parcel of this anarchism, testified by the popularity of Georges Sorel's *Réflexions sur la violence* (published as a serial in 1906, as a book in 1908). In general, there is a consensus among literary and other historians about the importance of violent and subversive fantasies, as well as the "virility cult," among the avant-garde, even in Valentine de Saint-Point's *Manifeste de la femme futuriste*, in which she tries to define a feminine Futurism.[189] Within the space of two decades, these writers, together with their foreign colleagues, developed a grammar of the automotive adventure which co-constituted the automobile culture of this early phase, as may be deduced from the silent movies. On its turn, early automobilism was very receptive to this anarchic behavioral mode, as the swarm-like character of pioneer motoring, its almost anti-systemic ideology, provided a material substrate not to be found in this intensity in other mobility modes. Quoting one literary exponent of the Dutch "1880s movement" (*Tachtigers*), Albert Verwey, these writer-drivers were in for "sensation," in all possible connotations of the word. Impressionism, indeed, seemed the preferred strategy for literati in this first phase to deal with "the atomization of perception [through an] atomization of stylistic means," despite criticism by the Futurists and other modernist artists (often from an urban viewpoint) of the Impressionists' aesthetics.[190]

It is remarkable that the literary and other artistic and less artistic representations of so different subcultures were so much alike.[191] Adventure is the essence of fiction, French literary historian Jean-Yves Tadié writes at the very beginning of his study of the adventure novel.[192] We can now add: it is also the essence of the car. The motivation was common, but the practices differed, and they aggregated, as we will see in chapter 4, into two main Atlantic subcultures of automotive adventure.

In the concluding pages of this chapter, we will further qualify this adventure as depicted in the body of autopoetic utterances presented here along the three levels of analysis explained at the start of this chapter: thematic content, symbolic imagery, and intermedial affinity.

First Level: Content Analysis

For the first level, we find four thematic clusters that can be characterized differently, depending on the perspective one chooses (table 2.1).

Table 2.1. Taxonomy of Automotive Adventures

Function	Adventure	Context	Subtheme
Driving	Functional	Car	Technology, skill, chauffeur
Gliding	Body/Erotics	Road	Flight, escape, unrest
Touring	Spatial/Colonial	Scapes/Other	Foreground, sublime, past and present, conquest
Racing	Temporal	Acceleration	Speed change, violence, aggression, angst-lust

The grouping in four thematic clusters is, of course, subjective, and other analysts may use different hierarchies and partitioning into groups (such as gender), and no doubt a frequency analysis of keywords could have supported the importance of the chosen terms, but it is claimed that the subthemes cover the thematic spectrum of the sources rather well. Other scholars, indeed, have proposed different thematic taxonomies, such as Kimberley Healey's analysis of female French travel writing through the concepts of "self," "time," "space" and "body," or Dorit Müller's "correlating fields of meaning" of "speed," "violence" and "life."[193] Such taxonomies, however, are not concrete enough for our 'cultural history of car technology' (see chapter 1). Taking as a criterion the environment the driver relates to (the 'context' in table 2.1), we distinguish between the car itself, the road, the landscape (or scapes in general, including cityscapes, but also smellscapes, etc.) and the Other, drawing—in this order—ever wider contextual circles around the car's interior and its occupants. In other words: we follow the author-drivers' perception of being, each and every one of them, at the center of the automotive world, if not the world per se. The temporal adventure should be seen in a context of promised acceleration of society, an issue we will come back to in chapter 4.

Focusing upon the functions realized by these occupants, the four environments coincide with the social, but also with bodily practices of driving, gliding, touring, and racing, respectively, either as driver or as passenger, or, for that matter, as nonuser who experiences the effects of these practices. The first, the third, and the fourth clusters are shorthand for the three well-known adventurous aspects of the emerging car culture: the skill in handling, maintaining, and repairing the car (the functional adventure), the spatial adventure of navigating and perceiving, and the temporal adventure of high speed, racing, and violence. We have seen that a fourth adventurous aspect can be added: the aggressive surveying of other lands and people. The second cluster could then appear as the fifth aspect of the automotive adventure, not previously recognized to its full extent: the bodily, eroticized movement within the automotive shell. The sixth adventurous aspect, the production of the past when driving into the countryside, has been subsumed under the second cluster (gliding). A more detailed analysis of subthemes can put this multidimensional adventure, as it emerged and developed over the course of a quarter century, into perspective.

Technical skills
Autopoetics was a medium that did not avoid technology. The car's technology formed a part of the narrative as a matter of course: several authors included a brief explanation of the car's inner workings or at least showed the reader that they knew what the mechanism was all about. In the very beginning this was, of course, nothing more than a passport to an audience of knowledgeable users, but soon the ability to handle the car became a status-related element in the social fabric of

early automobile culture. Passengers and some nonusers (on the street, as readers of juvenile literature or as moviegoers) projected feelings of admiration or eroticism onto the driver. This became confusing (and full of social and erotic tensions) if the owner delegated the skill (partially) to a chauffeur, a sure trigger for playing with class and gender roles.

The relationship between the occupants and the car's technology also generated cyborg-like subthemes, such as the anthropomorphizing of the car and its components, and the making of the car into an actor in the narrative. Autopoetics thus acquired a new type of acting literary persona: the machine (on wheels) driving into (and within) the garden. This machine functioned as a prosthesis, an "extended self" connecting mobility with consumption.[194]

One set of skills was not delegated to a chauffeur, however: erotic and sexual skills were developed often in parallel with autotechnical and driving skills. Autopoetics formed an arena where such skills could be practiced (or mentally prepared and rehearsed). In the case of male drivers these skills often acquired an aggressive flavor reminiscent of the conquest of nature: the machine and their desirable occupants had to be domesticated as much as the garden they were moving in.

Flight
The relation to the infrastructure in this phase was not expressed through systemic imagery of traffic flows and networks, but through a direct bodily feeling of the road, a feeling for which a surprisingly large number of authors and filmmakers used the concept of 'flight,' of gliding or being suspended in a womb. They thus used an old metaphor that depicted mobile vehicles as projectiles, such as the actress Anne Kemble who enjoyed her train ride in 1829 as a "sensation of flying," while others had used the "*impérial*" on top of the diligence to produce this experience.[195] In the case of male car occupants this regressive emotion was apparently so powerful that it tended to deny real, often poor, road conditions (although Desportes suggests that the flight sensation was mainly the result of the smooth propulsion as opposed to the shocks resulting from the steps of animal traction). This made the car trip at the same time into a trip into the Self, creating a strange confusion of inside and outside, of subject and object, of "moving and being moved."[196] This confusion, solidly based upon the peculiarity of car traveling, which can be characterized as the mobility of an immobile body, not only generated orgiastic feelings of happiness, it also seemed to lock the male body into the shell of the car through soft vibrations and other therapeutic, anti-neurasthenic dependencies that may explain the remarkable converting effect of the first car trip upon novices: once experienced, one is

hooked.[197] Authors used metaphors like disease, addiction, ozone flush, rush, rapture, horror, and hallucination to express this phenomenon.[198]

The flight metaphor also related to the possibility of escape that the car enabled: escape quite literally from the earth and its tedium (as in the flying fantasies of Marinetti, who saw the car as an airplane without wings, or a devil, as depicted by Bierbaum and Corelli) or from urban decadence (as in the tranquil pre-Futurist narratives by Buysse, Grahame, or Dreiser), or from straightforward, mundane experience itself opening the main character up for the sublime (as in Proust's caffeinated 'trips') or from women. This metaphor was all the more attractive to literati as, according to the critic Lionel Trilling, "the desire to escape society altogether is the chief modern element in literature."[199] At the same time it kept reminding those in flight of the "weightless culture of material comfort and spiritual blandness," which, according to T. Jackson Lear's analysis of turn-of-the-century American culture, bred "weightless persons who longed for intense experience to give some definition, some distinct outline and substance to their vaporous lives. This sense of unreality has become part of the hidden agenda of modernization."[200] And because they sought to escape weightlessness through weightlessness, they had to repeat the escape over and over again, just as with the smoking of a cigarette and other fleeting consumptions.

A third element of the flight image was the restlessness it created in the occupants: the fleeting superficiality of the car seemed an invitation to speed up, even in the biedermeier case of Bierbaum, and it generally produced a nervousness that incited traveling ahead and not staying too long in one place (as Margaret Winter experienced). In other words, the interplay between the vibrations of the engine, the shocks from the road surface transferred though the suspension and springing system, and the human bodies in motion generated powerful fantasies of independence, happiness, power, personal sovereignty, and freedom (experienced by women often as independence[201]). Well before the car became a part of an 'automobile system' in the next phase, the 'freedom of the car' was solidly embedded at the level of the individual body and in the pleasant sharing of this corporeal experience with traveling companions. It is social practice theory (inviting us to focus upon the 'doings and sayings'; see introduction) that helped reveal this condition.

As we have seen, this experience already existed in bicycle culture (which depicted biking women as butterflies[202]), and also train travelers knew it, but the addition of an engine as a 'mechanical multiplier' under personal control gave the flight experience in the car a special flavor, a crucial phantasmagoric power.[203] The motorcycle (largely absent in the narratives of this period) thus intensified the balancing experience of the bicycle, while the car, through the addition of two wheels, lost this very potent cyborg experience of being one with the machine through the moving of the entire body and replaced it with a stable flight emotion and a more passive massage (as Bierbaum found out), enabling a comparison with real flight, in an airplane. The experience was also related to other senses, as we saw in Proust's case: especially sound and music seem to have established connections between feelings of floating and the fourth dimension of time.[204] It took two decades before all the elements of this mechanism had acquired their respective roles, but from then on it formed a permanent source of intermediality between car driving and reflective writing, so much so that Gilles Deleuze much later could claim that "to flee is to produce the real, to create life, to find a weapon." This weapon, "created in flight from the self," is the travel text, as the scholar who quotes Deleuze concludes.[205]

Touring the landscape
The navigational, spatial element of the automotive adventure had a strong intermodal character from its very beginnings: drivers experienced the car trip as a reconquering of the foreground lost since the emergence of the supposed archenemy, the train. And yet, this new function of being able to stop wherever one wanted, and to make the remotest part of the landscape into a foreground by simply turning the steering wheel and driving to that part of the road network originally laid there for other purposes, caused anxieties and controversies: several authors claimed that motoring was the 'wrong' way to experience the landscape, and that the 'trivial motorist' saw nothing because of his preoccupations with the technology and the car's high speed (as Wharton said), indicating that here a truly new sensory technique was in the making. High speed, moreover, leads to "a sublime immobility and a contemplative state," according to Jean Baudrillard. At high speed "there's a presumption of eternity."[206]

Many authors described the landscape as a painting, or sightseeing as practicing poetry, while they often used qualifications such as "picturesque" (Mirbeau; *malerisch* in Winter and Bierbaum) to connect such experiences to the discourse of the sublime. This was a new type of sublime from the panoramic way of experiencing the landscape (Schivelbusch), being substituted by (or perhaps better: expanded with) a filmic style with a rapid sequence of images (Desportes). This, according to some authors, made the car experience superficial and fleeting. However this may be, the spatial adventure may seem dominated by vision (even if several authors pointed to the loss of vision through high speed), but the car enabled the entire body and its five senses to participate: sound (the organ of transmission in Proust) was as much a part of this experience as smell (Kipling; including the bad breath of the Other, in Konody; the exhaust fumes were for the nonusers along the roads) and the haptic sensation of the vibrating body. Remarkably, nobody ate in the car in these texts (although Dreiser may have consumed beer), but taste (if not used to support smell) was as active a part of the spatial adventure whenever the party stopped to have a picnic, or during the opulently described lunches and dinners (Williamson).

However experienced, the car was *in* the landscape, unmediated, as if it were moving through poetry or within the novel itself. Motorists experienced the landscape as a painting or a novel, setting in motion a blurring of the distinction between fiction and reality that formed another basis for the parallelism between car driving and reading, between producing the landscape experience and producing the practice of writing (Wharton, Williamson), between traveling and narrating. The written representation was highly impressionistic, characterized by improvisation, a seemingly spontaneous experience preferably put in the form of a diary or letters: the images followed each other without any inner relationship, their appearance dictated by the tempo of the traveler.[207] Paradoxically, even nonusers could experience the car trip, as we saw in the American Tim Hurst Tours. The intermedial affinity between movies and motoring made movie watching for such audiences into "an education in the logic of driving."[208] In other words, where in previous studies of early automobilism the motor bus has been given the role of an appetizer for those who could not afford a car,[209] the 'appetizing' started much earlier: through the reading of paintings, travelogues, and movies that had so many structural similarities in common with the car-driving experience.[210]

The spatial adventure also had a clearly anti-urban character (Buysse), both in rejecting the filth and nervousness of urban culture (at least for pre-Futurist authors) and in projecting fantasies of health and recovery onto 'nature.' But the car enabled a new type of nature experience, driven by variation and threatened by monotony (Winter), prefiguring the commodification of the landscape and affording a glance (Larsen) rather than a gaze (Urry). This experience was, like the other dimensions, heavily gendered, given the depiction of the landscape as a woman, or the taking of villages unawares (Wharton) or even raping them (Proust).[211] It also had a clear colonial, imperialist connotation, as the car helped invade 'exotic' landscapes and bring them into the orbit of the motorizing cultures.[212] Space became "conceived in terms of motoring needs."[213] Here also lies one of the possible connections to the world of sports and some of its aggressive subcultures, such as hunting, the "peacetime equivalent of the excitement connected with killing humans in times of war."[214]

The trip into the countryside, finally, also reshuffled the relation between the present, the past, and the future. The future was always present as a bundle of powerful expectations, making the dream about the trip as much a part of the spatial adventure as its actual execution. This formed the very cognitive and perceptual basis for the parallelism between the touring experience and the practice of reflective writing. The relation with the past was even more powerful: several authors emphasized the recovery of older experiences by modern means (Bierbaum, Wharton; Proust's entire artistic program), while Wharton, Konody, and many others used the car to view remote landscapes, architectural monuments, and 'backward' people: the tour traveled into the country's past, even if in the phase under study a shift can be observed from an interest in the past to an interest in the self, from exoticism to exporting modernity to the 'periphery.'

So, especially in this period the car's relation to the past was controversial: Proust used the car as a trigger of memories, Wharton as a magic vehicle to explore the past, but Mirbeau and Marinetti saw in the car a way to get rid of the past, to lose sight of it through the trip's superficiality. There seemed two kinds of modernities created here: a conservative modernity taken up by the Touring Clubs, and a flight forward into modernity as practiced by the Automobile Clubs: a middle-class, tranquil attitude, and an aristocratic, high-bourgeois attitude. It is remarkable that in literature the former occurred first, and then was expanded with (rather than substituted by) the latter. This has a technical equivalent: from the beginning of the first decade the cheap car (*voiturette*) was overtaken by the heavy closed big-family car, only to return in another form (the miniaturized car for owners of moderate means, as we saw in the previous chapter) after 1908.

And yet, the easy opposition between aggressive and modest use (and often by implication: between masculine and feminine, or male and female use) suggests a dichotomy that is analytically not very productive. "In literary terms," narrative psychologist Karl Scheibe concludes, "Don Quixote without Sancho Panza as counterbalance is not a good story."[215] Instead, the detailed analysis of Bierbaum's work showed how beneath the polished lacquer of the biedermeier similar aggressive fantasies were hidden that also had to find expression in automotive behavior. Similar conclusions could be drawn about the dichotomy between female and male adventurousness (of the 1700 books on foreign travel published before 1900 in the United States alone, more than one-tenth were written by women, many of them carrying the word 'adventure' in the title[216]), but we will postpone our assessment on this topic until chapter 5, once we have developed an overview of a longer period in which women had become more outspoken about their relationship to automobilism.

Speed and aggression
Speed was evoked in every source we have analyzed as quintessential or at least co-constitutive of the automotive adventure, and as such it represents a crucial explanatory factor for the emergence of automobilism. Writers interpreted speed (a relative judgment) as high speed, varying between 25 and somewhere well below 100 km/h, something one could hardly avoid while driving (Bierbaum), while some described it as ecstatic (Williamson), dangerous to nonusers, including animals (Mirbeau), leading to superficial perception of the environment (Gide), triggering memory (Proust) or erasing it (Mirbeau: *"le charme amnésiaque"*), or addictive (Grahame). Speed, in other words, is so ubiquitous in our set of examples that it functions as synonymous with driving per se ("we raced towards Salon," says Williamson; *"in sausendem Tempo vorwärts,"* exclaims Winter). Obeying speed limits made occupants bad tempered (Winter) or even led to social unrest when an entire automobile club opposed a speed limit as we have seen in the previous chapter. Speed had become such a universal catchword that it had already lost much of its explanatory power in this period, or so it seems, a situation that has not changed much in the analyses of historians, literary critics, and social scientists. "Speed," one recent commentator concludes, "intimately woven into a new paradigm of the modern subject's nexus of desires, becomes the new opiate and the new (after)taste of movement as power."[217]

Speed has often been enjoyed as maximum speed or speed records, which constitutes the connection to the world of sports. "The desire to experience speed, note, is not generated by a visible need," a recent study (announced as *A Speed Handbook*) concludes

> it is a desire without a visible object per se in the world of produced consumer goods; it is a desire that can only try and try again to consummate itself in ever faster, ever more desperate, and ever more dangerous speeds—a desire ... that submits itself to violence and a kind of torture. One might therefore see it as a new twist on older desires that, in its very strangeness, novelty, and excessive supplementarity, could be a force dangerous to the status quo of capitalist progress itself. Critics such as [Paul] Virilio would reply that that is why speed is so firmly policed, legislated, and controlled by the modern state.[218]

And yet, we know that the car's speed, when compared to other mobility modes, was not particularly high, and certainly not higher than that of the train. So, if speed as such cannot easily function as explanation, how should we redefine it to get a grip on its effects? As "private speed" (*vitesse privé*), speed controlled by the Self, as another observer concluded?[219] That was known, as we have seen, by bicyclists before, and was in fact 'discovered' on horseback, although animal muscle limited this to a far greater extent than engines. Thus it seems that 'speed' does not explain much, since it was a synonym for 'modern' and apparently did not require much further argumentation: it was an abstraction of something that has to be specified in order to become useful.

So, are we talking of speed as something 'objective'? Then it would probably have appeared as an indication of timesaving, as part of a socio-economic explanation. But the car did not function as a utilitarian machine in this period. And even if it did, one should compare average speeds (or worse: 'schedule speed,' a railway term including stops; in the case of the car: including tire defects, for instance), but the author-drivers invariably deal with maximum speeds, and they have never, even today, abandoned that misleading habit: potentiality and expectation take the place of real-world speeds.

The subjective experience of speed, then? As part of a psycho-cultural explanation? It is hard to experience speed. A high absolute speed is something that is part of the experience of the nonuser, such as the onlooker at races. Also, every engineer knows that, theoretically, bodies cannot experience speed, unless as speed *difference*, as relative speed, or, more specifically, speed *change*.[220] If we analyze this in a more detailed way, we observe that, when driving at constant speed, speed difference occurs in an open car, as friction with the surrounding air, a situation that has been often evoked as the source of the healthy effect of car driving. It was the type of speed experience that motorcycle drivers still produce, but here the thrill is enhanced by the balancing practice introduced by bicyclists. Speed difference (although not in the form of forces upon the body) also provokes exhilaration vis-à-vis another car on the road: going faster than someone else generates feelings of power and social distinction (Winter).

Once the car has become closed, speed cannot be felt, except as speed change, in two situations. At a bump in the road the body accelerates upward, followed by a sensation of being suspended in the air and a fall, moderated by the springs and dampers, just not far and fast enough to become seasick for most people. This is the corporeal basis for the flight experience. At the end of the twentieth century the German psychologist Rainer Schönhammer rediscovered this as a constituent of the driving thrill through interviews with drivers and passengers. His respondents reported a gliding experience, which Schönhammer related to the child's experience during birth. He also revealed the "suffering of the passenger'" at not being in control.[221] Earlier, in the 1950s, the American psychologist Michael Balint connected high speed to thrill (and introduced the term "angstlust" to emphasize the pleasurable sides of fear; the Williamsons talk of "fearful joy"). "Speed," a later commentator on Balint interprets, "is not only thrilling in itself, once it has been sufficiently accelerated, it also enables us to enter exposed and unfamiliar situations, far removed from the zones of safety and normality." Balint called such "avid thrill-seekers" "philobats" and he observed an "element of aggressiveness" in all their activities, whereas the "ocnophil... prefers to clutch at something firm when his security is in danger (and is) only at ease in the state of stabile security."[222] The car, we have to conclude, is a home for philobats and ocnophils alike.

The second type of speed change occurs when taking a corner at constant speed. Car occupants feel it as a centrifugal force on the body. The same is true when driving in a straight line at the moment of accelerating or decelerating; the latter (braking) especially has been a source of wonder among early motorists. Speed can be seen, but speed change can be felt: the visual bias in recent modernity research has closed our eyes to the bodily grounding of the car-trip experience.

Speed, then, seen as speed differences and speed changes, is directly related to the skill in handling the car, because, if these changes become too large, death or injury lurks around the corner. In other words, car occupants link speed change with danger, associating it with intrusion and violence (on one's own body, on the Other's body, on the environment in general) and this seems to have been recognized immediately upon the car's appearance as something belonging to the realm of the man. A recent analysis of "the pleasures and joy of technology" indicates "that the continuing reproduction of the car as a masculine technology also involves the production of men," a conclusion also true for the period under analysis here. The car, then, functioned as an anti-neurasthenic therapy in an age governed, according to Henry Adams, by a "law of acceleration," keeping men from sliding into passivity, which was "perilously close to the feminine condition."[223]

Thus, the risk of car driving evolved into a romantic fear-lust of/for danger seen as anti-nervousness, fascination for the power of the machine transferred into societal power and the thrill of taking part in societal acceleration and being among the avant-garde of modernity. It was also felt as satisfaction of being in the 'fast lane' of society, as a source of status by being able to 'afford' speed, connected to feelings of superiority over non-users, especially in 'backward' provinces. Around the early car a "covert culture" (Leo Marx) of aggression and violence was constructed, celebrated by Futurists, questioned by Expressionists, but enjoyed, secretly, by the Bierbaums among the early motorists and, if we are to believe Alice Williamson, by many women as well. All these utterances formed part of a culture made up of "an urge to buy, a lust for travel, ... an overactivity in business matters and a love of risk," which Joachim Radkau explains (referring to Germany, but it can generally be applied to neighboring countries and across the Atlantic) as emerging from "a mixture of a light mania with nervousness" evolving into a "widespread permanent condition, of which the stability was related to its confusing mixture of lust and suffering."[224]

The aggression resulting from this revealed itself in the most 'tranquil' samples of our body of literature: from the revolvers carried by Winter and used by Demolder to the tongue-in-cheek account of the number of animals killed (Mirbeau) and the stereotyping of the Other (Konody). By the dramatis personae themselves it is often seen as a response to the violence of the 'mob.' Indeed, the motorists' violence against man and animal, and the verbal violence of the avant-garde, can both be interpreted as the flipside of working class hooliganism. "In the end, the driver's sense of sovereign mastery and the bystander's perception of inhuman arrogance are two sides of the same coin. The abhorrence of cars is inseparable from their appeal."[225]

The discourse on speed also points to the fundamental continuity between the horse-and-bicycle worlds and the automotive world. Although this chapter resulted in enough particularities of the automotive adventure to enable us to explain the thrill many users and nonusers experienced about the new contraption, Bayla Singer's relativism remains important as a warning against too easy an emphasis on the innovative character of automobilism: "The enclosed portion of an automobile is not essentially different from the interior of a carriage, and the sexuality of a male chauffeur is not inherently different from that of a male coachman, gardener, or butler. It surely cannot be claimed that the sexual suggestiveness of the powerful automotive machine exceeds that of a high-spirited stallion."[226] Clay McShane, who unearthed several automotive phenomena among horse riders and lovers before the car (such as carriage racing, the 'pumping up' of horsepower, and even the introduction of exit lanes on parkways) would agree.[227]

Second Level: The Car as Shell

The second level of our analysis deals with the symbolic web spun around the car in this early phase, a phase in which the car already became what German cultural historians call a "collective symbol," or as semiologists would say, a "complex signifier."[228] Thus, in this phase, the car is described as either a "monster" or a "woman" to be tamed, a "*bête merveilleuse*" (Kistemaeckers, Mirbeau). It has been praised as a liberator (of pioneering men, of women; Marinetti, Williamson), as a unifier of mankind and compressor of time and space (Bierbaum, Proust), as a vehicle of progress and restorer of the past (idem), while the car trip has been presented as a Quest of the Self (Dreiser) or as something standing for life itself (Williamson). In chapter 5 we will observe how the metaphorical relationship turned upside down during the interwar years, as the ubiquity of the car enabled people to use the car and its components as signifiers for societal phenomena, while in this early phase it was the car (not society) that was adorned with epithets. In this period, some were anticipating this later use, such as a character in one of Edna Ferber's novels that possessed "a thirty-horsepower mind. I keep it running on high all the time, with the muffler cut out, and you can hear me coming for miles." Another character thought himself to be an "original self-starter," but found out that someone else "had to get out and crank me every few miles." Also, car lover Edith Wharton compared the "deteriorating," after the war, of works of fiction to what happens with "motor car tires; they have, like millions of other things, been ... overstrained and knocked about." In *A Six-Cylinder Courtship* the protagonist has problems with his "mental clutch" and his "transmission-gear of speech" when approached by the female character of the novel.[229]

The most potent symbols however grew from what I would like to call the car as a shell, in two chains of connotations. The car is first and foremost presented as a protective, but open space, like one half of a (clam) shell. The term itself does not appear in the sources (it is proposed here as an umbrella term for a cluster of images)[230]; in fact it is the French philosopher Gaston Bachelard who, in his *Poetics of Space* (1958), developed a "dialectics" of the shell, resulting, according to the American literary historian and poet Priscilla Denby, in opposing pairs such as "hard/soft, movement/stationariness, life/death, retreat/escape, little/big (for big things can emerge from shells) and perhaps most particularly, the very basic dichotomy of inside/outside." In the 1970s Denby interviewed a great number of people about their experiences in and around the car. Trying to historicize her results, she interprets the car as a vehicle that evolved into a "portable territory," bringing "the animalness in us to the surface" as we do not feel claustrophobic inside the car. On the contrary: "this feeling of enclosure is deemed rather pleasurable." The car as shell leads to "perceptual confusion" as the boundaries between "inside" and "outside" are blurred, as "we physically occupy a bigger space than our own body, and that space is still part of a much larger territory," an effect we also observed when analyzing Wharton's travel account.[231]

Secondly, the car as a shell has also been described in our examples as a beetle (a *scarabée*; Adam) and as a grenade (Kistemaeckers), in both cases meant to deploy violence to the outside world, because the first is protection to enable the second. The metaphor of the projectile has a long tradition: Schivelbusch points at the experience of train passengers of being "shot" through the air, thereby at the same time generating the flight metaphor.[232] A British eighteenth-century pedestrian who decided to take the coach used the same image, and saw places pass by the window like "meteors."[233]

From the sources (e.g., Winter) we know that the open cars could be closed with canvas hoods when the weather turned bad, and even when this was not the case, early motorists protected themselves against the cold by hiding in furs and leather, goggles and veils, "making drivers and passengers alike into illusory figures of a rather horrifying fiction."[234] Meanwhile we have come far from Leblanc's ecstatic eulogy of the bicyclist's own body as a fully sensitive sensory center of which only the skin forms the border with the environment—a supersensorium the motorist tried to recreate, but finding himself caught in a capsule and liking it.[235] Altogether, the car appears as an iron-clad offensive weapon geared to conquer hitherto unknown territory (and, as part of the same imagery, as a defensive protective hull against the feared aggression of the other), and as such this symbolic complex forms the metaphorical basis for all those travel accounts that cross continents, visit the European periphery (Konody), survey the American West, or even visit, on a Sunday spin, the countryside around town.[236] From this perspective the initial resistance to the car took the form of a covert civil or class war that did not last long, either because the rioters became motorists themselves (as happened with the farmers in the United States), or because speed-control systems appeared in the form of police traps and club campaigns against road hogs.

Third Level: Intermedial Affinity

The third level of analysis addresses the structural analogies between the producing of car experiences and experiences of and with other media, including writing (reading), singing (hearing), and filming (watching). We have come across several examples of intermedial affinity, such as the use of a filmic style in the application of rapid images, short and hasty sentences, or the description of the car as a perception machine, a rolling camera, of which the individual images remain visible as individual stills: "Images did not appear to dissolve into continuous flow, but retained their individual pictorial integrity in rapid succession."[237] Film, in other words, was loved because of its "capacity to capture the accidental" (in all senses of the word, we would like to add).[238]

As a matter of fact, the very act of writing travelogues while making a trip is an exercise of mutually adjusting "wheel and pen."[239] Jonas Larsen has opened the possibility of a taxonomy of these affinities, introducing an aesthetic affinity.[240] From a very general perspective: the *l'art pour l'art* of pre-1900 estheticism found its automotive equivalent in the trips without a destination, where the journey itself was savored. The artists' belief in the autonomy of the work of art rhymed with the shell metaphor and its 'anarchic' meandering through open space, while even the struggles between modernism and anti-modernism among the literary avant-gardes resonated with the motoring avant-garde and the resistance against it. Perhaps belletristic literature (and art in general) is not so powerful as a historical source because of some special inherent quality; rather, it is useful because artists and car drivers belonged to the same group. Artists were not so much 'better observers' than experts of experience, they were an eloquent and celebrated part of the early user culture themselves and co-shaped its culture. Differently put: literary (and other artistic) utterances have to be historically contextualized in order to be effective as historical source.

Together with the blurring of the boundaries between inside and outside the examples of intermediality in the previous pages give the impression on several occasions (such as Wharton) of the car trip becoming a part of the fictional narrative itself, of an interlacing of fiction and reality. A recent analysis (quoting Virilio) has called the car a "landscape simulator" (stressing the parallelism between the means of transport and communication) and an "illusion machine."[241] Another scholar places the tourist trip in the same league as literature, film, and visual arts as a way "to experience fictitious space."[242] And yet other analysts emphasize the composing function of the driver: "The technical perfection of the male mechanism, the plan of the moving sculpture, are visited, consumed by the pilot, somewhat like the musician actualizes the music, makes Mozart alive every time he plays, or like the reader actualizes the literary text, and every reading of Stendhal or Proust revives Stendhal and Proust."[243] Conversely, literary scholars may benefit from the same scholars' insight that the car is not so much thematically of interest for literary analysis, but because it has a "structuring role of the novel.... Quite quickly the car has become an object loaded with phantasms, no doubt because it modifies our relationship with space and time as it enables us to enter the world of speed; it thus converges with the mythical course (*coursier*), those opening up the roads to paradise, but also those pulling us in an infernal and tragic cavalcade."[244]

Because of the intermedial affinity of car trip and writing/reading, the motorist, then, may feel like an artist in the very act of driving, taking part in a collectively organized sublime experience. If one looks for the most basic explanation of the general and surprising attraction of the car, it is here: every driver a poet, through the democratization of the sublime.[245] Proustian inversion cements this desire into a material reality: driving the car produces a narcissistic, static center, while the environment starts to move, with rows of houses separating in front of the occupants, trees speeding along, wheat fields waving goodbye. "The writer is a body in the condition of transfer," Charles Grivel concludes. "Writing is transference: the body becomes something it was not or no longer is—on the one hand it leads us back to ourselves; on the other is helps us break out.... The shell is also the heart."[246] In a similar vein, Tim Armstrong has investigated "the reciprocal relations between body and writing," revealing in the process how writers transformed "their own literal and literary corpus." This affinity between writing and driving as "prosthetic" is more than a mere word play. Karl Marx observed a shift from the machine as a tool (an extension of the worker) to a motor (in which the worker was reduced to "a living appendage of the machine"). Through the process of becoming a "living appendage" of the car the driver could evolve into (the illusion of) a 'manager' of his own individual mobility. For Freud, man had become "a kind of prosthetic God." [247] The prosthetic activity, the cyborg-like experience was acted out as adventure, because "the form of human activity known as adventure has a central role to play in the construction and development of life stories, and ... life stories, in turn, are the major supports for human identities." The narrative psychologist who proposes this conclusion, sees adventure as a Western idea "strongly associated with the role of the male," an association "largely cultural in origin"—indeed, "horrors make the fascination." For Georg Simmel, adventure is shorthand for life and thus not a privilege of a vanguard, although he observed a "deep relationship between the adventurer and the artist, and perhaps also the inclination of the artist to engage in adventure." Adventure is the characteristic of "unhistorical man," a being living in the now (*Gegenwartswesen*).[248]

Our 'grammar of automobile adventure' thus results in a corporeal, eroticized, expansionist (imperialistic), and highly exciting experience, in space, in time, and in technology, heavily gendered as an aggressive, violent flight in and of a shell. It has been shouted from the beginning by the likes of Toad and Marinetti, and whispered by Bierbaum, Winter, and Williamson. And it will explode, literally, once 'adventure' and 'war' got increasingly intertwined during the First World War.

Notes

1. Rudyard Kipling, "Carmen Circulare," in "The Muse Among The Motors 1900–1930," in M. M. Kaye (ed.), *Rudyard Kipling: The Complete Verse* (London: Kyle Cathie Ltd., 1996), 555–581, here 557–558.
2. *New York Times* (2 May 1912); http://en.wikipedia.org/wiki/Bonnot_Gang. For Leblanc's story, see Maurice Leblanc, "The Most Amazing True Story of Crime Ever Told: 'The Auto-Bandits of Paris,' by the Author of 'Arsene Lupin'; Sent by Cable to the New York Times," *New York Times* (5 May 1912), http://query.nytimes.com/mem/archive-free/pdf?res=9F0DE4DC153CE633A25756C0A9639C946396D6CF (last accessed 15 March 2010). Richard Parry, *The Bonnot Gang* (n.p. [London]: Rebel Press, 1987), 5, 75, 114–115, 135, 166. Conan Doyle and 30,000: Philipp Blom, *The Vertigo Years: Europe, 1900–1914* (New York: Basic Books, 2008), 373.
3. Otto Julius Bierbaum, *Eine empfindsame Reise im Automobil von Berlin nach Sorrent und zurück an den Rhein: In Briefen an Freunde beschrieben* (Munich: Albert Langen—Georg Müller Verlag, 1979; reprint of 1903), 212, 216.
4. Gijs Mom, "Civilized adventure as a remedy for nervous times: Early automobilism and fin the siècle culture," *History of Technology* 23 (2001), 157–190.
5. Paul Fussell, *The Great War and Modern Memory* (Oxford: Oxford University Press, 2000; first ed. 1975), 157–158. Fussell emphasizes that the latter qualification does not hold for the United States as this country "has always done very well without a consciousness of a national literary canon" (ibid., 158). Helen Carr observed that "a remarkable number of novelists and poets were *travelling* writers" for the period between 1880 and 1940 in her "Modernism and travel (1880–1940)," in Peter Hulme and Tim Youngs (eds.), *The Cambridge Companion to Travel Writing* (Cambridge: Cambridge University Press, 2002), 70–86, here 73 (italics in original).
6. The latter phrase is the title of an early one-minute silent film (see later in this chapter).
7. Mobility history proper has not (yet) developed such a canon, as it has not been engaged in a 'cultural turn' (see introduction) long enough for it to crystallize, although some national icons have meanwhile clearly been identified, such as Otto Bierbaum in Germany, Rudyard Kipling in the United Kingdom, Octave Mirbeau and Marcel Proust in France, Cyriel Buysse in Belgium, Theodore Dreiser in the United States, and Louis Couperus in the Netherlands. Most of them are mentioned in overviews by literary and cultural historians, such as Jens Peter Becker, *Das Automobil und die amerikanische Kultur* (Trier: Wissenschaftliche Verlag Trier, 1989), Siegfried Reinecke, *Autosymbolik in Journalismus, Literatur und Film: Struktural-funktionale Analysen vom Beginn der Motorisierung bis zur Gegenwart* (Bochum: Universitätsverlag Dr. N. Brockmeyer, 1992), Roger N. Casey, *Textual Vehicles: The Automobile in American Literature* (New York and London: Garland Publishing, 1997), Cynthia Colomb Dettelbach, *In the Driver's Seat: The Automobile in American Literature and Popular Culture* (Westport, CT, 1976), and Ronald Primeau, *Romance of the Road: The Literature of the American Highway* (Bowling Green, OH: Bowling Green State University Popular Press, 1996).
8. A group of French writers "right on the heels of the last symbolists" was identified earlier by Henri Mitterand as "travelers" in his *La Littérature française du XXe*

siècle, quoted by Kimberley J. Healey, *The Modernist Traveler: French Detours, 1900–1930* (Lincoln and London: University of Nebraska Press, 2003), 3. In this group are Apollinaire, Larbaud, Cendrars, Segalen, and some others. The limited overlap with my group results from the fact that my selection criterion is not traveling, but "autopoetics" as defined later in the text.

9. Cecilia Tichi, *Shifting Gears: Technology, Literature, Culture in Modernist America* (Chapel Hill and London: The University of North Carolina Press, 1987), 18, calls trees, animals, and engines "structural analogues," an observation with "far reaching implications for imaginative literature." Now, novels were conceived "as a designed structure of component parts." Analogy relation (*Analogiebeziehung*) in Claudia Lieb, *Crash: Der Unfall der Moderne* (Bielefeld: Aisthesis Verlag, 2009), 14. Another way to distinguish between levels of the literary text is proposed by Henry S. Turner, "Lessons from Literature for the Historian of Science (and Vice Versa): Reflections on 'Form,'" *Isis* 101 (2010), 578–589, here 580–581, who distinguishes between stylistic, structural, material, and social notions of form.

10. Mary Baine Campbell, "Travel Writing and its Theory," in Peter Hulme and Tim Youngs (eds.), *The Cambridge Companion to Travel Writing* (Cambridge: Cambridge University Press, 2002), 261–278, here 262.

11. For an example of discourse analysis applied to mobility history, see Jürgen Link, "'Einfluss des Fliegens!—Auf den Stil selbst!' Diskursanalyse des Ballonsymbols," in Jürgen Link and Wulf Wülfing (eds.), *Bewegung und Stillstand in Metaphern und Mythen: Fallstudien zum Verhältnis von elementarem Wissen und Literatur im 19. Jahrhundert* (Stuttgart: Klett-Cotta, 1984), 149–163; for an analysis of car symbols in German literature, see Jürgen Link and Siegfried Reinecke, "'Autofahren ist wie das Leben': Metamorphosen des Autosymbols in der deutschen Literatur," in Harro Segeberg (ed.), *Technik in der Literatur: Ein Forschungsüberblick und zwölf Aufsätze* (Frankfurt: Suhrkamp Verlag, 1987), 436–482. See also W. Dale Dannefer, "Driving and Symbolic Interaction," *Sociological Inquiry* 47, no. 1 (1977), 33–38.

12. See, for instance, Harro Segeberg (ed.), *Die Mobilisierung des Sehens: Zur Vor- und Frühgeschichte des Films in Literatur und Kunst* (Munich: Wilhelm Fink, 1996); and for the Interbellum: Harro Segeberg, *Literatur im Medienzeitalter: Literatur, Technik und Medien seit 1914* (Darmstadt: Wissenschaftliche Buchgesellschaft, 2003). I thank Martin Knoll, Technical University Darmstadt and fellow in 2009–2010 at the Rachel Carson Center in Munich, for pointing out this connection to me. For an example of affinity between language and the telegraph, see John Durham Peters, "Technology and Ideology: The Case of the Telegraph Revisited," in Jeremy Packer and Craig Robertson (eds.), *Thinking with James Carey: Essays on Communications, Transportation, History* (New York: Peter Lang, 2006), 137–155, here 145–147: "Words ... are like telegraph signals," French linguist Michel Bréal opined.

13. Charles Grivel, "Automobile et cinéma I," in Michel Bouvier, Michel Lartouce, and Lucie Roy (eds.), *Cinéma: acte et présence* (Quebec: Éditions Nota Bene, 1999), 63–85, here 63, 68; Charles Grivel, "Automobil: Zur Genealogie subjektivischer Maschinen," in Martin Stingelin and Wolfgang Scherer (eds.), *HardWar/SoftWar: Krieg und Medien 1914 bis 1945* (Munich: Wilhelm Fink Verlag, 1991), 171–196, here 182. The double entendre is lost in the translation: a "*Schreibmaschine*" is not only a machine to write, but also a typing machine.

14. Jennifer Shepherd, "The British Press and Turn-of-the-Century Developments in the Motoring Movement," *Victorian Periodicals Review* 38, no. 4 (Winter 2005), 379–391, here 381–382.
15. Wolfgang Schivelbusch, *Geschichte der Eisenbahnreise: zur Industrialisierung von Raum und Zeit im 19. Jahrhundert* (Munich and Vienna, 1977), 59–62. For an English translation, see Wolfgang Schivelbusch, *The Railway Journey: The Industrialization of Time and Space in the Nineteenth Century* (Berkeley: University of Calfornia Press, 1986).
16. Schivelbusch, *Geschichte der Eisenbahnreise*, 59 (inversion), 47, 84–105.
17. Rainer Schönhammer, *In Bewegung: Zur Psychologie der Fortbewegung* (Munich, 1991), 102–111 (horse carriage); Lynne Kirby, *Parallel Tracks: The Railroad and Silent Cinema* (Exeter: University of Exeter Press, 1997), 43–44. The diorama worked with large transparencies through which the light was manipulated creating visual illusions. For a critique of Schivelbusch's "extreme sort of phenomenalism" see George Revill, "Perception, Reception and Representation: Wolfgang Schivelbusch and the Cultural History of Travel and Transport," in Peter Norton, Gijs Mom, Liz Millward, and Mathieu Flonneau (eds.), *Mobility in History: Reviews and Reflections (T²M Yearbook 2012)* (Neuchâtel: Alphil, 2011), 31–48, here 43.
18. Marc Desportes, *Paysages en mouvement: Transports et perception de l'espace XVIIIe–XXe siècle* (n.p. [Paris]: Gallimard, 2005), 155, 266; Ponts et Chaussée: review of *Paysages en mouvement* in *HOST, Journal of History of Science and Technology* 1 (Summer 2007), http://johost.eu/?oid=42&act=&area=10&ri=1& itid= (last accessed 9 May 2008).
19. Desportes, *Paysages en mouvement*, 60.
20. See, for instance, Judith Adler, "Origins of Sightseeing," *Annals of Tourism Research* 16 (1989), 7–29, here 10. Adler points at the shaping of "an internationally shared canon of observational method" from as far back as the sixteenth century, with books like *Brief Instructions for Making Observations in All Parts of the World* (Woodward, 1696) (ibid., 16).
21. Desportes, *Paysages en mouvement*, 173–174, 260–271, 241 (aura), 244–245 (traveling camera); on the landscape's loss of aura, see also Schivelbusch, *Geschichte der Eisenbahnreise*, 42.
22. Jonas Larsen, "Tourism Mobilities and the Travel Glance: Experiences of Being on the Move," *Scandinavian Journal of Hospitality and Tourism* 1, no. 2 (2001), 80–98, here 87 (preference), 91 (fleeting and disappear), 92 (aesthetic affinity).
23. Schivelbusch, *Geschichte der Eisenbahnreise*, 47 (my translation, italics in original).
24. Hobbes quoted in Tim Cresswell, *On the Move: Mobility in the Modern Western World* (New York and London: Routledge, 2006), 14; honey bee: William Morton Wheeler, *Emergent Evolution and the Development of Societies* (New York: W.W. Norton & Company, 1928), 37. In "Auf der Terrasse des Café Josty" (At the terrace of café Josty), by the German Expressionist Paul Boldt, the swarm is observed among pedestrians: "The people run on the asphalt / Busy like ants." Sabina Becker, *Urbanität und Moderne: Studien zur Grossstadtwahrnehmung in der deutschen Literatur 1900–1930* (St. Ingbert: Werner J. Röhrig Verlag, 1993), 219. Newton: Schivelbusch, *Geschichte der Eisenbahnreise*, 38.
25. Leo Marx, *The Machine in the Garden: Technology and the Pastoral Idea in America* (Oxford and New York: Oxford University Press, 2000; first ed. 1964); Schivelbusch, *Geschichte der Eisenbahnreise*, 86 (Marx).

26. Kurt Möser, *Geschichte des Autos* (Frankfurt and New York: Campus Verlag, 2002); Catherine Bertho Lavenir, *La roue et le stylo: Comment nous sommes devenus touristes* (Paris: Editions Odile Jacob, April 1999); Wolfgang Sachs, *Die Liebe zum Automobil: Ein Rückblick in die Geschichte unserer Wünsche* (Reinbeck bei Hamburg: Rowohlt Verlag, 1984); also as: *For the Love of the Automobile: Looking back into the History of Our Desires*, trans. Don Reneau (Berkeley, Los Angeles, and London: University of California Press, 1992).
27. Edmund Russell, *Evolutionary History: Uniting History and Biology to Understand Life on Earth* (Cambridge and New York: Cambridge University Press, 2011), 135; Edmund Russell, "Introduction: The Garden in the Machine: Toward an Evolutionary History of Technology," in Susan R. Schrepner and Philip Scranton (eds.), *Industrializing Organisms: Introducing Evolutionary History* (New York and London: Routledge, 2004), 1–16.
28. Marx, *The Machine in the Garden*, 367 (industrialism), 369 (ornithologist), 4–5, 10–11, 25, 27, 193 (meaning); Hawthorne: 11–14. Marx confesses in an "Afterword" to his analysis, that he initially intended to analyze what this author actually did as a Master's thesis, among a group of "literary sociologists": left-wing, "proletarian" novels from the 1930s, but decided against it because "the more predictable and formulaic, the less interesting, they seemed" (ibid., 370). He is right. I know, because I did the analysis: projectgroep 'literatuursociologie' 1 (= Marie-José Buck, Jean Lemmens, Gijs Mom, Kees van de Ven, Harrie Vleerlaag and Marcel Weijers), "Links Richten tussen partij en arbeidersstrijd: materiaal voor een theorie over de verhouding tussen literatuur en arbeidersstrijd" (Nijmegen, July 1975), 2 vols.
29. Marx, *The Machine in the Garden*, 219 (note), 238 (elites), 290–291 (narcissistic), 356–359 (Gatsby), 375 (celebratory). Peter J. Schmitt, *Back to Nature: The Arcadian Myth in Urban America* (New York: Oxford University Press, 1969), xii, presents his book as the "popular culture" counterpart of Marx's study.
30. Marx, *The Machine in the Garden*, 370.
31. Ellen Gruber Garvey, "Reframing the Bicycle: Advertising-Supported Magazines and Scorching Women," *American Quarterly* 47, no. 1 (March 1995), 66–101. I thank André van Oudvorst, Free University Brussels, for suggesting the metaphor of alternative scenarios to me. In her analysis of "urban domesticity" on the basis of American literature, Betsy Klimasmith, *At Home in the City: Urban Domesticity in American Literature and Culture, 1850–1930* (Durham: University of New Hampshire Press, Published by the University Press of New England, 2005), 221, also posits that "the novel form narrates the bodily experience of a theoretical 'What if?' as it imagines how characters' subjectivities develop in relation to the spaces they inhabit."
32. Sara Danius, "The Aesthetics of the Windshield: Proust and the Modernist Rhetoric of Speed," *Modernism/Modernity* 8, no. 1 (2001), 99–126, here 118.
33. Lawrence Buell, *The Future of Environmental Criticism: Environmental Crisis and Literary Imagination* (Malden, MA, and Oxford: Blackwell Publishing, 2005), 55. Buell on his turn borrowed the concept from Angus Fletcher. I thank Agnes Kneitz, Rachel Carson Center, Munich, for bringing this source to my attention, and her and Martin Knoll, University of Darmstadt and Rachel Carson Center, for discussing the usefulness of this concept in a 'reading group' at the Carson Center in the spring of 2010.
34. Healey, *The Modernist Traveler*, 13.

35. Ibid., 16. For a recent example of the self-quest approach, see Healey (ibid., 9), who is "primarily interested in writers who asked questions about themselves, of their destinations, and of their writing practice." The journey itself is lacking from her interests, except for the period when transcontinental aviation emerged. For Healey, travelogues "are tales of stasis, self-examination, and psychic immobility. In this manner they are modernist and metatextual" (ibid., 2). We will see later that autopoetics exactly combine mobility and immobility, and speed and stasis, because of the 'cocooning effect.'

36. Carr, "Modernism and travel (1880–1940)," 75.

37. See, for instance, Carr, "Modernism and travel (1880–1940)," who manages not to mention the automobile one single time when dealing with modernist authors. She also claims that "leisure travel" stops in war. For an analysis of automotive literature as one long 'Quest of the Self,' see Primeau, *Romance of the Road*, for instance: 19.

38. Volker Klotz, *Abenteuer-Romane: Sue—Dumas—Ferry—Retcliffe—May—Verne* (Munich and Vienna: Carl Hanser Verlag, 1979), 217.

39. Kevin O'Neill, *André Gide and the Roman d'Aventure: The History of a Literary Idea in France* (Sydney: Sydney University Press, 1969), 12. In this book, I will use the present tense when dealing with fictional 'reality.'

40. Jean-Yves Tadié, *Le roman d'aventures* (Paris: Presses Universitaires de France, 1982), 5–6; Kevin L. Cope and Alexander Pettit, "Foreword," in Serge Soupel, Kevin L. Cope, and Alexander Pettit (with Laura Thomason Wood) (eds.), *Adventure: An Eighteenth-Century Idiom; Essays on the Daring and the Bold as a Pre-Modern Medium* (New York: AMS Press, 2009), xi–xx, here xiv (aesthetic), xvi (forests), xviii (travel).

41. Freud: Patrick Laviolette, *Extreme Landscapes of Leisure: Not a Hap-Hazardous Sport* (Farnham and Burlington: Ashgate, 2011), 5; Richard Phillips, *Mapping Men and Empire: A Geography of Adventure* (London and New York: Routledge, 1997), 88 (Orwell), 99–100 (*rite de passage*; italics in original), 101 (anarchic, quoted from Zweig, see n. 45).

42. Herfried Münkler, "Strategien der Sicherung: Welten der Sicherheit und Kulturen des Risikos: Theoretische Perspektiven," in Herfried Münkler, Matthias Bohlender, and Sabine Meurer (eds.), *Sicherheit und Risiko: Über den Umgang mit Gefahr im 21. Jahrhundert* (Bielefeld: transcript Verlag, 2010), 11–34, here 21 (n. 15).

43. Tadié, *Le roman d'aventures*, 189.

44. Georg Simmel, "Das Abenteuer (Zur philosophischen Psychologie)," in Simmel, *Philosophische Kultur: Über das Abenteuer; die Geschlechter und die Krise der Moderne; Gesammelte Essays, Mit ein Vorwort von Jürgen Habermas* (Berlin: Verlag Klaus Wagenbach, 1986), 25–38, here 27 (*Beziehung*), 32 (love), 35 (form), 37 (universal).

45. Paul Zweig, *The Adventurer* (London: Basic Books, 1974), 208–213, 221 (repetition), 223–225 (challenge).

46. Martin Green, *Seven Types of Adventure Tale: An Etiology of a Major Genre* (University Park: The Pennsylvania State University Press, 1991), 206 (cradle), 208–209 (capitalism); the relationship between adventure novel and capitalism was analyzed by Michael Nerlich, *Ideology of Adventure: Studies in Modern Consciousness, 1100–1750*, volume 1 (Minneapolis: University of Minneapolis Press, 1987); O'Neill, *André Gide and the Roman d'Aventure*, 29 (literati), 55 (*génie naïf*), 62 (*inquiétude*), 61–62 (what comes next), 63 n. 13 (unknown; italics in original).

47. Hélène Dachez, "The Adventure of Sense(s) in Richardson's *Clarissa*," in Soupel, Cope, and Pettit (eds.), *Adventure*, 41–62, here 59.
48. Zweig, *The Adventurer*, 227 (internalized), 229 (pulp magazines), 249 (transcendence).
49. Sheila Rowbotham, *Dreamers of a New Day: Women Who Invented the Twentieth Century* (London and New York: Verso, 2010), 2: "The bicycle became the symbol of (the) self-propelled female vanguard contesting physical and psychological spaces."
50. Émile Zola, *Les trois villes: Paris; Édition présentée, établie et annotée par Jacques Noiray* (n.p. [Paris]: Gallimard, 2002), 448 (birds), 449 (velvet), 451 (fan). I thank Wulfhard Stahl (librarian, World Trade Institute, Bern) for bringing this source to my attention.
51. Willa Cather, "Tommy, the Unsentimental," in *Willa Cather's Collected Short Fiction, 1892–1912; Introduction by Mildred R. Bennett* (Lincoln: University of Nebraska Press, 1965; first ed. 1965), 473–480, here 478; Nanette Hope Graf, "The Evolution of Willa Cather's Judgment of the Machine and the Machine Age in her Fiction" (unpublished doctoral dissertation; University of Nebraska, 1991), 15, 35, 39–46; Garvey, "Reframing the Bicycle," 92–94.
52. Sherwood Anderson, *Winesburg, Ohio; With an Introduction by Malcolm Cowley* (New York and London: Penguin Books, 1960; first ed. 1919), 39–48, here 42 (adventurer), 45 (quote). The story is called "Mother."
53. Jerome K. Jerome, *Three Men on the Bummel* (n.p.: Penguin Popular Classics, n.d.; first ed. 1900).
54. Ibid., 66, 74.
55. Frances E. Willard, *How I Learned to Ride the Bicycle: Reflections of an Influential 19th Century Woman (Introduction by Edith Mayo; Edited by Carol O'Hare)* (Sunnyvale, CA: Fair Oaks Publishing, 1991; rev. ed. of *A Wheel Within a Wheel*, 1895); H.G. Wells, *The Wheels of Change* (London: Faber and Faber, 2008; reprint of 1896 edition).
56. Bettina L. Knapp, *Machine, Metaphor, and the Writer: A Jungian View* (University Park and London: The Pennsylvania State University Press, 1989), chapter 1 (14–28).
57. Mark Polizotti, "Patabiographical: The portrait of a louche French author and his kingly alter ego," *Bookforum* (December/Januray 2012), 36–37, here 36 (review of Alastair Brotchie, *Alfred Jarry: A Pataphysical Life*).
58. Sarah Wintle, "Horses, Bikes and Automobiles: New Woman on the Move," in Angelique Richardson and Chris Willis (eds.), *The New Woman in Fiction and Fact: Fin-de-Siècle Feminisms* (Houndsmill and New York: Palgrave MacMillan in association with Institute for English Studies, School of Advanced Study, University of London, 2001), 66–78, here 77. This analysis not only contains several factual errors about the novel, its conclusion is also questionable: it neglects the intermediary role of women in the masculine game of automobilism.
59. [Mary] Kennard, *The Motor Maniac: A Novel* (London: Hutchinson & Co., 1902), 137 (flying and gliding), 172 (train), 190 (coil), 194 (motorcycle), 199 (glued to the road), 242 (distinguishable).
60. Shepherd, "The British Press and Turn-of-the-Century Developments in the Motoring Movement," 382.
61. Race to doctor: Rudyard Kipling, "'They' The Return of the Children," in Kipling, *Traffics and Discoveries* (n.p.: Kessinger Publishing's Rare Prints, n.d. [1904]; reprint), 205–249, here 218; skill: Kipling, "Steam Tactics: The Necessitarian," in

Kipling, *Traffics and Discoveries*, 112–161, here 137. Angus Wilson, *The Strange Ride of Rudyard Kipling: His Life and Works* (London: Pimlico, 1977), photo 17 (Locomobile). For an enumeration of Kipling's cars, see the remarks by Alastair Wilson at http://www.kipling.org.uk/rg_steamtactics_kipearly.htm (last accessed 25 February 2010), and Derek Jewell, *Man & Motor: The 20th Century Love Affair* (London: Hodder and Stoughton, n.d. [1966]), 102.
62. Kipling, "The Muse Among The Motors 1900–1930."
63. On early vibration therapy against nervousness and Zander, see Joachim Radkau, *Das Zeitalter der Nervosität: Deutschland zwischen Bismarck und Hitler* (Munich and Vienna, 1998), 27–49, and Mom, *The Electric Vehicle*, 38. In Guy de Maupassant's novel *Mont-Oriol* (1887) an unscrupulous doctor in a newly built spa founds an "Institut de Gymnastique Automotrice" in which patients are immobilized in a "swivel chair" (*fauteuil à bascule*) which simulates the four "natural exercises": walking, horse riding, swimming, and canoeing. Larry Duffy, *Le Grand Transit Moderne: Mobility, Modernity and French Naturalist Fiction* (Amsterdam and New York: Rodopi, 2005). I thank Saurabh Arora (Eindhoven University of Technology) for bringing this source to my attention.
64. Bierbaum, *Eine empfindsame Reise im Automobil*, 21 (addictive), 44 (engine), 80 (*Rhythmus*).
65. Laurence Sterne, *A Sentimental Journey Through France and Italy* (transcribed from the 1892 George Bell and Son edition by David Price), http://www.gutenberg.org/cache/epub/804/pg804.txt (last accessed 4 July 2011), 15 (adventures), 41–42 (starling story), 51 (devil).
66. "Philister contra Automobil," in Otto Julius Bierbaum, *Mit der Kraft: Automobilia von Otto Julius Bierbaum* (Berlin: Bard Marquart, 1906), 325.
67. "Das höllische Automobil: Ein Märchen für sämtliche Alters- und Rangklassen nach einer Idee Alf Bachmanns," in Bierbaum, *Mit der Kraft*, 283–311, here 302–303. This story first appeared separately as Otto Julius Bierbaum, *Das höllische Automobil: Novellen* (Vienna and Leipzig: Wiener Verlag, 1905), 17–61. See for other travel stories, such as "Eine kleine Herbstreise im Automobil" and "Kleine Reise," his *Die Yankeedoodle-Fahrt und andere Reisegeschichten: Neue Beiträge zur Kunst des Reisens* (Munich: Georg Müller, 1910). Dushan Stankovich, *Otto Julius Bierbaum—eine Werkmonographie* (Bern and Frankfurt: Verlag Herbert Lang, 1971), 52.
68. Dorit Müller, *Gefährliche Fahrten: Das Automobil in Literatur und Film um 1900* (Würzburg: Königshausen & Neumann, 2004), 45–46 (my translation). Thomas Mann, *Tristan, Tonio Kröger, Der Tod in Venedig, Mario und der Zauberer, Meistererzählungen; Königliche Hoheit, Roman* (Zürich: Manesse Verlag, 2004; Nachwort von Hans Wysling).
69. Léon Bloy, *Journal I, 1892–1907 (Édition établie, présentée et annotée par Pierre Glaudes)* (Paris: Robert Laffont, 1999), 592 (crétinisme), 606–607 (car trip); Pierre Glaudes, "Avant-propos," in ibid., I–VI, here VI (Céline); Pierre Glaudes, "Introduction générale," in ibid., VII–LXXV, here XLI (devil), XLII (stinks), LV (Bonnot); assassin quoted in Marc Camiolo, "Anti-convivialité automobile," *Revue d'Allemagne et des Pays de langue allemande* 43, no. 1 (2011), 119–129, here 127.
70. Quoted in Michael Adams, *The Great Adventure: Male Desire and the Coming of World War I* (Bloomington and Indianapolis: Indiana University Press, 1990), 74.
71. Oliver Stallybrass, "Editor's Introduction," in E. M. Forster, *Howards End*, edited by Oliver Stallybrass (London and New York: Penguin Books, 2000; first ed. 1910), 7–17, here 10–11; Rob Doll, "An Interpretation [of] E.M. Forster's *Howards*

End" (2000), http://www.emforster.info/pages/howardsend.html (last accessed 27 July 2009). Quotation: E. M. Forster, *Howards End*, introduction and notes by David Lodge (London and New York: Penguin Books, 2000), 92.

72. Quoted in Andrew Gavin Altman, "Motoring for the Million? Cars and Class in Pre-1952 Britain" (unpublished PhD thesis, Boston College, Graduate School of Arts & Sciences, 3 April 1997), 160–161.

73. Evelyn Cobley, *Modernism and the Culture of Efficiency: Ideology and Fiction* (Toronto, Buffalo, NY, and London: University of Toronto Press, 2009), chapter 9: "Efficiency and Its Alternatives: E.M. Forster's Howards End," 246–281, here 246 (British), 249 (speed), 250 (heaved). For the following see: Kenneth Grahame, *The Wind in the Willows* (New York: Dell, 1969) 22 (well-to-do Toad), 33 (poetry of motion), 61 (skill), 104 (detoxication), 111 (conversion), 172 (Rat), 195 (large quote; addiction), agency of artefacts (examples): 80, 135, 167, 176, 181. Bank of England and suicide: http://www.kennethgrahamesociety.net/biography.htm (accessed 27 July 2011). Road hog, medieval and rural self-image: Altman, "Motoring for the Million?" 164-165, 171.

74. Marie Corelli, *The Devil's Motor: A Fantasy* (London: Hodder & Stoughton, n.d. [1910]); David Burgess-Wise, "Fiction of the Motor Car," *Automobile Quarterly* 31, no. 1 (1993), 89–101, here 101; feminist and not well sold: Teresa Ransom, *The Mysterious Miss Marie Corelli: Queen of Victorian Bestsellers* (Stroud: Sutton Publishing, 1999), 2; melodramatic: http://en.wikipedia.org/wiki/Marie_Corelli (last accessed 9 March 2010); he-females, Haggard, and religious: Jessica Amanda Salmonson, "Marie Corelli and her Occult Tales," at http://www.victorianweb.org/authors/corelli/salmonson1.html (last accessed 9 March 2010); New Woman: Robert Dixon, *Writing the Colonial Adventure: Race, Gender and Nation in the Anglo-Australian Popular Fiction, 1875–1914* (Cambridge: Cambridge University Press, 1995), 83.

75. Quoted in Marco Modenesi, "Locomotions nouvelles: Automobiles et écrivains à la fin du XIXe siècle," in Frédéric Monneyron and Joël Thomas (eds.), *Automobile et Littérature* (n.p.: Presses Universitaires de Perpignan, 2005), 25–36, here 32–34 (quote on 34).

76. Ransom, *The Mysterious Miss Marie Corelli*, 180.

77. Christa Zorn, *Vernon Lee: Aesthetics, History, and the Victorian Female Intellectual* (Anthens: Ohio University Press, 2003), xvi, 3.

78. Vernon Lee, "The Motor-car and the Genius of Places," in Lee, *The Enchanted Woods and Other Essays on the Genius of Places* (London and New York: The Bodley Head, 1905), 95–113, here 95–98, 102 (unreality), 103 (grievances), 107 (dust; italics in original); http://en.wikipedia.org/wiki/Vernon_Lee (last accessed 12 March 2010); transition and rhime: Zorn, *Vernon Lee*, xi, xviii; wastefulness: Vernon Lee, "Gospels of Anarchy," in Andrea Broomfield and Sally Mitchell (eds.), *Prose by Victorian Women: An Anthology* (New York and London: Garland Publishing, 1996), 711–729, here 719.

79. For a fragment, see Eugen Diesel, "Autoreise 1905," in Rüdiger Kremer and Wolfgang Rumpf, *Baby, won't you drive my car? Dichter am Steuer* (Frankfurt: Fischer Taschenbuch Verlag, 1999), 31–44.

80. P. G. Konody, *Through the Alps to the Apennines* (London: Kegan Paul, Trench, Trübner & Co, 1911; reprint 2009 Kessinger Publishing's Rare Prints), 1 (magic carpet), 2 (steam car), 3 (participants), 7 (racing quote), 24 (pictureesque), 34 (competition), 37 (anti-train), 48 (sublime), 51 (bad breath), 79 (impressionistic style), 107 (modest resistance); Francis Miltoun, *The Automobilist Abroad*

(London: Brown, Langham & Company, 1907), 301 (paradise); Miltoun, *Italian Highways and Byways from a Motor Car* (London: Hodder & Stoughton, 1909), 3 (pilgrim), 11 (real Italy quote), 12 (kinetoscope), 20 (pennies); Charles L. Freeston, *Die Hochstrassen der Alpen: Ein Automobilführer zum Befahren von über einhundert Gebirgspassen* (Berlin: Richard Carl Schmidt & Co., 1911).

81. Eugène Demolder, *L'Espagne en Auto: Impressions de voyage* (Paris: Société du Mercure de France, 1906; first ed. 1906), 240 (my translation).
82. Demolder, *L'Espagne en Auto*, 7 (summer), 10–11 (long quotation), 13 (bowels), 43 (braking), 65 (salon), 67 (skills), 72 (palette), 82 and 87 (tire defects), 188 (nothing new), 241 (anthropomorphizing), 253 (*vitesse*), 289 (defects), 290 (burn). http://en.wikipedia.org/wiki/Th%C3%A9ophile_Gautier (last accessed 13 March 2010).
83. Demolder, *L'Espagne en Auto*, 73 (mob), 75 (faun), 111 (guide), 185 (gypsies), 263–264 (gunpoint), 280–281 (adventure).
84. Pierre de Crawhez, *Les Grands Itinéraires en Automobile à travers l'Europe, l'Algérie et la Tunisie: Annuaire de l'A.C.N.L.* (Namen: Bertrand, 1906).
85. Dan Fellows Platt, *Through Italy with Car and Camera* (New York and London: G. P. Putnam's Sons, 1908).
86. Hans Peter Hahn, "The Appropriation of Bicycles in Africa: Pragmatic Approaches to Sustainability," paper presented at the workshop Re/Cycling Histories: Users and the Paths to Sustainability in Everyday Life, Rachel Carson Center (Munich, 27–29 May 2011), 3; Segalen and Cendrars: Healey, *The Modernist Traveler*, chapters 1 and 2 (17–77). Expansionist: Henning Eichberg, "'Join the army and see the world': Krieg als Touristik—Tourismus als Krieg," in Dieter Kramer and Ronald Lutz (eds.), *Reisen und Alltag: Beiträge zur kulturwissenschaftlichen Tourismusforschung* (Frankfurt: Institut für Kulturanthropologie und Europäische Ethnologie der Universität Frankfurt am Main, 1992), 207–228, here 211; exoticism: Peter J. Brenner, "Schwierige Reisen: Wandlungen des Reiseberichts in Deutschland 1918–1945," in Brenner, *Reisekultur in Deutschland: Von der Weimarer Republik zum "Dritten Reich"* (Tübingen: Max Niemeyer Verlag, 1997), 127–176, 127; empire: Dixon, *Writing the Colonial Adventure*, 3 (Empire), 198 (guns).
87. Winter was the only daughter of a leather manufacturer and she married the son of a paper manufacturer who at the death of her father took over the leather factory. Bernd Utermöhlen, "Margarete Winter—eine Automobilistin aus Buxtehude," *Technik und Gesellschaft* 10 (1999), 271–279, here 271.
88. [Margarete Winter], *5000 Kilometer Autofahrt ohne Chauffeur, Buxtehude 1905* (Stuttgart: Uhlandsche Buchdruckerei, n.d. [1905]), 9 (train), 14 (inversion), 21 (no time for nature), 29 (flight), 34 (change), 44, 55 (nomadic community), 57, 80–81 (unrest), 64 (stereotyping), 94 (per minute), 98 (*au courant*), 99 (conquest), 117, 133–136 (Belgian and Dutch hostility), 127 (racing), 139 (advice), 141 (no aristocrat), 149 (Frisians), 150 (*Herrenmenschen*), 151 (Opel).
89. Mary Suzanne Schriber, *Writing Home: American Women Abroad 1830–1920* (Charlottesville and London: University Press of Virginia, 1997), 191 (un-American), 195 (scholar-adventurer), 196 (two adventures). Wharton's reluctance to mention her means of travel can be related to her biographer Hermione Lee's remark about her "reserve and concealment [which] are everywhere in her fiction." Hermione Lee, *Edith Wharton* (New York: Alfred A. Knopf, 2007), 11.
90. For a similar conclusion, see Duffy, *The Speed Handbook*, 149.

91. Edith Wharton, *A Backward Glance* (New York and Berlin: Globusz Publishing, n.d.; first ed. New York and London: Appleton-Century, 1934), http://www.globusz.com/ebooks/BackardGlace/index.htm (last accessed 22 June 2009); on Henry James, including his participation in the 1907 trip through France, see the conclusion, p. 3; name of Charles Cook, the chauffeur, in chapter 10, p. 13; Wharton, *A Motor-Flight Through France* (London and Basingstoke: Picador/Macmillan, 1995; Picador Travel Classics; reprint of 1908, with an introduction by Julian Barnes), 3 (decorum), 4 (servants), 17 (first sentences), 31 (poetry), 99 (road), 107 (Pau), 115 (climbing), 123 (cross), 127 (sublime), 135 (Virgilian), 147 (observer), 151 (pilgrim), 154–157 ("one"), 162 (prose); Sarah Bird Wright, *Edith Wharton's Travel Writing: The Making of a Connoisseur* (Houndsmill and London: MacMillan Press, 1997), 76–77 (1904–1907; Panhard-Levassor), 82 (divorce), 113–155 (8 travelogues), 117 (Spain), 146 (*bon movement*; italics in original); Alan Price, *The End of the Age of Innocence: Edith Wharton and the First World War* (New York: St. Martin's Press, 1996), 132 (adventure).
92. Julian Barnes, "Introduction by Julian Barnes," in Wharton, *A Motor-Flight Through France*, 1–14, here 5, 7. Wharton, *A Backward Glance* (Wharton's account of the open car becoming closed is in chapter 6, p. 12). For the passage on the peasant faces, see Wharton, *A Motor-Flight Through France*, 90. Opposition: ibid., 70.
93. Wharton, *A Motor-Flight Through France*, 29 (inversion), 44 (*hôtellerie*), 83 (foreground), 102 (stations). A comparable conclusion also in Gianfranca Balestra, "Edith Wharton, Henry James, and 'the Proper Vehicle of Passion,'" in Francesca Bisutti De Riz and Rosella Mamoli Zorzi, in cooperation with Marina Coslovi (eds.), *Technology and the American Imagination: An Ungoing Challenge: A.I.S.N.A. Associazione Italiana di Studi Nord-Americani, Atti del Dodicesimo Convegno Biennale, Università di Venezia, 28–30 ottobre 1993* (Venezia: Edizioni Supernova, 1994), 595–604, here 597, where the automobile in Wharton's texts is analyzed as "an extension of the body" allowing "immersion in the landscape rather than mere spectatorship from a window."
94. Sidonie Smith, *Moving Lives: Twentieth-Century Women's Travel Writing* (Minneapolis and London: University of Minnesota Press, 2001), 180–181; rush quoted in Schriber, *Writing Home*, 204.
95. Balestra, "Edith Wharton, Henry James, and 'the Proper Vehicle of Passion,'" 596–597; attack quoted by Schriber, *Writing Home*, 197.
96. James B. Wolf, "Imperial Integration on Wheels: The Car, the British and the Cape-to-Cairo Route," in Robert Giddings (ed.), *Literature and Imperialism* (Houndsmill and London: MacMillan, 1991), 112–127, here 113.
97. Bertho Lavenir, *La roue et le stylo*, 52–53, 76–77, 205, 222. Bertho Lavenir lays the turning point from an elite to a cooperative car movement in France in 1905–1906.
98. Leo Faust, 'Het ontwaken,' *De Auto* (1909), 15–17, 35–39 (quotation on 38; my italics: note the gender of the engine). Neologisms and quotations in Alphonse van Lier, 'Een rit voor Geluk,' *De Auto* (1909), 64–67, 90–93; F. Hageman, 'De Truc,' *De Auto* (1909), 119–121, 146–148 (quotation on 146), 171–172, 194–196; J. M., 'De souplesse van den automobielmotor,' *De Auto* (1908), 918–921, here 919; last quotation: *De Auto* (1910), 391.
99. Carter, *The Proustian Quest*, 5 (father), 134–137 (quotations: 134, 136, 137); Ernst Robert Curtius, *Marcel Proust* (Berlin and Frankfurt: Suhrkamp Verlag, 1952), 7, 9–10.

100. Danius, *The Senses of Modernism*, 127.
101. Carter, *The Proustian Quest*, 32 (rape), 161 (brothel). On one occasion "caged, famished rats would be brought in and Proust, watching the bloody spectacle of the rats clawing and biting each other, would succeed in reaching a climax."
102. Curtius, *Marcel Proust*, 83, 86 ("*bürgerliches Patriziat*").
103. Schivelbusch, *Geschichte der Eisenbahnreise*, 41 (note) (my translation).
104. Wolfram Nitsch, "Fantasmes d'essence: Les automobiles de Proust à travers l'histoire du texte," in Rainer Warning and Jean Milly (eds.), *Marcel Proust: Écrire sans fin* (Paris: CNRS Éditions, 1996), 125–141, here 134–135, 125 (*hélas*), 139 (three phases); André Gide, "Journal sans dates," *La Nouvelle Revue française* no. 21 (1 September 1910), 336–341, here 340–341.
105. For the quotation I have consulted a very recent Dutch translation of the Swann part: Proust, *Op zoek naar de verloren tijd*, 248–249. See also Henri Bonnet, *Marcel Proust de 1907 à 1914: Bibliographie complémentaire (II); Index général des bibliographies (I et II) et une étude: Du Côté de chez Swann dans A la Recherche du Temps Perdu* (Paris: A. G. Nizet, 1976), 133. On Proust's treatment of the senses, especially the "dissociation of the eye and the ear," see also Sara Danius, "Orpheus and the Machine: Proust as Theorist of Technological Change, and the Case of Joyce," *Forum for Modern Language Studies* 37, no. 2 (2001), 127–140, and Jacques Nantet, "Marcel Proust et la vision cinématographique," *La revue des lettres modernes: histoire des idées et des littératures* 5, nos. 36–38 (1958), 307–313. Biographer: Claude Mauriac, *Marcel Proust in Selbstzeugnissen und Bilddokumenten* (Hamburg: Rowohlt Taschenbuch Verlag, 1976), 32.
106. Edith Wharton complained in her autobiography that Proust was only interested in the company of dukes and duchesses. Wharton, *A Backward Glance*, chapter 12, p. 15. The Irish modernist James Joyce, too, dedicated a short story "After the Race" (in *Dubliners* [1914]) to the Gordon Bennett race of 1903 (quoted in Andrew Thacker, "Traffic, gender, modernism," *The Sociological Review* 54, no. s1 [October 2006], 175–189, here 178).
107. Maurice Leblanc, *Voici des ailes: roman* (Paris: Phébus, 1999; original ed. 1898), 21; folk hero: Blom, *The Vertigo Years*, 384; lowbrow: Jean-Yves Tadié (ed.), *La littérature française: dynamique & histoire*, II (Paris: Gallimard, 2007), 576; Jacques Derouard, *Maurice Leblanc: Arsène Lupin malgré lui* (n.p.: Librairie Séguier, 1989), 200–201 (bicycle quotations), 211 (eight editions).
108. Leblanc, *Voici des ailes*, 26 (Métivet), 26, 51, 86, 109, 117 (my translation).
109. Maurice Leblanc, *Gueule-Rouge, 80-Chevaux* (Paris: Société d'éditions littéraires et artistiques, 1904; first ed. 1904), 5, 9; Kistemaeckers: Derouard, *Maurice Leblanc*, 249.
110. Andreas Braun, *Tempo, Tempo! Eine Kunst- und Kulturgeschichte der Geschwindigkeit im 19. Jahrhundert* (Frankfurt: Anabas-Verlag, 2001), 26. The picture (on p. 27) is utterly sadistic and sexist, as the running car at full speed must meanwhile have heated the radiator supporting the nude to near skin-burning temperature. This sadistic aspect of early car-related art is not included in the analysis by Lynda Nead, "Paintings, Films and Fast Cars: A Case Study of Hubert von Herkomer," *Art History* 25, no. 2 (April 2002), 240–255.
111. Derouard, *Maurice Leblanc*, 250; Bettina Knapp, *Maurice Maeterlinck* (Boston: Twayne Publishers, 1975), 129, 135–136; Nitsch, "Fantasmes d'essence," 136; Maurice Maeterlinck, " En automobile," in Maeterlinck, *Le Double jardin* (Paris: Bibliothèque-Charpentier, 1904), 51–65, here: 59, 60, 62 (quotations), 63

(prison), 64 (uniter); for an English translation, see Maurice Maeterlinck, "On a Motor-Car," in Maeterlinck, *The Double Garden* (London: George Allen & Company, 1915), 139–152; Pierre Michel, *Un moderne: Octave Mirbeau* (Cazaubon: Eurédit, 2004), 281. On Maeterlinck's boxing enthusiasm, see his short story "Éloge de la boxe," in Maeterlinck, *L'Intelligence des fleurs* (Paris: Fasquelle, 1907), 183–187.

112. Maurice Maeterlinck, "L'Accident," in Maeterlinck, *L'Intelligence des fleurs*, 237–253, here 239.
113. Joris van Parys, *Het leven, niets dan het leven: Cyriel Buysse en zijn tijd* (Antwerp and Amsterdam: Houtekiet/Atlas, 2007), 385 (two cars), 587 (Maupassant), 714 (marriage); Luc van Doorslaer, "Cyriel Buysse en de automobiel," in Cyriel Buysse, *Reizen van toen: Met de automobiel door Frankrijk*, edited and introduced by Luc van Doorslaer (Antwerpen and Amsterdam: Manteau, 1992), 7–22, here 18–19.
114. Buysse, *Reizen van toen*, 25 (my translation).
115. Ibid., 91; Cyriel Buysse, "Per auto," in ibid., 111–199, here 176; Cyriel Buysse, "De laatste ronde," in ibid., 203–260, here 217; congestion: van Parys, *Het leven, niets dan het leven*, 603. For examples of Buysse's anti-Americanism, see: ibid., 528, 699.
116. Henry Kistemaeckers, *Monsieur Dupont chauffeur: Nouveau roman comique de l'automobilisme* (Paris: Bibliothèque-Charpentier, 1908) (troisième mille), 13 (noise), 17 (smell), 19 (organ), 45 (piano), 129 (painter), 173 (petit-bourgeois), 253 (salon), 255 (elite bashing), 259 (grenade), 292–293 (train and Americans), 277 (passion), 377 (vulgar). Defective composition: there is a not very functional breach of style (probably the result of pre-publication in a magazine) in the main story which gave the book its title: on p. 80 the omniscient narrator interrupts his story for eight (fictional) years and then takes on a much more active role.
117. Adam quoted in Mathieu Flonneau, *Les cultures du volant: Essai sur les mondes de l'automobilisme XXe–XXIe siècles* (Paris: Éditions Autrement, 2008), 65–66; http://de.wikipedia.org/wiki/Paul_Adam (last accessed 12 March 2010); Tristan Bernard, *Les veillées du chauffeur* (Paris: Société d'éditions littéraires et artistiques, Librairie Paul Ollendorf, 1909); http://www.answers.com/topic/tristan-bernard (last accessed 2 July 2009).
118. See the following short stories from Tristan Bernard, *Sur les grands chemins* (Paris: Librairie Paul Ollendorff, n.d.; first ed. 1911): "Un mari chauffeur" (63–68), "Chiffons...." (82–87; quotation on 82), "Joyeux piétons" (174–179), "Nous roulons...." (180–186), "Les notes de routes" (214–220), "Avares spécialisés" (264–269). Defects: Charles Grivel, "D'un écran automobile," in Jochen Mecke and Volker Roloff (eds.), *Kino-/(Ro)Mania: Intermedialität zwischen Film und Literatur* (Tübingen: Stauffenburg Verlag, 1999), 47–77, here 55.
119. Alice M. Williamson, *The Inky Way* (London: Chapman & Hall, 1931), 20, 31 (success), 42 (editors), 54–55 ("lots of money and make it soon"; seven thousand words a day), 25–26 (Harmsworth quote). On Charlie Williamson: Burgess-Wise, "Fiction of the Motor Car," 93.
120. Compare the photo of the dog in Williamson, *The Inky Way*, facing page 114, and in C. N. Williamson and A. M. Williamson, *The Botor Chaperon* (London: Methuen, 1906; reprint BiblioLife, n.d.), facing page 85. A *botor* was a motorboat.
121. Burgess-Wise, "Fiction of the Motor Car," 94. For a Dutch translation of *The Lighting Conductor*, see C. N. Williamson and A. M. Williamson, *De kranige chauf-*

feur: wonderlijke avonturen van een motorwagen, transl. Ms. Van Heuvelinck (Utrecht: H. Honig, 1913). Spyker: Fons Alkemade, "De auto in de Nederlandse literatuur en speelfilm," report for Foundation for the History of Technology, Eindhoven, The Netherlands (Eindhoven: Nederlands Centrum voor Autohistorische Documentatie NCAD, August/September 2000), 7. Broadway: Casey, *Textual Vehicles*, 23.
122. Williamson and Williamson, *The Botor Chaperon*, 167–170; Tibe: 111 (italics in original); C. N. Williamson and A. M. Williamson, *The Lightning Conductor: The Strange Adventures of a Motor-Car* (New York: Henry Holt and Company, 1903; eleventh impression, revised and enlarged), 48–50.
123. Williamson and Williamson, *The Lightning Conductor*, 13 (italics in original).
124. C. N. Williamson and A. M. Williamson, *The Lightning Conductor Discovers America* (Garden City, NY: Doubleday Page & Company, 1916); C. N. Williamson and A. M. Williamson, *The Lightning Conductress* (London: Methuen, 1916); for a Dutch translation, see C. N. Williamson and A. M. Williamson, *De kranige chauffeuse: vervolg op (De kranige chauffeur)*, transl. mevr. J. P. Wesselink-Van Rossum (Rijswijk: Blankwaardt & Schoonhoven, n.d.); Alice Williamson, *The Lightning Conductor Comes Back* (London: Chapman & Hall, 1933).
125. C. N. Williamson and A .M. Williamson, *The Princess Passes: A Romance of a Motor-Car* (New York: Henry Holt & Company, 1905; first ed. 1903[?]), 23 (well-being), 35 (understanding), 38–39 (gliding), 40 (full speed), 78 (workshop), 323 (petrol).
126. Williamson, *The Inky Way*, 117, 120 (German tourists), 210–211 (Maeterlinck and Keynes). Charlie also brought into the writing team her husband's "advice as to motor-stuff" (ibid., 365). Alice is quite vague about the causes of Charlie's disappearance from her life, although she refers to his two years of illness at the end of the war that forced her to write her novels during the night. Alone, she finds it hard to be accepted as the real author of their novels (ibid., 364, 366). Whether this is connected with her suicide in 1933, two years after the publication of her autobiography, is not known. Burgess-Wise, "Fiction of the Motor Car," 95.
127. Edward Salisbury Field, *A Six-Cylinder Courtship* (New York: Grosset & Dunlap, 1907), 100 (fellow), 111–112 (italics in original), 126 (child). For an analysis of this novel, see Wolfgang Munzinger, *Das Automobil als heimliche Romanfigur: Das Bild des Autos und der Technik in der nordamerikanischen Literatur von der Jahrhundertwende bis nach dem 2. Weltkrieg* (Hamburg: LIT Verlag, 1997), 73–76
128. http://en.wikipedia.org/wiki/Tom_Swift; Berger, "Popular Culture and Technology in the Twentieth Century," 389; Jane S. Smith, "Plucky Little Ladies and Stout-Hearted Chums: Serial Novels for Girls, 1900–1920," in Jack Salzman (ed.), *Prospects: An Annual of American Cultural Studies, Volume Three* (New York: Burt Franklin & Co., 1977), 155–174, here 155–156, 167, 170; Deborah Clarke, *Driving Women: Fiction and Automobile Culture in Twentieth-Century America* (Baltimore, MD: The Johns Hopkins University Press, 2007), 27; Kathleen Franz, *Tinkering: Consumers Reinvent the Early Automobile* (Philadelphia: University of Pennsylvania Press, 2005), 44–50. See also Deborah Clarke, "Anxiously Popular: Women and the Automobile Culture of the Early 20th Century," paper delivered at the Society of Automotive History conference, Los Angeles (March 2000). For Motor Boys novels on the war, for instance, see Clarence Young, *Ned, Bob and Jerry in the Army: Or, The Motor Boys as Volunteers* (New York: Cupples & Leon Company, 1918); Clarence Young, *Ned, Bob and Jerry on the*

Firing Line: Or, The Motor Boys Fighting for Uncle Sam (n.p., n.d.; reprint Milton Keynes: Dodo Press, 2009); Clarence Young, *Ned, Bob and Jerry Bound for Home: Or, The Motor Boys on the Wrecked Troopship* (New York: Cupples & Leon Company, 1920). Big fight: Edna Brooks, *The Khaki Girls Behind the Lines: Or, Driving With the Ambulance Corps* (New York: Cupples & Leon Company, 1918), 56; Edna Brooks, *The Khaki Girls of the Motor Corps: Or, Finding Their Place in the Big War* (New York: Cupples & Leon Company, 1918).

129. Zorn, *Vernon Lee*, 22. For an analysis of the Motor Boys series and a comparison with the Motor Girls series, see Clay McShane, *Down the Asphalt Path: The Automobile and the American City* (New York: Columbia University Press, 1994), 144–147, 168–171 (liberating: 168). For a similar statement that the car had less to offer to women than to men (and less than the bicycle and the horse), see Sarah Wintle, "Horses, Bikes and Automobiles: New Woman on the Move," in Angelique Richardson and Chris Willis (eds.), *The New Woman in Fiction and Fact: Fin-de-Siècle Feminisms* (Houndsmill and New York: Palgrave MacMillan, in association with Institute for English Studies, School of Advanced Study, University of London, 2001), 66–78, here 78.

130. Molly W. Berger, "Popular Culture and Technology in the Twentieth Century," in Carroll Pursell (ed.), *A Companion to American Technology* (Malden, MA, and Oxford: Blackwell Publishing, 2005), 385–405, here 387, 388–389.

131. http://en.wikipedia.org/wiki/Tom_Swift. For such a Brechtian interruption, see Victor Appleton, *Tom Swift and his Electric Runabout OR The Speediest Car on the Road* (New York: Grosset & Dunlap, 1910), 8. Klaus Benesch "Our Bikes Are Us: Speed, Motorcycles and the American Tradition of a 'Democratic' Technology," *International Journal of Motorcycle Studies* 6, no. 1 (Spring 2010), characterizes the first Tom Swift issue on the motorcycle as expressing "the values of republican America: utilitarianism, self-reliance, antitrust sentiments, democratic technology, a fixation on the nuclear family and, finally, the ideal of the common man."

132. Speed-addict: Benesch "Our Bikes Are Us." Appleton, *Tom Swift and his Electric Runabout*, 5, 13, 19, 23 (tubular structure), 37 (quick recharging), 104 (wonder), 109 (silent), 210 (128 km/h). On Edison's nickel-iron battery as a "wonder battery," see Gijs Mom, "Inventing the Miracle Battery: Thomas Edison and the Electric Vehicle," in *History of Technology* 20 (1998), 17–45.

133. Schmitt, *Back to Nature*, 140. Schmitt analyzes children's fiction in two chapters (125–140).

134. Sammy Kent Brooks, "The Motorcycle in American Culture: From Conception to 1935" (unpublished dissertation, George Washington University, Washington, DC, 1975), 197 (Motorcycle Chums and Tarzan), 203 (*Stormbound* and played no part).

135. 1898: Enda McCaffrey, "*La 628-E8*: La voiture, le progrès et la postmodernité," *Cahiers Octave Mirbeau* 6 (1999), 122–141, here 124; Guillaume Apolinnaire, "Zone, recueil Alcools" http://www.bac-facile.fr/commentaires/1064-guillaume-apollinaire-alcools-zone.html (last accessed 20 November 2009); English translation by Donald Revell, http://www.poets.org/viewmedia.php/prmMID/19454 (last accessed 19 November 2009); the quotation from *Calligrammes* is from the translation of Donald Revell: [Guillaume Apollinaire], *The Self-Dismembered Man: Selected Later Poems of Guillaume Apollinaire*, translated by Donald Revell (Middletown, CT: Wesleyan University Press, 2004), 73. For an analysis of "Zone," see Healey, *The Modernist Traveler*, 82–85.

136. Reginald P. Carr, *Anarchism in France: The Case of Octave Mirbeau* (Manchester: Manchester University Press, 1977), 135; Michel and Nivet, *Octave Mirbeau*, 766-769; Jean Grave: Robert Ziegler, *The Nothing Machine: The Fiction of Octave Mirbeau* (Amsterdam and New York: Rodopi, 2007), 8. According to Michel, *Un moderne*, 283, Mirbeau's first car trip took place in 1902.
137. Ziegler, *The Nothing Machine*, 117, 133. On Mirbeau's violent anarchism, see also Cécile Barraud, "Octave Mirbeau, un 'batteur d'âmes' à l'horizon de la Revue Blanche," *Cahiers Octave Mirbeau* 15 (March 2008), 92-101.
138. Michel, "Maeterlinck et *La 628-E8*," 240, 247; launch: Michel, *Un moderne*, 281.
139. "L'Auto' et les artistes," *Les Annales politiques et littéraires* no. 1274 (24 November 1907), 491-493. Other contributing artists were Romain Coolus, Pierre Loti, Henry Kistemaeckers, Michel Corday, Yvette Guilbert, Michel Provins, Myriam Harry-Perrault, and Édouard Detaille.
140. Hiroya Sakamoto, "La genèse des 'littératures automobiles': Histoire d'une polémique en 1907 et au-delà," *La voix du regard* no. 19 (2006/2007; Dossier: en voiture!), 31-42, here 35 (my translation); the passus quoted does not appear in the English translation (see note 142).
141. On the mixture of reality and imagination in *La 628-E8*, see Pierre Michel, "Introduction," in Octave Mirbeau, *Oeuvre romanesque*, vol. 3 (Paris, Buchet, and Chastel: Pierre Zech/Société Octave Mirbeau, 2001), 260-277, here 271. On Bonnard and his circle of friends around the anarchistic journal *La Revue Blanche*, see also Beatrice Lavarini, "Bonnard und seine Liebe zum Automobil," in Gert Schmidt (ed.), "Automobil und Kultur: Nürnberger SFZ-Kolloquien 1999 und 2000," Schriftenreihe des Sozialwissenschaftlichen Forschungszentrums der Friedrich-Alexander-Universität Erlangen-Nürnberg, vol. 8 (Nürnberg, 2001), 186-202, here 192.
142. [Octave Mirbeau], *Sketches of a Journey: Travels in an early motorcar. From Octave Mirbeau's journal 'La 628-E8.'* With illustrations by Pierre Bonnard (London: Philip Wilson Publishers, in association with Richard Nathanson, 1989); Michel, "Introduction," 272 (quotes from Gide and Larbaud; my translation); Ziegler, *The Nothing Machine*, 174-175 (novel); atomisation: Roy-Reverzy, "*La 628-E8* ou la mort du roman," 263 (italics in original).
143. Conquest: Samuel Lair, *Mirbeau, l'iconoclaste* (Paris: L'Harmattan, 2008), 247; Paul Adam, *Le troupeau de Clarisse* (Paris: Ernest Flammarion, n.d.), 15 (scarabée), 183 (killing). Valery Larbaud, *A.O. Barnabooth: Son journal intime* (Paris: Éditions de la Nouvelle Revue Française, 1932[26]), 247-249. Ziegler, *The Nothing Machine*, 10-11 (dehumanized), 14, 173 (tableaux), 174-175 (hallucination), 182 (fictional narrative); car-author: McCaffrey, "*La 628-E8*: La voiture, le progrès et la postmodernité," 127.
144. Jesuit: Michel, *Un moderne*, 277.
145. Anne-Cécile Pottier-Thoby, "*La 628-E8*, opus futuriste?" *Cahiers Octave Mirbeau* 8 (2001), 106-120, here 117; Mirbeau anticipating Marinetti's worldview: Ziegler, *The Nothing Machine*, 198.
146. Müller, *Gefährliche Fahrten*, 53, 55.
147. "Épilogue: La Mort tient le volant...," in F. T.Marinetti, *La Ville charnelle* (Paris: E. Sansot, 1908), 221-229, here 224, 228, 229.
148. "À Mon Pégase," in ibid., 169-172, here 172 (my translation).
149. Rolf Parr, "Tacho. km/h. Kurve. Körper: Erich Maria Remarques journalistische und kunstliterarische Autofahrten," in Thomas F. Schneider (ed.), *Erich Maria*

Remarque: Leben, Werk und weltweite Wirkung (Osnabrück: Universitätsverlag Rasch, 1998), 69–90, here 84.
150. Thacker, "Traffic, gender, modernism," 177.
151. Gabriele D'Annunzio, *Forse che si forse che no* (Milan: Presso I Fratelli Treves, 1910). My translation from the German (*Vielleicht, vielleicht auch nicht,* 1910) as quoted in Müller, *Gefährliche Fahrten,* 60.
152. Joan Neuberger, *Hooliganism: Crime, Culture, and Power in St. Petersburg, 1900–1914* (Berkeley, Los Angeles, and London: University of California Press, 1993), 146.
153. Lewis H. Siegelbaum, *Cars for Comrades: The Life of the Soviet Automobile* (Itahaca, NY, and London: Cornell University Press, 2008), 177.
154. On *Frühexpressionismus* and the car, see Becker, *Urbanität und Moderne,* 164–222.
155. Peter Demetz, *Worte in Freiheit: Der italienische Futurismus und die deutsche literarische Avantarde (1912–1934); Mit einer ausführlichen Dokumentation* (Munich: R. Piper, 1990), 75.
156. Marie Holzer, "Das Automobbil," *Die Aktion* 2, no. 34 (21 August 1912), columns 1072–1073, here 1072; Hans Flesch von Bruningen, "Der Satan," *Die Aktion* 4, no. 30 (25 July 1914), columns 658–670, here 660; and Flesch von Bruningen, "Totentanz," *Die Aktion* 4, no. 30 (25 July 1914), columns 656–657, here 656 (all English translations are mine).
157. Gottfried Kölwel, "So stand ich vor dem Sterben ... ," *Die Aktion* 4, no. 12 (21 March 1914), columns 258–259; Oskar Kanehl, "Schrei eines Überfahrenen," *Die Aktion* 4, no. 6 (7 February 1914), columns 119–120.
158. Müller, *Gefährliche Fahrten,* 82–83.
159. Catherine Bertho Lavenir, "Autos contre piétons: la guerre est déclarée," *L'Histoire* no. 230 (March 1999), 80–85, here 83.
160. Müller, *Gefährliche Fahrten,* 115; shift: 150, 271.
161. Carl Einstein, *Bebuquin* (ed. Erich Kleinschmidt) (Stuttgart: Philipp Reclam, June 2008); Müller, *Gefährliche Fahrten,* 121, quotation on 118.
162. Harro Segeberg, "Literarische Kino-Ästhetik: Ansichten der Kino-Debatte," in Corinna Müller and Harro Segeberg (eds.), *Die Modellierung des Kinofilms: Zur Geschichte des Kinoprogramms zwischen Kurzfilm und Langfilm 1905/06–1918* (Munich: Fink, 1998), 193–219, here 215–216.
163. Matthias Luserke-Jaqui, "Carl Einstein: *Bebuquin* (1912) als Anti-Prometheus oder Plädoyer für das 'zerschlagene Wort,'" in Matthias Luserke-Jaqui, in cooperation with Monika Lippke (ed.), *Deutschsprachige Romane der klassischen Moderne* (Berlin and New York: Walter de Gruyter, 2008), 110–127, here 115, 112.
164. "Einleitung," in Paul Raabe (ed.), *Ich schneide die Zeit aus: Expressionismus und Politik in Franz Pfemferts "Aktion"* (Munich: Taschenbuch-Verlag, 1964), 7–15, here 14. René Schickele, *Benkal der Frauentröster* (Berlin: Paul Cassirer, n.d.; 'Zweites bis viertes Tausend,' first ed. 1913), 109–111. On Schickele, see Müller, *Gefährliche Fahrten,* 108–113, and Eric Robertson, *Writing Between The Lines: René Schickele, 'Citoyen français, deutscher Dichter' (1883–1940)* (Amsterdam and Atlanta, GA: Rodopi, 1995).
165. Reinhard Bendix, "Tradition and Modernity Reconsidered," *Comparative Studies in Society and History* 9, no. 3 (1967), 292–346, here 343.
166. Julian Smith, "A Runaway Match: The Automobile in the American Film, 1900–1920," *Michigan Quarterly Review* 19/20 (Fall 1980/Winter 1981), 574–587, here 574–575.
167. Smith, "A Runaway Match," 578.

168. James Leo Cahill, "How It Feels to Be Run Over: Early Film Accidents," *Discourse* 30, no. 3 (Fall 2008), 289–316, here 298.
169. Smith, "A Runaway Match," 584 (happy accident); United Kingdom: Cahill, "How It Feels to Be Run Over," 295; on French silent movies with the accident motif, see Maurice Girard, *L'automobile fait son cinema* (n.p. [Paris]: Du May, n.d.), 32–36.
170. Karen Beckman, *Crash: Cinema and the Politics of Speed and Stasis* (Durham and London: Duke University Press, 2010), 29.
171. Cahill, "How It Feels to Be Run Over," 291 (repressed), 292 (mechanics), 293 (Paris); Méliès: Bryony Dixon, *100 Silent Films* (Houndsmill and New York: Palgrave Macmillan, 2011), 229.
172. All quoted in Lieb, *Crash: Der Unfall der Moderne*, 218–219.
173. http://www.images.ch/publications/martini.html (last accessed 15 December 2009).
174. McShane, Down the Asphalt Path, 143–144; Dave Mann and Ron Main, Races, Chases & Crashes: A Complete Guide to Car Movies and Biker Flicks (Osceola, WI: Motorbooks International, 1994), 97 (Oldfield); Dashed to Death: Joseph Anthony Interrante, "A Movable Feast: The Automobile and the Spatial Transformation of American Culture 1890–1940" (PhD thesis, Department of History, Harvard University, Cambridge, MA, June 1983), 251.
175. Müller, *Gefährliche Fahrten*, 160 (phantom rides), 194 (stenogram), 225–227 (Hepworth), 217 (Méliès); Dorit Müller, "Transfers between Media and Mobility: Automobilism, Early Cinema and Literature, 1900–1920," *Transfers* 1, no. 1 (Spring 2011), 53–75; Cecil Hepworth, "How It Feels To Be Run Over," in *The Movies Begin: A Treasure of Early Cinema 1894–1913*, vol. 3 (New York: Kino Video, 2002). Chaplin: Marjorie Perloff, *The Futurist Moment: Avant-Garde, Avant Guerre, and the Language of Rupture* (Chicago and London: The University of Chicago Press, 1986), 14; Mann and Main, *Races, Chases & Crashes*, 6 (Keystone). The Internet Movie Database (IMDB) is the standard source for most of the films mentioned, and many can be seen in part on YouTube. For the Keystone Kops, for instance, see http://www.youtube.com/watch?v=Sx 9Tovevx6w&feature=related (last accessed 27 January 2012).
176. Michael McGerr, *A Fierce Discontent: The Rise and Fall of the Progressive Movement in America, 1870–1920* (Oxford and New York: Oxford University Press, 2003), 259–261.
177. Rick Altman, "From Lecturer's Prop to Industrial Product: The Early History of Travel Films," in Jeffrey Ruoff (ed.), *Virtual Voyages: Cinema and Travel* (Durham and London: Duke University Press, 2006), 61–76, here 65, 67, 73.
178. Jennifer Lynn Peterson, "'The Nation's First Playground': Travel Films and the American West, 1895–1920," in Ruoff (ed.), *Virtual Voyages*, 79–98, here 91.
179. Nancy Koppelman, "One for the Road: Mobility in American Life, 1787–1905" (unpublished doctoral dissertation, Emory University, 1999), 339.
180. McShane, *Down the Asphalt Path*, 141–142, 170 (Mills). The car song, of course, was an international phenomenon. See, for instance, the German hit "*Schorschl, ach kauf mir doch ein Automobil*" (Honey, please buy me an automobile) in which, of course, an accident happens, from which the lovers come away completely unharmed. Hermann Glaser, *Das Automobil: Eine Kulturgeschichte in Bildern* (Munich: Verlag C.H. Beck, 1986), 23.
181. *The Magnificent Ambersons* was made into a movie in 1942, featuring Orson Welles; see André Waardenburg, "Meesterwerk van Welles," *NRC Handelblad* (21 and 22 January 2012), 29. Other novels that look back from the second phase

into the first are Céline's *Death on Credit* (see note 236); Paul Morand's *1900*, in which he describes his fascination for car races; and Oskar Maria Graf's *Das Leben meiner Mutter* (1978, The Life of My Mother), in which a pedestrian's view is given on the "noisy monsters (which) stank after petrol and swirled thick clouds of dust. Their occupants look no less ugly, indeed almost dangerous.... No wonder that we avoided such contraptions and followed them in fear and full of hostility from afar." Quoted in Matthias Bickenbach and Michael Stolzke, "Schrott: Bilder aus der Geschwindigkeitsfabrik; Eine fragmentarische Kulturgeschichte des Autounfalls," http://www.textur.com/schrott/ (last accessed 26 April 2011), 5. Paul Moran, "Teuf-Teuf," in Moran, *1900*, quoted in Birgit Winterberg, *Literatur und Technik: Aspekte des technisch-industriellen Fortschritts im Werk Paul Morands* (Frankfurt, Bern, New York, and Paris: Peter Lang, 1991), 145.

182. Quoted in Brian Ladd, *Autophobia: Love and Hate in the Automotive Age* (Chicago and London: University of Chicago Press, 2008), 71.
183. Munzinger, *Das Automobil als heimliche Romanfigur*, 94 (speed), 96 (grease), 97 (rolled over), 98 (mind), 99 (citizenship and belligerence). I consulted Booth Tarkington, *The Magnificent Ambersons* (1918), http://etext.virginia.edu/etcbin/ toccer-new2?id=TarMagn.sgm&images=images/mode... (last accessed 22 April 2011).
184. Graf, "The Evolution of Willa Cather's Judgment of the Machine and the Machine Age in her Fiction," 123–136 (quote on 133), 8 (*McClure's Magazine*), 42 (superwoman). Willa Cather, "The Bohemian Girl" (1912), in *Willa Cather's Collected Short Fiction, 1892–1912*, 3–41, here 4 (Old Lady), 5 (all other quotes).
185. Quoted in Munzinger, *Das Automobil als heimliche Romanfigur*, 42.
186. Quoted in ibid., 51–57 (*Sister Carrie*: 54; *The Titan*: 55; current: 56).
187. Richard Lingeman, *Theodore Dreiser: An American Journey 1908–1945* (New York: G. P. Putnam's Sons, 1990), 58 (advocate), 70–71 (Renaissance), 86 (underside).
188. Theodore Dreiser, *A Traveler at Forty* (Urbana and Chicago: University of Illinois Press, 2004; first ed. 1913); Theodore Dreiser, *A Hoosier Holiday*, with illustrations by Franklin Booth and an introduction by Douglas Brinkley (Bloomington and Indianapolis: Indiana University Press, 1997; first ed. 1916), 29 (conversation), 34 (beer bottles), 40 (quotation Victorian), 53 (chauffeur), 61, 83–84 (last quote), 95 (technology), 100 (middle class), 117 (quotation motorcycle), 133 (farmers), 154 (quotation sensory response), 155 (camera), 204 (slum; quotation etching); Lingeman, *Theodore Dreiser: An American Journey*, 34 (bohème), 116 (Pathfinder), 165–167 (radical).
189. Arthur Redding, *Raids on Human Consciousness: Writing, Anarchism, and Violence*, Cultural Frames, Framing Culture Series (Columbia: University of South Carolina Press, 1998), 80; Hanno Ehrlicher, *Die Kunst der Zerstörung: Gewaltphantasien und Manifestationspraktiken europäischer Avantgarden* (Berlin: Akademie Verlag, 2001), 53 (Sorel), 133 (virility cult), 135 (Saint-Point).
190. Becker, *Urbanität und Moderne*, 54.
191. Despite their basic similarities, Healey, *The Modernist Traveler*, 109–110, makes a difference between French travel writing on the one hand and British and German on the other, because in the latter countries "sports, dance, and health were all emphasized in nascent cults of the body," while France "seems not to know what to do with the material repository of soul, mind, and intellect." At the same time, Healy acknowledges the limits of such generalizations: "women travelers cannot be generalized as a type" (ibid., 121).

192. Tadié, *Le roman d'aventures*, 5.
193. Healey, *The Modernist Traveler*, 14; Müller, *Gefährliche Fahrten*, 269.
194. Russell Belk, "Possessions and the Extended Self," *Journal of Consumer Research* 15, no. 2 (September 1988), 139–168.
195. Diane Drummond, "The impact of the railway on the lives of women in the nineteenth-century city," in Ralf Roth and Marie-Noëlle Polino (eds.), *The City and the Railway in Europe* (Aldershot: Ashgate, 2003), 237–255, here 255; *impérial*: Jeffrey T. Schnapp, "Crash (Speed as Engine of Individuation)," *Modernism/Modernity* 6, no. 1 (1999), 1–49, here 11.
196. Jörg Beckmann, "Mobility and Safety," *Theory, Culture & Society* 21, no. 4/5 (2004), 81–100, here 85.
197. Desportes, *Paysages en mouvement*, 238.
198. Schnapp, "Crash (Speed as Engine of Individuation)," 11.
199. Arthur Redding, "The Dream Life of Political Violence: Georges Sorel, Emma Goldman, and the Modern Imagination," *Modernism/Modernity* 2, no. 2 (1995), 1–16, here 10; mobility as flight from women in Beckman, *Crash*, 19.
200. T. J. Jackson Lears, *No Place of Grace: Antimodernism and the Transformation of American Culture 1880–1920* (Chicago and London: The University of Chicago Press, 1984), 32.
201. For an analysis of the concepts of freedom and independence in women travelogues, see, for instance, Beth Kraig, "Woman At The Wheel: A History Of Women And the Automobile In America" (unpublished doctoral dissertation, University of Washington, 1987), chapter 7 (239–292).
202. Petra Naumann-Winter, "'Das Radfahren der Damen': Bildbetrachtungen zum Diskurs über Modernisierung und Technisierung um 1900," in Christel Köhle-Hezinger, Martin Scharfe, and Rolf Wilhelm Brednich (eds.), *Männlich. Weiblich. Zur Bedeutung der Kategorie Geschlecht in der Kultur: 31. Kongress der Deutschen Gesellschaft für Volkskunde, Marburg 1997* (New York, Munich, and Berlin: Waxmann, 1999), 430–443, here 435.
203. For the tramway as a predecessor of the flight experience, see Desportes, *Paysages en mouvement*, 247; for the experience of "floating" above the ground during horse dressage, see Rhys Evans and Alexandra Franklin, "Equine Beats: Unique Rhythms (and Floating Harmony) of Horses and Riders," in Tim Edensor (ed.), *Geographies of Rhythm: Nature, Place, Mobilities and Bodies* (Farnham and Burlington: Ashgate, 2010), 173–185, here 175.
204. Hans-Joachim Braun and Elena Ungeheuer, "Formt die Technik die Musik? Montageästhetik zu Beginn des 20. Jahrhunderts," *Berichte zur Wissenschaftsgeschichte* 31, no. 3 (2008), 211–225, here 220.
205. Healey, *The Modernist Traveler*, 114.
206. Baudrillard quoted in Nead, "Paintings, Films and Fast Cars," 253.
207. Wolfgang Reif, "Exotismus im Reisebericht des frühen 20. Jahrhunderts," in Peter J. Brenner (ed.), *Der Reisebericht: Die Entwicklung einer Gattung in der deutschen Literatur* (Frankfurt: Suhrkamp Verlag, 1989), 434–462, here 441–442.
208. David Gartman, *Auto Opium: A Social History of American Automobile Design* (New York and London: Routledge, 1994), 36; Duffy, *The Speed Handbook*, 136.
209. See, for instance, Gijs Mom, "Mobility for Pleasure: A Look at the Underside of Dutch Diffusion Curves (1920–1940)," *TST Revista de Historia: Transportes, Servicios y Telecomunicaciones* no. 12 (June 2007), 30–68.
210. For an analysis of the affinity between production of the car and destruction of the animal body (in abattoirs) and the relation with film, see Nicole Shukin,

Animal Capital: Rendering Life in Biopolitical Times (Minneapolis and London: University of Minnesota Press, 2009), chapter 2 (87–130).
211. According to Carroll Purcell, "American Boys, their Books, and the Romance of Modern Transport," paper presented at a workshop on 'gendered mobility,' University of Wollongong (14–15 December 2010), 11, the male relationship with nature "lies very close to the suggestion of rape."
212. For the relationship between speed and imperialism, see Duffy, *The Speed Handbook*, 40–44.
213. Henri Lefebvre, *Everyday Life in the Modern World* (New Brunswick, NJ, and London: Transaction Books, 1984), 100.
214. Norbert Elias and Eric Dunning, *Quest for Excitement: Sport and Leisure in the Civilizing Process* (Oxford and New York: Basil Blackwell, 1986), 161.
215. Karl E. Scheibe, "Self-Narratives and Adventure," in Theodore R. Sarbin (ed.), *Narrative Psychology: The Storied Nature of Human Conduct* (New York, Westport, CT, and London: Praeger, 1986), 129–151, here 133.
216. Schriber, *Writing Home*, 47, 59.
217. Duffy, *The Speed Handbook*, 35; McCaffrey, "*La 628-E8:* La voiture, le progrès et la postmodernité," 137.
218. Duffy, *The Speed Handbook*, 35.
219. Christophe Studeny, *L'invention de la vitesse: France, XVIIIe–XXe siècle* (n.p. [Paris]: Gallimard, 1995), 291–337.
220. A similar argument is formulated by Paul Virilio, a former physicist: boredom is countered "by precisely the one variation the motor is capable of: *acceleration*"; Virilio, *The Art of the Motor* (Minneapolis and London: University of Minnesota Press, 1998), 88 (italics in original).
221. Schönhammer, *In Bewegung*.
222. Peter Wollen, "The Crowd Roars: Suspense and the Cinema," in Jeremy Millar and Michiel Schwarz (eds.), *Speed: Visions of an Accelerated Age* (London, Guelph, and Amsterdam: The Photographers' Gallery / Trustees of the Whitechapel Art Gallery / Macdonald Stewart Art Centre / Netherlands Design Institute, n.d. [1998]), 77–86, here 77 (high speed), 79 (philobats), 86 (ocnophil). Angstlust: Karl-Heinz Bette, *X-treme: Zur Soziologie des Abenteuer- und Risikosports* (Bielefeld: transcript Verlag, 2004), 19; Williamson and Williamson, *The Princess Passes*, 68.
223. Catharina Landström, "A Gendered Economy of Pleasure: Representations of Cars and Humans in Motoring Magazines," *Science Studies* 19, no. 2 (2006), 31–53, here 33; perilously: Janet Oppenheim, *"Shattered Nerves": Doctors, Patients, and Depression in Victorian England* (New York and Oxford: Oxford University Press, 1991), 141; Adams quoted (in German: *Gesetz der Beschleunigung*) in Radkau, *Das Zeitalter der Nervosität*, 190–191.
224. Leo Marx, *The Pilot and the Passenger: Essays on Literature, Technology, and the Culture in the United States* (New York and Oxford: Oxford University Press, 1988). I thank Kurt Möser for bringing this study to my attention. According to Marx, literature could unveil this covert culture. Radkau, *Das Zeitalter der Nervosität*, 256.
225. Neuberger, *Hooliganism*, 142–152; Brian Ladd, *Autophobia: Love and Hate in the Automotive Age* (Chicago and London: University of Chicago Press, 2008), 177.
226. Bayla Singer, "Automobiles and Femininity," *Research in Philosophy and Technology*, vol. 13: Technology and Feminism (Greenwich, CT: JAI Press, 1993), 31–42, here 35. For an anthropology of "accelerated speed" (arguing that speed

fascination is much older than the bicycle and the car), see also T. Schnapp, "Crash (Speed as Engine of Individuation)."
227. Clay McShane and Joel A. Tarr, *The Horse in the City: Living Machines in the Nineteenth Century* (Baltimore, MD: The Johns Hopkins University Press, 2007); for the exit lane, see photo 2.4 in McShane, *Down the Asphalt Path*.
228. Naumann-Winter, "'Das Radfahren der Damen,'" 430.
229. Price, *The End of the Age of Innocence*, xiv; Ferber quoted by Tichi, *Shifting Gears*, 28; Field, *A Six-Cylinder Courtship*, 8.
230. Cultural historian Raymond Williams used the shell metaphor to characterize the "mobile privatization" of television, thus providing an intermedial bridge between communication and mobility. Williams, *Television* (1989), quoted in Shaun Moores, "Television, Geography and 'Mobile Privatization,'" *European Journal of Communication* 8 (1993), 365–379, here 376.
231. Priscilla Denby, "The Self Discovered: The Car in American Folklore and Literature" (unpublished dissertation, Indiana University, May 1981), 99–102.
232. Schivelbusch quoted in Larsen, "Tourism Mobilities and the Travel Glance," 82–83.
233. Robin Jarvis, *Romantic Writing and Pedestrian Travel* (Houndsmill, London, and New York: MacMillan Press / St. Martin's Press, 1997), 22.
234. Samuel Lair, "La 628-E8, 'le nouveau jouet de Mirbeau,'" *Cahiers Octave Mirbeau* 15 (March 2008), 54–67, here 60.
235. Leblanc, *Voici des ailes*, 68.
236. Not always as an offensive weapon, at least not in retrospective, as is testified by the sour-hilarious way in which Louis-Ferdinand Céline described perhaps the first family outing in a car in Europe in his *Mort à crédit*. "Our adventure," as Céline calls it, took place before the war in a one-cylinder tricycle from uncle Édouard, behind which father cycles to pick up the components that drop from the vehicle. For uncle, "the air was surely beneficial, but the automobile he finds irritating." Louis-Ferdinand Céline, *Mort à crédit: Roman* (Paris: Denoël et Steele, 1936), 80–82 (81: adventure; 82: uncle). For an English translation, see Louis-Ferdinand Céline, *Death on Credit: Translated by Ralph Manheim* (London: John Calder, 1989); the trip is on 71–73.
237. Nead, "Paintings, Films and Fast Cars," 254.
238. Cahill, "How It Feels to Be Run Over," 290.
239. For an analysis of the "mediology of the journey" in France, Italy, and Belgium, see Bertho Lavenir, *La roue et le stylo*.
240. Larsen, "Tourism Mobilities and the Travel Glance."
241. Jürgen Gunia, "Extreme Diskurse: Anmerkungen zur Kritik medialer Beschleunigung bei Günther Anders und Paul Virilio," in Leonhard Fuest and Jörg Löffler (eds.), *Diskurse des Extremen: Über Extremismus und Radikalität in Theorie, Literatur und Medien* (Würzburg: Königshausen & Neumann, 2005), 173–188, here 182, 179.
242. Christoph Hennig quoted in Anne-Katrin Ebert, "Ein Ding der Nation? Das Fahrrad in Deutschland und den Niederlanden, 1880–1940. Eine vergleichende Konsumgeschichte" (doctoral dissertation, Universität Bielefeld, Fakultät für Geschichtswissenschaft, Philosophie und Theologie, January 2009), 239.
243. Frédéric Monneyron and Joël Thomas, *L'Automobile: Un imaginaire contemporain* (Paris: Éditions Imago, 2006), 177.
244. Frédéric Monneyron and Joël Thomas, "Introduction," in Monneyron and Thomas (eds.), *Automobile et Littérature*, 7–11, here 8.

245. On the anthropological roots of my image of the car driver as poet, see Tim Ingold, *The Perception of the Environment: Essays on Livelihood, Dwelling and Skill* (London and New York: Routledge, 2000), chapter 23 ("The Poetics of Tool Use").
246. Charles Grivel, "Travel Writing," in Hans Ulrich Gumbrecht and K. Ludwig Pfeiffer (eds.), *Materialities of Communication* (Stanford, CA: Stanford University Press, 1994), 242–257, here 254.
247. Tim Armstrong, *Modernism, Technology, and the Body: A Cultural Study* (Cambridge: Cambridge University Press, 1998), 7 (reciprocal), 8 (corpus), appendage (79). God: Robert Dixon, *Prosthetic Gods: Travel, Representation and Colonial Governance* (St Lucia QLD: University of Queensland Press in association with the API Network, 2001), 14.
248. Scheibe, "Self-Narratives and Adventure," 130; Simmel, "Das Abenteuer (Zur philosophischen Psychologie)," 27.

CHAPTER 3

Driving on Aggression
The First World War and the Systems Approach to the Car

> "We arrived in Paris
> at the very moment of the mobilization
> we realized my comrade and I
> that the little car had driven us to a new epoch
> and though we were full-grown men already
> we'd just been born."—Guillaume Apollinaire[1]

When, on the night of 31 August 1914, the French avant-garde poet Guillaume Apollinaire entered Paris in a "little car" (that is, of course, if we may assume that the fiction of the poem corresponds with a similar event in Paris) he entered the war. The epithet "little" is significant here: the "new epoch" was one of a new type of car, ready for a new gamut of users. Like the car manufacturers, the poet also experimented with his production methods, as the middle lines of the poem are shaped in the form of a car (seats, chassis, wheels; there is even a suggestion of a road, as if he senses the importance of a systems approach to the car). The silent night ride, these lines disclose, reveals the unreliability of the small car's technology: the trip starts with the loss of three lights, while the line forming the road alerts us that the three passengers had to stop three times "to change flat tires."[2]

This chapter investigates the connection, suggested in Apollinaire's poem, between the use of the unreliable petrol car and its entering into a war. In the historiography of mobility the First World War functions indeed as a caesura in the spread of motorization. Whereas before the war road motorization was largely an affair for the moneyed elite, or so the interpretation goes, the crying urgent military need for large numbers of both passenger cars and trucks and the subsequent spread of driver experience in these vehicles established a largely utilitarian, middle-class user culture as soon as hostilities were over, as if the new users had simply been waiting, money in hand, for the car to become reliable and cheap. Although in many respects older contemporaries were stunned by the changes brought about by postwar society, this chapter emphasizes the continuity, rather than the breach, with both the prewar and the postwar periods arguing that the war's influence, while indeed crucial, was catalytic rather than interruptive, stimulating existing trends rather than creating new ones, such as the relationship between car clubs, the military, and aggression, and the logistics of fleet management within the army as we will explain in subsequent sections in this chapter. Most of the phenomena that made up the mobility spectrum in the 1920s were already observed before the war, but their occurrence was multiplied so greatly that one can speak of a new quality. A first massive, pervasive wave of motorization created huge opportunities for the middle classes and equally huge problems for those responsible for traditional infrastructures such as the railways, as we will see in the following chapters.

Here, though, we will analyze four aspects of this catalytic process, four 'war trophies,' as it were: the breakthrough of the truck; the systemic approach of motorized road mobility expressed in logistics and fleet management; the handling of technical unreliability and the emergence of a service concept; and the new form of motorcycle and car tourism and travel called 'thanatourism.' In other words, our analysis will shift toward a more systemic approach. Whereas, in line with the existing historiography, the war enhanced the emergence of a new type of users that had begun around 1910; it also, surprisingly and less well recognized, reinforced some violent traits of elite automobilism from before 1910. Therefore, we start this chapter with a more detailed investigation of the relationship in general between the military and the automobile, before and during the war, against a background of raging discourses on masculinity, aggression, and violence.

Preparing for War: Clubs, the Military, and Aggression

To establish to the full extent the impact of the military on early automobilism, one needs to look beyond motorization and include the earliest form of mechanically based individual mobility, the bicycle movement. Armies in the Atlantic world gradually recognized the potential of this new vehicle, which was far more than just the simple mechanical replacement of the horse. Apart from the euphoric experience of nature and physical health and hygiene that the bicycle enabled, the function of sport was also to internalize social standards while expressing and channeling individual aggression by means of group discipline. The increasing interest of the German army in the bicycle sport, for instance, ran parallel to a shift of emphasis within the bicycle movement from an internationalist-cosmopolitan to a patriotic-nationalist attitude.[3]

When the automobile gradually started to displace the bicycle as the ultimate sports vehicle, the emphatic presence of military men and the consciousness of serving a national interest by participating in automobile sport were part and parcel of the new movement. For instance, the proceeds of the famous thousand-mile trial in England in 1900, sponsored by newspaper magnate Alfred Harmsworth, were used to help finance the South African War.[4] In the United States, the relationship between early automobilism and the military was also quite common at the individual level. Both at the *Chicago Times Herald* race on 28 November 1895 and during the race organized by *Cosmopolitan Magazine* in New York City on 30 May of the following year, a military man (in the latter case flanked by millionaire John Jacob Astor, among others) chaired the jury.[5]

In a 1901 American overview of early automobile tests by armies in Europe, experiments in Germany, Austria, Italy, France, Russia, Belgium, Switzerland, and Norway were reported.[6] The German army compared steam, gasoline, and electric propulsion, rejecting the latter alternatives because of a lack of speed, distance, and hill-climbing capability. Russia boasted to be the first with its experiments with steam traction engines, used for transport in the Russo-Turkish war in 1870. And France sent a dozen vehicles to its military academies (the propulsion type was not specified). It seems that the Boer War fueled this interest in motorization. Not mentioned in the American overview, the British army set up a Mechanical Transport Committee in 1901, following its successful deployment of steam traction engines during this war, two years before the German army founded a "Self-Propelled Commando" (*Selbstfahrer-Kommando*) within the experimental department of the Communication Troops (*Verkehrstruppen*). France followed in 1905 with a *Commission Militaire des Transports par Automobile*. By that date all European countries had started to take measures indicating the willingness to include motorization into rearmament planning.[7]

In the battle between the propulsion systems, the armies soon favored the gasoline-propulsion alternative, although not without a lot of discussion, remarkably enough mostly in the United States. The reason for this was the lack of reliability of the gasoline option, although there was some positive interest in hybrid (gasoline-electric) propulsion among the military as well. For instance, in 1905 the French Automobile Club organized a heavy truck contest with a separate category for military trucks. The majority had gasoline traction, but Louis Antoine

Kriéger (1868–1951), one of the pioneers of electric automobile traction in France and creator of the most widespread electric vehicle type before WWI in Europe, introduced a hybrid truck equipped with a big searchlight.[8] When the American army bought three cars in 1899 "for the use of officers" it added: "Each is equipped so that a mule may be hitched to it should it refuse to run."[9]

The reluctance, observed by contemporaries and historians alike, of the military authorities to motorize seems to have been less a token of conservatism than a hesitance to invest large sums in a technology that was in a turbulent technological flux, for a war that most military planners expected to be a short and offensive one.[10] Clearly, military expectations were less optimistic than those of the civil elite, despite their close relationship. For the military the question of motorization was part of a much wider 'battle of the systems' between a road-oriented strategy and one based on railways, a battle occurring one to two decades before a similar one raging in civil society, as we will see in chapter 7. One historian has called the First World War "*the* railroad war in history" and another even sees the race to expand the railway network into a system directed at the neighbors' borders as the ultimate cause of the war itself. Indeed, Germany had systematically developed the railroad network from a military-strategic point of view (including the construction of lines for purely military reasons) and by the outbreak of the war a fully functional network with a large capacity was in place in the western part of the country. Since the Austro-Prussian and Franco-Prussian wars of 1866 and 1870 respectively, both governed strategically by mobilization and concentration (the distribution of mobilized men and goods to the fronts), this resulted in a "silent revolution in transport," which trebled the European track length to about 300,000 km, the result of a catch-up by France that seriously challenged Germany's prior "railway superiority" of 1870. But alongside the expansion of the infrastructure and the increasing performance of the rolling stock, the real basis of the close connection between railway development and war was the logistic aspect, where "Germany set the norm" for cooperation between railway and army planners. In general, modern transport logistics as a scientific practice has clear military roots (as well as roots in mathematics), going back to Von Clausewitz in the early nineteenth century and referring to the necessity of mediating between industrial society and the war effort. The first director-general of transport for the British armies was Sir Eric Geddes, who mobilized "an influx of civilian experts [who] imported management tools such as statistical forecasting." And although the deterministic opinion about the railway expansion as the cause for the war is not shared by many historians, its role was crucial:

"Had it not been for the railway construction and planning executed over previous decades Germany could not have contemplated risking a war on two fronts [France and Russia] and France could not have contemplated risking a re-run of the 1870 disaster," while "French railway superiority contributed powerfully to denying Germany victory in a short war."[11]

In this infrastructural stalemate the automobile got its chance. For Germany, it was the lower echelons of the troop officers (*Truppenoffiziere*) who provided the impulse for motorization, but this was not the general European picture. In the Netherlands the automobile club had to be explicitly invited by a military "Autotractie-Commissie" to set up a volunteer member-motorists group after the example of the German, Austrian, and British automobile clubs. The "Vrijwillig Militair Automobiel Korps" (VMAK, Volunteer Military Automobile Corps) founded in 1913 alongside an already existing "Vrijwillig Motorwielrijders Korps" for motorcycle owners consisted exclusively of members of the Dutch automobile club, which earned itself the epithet *Royal* with that move. As one of the many examples of the fundamental continuity between prewar motor sport and car deployment during military actions, the eighty-four largely noble and *haut-bourgeois* VMAK members enjoyed free petrol and tires when driving military officers around during maneuvers.[12]

The army's main interest concerned the heavy-duty transportation vehicle, but it also showed a lively interest in the passenger car. The "Kaiserpreis" (Emperor's prize) contest, for instance, held annually in Germany from 1905, was explicitly organized as a series of "suitability tests for military application" of the car.[13] In Germany, the early integration of military and industrial interests was also expressed by the presence of high-ranking officers in the supervisory boards of many automobile enterprises. Senior military officers were as a matter of course part of the aristocratic and bourgeois ruling elite.

Military interests may well have stimulated an early interest in utilitarianism. In the Netherlands, for instance, the paralysis of national transport by the famous Railway Strike in 1903 (see chapter 1) was immediately countered by an offer from the Automobile Club to provide the army and the police with their cars. Through telegraphic messages to its members, twenty-six cars in Amsterdam and The Hague were immediately available. One of the club members, Peugeot importer Verwey & Lugard, boasted afterward in a brochure that the strike had been "one big reliability trial." And the magazine of the Dutch Touring Club ANWB observed in a somewhat startled tone that opponents of the car had become friends and "even the authorities, who tried to oppress automobilism by harsh laws, have sought their rescue with the automobile."[14] According to the magazine, these authorities praised the fastest cars above all. Even the bourgeois highbrow newspaper *Handelsblad*, which up to then had derided the parvenu character of the new vehicle and come out in favor of the much quieter electric vehicle, was full of praise about "the automobile savior."[15]

It is, in view of the subsequent development of automobilism, remarkable that the flexible, 'anarchistic' automobile was successfully used here against a centrally controlled, but paralyzed railway system, prefiguring a movement on a much larger scale during the interwar years, as we will see in chapter 7. Former opponents now seemed more than willing to pay the price in terms of noise, smell, safety problems, and technical unreliability. For the military this seemingly systemic weakness of the car had to be countered by a rigorous, centralized fleet management, forced upon the "total anarchy" (as one French lieutenant called it) of individually controlled cars.[16] This contrasted with the railways, where management had grown up more or less 'naturally' out of the common use of a linear rail infrastructure and hence could quite easily adapt to military requirements. The importance of this 'systemic enforcement' on a basically 'anarchic' car culture in all Atlantic countries can hardly be overstated.

To bring this about, however, car owners had to cooperate, to volunteer in the co-construction of an explosive cocktail of militarism, aggression, and lust for violence that was spun around the new vehicles by men who equated 'adventure' with 'not yet dead.' We do not need an excursion into colonial travelogues to substantiate this: even on a quiet touring trip through the countryside the military metaphors inspired by the association with a conquest sprang easily to mind, as in the case of the British pioneer Sir Francis Jeune: "To many of us come all the pleasures and excitement of exploration. . . . I believe that the Duke of Wellington used to say that the best general was the man who knew what was on the other side of a hill. We are all of us in that sense qualifying to be generals now."[17] The German cultural historian Kurt Möser has exposed the close relationship between the early 'sporting' automobile culture and the military culture of honor, manliness, disgust for the decadent bourgeois society, and contempt of death, while the Swiss economic historian Christoph Merki has observed preindustrial, chivalrous traits in early car culture. In the previous chapter, we have given several examples of masculine, aggressive fantasies among autopoetic writers.[18]

Adventure, in this context, acquired a deadly face. Peter Gay, in his monumental, Freudian analysis of nineteenth-century "bourgeois experience," approaches aggression as an anthropological constant, a universal trait as much as its "instinctual ally and adversary," sexuality. Although Gay cites the psychoanalyst David Rapaport as saying that "the origin and development of *aggressive drives* is still unresolved," the philosopher Slavoj Žižek, likewise inspired by Freud (and Lacan), locates "the ultimate cause of violence in the fear of the Neighbour." While some analysts define violence quite narrowly as an act *intended* to harm the other materially or physically (such as the Dutch philosopher Achterhuis, who emphatically objects to the easy application of the concept of 'structural' or 'systemic violence'), both Gay and Žižek give the concept a much wider meaning. For Gay, "aggressive acts ... are not all primitive pugilism, wanton cruelty, or routine murder. They range across a broad spectrum of verbal and physical expression, from confident self-advertisement to permissible mayhem, from sly malice to sadistic torture. They emerge as words and gestures—less fatal, to be sure, than physical violence, but little less unmistakable." His definition goes so far as to include nearly all innovative practices: "Solving a tantalizing puzzle, climbing an unclimbable mountain, gaining proficiency in an obscure tongue, inventing a labor-saving device, are all in their way aggressive acts.... I cannot repeat too often that the invention of tools to extract and refine natural resources, the establishment and improvement of rapid communications, the development of techniques easing the exchange of goods and services, the founding of institutions to distribute necessities and, for those better situated, comforts and luxuries, all exemplify the kind of 'attack' on the world in which the nineteenth century was preeminent."[19]

Žižek distinguishes between three types of violence, subjective, objective, and symbolic. While the first can be seen, the other two are "systemic," part of the normal functioning of society, invisible "since (they) sustain ... the very zero-level standard against which we perceive something as subjectively violent." In this he is following a well-known distinction in sociology between "exceptionalist" or "irrationalist" theories of violence, which analyze violence as "a spontaneous or deliberate outbreak or departure from the normative rules," and "structural" or "systemic" theories, which consider violence to be "an integral, vitiating ground of any dynamic system." A recent textbook aimed at "controlling violence" defines aggression as "behavior that violates the personal rights of others, with an emphasis on physical violence or the threat thereof," whereas an expert in the trade of military killing sees the history of military training as a sequence of diminishing an "innate resistance to killing."[20] Literary and cultural theorist Raymond Williams lists as many as seven "senses" of the word "violence," varying from physical assault, the use of physical force in general, to threat, unruly behavior, and even intensity or vehemence, "violence on television," and "to wrench for meaning or significance."[21]

For our purpose, such a violence discourse has to be historicized. Such a historicization would no doubt bring George Sorel to the fore, whose theories of anarcho-syndicalist violence we connected with the 'anarchist' group of French-Belgian writer-motorists in the previous chapter. Sorel, an engineer before he turned to philosophy, saw violence as a creative power (of the proletariat, against the petty-bourgeois and society in general), but was interested in revitalizing "old heroisms" such as thrift, work, and family. His "myth" of the general strike should not so much lead to a communist state, but would have to put the proletariat in "a state of permanent alertness and preparedness," a similar state as intellectuals, after the war, would strife for (see chapter 6), and which we will connect to the skill of car driving.[22]

Peter Gay starts his analysis with a chapter on the German "Mensur" (the student fencing duel preferably resulting in a scar in the face) as an example of the "cultivation of hatred." The crucial concept here (and one reminiscent of Norbert Elias) is "cultivation." The Mensur is "an exercise in aggression checked by accepted rules," and one wonders whether a similar analysis for the beginning of the twentieth century should not instead have chosen the automobile as the instrument capable of cultivating aggression and violence.[23] For the period he studied Gay distinguishes between three "alibis" for aggression: competition, the construction of the Other, and the cult of manliness. Remarkably, he mentions Otto Julius Bierbaum, the (upon closer look, as we saw, not so) 'quiet motorist' we met in the previous chapter, as someone who expressed a "virile contempt for the spineless bourgeoisie," a class that to him was "the embodiment of all that is mediocre," blind to the fact that "danger is the spice of life." Manliness, according to Gay, "was a potentially volatile compound of desperate restraints and ferocious desires barely held in check, the whole ready to burst into flame if combined with other alibis for aggression."[24]

With this catalogue of expressions and forms of aggression and violence it is not very hard to see the fundamental historicity and the limits of the declared universality of these terms. The field of men's studies has unearthed a "plurality of masculinities" in modernity (including definitions relevant to nonwhites and queers), but even the "hegemonic manliness concept" developed within this discipline has to concede that the masculinities of the turn of the century were not only heavily militarized, but also depended to a large extent on a rediscovery of the body.[25] The quintessential 'man' in this period was Theodore Roosevelt, who mobilized all three of Gay's alibis to develop, from a sickly boy, into an advocate of the "strenuous life," an obsessive killer of animals and a "rough rider" in the Spanish-American War and, during his presidency from 1901–1909, he proved to be a racist and aggressive imperialist.[26] Some propose to distinguish between an American and a European man, the latter also shaped by a collective experience of manliness production, in the "homosocial" ("the seeking, enjoyment and/or preference for the company of the same sex") group with "sacrifice" as a central category to indicate the price to pay for "obeying the norms of militarised masculinity."[27]

A multitude of studies has fine-tuned the connection between this type of turn-of-the-century manliness and sports, sexuality, and violence. These studies are important because they stress the collectivity of the masculine experience, male bonding in games and sports, and remarkable anti-commercialism (the emphasis on honor, and amateurism), anti-individualism (the readiness to sacrifice oneself for the group), and anti-urban sentiments (in a rediscovery of 'nature') they generated. The effect was to defeminize masculine culture: "Aggression was not just a rupture but a relationship, serving men who could not express mutual affection more openly and vulnerably." For such men, the British under-secretary for war opined in 1896, "the terrible things in war are not always terrible; the nauseating things do not always sicken. On the contrary, it is even these which sometimes lifts the soul to heights from which they become invisible." He found that for many men "danger is a strong wine" and battle "the equivalent of motherhood for women."[28] During the following world war, those that could not cope with the "shell shock" were diagnosed as suffering from "male hysteria," a diagnosis "with a powerful negative stigma and enduring feminine associations." The giant masculine community at the front was the healthy society, so shell-shock sufferers had to be kept away from the (feminized) home front.[29] But on the home front, too, chivalrous attitudes, as a typical revival of old and persistent imagery within a modern setting, were propagated and popularized, for instance among the British Boy Scouts, who could "kill Germans by proxy" by freeing soldiers from the Coast Guard for the fighting line. The search for something to replace the horseman in war may well have stimulated motorists to volunteer for supporting roles and the military to seriously consider the deployment of the automobile, once, by the end of the first decade, it seemed reliable enough to stand on its own. Motorists could now fantasize about being knight-errants instead of urban flâneurs.[30] And they did so, in a massive way.

Preparing for War: Organizing Mobility

In 1907 the German army institutionalized the *Selbstfahrer-Kommando*, renaming it the *Kraftfahr-Abteilung* (Motor Vehicle Department). That happened in the same year that the concept of a military truck (*"Armeelastzug,"* with trailer) was anticipated. A year later an effort to enforced motorization was started following a subsidy scheme. The production numbers of these *"Subventionslastwagen,"* standardized 9-ton trucks with a maximum net mass of 4.5 tons, were not spectacular, however. In 1909 the scheme covered 2004 vehicles, not even 9 percent of the German civilian motorized fleet, and until the end of March 1914 only 917 trucks in total were subsidized. But their symbolic value was of much greater significance, as it entailed a new recognition of the (gasoline) automobile as a heavy-duty utility vehicle. The scheme, granting a subsidy if the vehicle met certain standards and if the owner was prepared to hand it over to the army in case of war, assumed a life span of five years. The AEG subsidiary and automobile manufacturer NAG supplied the largest share of the annual numbers manufactured, but its competitor Namag also concentrated on the military application of its products at an early stage: during the First World War, the Namag subsidiary Hansa-Lloyd grew into a large supplier of the German army by producing two hundred (gasoline) trucks a month. Ironically, both producers were among the major German electric vehicle manufacturers before the war. Because most of the trucks were deployed by breweries, a military history of early army motorization concludes that the army helped motorize the German brewing sector. In 1911 the *Kraftfahr-Abteilung* was made into a true *Kraftfahr-Bataillon*. France followed Germany with a subsidy scheme in 1910, Great Britain in 1912.[31]

Around 1908, and only when interest among European armies in the automobile had visibly increased, the US Army started a long-lasting internal discussion about the possible benefits of motorization. Like its European counterparts, the US Army had also started experiments at an early stage. Here the preference was clearly in favor of electric traction, as the first three vehicles purchased by the Signal Corps in 1899 were electric. But, as in Europe, they were rejected for military purposes after extensive testing in 1900 and 1901 because of the difficulty of recharging in the field. Later tests on steam and gasoline automobiles led to the conclusion that the latter were superior because of their lower fuel and water consumption, but steam propulsion was not rejected for trucks. Although the higher echelons of the War Department did not show much interest in motorization, some individual branches (such as the Corps of Engineers and the Ordnance Department's Board of Ordnance and Fortifications) continued experiments during 1903 and 1904 on a limited scale. And although the relatively small American army had been flooded with offers from the industry to construct cars and trucks for military use since 1896, all attempts in this direction failed during the whole of the first decade of the twentieth century and no subsidy program comparable to those in Europe was ever developed. Nor was the plan to establish a Volunteer Motor Corps ever realized.[32]

Part of the explanation for this can be found in the policy shift in 1905 to use a standard automobile design instead of specialized commercial products, a policy that lasted until 1916. By July 1907, the Quartermaster Department had bought and put into service about twenty motorized vehicles (sixteen passenger cars, two ambulances and two trucks—the latter two both steam propelled), but the first extensive report on this topic clearly explained the hesitancy of the American army: the automobile was only applicable for transport of persons and goods "over city streets or well-kept roads," and thus did not recommend itself to replace "any of the present means of Army transportation." At the end of 1908 the Quartermaster Department and the Signal Corps owned thirty of the thirty-two automobiles in the Army, most of them passenger cars.[33]

In 1909 there was an increase in military attaché's reports received on the use of automobiles in European armies. At the same time, automobile technology had improved to such a degree that both passenger cars and trucks were deemed fit to negotiate road conditions much harsher in the United States compared with Europe. It seems that shortly after 1908 the confidence in at least a minimal reliability had spread far enough among the specialists within the American army to have them seriously consider deploying motor vehicles on a more than experimental scale. This is consistent with the shift toward a new type of user around this year, as we have seen in chapter 1. But the vehicles were used for supply functions only, whereas in Europe the transportation of high-ranking officers was also implemented.[34]

From 1912 onward, the Quartermaster Department seriously and systematically started to study and institute post and field motorization, for instance by instigating a cross-country test with four trucks in that year. But as late as 1915, only fifty vehicles were to be found in the Quartermaster Depots.[35] Norman Cary, who performed the most detailed study of the US Army's motorization at the level of the artifacts, explains this by a "misreading [by Army authorities] of the European practical experience." Pointing at the still despicable roads in the American countryside (and even many towns) an analysis from as late as 1916 concluded that "notwithstanding these fine roads in France, the combat and field train of combat units in their entirety, as well as a large portion of their corps train (our division trains) remain animal-drawn. There seems to be no doubt that when we consider the road conditions in our possible theater of operations we will not be able to change to the motor truck until a much later date than the European Army." Warfare in Europe had by then already immobilized itself into a static trench war, whereas the probability of Mexico (rather than Germany) being the future American enemy would necessitate a very mobile campaign. The European experience, Cary argues, "confirmed the view of the more conservative in American military circles, most particularly that the mule and the horse would always remain the sinews of Army supply."[36]

Mobilization, Immobility, Remobilization: Aggression, Violence, and Atrocities

Meanwhile, on 4 August 1914, the German army invaded neutral Belgium after a mobilization by the railways that ran like clockwork. It all started with a pedestrian killing an archduke in a car in Sarajevo. This mobilization was the prelude to an orgy of violence and mobility, a destructive combination that soon took an entire continent in its grip. At the same time, it was an intermedial event never witnessed before, in terms of trains deployed, telephones and telegrams used to support a "precipitous diplomacy" just before the war, and cars. Large numbers among national populations were elated by "the deliverance from pre-war sterility," "a liberation from vulgarity, restrictions and conventions," and a "vacation from life." We have heard this all before, in the journals of the auto clubs.[37]

While the French railways deployed seven thousand trains to transport the soldiers to the 760-km-long front (at the junction of Troyes one could see a train pass every four minutes), the German railways used eleven thousand to bring three million men and one million horses into Belgium. With a speed of 20 to 30 km/h and with jolly panels announcing "To Berlin" on the French or "Excursion to Paris" on the German trains, the rush to the front released anxieties, expectations and the hopes for a "liberation of Europe" that had been building up over the course of at least a decade, if not more: similar accounts abound of the Franco-Prussian war.[38]

Long arrays of quotations have been produced to convey this enthusiasm, from generals, to well-known national writers, to the cheers in the street. For all of them, not to be repeated, the Reverend Bishop of London A. F. Winnington-Ingram's speech can stand: "Kill Germans! Kill them, not for the sake of killing, but to save the world. . . . Kill the good as well as the bad ... kill the young men as well as the old ... kill those who have shown kindness to our wounded as well as those fiends who crucified the Canadian sergeant. ... As I have said a thousand times, I look upon it as a war of purity, I look upon everybody who dies in it as a martyr."[39] Recent scholarship has emphasized that the enthusiasm to go to war was not universal, however.[40] In Italy, for instance, businessmen, workers, and peasants alike were hostile to the war, but the upper classes, and especially the intellectuals, were thrilled, most spectacularly of all the Futurists, supported by students, large portions of the urban middle classes, and a part of the industrial elite. The "warrior-poet" Gabriele D'Annunzio, macho, womanizer, and "welcomed [by his followers] almost as a second Messiah," started a campaign for war, supported by the "agitator for war" Marinetti, both eager to realize their "Ode to Violence" (referring to the title of a 1912 poem by Enrico Cardile from 1912). Most avant-garde artists in Britain (the "Vorticist" Ezra Pound), Germany (the Expressionist Johannes R. Becher), France (Apollinaire), and Russia (the "Suprematist" Malevich), however, did not follow the Italian example. In the United Kingdom, it seems that especially the lower ranks of the white collars and the higher echelons of the workers were enthusiastic about the war. During the first months of the war millions of volunteers were registered as soldiers on both sides (two million in the United Kingdom alone) and several scholars have observed that this willingness to sacrifice one's own life for the good of the cause did not disappear once the expected *Blitzkrieg* had been frozen into a *Sitzkieg*, even when from 22 April 1915 German gas attacks (an example followed in September by the British army) were added to the futile strategies to remobilize the hostilities. In the United States, too, a majority of the population supported the decision to go to war. In Russia only the Bolsheviks were opposed. In the end, 12 percent of British, 15 percent of German, and 16 percent of French soldiers died (about four million), while total casualties, including civilians, amounted to fifteen million, with twenty million injured (also victimized by the war were half a million dead horses). Smaller, weaker countries lost a higher ratio of their soldiers: Serbia one-third, Turkey and Romania a quarter, and Bulgaria one-fifth. In early 1917 a total of fifteen million men were mobilized, a gigantic male 'society' with an "atmosphere of brotherhood" ready to kill and to indulge in a feeling of community that could for a while erase former class distinctions and diffuse social and cultural values into new "national cultures." During the Battle of the Somme, the British general Douglas Haig had his engineers construct a dense telephone network, roads to the British trenches, field hospitals and an entertainment infrastructure for sports, film, pubs and "other amusement."

Keeping this society functioning cost one-third of the UK national budget; Germany spent half of its budget to keep its own 'society' running. "Even if we perish as a result of this," the future chief of staff Erich von Falkenhayn said to Chancellor Theobald von Bethmann Hollweg, "it was beautiful." Likewise, the American Expeditionary Forces organized tennis tournaments, horse shows, and golf contests, and bought 10,000 baseballs, 2000 footballs, 1800 soccer balls, 1500 basketballs, and 600 sets of boxing gloves in January 1919 in Paris, all to prevent its soldiers to "go Bolshevist," as John J. Pershing and his generals feared.[41]

Hardly had they arrived on foreign territory (during the entire war the armed conflict took place *outside* Germany) German troops, on foot, bike, and horseback, started the "August massacres," a week-long frenzy of mass killing (mostly in the towns of Aarschot, Andenne, Tamines, Dinant, and Leuven), rural population displacement into "concentration camps," and infrastructure destruction (25,000 homes and other buildings destroyed in 837 communities). All this was carried out under the pretext that snipers had been shooting at the invading army. By the end of that month 6000 Belgian civilians had been killed, "the equivalent of about 230,000 Americans today," as an angry American scholar asserts, reminding us of the largely forgotten atrocities from the beginning of the war. Twenty percent of the Belgian population—1.5 million people—fled the country, a million to neutral neighboring Netherlands, where they were interned in camps.[42]

This all-too-brief account of only one episode in a war lasting four years illustrates the mobility frenzy of an entire continent, a dynamism in which the automobile played a modest role, but one which was invested with a great deal of symbolism. 'Modern warfare,' with its typical mix of high-tech violence against the entirety of the enemy's culture and civilians on the one side, and preindustrial, 'mediaeval' cruelty on the other, had already been rehearsed during the first Balkan War in 1912, and after August 1914 was not only visible on the Western front: the German and Austro-Hungarian atrocities against the Italian population, the 1915 Armenian genocide by the Turkish army, and the "explosion of barbaric joy" when the special Italian assault units (*arditi*, with their daggers, mostly recruited from the countryside) attacked the Austrian-Hungarian population show that the mobilization of entire nations into the war was accompanied by a wish for national purity and a hatred of the 'Foreigner.' In Germany, however, this clash of the new upon the old was especially fierce. Several observers have emphasized that German cruelty deployed violence "less dampened" than elsewhere. The First World War, the first "total war," has recently been described as a cultural war. Not only did the mobilization of even the remotest corners of every nation lead to a rapid diffusion of 'modern' attitudes and behavior, but contemporaries also spoke of a widespread euphoric feeling of 'community,' of the sharing of a common goal among peoples that before had been each other's adversaries. The flipside was an increased fear of spies and strangers, and of the 'Foreign' in general, not the best basis for restrained behavior when entire armies traveled abroad, as was the case with the German army. "Germany needed war in order to purge society, especially the lower orders, of their addiction to personal possessions, pleasure-seeking, and their demand for rights," was the broad feeling among dominant portions of the ruling classes. "British militarists argued in very similar terms that modern urban society was decadent, and that military service would be healthy for the body politic. ... In other words, militarization of society for war [was necessary] in order to turn the clock back, restore traditional gender roles, and reverse urbanization and modernization." Ironically, this was to be accomplished by high-tech means, the telegraph, aerial reconnaissance by airplanes, and the massive deployment of the automobile. In other words, the emergence of the car cannot be readily understood without a reference to this ambiguity between novelty and nostalgia, new and old.[43]

The "waiting for the war" was an international phenomenon.[44] At the outbreak of the war, Germany (where 1,851 trucks had been manufactured compared to 12,400 passenger cars in 1913) appeared to be at a considerable material disadvantage in comparison to the Allies. During the German offensive in Belgium and France, a few hundred trucks were deployed, but vehicle-fleet management was badly organized and 60 percent of the trucks had broken down before the start of the battle of the Marne, the largest battle since Waterloo (1815). The catalyzing effect of the mechanized war on the stock of army vehicles was also true for Germany, however. Whereas the German army had at their disposal 9738 trucks and 4000 staff cars, ambulances, and motorcycles at the outbreak of the world war, by the end of the war 25,000 trucks, 12,000 staff cars, 3200 ambulances, and 5400 motorcycles were registered as German army vehicles.[45]

Initially, the commanding staff of the other armies also thought that railroads and horse traction would form the backbone of the warfare.[46] The British army, for example, sent 60,000 horses and only 1200 trucks to the front in August 1914. This number was quickly increased, at first by impounding city buses (the famous London double-deckers, sometimes used with their original destinations still on display at the front), passenger cars, and extra trucks, but soon by increasing national production. By the end of the war there were 32,000 British trucks in France, of which 800 were steam powered. The French had exactly 220 vehicles when hostilities started: 50 tractors, 91 trucks, 31 ambulances, and 38 passenger cars, one-third of the latter equipped with a machine gun. At the end of August 1914 requisition and confiscation had brought in an additional 6000 vehicles, but at the end of the war the French army possessed nearly 100,000 motorized vehicles, including 54,000 trucks (Berliet, Renault, Peugeot), while Italy supplied almost 46,000, most of them manufactured by Fiat. In the first half of 1915 alone, England and France bought 26,000 automobiles in the United States. When the United States entered the war on 6 April 1917, they had to buy large numbers of cars and trucks from European manufacturers as well, just as they used other European—mainly French—equipment, such as airplanes, cannons, and even tanks. The latter, the ultimate hope of all those military planners who were still dreaming of an offensive war, were deployed by the British during the final phase of the battle of the Somme (15 September 1916). It was only in the fall of 1917 that the potential of a massive tank deployment to get an immobilized war moving again became visible during the "Tank Battle of Cambrai," with 476 tanks, 216 in the front line. More than half of them were hit and broke down during the first day, but a recent study suggests that it was the tanks that "loomed large in German military justifications for the decision to seek an armistice," even if their actual contribution to the Allied war effort was "useful rather than essential." Tellingly, the first prototypes were called "Mother," equipped with six machine guns, while the later "male versions" had two rapid-fire guns.[47] Historian Modris Eksteins sees the tanks, "the only invention by the allies of any importance to the trench war," as a failure because of the British war planners' chivalrous and Victorian reluctance to use the element of surprise; instead they tested the new weapon in a series of smaller battles.[48]

A count undertaken by the Japanese (!) government at the moment of the American entry into the war showed that the Allied powers had 170,000 motorized vehicles at their disposal and Germany and its allies had 130,000.[49] Yet, one cannot maintain that horse substitution had been completed by the motorization of the nations at war. On the contrary: over the course of the entire war, Great Britain shipped more tons of animal fodder to the front than of both ammunition and soldiers' rations. As so often, innovation does not follow the path of substitution but of accumulation of new upon old, where new incites the old to catch up and to counter a threatening replacement.

War Trophies: The Truck, Logistics, and Maintenance

From the viewpoint of the individual motorist, it is understandable that on the basis of such largely quantitative information at the level of the artifacts, transport historiography has concluded that the war and its preparation in the Western industrialized world did not have much impact on the automobile. But seen from a different perspective, that of the automobile system, this judgment has to be corrected. Whether based on horse traction or on the high-tech planes, trucks, passenger cars, and tanks, the irony (if not the tragedy) of this war from a mobility-historical perspective was its participants' belief in speed as the ultimate strategy for lethal warfare, whereas within a couple of months a three-year stagnant trench war had been established, one which was certainly no less lethal. Alfred von Schlieffen's offensive plan to quickly surprise the French and then run to the east to defeat Russia was fully based on this expectation of the lethal character of speed.[50]

Another illustration of the continuity between prewar experiences and the war itself can be found in the punitive expedition from May to August 1916 against Pancho Villa in Mexico, which, in the words of the military historian Marc Blackburn, was "a laboratory to test the capabilities of motor trucks in an operational setting." When access to the Mexican Northwestern Railroad was denied by President Venustiano Carranza, the Quartermaster Corps purchased 55 motor trucks from the White Motor Company and the Jeffery Company, both steam propelled and both winners of the cross-country test in 1912. By June 1916 campaign leader general John J. Pershing (who would go on to play a leading role in the American presence at the European theater of war only a year later) had bought a total of 588 motor trucks (including 67 specialized vehicles), 75 passenger cars, and 61 motorcycles. But the inexperience of men and officers with the new vehicles and the fact that the vehicles were much less reliable than expected made a special maintenance depot necessary.[51] For the military, the main result of this campaign was the reevaluation of the need for a standard motor truck, given the fact that the trucks had been produced by thirteen different manufacturers.[52] With hindsight, however, the recognition that operating a fleet of gasoline vehicles was not possible without a vast infrastructure of maintenance and service was no less important.

A second demonstration of the crucial importance of the systems approach took place at the Western front, shortly before the United States entered into the war. The logistical triumphs based on the automobile and set up around the so-called race to the sea in the fall of 1914, and the battles of Verdun and the river Somme in 1916 can be analyzed as an answer to all that went wrong during the famous campaign of the "taxis de la Marne" on 6–8 September 1914. In the latter case especially, the role of the media was much more important than the vehicles' actual impact on the course of events. The taxis became famous because of the seemingly improvised transport that the military governor of Paris, General Joseph Gallieni, had to organize for fresh troops to hold the German advance toward Paris.[53] In reality, all vehicles in Paris, horse drawn or motor propelled, had been requisitioned a week earlier, with the exception of some very specific types, such as those for milk delivery. At the end of August, the 10,000 automobiles requisitioned from across the nation (including more than 1000 motor buses, 2500 passenger cars, and 6000 trucks) were used as a reservoir for a massive retreat from and an evacuation of the city of Reims, mostly by motor bus (most of the other buses, by the way, some 700, were used to transport fresh meat). Of the Parisian vehicles only 250 were kept as "ministerial reserve," while another 200 were sent to Bordeaux along with the government. This operation ended in chaos, as the military drivers were not prepared to be disciplined into convoys because of a lack of officers. The 10,000 Parisian taxis were mostly out of order because their drivers had been called up. Of the 3000 still in service, driven by chauffeurs aged between 55 and 65, 150 were kept as a "permanent reserve" in three large private taxicab garages, but 1200 cabs were needed for the transport of parts for the infantry of the sixty-second division—six thousand men, who had disembarked at a railway station near Paris after a four-day trip from the eastern border. The subsequent "hunt for taxis" on the streets of Paris (often by *gendarmes* who left the passengers to walk to their destination) and their collection in the village of Gagny was as hilarious as any prewar automobile event: the taxi drivers, once requisitioned and promised their normal reimbursement for the distance traveled, started to race against each other on the wide Parisian boulevards having first lowered the metallic flags signaling that they were occupied. At the collecting area, they created large jams; meanwhile, the local military had not received any orders. Another group did not manage to leave Paris because the "*porte*" was closed and the gatekeepers refused to let them through. Orders, given by telephone, were so vague that further chaos ensued. After the six thousand soldiers, in groups of three to five men with their weapons, were delivered at their destination 50 km away, the drivers were reimbursed in Paris, some of them showing 120 to 200 km on their taximeters. Cars that were overturned or had to stop because of broken wheels or other major problems were simply left behind and it took days to recover them. Some drivers only came back several days later, "their wives worried," because they were "seduced by the adventure and the curiosity which is so strong among Parisians, . . . and had offered their help at the organization of the frontlines." Others transported the wounded back to Paris. Later that month, less well-known convoys of 50 to 150 taxis were deployed to "feed" other battles.[54]

All in all, five battalions of 800 men each were thus transported over a distance of about 50 km, while the bulk of the 150,000 men of the sixth army had been brought there earlier by train. In his memoirs published a decade later, Gallieni conceded that "one has exaggerated somewhat the importance of the taxis, but in the end it was a good idea, although very simple."[55] And although a recent study, again, called the event "militarily … insignificant," its symbolic role was all the more important, as so many events in the early history of motoring. Many of the cabs were of the Renault Type AG 1 owned by the "G7" company; developed shortly after 1905, they heralded the beginning of the end of the electric taxicab. Before starting its career as a taxicab, this Renault car was subjected to several months of testing, leading to considerable structural changes in the car's design characterized by *"une interchangeabilité absolue"* (absolute interchangeability) and a simplification of its handling characteristics.[56]

Something that had a much greater impact on the course of the hostilities, however, was the subsequent "race to the sea" (*course à la mer*), a largely forgotten effort by the opposing armies to break through the enemy lines and stretch the front right up to the Belgian Channel beaches over the course of a month. This 'race' started on 14 September 1914 with the battle of the Aisne River and included the notorious battle of the Yser (16–31 October) and the first battle of Ypres (21 October to 11 November) with *"lieux de mémoire"* (places of remembrance) such as Passchendaele and Dixmuide. Meanwhile, Antwerp had fallen (on 10 October). From a French perspective, this sequence of battles is also a better candidate for attention than the counteroffensive at the Marne in September, because the latter was on the initiative of an individual (Governor Gallieni), whereas the former was fully organized and executed by the *Direction des Services Automobiles* (DSA) under the directorship of Aimé Doumenc (as a "road commissioner" or a "director of transport"). Between September and November 1914, DSA transported 350,000 men in automobiles through the northwestern part of France and the southwestern part of Belgium where no railways existed.[57]

Two years later, the logistical preparations for the counterattack at Verdun were similar feats of 'transportation engineering.' This was something that had been practiced before the war on a much smaller scale by the managers of (mostly electric) taxicab fleets, but which at this time and on this scale could only have been realized by the military. The narrow-gauge railroad, "Le Petit Meusien," between Bar-le-Duc and Verdun only allowed a flow of goods of 800 tons a day, ten times less than what was needed. Faced with this infrastructural bottleneck (a "reverse salient" in the literally Hughesian sense[58]) the French army, under the leadership of a true "system builder," Captain Aimé Doumenc, decided to use the departmental road from Bar-le-Duc. This road was 7 m wide and 75 km long, muddy in the fall, icy and slippery in the winter. For ten months, 24 hours a day, between 9000 and 11,500 vehicles, including 3500 trucks, 200 buses, 500 tractors, 800 ambulances, 2000 passenger cars, and an unknown number of motorcycles, drove to and fro, one every fourteen seconds, transporting 90,000 men and 50,000 tons of material per week, at an average speed of 15–20 km/h. At certain times of the days the interval between vehicles was as little as five seconds. On 21 February 1916, when the German army started its attack at 4:45 PM, the road was swept clean in four hours by a special commission (a "*commission régulatrice*") housed in Bar-le-Duc (and commanding 300 officers and 8500 men) and controlling an area in which "every isolated vehicle, every displacement, every embarkement, every convoy [was] subsumed under one single authority." In total, 2.4 million men and 2 million tons of freight were transported during eleven months, the last six under the certainty that the French army dominated the battle scene. To counteract the rapid deterioration of the road bed, 700,000 tons of limestone were spread on this "*Voie Sacrée*" (Sacred Road) and 1200 workers (some of them German prisoners of war) shoveled stones under the trucks' massive rubber tires so that these subsequently crushed them, thus paving the road while using it, but destroying the tires in the process.[59]

It was forbidden to overtake or to stop; defective vehicles were driven to one side and repaired by groups of mechanics or simply shoved into the ditch. Later, another logistical innovation was introduced (the "*créneaux*"): vehicles were ordered to stay together in groups of eight or ten, leaving fifty meters between them so that passenger cars and ambulances could overtake and a flow of two speeds could be maintained in one direction. It was the conveyor belt of the mechanized war: at the end of February 1916, Verdun saw gasoline traction in the shape of the truck celebrate its victory. This happened not so much because of a sudden new reliability in the cars and trucks themselves, but because of the centrally regulated traffic flow, the dream of traffic engineers long before they were in a position to develop it in a civilian setting. This central control simply had to be managed more rigorously to compensate for the artifact's unreliability—regulation worked by streamlining the flow's organization and by constantly maintaining its underlying infrastructure. The auto service did for the war what the mechanic did for the car: compensate logistically for what the technology proper could not achieve. In other words, technology, in its development, may be endowed with a certain relative autonomy, but in the end there will always be a social and organizational solution to every technical bottleneck that occurs. In Verdun, it was the role of the road infrastructure that appeared to be the most crucial: the German army in its planning had refrained from such maneuvers as they thought that the roadbed would never sustain such a constant destructive load. They were right, but technology is more than functionality, as we now know. It is also important to be aware of the fact that the motto "*Dégager la route*" (free the road) was maintained in a rigorous and wasteful way: defective cars were shoved aside. In this sense, the constant reiteration in postwar military analyses of the "efficient" character of the automobile deployment has a certain "military," non-economical, and romantic tint to it. According to some contemporaries, especially those responsible for the much more impressive railroad logistics (or historians writing in their tradition), the Sacred Road "was really nothing more than an improvisation which should not have been necessary" if the military authorities had not neglected "to provide Verdun with an alternative railway when the existing line came within range of German shellfire." But the authorities did not, and their neglect is no coincidence: the mazes of the railroad network, some authorities started to realize, could much easier be filled by the 'flexible' motorcycle, car, and truck, thus prefiguring the coordination crisis we will be dealing with in chapter 7. So, the counterattack at the Somme was, according to a high-ranking French officer, "the triumph of method and order," of "order and speed," enough to change public opinion which "at first was quite unfavorable towards automobilists." Part of this order was the rigorous separation of the truck flow from all other vehicles, which made some historians see in the *Voie Sacrée* the first freeway in history. Other contemporaries (among them General Pétain, who took over command at the end of the battle and unjustly claimed to have been the initiator of this feat) called the operation "*la Noria.*" This referred to a water wheel as a metaphor for the "carousel" or "endless chain" of a constantly rotating reservoir of men and ammunition. One of the military histories of the battles compared the war machine with the flow of blood in the veins of an organism, a metaphor that after the war became very powerful in the hands of the fledgling profession of traffic engineering. In fact, there were several similar events during the war, such as the road between Amiens and Bray-sur-Somme. The *Voie Sacrée* was also a media construction: during almost a year of its existence, press trips were organized, resulting in newspaper stories that not only allowed motorists to "regain the favor of the population," but also depicted "the soldiers of the automobile" as real heroes. Just like so many other elements of the Great War, such as the "futility myth" (that the entire carnage was in vain) and the "great casualty myth" (that the onslaught was quantitatively extraordinary), the French logistic genius soon turned into a myth as well. High-ranking German officers studied the organization of the Sacred Road after the war "to prepare for the next war."[60]

Founded just before the outbreak of the war following a preparatory period lasting nearly a decade (in which detailed plans were developed, and even, in 1913 and 1914, longevity tests performed following the example of similar British tests), the *Direction des Services Automobiles* DSA started its remarkable career with the requisition of a part of the French passenger car fleet (then about 80,000 vehicles strong). To this purpose it set up a number of local assessment commissions and two "*Grands Parcs Automobile de Réserve*," the main camp in Vincennes near Paris, the "annex" near Lyon. There, nearly 14,000 cars and trucks were assembled and repaired during the first month of the war.[61]

The emergence of the systems approach in France can quite accurately be pinpointed: it occurred at the end of 1915 and the beginning of 1916, the period that saw the emergence of a rigorous centralized structure, which was however dependent on "the initiative of every one of its members." This dual, Foucaultian structure of rigorous centralism together with voluntary cooperation between the centralized was cleverly expressed in the "*insignes*," the markings of each unit that were artistically painted on the vehicles, creating a personal adherence to the small group rather than the control center.[62] It is this 'traffic pattern' of the (externally, and as a response also internally) 'controlled swarm' that we see reappear after the war in the minds of traffic engineers, when they start racking their brains over how to solve the anarchism and lethality of automobile traffic. Nowadays, military analysts study this kind of movement in the hopes of developing a strategy of 'swarming' against guerilla and terrorist threats.[63]

After two years the DSA commanded 13,000 passenger cars and nearly 31,000 trucks distributed across 846 "sections," in which 1500 officers and 75,000 soldiers—among them 7200 mechanics ("*ouvriers*")—were responsible for the proper running of this enormous fleet. By the end of the war, nearly 4000 women were involved in this work, most of them active in rear positions (not in the armies). In less than four years, the DSA personnel had quadrupled and by 1918 the DSA controlled 102,412 vehicles (including 314 tanks), more than half of these trucks, and one-third purchased outside France, mostly in Italy and the United States. They were repaired by using spare parts from a central depot in Paris (where a thousand women also worked alongside two thousand men), in eight "army parks" (close to the different armies) and ten central "revision camps" (later called "repair camps") spread over French territory.

While the "army camps" were organized as garages (on average one worker delivered one repaired vehicle per month), the "repair camps" were set up as true factories, in which the different repair sequences such as dismantling, body work, and engines were taken care of in separate *ateliers*. Rigorous statistics were kept about the *"rendement"* (efficiency) of the entire fleet management, resulting, for instance, in the conclusion that 3 percent of the vehicles per month could not be salvaged. On average, at any given time, 12,000 vehicles were under repair or under revision, and 323 vehicles per day were 'produced' by this gigantic maintenance machine. The active vehicles were transporting one million tons of freight per year by the end of the war, and nearly an equal number of men, and they were using more than two-thirds of the entire national petrol consumption.[64] In other words, the personnel of the DSA were professionals, not volunteers like the Volunteer Military Corps in Germany and the Netherlands. Tellingly, France had no experience of a volunteer corps before the war. Instead, Doumenc and his DSA analyzed what went wrong at the Marne. The long list of 'unprofessional' points included vehicles that were empty for half of their distance, the fact that they did not run at exactly the same speed, and that no road police was provided, nor crossings kept free from other traffic.[65]

When the United States entered the war in November 1917, their fleet management and repair system was modeled on that of the decentralized British, but they soon switched to the French centralized approach (and had their soldiers instructed by the French), as did the British. The first 120 American trucks (of the 50,000 vehicles Pershing thought to be needed) arrived in France in June 1917. Only at the end of the war, by late April 1918 (when America had two million men serving in France), was procurement of material, storage, and transport brought under the command of "the General Staff Division of Purchase, Storage and Traffic," whereas hitherto there had been "chaos and disorganization of the war effort at home" and particularly "ordnance business in chaos." Captain Doumenc's American counterpart, Clarence C. Williams, was instrumental in "cleaning up the ordnance mess": "If the fighting men want elephants, we get them elephants."[66]

Why, then, did the taxis at the Marne attract so much attention, whereas the railroad management of the Germans and the silent DSA logistics of trucks were at least as effective and impressive, and yet have since been forgotten? In the end, both Verdun and several other battles were basically 'railway battles' in that the bulk of men and material was brought to the battlefields by railroads; indeed, in Verdun, the railroads had been especially constructed for the purpose by the German army. If Germany had won, we would now be awed over the offensive against

Verdun, when in three weeks 860 freight trains and 550 troop trains brought soldiers and their goods as close as possible to the frontline, the last distance covered mostly by horse traction, as that was considered by far the most efficient option, more so than the truck that caused congestion directly behind the lines. Some military historians claim that this was exactly the reason why the Germans lost: they were logistically outperformed in the Allied "*Materialschlachten*" of the Somme offensives. All in all, the impressive logistic events made one issue clear: had the immobility of the fixed frontlines not occurred, the provision of men and material would have stalled much more often than was the case during the German campaign in the East in 1915 (where the armies had to make a stop, waiting for "*Nachschub*") or, in general, when the offensive from an immobilized position at the Western front went too fast.[67] The American army, the most heavily motorized of all, experienced congestion the most, already starting at the docks on the American East Coast.[68]

A factor that undoubtedly played a role in the myth formation around the *Taxis de la Marne* was the underdog position of the Parisians who, as the story went, cleverly outwitted the Germans by means of improvisation—the same reason that German accounts of automobile deployment at the front (emphasizing last-minute assignments, just-in-time delivery of important materials, and other elements that constitute 'adventure') were so popular in Germany. Indeed, it was the 'adventurous' and 'anarchistic' character of the Marne taxis, the 'free men' (and elderly men at that) behind the wheel and driving through the night that fanned the flames of fantasy. It had the same adventurous attraction as a 'spectator sport.'

On the other hand, the largely hidden effectiveness of the Verdun and Somme events can only be understood if we accept that it amounted to nothing less than a scientific feat to discipline the 'swarm' of individual vehicle drivers into one single 'body' for one clear purpose: to get the men and the material there on time.

War Trophies: Thanatourism and Other Adventures

The second realm in which mobility became intertwined with the open violence of the war was personal travel, and especially travel for its own sake. In this case, the continuity with the prewar tourist experience is expressed in the neologism 'thanatourism,' suggesting a tradition that went well beyond the experience of motorization alone. The term conveys a double meaning: it covers battlefield tourism in the sense of people traveling to places where battles had raged (or were raging), and it covers the traveling of soldiers to these places to engage in battle.[69]

During the nineteenth century, when British travelers were in the process of shaping tourism, a train and carriage trip to the battlefields of Waterloo, organized by Thomas Cook, was highlight on the list of attractions. Cook also set up tours to the Boer War battlefields, and these enjoyed such popularity that the high commissioner complained about the nuisance caused by tourists in the war zone. That incited *Punch* to comment: "And almost ere the guns are dumb, / The picnickers' champagne will pop / Upon the plains of Spion Kop." During the Franco-Prussian war, the *Observer* had published a leading article that made the connection between the curiosity for cruelty and the picturesque. Next to the 'technical sublime' (David Nye) apparently there is also a 'sublime of violence.' War tourism, in other words, was shaped and practiced long before the emergence of the automobile, but it was expanded beyond all expectation during and after WWI. Directly after the war, the Belgian automobile club decided to rename its annual car festival in Ostend the *Rallye des Champs de Bataille* (Battlefields Rally) and combine the contest with visits to the Yser battlefields, while the Touring Club organized coach trips to "the devastated regions," backed by a national government eager to capitalize on its recent destruction. Michelin published its guide of the Marne and other war locations in 1917 (that is before the end of the war), and by 1920 it was easily outselling, with 950,000 copies, Henri Barbusse's famous anti-war novel *Le Feu*.[70]

Even during the war tourism in Luxemburg had blossomed. In the Netherlands, neutral during the war, mobile life initially continued as if nothing had happened, until car use was prohibited in December 1917 because a ship carrying petrol was hit by a torpedo off the Dutch coast. Mobilization in 1914 (177,000 men, 6600 horses, and thousands of wagons, guns, and marines to the harbors) had been rehearsed by the railway companies and, as in Germany, it was executed with clockwork precision. After mobilization, the trains remained overfull with military personnel on leave, some 60,000 to 80,000 Germans expelled from Belgium, and thousands of American tourists who had been stranded in Switzerland and Germany on their way back to the ocean liners in Rotterdam. In addition, there were one million Belgian refugees who fled between August and October. Towns now started to experience their first "traffic complaints." The Dutch equivalent of the Bois de Boulogne, the Scheveningen boulevard along the coast near The Hague, developed into an "eldorado of motoring" made unsafe by "parvenu-type motor bolshevists" with earrings and in the company of "yelling, painted ladies." Some of the automobile club members, fully absorbed in their picturesque dream world, tried to realize a sport of "grand tourism," but the club journal warned that the German authorities in Belgium had the habit of requisitioning cars from curious Dutchmen. For ordinary club members, the only solace its journal could offer were translated stories about masculine risk-taking and the fascination for speed, aggression, and violence, such as "an automobile contest with the enemy" (in which chauffeurs deployed their driving skills while "flying over the road"), or the "petrol adventure" by a German soldier in Poland. In addition, there was a made-up travel account of an ammunition convoy behind the front lines, and a "tank history" in which the adventures were described in the style of an exciting youth novel. With no war to fight, Dutch army officers continued to participate in car and motorcycle races.[71]

In the German journal *Der Motorfahrer* (The Motorist) the double continuity in terms of sport ("the cold-blooded and rashly daring spirit of the sportsman") and the writing of travel accounts (we chose "the more picturesque road via Eupen") can be observed during the entire war, despite the omitted place names and other information that fell victim to the censor. Adventurous it was, no doubt: "We were intoxicated," including the chilling account of a patrolling trip literally "over dead bodies.... All full of blood, dripping from the wheels and mudguards." One motorcyclist experienced his war as a rally, while another celebrated the unity with his machine. Others marveled at being "only modest parts of an enormous whole. An enormous, impressive whole." Being a "volunteer fast driver" (*Schnellfahrer*, abbreviated to "S.F.") was a special honor: they shot spying "vagabonds," after they had forced them to dig their own grave. Spreading glass and other sharp objects on the roads (which had been done during a general craze about spies near the German border) was not needed anymore. For motorists eager to deploy high speed the war enabled what was not possible in peacetime: get the people, and even the chickens, from the road. The central car depots resembled "an impressive conference on car sports. ... For everyone who knows the 'struggle against the car' from his own experience, it is satisfying to hear how gratefully people acknowledge the car's performance."[72]

What the war brought about, was a radical extension of the traveling experience to other sections of the populations. For Möser, who initiated the analysis of the car-war connection among historians of mobility, the first occasion a European worker was likely to experience car travel was as an injured soldier, transported by middle-class or elite volunteers in their cars-turned-ambulances.[73] But the war did more to the soldiers: hammered into speechless and senseless machines by the atrocities, the cruelty, the suffering in their cold, wet, and filthy trenches under the constant noise of explosions and machine-gun fire and the threat of poison gas clouds, they started to produce poems. Some 1.5 million were written in the first war month of August alone, according to an estimate from 1920. That is, 50,000 poems per day, of which about 100 per day ended up in print.[74] While historian Modris Eksteins sees the men live "according to their reflexes" and "functioning on their instincts," this remarkable flood of poems rather testified to the widespread bellicosity in Europe, as can be taken from this English "demented exhortation" (which ends, as the analyst states, after "half a dozen breathless pages of necrophile ranting"):

> Burn, burn,
> set fire to this world till it becomes a sun.
> Devastate smash destroy,
> Go forth, go forth, oh lovely human flail,
> be plague earthquake and hurricane.
> Make a red spring
> of blood and martyrdom
> bloom from this old earth,
> and life be like a flame.
> Long live war![75]

Did Europe really need a Marinetti, one wonders when reading such poetry? Because of the unspeakable nature of their feelings, the millions of poets used existing linguistic structures to nonetheless express them on paper.

Indeed, recent studies have shown how sporting discourse provided an important linguistic and experiential reservoir that tended to describe war as a contest (the war as a "world devastating Olympiad," as a German general saw it, compensation for the canceled Olympics planned for Berlin in 1916), a competition of able-bodied men. Such poetry, written by former members of the *Turnvereine* (Athletics Clubs), initially idealized dying, and then mobilized the bodily experiences of war to evoke the desire to destroy, especially those "unfit for life": "We, chosen to stand here at the Front / Are all a part of that vast hand / Which brutally crushes all that is not worthy / And gives the course of history new glory."[76] War is not just about mobilizing people in one splendid organizational stroke: long before the trains started rolling, a slow and steady psychological and cultural mobilization had been taking place, and it is certainly not a coincidence that this cultural mobilization had the same duality of old and new, the fascination for high-tech mobility and enthusiasm for preindustrial values, as we saw earlier as pervading the entire culture. This war was, in other words, not only a war about territory, it was also a war about the power of words: the urge to write one's experiences down in the hundreds of thousands of poems was motivated by the desire of "finding and communicating 'meaning.'" One analyst has likened what these soldiers did to the "loud singing of the child sent into the dark cellar."[77] Culture here should be understood as cultural *praxis*: after the war, German soccer was infested by a "never before witnessed degree of fouls and manifestations of rudeness," which could hardly be brought back to normal proportions.[78]

Another recent study sums up an entire tradition of scholarship in claiming that travel, tourism, pilgrimage, and going to war have much more in common than is often realized. The quintessential tourist experience of departing in order to come back, the building up of expectations before departure, the openness to experiencing the sublime in nature, and even, for the common soldier, the parallels between the hotel and the barracks disqualify the usual statement of tourism research that "during wartime, tourism simply stops." Tourism did not stop, but the fledgling middle-class travel movement certainly collapsed. The war instead forced a shift of the traveling culture toward the lower middle classes and the poor, as drivers, but mostly as "freight" to be delivered to the frontlines by truck, and to be collected from there by ambulance. Falling back on a stable narrative of tourist travel for these soldiers went far beyond using travel writing as a reservoir of stories: classical travel literature had brought to them "the dialectical experience of alterity and identity, the tradition of the authentic as well as literary simplicity as stylistic ideal," that is to say, discourses, experiences, and techniques that had been practiced before the war in the *Wandervogel* and Rambling movements. It was practiced by the Boy Scouts too with their "functionalizing of youth's lust" through the singing of war songs and militaristic plays, and for the more literary-inclined soldiers in "the feeling of casting off chains and starting anew" (*Auf- und Ausbruchstimmung*) of German Expressionism. "Off into Liberty" was the slogan for intellectuals during the first months of the war, and the Expressionists added their evocations of community, vitality, and "collective adventure." The description of "big adventure," then, was an implicit comment on prewar experiences in all Western countries, of societal disintegration (the dissolution of the boundary between elites and proletariat, between industrialization and backwardness, between center and periphery) and the "sluggishness, boredom and senselessness of the existence ... especially (among) young people." Adventure, it seems, was mobilized as a violent undertaking with the aim of revitalizing an uneventful life.[79]

We know about the tourists' experiences because, as we have seen in the previous chapter, they occur simultaneously with and are co-constituted by the production of graphic representations, photographs, letters to the family back home, poems, and other forms of literary appropriation. We have already mentioned the tourist jokes on the trains going to the front: "To the summer freshness," "France via Belgium! Sleeping car to Paris," and "See you on the boulevard" were some of the slogans soldiers coined playing on the parallelism of travel and going to war, as if they both generated the same kind of excitement.[80] And perhaps they did. War, in other words, was for most nonwealthy people the first and only way to travel into unknown territory, and to experience the Other, the Foreigner—as an Enemy in this case. Their letters and poems functioned to stabilize the home front, as they stylized their daily experiences into linguistic allegories of sport and war. The experiences were of preparing for, performing, and working through conflicts. Sport and travel were the two big symbolic reservoirs available to the men in the trenches. The military on both sides of the conflict soon realized the importance of propaganda—huge press departments were set up (in Germany in 1917, hundreds of officers, journalists, caricaturists, painters, photographers, and technicians populated the *Kriegspresseamt*, the much bigger successor to the *Nachrichtenabteilung IIIb* from the beginning of the war) and well-known authors were invited to visit the "war theatre."[81] It is no wonder then that apologetic literature was far more sizable than that critical of the war. An analysis of the list of bestsellers from interwar Germany, an "enormous sense-making machine" (including a publishing infrastructure) with 'authenticity' as its central claim, results in the conclusion that readers were primarily interested in "war as adventure," including, one needs to add, automotive adventure. This may be obscured by the immense popularity of Rainer Maria Remarque's anti-war novel *Im Westen nichts Neues* (published in 1929, and selling 900,000 copies, making it the third bestselling interwar title), but in the number of titles apologetic (and "embedded," to borrow a phrase from the Gulf War) literature was by far in the majority.[82] Besides, a recent study on French WWI fiction refers to Remarque's novel to argue that it may as well have conveyed the pleasure of violence and thus its success is in no way an unquestionable testimony of postwar pacifism.[83]

In September 1914, Dutch journalist Tom Schilperoort managed to go to the war front by bicycle, and report on it in the touring club's magazine; the Belgian writer Stijn Streuvels did the same ("for fear that it was over before we had seen anything") and his colleague Cyriel Buysse published in a Dutch newspaper his accounts of a train trip to the Somme battle in the Summer of 1916 *"Achter het front"* (Behind the front lines) and *"Reizen in oorlogstijd"* (Travel in times of war). It resulted in a unique look from the noncombatant's perspective of battle logistics:

An automobile, a big gray truck, amidst a blown up, gray dust cloud! Fast and droning it speeds by, immediately followed by a second, a third, a fourth, by an innumerable and endless row of automobiles ... and while our train waits snorting to let that hurricane rage past, a second automobile-storm arrives from the opposite direction and then it seems to become one wild chaos, some infernal affair of dark animals dashing up against each other in clouds of smoke and flashes of fire, with here and there men clenched as demons at the wheel of the wild monsters [where have we heard this before? Remember Marie Corelli, chapter 2]. All that runs to and from the Somme, from the destruction or towards it, with the grenades and cartridges of death, or with the dead and injured of the battlefield, as in one horribly wild Bacchanal of world cataclysm through the soft wealth of quiet-idyllic landscape, where the field workers labor and where high in the sky the larks sing.

Buysse's prose forms his passport to the position of official war correspondent, complete with a fast passenger car (a *"Panhard-de-guerre"*), a chauffeur, and a captain as guide, and results in disgusted-fascinated observations like when he sees a train, full of blood, being cleaned: "That's how it works! The human slaughter cattle is withdrawn from the battle after the slaughter; the animal-cattle is driven towards it, upon the train's return. This is how it can last. This is how the blood can continue to flow, as it is replenished by fresh blood."[84] War, Buysse seems to testify, is a forced and cruel synthesis of the pastoral and the machine, not of the machine in the garden, but the garden as the machine. It was, it seems, the masculine combination of fascination and disgust for violence, which after the war was disqualified by Willa Cather's novel *One of Ours,* among others (see later in this section and also in chapter 6).

On the other hand, one should not exaggerate this: as late as the beautiful summer of 1916, soldiers were writing home from the trenches comparing their stay with a camping excursion: "We walk through the trenches, the air is fresh, the sun shines."[85] When the elitist and prestigious American Volunteer Motor Ambulance Corps entered the war scene (their cars a personal gift and each member prepared to drive him- or herself, as they not only paid for their own salary but also for the upkeep of their cars), French military planners complained that "the foreign personnel is not military enough, [and] too voluntary. The drivers tend to behave as sportsmen."[86] Given what we have seen before, the planners were not complaining about the fact that they saw war as sport, but that they showed it too openly, that they didn't hide it behind the cold, manly face of utilitarianism. That was certainly the case for Gertrude Stein and Alice Toklas, who were treated with suspicion by a French officer near Nîmes when they were driving around with medical supplies in their new Ford. Stein had ordered the car in the United States upon advice from Isabel Lanthrop, "wearing a pink dress with pearls and a garden-party hat that belied her efficiency" (as one of Stein's biographers muses), the head of the American Fund for French Wounded, a privately funded volunteer organization. Upon arrival in France, the car was rebuilt into a truck, while Stein had meanwhile taken driving lessons from a painter-friend in his Renault cab (which he drove during the war to earn money). Stein had an adventurous style of driving, and she refused to drive in reverse, for instance when she had to park her car. They called the car "Auntie" after an aunt who "always behaved admirably in emergencies and behaved fairly well most times if she was properly flattered," and the couple undertook their first long trip in the spring of 1917, which they made into "a gastronomical tour" as well. "We had a few adventures," Stein remembered. "We drove by day and we drove by night and in very lonely parts of France and we always stopped and gave a lift to any soldier, and never had we any but the most pleasant experiences with these soldiers." Stein never repaired the car herself, but knew how to convince others to take care of it, while she herself meanwhile wrote poetry. They visited Verdun as tourists and enjoyed, as one historian concludes, a "perverse exhilaration" from their trips.[87]

Similar experiences have been reported by other Americans closer to the frontline. Accompanied by an angry letter from former President Theodore Roosevelt, who accused the American government of a lack of "manliness" as it did not declare war and failed to protect its own citizens in Europe, wealthy American volunteers, mostly "college men who have come over from the United States to 'do their bit' for France and see the war at the same time," were integrated in the *Service des Autos* and transported the wounded. Meanwhile, they found time to describe how they "rode along the Meuse, through a beautiful country where the snow-covered hills, with their sky-lines of carefully pruned French trees, made me think of masterpieces of Japanese art." The temporal adventure was satisfied, too: "There is a great rivalry between the men of the several sections in matter of speed and load. . . . The American product has the record for speed, which is, however, offset by its frequent need of repair." The latter was done in well-equipped workshops. "Except for our experiences of the road, there was little romance in the daily routine. True, we were under shell fire, and had to sleep in our cars or in a much-inhabited hayloft, and eat in a little inn, half farmhouse and half stable, where the food was none too good and the cooking none too clean, but we all realized that the men in the trenches would have made of such conditions a luxurious paradise, so that kept us from thinking of it as anything more than a rather strenuous 'camping out.'" Literary historian Paul Fussell's conclusion about the war's symbolic status as "the ultimate anti-pastoral" should be relativized.[88]

Edith Wharton's account of her trips to the front during the first fourteen months of the war certainly does not comply with Fussell's characteristics. Giving the means of travel (an open car) much more attention than in her prewar *Motor-Flight Through France* (see chapter 2), her descriptions in *Fighting France* (1915) of the thrill of (watching the) fighting intertwine with the same literary indulgence in nature and its smells and sounds as in her prewar travelogue. The adventure (defined as a venture into the "unknown") is everywhere: on two pages alone the car "flew," "spun(s)," "hurr(ies) on," and "dash(es) through" on her way to a guided tour ("our sightseeing," she calls it) of the frontline in the Vosges, as an interlude on her visit to frontline hospitals. Continuity with the prewar motorized mobility is everywhere, for instance right at the start when she sees Paris empty itself of inhabitants traveling to the army while another army, of refugees, comes in, joining "the great army of tourists who were the only invaders [Paris] had seen for nearly half a century."[89] That same war she wrote for *Scribner's Magazine* (and published in 1918 as *The Marne: A Tale of War*) of her visits to the front undertaken upon request of the Red Cross and upon special permission by General Joffre himself, a series of "dispatches" meant to incite her country (and her friend Roosevelt) to interventionism in the war. "I don't know anything ghastlier & more idiotic than 'doing' hospitals en touriste, like museums!" she wrote to a friend, but her appearance at the Verdun front in February 1915 caused quite a stir nonetheless, as one of the first 'embedded' war correspondents, and female at that. *The Marne* tells of an American ambulance driver, Troy Belknap, whose family donated the money for his vehicle, who is stunned by the endless flow of "huge grey motor-trucks, limousines, torpedoes [cars with a torpedo-shaped body], motor-cycles, long trains of artillery, army kitchens, supply wagons." A truck collides with his ambulance, after which he decides to go to the front and take part in the trench battles: "and he had really been in the action!"[90]

Other countries also saw their famous writers travel to the front. From the United Kingdom came Rudyard Kipling and Arthur Conan Doyle. One of the most remarkable accounts is given by the popular British writer (and former suffragist) May Sinclair (pseudonym of Mary Amelia St. Clair), who in 1915 published a novel in which the autobiographical frustration of being sent away from the Belgian front because of her clumsy eagerness to indulge in adventure is hardly hidden from her *Journal of Impressions in Belgium*. Sinclair's fame was used by the initiator of the Motor Field Ambulance Corps, Hector Munro, to raise financial support. Her novels (she wrote several on the topic) hardly deal with the actual war practice itself, but all the more about the "almost orgastic" experiences of thrill and the proximity of danger.[91] In *Journal of Impressions,* with many passages written in the 'you' form, a group of voluntary nurses, led by a man but able to drive and maintain their (Daimler) ambulances themselves, indulge in adventurous trips, "drunk, very slightly drunk with the speed of the car," leading to "ecstasy," steadily mounting "thrill," and further "desire ... to get into the greatest possible danger—and to get out of it," meanwhile "accused bitterly of sightseeing." The thrill experience is very precisely described: it is a "steadily mounting thrill which is not excitement, or anything in the least like excitement, because of its extreme quietness. This thrill is apt to cheat you by stopping short of the ecstasy it seems to promise ... it became intense happiness." But there are also other thrills, such as the one experienced during the flight from Antwerp or Ghent, when the German armies conquered those cities (adding a new dimension to the concept of flight; see chapter 2). The group initially still has "the appearance of a disorganized Cook's tourist party" and experiences "some tremendous but wholly visionary adventure" of sighting German guns, an "unreal scene, the most interesting part, say, of a successful cinematograph show." The "thirst for experience, or for adventure, or for glory, or for the thrill" can hardly be quenched. But unlike Edith Wharton, for Sinclair's protagonist, in a passage where the "you" shifts into an "I," the adventure "lacks every element of surprise. It is simply what I came out for." The "perilous adventure" is appreciated because of "the fun of the thing," but it is also a self-test. Adventurous trips are high speed ("we drove very fast" in "sheer excitement of the rush through the danger zone") and if not, they are "a tame affair." The continuity with prewar motoring experiences is very obvious indeed: a visit to "Belgian batteries at work" is compared to a "stop at Olympia and have a look at the Motor Show on our way to Richmond." Good drivers are admired because of their cyborg-like unity with the car. The car, indeed, is a shell, a protective capsule against the shells (grenades) outside. Meanwhile, the women drivers are constantly corrected or called back by the men, and after only a couple of weeks, the protagonist is sent back home with a trick, to raise new funding for the Motor Field Ambulance Corps. The exciting mixture of awe and fear, the hyper-experience, had become too much.[92]

Helen Zenna Smith's anti-war novel *Not So Quiet*... was meant as a "feminist reprise" to Remarke's success. A pseudonym of Evadne Price (1896–1985), "a very successful free-lance journalist (and) author of several hundred paperback romances," her first novel was an immediate success. Although published in 1930 as part of a wave of war novels (translated in French and Dutch), and although she never went to the war zone herself, her "semi-biographical" story is based upon the diary of ambulance driver Winifred Young, told in the novel by protagonist Helen Z. Smith. Her modernist style as well as her emphasis on "efficiency" and her neglect of the racing adventure (women drivers are more "thoughtful") belong in the next phase, but her description of the war as a bodily experienced hell, the preference of the protagonist for the dirt of the machine over the dirt of the human body of the casualties, as well as the emphasis on "driving sense" and related cyborg-like experiences add to our knowledge of the relationship between automotive mobility and the war. This is a novel about the acoustic sense: the deafening roar of the war inspired its title. Again, women's technical dexterity is emphasized, but also the female war raging behind the front lines in the hospitals, in this case a struggle about the expulsion of a lesbian from the midst of the women drivers, who is sent back home after being stealthily accused of an amoral attitude. Although feminist analyst Jane Marcus criticized the novel as an effort "to clear the volunteers of the charge of lesbianism," Smith's comment on woman's role in the carnage is no less convincing:

> Doesn't it make you sick? Slack as much as you can, drive your bus as cruelly as you like, crash your gears to hell, muck your engine till it's in the mechanic's hands half its time, jolt the guts out of your wounded, shirk as much as you can without actively coming up against the powers-that-be—and you won't be sent home. But one hint of immorality and back you go to England in disgrace as fast as the packet can take you.... Personally, if I were choosing women to drive heavy ambulances their moral characters wouldn't worry me. It would be "Are you a first-class driver?" not "Are you a first-class virgin?" The biggest harlot or the biggest saint... what the hell does it matter as long as they put up a decent performance behind the steering wheel and can keep their engines clean?

The novel is less about the temporal (racing) adventure, as is mostly the case, but more about the "erotics of war," including the flirt with the men, and including the personal unity with the "bus" (the ambulance). "Brake off. Clutch out. Gear. Gas. I am not doing it—it's doing itself somehow."

Marcus, in her comments, emphasizes the importance of the prewar suffrage campaign (the women's movement "practically destroyed" by the war) as "the training ground for ambulance drivers and... nurses in World War I. Bravery, physical courage, chivalry, group solidarity, strategic planning, honor—women learned these skills in the streets and gaols of London." At the same time, Smith knows why the power that be opted for 'classy' women: "Why should they want this class to do the work of strong navies on the cars, in addition to the work of scullery maids under conditions no scullery-maid would tolerate for a day? Possibly this is because this is the only class that suffers in silence, that scorns to carry tales. We are such cowards."[93] Who are the 'we,' then?

> There may be an odd few who enlisted in a patriotic spirit—I haven't met any, personally. Girls who were curious, yes; girls who were bored stiff with home (like myself) and had no idea of what they were coming to, yes; man-hunters like The B.F. [one of the drivers]; man-mad women, semi-nymphomaniacs like [another ambulance driver] Thrumms, who was caught love-making in an ambulance and booted back p.d.q. to England, yes; megalomaniacs like Commandant who hope "bossing the show" and have seized on this great chance like hungry vultures, yes; girls to whom danger is the breath of life, yes; but my observation leads me to the conclusion that all the flag-waggers are comfortably at home and intend to stay there.[94]

Much closer to the end of the war (1918) Enid Bagnold's (1889–1981) *A Diary Without Dates* appeared, an early novel critical on the conditions in a war hospital (from where she was dismissed as a result), while two years later she published a novel based upon her own experiences as an ambulance driver in the Voluntary Aid Detachments (VADs). This remarkable novel is one about car use and especially car maintenance and repairing skills, the constant struggle to keep the ambulance running, a nice example of our thesis that the war introduced the systems perspective in automotive mobility. "She passed the morning in the garage working on the Renault, cleaning her, oiling her." Maintenance requires listening skills: "the engine began to knock. She saw [the driver] Foss's head tilt a little sideways, like a keen dog who is listening."[95] No less controversial was Mary Borden's *The Forbidden Zone*, written in 1929 as part of the anti-war novel wave, but based on her experiences with a mobile hospital she equipped and financed herself. Daughter of a wealthy Chicago family, having completed a world tour during the second half of the first decade of the new century, involved in the London suffragette movement, and connected to the British literary avant-garde (Vorticism), she would again set up a mobile hospital during the Second World War.[96]

From the United States it was the future Nobel Prize winner Ernest Hemingway who served as a Red Cross volunteer (an experience he would incorporate in his *A Farewell to Arms*) and in his diary called the war an adventure, writing home that he "has seen all the sights, the Champs Elysees, the Tuileries and the Arc de Triumph."[97] The enthusiasm to take part in the war was so great that even Henry Ford, when he proposed to send a Peace Ship to Europe in 1915, was depicted in the press as a clown.[98] At the home front, the American Progressive movement backed President Woodrow Wilson's decision to go to war as a "model of middle-class utopia." One reformer found, "war necessitates organization, system, routine, and discipline," and another exclaimed, "long live social control." But already in 1917 a like-minded journalist observed in Minneapolis "a whole common people rolling carelessly and extravagantly up and down these streets in automobiles, crowding insipid 'movie' shows by the tens of thousands.... All overdressed! All overeating! All overspending.... We need trouble and stress!"[99]

Among the accounts of the German writers, Anton Fendrich's *Mit dem Auto an der Front* (1915, To the front by car), which sold 90,000 copies (not counting an unknown number in Dutch translation), and Paul Grabein's *Im Auto durch Feindesland* (1916, Through enemy country by car) are investigated here on their reflection (and refraction) of the connection between violence and mobility under war conditions. Both authors were well known, Grabein (1869–1945) as an author of several *Unterhaltungsromane* (entertainment novels), Fendrich (1868–1949) as a social democrat and SPD member who had written several books on rambling and nature, and who evolved into a fierce proponent of the "integration" of the social-democratic masses into the *Volksgemeinschaft*. More than Grabein, Fendrich sees the war as a unique chance for the rebirth of the nation: in the bombastic parlance typical for the time, he suggests that "only tremendous shocks were able to take from people their silly fear of death."[100] Both books include the "narrative-descriptive double-face of the genre" of travel writing, consisting of "a dynamic-narrative description of arrival and departure, and the static-descriptive representation of landscape and culture of the travelled-through land." The well-tried pattern of "the narrative and thematic structure of the travel account (departure—experience of the strange—return) ... completes the sense-making of (the efforts to project meaning into) the accounts of the war theatre, which travels the minefield of unspeakability and propaganda, caught between being silent and the urge to speak." That, indeed, is what strikes the reader when entering their fictional worlds: that they establish continuity with prewar experiences, describing their high-vitalist adventures as experienced on the home front's behalf. Although a satisfactory theory about how the unspeakable is transformed into text is still lacking, it is quite clear why the war travel accounts are such an interesting source, especially for mobility history: they focus on "the unexpected, the remarkable, the challenged," that is to say, the strange.[101]

Both convey their 'authenticity' through their bearing witness to well-known battles, heightening their credibility by calling the war a "terrible destroyer" (*Vernichter*) and even including some allusions to atrocities: Grabein's defense of the burning down of entire villages because of the presence of Belgian snipers is followed by his dwelling on the German "benevolence" toward the Belgian and French population, mentioning in passing civilians on the run from the violence, and describing German soldiers' glee whenever they can burst into action. Fendrich shares his fun with his readers at the sight of coughing Frenchmen overwhelmed by poison gas, "bayonets fixed and knives between their teeth." Both books are examples of thanatourism, drawing on prewar tourist conventions, including the meeting with the Other, in two ways: as members of the lower echelons of their own country and as 'foreigners' abroad. Ordinary German soldiers, pedestrians at the frontline, stare in awe at Grabein's female companion in the car while they search for his passenger's husband, a physician. Somewhat later he drives along a "romantic" road to a French chateau, "looted by the British and ruined by the French." The Other as Foreigner, as Enemy, is abundantly present in these stories full of racism (blacks as "pigs" (*Schweine*), "half humans," or even "half-apes") and condescending stereotypes (the French are dirty; Belgians are children)—nothing new, nor anything specifically German, just rather harshly formulated.[102]

"When one comes into the field," Fendrich muses, "one's heart first turns to the trenches. That is so natural. Abundant thankfulness, followed by a kind of timid curiosity that would like to look behind and beyond the trenches, and finally the awe often mixed with a bit of envy, all this grips us when we, in the middle of a field, at the border of a cornfield or on the road, see those white crosses. The envy is directed to those who are ennobled here by another death distinct to the one we will all once have to suffer at home in our lonely struggle. They however have their souls inspired, carried along by the rushing vitality of the struggle for the fatherland." Fendrich has this contemplation followed by a description of a rhododendron ("what a pity it doesn't blossom yet!") suddenly interrupted by the shimmering of a white cross through the dark green bushes, thus continuing prewar nature description mixed (literally, in the depicted reality itself) with emblems of vitalist and honorable life in war.[103] 'Nature' is depicted as violated by war (trees scarred by grenade fire, or removed to provide a better line of fire, forests decimated for fuel in the homes), or as something to be 'cleansed' from enemies, or as "thick air" (an expression referring to the smoke-filled atmosphere during battle).[104] "It is still early in the morning," Grabein starts his account of his entrance into France. "From the freshly broken ground a spicy smell of soil emanates. From the north the wind caresses the stubble fields and carries sea air. Nothing here commemorates the bloody struggles farther down the road. The sun laughs in the deep-blue sky. This morning is like a joyful holiday. But then it races past us, a field-gray rakishly-shaped car. Three members of the general staff in it with faces full of masculine sternness, great intelligence and cold energy." The car is equipped with a knife-like steel contraption running over its entire length, to cut ropes across the road.[105]

Against this war-torn nature both narratives give us stories of masculine adventure. Both speak admiringly of a 'hard' and 'cold' attitude, and are dismissive of the cowardice of feminized intellectuals and "*Spiesser*" (philistines) and of women and sentimentality. Adventure is clearly a car adventure, most explicitly in the case of Grabein, who, as leader of a convoy of volunteer motorists, receives adventurous assignments, from getting demolition teams to the front to the transport of soldiers and wounded, as part of the conveyor belt of mechanized war. Such trips, whenever possible, are accomplished at "a crazy speed" or in a "roaring tempo" at full throttle ("*mit Vollgas*") and sometimes even an open exhaust, described using vocabulary that evokes the "romantic warrior" and the "wild-romantic" experience. As here (reminding us of Corelli again): "Forward, with the highest possible speed! and with a fully opened valve in the exhaust pipe under a hellish clatter the cars storm forward." The volunteers Grabein meets are "members of better families, even owner-drivers (*Herrenfahrer*), who have put on a soldier's uniform to show their enthusiasm for the war and also their lust for adventure." The senses are strained to the limit amid the cacophony and stench of war, the car's exhaust noise competing with exploding bombs or sometimes operating in complete silence when a covert operation has to be accomplished. "Bruges is a town dating from the high-Gothic period," Fendrich starts a chapter as if he is about to engage in a peaceful, static rendering of the monumental Belgian town, and indeed the opening sentence is followed by a description of the landscape with all the paraphernalia of the prewar travel account, even making use of Proust's inversion technique, when "white crosses rush past right and left of the road."[106]

Although it is a bit of a stretch to generalize from two books in the midst of an avalanche of publications, there are some conclusions that can be usefully drawn. Contextualizing this wartime dynamism of the privileged, motorized few means contrasting it with the equally modern, but static unification with the machine that took place in the trenches.[107] Duty and loyalty, combined with sensory overload, generated a symbiosis of man and machine, but we need the dynamism of the car drivers to communicate the soldiers' heroism to the home front (and to us). In his exemplary search for the motivations behind this absurdist theater, Alan Kramer quotes the French Jesuit writer Pierre Teilhard de Chardin, who tells of "the unforgettable experience at the front," which, "to my mind, is an immense freedom." At the same time, the feeling of solidarity with fallen comrades was a major basis for postwar tourism. To fuel this tourism, a willingness to experience the unexpected (core of the adventure, as we have seen in the previous chapter) had to be developed, a willingness that can be illustrated by a letter written by a volunteer captain and Hamburg businessman, who relates his trench experiences to an artist-friend, describing "a kind of fatalistic death-wish which he could not discuss even with his wife.... Since that time I feel the need—keep this to yourself—to expose myself to greater danger."[108] All this is in great contrast with the later anti-war novels, also in Germany, for instance in Adrienne Tomas's (pseudonym of Hertha Strauch, 1897–1980) *Die Katrin wird Soldat* (1930, Katrin becomes a soldier) where the life behind the front (in the Lorraine city of Metz), based on her own experiences, is depicted. Katrin become a "half soldier" through her work for the German Red Cross, a witness of the conveyor belt of the war in the form of train after train full of dead and wounded, and fresh soldiers in the other direction. What strikes the reader focused on mobility in this novel is the surge of automobile traffic in a midsize provincial town because of the war: "Military cars race though the town, honk deafening and act important as if they are the fire brigade."[109] The war was a multisensorial mobility experience, indeed.

In dealing with the unspeakable, such is the main conclusion of our analysis, pedestrian-soldiers and motorists alike sought support in continuity with all that was familiar from the prewar period, despite declarations to the contrary about the new vitalist life in the trenches. The continuity is even confusing, as Grabein finds when he describes how a "real garrison life" develops around him and that the only thing keeping him from thinking that he is in a "peaceful provincial town" is the rumble of the guns in the distance.[110] Constantly recurring elements are the concepts of 'adventure,' a concentrated experience of bodily and psychological thrill, and 'tourism,' a prolonged experience of sightseeing and nature's sublime, interspersed with much more violent images than during peacetime tourism. The sublime, in fact, has two sides: on the one hand, thanatourists were in awe of the unspeakable beauty of larger-than-life nature; on the other hand this awe was intermingled with a fascination for destruction. In the thousands of photographs German soldiers took and sent home "damaged and destroyed churches were a favorite motif." Often, the two aspects of the sublime were combined, leading to an "aesthetization of destruction": how else can we explain the fascination for a landscape that had been subjected to a German "scorched-earth policy" (after March 1917, inspired by the Russian strategy from 1915), or for the landscape of La Malmaison/Laffaux after the French conquest in 1917, where a single week saw the dropping of eight tons of shells per meter of frontline?[111]

In 1916 some artists in neutral Switzerland, soon followed by contemporaries in Berlin, turned the unspeakable absurdity of the war into linguistic nonsense. They called themselves Dada: "The battle is our house of joy," Hugo Ball wrote, and co-founder Richard Huelsenbeck told the audience of his Cabaret Voltaire in Zürich that Dada "wanted to end the war with nothing. We don't have a goal, we do not even intend to entertain you."[112] Dada (characterized by Huelsenbeck as "the screaming of the brakes") was just the absurdist outcome of an earlier Expressionist play with violence and art "racing automobiles" (Johannes Becher), and aerial warfare. Franz Behrens, for instance, brings "expressionists" and "artillerists" together in one poem, and others mimic his linguistic analogy through the use of the "radically explosive, telegram-short word artillery." In the sense of our three-level literary analysis, the war provoked a bacchanal of 'affinity' with novels and poetry. Even Thomas Mann had, upon the start of the hostilities, included the soldier in the creative avant-garde.[113] Especially racing (*"Mit Vollgas auf den Feind!"* [At full throttle against the enemy!], as an anonymous story in the *Allgemeine Automobil-Zeitung* formulated) formed one of the thematic continuities between prewar and war experiences.[114]

Just as dependent on continuity with prewar circumstances was the situation in Russia, where Germany deployed fewer cars because of the poor road conditions.[115] From the Russian side, however, we have a unique report by Viktor Shklovskij. In *A Sentimental Journey*, he not only relates his *Memoirs, 1917–1922* (as the subtitle ran) to Otto Bierbaum's early car experiences (who, as we saw, also used Laurence Sterne's phrasing), he also develops a literary theory about "retardation" and "estrangement" (a precursor of Brecht's aleination) as a way of reading Sterne. Shklovskij started his remarkable career (before he as a Social Revolutionary had to flee from the Bolsheviks and ended up in Berlin, as we will see in chapter 6), "as an instructor in a school for armored-car personnel." His descriptions of how he managed the car fleet (mostly American Packards and Locomobiles) while at the same time working on his literary theory and giving lectures in Gorky's Institute of Art History in Petersburg are littered with adventures: races along the Eastern front ("A shell fell in front of the radiator of our car; I think they had aimed at the cloud of dust"), the seizing of armored cars during the Revolution ("the car's forty to sixty horsepower makes an adventurer out of a man"), the use of the car as a deadly weapon ("why did you want to go through with this [shooting at an enemy]? We could have run him down in the car"); in short: "bombs and cars change a man's character."[116] In other words, Shklovskij is the living proof of the crucial role of 'affinity,' as he produced a new, 'revolutionary' car experience in conjunction with the production of literary theory based on an eighteenth-century travelogue, and written in a style as innovative as the technology of his vehicle.

Stylistically more traditional was American writer Willa Cather's *One of Ours*, which appeared after the war and won her the Pulitzer Prize. Her coverage of the war was controversial because she emphasized exactly this continuity between war and peace, rather than its atrocities. Whereas Hemingway (*A Farewell to Arms*, 1929), Dos Passos (*Three Soldiers*, 1921), and e. e. cummings (*the enormous room*, 1922) indulged in depictions of cruelty and shock, Cather (who did her research by interviewing boys from her Nebraska hometown and visited France, including Verdun, immediately after the war) followed protagonist Claude and his fellow farm boys to France, where they experienced "the tingling sense of ever-widening freedom" and concluded that a war with no fighting "was no sport." Claude "was enjoying himself all the while and didn't want to be safe anywhere. . . . Life had never seemed so tempting as it did here and now." And defensively: "anyway, he wasn't a tourist. He was here on legitimate business." One of the characters remarks "drily" that the war is "a costly way of providing adventure for the young." Indeed, the fifth section in *One of Ours* "is more travelogue of the French countryside than war chronicle," as one of the analysts concluded, a picture Cather's fellow-writers could not appreciate.[117]

Conclusions

Those war narratives issuing from the perspective of the car interior provide us with a counterimage to the immobile pedestrianism of the trenches, a dualism that was only overcome when the frontlines started to move again on 21 March 1918, first with a German offensive and then with an Allied counteroffensive, only to grind to a halt once again with the occupation of Germany. There, meanwhile, Chancellor Bethmann Hollweg was replaced by a military dictatorship and a left-wing insurgency threatened to put an end to middle-class society, followed soon by a right-wing countermovement, both sides using trucks, not only as soldier buses but also mimicking "the tightly-packed troops of the 'trucked' raiding patrols." The Italian fascists did the same, following a strike in which 100,000 workers in Turin threatened to take over the city, and so did the revolutionaries in Russia. French troops started mutinies, stopping trucks shouting "Long live peace! Down with the war!"[118]

When Europe stumbled into the 1920s, ready to witness a first massive peace-time wave of motorization, looking back at the war made clear that many things had changed drastically, even if they were in fact intensifications, extensions, and expansions of phenomena present before the war. One of these was the interventionism of the national state into the daily life of its citizens: central government budgets increased greatly to finance a welfare state in the making, leading to the redistribution of wealth that made the wave of motorization possible, and also to unprecedented tax increases.[119]

A second change was the involvement of women in the public sphere, a movement toward general voting rights in which the automobile played a special role. In the United States, for instance, the suffragette movement during the war was hardly conceivable without the transcontinental car trips, as we have seen in the previous chapter. In Europe, after the war, similar developments were preceded by a massive involvement of women on the home front as well as at war, often behind the steering wheel of an ambulance or a truck. In the United Kingdom, the number of women participating in paid work increased from 6 million to 7.3 million (37 percent of the working force in 1918) during the war years, while in France the proportion of women working increased from 38 to 46 percent.[120] At the outbreak of the war, "several flourishing ladies' motor businesses in London and the provinces" could be counted, one of them bearing the slogan "Women Trained by Women," while the Scottish car manufacturer Arrol-Johnston started an "ultramodern factory" in the Scottish village of Tongland run by sixty women (with two male engineering instructors and two female supervisors), "a paradise for the woman engineer." At its height, two hundred women were employed, proving that "girls can be boys, as far as mechanics are concerned."[121] In the United Kingdom, over one hundred thousand women were "enrolled" (not "enlisted," as they did not gain military status but kept being treated as civilians) in so-called auxiliary services between 1917 and 1919. Five thousand of them were in the Women's Legion Motor Transport Section, and the majority (fifty-seven thousand) in the Women's Army Auxiliary Corps (WAAC) where they worked behind desks or in kitchens, or as gardeners in cemeteries. When rumors emerged about the "immoral" behavior of these "wonderful little Valkyries in knickerbockers" and when newspaper stories appeared about the killing of eight women during the German spring 1918 offensive, an indignant WAAC leader set up a press conference stating that women were as entitled to being killed as men.[122] The iconic role of such women was especially visible in the case of "Elsie and Mairi" (no chance that heroic men would have been called by their given names). Elsie Knocker and Mairi Gooden-Chisholm were members of the Gypsy Motor Cycle Club at a time when the United Kingdom counted only fifty woman motorcyclists; they volunteered, when war broke out, in the Flying Ambulance Corps. Their adventures were already made famous in a book written during the war by Geraldine Mitton, and the "heroic Scottish ladies," "the most photographed women of the First World War," became newsworthy, shoring up the patriotic mood at the home front and leading to the granting of a Military Medal, something that made it into French and Belgian newspapers.[123]

At least 16,500 women worked overseas in the American Expeditionary Force, most of them lower-middle class as nurses, in canteens and offices, and as drivers of ambulances and other automobiles, the latter especially in the Red Cross Motor Corps. This gigantic logistic machine, employing twelve thousand men and women in 1918, consisted of more than six hundred vehicles, more than half trucks, maintained in two large garages in Paris by skilled mechanics. Although most women worked far from the front, several had tasks as dangerous as men, which they liked at least as much: typically, the requests to be put at such places near the front outnumbered the actual positions by a factor of ten. Some explained their eagerness because they felt themselves to be "an adventuress" who "wanted to travel, to see a lot of life." After the war, several ex-servicewomen became tour guides in France, England, and Italy.[124] It was the (upper-) middle-class woman, however, who got the most attention from contemporaries and historians: "a woman of culture who was familiar with both the French language and the intricacies of auto mechanics. Drivers usually performed their own maintenance work, including oil changes, small repairs, and cleaning.... Despite the hectic schedule, though, many volunteers enjoyed their days in Paris and the excitement of driving for the war." Their supporting role was impressive, though: the US Women Motor Corps served nine hundred hospitals spread over the territory of France.[125]

However, the expansion of 'motordom' during the war did not precondition a physical involvement in battles, as the example of the neutral Netherlands shows. Without an automobile industry of any significance, the Dutch army had set up seven automobile companies, maintained in two large workshops in Delft and Schiedam (near Rotterdam) where the "*Motordienst*" (the Car Service Department) also had a mobile workshop designed by one of the major motoring publicists from the prewar period. According to the journal of the Dutch auto club, both in France and England "an army of new automobilists" was formed, standing at 2.5 and 3.5 million respectively. These were drivers who were not, as before the war, trained to drive one single car, but who had become universalists, able to drive both passenger cars and trucks. By 1918, even in the neutral Netherlands about three thousand soldiers had been trained as drivers, several of them in possession of a driver's license, a novelty in Dutch automobilism.[126] Likewise, in Switzerland, another small, neutral country, the army had by the end of the war a *"Service automobile"* of 2344 motor vehicles at its disposal, including 240 motorcycles, most of them with sidecars, 800 passenger cars, and 1200 trucks, an impressive increase from the 40 held by the *"Corps des automobilistes volontaires de 1914."*[127] Möser estimates the number of Germans who received a car driver's course from the army during the war at 150,000.[128] Perhaps the nonmaterial consequences of the war were even more important. In the United States a Forest Service employee opined that the war "introduced a large number of the male population of the country to outdoor life and physical exercise in the open."[129] Likewise, the son of the famous General Doumenc claimed at a conference in 1981, using the notes of his father, that "a considerable part of the professionals (*'cadres'*) in the car industry until the 1950s were educated in the repair parks of the automobile service for the war of 1914–1918."[130]

Whether the First World War was in fact a 'railroad war' or not, in the (American) perception what had happened during 1917 and 1918 was quite clear: it was (in the words of the British Lord Curzon) "a victory of allied motors over German railroads."[131]

Notes

1. Guillaume Apollinaire, "La Petite Auto/The Little Car," in Donald Revell (transl.), *The Self-Dismembered Man: Selected Later Poems by Guillaume Apollinaire* (Middletown, CT: Wesleyen University Press, 2004), 68–75, here 75.
2. Apollinaire, "La Petite Auto/The Little Car," 73.
3. Rüdiger Rabenstein, *Radsport und Gesellschaft: ihre sozialgeschichtlichen Zusammenhänge in der Zeit von 1867 bis 1914* (Hildesheim, Munich, and Zurich, 1991), 249.
4. Immo Sievers, *AutoCars: Die Beziehungen zwischen der englischen und der deutschen Automobilindustrie vor dem Ersten Weltkrieg* (Frankfurt, Berlin, Bern, New York, Paris, and Vienna, 1995), 183.
5. Friedrich Schildberger, "Die Entstehung der Automobilindustrie in den Vereinigten Staaten" (manuscript Daimler Benz archives, Stuttgart), 87–88; Paul A. C. Koistinen, *Mobilizing for Modern War: The Political Economy of American Warfare, 1865–1919* (Lawrence, KS, 1997), 95.

6. The overview was based on "a recent issue of *Armee und Marine*"; see "Military Automobiles," *Horseless Age* 8, no. 15 (10 July 1901), 335. The following pages are based upon Gijs Mom and David Kirsch, "The Holy Road to the Automobile System: Interactions between the Military and Early Automobilism, 1898–1920," unpublished paper presented at the annual SHOT Conference in San Jose, California (4–7 October 2001). I thank the participants to the session in which the paper was presented for their helpful comments.
7. British army: Norman Miller Cary Jr., "The Use of the Motor Vehicle in the United States Army, 1899–1939" (unpublished dissertation, University of Georgia, 1980), 6; also see: E. S. Shrapnell-Smith, "Five decades of commercial road transport with inferences about its future," *Journal of the Institute of Transport* (February–March 1946), 214–229, here 214; *Selbstfahrer*: Dieter Storz, *Kriegsbild und Rüstung vor 1914: Europäische Landstreitkräfte vor dem Ersten Weltkrieg* (Herford, Berlin, and Bonn: Verlag E. S. Mittler & Sohn, 1992), 353; France: Rémy Porte, *La Direction des Services Automobiles des Armées et la Motorisation des Armées Françaises (1914–1919)* (Panazol: Charles Lavauzelle, 2004), 9.
8. "Concours de véhicules automobiles industriels," *Locomotion Automobile* (27 July 1905), 28–30; "Le projecteur Kriéger," *Locomotion Automobile* (28 September 1905), 165.
9. Mark Adams, "The Automobile—A Luxury Becomes a Necessity," in Walton Hamilton et al., *Price and Price Policies* (New York and London: McGraw-Hill, 1938), 27–81, here 29 (n. 1).
10. Storz, *Kriegsbild und Rüstung vor 1914*, 353.
11. Winfried Baumgart, "Eisenbahnen und Kriegsführung in der Geschichte," *Technikgeschichte* 38 (1971) no. 3, 191–219, here 197, 200, 210 (railroad war, italics in original; my translation); David Stevenson, "War by timetable? The railway race before 1914," *Past and Present* 162 (1999), 161–194, here 169 (trebled and silent revolution), 171 (norm), 183 (superiority), 190, 193; A. J. P. Taylor, *War by Time-Table: How the First World War Began* (London and New York: Macdonald, 1969); Geddes: David Stevenson, *With Our Backs to the Wall; Victory and Defeat in 1918* (London/New York: Allen Lane/Penguin, 2011) 226; Von Clausewitz: Stephan Rammler, *Mobilität in der Moderne: Geschichte und Theorie der Verkehrssoziologie* (Berlin: edition sigma, 2001), 70; Hervé Mathe, "Problème de la conduit de la démarche logistique: le concept de logistique dans une histoire contemporaine de la pensée sur la guerre," in Gérard Canini (ed.), *Les fronts invisibles: Nourrir—fournir—soigner; Actes du colloque international sur La logique des armées au combat pendant la Première Guerre mondiale organisé à Verdun les 6, 7, 8 juin 1980 sous la présidence du général d'Armée F. Gambiez, membre de l'Institut* (Nancy: Presses Universitaires de Nancy, 1984), 21–39, here 22–23; Horst Rohde, "Facktoren [sic] der deutschen Logistik im Ersten Weltkrieg," in ibid., 103–122, here 103.
12. Kurt Möser, *Geschichte des Autos* (Frankfurt and New York: Campus Verlag, 2002), 124–125; W. Zweerts de Jong, *De geschiedenis van het Vrijwillig Militair Automobiel Korps, naar verschillende gegevens verzameld en bewerkt* (Haarlem: J. A. Boom, 1918), 9–10, 18–19, 24, 32, 114–115 (free petrol and tyres); *Gedenkboek van het 25-jarig bestaan der Koninklijke Nederlandsche Automobiel Club, 1898–3 juli–1923* (Haarlem, 1923), 78–80.
13. Gerhard Horras, *Die Entwicklung des deutschen Automobilmarktes bis 1914* (Munich, 1982), 285.

14. *De Kampioen* (6 February 1903) (quoted in Fons Alkemade, *Het beeld van de auto 1896–1921: verslag van een speurtocht door Nederlandse collecties* (Deventer 1996) 37), my translation.
15. Ibid. Not all cars were deployed in a manner loyal to the state powers. Also the anarchist Domela Nieuwenhuis and the poet and radical leader of the socialist movement Herman Gorter used bicycles and automobiles during the strike. A. J. C. Rüter, *De spoorwegstakingen van 1903: Een spiegel der arbeidersbeweging in Nederland* (Leiden: E. J. Brill, 1935), 291 (note 5). *Gedenkboek der Werkstakingen van 1903 (Actestukken der Samenzwering), Deel I: Voorspel. Eerste en Tweede Periode* (Wageningen: N. V. Drukkerij "Vada," 1903), 353; Rüter, *Deel II: Derde en Vierde Periode, Naspel, Bijlage en Register* (Wageningen: N. V. Drukkerij "Vada," 1904), 553.
16. Anne-Marie Mans, "Le Service Automobile vu sous l'angle du soutien, 1913–1938," *Revue historique des armées* no. 3 (1980), 37–52, here 42.
17. Sir Francis Jeune, "The Charms of Driving in Motors," in Alfred C. Harmsworth, *Motors and Motor-Driving* (London and Bombay: Longmans, Green, and CO., 1902), 341–345, here 341–343.
18. Möser, *Geschichte des Autos*. Christoph Maria Merki, *Der holprige Siegeszug des Automobils 1895–1930: Zur Motorisierung des Strassenverkehrs in Frankreich, Deutschland und der Schweiz* (Vienna, Cologne, and Weimar: Böhlau, 2002), 257.
19. Peter Gay, *The Bourgeois Experience, Victoria to Freud*, vol. 3: The Cultivation of Hatred (New York and London: W. W. Norton Company, 1993), 4 (constant), 7 (ally), 115 (manliness), 532 (Rapaport; italics in original), 534 (puzzle), 535 (repeat); Slavoj Žižek, *Violence: Six Sideways Reflections* (New York: Picador, 2008), 206. On Freud's struggle with aggression, see Samuel B. Kutash, "Psychoanalytic Theories of Aggression," in Irwin L. Kutash et al., *Violence: Perspectives on Murder and Aggression* (San Francisco, Washington, DC, and London: Josey-Bass Publishers, 1978), 7–28. Freud concluded ultimately that "man possesses an aggressive drive from within" which he coupled with the sexual drive. Tellingly, after the First World War he developed a different theory, based on a coupling of life and death instincts (Eros and Thanatos) (ibid., 8); Hans Achterhuis, *Met alle geweld: Een filosofische zoektocht* (Rotterdam: Lemniscaat, 2008), 226. Achterhuis acknowledges the existence of a mild form of 'structural violence' (ibid., 227).
20. Žižek, *Violence*, 1–2; Arthur Redding, *Raids on Human Consciousness: Writing, Anarchism, and Violence* (Columbia: University of South Carolina Press, 1998), 4–5 (the first set of theories are called "structuralist-functionalist" in sociology, but Redding coined the names "exceptionalist" and "irrationalist" for them); Eli Osman and Ching-tse Lee, "Sociological Theories of Aggression," in Kutash et al., *Violence*, 58–73, here 60; Dave Grossman, *On Killing: The Psychological Cost of Learning to Kill in War and Society* (New York, Boston, and London: Back Bay Books / Little, Brown and Company, 2009), 13, 134–135 (Grossman is a former paratrooper and a psychology professor at West Point). Grossman expresses the relationship between violence and sexuality in the subtle statement that the carrying of a gun is "like having a permanent hard-on."
21. Raymond Williams, *Keywords: A Vocabulary of Culture and Society* (New York: Oxford University Press, 1985; revised edition), 329–331.
22. James H. Meisel, "Georges Sorel's Last Myth," *The Journal of Politics* 12, no. 1 (February 1950), 52–65, here 52; creative power: "Sorel, Georges" (*The Co-*

lumbia Encyclopedia, 6th edition, 2001–2007; http://www.bartleby.com/65/so/ Sorel-Ge.html, last accessed 20 January 2008).
23. Gay, *The Cultivation of Hatred*, 9–33.
24. Ibid., 35 (alibis), 97 (Bierbaum).
25. Ulrike Brunotte and Rainer Herrn, "Statt einer Einleitung: Männlichkeit und Moderne—Pathosformeln, Wissenskulturen, Diskurse," in Brunotte and Herrn (eds.), *Männlichkeiten und Moderne: Geschlecht in den Wissenskulturen um 1900* (Bielefeld: transcript Verlag, 2008), 9–23, here 10; hegemonic: Wolfgang Schmale, *Geschichte der Männlichkeit in Europa (1450–2000)* (Vienna, Cologne, and Weimar: Böhlau Verlag, 2003), 153, 195; Christa Hämmerle, "Zur Relevanz des Connell'schen Konzept hegemonialer Männlichkeit für 'Militär und Männlichkeit/en in der Habsburgermonarchie (1868–1914/18),'" in Martin Dinges (ed.), *Männer—Macht—Körper: Hegemoniale Männlichkeiten vom Mittelalter bis heute* (Frankfurt and New York: Campus Verlag, 2005), 103–121, here 104–105, 109.
26. See, for instance, Douglas Brinkley, *The Wilderness Warrior: Theodore Roosevelt and the Crusade for America* (New York: Harper, 2009), passim. See also Gay, *The Cultivation of Hatred*, 116–127, and Gail Bederman, *Manliness & Civilization: A Cultural History of Gender and Race in the United States, 1880–1917* (Chicago and London: The University of Chicago Press, 1995), 170–215.
27. Jürgen Martschukat and Olaf Stieglitz, *Geschichte der Männlichkeiten* (Frankfurt and New York: Campus Verlag, 2008), 113–115 (quotation on 115); Stefan Dudink, Karen Hagemann, and John Tosh (eds.), *Masculinities in Politics and War: Gendering Modern History* (Manchester and New York: Manchester University Press, 2004), 32.
28. Michael C. C. Adams, *The Great Adventure: Male Desire and the Coming of World War I* (Bloomington and Indianapolis: Indiana University Press, 1990), 7, 37 (male bonding), 45 (rupture), 48 (motherhood).
29. Paul Lerner, *Hysterical Men: War, Psychiatry, and the Politics of Trauma in Germany, 1890–1930* (Ithaca, NY, and London: Cornell University Press, 2003), 62 (quotation), 156.
30. Allen J. Frantzen, *Bloody Good: Chivalry, Sacifice, and the Great War* (Chicago and London: The University of Chicago Press, 2004); knight-errant: Adams, *The Great Adventure*, 65.
31. Olaf von Fersen (ed.), *Ein Jahrhundert Automobil-technik: Nutzfahrzeuge* (Düsseldorf, 1987), 18; "10 Jahre NAG," *Stahlrad und Automobil* 27, no. 1 (7 January 1912), 12–17; Sievers, *AutoCars*, 214–216; Storz, *Kriegsbild und Rüstung vor 1914*, 353.
32. Cary, "The Use of the Motor Vehicle in the United States Army," 11–21, 94.
33. Ibid., 29–30. In 1908 the US Army bought two additional electric buses (ibid., 32).
34. Ibid., 33, 39–43, 49.
35. Marc K. Blackburn, "A New Form of Transportation: The Quartermaster Corps and Standardization of the United States Army's Motor Trucks 1907–1939" (dissertation, Temple University, August 1992), 16, 21–22.
36. Cary, "The Use of the Motor Vehicle in the United States Army," 52 (quotations on 79, 81).
37. Wolfgang J. Mommsen, "Conclusion," in Jean-Jacques Becker et al., *Guerre et cultures 1914–1918* (Paris: Armand Colin, 1994), 429–442, here 431; Modris Eksteins, *Tanz über Graben: Die Geburt der Moderne und der Erste Weltkrieg*

(Reinbek bei Hamburg: Rowohlt Verlag, 1990), 147; Sarajevo: Taylor, *War by Time-Table*, 51–59 (I thank Clay McShane for suggesting the Sarajevo attack to me); "precipitous diplomacy," a quote from Stephen Kern, *The Culture of Time and Space*, cited in Kristen Whissel, *Picturing American Modernity: Traffic, Technology, and the Silent Cinema* (Durham, NC, and London: Duke University Press, 2008), 224.

38. John Westwood, *Railways at War* (London: Osprey, 1980), 134. 760 km: Stevenson, *With Our Backs to the Wall*, 8. Estimates about the number of train wagons deployed during mobilization vary widely between scholars. Franco-Prussian war: Bart Stol, "Iedereen wilde oorlog in 1870," *Historisch Nieuwsblad* (September 2007), 45–49, here 46.

39. Stanley Cooperman, *World War I and the American Novel* (Baltimore, MD: The Johns Hopkins Press, 1967), 99–100.

40. For the following paragraphs, see Mark Thompson, *The White War: Life and Death on the Italian Front 1915–1919* (London: Faber and Faber, 2008), 39ff.; see also Angelo Bazzanella, "Die Stimme der Illiteraten: Volk und Krieg in Italien 1915–1918," in Klaus Vondung (ed.), *Kriegserlebnis: Der Erste Weltkrieg in der literarischen Gestaltung und symbolischen Deutung der Nationen* (Göttingen: Vandenhoeck & Ruprecht, 1980), 334–351.

41. Gas attacks and brotherhood: Eksteins, *Tanz über Graben*, 246–247, 348; United States: Manfred Henningsen, "Das amerikanische Selbstverständnis und die Erfahrung des Grossen Kriegs," in Vondung (ed.), *Kriegserlebnis*, 368–386; Kramer, *Dynamic of Destruction*, 94 (last quotation), 105 (fifteen million mobilized), 107 (Russia), 121 (Italy), 165–175 (Futurists), 251 (fifteen million deaths); horses: Leo van Bergen, *Before My Helpless Sight: Suffering, Dying and Military Medicine on the Western Front, 1914–1918* (Franham and Burlington, VT: Ashgate, 2009), 26; two million: Karsten Alnæs, *De geschiedenis van Europa, 1900–heden: Onbehagen* (Amsterdam and Antwerpen: Ambo/Standaard Uitgeverij, 2007), 166–167; weaker countries: James J. Sheehan, *The Monopoly of Violence: Why Europeans Hate Going to War* (London: Faber and Faber, 2008), 100; more than half: Bernard Wasserstein, *Barbarij en beschaving: Een geschiedenis van Europa in onze tijd* (n.p. [Amsterdam]: Nieuw Amsterdam, n.d. [2008]), 95; amusement: Christiane Eisenberg, "Geselligkeit im Kaiserreich und in der Weimarer Republik: Das Beispiel des Sports," in Werner Plumpe and Jörg Lesczenski (eds.), *Bürgertum und Bürgerlichkeit zwischen Kaiserreich und Nationalsozialismus* (Mainz: Verlag Philipp von Zabern, 2009), 95–106, here 104. For a relativization of 'war enthusiasm,' see: Niall Ferguson, *The Pity of War* (New York: Basic Books, 1999), 175–186. Enthusiasm of workers and the middle class: Alan A. Jackson, *The Middle Classes 1900–1950* (Nairn: David St. John Thomas Publisher, 1991), 321; soccer balls: Mark A. Robison, "Recreation in World War I and the Practice of Play in *One of Ours*," *Cather Studies* 6 (2006), 1–22, here 6 (http://cather.unl.edu/cs006_robison.html, last accessed 18 September 2011).

42. Alnæs, *De geschiedenis van Europa*, 121; Jeff Lipkes, *Rehearsals: The German Army in Belgium, August 1914* (Leuven: Leuven University Press, 2007), 13 (20 percent), 17, 565; Alan Kramer, *Dynamic of Destruction: Culture and Mass Killing in the First World War* (Oxford and New York: Oxford University Press, 2007), 7; one million: Gijs Mom and Ruud Filarski, *Van transport naar mobiliteit: De mobiliteitsexplosie (1895–2005)* (Zutphen: Walburg Pers, 2008), 107: most of the Belgian returned to their country after a year, but 100,000 stayed in exile for the remainder of the war.

43. Kramer, *Dynamic of Destruction,* 55 and 127–129 (Italy), 135 (Balkan), 139 (cultural war), 149 (Armenian genocide), 151 (the "foreign"), 153 (mobilization of entire society), 160 (quotation); clash and dampened: Eksteins, *Tanz über Graben,* 107, 133; total war: Lipkes, *Rehearsals,* 17; community: Alnæs, *De geschiedenis van Europa,* 115.
44. Jost Dülffer, "Attentes allemandes, attentes italiennes: Préfigurations de la guerre en Allemagne avant 1914," in Becker et al., *Guerre et cultures 1914–1918,* 65–78, here 73 and passim.
45. James M. Laux, *The European Automobile Industry* (New York: Twayne, 1992), 33, 45–49; Laux, "Trucks in the West during the First World War," *The Journal of Transport History* 6, no. 2 (September 1985), 64–70; Helmut Otto, "Die Herausbildung des Kraftfahrwesens im deutschen Heer bis 1914," *Militärgeschichte,* 3/1989, 227–236; Kurt Möser, "World War One and the Creation of Desire for Cars in Germany," in Susan Strasser, Charles McGovern, and Matthias Judt, *Getting and Spending: European and American Consumer Societies in the Twentieth Century* (Cambridge: Cambridge University Press, 1998), 195–222; largest battle: Holger H. Herwig, *The Marne, 1914: The Opening of World War I and the Battle That Changed the World* (New York: Random House, 2009), xii.
46. See, for instance, Baumgart, "Eisenbahnen und Kriegsführung in der Geschichte."
47. T. C. Barker, "The Spread of Motor Vehicles before 1914," in Charles P. Kindleberger and Guido di Tella (eds.), *Economics in the Long View: Essays in Honor of W. W. Rostow;* vol. 2: Applications and Cases, Part I (London and Basingstoke, 1982), 149–167, here 156; French statistics 1914 and 1918: [Aimé] Doumenc, *Les transports automobiles sur le front français 1914–1918* (Paris: Plon-Nourrit et Cie., 1920); André Duvignac, *Histoire de l'armée motorisée* (Paris: Imprimerie Nationale, 1947), 179; use of French equipment: Porte, *La Direction des Services Automobiles des Armées,* 232. For a request from Renault to the French government to boycott Ford because of its preference "for our enemies," see ibid., 104. The French car industry profited extensively from the war situation, so much so that a military accountancy check revealed price increases of between 15 and 40 percent. 476: Stevenson, *With Our Backs to the Wall,* 27; mother: Möser, *Geschichte des Autos,* 133; useful: Stevenson, *With Our Backs to the Wall,* 211–222 (esp. 212, 222).
48. Eksteins, *Tanz über Graben,* 253.
49. Japanese: "Enormous use of motor vehicles in war," *NELA Bulletin* (October 1917), 764; according to Porte, *La Direction des Services Automobiles des Armées,* 294, the French armies had 95,000, the British 45,000 (including 800 caterpillar vehicles), and the Americans 40,000.
50. Alnæs, *De geschiedenis van Europa,* 252.
51. Blackburn, "A New Form of Transportation," 37–43 (quotation: 37); for the punitive expedition, see also: Cary, "The Use of the Motor Vehicle in the United States Army," 96 ff.
52. Blackburn, "A New Form of Transportation," 42–43.
53. J. Caloni, *Comment Verdun fut sauvé* (Paris: Étienne Chiron, 1924).
54. Henri Carré, *La véritable histoire des taxis de la Marne (6, 7 et 8 Septembre 1914)* (Paris: Librairie Chapelot, 1921), 19–20, 27, 30–35, 55–57, 87–93; Doumenc, *Les transports automobiles sur le front français,* 4–6.
55. Porte, *La Direction des Services Automobiles des Armées,* 156–157 (Gallieni: 157); Wasserstein, *Barbarij en beschaving,* 74.
56. Pol Ravigneaux, "Les nouveaux fiacres automobiles," *La vie automobile* (19 De-

cember 1905), 801–803; Claude Rouxel, *La grande histoire des taxis français 1898–1988* (Pontoise, 1989), 38, 112–113; René Bellu, *Toutes les Renault* (Paris, 1979), 25, 112–113; Mom, *The Electric Vehicle*, 132–133; insignificant: Herwig, *The Marne*, 262.

57. Porte, *La Direction des Services Automobiles des Armées*, 159–163.
58. For the concept of "reverse salient," see Thomas P. Hughes, *Networks of Power: Electrification in Western Society, 1880–1930* (Baltimore, MD, and London: Johns Hopkins University Press, 1983). Hughes borrowed the phrase from the military.
59. Paul Heuzé, *La Voie Sacrée: Le Service automobile à Verdun (Février–Août 1916)* (Paris: La Renaissance du Livre, 1919), 36, 42 (motorcycles), 52 (1,200 workers); see also "La Voie Sacrée," *La lettre de la fondation de l'automobile Marius Berliet* no. 94 (July–August 2001), 9.
60. Porte, *La Direction des Services Automobiles des Armées*, 173–186, quotations: 178, 185; German scepticism: Baumgart, "Eisenbahnen und Kriegsführung in der Geschichte," 213; prepare: Möser, *Geschichte des Autos*, 129; veins: Doumenc, *Les transports automobiles sur le front françois*, 120; Heuzé, *La Voie Sacrée*, 2 (unfavorable), 9 (order and speed); Westwood, *Railways at War*, 145; Amiens: Georges Reverdy, *Les routes de France du XXe siècle 1900–1951* (Paris: Presses de l'école nationale des ponts et chaussées, 2007), 65; myths: Barbara Korte, Ralf Schneider, and Claudia Sternberg, "Einleitung und Grundlegung," in Korte, Schneider, and Sternberg (eds.), *Der Erste Weltkrieg und die Mediendiskurse der Erinnerung in Grossbritannien: Autobiographie—Roman—Film (1919–1999)* (Würzburg: Königshausen & Neumann, 2005), 1–32, here 29.
61. Porte, *La Direction des Services Automobiles des Armées*, 9–23; Heuzé, *La Voie Sacrée*, 7.
62. Heuzé, *La Voie Sacrée*, 16–17, 38 (quotation; italics in original), 41 (insigne).
63. See, for instance, John Arquilla and David Ronfeldt, *Swarming & the Future of Conflict* (Rand National Defense Research Institute, n.d.).
64. Porte, *La Direction des Services Automobiles des Armées*, 58, 77, 83, 87, 110, 132–133, 196, 207, 219; Heuzé, *La Voie Sacrée*, 55. One thousand women: Mans, "Le Service Automobile vu sous l'angle du soutien, 1913–1938," 41. Mans explains that during the entire war there existed two separate "services automobiles" in the French army, one central (dealt with by Doumenc in the text) and one belonging to the Artillery (ibid., 43). On the development of the Renault tank, see Gilbert Hatry, "Les rapports gouvernement, armée, industrie privée pendant la Première Guerre mondiale: le cas des usines Renault," in Canini (ed.), *Les fronts invisibles*, 171–188.
65. Porte, *La Direction des Services Automobiles des Armées*, 157. The best treatment of the maintenance and repair activities of the French army during WWI can be found in Doumenc, *Les transports automobiles sur le front français*, 126–139.
66. Porte, *La Direction des Services Automobiles des Armées*, 222–223; last quotation: Daniel R. Beaver, "The Problem of American Military Supply, 1890–1920," in Benjamin Franklin Cooling (ed.), *War, Business, and American Society: Historical Perspectives on the Military-Industrial Complex* (Port Washington, NY, and London: Kennikat Press, 1977), 73–92, here 79, 83, 87; two million: Westwood, *Railways at War*, 156; 120: Marc K. Blackburn, *The United States Army and the Motor Truck: A Case Study in Standardization* (Westport, CT, and London: Greenwood Press, 1996), 32.

67. Alnæs, *De geschiedenis van Europa,* 131, 155; Rohde, "Facktoren [sic] der deutschen Logistik im Ersten Weltkrieg," 106 (*Materialschlacht*), 117–119 (cadres). German shortage of trucks: Daniel R. Beaver, "Politics and Policy: The War Department Motorization and Standardization Program for Wheeled Transport Vehicles, 1920–1940," *Military Affairs* 47, no. 3 (October 1983), 101–108, here 101.
68. Stevenson, *With Our Backs to the Wall,* 242.
69. Rudy Koshar, *German Travel Cultures* (Oxford and New York: Berg, 2000), 68–69.
70. David W. Lloyd, *Battlefield Tourism: Pilgrimage and the Commemoration of the Great War in Britain, Australia and Canada, 1919–1939* (Oxford and New York: Berg, 1998), 20–21 (poem: 21). "Thanatourism" has been introduced as a concept by A. V. Seaton in 1999 in an analysis of Waterloo: Charlotte Heymel, *Touristen an der Front: Das Kriegserlebnis 1914–1918 als Reiseerfahrung in zeitgenössischen Reiseberichten* (Berlin: Lit Verlag Dr. W. Hopf, 2007), 17, 335 (Michelin). See also Susanne Brandt, "Le voyage aux champs de bataille," *Vingtième Siècle; Revue d'histoire* 41, no. 1 (1994), 18–22, here 20; Donald Weber, "Automobilisering en de overheid in België vóór 1940: Besluitvormingsprocessen bij de ontwikkeling van een conflictbeheersingssysteem" (dissertation, University of Ghent, 2008), 321.
71. Mom and Filarski, *De mobiliteitsexplosie (1895–2005),* 107, 110; Luxemburg: Heymel, *Touristen an der Front,* 73.
72. Rssbch, "Der nicht felddiensttaugliche Kraftfahrer zu Kriegszeiten," *Der Motorfahrer* 12, no. 33 (15 August 1914), 5–9, here 8 (cold-blooded); Wa.O., "Strassen frei!" *Der Motorfahrer* 12, no. 34 (22 August 1914), 6; Friedrich Baumann, "Eine Patrouillenfahrt im Oberelsass," *Der Motorfahrer* 12, no. 38 (19 September 1914), 7 (dripping); Ebel, "Meine Fahrt nach Frankreich," *Der Motorfahrer* 12, no. 47 (21 November 1914), 3–4, here 3; e., "Der Motor im Kriegsdienst," Ibid., 5 (conference and acknowledgment); F. R. Schmaltz, "Kriegserlebnisse eines Motorradfahrers," *Der Motorfahrer* 12, no. 50 (12 December 1914), 5–6, here 5; Wilhelm Hoerschgen, "Kriegs- und Schlachtenbericht," *Der Motorfahrer* 12, no. 51 (19 December 1914), 3–4 (*Schnellfahrer*).
73. Möser, *Geschichte des Autos,* 129.
74. Thomas F. Schneider, "Die Wiederkehr der Kriege in der Literatur: Voraussetzungen und Funktionen 'pazifistischer' und 'bellizistischer' Kriegsliteratur vom Ersten Weltkrieg bis zum Dritten Golfkrieg," *Osnabrücker Jahrbuch Frieden und Wissenschaft* 12 (2005), 201–221, here 203; speechless: Heymel, *Touristen an der Front,* 193; senseless machines: Wilhelm Deist, "La mort et le deuil: La 'moral' des troupes allemandes sur le front occidental à la fin de l'année 1916," in Becker et al., *Guerre et cultures 1914–1918,* 91–102, here 101.
75. Thompson, *The White War,* 181; Eksteins, *Tanz über Graben,* 261.
76. Peter Tauber, "Der Krieg als 'welterschütternde Olympiade': Der Sport als Allegorie für den Krieg in Briefen und Gedichten des Ersten Weltkrieges," *Militärgeschichtliche Zeitschrift* 66, no. 2 (2007), 309–330, here 326 (I thank Katie Ritson, Rachel Carson Centre, Munich, for her translation).
77. Klaus Vondung, "Einleitung: Propaganda oder Sinndeutung?" in Vondung (ed.), *Kriegserlebnis,* 11–37, here 13, 17–18 (quotations).
78. Eisenberg, "Geselligkeit im Kaiserreich und in der Weimarer Republik," 104.
79. Heymel, *Touristen an der Front,* 15 (theologicians), 35 (hotel), 72 (tourism stops), 24 (dialectical), 45–47 (Boy Scouts), 54 (expressionists); last quotation: Vondung, "Einleitung," 18–19.

80. Heymel, *Touristen an der Front*, 55.
81. Ibid., 188–189
82. Schneider, "Die Wiederkehr der Kriege in der Literatur," 202, 205 (quotation).
83. Pierre Schoentjes, *Fictions de la Grande Guerre: Variations littéraires sur 14–18* (Paris: Éditions Classiques Garnier, 2009), 252 (pleasure of violence: 120).
84. Joris van Parys, *Het leven, niets dan het leven: Cyriel Buysse en zijn tijd* (Antwerp and Amsterdam: Houtekiet/Atlas, 2007), 465 (Streuvels), 507 (newspaper), 512, 514 (quotations); Schilperoort: Anne-Katrin Ebert, "Ein Ding der Nation? Das Fahrrad in Deutschland und den Niederlanden, 1880–1940. Eine vergleichende Konsumgeschichte" (doctoral dissertation, Universität Bielefeld, Fakultät für Geschichtswissenschaft, Philosophie und Theologie, January 2009), 343
85. Eksteins, *Tanz über Graben*, 228.
86. Porte, *La Direction des Services Automobiles des Armées*, 226–227.
87. [Alice B. Toklas], *The Autobiography of Alice B. Toklas* (New York: Harcourt, Brace and Company, 1933), 207 (American Fund), 210 (truck body), 213–214 (adventures quote), 223 (officer); James R. Mellow, *Charmed Circle: Gertrude Stein & Company* (New York and Washington, DC: Praeger, 1974; first ed. 1974), 226–237 (quote on Auntie on 228); perverse: Mary Suzanne Schriber, *Writing Home: American Women Abroad 1830–1920* (Charlottesville and London: University Press of Virginia, 1997), 205.
88. Frank Hoyt Gailor, "An American Ambulance in the Verdun Attack" (originally published in *Cornhill Magazine*, July 1916), together with "An Introduction by Theodore Roosevelt" included in Abram C. Platt, *Friends of France: The Field Service of the American Ambulance Described by Its Members* (Boston, 1916), http://www.vlib.us/medical/FriendsFrance/ff02.htm (last accessed 26 April 2010); quotes on 2 (manliness), 12 (do their bit), 13 (Meuse), 14 (rivalry), 15 (romance); Paul Fussell, *The Great War and Modern Memory* (Oxford: Oxford University Press, 2000; first ed. 1975), 231 (anti-pastoral). For a recent critique of Fussell as a literary historian who privileges highbrow literature as "providing a predestined access to the experience of this war," whereas middlebrow and lowbrow texts produce alternative narratives conveying a more positive image, see Barbara Korte, "Erfahrungsgeschichte und die 'Quelle' Literatur: Zur Relevanz genretheoretischer Reflexion am Beispiel der britischen Literatur des Ersten Weltkriegs," in Jan Kusber et al. (eds.), *Historische Kulturwissenschaften: Positionen, Praktiken und Perspektiven* (Bielefeld: transcript Verlag, 2010), 143–159, here 148–154 (quotation: 148).
89. Edith Wharton, *Fighting France: From Dunkerque to Belfort* (New York: Charles Scribner's Sons, 1919; first ed. 1915), 7 (army), 75 (hospital visits), 95–96 (landscape description), 122–123 (dash etc), 139 (open car), 191 (sightseeing), 206 (frontline Vosges).
90. Kate McLoughlin, "Edith Wharton, War Correspondent," *Edith Wharton Review* 21, no. 2 (Fall 2005), 1–10, here 5–6; Edith Wharton, *The Marne: A Tale of the War* (London: MacMillan, 1918), 17 (donated), 93 (motor-trucks), 110 (accident), 120 (to the front), 135 (action).
91. Suzanne Raitt, "May Sinclair and the First World War," *Ideas* 6, no. 2 (1999), two parts, http://nationalhumanitiescenter.org/ideasv62/raitt.htm (last accessed 5 March 2012); Rosa Maria Bracco, *Merchants of Hope: British Middlebrow Writers and the First World War, 1919–1939* (Providence, RI, and Oxford: Berg, 1993), 44.

92. May Sinclair, *A Journal of Impressions in Belgium* (New York: The Macmillan Company, 1915), http://www.archive.org/details/journalofimpress00sinciala (last accessed 2 March 2012), 8 (sightseeing), 12 (drunk), 13 (danger), 21 (Cook), 27 (cinematograph), 35 (perilous), 52 (shell), 11 (Daimler), 69 (thirst), 71 (woman called back), 115 (no surprise), 120 (flight from Antwerp), 132 (fun), 135 (self-test), 148 (Motor Show), 151 (very fast; rush), 168 (happiness), 173 (tame affair), 208 (cyborg), 249 (flight from Ghent), 291 (Corps).
93. Helen Zenna Smith, *Not so Quiet... Stepdaughters of War* (New York: The Feminist Press at the City University of New York, 1989; first ed. 1930), 33 (bodily hell), 50 (cowards), 55 (inefficient), 59 (dirt, thoughtful), 61 (dexterity), 75 (driving sense), 113 (female war), 126 (virgin), 127 (cyborg experience), 145 (erotic adventure), 151 (cyborg: "my steering wheel"), 154 (last quotation); Jane Marcus, "Afterword: Corpus/Corps/Corpse: Writing the Body in/at War," in ibid., 241–300, here 242 (erotics of war), 257 (suffrage), 263 (romances), 266 (Winifred Young), 271 (acoustic), 285 (lesbianism).
94. Smith, *Not so Quiet*, 135.
95. Enid Bagnold, *A Diary Without Dates,* with a new introduction by Monica Dickens (London: Virago/William Heinemann, 1978; first ed. 1918); Enid Bagnold, *The Happy Foreigner* (Middlesex: The Echo Library, 2007; reprint of 1920), 4, 34, 95 (dog), 119 (garage).
96. Mary Borden, *The Forbidden Zone* (London: William Heineman, 1929); Marcia Phillips McGowan, "A Nearer Approach to the Truth: Mary Borden's *Journey Down a Blind Alley*," *War, Literature and the Arts* (n.d.), http://wlajournal.com/23_1/images/mcgowan.pdf (last accessed 30 March 2013); Jane Conway, *Mary Borden: A Woman of Two Wars* (n.p.: Munday Books, 2010), 23–36.
97. Heymel, *Touristen an der Front*, 331.
98. Celia M. Kingsbury, "'Squeezed into an Unnatural Shape': Bayliss Wheeler and the Element of Control in *One of Ours*," *Cather Studies* 6 (2006), 1–14, here 3, http://cather.unl.edu/cs006_kingsbury.html (last accessed 18 September 2011).
99. Michael McGerr, *A Fierce Discontent: The Rise and Fall of the Progressive Movement in America, 1870–1920* (Oxford and New York: Oxford University Press, 2003), 281–282.
100. http://de.wikipedia.org/wiki/Anton_Fendrich (last accessed 19 November 2009), and for Grabein, http://www.literaturport.de/index.php?id=26&no_cache =1&user_autorenlexikonfrontend_pi1 percent5Bal_aid percent5D=3090&user_ autorenlexikonfrontend_pi1 percent5Bal_opt percent5D=1 (last accessed 20 January 2010); Anton Fendrich, *Mit dem Auto an der Front: Kriegserlebnisse* (Stuttgart: Franckh'sche Verlagshandlung, n.d.; 51.–55. Tausend; first ed. 1915), 150; Paul Grabein, *Im Auto durch Feindesland: Sechs Monate im Autopark der Obersten Heeresleitung* (Berlin and Vienna: Verlag Ullstein & CO, 1916).
101. Heymel, *Touristen an der Front*, 173, 215, 317–318 (my translation).
102. Grabein, *Im Auto durch Feindesland*, 88, 98, 120, 145, 148, 183, 185 (*Vernichter*); Fendrich, *Mit dem Auto an der Front*, 16 (Belgians), 89 (gas attack), 98, 105, 107, 127.
103. Fendrich, *Mit dem Auto an der Front*, 30–31.
104. Grabein, *Im Auto durch Feindesland*, 55, 57, 68, 83, 96, 122, 147, 239.
105. Ibid., 50. A photograph of such a car can be seen in Duvignac, *Histoire de l'armée motorisée*, 267.
106. Fendrich, *Mit dem Auto an der Front*, 16–17 (tempo), 19 (racing), 32 (inversion), 40 (Bruges), 49 (crazy speed), 96 and 98 (senses); Grabein, *Im Auto durch*

Feindesland, 92 (cold quiet), 205 (cold blood), adventure lust: 72, quotation: 203, transporting wounded soldiers: 228.
107. Deist, "La mort et le deuil,"101.
108. Kramer, *Dynamic of Destruction,* 241, 255 (sensory overload), 256–257 (last quotations; the first quote is from Kramer himself).
109. Adrienne Thomas, *Die Katrin wird Soldat; Ein Roman aus Elsass-Lothringen* (Amsterdam: Allert de Lange, 1936) (225. Tausend; first ed.: 1930), 121 (honk), 182 (half-soldier); Rebecca Biener, "Die literarische Verteidigung des kleinen Glücks am Beispiel der Autorin Adrienne Thomas" (dissertation, Universität-Gesamthochschule Siegen, Sprach- und Literaturwissenschaften, January 2005), http://dokumentix.ub.uni-siegen.de/opus/volltexte/2007/286/pdf/biener.pdf (last accessed 2 March 2013).
110. Grabein, *Im Auto durch Feindesland,* 222, 236.
111. Kramer, *Dynamic of Destruction,* 224, 226, 234 (churches).
112. Alnæs, *De geschiedenis van Europa,* 340 (Ball and Hülsenbeck); Eksteins, *Tanz über Graben,* 317 (house of joy); on Dada and violence, see Hanno Ehrlicher, *Die Kunst der Zerstörung: Gewaltphantasien und Manifestationspraktiken europäischer Avantgarden* (Berlin: Akademie Verlag, 2001), chapter 4.
113. Joseph Vogl, "Krieg und expressionistische Literatur," in York-Gothart Mix (ed.), *Naturalismus, Fin de siècle, Expressionismus 1890–1918* (Munich: Deutsche Taschenbuch Verlag, May 2000), 555–565, here 558 (Mann), 561 (Behrens); brakes: Harro Segeberg, *Literatur im Medienzeitalter: Literatur, Technik und Medien seit 1914* (Darmstadt: Wissenschaftliche Buchgesellschaft, 2003), 23 (word artillery), 31 (brakes).
114. *Vollgas,* quoted in Dorit Müller, *Gefährliche Fahrten: Das Automobil in Literatur und Film um 1900* (Würzburg: Königshausen & Neumann, 2004), 128.
115. Meyer, "Kraftwagen und Krafträder im Feldzuge 1914," *Der Motorfahrer* 12, no. 44 (31 October 1914), 4–5, here 5.
116. Viktor Shklovsky, *A Sentimental Journey: Memoirs, 1917–1922,* translation and literary introduction by Richard Sheldon; historical introduction by Sidney Monas (n.p.: Dalkey Archive Press, 2004; first ed. 1923), 35 (race), 44 (cloud), 136 (seizing), 137 (horsepower), 167 (Packards), 195 (weapon), 210 (bombs), 233 (retardation), 234 (lectures); Richard Sheldon, "Making Armored Cars and Novels: A Literary Introduction," in ibid., ix–xxv, here xi (instructor), xvi (estrangement).
117. Willa Cather, *One of Ours* (New York: Vintage Books, 1991; first ed. 1922), 246 (tingling), 251 (enjoying), 327 (legitimate), 339 (adventure), 345 (no sport); war chronicle: Robison, "Recreation in World War I and the Practice of Play in *One of Ours.*" For the controversy around Cather's novel after the war, see Hermione Lee, *Willa Cather: Double Lives* (New York: Pantheon Books, 1989), 167–173.
118. Möser, *Geschichte des Autos,* 128; Kramer, *Dynamic of Destruction,* 23, 131; Porte, *La Direction des Services Automobiles des Armées,* 210, 255 (quotation).
119. Wasserstein, *Barbarij en beschaving,* 95.
120. Möser, "World War One and the Creation of Desire for Cars in Germany," 205; Wasserstein, *Barbarij en beschaving,* 94.
121. Georgine Clarsen, *Eat My Dust: Early Women Motorists* (Baltimore, MD: Johns Hopkins University Press, 2008), 32, 47–51 (factory), 53 (quotation).
122. David Mitchell, *Women on the Warpath: The Story of the Women of the First World War* (London: Jonathan Cape, 1966), 128 (Valkyries), 221–223 (WAAC), 227 (cemeteries), 228 (press conference).

123. Diane Atkinson, *Elsie and Mairi Go To War: Two Extraordinary Women on the Western Front* (London: Preface Publishing, 2009), 2–3, 22–23, 119, 147 (Scottish ladies), 172, 179, last quote under photo on second unnumbered page opposite 184.
124. "Automobile Transportation in France," *The Club Journal* 10, no. 1 (June 1918), 18; Susan Zeiger, *In Uncle Sam's Service: Women Workers with the American Expeditionary Force, 1917–1919* (Ithaca, NY, and London: Cornell University Press, 1999), 2 (16,500), 49 (adventuress), 129 (as dangerous), 167 (tour guide). Twelve thousand: Lettie Gavin, *American Women in World War 1: They Also Served* (Niwot: University Press of Colorado, 1997), 202 (for a photo see Ibid.).
125. Kimberly Chuppa-Cornell, "The U.S. Women's Motor Corps in France, 1914–1921," *The Historian* 56, no. 3 (Spring 1994), 465–476, here 468, 470.
126. Mom and Filarski, *De mobiliteitsexplosie (1895–2005)*, 114–115.
127. Jean-Philippe Matthey, "Un aspect de la motorisation dans le canton de Vaud: la question de l'interdiction dominicale de la circulation automobile (1906–1923)" (unpublished Master's thesis, Université de Lausanne, Faculté des Lettres, 1992), 55.
128. Kurt Möser, "'Der Kampf des Automobilisten mit der Maschine'- Eine Skizze der Vermittlung der Autotechnik und des Fahrenlernens im 20. Jahrhundert," in Lars Bluma, Karl Pichol, and Wolfhard Weber (eds.), *Technikvermittlung und Technikpopularisierung—Historische und didaktische Perspektiven* (Münster, New York, Munich, and Berlin: Waxmann, 2004), 89–102, here 100.
129. Quoted in Paul S. Sutter, *Driven Wild: How the Fight against Automobiles Launched the Modern Wilderness Movement* (Seattle and London: University of Washington Press, 2002), 25–26.
130. André Doumenc, "Les transports automobiles pendant la guerre de 1914–1918," in Canini (ed.), *Les fronts invisibles*, 371–380, here 375.
131. "Plan Sacred Highway," *American Motorist* 11, no. 4 (May 1919), 14.

PART II

Persistence (1918–1940)

CHAPTER 4

"Why Apologize for Pleasure?"
Consuming the Car in Boom and Bust

"Once a motorist, always a motorist."[1]

In December 1927 *The American Magazine* published a contribution written by department store truck driver Archie Chadbourne, accompanied by a picture of his wife and two daughters, without mentioning the area he was living, as if to emphasize the typicality of their case. Buying their clothes at the Salvation Army store, they belonged to the "respectable poor," he wrote, but their shortlist of expenditures also included "auto upkeep." That sum represented 4 percent of his monthly income of a hundred dollars, not much less than home economists calculated for the American average. Chadbourne spent this amount on a sixty-dollar second-hand car, for which he had to spend "five dollars down and five dollars a month." The Chadbournes were nonetheless optimistic spenders: "We're going to live in this world but once, so we must keep up with the crowd, no matter the cost." That's why they bought a second-hand phonograph, too, on payments, followed by a radio. "We have a car to take us away from home. Now we'll get a radio to keep us home." Chadbourne bought his tires at a scrap yard. The car's upkeep "costs a little more money than we anticipated [because] we've used it more than we thought we would. But, oh, boy! the pleasure we've had." For Chadbourne, getting in debt was "the only adventure a poor man can count on."[2] The article, whether doctored or not (the following month's contribution announced the tale of an orphan "backed up by facts out of the author's own life"), emphasized the respectability of the urban poor: spending on expensive, durable commodities was such a pervasive dream that even the poor had to be part of it, at whatever cost. "Why apologize for pleasure?" was the rhetorical question *Automotive Industries* asked its readers in 1924.[3]

In this chapter, the first of a second group of four, we will investigate the emergence and consequences of the first massive wave of car ownership in the Atlantic world and the reformulation of the automotive adventure this entailed during the interwar years. Different from what historiography suggests, we will approach the car ("that always up-to-date icon of middle-class modernity"[4]) as a commodity rather than a transportation device. Although the automobile's profile of use (its functional spectrum) in this phase cannot be thought of without the transportation function in mind, the unique characteristics of the car culture are only understandable, such is the claim of this chapter, by embedding it in a pervasive consumption culture. It is further claimed that the car should be treated not only (or perhaps not so much) as a status symbol, but also (and perhaps especially so) as a tool to consume space. In other words: not only the artifact 'car' is analyzed as a commodity, it is the commodification of the 'car system' that sets the automobile apart from other commodities in that the car *multiplies* consumption opportunities. We are especially interested in the question of whether the automotive adventure in this second period was still mainly fed by touring and other pleasurable uses (as described in chapter 1 for the previous period), or whether this type of use was replaced by more utilitarian forms of use, mostly commuting, as is claimed by historiographic consensus.

In our search for answers to these questions we will observe the shift of the center of gravity of the Atlantic automotive culture from Europe to the United States, a process that already had started before the war. This coincided with several other shifts, such as that from the car (both technically and culturally) as 'shell' (as we have characterized it in chapter 2) to the car as 'capsule': from an open tourer it evolved into a closed-bodied family car usable all year round in the city, but most of all in the countryside, on ever-longer trips. At the same time, the 'automotive subject' shifted from the small, elite situational group (the party of friends, the extended family, the couple in love) to the (lower-) middle-class nuclear family. This encapsulation of the family will be studied in detail in chapter 5.

We will also witness a shift in the way the car was seen by central government, industry, and car and touring clubs alike: whereas in the previous period the car was perceived as a seemingly autonomous artifact providing the motorist with an individualized pleasure quite distinct from the discipline of the train, now the automobile was taken up in a system of maintenance by a service infrastructure, and of registration and taxation by a bureaucracy on several governmental levels. Furthermore, it could no longer function without an extensive road and fuel system that covered the entire nation, as we will see in chapter 7.

And as in the previous phase, we will also dedicate a special chapter (chapter 6) to the reformulation of the adventure on the basis of literary sources and sources from related media. There, we will investigate another shift, from speed to space, from temporal to geographical aspects of the automotive adventure. Together, the four following chapters will enable us to characterize, in the concluding chapter, the 'Atlantic automobile,' at the moment of the outbreak of WWII.

In this chapter, we will first concentrate on figures and try to reconstruct the diffusion of the car throughout the North-Atlantic realm. Then we will focus on Europe and ask whether the car's diffusion on that continent was an example of 'Americanization.' The answer to that question enables us to analyze the use profile of the 'Atlantic automobile,' and especially address the question about the share of commuting. The next section investigates the trend to long-range use for pleasure and tourism on both continents. We will close this chapter by emphasizing the evolutionary aspect of this development, as well as the role of expectations.

The Car as Commodity: Its Spread Among the Atlantic Middle Class

Some 250,000 American soldiers per month were rushed onto the war scene (bringing with them supplies of corned beef, biscuits, wine, fur jackets, and fresh underwear, things German soldiers—whenever they stumbled upon an Allied depot—could only dream of) before the guns fell silent on the European battlefields on 11 November 1918. This was in the middle of a revolutionary and counter-revolutionary wave of violence that had not only overthrown the autocratic Russian regime but also led to the setting up of soviets in towns as dispersed as Turin, Munich, Vienna, Glasgow, and even Seattle. Quite a few women took part in the riots: "They were not always the ones throwing the stones, but they organized the protests and encouraged or directed their adolescent sons." During the Russian Civil War ten million died, more than five times the number during WWI. In Italy, the *Arditi*, privately organized fighter groups of mostly middle-class background, most of them former soldiers and led by Futurists and other car lovers, initiated a wave of fascist violence against peasant and worker uprisings in Turin and Tuscany, while the mini-imperialist but at the same time anarcho-syndicalist occupation of Fiume (current-day Rijeka) under the leadership of Futurist D'Annunzio challenged the peace treaties of London and Versailles. The United States had its own anti-radical Red Scare including a revived Ku Klux Klan targeting African Americans, Jews, and Catholics. The new president Harding promised "normalcy" and "a return to a more individualistic society."[5]

"Everyone knows, everyone feels that this peace is only a poor covering thrown over ambitions unsatisfied, hatred more vigorous than ever, national anger still smouldering." This observation was not made by a belligerent Hitler or Mussolini, but by a captain of the French army taken prisoner in Verdun, Charles de Gaulle, an observation made in a country that was relatively free of postwar violence. In Germany however, the general "passionate longing for radical change" fueled uprisings in Kiel, Hamburg, Berlin, and Munich, which were answered by a group of twenty to forty thousand right-wing fighters, young middle-class men "addicted to violence" and gathered in the *Freikorps,* secretly supported by the state until they too tried a coup, the Kapp *putsch* of March 1920. In all these struggles technology, and especially automotive technology in the form of cars and mostly trucks, played an important supportive role.

Belgium and northern France were completely destroyed. After the German political elite had grudgingly accepted the obligation to make reparation payments, real wages would not reach prewar levels until 1928. Although the growth rates of the gross national product were similar in Germany and France, in Germany this was based on capital goods made by large export-oriented conglomerates of iron and steel producers; in France, with an exceptionally low birth rate, an unusually high rate of women working outside the home and a broad middle class, an eager internal market was starting to absorb consumer goods produced by middle-sized manufacturers. And although the United Kingdom mourned its 'lost generation,' they had half as many deaths as France, the survivors being ready to indulge in mobile leisure as soon as their increasing income (and decreasing car prices) would allow.[6]

Indeed, in most European countries the 1920s showed a spectacular wave of large-scale car diffusion among a new layer of middle-class owners, thus accelerating a process started well before the war. The United Kingdom and France were most eager, and, as they, together with the United States, had won the "technology race" during the war; they were indeed able to motorize the first.[7] But behind them a broad band of diffusion rates showed large differences among the countries—Italy, Germany, and the Netherlands at the bottom of the scale, Belgium and Sweden in the middle range, while Denmark initially followed France and the United Kingdom but during the 1930s was falling back together with other Scandinavian countries as Germany started to catch up (figure 4.1).

In France, car *use* during the interwar years grew more then tenfold, from 1640 (1921) to 18,080 (1939) vehicle-kilometers per year, a spatial expansion that mainly took place in a broad area around Paris and in

"Why Apologize for Pleasure?"

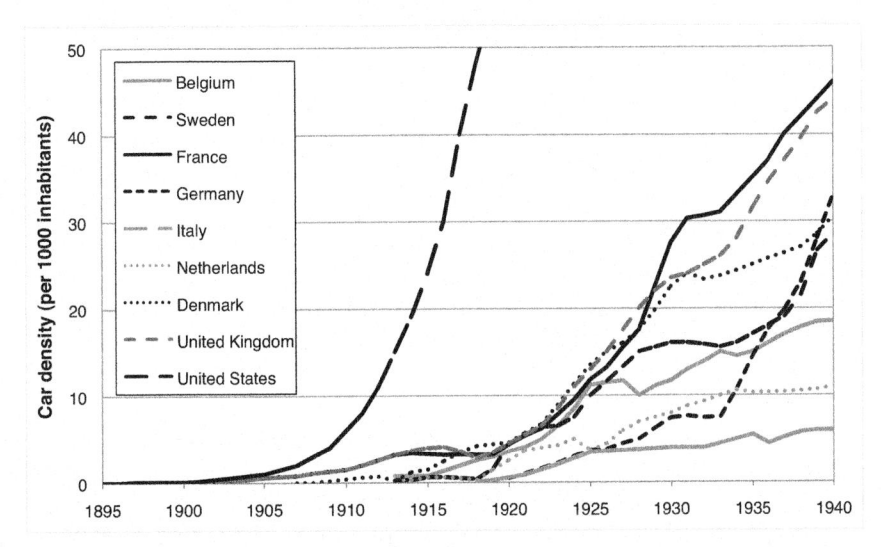

Figure 4.1. Car densities (per one thousand inhabitants) in selected countries until 1940.
Source: See figure 1.2

the Bordeaux region, as well as a wealthy corridor in-between.[8] The highest number of car *purchases* per capita could also be found around Paris. But focusing on cars can be misleading: the countryside middle classes motorized predominantly on the motorcycle, which showed the highest densities in the north- and southeast. These must have been used mainly for local needs, as their presence was hardly expressed in the road censuses measuring mainly interurban traffic. Only during the 1930s did the countryside take the lead, but we should be careful what we mean by that: in 1932, two-thirds of car owners lived in villages or small rural towns, while farmers only owned 3 percent of all registered cars. Shortly thereafter the second-hand market was 2.3 times as big as the new car market. That was nearly as much as in the United States (as we will see later), but only 14 percent of the cars were bought on credit.[9] Neighboring Switzerland, going through a comparable diffusion wave, only had 28 percent of its cars as second-hand as late as 1938, probably the result of a higher middle-class income.[10]

In Germany, a correspondent on an American car journal found hardly any cars on the roads in 1920. And when the ban on the use of automobiles was alleviated a year later (and the German car and motorcycle club ADAC referred to a "right on pleasure" for its members), rampant inflation resulted in a redistribution of income to a small elite of industrial and commercial bourgeois (*Wirtschaftsbürger*) and those who had gathered a fortune during the war, while the traditional elite (the *Bildungsbürger*) lost most of their money. As a response, the lower middle classes (encompassing the so-called *Angestellte* and the technical intelligentsia) and higher echelons of the workers started to buy motorcycles in numbers unequaled in other European countries. On average, two-thirds of the vehicles driving on the German roads were motorcycles. Forty-four percent of all cars were registered in large cities at the end of the 1920s, whereas three-quarters of new cars were owned by tradesmen, free professions, and government institutions, and a second-hand market had hardly developed. By the beginning of the 1930s, however, also in Germany the second-hand market amounted to 73 percent of the total.[11]

Despite the efforts in the historiography to emphasize the national exceptionalisms rather than the European commonalities, or to attune every scholarly judgment to what happened on the other side of the ocean, Germany too witnessed a brief but spectacular phase of motorization until the end of the 1920s: from around 1924 the membership of the middle-class club ADAC expanded, after a lull during the first half of the 1920s, with the same enthusiasm as in other countries (figure 4.2). The expansion must have been caused, at least to a large part, by motorcycle owners, although in Germany too (as in France) their number expanded the fastest outside the big cities. An explosion of motorized bicycles (*Kleinkrafträder*, exempt from taxation since April 1928 if

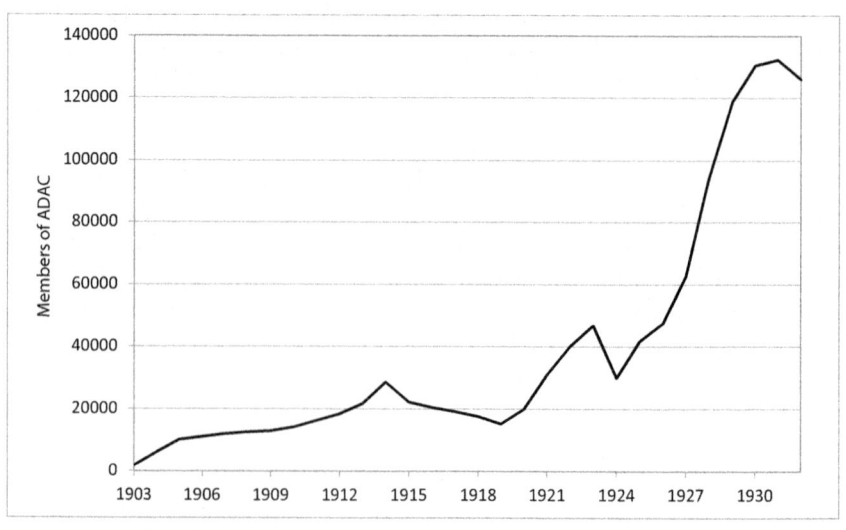

Figure 4.2. Members of ADAC until 1932.
Source: Seherr-Thoss, 27

their cylinder volume was less than 200 cm^3) took place, just as in the Netherlands.[12] In 1939, on the *Autobahn* between Munich and Salzburg and between the large cities and the *Ostsee* in the north, 80 percent or more of the counted traffic were motorcyclists. ADAC became not only the largest motoring association of Germany (with one-third of its members having a motorcycle, two-thirds a car), but also of the entire continent, although the Dutch ANWB in relation to the population or to the motorized parts of it, was much larger and influential, while the British Automobile Association (AA) was in absolute figures three times as large as ADAC.[13]

When motorization stagnated around 1930, ADAC radicalized into a fighting club, organizing its members into a "united front" of "consumers," emphasizing that its "unity and tight organization" was the secret of its success.[14] In 1931, when Germans began taking their vehicles out of registration in massive numbers, it started an aggressive campaign (*Abwehrkampf*, a defensive struggle) against further tax increases, demanding attention to the "catastrophic condition of the car industry" and arguing that Germany had the highest taxes on cars on the continent.[15] ADAC slowly increased the pressure on the national government (declaring, just like ANWB in the Netherlands, that it could hardly restrain its members from taking more radical actions). This included objections to the "injustice and oppression" and declarations that "we won't cooperate any more."[16]

When Hitler in 1933 started his program of enforced motorization based on the construction of a megalomanic *Autobahn* network and the lifting of taxes on new cars, "a sigh of relief went through the ranks of motorists," as a chronicler of ADAC remarked after WWII. For Hitler the car was a "conciliatory tool [to alleviate] class contradictions." His tax reduction scheme was first tried out in the German enclave Danzig (the later Gdansk) in Poland, where it led to an immediate revival of car diffusion; in fact, in October 1932 as a result of a changing basic economic trend, well before Hitler's rise to power, car purchases were higher than the previous year for the first time: the increase in car sales preceded the rise of the national production index. It seems that the subsequent trebling of Germany's car fleet, starting in March 1933, was mainly caused by a shift from motorcycles to cars, while foreign makes, and especially Opel (owned by General Motors), also benefited from the tax exemption.[17] Indeed, expressed in passenger-kilometers, cars started to outperform motorcycles from 1935 on.[18]

By then, however, the spectacular increase of ADAC's membership was over, until the end of the 1930s, when the touring club of occupied Austria was merged with the German, resulting in an organization of 200,000 members.[19] Meanwhile, ADAC was "assimilated" (*gleichgeschaltet*) into a national DDAC under the leadership of Adolf Hühnlein, a WWI and *Freikorps* veteran, leading SA figure and leader of the national socialist paramilitary motor vehicle corps NSKK. ADAC was not destroyed (according to one historian because Hitler himself had been a member), but the car clubs were forced to merge with DDAC. The confiscation of the assets of the clubs by the Nazis was outright plunder: bicycles from the social-democratic bicycle factory were distributed among members of the police and the *Hitlerjugend*, while motorcycles from communist and social-democratic groups were given to the motor divisions of the SA and SS. Later, part of the stolen money was used to buy motorized bicycles for the *Hitlerjugend*. Although this led to new groups of Germans getting involved in motorization, the "*Motor-HJ*" (the motor division of *Hitlerjugend*) failed to satisfy most of its eager members' desire to own a motorized vehicle. In 1937, one thousand light motorcycles and sixty-three cars had to be used by fifty thousand members; two years later, the number of motorcycles had doubled, but the membership had also doubled. Instead, the organization became a place for many German youth to acquire a driver's license, continuing the German tradition of militarized driving education started during WWI. Cars and trucks *were* given (mainly through sponsoring by the industry) to the Motor Sport Schools (*Motorsportschulen*), however, where more than 180,000 drivers were trained before the war broke out. In the end, the Nazis failed to realize their promise to increase the German vehicle fleet by 3 to 4 million, although the trebling of the fleet (to 3.4 million between 1933 and 1939) was still spectacular, helping to trigger economic recovery.[20]

The mix of car and motorcycle densities per European country gives a helpful indication of the purchasing power of the respective middle classes in an overall trend of spectacular motorization (and warns us against focusing upon cars alone): as a combination of figures 4.1 and 4.3 (covering car and motorcycle densities respectively), figure 4.4 gives an idea about the partial compensation of car diffusion by an expansion of the motorcycle market.

Such statistics suggest that what the United States did by car, Europe did by car and motorcycle. And yet, these two-wheeler diffusion waves do not compensate fully for the 'missing cars': in Germany, for instance, it is quite clear that, even if one combines the car and motorcycle densities under one heading of 'road motorization' it still does not reach the combined densities for the United Kingdom, or France. Figure 4.4 shows how France, from the middle of the period investigated here, took the European lead as a modern, motorized country without much help

"Why Apologize for Pleasure?"   293

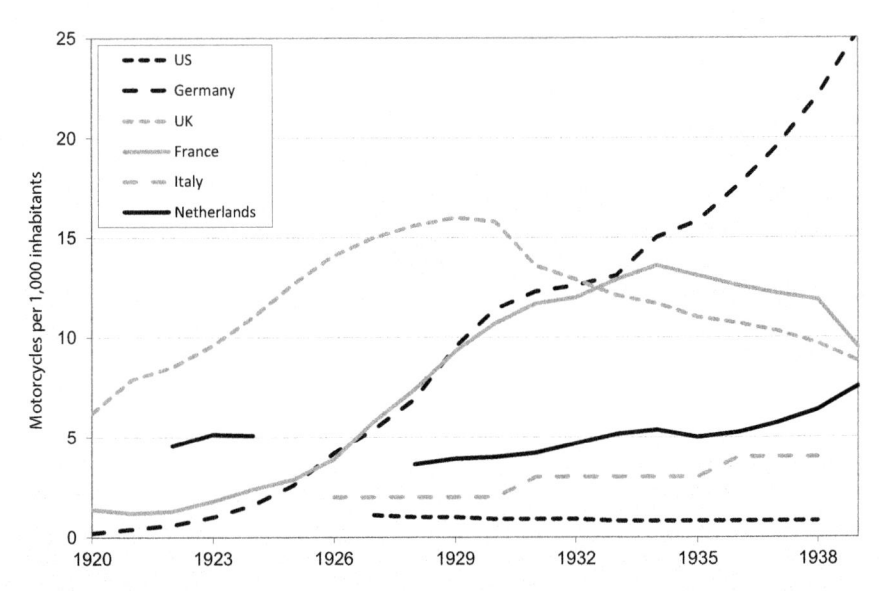

Figure 4.3. Motorcycle densities in the United States and some European countries, 1920–1939.
Source: Flik, 289; NL: CBS

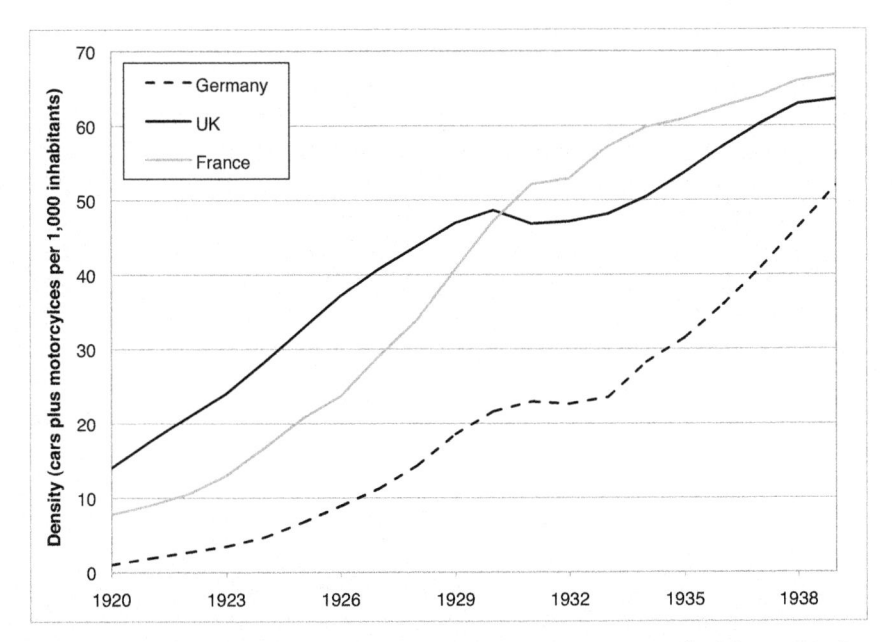

Figure 4.4. Motorization (car and motorcycle densities totaled) of Germany, the United Kingdom, and France during the interwar years.

from a motorcycle market, probably because most of the car's spread was in small towns and villages. Germany, indeed, began to catch up during the Nazi reign, but was still about one-fifth less car dense by the outbreak of the war, probably as a result of the high taxes on fuel that partially compensated the car tax advantages. In 1928 Austria had the largest motorcycle population in relation to the automobile (twice as big), followed by Germany (1.3 times larger) and the Netherlands (about the same). All other countries had smaller motorcycle populations (80 percent in the United Kingdom, 67 percent in Italy, 27 percent in France), with the smallest by far in the United States (with only 7 percent), although in absolute figures the United States was still ranked fourth among motorcycle countries. There, the war was a boost to the four-wheelers, but the deathblow to the motorcycle industry was the Model T, which was "cheaper than a good motoryle and a sidecar." America's emphasis on 'comfort' (see the next chapter) seems to have decimated the motorcycle, while the relative decline of "unattached males" (because of the relative increase of marriages and of residential stability) meant a further blow to its attractiveness. With a declining production during the 1920s (leading to a halving of the motorcycle registrations of which about half was exported to other countries where they were absorbed as luxury bikes) the United States lost its fourth place to Italy in 1933. In 1930, there were 108,000 motorcycles registered, against 23 million cars. By then, motorcycle clubs had become "more social," among other reasons because of a modest influx of women.[21]

The British motorcycle industry grew into the world's largest during these years. It was overtaken by Germany by the end of the 1920s, which then produced nearly 40 percent of the world's motorcycles (against less than 30 percent in the United Kingdom). The "small family car" Austin Seven, 290,000 of which were produced between 1922 and 1935, was explicitly aimed at 'upgrading' the motorcyclist. It was not until around 1925—the Dutch national statistics bureau was reorganizing then, so data are lacking for these years—that the motorcycle in the Netherlands was overtaken, in absolute registration numbers, by the car.[22] The same happened in the United Kingdom, but in devastated Germany, as we have seen, the motorcycle dominated well into the 1930s, when more than half of all motorized vehicles were motorcycles.[23] By then, Germany was not only the largest producer of motorcycles (with DKW as the largest motorcycle factory in the world), but probably also of light cars, although the launch of a "European Volkswagen" failed. In 1938, workers formed exactly 1.2 percent of car buyers in Germany.[24] German motorcycles in 1938 were 49 percent owned by workers (but its motorcycle clubs were predominantly middle class), raising the interesting question of whether motorization did not penetrate deeper (in terms of class) into Germany than in any other country.[25]

From a user perspective, however (measured in passenger-kilometer), the car dominated during the entire period, as figure 4.5 shows, except for the period between 1930 and 1934 when car use experienced a sudden dip, and motorcycle use did not.[26]

The enthusiastic embrace of the car in rural small towns is also the secret behind the miracle that happened on the other side of the ocean. Indeed, after 1918 "the world was no longer Eurocentric." In the three decades since the turn of the century manufactured goods increased by 150 percent in the United States, two and a half times more than the population growth of 65 percent. Relative prices of durable goods declined by one-quarter during the 1920s and did not return to prewar levels until after the next war (in constant prices of 1982); car prices decreased by 40 percent in the 1920s alone, gasoline prices by one-third.[27] Real wages increased nearly 40 percent during the first half of the 1920s.[28] And although by the end of the 1920s only 40 percent of American families earned more than the threshold income of $2000 considered necessary to purchase basic 'necessities,' an aggressive form of marketing accompanied by a generous credit system "induced people to buy more than they could afford," as several concerned contemporaries warned.[29] Later consumption historians emphasized the "middle class type" of this consumption, which especially becomes clear in the Ger-

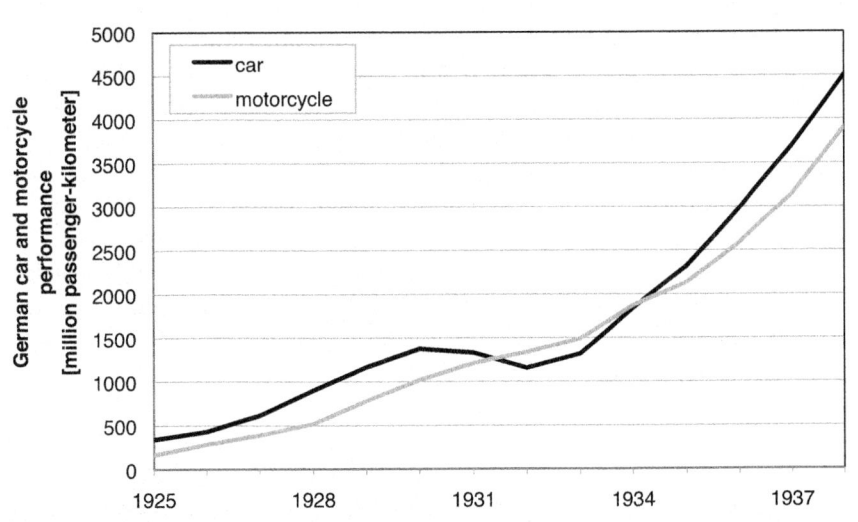

Figure 4.5. German car and motorcycle performance (in million passenger-kilometers).
Source: Hoffmann 693, table 186

man case, where *Angestellte* (white-collar employees) were involved in a struggle to distinguish themselves from the elite 'above' and the workers 'below': "Instead of spending more for housing and education, they were buying entertainment and insurance." As "consumer innovators," they were shaping a specific form of domesticated hedonism: fun, but with an insurance policy.[30]

The American discourse about consumption became so fierce that it is hard to keep fact from expectation. Secretary of Commerce (and in 1928 elected president) Herbert Hoover expressed some concern in 1925 that consumption would keep pace with production as America still lacked "the basic data of distribution."[31] Yet, in less than 1.5 decades around WWI American households had not only shifted their expenditures within their budget from savings to durable goods (the very trigger of consumerism), but within the latter they also shifted from furniture to the car, a shift completed by 1924. Indeed, the "transport revolution" ignited and fueled the twentieth-century consumption society in more than one way. It first and foremost made commodities so much cheaper because of a spectacular decrease in transport costs. And then it stimulated what has been labeled "dribble spending," as we will see later. Cars not only competed with shoes (which increased in longevity and hence sold less) but also with clothes in general. In 1922 car ownership reached eleven million, prompting some observers to emphasize that families should buy a home first (a "primal instinct of us all," Hoover again), and then a car.[32]

In the 1920s cars (and their parts) overtook furniture in the average American household budget in an orgy of consumerism, after they had displaced (and left behind at a much lower level) expenditures for horse-drawn vehicles a decade previously (figure 4.6). By the 1930s, the car's share in the total expenditures for 'major consumer durables' rose to more than 50 percent. In 1922 *Literary Digest* characterized the "typical purchaser" of a car (based on thousands of installment transactions of a financing company in Cleveland) as follows: he was a married man of thirty-three with a bank account and a life insurance policy, who bought a $1400 car paying half of that amount down and paying the balance at a monthly rate of $100, while his monthly income was $350 and he owned an equity in real estate of $5000. His personal property was worth about $2000 and he had bought a car before, also on the installment plan. The "trait" the magazine did not mention was that this man was white.[33] Since 1917 the number of Americans earning between $1000 and $3000 per year had doubled (to 4.6 million in 1923), the group that was considered to be especially interested in cars.[34] In 1934 the Massachusetts Highway Department characterized the aver-

"Why Apologize for Pleasure?" 297

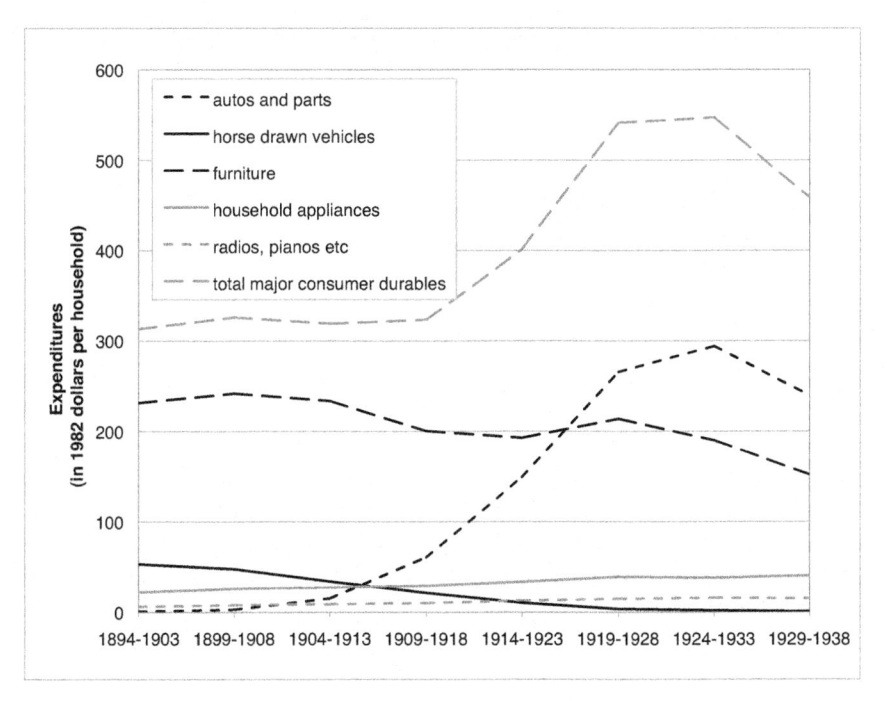

Figure 4.6. Average household expenditures in the United States per overlapping decades.
Source: Olney, 17, table 2.2a

age motorist (on the basis of a survey of twelve thousand) as: "over 90 percent male, most between 21 and 44 years old, and with at least five years of driving experience."[35]

By then, cars formed half of the total of credit outstanding (grown fivefold between 1922 and 1929), followed by furniture (19 percent) and pianos (7 percent).[36] Within the interwar period, 1929 formed the absolute peak year of consumerism, a year in which also the advertising budget of American business reached its maximum before WWII: in that year, when for the first time more than half of American families owned a car, one-quarter of them bought one (15 percent on credit).[37] By the end of the decade the average working-class family probably had about a 30 percent chance of owning a car, and an unskilled worker much less than that. On the other hand, the average nonfarm family had a 50 percent chance of owning a car. Another analysis of family car ownership gives for 1918 a penetration in laborer families of 9 percent, among wage earners 13 percent, and among salaried families 31 percent, while these percentages for 1935 were 33, 43, and 80, respectively. Only 11 percent of black families in that year owned a car, and they spent less on gasoline than white families did.[38]

Average and aggregate data, however, can be quite misleading in view of the enormous variety of diffusion rates across geographic space, rate of urbanization, income, gender and race differences, and cultural preferences. For instance, car densities in metropolitan New York differed as much as a factor of five between the lowest (Manhattan, with its well-developed public transport and its high costs of car upkeep) and the highest value (the county of Monmouth, NJ).[39] And nearly half the privileged Ford employees in Detroit at the beginning of the 1930s possessed a car, against nearly all of Iowa farmers.[40] In fact, classical econometric diffusion analysis failed to explain the car's spread: it could explain the spread of expensive cars such as Pierce Arrow and Oldsmobile to a large extent on the basis of seven variables (from income through congestion to the presence of blacks), but Ford's popularity escaped their analysis, as the variables could explain only a quarter of its spread.[41] The best basic analysis of the variety of car enthusiasm remains a hardly noticed dissertation by Adrian Jarvis who concluded that car densities (in cars per capita, calculated for five 'turning-point years') were soon highest in less densely populated regions in the Midwest and the Far West, while the South lagged behind during the entire period (figure 4.7).[42]

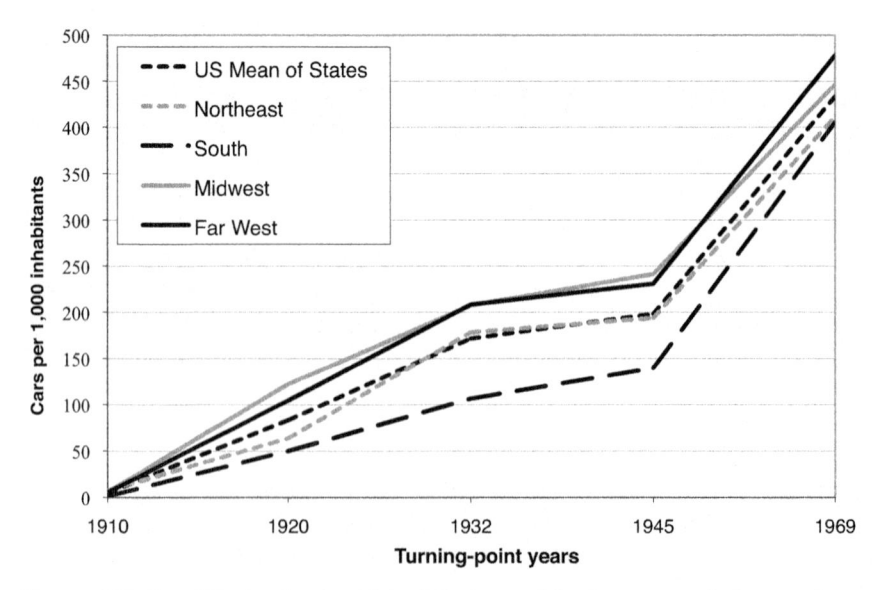

Figure 4.7. Car diffusion in the United States and in the country's four regions for five turning-point years. Note: the x-axis is not linear.
Source: Jarvis, 182–183

Who were these pioneer middle-class car consumers? Many farmers during the agricultural crises of the 1920s and 1930s kept using the cars they had bought during the boom years of the war, but in Minnesota prosperous farmers increased their average car ownership further from 57 percent in 1920 to 92 percent by the end of the decade; 40 percent owned a Ford, while 35 percent had two cars. Dutch and Italian immigrants, formerly farmers, in predominantly rural Wayne County, New York, showed an "exceptionally high per-capita automobile ownership." Other surveys of farm car ownership in the states of New York, Ohio, Iowa, and Wisconsin show similar penetration rates of 80 to 95 percent, but a 1928–1929 survey on 368 farm families in Allegany, Chautauqua, Niagara, and Yates counties in New York State with an average annual household expenditure of less than $1000 did not even mention cars, perhaps because this region abounded in interurban trolleys. Seventy-four dairy-farm households (of which 60, or 83 percent, had cars) in Vermont, however, with a similar annual average amount of "family expenditures," spent 4 percent of this amount on cars, plus 2 percent on recreation (including car trips) during the second half of the 1920s, but car purchase was considered to be part of the "business" portion of the gross income. While farm ownership of cars increased during the 1920s (from 30 to 60 percent), it stabilized in the 1930s (in parallel with the general near-stabilization during this period) and even started to compete with the telephone, ownership of which decreased during the entire period from 40 to 25 percent: remarkably (and uniquely so for the United States, as far as this is known), the car competed with the telephone in the household budget during the crisis, a true 'intermedial' phenomenon. By then, farmers owned nearly one-quarter of the cars registered in the United States, but farm women used the telephone more than the car (less than half of them had a driver's license by 1937).[43]

As small family farms were rapidly vanishing from the American countryside, car densities increased with the size of the villages. Indeed, "small rural towns" were the gravity centers of the car-buying frenzy, as President Herbert Hoover's Commission on Recent Social Trends, installed at the end of 1929, observed.[44] In other words, whether rural or metropolitan, it was the *urban* car culture that formed the real driver of the interwar American expansion of automobilism: the census of the 1920s showed for the first time more inhabitants in cities than on farms (in terms of families, the dominant automotive subjects in this phase, the urban share was, with more than 58 percent, even higher), a fact that did not go unnoticed at the car manufacturers' headquarters where concerns about the impending "saturation crisis" were mounting. It was automobility historian James Flink who claimed that this crisis was a major factor triggering the Depression. Whereas the 1910s had been the "golden years" of car diffusion among American farmers, the "saturation crisis" of 1923–1924 made the urban middle class a renewed target of the car industry.[45] And although the expansion toward "new owners" during the interwar phase was erratic and first declined and then increased again (a rebound of the depression or a new layer of buyers in the lower middle classes?), figure 4.8 indicates a repeated push toward new users during the entire period. Most of these cars were in the cheapest segment (less than $500), while about 95 percent of new cars cost less than $750 in 1934, about half the annual wage of a worker. In 1935 for every new car, 1.5 used cars were sold; they overtook new cars in annual sales numbers in 1925, and their sales reached a peak in 1938, when their volume was 2.5 times that of new cars. Around 60 percent of all used and new cars were bought on installment credit. Remarkably, this set a threshold for the lowest price on the second-hand market, strengthening the general trend toward a middle-class car culture; the only way, now, to keep the mills of car consumption running was not lowering the price further, but selling new cars every year, as we will see in the next chapter. By the end of the 1920s there were 23 million cars registered owned by three-quarters of the American households.[46]

Many midwestern farmers had meanwhile become as sophisticated and "streetwise" as urbanites, acting as entrepreneurs whose families decreased in size at the same pace as in cities, as a survey of 140 villages in 1930–1932 testified. Contemporary sociologists also emphasized that

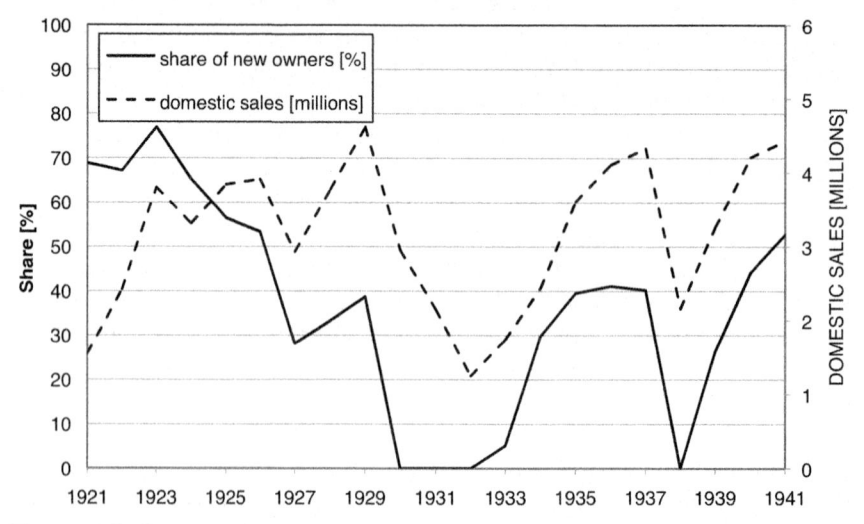

Figure 4.8. Share of new car owners in domestic sales in the United States, 1921–1941.
Source: St. Clair, 125, table 4.1

it was "less and less possible to portray sharp contrasts between 'the rural' and 'the urban' family because of greater heterogeneity within both rural and urban groups." This suggests that it is perhaps better to characterize American automobilism in this period as an 'urbanizing' culture rather than an 'urban' culture, so as to include many of the entrepreneurial farmers in this culture.[47] By the end of the 1920s, for the 30 million people living on farms and the 32 million living in towns of fewer than ten thousand inhabitants (even if we know that in American political geography almost all land outside cities is organized in towns[48]), it was the car (and related media such as the telephone, and newspapers) that helped homogenize the countryside into an urbanized culture. The other part of this homogenization was performed in the cities, where a movement in the opposite direction took place during this period: while the city centers had precipitously grown during the two decades before 1920, they stabilized thereafter and the city rings started to grow until well after the Second World War.[49] Indeed, the importance of 'suburbanization' to explain the persistence of the American car culture, a phenomenon which from 1930 led to a decline of the city population, cannot be stressed enough.[50] While in 1920, 9 percent of Americans lived in suburbs, a decade later this had risen to 14 percent. Instead of the hub-and-spoke shape of the streetcar cities, a "network" shape tuned to the car emerged. In the suburbs, Leo Marx's 'middle landscape' took shape: a replication of "small town America on a mass scale. Into existence came the automobile-centered, mass-produced suburb with its landscape of single-family tract houses, highway shopping, and single-story school buildings, because of its high income structure the very epicenter of the consumption society. Symbolically fused to it were images of family life, domesticity, safety, and the innocence of childhood." A decidedly male-oriented middle-class family culture emphasized family vacations and organized sports, where women did domestic work and acted as "Mrs. Consumer." Most suburbanites were young, white, economically homogenous in the sense that the poor were excluded. "The suburban lifestyle in its full glory was the province of the young, white, educated, married couples with children. All others participated to a lesser extent."[51]

It was the small town, either in the form of a suburb or as an outgrowth of a village in the countryside, where American culture in general, and its car culture especially, took shape as a hybrid of modernization and a "bucolic note" that was ever present, a culture that was shared by a much more diverse part of the population (including workers) than often assumed. Indeed, Jarvis's analysis of average car densities at the state level indicates that American car diffusion sought a 'middle' in nearly everything: the spread of the car (per capita) was highest in states with a middle income of people living in states with a middle rate of urbanization (see figures 4.9a and b).

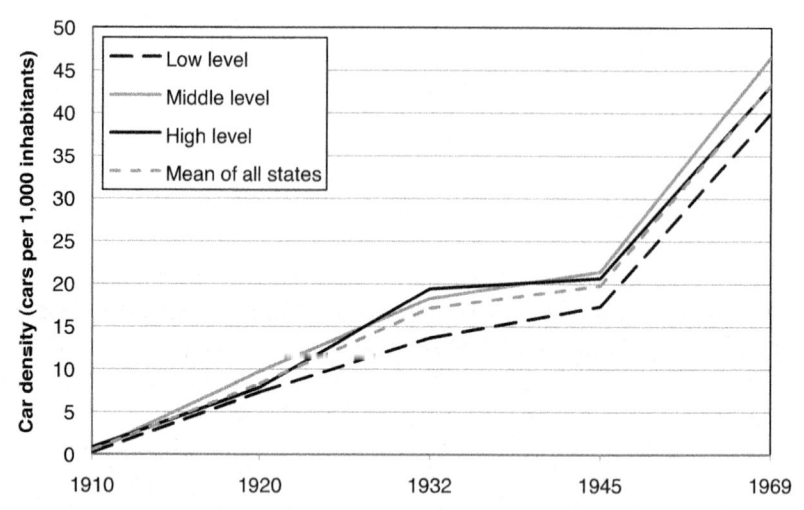

Figure 4.9a. Car density and urbanization levels.
Source: Jarvis, 235, table 36

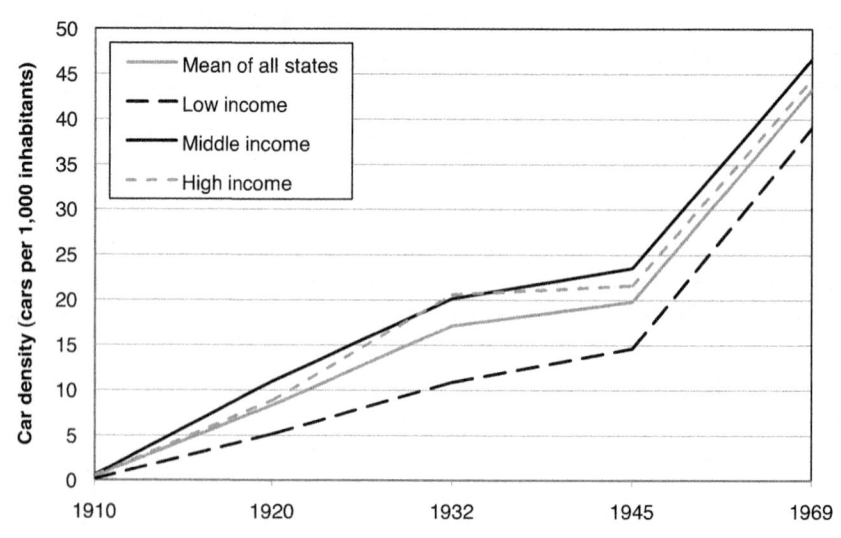

Figure 4.9b. Car densities in states with low, middle, and high income levels.
Source: Jarvis

It was "Middletown," analyzed by sociologists Robert and Helen Lynd in 1924–1925 in Muncie, Indiana, that could stand for all these middle-size towns with middle incomes, although ideological blindness prevented the Lynds from identifying the middle class as the real driver. Instead, they distinguished between a "working class" and a "business class," the latter also including what others were meanwhile identifying as the "middling classes." The Lynds' blindness for the middle class must have hindered several historians in their effort to analyze the American car culture along class lines. While it seemed that workers had a more 'technological' relationship to the car (and, one is inclined to say, remained more receptive to the openly adventurous side of its culture), the middle class preferred a more 'tamed' adventure. Deviants from this domesticated pattern, such as nomads, celebrated their mobility as a permanent state of being, whereas the middle-class family experienced mobility from a primarily sedantarist position. They became "enslaved" by their cars, as "a keen tramper and camper" in the United Kingdom opined. Or they used the "family machine," as an Australian car journal opined in 1928, to "bind ... the family together although it takes [it] far from home."[52] Other 'deviants' were the American youth who invented a new cocktail of mobilities in these years, varying from "tough dancing," via going to the movies and using the telephone, to using the car, all against a background of new courting techniques.[53] For members of the middle class, being in the middle between an elite and a working class not only meant that "fuzziness" and "ambiguity" belonged to their characteristics—if not political facelessness according to one of the most prominent students of this class, C. Wright Mills—but also that their existence as an "ideological construct" functioned as a neutralizer of class conflict.[54]

Car ownership in Middletown was quite large: two-thirds of all families owned one, 40 percent of them Fords. And although, according to one historian, the Lynds miscalculated the car ownership rate, they did acknowledge the existence of a middle class during their second round of research, when they returned to Muncie one decade later. But the damage was done: they portrayed the workers as would-be middle class, projecting their own preoccupation upon a group that had their own motives to aspire to becoming a car owner. The very popular analysis by the Lynds (with six editions sold in the first year of publication alone) showed convincingly how the car competed with houses (they pointed at the "not uncommon practice of mortgaging a home to buy an automobile") and even with clothes in the family budget. Half of their sample of working-class families owned cars (more than 40 percent of them owned a home as well) but for many of these families "even short trips" were out of the question, while only a couple of worker families enjoyed an annual holiday of one week.[55] The Lynds did not go further into this remarkable feat of owning without being able to use, but stories about the way cars were sold to families that could not afford any upkeep were well known, and not only among the 'respectable poor,' introduced at the beginning of this chapter.[56] "I am not a 'tightwad,'" a small-town banker confessed in 1925 in the *Atlantic Monthly,* "I am careful and I am thrifty—at least I was until I became a motorist ... First you want a car; then you conclude to buy it. Once bought, you must keep it running, for cars are useless standing in a garage. Therefore you spend and keep on spending, be the consequences what they may. You have only one alternative—to sell out; and this pride forbids. The result upon the individual is to break down his sense of values." The banker called Americans like him "car-poor" because for them "saving is impossible. The joy of security in the future is sacrificed for the pleasure of the moment."[57] It was this "pleasure of the moment," which made the car trip comparable to, say, a cigarette or a drink, which introduced the phenomenon of "dribble spending" into twentieth-century consumption culture, providing a never-ending impulse to repeated purchase.[58]

Another survey from the end of the period under investigation here leaves no doubt about the largely middle-class character of automobilism between the wars: as late as 1939, 60 percent of low-income groups (earning up to $500 per year) in villages made no expenditures for transport purposes at all.[59] The best analysis (and partial debunking) of the myth of general prosperity in the 1920s, undertaken by labor historian Frank Stricker in 1983, shows convincingly that unskilled workers gained little from the boom years between 1922 and 1929 and that one-third of Americans at the end of that period still lived under the poverty threshold of $1500 to $1600. For 1929 Stricker found a penetration rate among wage earners of 46 percent of households owning a car, taking into account an estimate of households possessing more than one car and a considerable amount of fleet ownership and other forms of business use, such as the use of cars by traveling salesmen.[60] Other sources are quite clear about the low penetration rates of car ownership among blacks ("One of the Groups Middletown Left Out"): halfway through the 1930s nonrelief black families reported car ownership for only 20 percent, while two-thirds of comparable white families (in terms of income) by then owned a car. This share was much lower in large cities, but that was true for all families.[61] Likewise, an overview of studies on the spread of the car among American workers concludes that "urban workers ... were unlikely to own an automobile in this period," while "workers in smaller cities were more likely to own automo-

"Why Apologize for Pleasure?"

biles, but even then their diffusion was fairly limited."[62] This is confirmed by automobility historian James Flink: "contrary to myth, it was not until the 1960s that the American working class really became automobile owners on a very large scale."[63]

This begs the question why workers did not motorize on a much larger scale, given the opportunities of sale on credit, the extensive second-hand market and the possibility of car sharing. The reason must be cultural rather than financial: the car in this phase evolved into a middle-class vehicle of distinction, against a working class "which was uncomfortably close to them materially." Like so many paraphernalia of bourgeois culture (such as the parlor in the house, the visiting of museums and the philharmonic, the collecting of souvenirs of foreign travel and the "paintings of deceased forebears"), the middle classes had appropriated the car and had started to redesign it (as an 'affordable family car') according to its culture, despite the slight decline of average nonfarm income during the 1920s.[64] In the United States, when compared to car use by "executive professionals" and "labor," white-collar families used the car the most for pleasure traveling, and the least for commuting (expressed as "average time expenditures" in minutes).[65] Their desires became supported by eager car manufacturers who in this phase started to experience market saturation, and discovered that the second-hand market and the credit system prevented the further lowering of the price for new cars from expanding the market. Right at this moment they started to introduce the rapid changes in automobile styles and the fine-tuned segmentation of the car market in classes. The turning point was 1927, when Ford's Model A was introduced and a period came to a close wherein a 'classless' America had learnt to drive.[66]

And yet, the surprisingly high penetration of the car in some unexpected pockets of society illustrates the eagerness of the American motorization process. In 1926, 65 percent of the "harvesters" in some Californian valleys came by car, with their entire family, driving from job to job.[67] A government overview from 1941 showed that lower-income groups did possess cars (black and white sharecroppers and black farmers between 25 and 50 percent), two-thirds of these being second hand, and even at higher income levels (up to $4000) one-quarter to one-third had second-hand cars. New cars were essentially to be found in large cities among high-income groups.[68]

Once bought, car ownership proved to be remarkably 'depression-proof': although car sales more than halved in 1929, and halved again in 1932, gasoline sales only dropped 4 percent between 1929 and 1933 (and even car registrations fell back by some 3 percent) leading the Lynds to the assumption that there may have been "somewhat less random driving about on summer evenings to cool off." And for those who had to abandon their possession, a Muncie newspaper consoled: "Many a family that has lost its car has found its soul," the second *it* referring to the family, not the car. Many lost their houses as well. They also witnessed a bicycle revival, with half a million produced nationally in 1935.[69] If one realizes that radio sales collapsed during the Depression (dropping 80 percent of its 1929 value) the conclusion seems clear: forced to chose, Americans preferred mobility above sessility.[70]

Indeed, the Lynds dedicated a dominant part of their analysis to consumption and leisure, especially the effect of mobility and communication on family life. Middletown was an eloquent example amid a host of family budget surveys that tried to capture the changes taking place in the ways Americans were applying "the philosophy of 'spending for prosperity'" or were indulging in "the Saturnalian orgy of consumers' goods," revealing at the same time persistent class divisions.[71] *Americans* should be read here as a "new middle class" of "college educated professionals and white collar and managerial workers," who were quickly dominating American culture, succeeding the diminishing entrepreneurial "old middle class" of the "self-employed." Leisure, for these groups, became "the center of character forming influences, of identification models," models that predominantly were masculine in nature.[72] Heavily debated and controversial among American social scientists of the period, white collars "have little in common occupationally. They can and do, in their social intercourse, talk at length about consumption problems—comparing car models, furniture, rugs, color combinations, the dress and achievements of children, etc., etc."[73] In other words, middle class was less about income but more about spending, and about self-image: according to a 1940 survey nearly 80 percent of Americans identified themselves as belonging to the middling people. Middle class was a "meaningfully meaningless category of identity." According to economic historian Michael Lind, the American middle class is the result of "economic engineering by means of laws and public policies."[74]

European Car Consumption and 'Americanization': Eagerness Compared

The sheer size of the American miracle should not make us blind to comparable developments on the other side of the Atlantic, nor make us jump too easily to conclusions varying upon the mantra of American exceptionalism, such as the diffusionist myth. Similar budget surveys in Europe among the new middle classes in Germany (*Angestellte*), Great Britain (clerks), and France (*employés salariés*) revealed comparable spending patterns. There were important differences among these nations: a recent study argues that in every country the "factor bundle" governing modernization, consisting of bureaucratization, industrialization, and democratization, was differently 'distributed,' the first being weak in the United Kingdom, the second in France, the third in Germany, but an ongoing study of the German "*Bürger*" has led to an ever louder plea to abandon the idea of a German exceptionalism. Nonetheless, the difference in real wages cannot be ignored: indexed at 100 for London in 1928, these wages were 179 in New York, but 66 in Berlin, 55 in Brussels, and 48 in Vienna.[75]

In highly urbanized United Kingdom, where as early as 1901, 80 percent of the population lived in towns, white collars formed nearly one-quarter of the employed population in 1931, whose living standards were also steadily increasing. Like in the United States, these groups were often not so much distinguishable from the better-paid echelons of the workers by income, but by the fact that they performed non-manual work based on "respectability" and outside appearance, and by a "lifestyle" characterized by "a strong belief in the importance of family and home." In 1911, the nearly one million male white collars formed one of the largest occupational groups in the British economy, more than half of them belonging to the commercial (nonindustrial) clerks and managers. Here, too, car purchase on credit spread quickly, while expenditures on "holiday, clubs, car and leisure" also represented about one-tenth of the family budget (workers spent about half of that share). But due to the higher taxes on cars and fuel along with the relatively lower real income levels, private car ownership was concentrated in the southeast of England, despite the halving of car prices between 1920 and 1927. Motorcycle ownership, reaching a peak in 1930, only started to decline during the following decade. "Clerks," belonging to the "lower middle class" and largely recruited from the working class (and, unlike in the United States, often unionized), formed 4.6 percent of the working population in 1911 and 7.8 percent in 1931, and were much less likely to own a car. Nearly half of the clerks in 1911 were younger than twenty-five, and two-thirds of them were female. The old and new middle classes together represented 23 percent in 1923 (29 percent in 1931) of the employed population, 40 percent of them female, representing 1.6 to 1.8 million women and not a few of them "new women" who helped decrease family size precipitously (to 2.4 in 1915) while 40 percent of them practiced birth control by 1919.[76]

A government report on expenditures of 1360 middle-class families in 1938–1939 showed that one-fifth of their income was spent on holidays and recreation, but a surprisingly low share of 0.5 to 1 percent of the family budget was spent on cars, motorcycles, and bicycles, the highest earners spending six times more than the lowest.[77] It seems that white-collar wages rose less than those of blue collars during the Hungry Thirties, but falling prices during the second half of the 1920s enabled white collars to spend more of their income on motorcycles and cars than they had previously done. "Motorists who make holidays with their cars," the journal *Autocar* conceded, "are not of the very rich class."[78] No wonder, then, that a recent, very detailed and quite original study of 12,439 journeys to work reported by 1834 Britons only resulted in 10 percent of all trips to work being done by car during the interwar years (and an additional 3.5 percent by motorcycle), while about 45 percent walked or cycled and about 35 percent used public transport (with an impressive share of 11 percent for the bus). This was not only a question of income: many European cities were flatter, more temperate, and denser than American cities, thus more easily allowing for walking. Only in 1930 a "period of unprecedented change" started (and lasted until 1970) in which motorized road transport, and especially the car, began to take the lead. Commuting distances had already started to increase before, as they grew between 1921 and 1966 by about 50 percent, starting from an average (for northwest England) of 2.4 km.[79] By then, British consumers spent a three- to four-times-higher share of their total expenditures on private transport than did the German consumers.[80]

Likewise, the German *Angestellte* represented a new, future-oriented middle layer of "agents of modernization," even if they appeared susceptible for right-wing influences, much more than their American colleagues, a situation that had led to a debate on the "extremism of the middle" among historians of Germany.[81] While a million of them in 1895 stood against nearly 13 million workers, by 1925 this had exploded to 3.6 million (against 14.4 million workers) only to be further increased to 4.7 million (against 17.5 million workers) on the eve of WWII. The sociologist Theodor Geiger calculated on the basis of the 1925 census that the "*neuer Mittelstand*" formed one of five occupational layers, representing 17.9 percent of the total population and about to surpass, in numbers, the old middle class and the independent professions.[82] In contrast to the more traditional government officials, the "*Beamte*," with whom they were often compared in budget surveys (and who were grouped together with the *Angestellte* under the label of "lower middle class" by Anglophone scholars[83]), their lifestyle was pleasure and recreation oriented, and they spent less (in share of total expenditures) on clothes, housing, and children than either the civil servants or the workers, while their fertility was exceptionally low: 1.5 children seems to have been a "class ideal." These relatively small families spent one-third of their income on food, and one-fifth on housing. Some surveys of the period claimed that they deliberately remained undernourished to enable them to spend more on the third, crucial item: clothing (10–12 percent). Of the remainder of the expenditures, taxes and (partially compulsory) social insurance formed a large part, while "recreation costs" ("*Erholung*") formed about 3 percent of the family budget, spent on vacation, trips, and recreation. Almost all of the employees surveyed enjoyed time off from work for holidays, but few of them could afford to go away on holidays. Nevertheless, these costs competed with the costs for children within the household budget, a phenomenon none of the American surveyors mentioned, although American fertility declined too. They ate in restaurants more, spent more on personal hygiene, invested in having a 'social life' and often occupied older houses in the city, not in the suburbs, probably in order to be close to entertainment (and perhaps because suburban land cost more than in the United States). Transport took around 3.5 percent of their budget; 47 percent of that amount was spent on commuting, and 53 percent on pleasure travel (*Reisen*). Around the middle of the Interbellum, 80 percent of their transport needs were satisfied by public transport. When budget surveys dealt with their pleasure transportation, the car was not mentioned. As for commuting, a majority walked to work (often more than an hour each way), while one-fifth took the bicycle. They started to motorize in the 1920s, well after the old middle classes had started in the previous period, but often by purchasing a motorcycle or a cyclecar or "*Kleinwagen.*" In one survey of 460 families of *Angestellte* from 1931 only four families used a motorcycle (and no one used a car) to drive to work. Families with children spent less on transport costs, as they often went on foot; they hardly ever took a vacation trip. The generally worsening economic situation during the interwar years puts the persistence of car and motorcycle ownership in a particular light, because of its compensatory function. Remarkably, many members of these groups improved their real income; this not only made them expand their savings but also their car ownership, which increased from 1929 onward. They spent less on savings than the civil servants (*Beamte*) and were less family oriented then these, and used the installment plan as enthusiastically as their American counterparts.[84] German workers at the end of the 1920s, on the other hand, used transportation mostly for commuting: two-thirds used public transport and one-third the bicycle.[85] In a survey from 1930 among more than five thousand young *Berliners* (many workers, but most of them lower middle class), fewer than one percent of the boys and more than double that share of girls owned a light (and sometimes a heavy) motorbike, with which they raced against each other, and which they maintained themselves. One-third of them roamed the countryside (probably on foot) and more than that took part in outings. Among worker youth the golden rule was: one hour driving, two hours tinkering. Some sources claim that this kind of use declined around 1930 when utilitarian use started to dominate.[86]

The 'Atlantic' character of the middle-class car culture was further enhanced when American manufacturers, faced with an impending domestic 'saturation crisis,' started to export their products while the home market was comfortably protected by high import tariffs. When European countries with important manufacturing facilities introduced or intensified their protectionist measures, American makes were assembled or produced locally (sometimes under 'foreign' names, such as Opel). American car exports equaled the car production of all other countries together and Europe accounted for one-third of the total sales of American makes in 1928 (from one-fifth four years earlier).[87]

Indeed, in mobility matters the interwar years saw a division of two types of countries in Europe: those with a domestic motor vehicle industry, and those without. While the former increased their import tax on cars and trucks considerably (33.3 percent in 1928 in the United Kingdom; 190 percent in 1932 in Italy; 40 percent in Germany and 45 percent in France, both in 1928 and 1932), most smaller countries levied percentages between 14 and 20 percent, with Belgium, Spain, Switzerland, and Norway as exceptions, with 30 percent (Belgium) to 110 percent (Spain 1932) as extremes. Nonetheless, Belgium (whose government more or less sold out its fledgling industry to the American imports) was the largest importer of American cars in Europe. Combined with different levels of purchasing power, during the 1930s a very unequal (and diverging) pattern of national diffusion arose, as figure 4.1 testifies. The same is true for the popularity of American imports, especially Fords, who, by the end of the 1920s had captured about one-quarter of the entire European market, about the same as the American market share in Germany. The Scandinavian countries all had a penetration ratio of American cars of more than 80 percent (Finland was extreme, with 90 to even 97 percent) in the period between 1922 and 1931. At the other extreme, France had been the most rigorous in protecting its domestic market with a share of American cars of only about 5 percent in the years around 1930.[88]

The Netherlands were between the European extremes, because of the dominance of French (and during the 1920s also Italian) makes. As a Dutch historian of design phrased it, the Dutch elite openly declared their orientation toward France, but they secretly admired America.[89] Although further research is necessary, the probably rather distinctive

Dutch statistics for this period (which enable researchers to study the spread of the car per province, per municipality, and per make) suggest that the Dutch countryside especially started to buy Ford's Model T in massive numbers, much more per capita than was the case in many other European countries, and certainly the larger ones.[90] On a tiny scale, Dutch would-be motorists mimicked the American midwestern farmer, making it, according to a contemporary German transport expert, "the most mobile country in Europe."[91] The much larger middle-class market in the big towns seems to have absorbed Ford cars in larger numbers when in 1931 an assembly factory in Amsterdam started production of Europeanized Fords, such as the Model Y and, later, the V8, a nice example of the fact that Americanization is not conceivable without a coincidental Europeanization. An American consular report from the period explained this by pointing at the high purchasing power of the Dutch middle classes, and another consular report connected this with the introduction of the sales system on credit.[92]

In the Netherlands the spread of the car during the Interbellum has been studied in detail, although an explicit connection, based on quantitative evidence, between middle class and car ownership has so far not been established.[93] A German observer stated in 1934 that by then all Dutchmen who could have afforded a car already had bought one, and five years later not more than 100,000 passenger cars were registered nationwide, equally spread over the country (to give an idea of its size: this represented half of the *annual* car market in Germany).[94] This saturation was also apparent in Switzerland with its wealthy middle class and, to a certain extent, also in the United Kingdom with its blossoming middle-class car culture. In the latter case, however, it seems that among the passenger car owners the 'professional and commercial' parts of the middle class were strongly represented and that in the 'classy culture' of the United Kingdom of the 1930s, pleasure motoring was still loaded with aristocratic connotations. According to an analysis of 1950, as late as 1939 about 40 percent of car sales were to business users and in 1938 still 500,000 potential car buyers did not own a car. By then, 20 percent of British households owned a car, but there were large regional differences, with the North having the lowest share, and the Midlands the highest (more than double the share of the North).[95] The modest Dutch diffusion (already revealing itself during the much less heavily taxed car culture before the First World War, as we saw in chapter 1) can perhaps be attributed to a culture of collective leisure practices related to the 'pillarized' structure of Dutch society, as well as by a highly urbanized spatial structure that indeed made the car less a 'necessity' than in less urbanized European countries.[96]

The Dutch bought especially those cars that, according to the classification of the tax revenue statistics, were just above the lowest horsepower rating. And although a Belgian study suggests a (negative) causal relationship between taxes and car ownership, and a Dutch study even explains the entire Dutch modesty in motorization out of government tax policy, this middle-class behavior of not opting for the cheapest cars tells us otherwise; there was, in many European countries, a clear threshold that many lower-middle-class members did not chose to cross even if they had the money to buy a second-hand car on credit.[97] An insight into this phenomenon cannot be found through the quantitative route. Studying the "underside of the diffusion curve" (the part of motorization played by the lower middle classes and the better-paid workers) may shed more light on this question, and it seems that, indeed, whereas in the United States we see car ownership without use (in Middletown, for instance, as shown previously), in Europe we witness use without ownership, in the motor bus. Use without ownership was also to be found in Germany, with its *Selbstfahrer-Union,* a kind of early car-sharing initiative meant "to enable those circles, for whom the possession of an own car is not realistic, the use of a vehicle at any time."[98]

The Dutch case also illustrates why one should be careful calling the motorcycle a would-be car for those who could not afford the latter. Although it cannot be denied that for many the motorcycle was a choice based on a lack of money, for many others the youthful and adventurous culture promised experiences the car did not seem to be able to. The low price of the Model T (about the same as a motorcycle-cum-sidecar) is an argument in favor of the 'would-be thesis,' while the prices of second-hand cars and the fact that in many 'motorcycle countries' such as the Netherlands and Germany people did not buy the cheapest cars are arguments in favor of the 'adventure thesis.'

The motorization of the Dutch countryside had increased so far by the end of the 1920s that the road censuses established an average overall dominance of the motorized vehicle over horse traction, a process that continued during the 1930s: the censuses showed a consistently higher growth rate of motorized road traffic in the non-Western less-urbanized Dutch provinces. This is, in principle, exactly according to the 'American model' of diffusion, with one big difference: whereas in the United States car densities in the less-populated states already during the 1920s and 1930s rose above those in the larger cities (because of the wider range of mobility alternatives in these cities and because of congestion), in the Netherlands car densities in the small towns only surpassed those of the large towns as late as the 1970s. The same is true for French car diffusion in this period: car densities in Paris and its environs remained the highest in the country until at least 1938.[99]

And although similar analyses from other European countries are not available,[100] this leads to the assumption of the existence of two different diffusion patterns, one characteristic for 'empty countries' (countries with a low average population density such as the United States, and, in Europe, say Sweden, Germany, Spain) and one for 'full countries' with a high urbanization rate (such as the Netherlands, and perhaps the United Kingdom and some other smaller European countries, such as Belgium). This would mean, of course, that the American way of motorization is not *the* model, but just one of the two possible diffusion models developed in industrialized countries during the first half of the twentieth century.[101] And although a 'gap' between European diffusion patterns (certainly those of the empty countries) and the American one cannot be denied, nor can it be denied that all relevant issues accompanying the diffusion did not show such a gap: the tax laws on cars before the First World War, the motorizing middle class, the alarming rise in traffic fatalities during the 1920s, the improvements in the road network, the decline in railway ridership, these systemic traits of the motorization process (many of which will be investigated in chapter 7) all happened at the same time on either side of the Atlantic. Moreover, the same saturation phenomena that were observed for the American market (and which was in part due to the car not being adopted on a massive scale by workers), also were visible in Europe: market saturation was observed in the Netherlands, in the United Kingdom, and even in Germany where in 1928 the number of those households that earned at least 8000 Reichsmark (considered to be the threshold that enabled access to car ownership) were the same as the number of car registrations.[102] In short, the 1930s opened the market for lower echelons of the middle classes, but not beyond.

The Car as Necessity: A Profile of Car Use in the Interbellum

Now that we have charted the car's quantitative spread among the middle class, we will try to get a grip on the Atlantic user culture, most particularly the car's commuter function as the basis for the idea that the car had become a 'necessity' for the average middle-class household. After all, whereas for the explanation of the *emergence* of automobilism the motorists' desire for adventure in all its regional and social subcultures would suffice (as we have argued in the previous chapters), for the *persistence* of automobilism some functional embeddedness in the wider society and its culture seems to be crucial. The debate about 'necessity,' accompanying the car's spread among the middle class in several Atlantic countries, can help us determine the main factors that shaped interwar car culture.

Use by Farmers

In the United States, the car boom started with a panic among the car proponents during WWI. The journal of the Automobile Club discussed the threat of carless Sundays (when "pleasure motoring" should be suspended) and worried whether the car industry would be fully geared to the war effort, questioning the wisdom of calling the car a pleasure vehicle. Indeed, it seems that the war started the substitution of "passenger car" for "pleasure car."[103] Although the consensus among historians of American rural motorization is that the car was used by farmers as "business with a lot of pleasure attached," a detailed federal survey from as late as 1939 in 64 farm counties, 140 villages, and 20 small cities up to 20,000 inhabitants gives a somewhat different picture.[104] Largely limited to white nonrelief and native-born families (except for some black families in the Southeast), and hence overestimating national car ownership, this survey clearly indicated that only 20–30 percent of the car's costs were related to farm business, a share which diminished to as low as 15–20 percent in villages and small cities. Combined use (business and "family use," the latter including commuting) was customary among 43 percent of farmer families and only among 20 percent of families living in villages. Among wealthy Minnesota farmers 18 percent reported more than 50 percent on average of their mileage as being "non-farm use"; they drove more than eight thousand miles per year.[105] This is confirmed by travel performance estimates from the US Department of Commerce, whose statistics suggest that "rural travel" (vehicle miles observed in rural areas) surpassed "urban travel" by 1940.[106] Given the lower motorization among farmers this indicated that urbanites and rural motorists were meanwhile both indulging in driving in the countryside, but for what purpose?

Anecdotal evidence suggests that farmers certainly did not shun the pleasurable sides of motorization, although the surveys cited earlier are contradictory. The Vermont farmers already mentioned, for instance, reported only in 1930 that they took one week off per year (often in chunks of some days), which they spent on "visiting friends or on automobile trips," the former no doubt not performed without the latter. "A few went camping, visited the seashore or the city, attended meetings," while their wives attended homemakers' camps. Some "took business trips or went hunting." In short, they hurried around in their car. "These farmers," *Motor Age* exclaimed already in 1915, "want touring cars, not roadsters."[107]

In 1926 Yellowstone National Park reported that the number of visitors with "agricultural pursuits" were more than double any other profession, such as salesmen, "professionals," merchants, teachers, and business proprietors and executives, clerks, and retired people. Nearly 60 percent of the 6500 readers of *Woman's World* indicated they were planning a holiday for 1930, while nearly half of those living on farms gave a similar answer. At that moment, rural ownership of all registered cars was 40 percent, equally divided between farms and villages.[108] And yet, when the Bureau on Public Roads published a report on traffic censuses in eleven states in 1930, farmers' use of cars differed crucially from village and city use because of its very high share of local trips (twenty to forty miles). This would mean that if farmers used their cars for pleasurable purposes, they mainly used it on short trips. This also confirms our earlier conclusion that the real core of the postwar wave was to be found in the small town and that much of the indistinction in the historiography is caused by confusion about what constitutes "rural," "agricultural," and "village."[109]

Commuting

After the start of rural motorization during the First World War, the urban equivalent of the farmers, the small shopkeepers, started to motorize as well. About them surprisingly little is known apart from the fact that they, too, were caught between jumping on the bandwagon of business expansion and increased efficiency on the one hand and having to make decisive shifts in the family budget on the other, here to allow for the purchase and maintenance of the car, even if bought second hand, on the installment plan.[110] Car owners, although a minority representing 14 percent of the population by 1923, managed to evolve into the very vanguard of a culture of consumption, with the nuclear family as its core, and located in the rapidly expanding suburbs. Indeed, as one leisure study halfway during the interwar period showed, more than 70 percent of suburban pleasure driving took place within the context of the family.[111]

Urban history abounds with analyses of competition between public transport and cars, but failed so far to shed some authoritative light on the user culture of the latter, especially not in a quantitative sense.[112] There is consensus, however, on the fact that the shape of American cities until the end of the 1920s was determined by travel time and that it was the railways, not the car, that had up to then driven that process.[113] Consensus exists also about the car enabling new, especially 'flexible' and 'individualist' user patterns, suggesting that what at the artifact or system level appeared as a substitution of technologies, was at the functional level a much more complicated and nonlinear process. Women, of whom at the start of the interwar phase about one-fifth drove (on farms probably a bit more, although they *owned* cars to a slightly lesser extent than their urban counterparts), seem to have been especially versatile in this respect, combining "the men's errands with their own." For comparison, in the French department of Puy de Dôme during the 1920s, not more than 1–5 percent of registered cars were owned by women, and even less took a driver's license exam.[114]

Modal split (the share of vehicle types in the movement of people) is not the same as functional split: streetcars were already used for commuting, shopping, and social and recreational trips as far back as the horse days of the 1850s and 1860s (and the scarce data available indicate that 'pleasurable' use dominated in those early days of the streetcar suburb), but the question of which of these functions were first taken over by the car cannot be answered on the basis of the available scholarship, although some statistics on the relative increase of commuting on Class I steam railroads in the United States suggest that it was 'pleasure' that was taken from the train first, by the car.[115]

It is impossible to assess exactly the role of commuting in early urban car use. Certainly, it was the middle class, led by business people (apart, perhaps, from such pockets of well-off workers, such as in Detroit or Flint, Michigan), who initiated this type of use, but until well into the 1930s public transport (in terms of the number of trips, *not* passenger-miles) dominated the urban mobility spectrum. As late as 1935, 80 percent of white working-class families reported using the streetcar for commuting, representing one-third of their transportation budget.[116] In 1929 only 8 percent of a sample of hundred employee families in the "Dynamic City" of Detroit (as the mayor called it) commuted regularly by car (17 percent of the forty-seven families owning a car), while another 13 percent (28 percent of the forty-seven car owners) commuted sometimes. In the other "vehicle city," Flint (Michigan), by 1936 two-thirds of the workers used the car to commute, more than half of them as single drivers. By contrast, a recent study of the "automobile suburb" city of Los Angeles (with Detroit belonging to the cities with the highest car density) claims that "while auto owners in the 1910s used their vehicles mostly for recreation, like weekend drives, by the 1920s daily commuting to work and shopping became commonplace," echoing the original surveys that did not distinguish between the latter two functions. Evidence from the struggle around parking bans in the central business districts of many cities during the 1920s suggests that many of the cars parked there were used for shopping rather than driving to work. One historian claims that shopping was "the most common reason" for getting behind the wheel: three-fifths of motorists in 1923 said that this was the case, while in 1940 this was true for 86 percent of 76,000 drivers surveyed. Hitchhiking was also quite common, especially among the young and among women, several of the latter going "back and forth just for the fun of riding in nifty cars," and indicating that even in such a 'serious' use as commuting car culture could be exciting.[117]

As most new suburbs by far in the interwar years did not have any form of public transport, the number of middle-class suburbanites gives an indication of the importance of commuting by car, although we know from circumstantial evidence (see chapter 6) that the car was used to bring commuters from their home in the suburbs to the nearest train station. In 1930, 2.2 million Americans and one decade later 15 million of them did not have access to public transport.[118] However, a suburban leisure study of Westchester County counted only one-tenth of the adult population as regular commuters in 1930, although it conceded that "in selected villages and districts as much as 29.4 percent of adults have been found to commute." Other suburbs around New York City reported commuter shares of between 8 and 22 percent; distance was not a criterion as they all spent 30 to 40 minutes on their commute.[119] Many workers, however, even if they possessed a car, could not afford to use it for commuting. One-third of Pittsburg mill workers owned a car, but only 12 percent commuted in it. Many of them maintained and repaired the car themselves, or engaged an affordable colleague from the neighborhood.[120]

Commuting was also not unknown among farmers: many of them went to the village to live, leaving only "a shed for tools and housing for accessories" at their farmland, which often was as far as fifty miles away.[121] It should be noted that this kind of observations was often based on common sense rather than statistics or other evidence: it is plausible but not concrete enough to allow answering simple questions that go beyond the obvious, such as: how much commuting was done by car, and did this new, 'necessary' function overwhelm the earlier pleasure function or not? Perhaps Joseph Interrante (referring to a "different relation between work and leisure" of farmers and suburbanites) comes nearest to the truth when he concludes that it was the "very separation between farm work and family/recreational activity which distinguished farmers from suburbanites who used their cars to commute to work."

This begs the question, however, how to explain the explosion of car ownership in 'middle towns' during the 1920s and, against this background, what to think of the widespread argument that, while the War Industries Board in 1918 called the automobile "among the least essential manufactures," it would quickly come to be seen as a necessity?[122]

Surveys on Use Profiles

The answer to this question lies hidden in some surveys on family expenditures in the mid-1930s. Apart from indicating that car ownership was virtually universal among families with the highest incomes (above $5000), decreasing to 20 percent in the income bracket between $750 and $1000, they showed that "the automobile is chiefly a vehicle for convenience and recreation" and that "occupational classification has little bearing on the prevalence of automobile ownership at given income levels."[123] In other words, if we can't convincingly show that commuting did not dominate, perhaps we can make it plausible that pleasurable use was very important.[124]

Remarkably, the Depression only slightly reduced the trend of increasing car touring, despite the devastating decrease in new car sales. By the beginning of the 1930s, American motorists consumed as much gasoline during the winter as during the summer.[125] During 1932 only 5 percent of the cars bought on the installment plan were repossessed, indicating that whatever happened, Americans clung to their car more than candy, bathtubs, and even clothes.[126] Circumstantial evidence, such as the revenues from US gasoline taxes, clearly shows that not only was 1928 the year of the steepest rise in car use, but also that an absolute decline took place only in 1931, followed by a quick recuperation in 1933 (figure 4.10). In 1934 auto touring was back at its 1930 level. This is quite different from circumstantial evidence from France, showing drivers license exam applications never recovering during the 1930s (figure 4.11).

In the United States, tourism hardly suffered from the depression, perhaps also because of a spectacular increase of trailers and "motor homes" that helped avoid the high costs of hotels and restaurants, or were even subsituting for houses: within two years of 1935 (when there were 25,000), 300,000 trailers roamed the roads, while four hundred companies together produced 100,000 per year, before the movement collapsed at the end of the 1930s. In short, the Depression confirmed leisure as a 'necessity' for many Americans, including the twelve to fifteen million unemployed (in 1932). Concerned social scientists of the period supported this by emphasizing that the "pursuit of happiness" through active leisure was a welcome antidote to social unrest. Several of them were amazed about the "resistance of motor transport to the

"Why Apologize for Pleasure?"

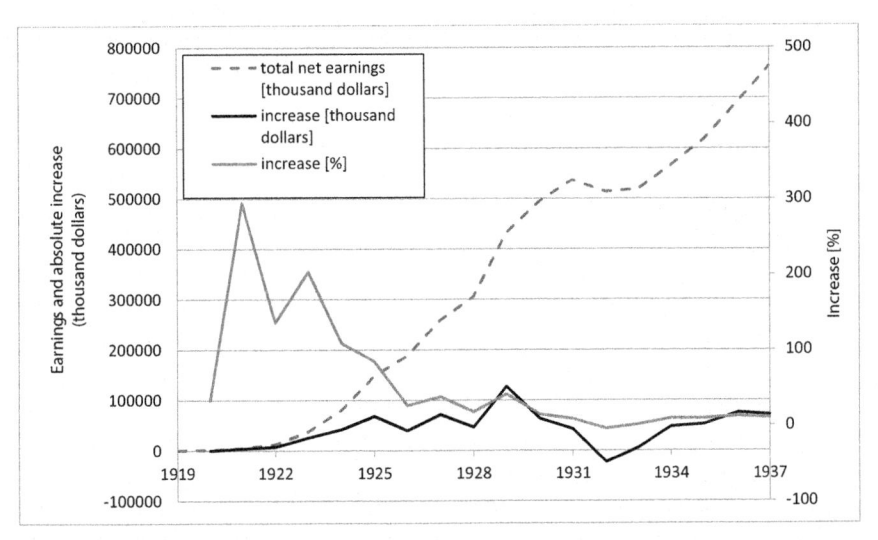

Figure 4.10. Net earnings on gasoline taxes in the United States, 1919–1937.
Source: Crawford, 6–7, table 3

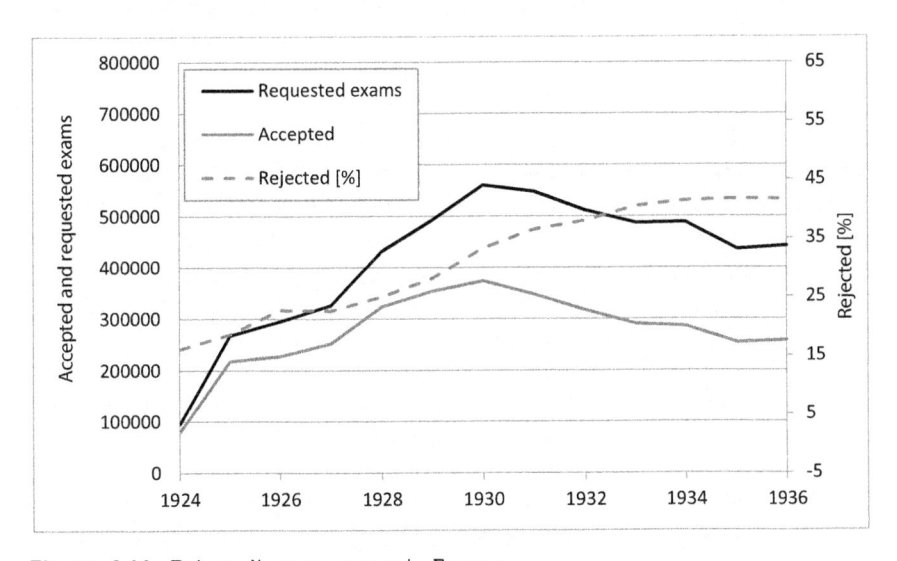

Figure 4.11. Drivers license exams in France.
Source: Report Bedaux et al., PIARC Conference, The Hague, 1938, 11, table 5

depression" as "one of the most remarkable features of the situation in this country," a situation which is explained by one student of the interwar years because of "high taxes, chronically low working-class wages, and unprecedented middle-class prosperity"—in other words, by the urge to distinction within the middle classes.[127] For the American middle class, indeed, the car was really a 'necessity' in more than one sense.

Analyses of the "house trailer movement" revealed that half of the interviewed Americans dreamt of owning a mobile home. They revealed, too, that many Americans "had originally bought their trailers to use on vacation tours but had since disposed of their permanent homes and furnishings in order to enjoy trailering the year-round." About one tenth (20,000) of "trailer dwellers" were "permanent." Such analyses make clear that the trailer movement was partly a form of subversive mobility, "a revolt, of our people against what they apparently feel to be a condition of oppression," as the journal *Trailer Travel* opined. According to Thomas Heany, who studied the movement in detail, "trailering in part 'solved' the problem of class conflict, by breaking up the working class into mobile, nuclear families," leading to the emergence of "a deeply conservative movement that emphasized the importance of small businesses and limited government intervention as well as middle-class values of independence and individual initiative," a movement that in the end (in 1937) was "undermined by government regulation and ultimately by the economic recession of the same year." But originally, it seemed to offer a "new way of life," "an alternative vision of a leisure-oriented society which differs dramatically from the consumer culture that evolved during the late 1930s."[128] Next to the abandoned jitney movement of two decades before (see chapter 7), this is the second example of subversive mobility that had to give in to mainstream petty-bourgeois family mobility.

At about the same time, the trailer movement also rolled over Europe. In 1934, for instance, the German writer Heinrich Hauser declared in his *Fahrten und Abenteuer im Wohnwagen* (Journeys and adventures in my mobile home) that their trailer satisfied a "double desire, fully contradictory so it seems: travelling and at the same time staying at home."[129]

In 1930 summer passenger car traffic on interstate highways exceeded winter traffic in eleven western states by 7 (Arizona) to 89 percent (Wyoming), much more than the also higher rates of truck traffic. On Sundays, 36–109 percent more passenger cars were on the road than the weekly average. This must have been pure leisure travel as shops were closed on Sundays. Interstate traffic formed 15 percent on average (5 percent in California and 38 percent in Arizona and Nevada) of intrastate traffic.[130] In Iowa halfway through the 1920s road use was also highest on Sundays and in the summer, while a decade later on the Bronx River Parkway in New York the traffic on Saturday and Sunday was two to three times as high as on weekdays. A Californian survey estimated the Saturday and Sunday traffic to be 42 percent of the weekly total. A detailed road census in Connecticut in 1922 revealed that 40 percent of the domestic cars stopped on the through-roads were used for business purposes, a share that decreased to 32 percent when only foreign cars (from other states) were checked. Business use was more frequent on local roads (48 percent). As a general average the traffic engineers calculated 44 percent of cars and 22 percent of passenger-miles in the state of Connecticut to be "business," defined by the drivers themselves at a moment that the real wave of long-range touring still had to come.[131]

It seems that the expansion of the road network not only stimulated the further popularization of domestic tourism, but also convinced the business user, who until then had been traveling in the train, to decide to buy a car. This was an international phenomenon and can be substantiated not only by the loss, by national (and often nationalized) railroad companies, of many first class and subscription clients, but also by sporadic road censuses indicating a dramatic increase in annual mileages per car. Seen with hindsight, the pure business deployment of the car only became manifest during a brief period of two decades, interrupted by the Second World War: as soon as postwar mass motorization emerged in the late 1950s, this type of car use receded into a minority position of perhaps 20 percent or less. Such a phenomenon has not only been observed for the Netherlands, a similar increase to extremely high mileages can be seen in the United Kingdom (18,000 km and more from 1933), an increase that had started about eight years before from a level of around 8000 km.[132] This should be placed against a background of a general explosion of mobility, for instance in the United States from about 800 km per American in 1920, to more than 3000 km in 1930, in the former year done by train, in the latter mostly by car (and 10 percent by train, much less than a decade before).[133]

Expressed in terms of passenger-miles and number of trips taken over an entire year, leisure use probably did not dominate over the more 'serious' sides of car use, although some incidental evidence on San Francisco in 1927 suggests that the two trip purposes of "shopping" and "restaurant" were about balanced with both other purposes of "business" and "banks" (all four trip goals had about an equal share of one-forth). If the United States did not differ from Europe in that business trips increased the average annual mileage figures, then such trips were probably not in the majority in 1938, when one-third of the interviewed Americans declared to drive 9000 miles per year, and another 26 percent between 5000 and 10,000 miles. For more than one-fifth of the respondents, the longest trip that year was longer than 1000 miles.[134] A post-WWII history of "buying habits" from the US Department of Labor stated, without giving evidence (and without indicating in what units this was measured), "that over 50 percent of all automobile use [before the war] was for social and recreational purposes." Although it is very likely that touring and tourism did dominate longer-range car use, in terms of passenger-miles this type of use was negligible compared to the other usage types, because its frequency was so much less than that of the other functions.[135] What is more important in this context, however, is the increased combined use of single car trips, an intensifying of what was already customary on streetcar trips: commuters used the trip back home to shop, or were visiting friends on their long-distance search for jobs. Or women used the car for shopping after they had brought their husband to the railway station for his commute.[136] The car occupancy rates of around 2 to 2.5 are an indication of this trend to use the car as an extremely flexible, door-to-door 'function combiner.'[137] Another indirect piece of evidence, the average annual mileage of about 6000 in 1923, 7000 miles a decade later, and a slightly increasing value around 9000 between 1936 and 1940, also points, for the United States, to the direction of a relative modest share of pure business use of the automobile.[138] American credit finance companies claimed that 30 percent of all vehicle sales were to businesses. If that is correct, a thought experiment shows that there were not many pure business passenger cars in the United States: if we assume that half of this percentage was bought as trucks or vans, and if we also assume that these were twice as expensive as passenger cars, and if we further assume that businesses bought only new passenger cars, which were twice as expensive as the average second-hand car, then, indeed, not many pure business passenger cars were part of the American car fleet. The present state of our knowledge can probably best be summarized by concluding that pleasurable car use represented about half of total use in an urban setting of short trips, while it dominated in larger trips (in both cases in number of trips or in passenger-miles). In fact, the popularity of the 'family car' provides quite convincing evidence that, whatever the actual use, purchasing motives were decisively influenced by the expectations of pleasure during the evening spin, the weekend trip, and, most of all, the holiday tour with the family.[139]

The Debate on 'Necessity'

It is therefore all the more surprising to find in the sources the constant expounding on the car as a utilitarian vehicle or as a necessity. Most of all, it was the automobile proponents who hammered on the idea, such as the car dealer who wrote in the *Washington Post* as early as 1922 that the car had meanwhile become "an essential rather than a luxury." The issue must have been crucial because every letter of the National Automobile Chamber of Commerce (NACC) carried a quote from President Warren G. Harding from 1921 that the car was now "an Indispensable Instrument in our Political, Social and Industrial Life."[140] The NACC also published a report showing that only 3 percent of the cars were bought for pleasure alone, but even the headquarters of the car lobby had to acknowledge that recreational and utilitarian functionalities had about an equal share in the use profile.[141] The pounding on the car as necessity was also directed against the railroads and some city planners who kept using the term 'pleasure' when referring to the automobile. Contemporary social scientists followed in their path, such as the Berkeley professor who denounced "comfortable conservatives [who] still regard [the car] darkly as a luxury."[142] On their turn they were followed by historians and other students of the American car.[143]

By the end of the 1930s even relief authorities had joined the chorus, although the Lynds, when they came back to Middletown a decade after their famous study, found several "business-class people [who] regard it as a scandal that some people on relief still manage to operate their cars." Although this perhaps may be explained away by pointing at the small shop owners who saw their very existence threatened by the increasing car use of their customers, this is more difficult to maintain for the labor leaders who complained that "many Middletown workers were more interested in figuring out how to get a couple of gallons of gas for the car than they were in labor's effort to organize."[144] No wonder that the Lynds considered the car a luxury, and the question of 'necessity' an issue of morality.[145] In other words, the discourse on 'necessity' was part of a general reflex among budget specialists and researchers to accept consumption while admonishing the middle class to pursue "non-luxurious 'abundant living'" as part of what they called a "higher life" in which cars and motor boats were included as a matter of course. The keyword for solving this paradox, and the reason why we will dedicate a separate chapter to this issue (see chapter 5), was "comfort," as a "middle ground between necessity and luxury."[146] As one specialist remarked in 1938, the car has become "a necessity to acceptability in middle-class society." Such specialists defined, in 1935, an idealized middle-class budget of $4000, of which 11 percent would be spent on transport, nearly double the actual share.[147] Consumption, such projections suggest, was as much about expectations and the thrill of getting 'more,' than about the actual 'use' of the commodity. This, Warren Susman concludes, was the American Dream: "It is not a question of whether such abundance was a real possibility. The significant issue is the belief that it was."[148] Or it was the *American Hunger,* the name of Richard Wright's fierce attack on consumerism: "The most valued pleasure of the people

I knew was a car."[149] This use, and these expectations, had to be justified, somehow.

The evidence so far suggests that utilitarian functions were often used as an alibi in a culture where 'pleasure' did have a mixed press.[150] Although this is especially true for the European car culture, as we will see later, in the United States as well the debate about the term 'pleasure car' reveals elements of a "utilitarian morality," inspired initially, perhaps, by the fear that the federal government would use this argument to follow European governments, who, one after the other, started to impose luxury taxes on cars. "Luxuries which are such in truth," *Automotive Industries* responded to "wiseheads" who made pleas for a luxury tax, "have been curtailed, but the automobile has long since become a utility of the first order, a necessity in our modern scheme of civilization. As such it can justly be taxed only as other necessities are taxed and in particular only as other means of transportation are taxed."[151]

Thus, the mantra of the car as necessity must be seen as part of a legitimation effort. At the ideological front, it was the Progressive movement that had to be rebuked: its campaigns against "the culture of pleasure" (prohibition, the National Board of Censorship on films, anti-divorce campaigns, jazz-playing blacks) were increasingly seen as an uphill battle against the unassailable enemy of "individualism." Already during the war, the War Industries Board did not succeed in curtailing all "inessential" industries. After the war "controlling civilians' impulses" appeared even more hopeless considering the utterances of "aristocratitis [which] manifested itself in a preoccupation with automobiles, oriental rugs, and fur coats."[152] But there was also a nonideological, very down-to-earth reason for the hesitation about car pleasures. A recent study of the surprisingly widespread urban resistance to the car in the early 1920s triggered, among others, by its dangerous aspects, supports this view: a Californian labor newspaper for instance described the tension between pedestrians and motorists in terms of a class war. A similar revival of an anti-car attitude can also be found in Europe during the early 1920s.[153] Ironically, the success of the spread of the American car provided its proponents with undeniable arguments. One of the car's effects on the countryside was the vanishing of the country doctor from the villages, isolating those without a car from what had been perceived as adequate medical care (with a doctor "in nearly every hamlet"). The same is true for small shops and local industry, whereas the explosion of suburbs was another powerful case in point.[154] In other words, car users created their own necessity by inducing a change of what a report on the "village merchant" from 1928 called "a combined railroad and horse geography." This consisted, for the forty-five villages (of less than 1600 inhabitants) in the state of Illinois surveyed by the report, of county seats positioned every twenty to thirty miles with villages in between (every four to six miles) that "served as supply stations for the farmers." In this geography, paved roads were built during the last seven years, and a decline of sales from small shops was observable, that had already started before the emergence of the car and was also related to a decline in the population of about 15 percent since 1900.[155]

Many observers, however, realized that some nuance was asked for, beyond the simple dichotomy between 'luxury' and 'necessity.' Writing in 1921, one such observer distinguished between urban and "agricultural districts." In the latter context, car purchase seemed "justified on economic grounds." In the city, "except in isolated cases, [the cars] exist only for convenience, flexibility in transportation and pleasure, and must properly be counted luxury."[156] Home economists tried to nuance the debate further, for instance by introducing "decencies" (objects used "to preserve [the individual's] existing rank in society") next to "luxuries" and "necessities," or by introducing two extra intermediary levels of "convenience" and "comfort" between a subsistence and a luxury level, implying that the necessity of the car was relative, and dependent on the standard of living.[157] But several social scientists kept emphasizing the mainly pleasurable character of the car, such as in the case of one of the follower-surveys of Middletown on "Plainville," pseudonym for Wheatland, Missouri, where the car was seen, according to the researcher, "as more a social convenience than a useful farm 'implement,' as are telephones ... and radios."[158] For the sociology professor Jesse Steiner, writing as late as 1933, the car was a "luxury" without doubt. Three years later, another commentator simply stated that "three-fourths of the driving [by car] in the United States is for pleasure," a condition that he, erroneously, called "motion without motive."[159] Nonetheless, it is revealing that among such social scientists 'pleasure' was not seen as a motive worthy of analysis. Even so, when the American government during the Second World War tried (in vain) to suppress the pleasure element of the passenger car, the Office of Defense Transportation calculated (rather conservatively, as it declared) that 43 percent of the average annual mileage was "non-essential."[160]

For home economist Hazel Kyrk, still influenced by a Progressivist morality, American exceptionalism revealed itself in a kind of "mob" culture in leisure, such as "yelling in unison" and seeking "outlets for superfluous energy" like chewing gum and rocking chairs. And clearly setting the new prosperous phase apart from the 'strenuous' Roosevelt prewar days, she emphasized: "To these may be added dancing, smoking and ... automobile-riding. None of these involve strenuous activity but each gives a sense and degree of motion or activity that is relaxing and soothing." It was, in the words of the later historian Martha Wolfenstein, a "fun morality."[161]

In one way or another, other Atlantic countries witnessed similar debates, boiling down to a recognition that car culture was in fact the touchstone of hedonistic consumption, which had became necessary to recuperate from strenuous work. In Germany, *Angestellte* considered their "cultural costs" to be as much a "necessity" (*zwangsläufig*) as their colleagues in the United States.[162] And Hitler could not stop pounding in his speeches on the car that had meanwhile become a "general good," a "utilitarian object" (*Gebrauchsobjekt*), or a "general traffic medium" (*allgemeines Verkehrsmittel*) to emphasize the promises of leisure joys in the Volkswagen. German car statistics are notoriously unreliable, according to contemporary analysts, but it seems that the initial preponderance of the 'old *Mittelstand*' among car buyers had increased the relative importance of commercially deployed cars. One historian even calls the German interwar car a "business car" (*Geschäftswagen*) for which it is "an open question to what extent it was used for leisure."[163] If we extend our analysis to the motorcycle, however, it seems that the mobility culture in terms of pleasure trips did not differ considerably from what we have seen in other countries. But even if we do not use the motorcycle to make Germany into a motorizing country like the others, the abundance of qualitative evidence in this and the following chapter makes convincingly clear that the clumsy German statistics were part of a 'necessity' debate that took on a particularly harsh tone, especially in Germany. The national statistics office only distinguished between three functionalities: taxis and buses, governmental use, and "commercial, professional and other purposes" (*gewerbliche, berufliche und andere Zwecke*). Pleasure wasn't even mentioned. Reiner Flik, who cites the national statistics on the *Gebrauchswagen* extensively, mentions how even the educated *Mittelstand* (the petite-bourgeoisie) during the 1920s asked for heavier taxes on cars to punish the parvenus with their "profiteer luxury cars."[164] The communists also rejected the car as luxury: they called the *Autobahnen* "pleasure roads" (*Vergnügungsstrassen*), but kept the possibility open for their own dream of "the time that every working man will possess a car."[165]

When Hitler came to power, however, he knew how his people moved around. For him, depicting the Weimar Republic as an era of restrictions was very important, against which he then could depict the new Nazi Reich as a country in which his people could enjoy a "yet-unknown happiness..., particularly on Sundays and holidays." "Travel," according to the German literary historian Hans Dieter Schäfer, was "not a luxury, but a 'national duty.'" But Hitler allowed himself more than his underlings: in the Nazi press the same pounding on the necessity mantra went on as in other countries: "Utility is our slogan ... The typically German and at the same time leading insight in the west is that a car is an economic tool and no luxury." Economy and simplicity were the terms the Nazi press used to distinguish itself from the Americans with their desire for comfort.[166]

Close reading of the crucial years of the journals of the touring club ADAC and the elite car club DAC reveals a split motorized universe in Germany: whereas the DAC's *Allgemeine Automobil-Zeitung* covered a trouble-free, cosmopolitan, and mundane indulgence in the automotive adventure, *ADAC Motorwelt* expressed a troublesome, threatened mobility culture, clearly on the political right, but modern. For historians of mobility, it makes a difference which of the two sources they use to reconstruct German interwar mobility culture. Especially ADAC knew how to present the pleasure of touring and racing as a national 'necessity.' The ADAC culture fit in with what has been described as the German *Angestellte* (the salary workers, especially the Protestant part in the smaller towns): nationalistic, anti-socialist, anti-Semitic, and anti-feminist. ADAC's club culture showed remarkable similarities with one of the important clerks' organizations, DHV, whose youth undertook journeys to the "*Völkerschlachtdenkmal*" (a monument commemorating Napoleon's defeat at Leipzig) and organized "*Grenzlandfahrten*" (trips through border regions) accompanied by a "glowing nationalism."[167]

Initially, just as in other countries, the number of motorsport events (with a clear emphasis on the motorcycle) exploded during the 1920s, so much so that ADAC had to rationalize. Remarkably, many of the German motorsport events were collective, such as the first "*ADAC-Fünfländer-Tourenfahrt*" (Touring journey in five countries) abroad, with 450 participants, a trip that was explicitly acknowledged to be a "highly political act" in the rivalry between European nations. No less remarkable was the "*Reichshuldigungsfahrt*" (trip celebrating the 'Reich'), which ADAC organized to celebrate Hindenburg's eightieth birthday on 2 October 1927: a true mobilization, complete with maps published in its journal, of twelve thousand ADAC members in and on three thousand cars and motorcycles from all corners of the Weimar republic. The same year saw a trip to the occupied Rhineland, where the local ADAC representative expressed the "loyalty of the Rhineland to the German Fatherland."[168]

No wonder that ADAC became an ardent supporter of the German car industry. Its leadership redefined national motorsport as a testbed for domestic cars (which it called "*Gebrauchswagen*," utilitarian cars) rather than for its members as drivers. And just like in the Netherlands it shaped itself into an ambassador for the nation, a second Ministry of

Transport as it were, keeping track of the nation's (car) cultural pulse.[169] A revealing conflict arose around the ADAC's flag, depicting the Empire's colors, which it refused to change into the republican colors black, red, and gold, and which led to secession of the Deutsche Auto-Club. In Germany being a motorist was a political act: in 1931 communists killed a man wearing an ADAC cap.[170]

During the remainder of the 1930s *ADAC Motorwelt* heightened its chauvinism one notch further and praised the "irreproachable good order" of the new members. Assembled in the three million–strong SA (or stormtroop), these new members were clearly considered to be recruited from the working classes. The peak of the euphoria of this new automotive adventure was a 2000 km trip to the Eastern part of Germany, to the battlefields of Tannenberg, with Hühnlein and Hitler and many other Nazis present. Two hundred thousand disciplined SA and SS members created a kind of social *Autobahn* closing off crossings along the entire route and keeping the seven million spectators from the road, so much so "that the vehicles could effectively race at a speed of 100 km/h through towns, yes even through large cities, without having to bother about any interruption whatsoever."[171] When it comes to flow as ordered and chaneled movement, fascists knew how to realize that. Nonetheless, it would be too far-fetched to explain the German hesitancy to motorize out of a desire for '*Gemeinschaft*' and a lack of individualism. The reason for the slow spread was basically much more mundane (a lack of income), but it remains remarkable that the less 'liberal' the country, the less dense its car population.

Summarizing, the debate about 'necessity' (both in history and by historians) has been confused with the issue of 'utilitarianism.' Likewise, 'pleasure' and 'business' have been opposed in an unhistorical way. Once we accept that calling the car a necessity is the result of an historical debate about consumption and leisure, we can try to study the 'functional split' (the use profile) of the car in a quantitative and qualitative sense. 'Pleasure,' then, can also be hidden in business use, as George Babbitt will argue in the next chapter. But middle-class families no doubt generated most of their mobile pleasure during the evening and the weekend spin, and the long-range tours across countries or into other countries.

Testing a Different Approach:
Migration, Mass Tourism, and the Family Car

Perhaps the new functionality of the car during the interwar phase can best be described in terms of migration. First, of course, there was suburbanization as a double form of migration, from within the original city and from the countryside. This migration created "metropolitan communities" expanding in concentric circles with erratic traffic flows as opposed to the star-like pattern from earlier streetcar-suburb years.[172]

Not less spectacular were the annual migratory flows of tourists, segregated and highly gendered, to the National Parks (visited in the early 1930s by half—16.75 million—of all cars in the country) and to the South and West, flows that were accompanied by counterflows of southerners looking for jobs and farmers prospecting for better land. In a situation where the United States restricted foreign migration as of 1924 (which lasted until 1965), the flow to the South involved four to five million car tourists per year during the first half of the 1920s. Average car occupancy was 3.6, far above the year-round average of 2.2 and indicating that the husband, wife, and 1.6 kids were all included (the average size of the family was 4.1 in 1930).[173] Here, the prewar adventurous use of the car was continued, first in a "roughed" form (taken over in less pleasurable circumstances by the poor during the Depression who often transformed the camping grounds in "squatter camps"), then in a softened pattern dominated by middle-class families, reflecting the ideology of the Progressive reformers.[174]

The Great Migration (started in WWI) brought about six million Americans to the North and the West, most of them white, but 1.6 million black. This "Southern Diaspora" was mainly a blue-collar phenomenon of Americans who could not take part in the "Southern Renaissance" of the 1920s and 1930s, but white salesmen, businessmen, and students, and black professionals were also among them. It is not known how they traveled (the best study mentions train trips as well as passenger cars and trucks), but the recollection of a farmer family that the countryside in the 1930s was in "a perpetual state of motion," indicated that Steinbeck's Joad family (see chapter 6) was certainly not an isolated case. The flight from the drought of the Great Plains alone was undertaken by almost 60,000 families, some 300,000 people between 1930 and 1938.[175]

Thirdly, large waves of restlessness stimulated multiple waves of short-range touring, on a regional scale during holidays, but also in the form of the evening spin, and the Sunday outing, the latter a new type of vacation the car helped shape: between 1921 and 1941 total car mileage on rural roads increased sixfold, mainly generated by urbanites. It is probably this type of short- and middle-range use that dominated (in terms of passenger-miles) extra-urban automobile culture. A survey in five states at the end of the 1920s revealed that the trip length of one-third to one-half of the cars did not exceed twenty miles, while one-half to two-thirds did not exceed fifty miles.[176] This touring culture became embedded in a prewar 'wandering' culture, in which 'hiking' was enhanced into a motorized hike. In fact, a recent study even makes it plausible that increased car use stimulated hiking on foot, which now could take place much deeper into 'wilderness,' thus having a much more severe ecological effect. This was a North-Atlantic phenomenon: in Europe, hiking emerged in Germany first, before it, for instance, experienced a boom in the 1930s in the United Kingdom, supported by a highly successful youth hostel movement, also imported from Germany.[177] According to the consensus among tourism historians, the "romantic gaze" of wandering culture would dominate tourism until far after WWII, when it around 1970 would be superseded by an event-oriented "collective gaze" aimed at the pleasure of just moving around.[178] But car touring clearly prefigured this collective gaze, at a much earlier date.

The economic impulse for a middle-class family to indulge in 'non-essential' car use was pretty obvious: from an early date, cost calculations by tourists showed that it was cheaper for a whole family to travel by car than by train, especially when a tent was carried along, and a lot of canned food.[179] The train-hotel-holiday system of the precar era may have been a part of a masculine, work-related culture, the subsequent car tourism culture provided instead a proving ground for a new, 'companionate' family ideal, where the presence of the father was used to restore Victorian values of "mutual, voluntarily given affections." The new middle-class family, "based on love criteria in mating, implicit fidelity growing from vital sexual compatibility, and a deep commitment to child-bearing and democratic child-rearing," became driven by a "less structured, more emotional interaction between husband and wife and between them and their children." At the same time middle-class families got "more separated from other social institutions," its size (of about 3.3 persons, clearly smaller than working class families) related to the educational level of the woman rather than the man. The "nuclear family," with its main function to "produce happiness," thus became a "shell," a "shield of privacy" or a "shelter of domesticity" less penetrated by "traffic" of people than in the nineteenth century, a unit of individuals wishing to be sexually and emotionally "free," a wish inspired, according to one student of the nuclear family, by the "capitalist marketplace." While "from society's standpoint the family became less important," new concepts such as the "practical family" (its members bargaining, compromising, accommodating) and the "managerial family" (the "domestic team" tied together by expert advice on child care, domestic relations and household management) tried to give the all-pervasive concept of the 'family' a new, up-to-date functionality.[180]

Warren Belasco, whose study of the car touring and camping movement from 1979 is still unsurpassed, points at the "summer ideal" of the "recreation-based family" as the counterpart of the Prohibition movement to "save the family." To this end, the "canned family" celebrated the car as "more homelike than home itself," a culture where the father and his kids enjoyed their leisure time, and where mother complained: "They have called it sport; I have known it was work." American middle-class families, another analyst concludes, invented "mobile domesticity" to combine the safety of the family with the pleasure of the outdoors, security with adventure, comfort with freedom.[181] Whether work or sport, investing their energies in "hedonizing the outdoors" and, in general, the redefinition of the family as both a consumption and suburban recreation unit at the base of "a new America" led to the "defection" of large parts of the middle class from the Progressive movement.[182]

Apart from being anti-Progressivist (or better perhaps: seeing oneself as the true Progressives), tourist families should be seen within a wider context of increased road building, the emergence of the vacation with pay and a general climate of stimuli of an "urge to travel," at the federal and the local level. While the Reclamation Bureau of the government reclaimed land to give to the two million soldiers coming back from the war in Europe, the National Park Service, founded in 1916 as a special bureau of the US Department of the Interior, started to develop land for the American masses to 'See America First.'[183] The annual statistics of the Service indicate that the movement was impressive (figure 4.12). In

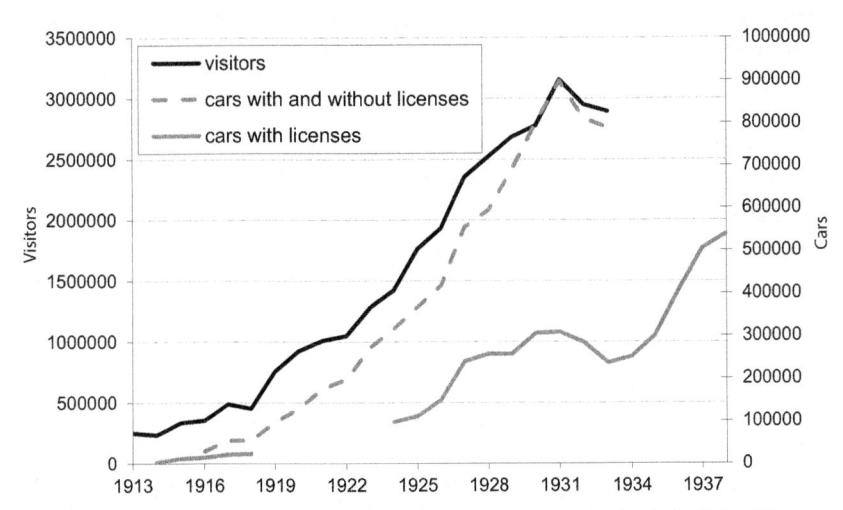

Figure 4.12. Visitors and cars visiting the national parks in the United States. *Source:* Annual report, Department of the Interior, National Park Service, 1917, 79–80; 1920, 101; 1928, 341–342; 1933, 194–196; 1936, 144; 1938, 40

1928, 11 of the 20 million registered cars took part in the summer tourist movement, carrying 44 million Americans, of whom 32 million stopped at a hotel and 12 million slept in auto camps or cottage camps. Halfway through the 1920s cars overtook railroads as the main travel mode to the parks, probably much earlier than for commuting. Whereas hotel tourists spent $7.70 a day during an average stay of ten days, campers spent $3.30 a day, during an average stay of thirty days. Of the tourists from nine most popular vacation regions that took part in "the business of selling scenery," only 7 percent were "home-staters."[184] Thirty-eight million motorists visited or passed through national forests in 1934, and thirteen million "stayed long enough to enjoy real recreation." Many of them stayed in camps or drove over roads and walked on trails built by the 150,000 unemployed of the Civilian Conservation Corps. A nice example of this federal support for building recreation infrastructure is the Going-to-the-Sun Highway in Glacier National Park, finished in 1933 with the sole purpose of offering a pure scenic experience.[185]

In the course of the 1920s, when states started to compete for tourists, a frenzy of road building took hold of the nation. In several states the Good Roads movements, founded to connect the rural population to the cities, were gradually taken over by local commercial interests who campaigned for road building to help redirect the flows of tourists.[186] The federal state started to build interstate roads around this time (the first Federal Highway Act dates from 1916), and the National Park Service even built a park-to-park highway (a gigantic motorists' playground of 3500 miles connecting a dozen parks), thus layering the park-to-park mobility upon the earlier farm-to-market transport.[187] This puts the building of roads (for tourism, for pleasure) in the United States during the interwar years in a different perspective.

Indeed, tourists started to speed up and prepare for longer trips, following the example set by the "transcontinentalists," car pioneers who cruised the hardly paved Lincoln Highway from East to West and whose numbers totaled twenty thousand by 1921 (compared to twelve a decade earlier). On average, tourists' trip lengths increased from 125 miles in 1916 to 300 in 1931 and 400 in 1936; the increase was probably strongest during the second half of the 1920s. There were large differences between states, though: Hoover's Research Committee on Social Trends gave 348 miles for the average touring trip length in Vermont, 370 miles for New Hampshire, and 667 for Ohio, urbanites driving farther than farmers and villagers.[188] The gradually increasing importance of long-range compared to short-range tourism can also be seen in the patronage of Yellowstone Park: in 1919 forty thousand people camped in the park, arriving there in ten thousand cars from forty-six states, but most long-distance tourists arrived by train; three years later only one-third of the visitors arrived by train, the rest by car (but only one-third of the visitors from New England came by car); again three years later, half of the New Englanders arrived by car.[189] It was this long-distance tourism that in fact created long-range automobilism, and that later appeared to be affected the most by the Depression.[190] Long-distance driving differed from railway riding in that it was often an accumulation of short-distance trips, as a clever Swedish tourist understood in 1924. Praising the car as a holiday vehicle, he found that there were "no distances to speak of, every place that turns up at the horizon can be reached in a couple of minutes."[191]

Contrary to Warren Belasco, however, close scrutiny of several automotive journals indicate that the first massive tourism wave was multilayered from the very beginning.[192] Instead of a sequence of different cultures (vulgar "car gypsying" in the wild and soon thereafter in municipal tent camps, followed by a tamed culture of "hoboing de luxe" in municipal cottage camps by the end of the 1920s and eclipsed by an institutionalized "motel tourism" in the 1930s), the hotels originally patronized by train travelers were soon occupied by upper-middle-class members of the American Automobile Association (AAA), some of whom also chose the "better class" of auto camps.[193]

Despite the loud celebration of camping 'democracy' (where social status could not be read from the clothes campers wore) the more adventurous readers of *Outing* chose the "non-rebellious anarchism" of the tents (made popular through the experience of millions of young men during the war) and the auto camps, continuing their tradition that had started as "horse-drawn camping" and "hiking." Most of these concepts provided "a middle-class, family-based model of outdoor recreation." Also, sociologists observed in the Dallas area in 1936 that three-quarters of tourist-camp patrons were not tourists at all, but local couples: some cabins were rented sixteen times per night.[194]

Belasco's analysis of the emergence of auto camping led to his conclusion that "gypsying" by car was a rather new phenomenon, and if it had a tradition, it should be placed in the middle-class habit of romanticizing simple countryside life. While this explanation is certainly valid for at least a part of the early auto campers, their behavior shows remarkable similarities with what in chapter 1 was called the 'functional side of the automobile adventure.' This adventure, manifesting itself in loud declarations by (mostly male) motorists that they loved tinkering and enjoyed repairing the frequent, but easily detectable defects, was an exclusive elite pastime originally, and had nothing to do with a middle class not having the means to repair their cars themselves. A technically informed history of auto camping should construct a direct line between the (often female) 'tinkering' with kitchen gear on the one side, and the tinkering with cars and tents on the other. This willingness to tinker, then, developed in a tamed form of 'roughing it,' a way, proponents of the American Wilderness movement declared, "to barber and manicure wild America as smartly as the modern girl." This kind of tourism, a recent study concludes, "requires a nature that is separate, distant, and exotic—a nature that one goes to see." The goal of tourism, then, is precisely "to acquire the experience of having done so."[195]

The American touring exodus seems to have started as a movement in which a flight from the city and from work was combined with a massive prospection tour for new space, new homes and new work. In spring and fall the majority consisted of "business tourists," if we are to believe the AAA journal *American Motorist*.[196] During the remainder of the year the AAA tourism culture became dominant, following an international pattern of a 'civilized' middle-class culture promoted by the automobile and touring clubs.[197] Like the touring and car clubs in Europe, AAA (in 1928 representing more than one thousand local automobile clubs of together nearly 2.5 million members) erected direction signs, published road maps and Touring Books, provided technical assistance along the road, gave advice in its journal as to the preferred itineraries, and nominated itself as the spokesperson to the automobile industry on technical matters, channeling the perceived functions their members fantasized about.[198] In 1930 alone (a year the AAA coined the slogan "farther, faster and more often"), it helped fifteen million Americans (about 15 percent of the total population) prepare for their trips by giving them planned itineraries; half a million went abroad. But it was not only urbanites who toured: the "Tin-can Tourists of the World" was described, in a British car magazine, as "not a hobo or 'down and out' association, but an organization of retired farmers of moderate means who own their little flivvers, Dodges, Buicks, Maxwells, and Chevrolets, and who every year take long land cruises in them." Fifteen hundred cities had car camping grounds, one thousand of them in California. Although in terms of car-miles tourism represented a minor activity, the two weeks of holidays in the early 1920s and the many weekends witnessed waves of millions of 'pleasure' seekers, which indeed gave the impression that America was constantly on the move. It was supported by the industry whose engineers defined the double utility of car driving as follows: "The physical pleasure of riding is in mild exercise and the sensation of moving through space, but the mental pleasure is in the changing scenes, the relief from customary sights and thoughts and from the extreme boredom of living with oneself." "Variety" became the

"Why Apologize for Pleasure?"

catchword of the movement, variety of scenery as much as variety of people, being open to surprise, including the surprise of small and easily repairable defects in the car.[199]

In Europe the interwar years showed a first massive wave of car tourism as well. In the Netherlands, a similar explosion of long-range tourism happened shortly after the First World War, albeit on a much smaller scale. There, the number of salary and wage earners enjoying a vacation doubled between 1920 and 1940 to 20 percent (representing one million). 'Collective contracts' (contracts between unions and employers for an entire sector) including vacation with pay increased in number, too, reaching its peak in 1931, in the midst of the Depression.[200] The gained leisure time was spent on international holidays on a massive way, perhaps more so than in other, larger, countries. The number of border documents issued by the national touring club ANWB increased tenfold during the interwar years. Although the figures are difficult to interpret (every European country had set its own rules and during the second half of the 1930s fewer documents were necessary because of liberalization of transborder traffic), an educated guess results in about 30 to 50 percent of the ANWB members taking part annually in international tourism, a quarter of them coming from Amsterdam. Most of them traveled by car, followed by bicyclists and motorcyclists (figure 4.13).

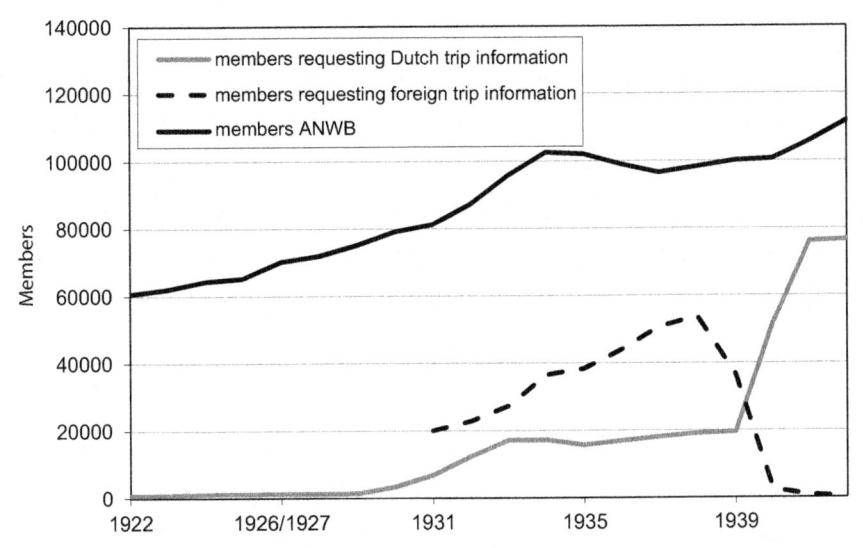

Figure 4.13. Trip information requested by ANWB members, the Netherlands, 1922–1942.
Source: Annual reports, ANWB

When between 1924 and 1931 the number of foreign cars in Germany doubled to 131,000, Dutch cars had gained a share of 23 percent, closely followed by Swiss car tourists. In 1933 by far the largest share of foreign cars counted in Germany were Dutch representing more than one-third of the entire Dutch passenger car fleet. Foreign cars made up one-tenth of the German vehicle performance, to which the Dutch were contributing in 10,000 cars (in 1926–1927). If one combines these figures with incidental evidence from the automobile and motorcycle clubs in the Netherlands, the 'organized motorists' (who formed about half of all registered motorists in the country) seem to have integrated long-range tourism as a normal pastime into their annual mileage in the course of the 1930s.[201]

The most convincing evidence, however, of Dutch car culture being driven by pleasure comes from a special census (including an interview with the passing motorists) on thirteen crucial bridges over the main rivers in the Netherlands in April 1935. Whereas during the first two weeks of the month (including two national holidays) 56 percent of the 100,000 passers-by declared to be on the road "for business, etc." and the remainder declared to be there for "tourism" purposes, during the following week business drivers formed 77 percent of the 91,000 vehicles. If one subtracts the 20–30 percent passing trucks, the conclusion is as clear as surprising: on work days half of the motorists were on the road for tourism purposes, whereas other censuses indicated that on Saturday and Sunday "recreative traffic" dominated.[202] A Belgian road census from 1938 confirms this: on Sunday three times more cars passed a counting point in the city of Overijse, and five times more motorcycles.[203]

In Germany, tourism was called an "economic necessity" (*volkswirtschaftliche Notwendigkeit*). Collective 'package tours' (*Pauschalreisen*) were developed attracting a majority of middle-class women. The 'new *Mittelstand*' dominated pleasure travel figures in this period, while working-class travelers occupied a share of 10 percent. Compared to the active population, however, participation in this *Volkstourismus* was modest. This was until the Nazi regime followed the example of the Italian fascists with their leisure organization *Opera Nazionale Dopolavoro* and managed to make the German *Kraft durch Freude* (KdF) into the world's largest travel organization, with almost 10 million holiday trips at its peak in 1937 and 700,000 cruises, but without the promised *Volkswagen*. KdF was dominated by '*Mittelstand*' Germans, and the 'travel fever' exploding between 1933 and 1938 was mostly enjoyed within the country.[204]

As in no other country on this scale, the German youth undertook a "rightward turn," as part of a general "shift to the right" of the *Mit-*

telstand since it had begun losing its self-confidence of being the "incarnation of social 'normalcy.'" Their leisure culture was dominated by violent and 'military' sports like hunting, shooting (which boomed because of the Versailles restrictions on army activities), horse riding and football (characterized during these years by "an almost endless succession of rudeness"). Surveys among sport teachers revealed that they observed "aggression" and related attitudes primarily in boys. From the mid-1920s the bicycle and car sports ranked high among sporting Germans' preferences, including mostly lower-middle-class men (only 6 percent of sport club members were women, if one neglects the 100 percent female rowing club) and hardly any workers. The new car users lacked especially "social driving skills," a worried observer from ADAC concluded.[205]

The entire KdF mobility (in terms of accommodations per night) only represented one-tenth of German travel between 1933 and 1939. The promise of a holiday (and a car) for every "*Volksgenosse*" (people's comrade) by the Nazis was based not on plans to increase income and purchasing power, but by breaking down the bourgeois privileges hitherto surrounding motorized leisure. "Boredom," one of the Nazi leaders declared, "leads to criminal thoughts." KdF's ideology was based on a collective idea of leisure activities, rooted in the "Friends of Nature" (*Naturfreunde*) and related subcultures that had started around the turn of the century, but the single traveler dominated (followed by couples, and only then by families with children), because the trips were too expensive for many workers and lower-middle-class families. The "collective gaze," observed on the basis of images as source and by some historians stylized into the paradigm of mass tourism, did not replace the "romantic gaze." Nonetheless, despite the difference in car density, the sale of refrigerators, radios, cameras, and camping gear "proceeded apace until 1939," just like in the United States. And although most historians conclude that even the Nazis did not succeed in advancing car ownership and family holidays beyond "middle-class tourism," there is probably no other example before WWII of institutionalized tourism with such a large participation (one-third for land travelers, less than one-fifth for cruises) of the lower strata of society, strata the KdF programs "sought . . . to make less proletarian in manner and deed." But the Nazis kept emphasizing the 'communal' character of their holiday programs, stimulating "intimate contact with local residents" through the organization of "dances and Heimat evenings," although, paradoxically, the promise of the Volkswagen (and the KdF excursions organized annually to the motor show in Berlin) hinted at a different, more individualistic part of their ideology. For a high-ranking representative of the German car industry, the car was "an individualistic form of transport of a new era, implying a rejection of state socialist, collectivist tendencies."[206]

Whereas the KdF trips were mostly done in trains and by bus (German railways offered discounts up to 75 percent; only in 1939 did KdF set up a division for "*Motorwandern*"), the Nazis *did* set up a motor vehicle organization to support and enable violent political actions to "conquer the street" (as propaganda minister Goebbels said) and the roads (in the countryside) as well. The insight among leading Nazis in the crucial role of cars and trucks goes back to the political turmoil right after the war (in 1922 the SA stormtroops set up their first transport division), but in 1931 they organized a *Nationalsozialistische Kraftfahrkorps* (NSKK, national socialist motor vehicle corps, its name reminiscent of the prewar volunteer automobile corps, see chapter 3) parallel to the existing motor divisions of SA and SS. Started with ten thousand members, NSKK grew to more than half a million members during WWII, while women (perhaps two thousand members during the war) were excluded from active membership. Ownership of a motor vehicle or even a driver's license was no condition to membership, however, and this membership did not reflect the average population distribution: one-third of the members (mostly recruited from the car industry and the garage sector, and mostly from an urbanized population) were workers (against more than half in the German population), whereas *Angestellte* and independent professions also formed one-third of the membership (against only one-quarter in the population). Peasants were very much underrepresented among the members. Only one-third of the members owned a motor vehicle and one-third did not even have a driver's license (until 1937, when light motorcycles of less than 250 cm^3 needed a special license). NSKK, in other words, was an association of car lovers. Even within the SA, only half of its 26,000 members in 1932 possessed a motor vehicle; about half of the members were mechanics in garages, workers in the car industry, and truck drivers. The SA motor division functioned as a waiting room for motorization, as its members followed courses in traffic rules, map reading, the use of small arms, and other forms of bodily violence useful at political manifestations, while their practice and propaganda trips were very popular. "The *Motor-SA* man must become one with his machine like the Huns with their horses," Hitler decreed. One of their songs expressed the adventurousness they were after in the style of the prewar futurists: "What rattles and rumbles and thunders along, / as if the devil was driving." While crossing the "acoustically occupied public space," these men could indulge in a "common nature experience, a spirit of comradeship, and day-trip cheerfulness." NSKK was much better organized than the social-democrat and communist motor divisions: among the 314,000 members (1930) of the social-democratic workers cycle and motorcycle league *"Solidarität"* there were 29,000 (mostly light) motorcycle owners who did communication work but they did not form a political fight club, and although communists received a 40 percent discount on expensive petrol at the filling stations of the German-Russian *Derop* oil company, their disgust for the 'luxurious' car prevented them from deploying this vehicle in political campaigns; instead they had cycle and motorcycle sections. Hitler's intermedial insight was crucial: "Without the motor vehicle, without the airplane and without the loudspeaker we would not have conquered Germany." Indeed, he covered 50,000 km by plane during his election campaign of 1932 alone. During that campaign he visited 150 manifestations and spoke to two to three million Germans (a "political propaganda which put even the American methods in the shadow") and his press office announced proudly that he had covered 1.5 million km by car between his release from prison to his seizure of power (1924–1933). He seemed to have never driven a car himself, but his "cauffeureska" belonged to his most intimate circles.[207] The car, he mused, "granted me the most beautiful hours of my life, I really have to confess so: people, landscapes, monuments!" Before they gained national power, the SA consciously avoided the use of the train as this could be easier controlled by the authorities. It meant that they used and adjusted the 'swarm character' of the car by enforcing military discipline and logistics upon their motorized troops. The noisy conquering of a village by a column of motorcycles preceded by a loudspeaker car impressed many Germans. In their "almost Sorelian cult of violence," tacitly accepted by many, they had created "a distinctly un-bourgeois counter-world," spreading fear and admiration at the same time as it took shape against a background of "bourgeois vigilantism" which had started as far back as the revolutionary phase right after the war.[208]

In France in the mid-1930s only about 5 to 10 percent of the people went on holidays, often in a collective setting (as 'social tourism'), in trains and buses. What in the United States were the local Chambers of Commerce were the *syndicats d'initiative* in France. But unlike the United States, and more like the Netherlands, the Touring Club de France was exceptionally active in promoting national and international tourism. The same was true, by the way, for the Belgian touring club: it would be a serious mistake to approach many Western-European touring clubs as car clubs.[209] When the Popular Front gave every employee and wage earner a two-week paid holiday in 1936 and a work week of forty hours, tourism started to become a national project actively supported by a *Bureau de Tourisme* with reduced rates for Sunday trips on the national railway network, especially from Paris; however, the "*Bon Dimanche*" tickets for a Sunday trip and the "Lagrange tickets," named after Undersecretary of State Léo Lagrange in Léon Blum's cabinet, with nearly a million sold in 1937, could not hide the fact that the majority of the French during the recession preferred to stay home. By then, about 15 percent of the French (and an equal share of the Germans) undertook a holiday trip of at least five days, against 40 percent of the British, with their higher real incomes and their unions, which were very active agents in the organization of collective holidays. Trying to "save liberalism by fusing it with mass politics," the French national leisure policy was explicitly set up to show an alternative to Nazi Germany and fascist Italy, and at the same time it was an effort to overcome the negative attitude among French working-class organizations toward the concept of leisure, which they thought had been "hijacked by American business managers." A recent analysis concludes that Lagrange tried "to reconcile liberal individualism and totalitarian effectiveness."[210]

In the United Kingdom, too, tourism started to emphasize to 'see your own country first' during the second half of the 1920s, and British tourists were joined by an increasing army of foreigners (318,000 in 1921 and 505,000 in 1938). They revived walking and cycling, also for workers who practiced an "associational culture which underpinned community-based social democracy," popular among a youth movement as well: bike sales reached a peak of 1.6 million in 1935. Youth hostel associations and Rambling Club membership varied from some thousands to some ten thousands, compared to seven million visitors to Blackpool in 1939, brought there by hundreds of special trains.[211]

Overall, the number of Europeans with paid vacation increased to between seventy and eighty million in 1939, nearly four times as many as in 1919, in the same order of magnitude as the sixty million Americans who effectively went on holidays. When by 1924 Germany had extended a paid holiday to 82 percent of its workers, only 13 percent of their British colleagues and one percent of their French enjoyed such a right. Often these workers had only a couple of days off, like in Germany, where only the *Angestellte* enjoyed a "travel relevant vacation" of a week or more. But the Germans largely stayed within their own country, held inside by quota for group tours, a 100 Reichsmark tariff on every transborder trip, and even a 1000 RM penalty on a trip into Austria, "a move that practically crippled the Austrian tourist industry." Like the See America First campaign, Germans were also advised 'to get acquainted with your own home country" (*lernt eure eigene Heimat kennen*). And yet, transborder tourism by Germans kept increasing (to twenty-five million days in 1936), and trips to the land of Mussolini, the

"First Motorist of Italy," were in fact promoted. Likewise, French international tourism also increased, for instance to its colonies Algeria and Tunisia, visited by 350,000 French in 1923 who wanted to satisfy their "desert fantasy." The French state built an extensive network of railways to enable this (at the same time preventing indigenous "obstinate vagabonds" from moving around and forcing them to settle), or the rails were simply taken along, in the caterpillar vehicles deployed by Citroën and Renault on their desert expeditions (see chapter 6), showing that the car driver, as an admiring general opined, not only would be freed from the rails, but would even cease being "a slave of the road." It shows that also the 'periphery' knew a lively tourism, although examples of domestic tourism, given the lack of road infrastructure and especially the lack of the middle class as main driving force, were few and far between, indeed. Meanwhile, vacation cultures under socialist, social democratic and fascist regimes converged to a common set of values on "social prestige, the role of vacations in familial and national unity, the return to and reconciliation with nature, and the regeneration of the body." The largest exodus took place during Pentecost, for instance in Germany where the new *Autobahnen* even prevented the traffic jams from forming during this top season. These new roads were pleasure roads with large seasonal fluctuations; the average per month in July was more than three times as high as in January.[212]

In the United States, although the working hours per week remained higher than in Europe (varying between "as low as" forty-four hours in some industries and as high as sixty hours in others), an average decline of about fourteen hours of work per week for industrial workers between 1914 and 1936 allowed a culture based on rural traditions to be surpassed by an urban culture of recreation governed by "the joys of outdoor life," as a study on "fringe benefits" concluded. In 1920, 40 percent of the white-collar workers had a paid vacation, a percentage that had doubled a decade later. Another study claims that at the beginning of the interwar years the spread of paid holidays was 10 percent of all wage earners, whereas a poll in 1939 revealed that about a quarter of the respondents enjoyed a paid vacation. Neither the unions, nor the workers themselves seemed to be much interested in getting this recognized by the employers. By the end of the period, though, this percentage had increased to 50, mainly imposed by the employers. The workers among them represented one-quarter of all wage earners.[213] By 1930 Americans spent about 15 percent of their national income on recreation, representing more than the entire national tax burden, but workers gave up their habit of "thrift over spending" more reluctantly "than optimistic merchandisers liked to believe."[214]

To relativize the effect of the American See America First campaign it is worthwhile to note that tourist traffic over the Canadian border had increased fivefold during the previous decade, while international tourism in general (from seaports and through the air) had nearly doubled between 1918 and 1931. In the 1920s bus tours were offered, enabling tourists to "See Europe from an armchair." Thousands of students roamed Europe in "two-month summer tours of ten to twelve countries." After Canada, Americans spent by far the most of their money in France, which during the Interbellum remained the world's largest tourist destination. France's main tourist income came from the United Kingdom (with more than a million tourists in 1929), followed by Spain, but the United States came in third for spending. While during the Depression the French tourist industry collapsed to one-fourth of its 1929 value, car tourism did not suffer in a similar way, however. Thus, in Europe especially, the crisis seems to have catapulted the car into a dominant position in the mobility spectrum.[215]

There is no doubt that international car tourism increased as much as general tourism, if not more, but the share in the total tourist flow was modest, with 5 percent for Britons going abroad in 1930, for instance. Car tourism to Italy increased nearly twentyfold during the decade before 1932, without any decrease during the Depression years, but they formed only about 10 percent of the tourists coming (and going again) by train. Car tourists into Norway in 1931 also formed one-tenth of the total tourist flow. Domestic tourism in Europe was largely carried by the train, although the habit of British tourists going by train to the holiday destination (often "nearby seaside resorts") and then "hire a car in their holiday area," as one historian observed (without giving any evidence), may not have been widespread on the continent. With a higher aggregate proportion of car use vis-à-vis other forms of mobility than in any other European country, in the United Kingdom the half million automobile club members in 1934 submitted 1.4 trip requests per member. A year later 150,000 families hired a caravan. Less than 2 percent of British motorists went abroad.[216]

By far the most detailed and accurate quantitative study of interwar car tourism has been performed for Switzerland, a major part of the Alpine "Playground of Europe" located between Geneva, Vienna, Munich, and Venice. A road census of 1936–1937 in all cantons but one gave 4.6 percent as the share of foreign cars in total Swiss road traffic. Nearly 90 percent of these foreign cars were driven by tourists from the three neighboring countries France, Germany, and Italy, the latter two doubling their contribution between 1935 and 1938, while collective travel from France even increased more than fivefold, no doubt because of

"Why Apologize for Pleasure?"

the Popular Front holiday campaigns. One-sixth of the number of all passenger car trips undertaken on Swiss soil was utilitarian in character according to this study (30 percent in local traffic and 12 percent of interurban traffic), one-third had a holiday purpose (8 percent of local traffic and 35 percent of interurban traffic) undertaken in the case of foreign traffic with a car occupancy of 3.66. Unfortunately the study did not try to estimate tourism's part in the total Swiss 'traffic performance' (in passenger-kilometers), but it did conclude that in nonlocal traffic the largest shares in terms of trips were taken up by domestic and international car tourism. Another study gives further evidence that Switzerland was the European paradise of motorists: nearly one-third of the 1.6 million foreign visitors in 1930 came by car. More persons came in by car than by train from 1933 onward, the result of an ongoing growth of car use and a sudden decline of train use from 1930, as figure 4.14 shows.[217]

Thus, while in all tourism statistics the year 1929 is presented as the indisputable peak year of the interwar phase, car travel in Switzerland seemed to continue to grow without much disruption. That leads us to conclude that motorists used the crisis of the 1930s to gain a crucial foothold in the modal split as well as in the minds of the people.[218] This happened, just as in the United States, on the basis of the nuclear family, and although for Europe the relation with car ownership has not

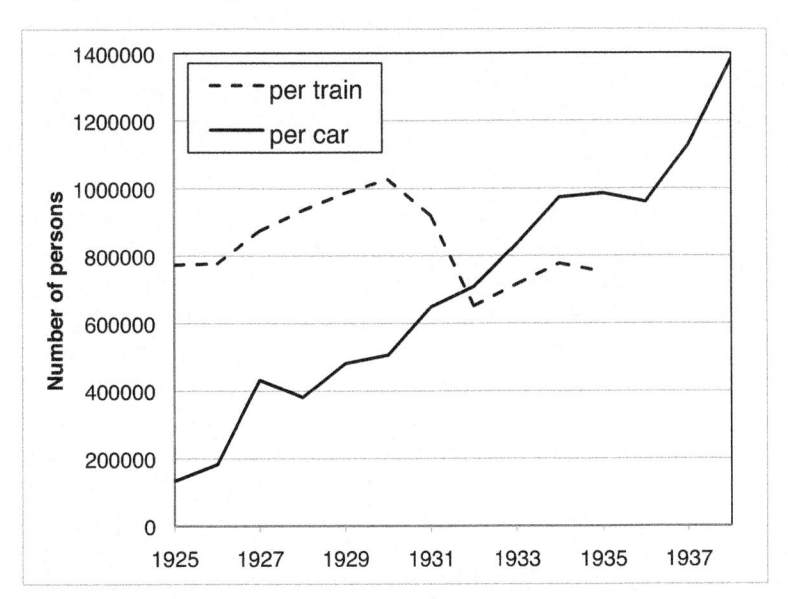

Figure 4.14. Persons traveling into Switzerland, 1925–1938.
Source: Gurtner, 229

been investigated, the family was emphasized there too. This was not a matter of course: the substitution of the traveling family for the traveling man as dominant traveling subject was not a linear process, but went through a transition phase of several decades with a broad spectrum of holiday-making styles.[219]

Conclusions

In the Russian satirical novel *The Golden Calf* (1931), by Il'ia Il'f and Evgenii Petrov, the hero passes the village of Novo-Zaits (New Rabbit) during a rally in his Antelope-Gnu car, and a man emerges from the cheering crowd shouting, "The iron steed is coming to replace the puny peasant horse! ... The automobile is not a luxury but a means of conveyance!" The hero shouts back, "Improve the roads! Merci for the reception!"[220] What was achieved in the West, and especially in the United States, with the tools of marketing, took the form of slogans in the East. But what was perhaps the most surprising was the distance the concept of 'the car as necessity' had meanwhile traveled around the Atlantic world. The car as a common vehicle, possessed or dreamt of, had become a universal phenomenon in this world, and, if we are to take for granted Il'f and Petrov's fiction, beyond.

In this chapter the car, as soon as its use spread toward the new middle-class families in the middle towns of the United States, acquired a 'functional mix' of rural and urban use, of utilitarian, business, and pleasurable applications, of touring, commuting, and shopping, in a highly dispersed and multilayered process full of nonlinearities and contradictions. In this messy process it appeared quite difficult to assess the role of business use and the commute, but in many cities the use of public transport (in number of trips, not passenger-kilometers) dominated until well in the 1930s, although its decline had started in the early 1920s. And although the number of suburbanites (out of reach of public transport) gives an indication of the importance of car commuting, we have seen enough counter-examples (women driving their husbands to the railway station, low shares of commuters and middle-class members in general in some suburbs) to question a simple linearity of public transport substitution. If we really want to assess the share of 'serious' mobility in early automotive culture, shopping perhaps will acquire a central (and gendered) role: suburbanites were at least as eager to shop in the city center as they were to commute at a moment when CBD businesses had not yet begun to follow their clientele into the suburbs. It is impossible to distinguish between 'serious shopping' and shopping for fun, and we don't need to: as long as we keep in mind that the driving itself was often perceived as a pleasurable practice, because of the psycho-physical satisfaction it provided (as we have analyzed for an earlier period in chapter 2 and we will analyze again for this period in chapter 6), car driving was seen as something directly rewarding (like the smoking of a cigarette) and shopping for fun, joyriding, the weekend spin, and the family holiday were the most obvious expressions of this pleasure. In an era without shopping malls, shopping took place at the same place as most commuters went to and came from: the central district.[221] In this chapter we dedicated more research effort to the (better-documented) tourism than to the shopping pleasures, an analysis that leads to the conclusion that it was this mobile pleasure that cemented the car into the fabric of American life. The term 'necessity,' which the contemporaries coined for this embedding, was at the same time defensive and programmatic: it tried to shield automotive culture from state interventionism and placed automobilism squarely into consumption society, as a 'tool' to consume and (re)produce spaces and scapes. We found that 'necessity' should not be confused with utilitarianism: it was the pleasure the car afforded that appeared necessary, from the New Deal all the way to Hitler, and even utilitarianism could not do without this pleasure. This pleasure had a clear individualist connotation, but there were also more collectively grounded experiences around the car, especially in Europe, especially in Germany: the *Volksgemeinschaft* was a term of the 'democratic middle' and not a 'fascist' term, originally.[222]

This chapter has shown how it was the family (as an 'individual' unit, a collective subject) that became the pioneer and backbone of a multiple-use automobilism, both in the United States and in Europe. This multiple use, to formulate the most general claim this chapter allows us to make, consisted for a quarter in pure pleasure, another quarter in pure business, while the remaining was an inextricable mix of pleasure, utility, and business use.[223] Much more important, therefore, is the mixed use of the car, which emerged shortly before the First World War in all North-Atlantic countries, started to dominate during the interwar period, and became the norm during the second half of the century. On both continents, the second half of the 1920s was the crucial phase in which the real breakthrough took place, although one should not forget that even in the United States a large minority of families (44 percent in 1927), most of them from the urban workers, did not have access to a car. That happened in a phase in which "60 percent of American families fell beneath (the) poverty line [of $2000]," a phase characterized by several observers as one of "rugged individualism" or, in the words of president Franklin Roosevelt, "individualism run wild."[224] Equally important is the conclusion that the pleasurable side of this mixed use was much more common than hitherto recognized, leading to the conclusion that the so-called utilitarian car is a historical myth, consciously manufactured by automobile interests and at the individual level functioning as an alibi for a cultural practice which as yet has to be investigated by historians in all its consequences. A part of this mixed use was the holiday trip and other forms of (car) tourism: in the entire North-Atlantic world, at least half of all motorists took part in this pleasurable form of mobility. Within this use profile, the family used the car to conquer long-range mobility. This is how the 'persistence' of Atlantic car culture came about, a persistence, though, under a constant simultaneous change. The author of a 1938 budget survey formulated it quite precisely: "An automobile, if no longer a badge of great distinction, has become at least a necessity to acceptability in middle-class society."[225]

This conclusion at first sight seems to be consistent with the only research done so far by transport scientists (and nonhistorians) into the long-term changes in people's "journeys" in the United Kingdom (mostly through interviews with elderly people). This study not only concludes that it is rather difficult to distinguish between trips as a "necessity" and trips for "leisure and pleasure," but also that the car did not provoke any spectacular behavioral change. The car substituted walking and cycling rather than public transport (which seems especially to apply to the United Kingdom). The switch to the car on an individual level was perhaps most dramatic in smaller towns (up to 100,000 inhabitants). Mobility, these researchers conclude, is mostly local, very stable and "instrumental ... designed to allow an individual to fulfill their [sic] everyday needs and obligations."[226]

Although interviewing people is notoriously biased toward the utilitarian side of mobility (and as such this UK study, too, fits into a long tradition of overemphasizing the 'serious side of mobility' among transport experts), it is nonetheless beneficial to relativize the often uncritically repeated determinism that the introduction of a revolutionary vehicle automatically generates revolutionary behavior. It is, indeed, much more plausible that the universal 'affordance' of the car allowed a broad array of practices which only changed quite incrementally and slowly, in the process influencing each other. The central argument, then, in the explanation of the car's diffusion is its role as a consumption item rather than as a transportation device, a commodity for the continuous production and consumption of 'adventure,' despite the smoke screen (and self-delusion) of contemporaries who kept pounding on the utilitarian turn of the car during the period under scrutiny here.[227] Predominantly used for the pleasurable purposes of leisure time, luxury taxation was more readily accepted in Europe than in the United States and prevented the car from diffusing beyond the better-paid echelons of the middle classes. The size of this group, the frontier between users and nonusers at the 'underside of the diffusion curve,' depended strongly on the balance between income (threatened, because of the economic crisis) and taxes (increasing, for the same reason) during the 1930s, which differed widely both within countries and between them. Superposed upon these multiple mobility modernities are some overarching differences between a European and American mobility culture.

This is not to deny the importance of the economic argument, but it is to argue that this argument can only be used to a certain extent: a largely economic explanation should first and foremost explain why the lower echelons of the middle classes or the higher echelons of the working classes did not use the extensively widespread credit sales system or did not purchase a second-hand car. As long as such an explanation is lacking, we conclude that the desire to possess and use a car was mostly a middle-class desire, as it fit in the Interbellum modernity discourse, and especially in the car's usefulness as a commodity in a leisure context. But at the individual level, there had to be a purchasing argument, which preferably should be based on utilitarian reasoning, and the discourse provided such arguments.

The motorization of Atlantic societies has been an evolutionary process rather than a revolution. The statistics gathered in this chapter are only one preliminary way to approach this phenomenon. The actual purchase of a car can be inspired by totally different impulses than statistics suggest. The constant pounding on the statistics by the contemporaries themselves worked as a self-fulfilling prophecy. Yes, only 10 percent of international tourist flows were transported in cars, but for contemporaries these cars seemed omnipresent, on roads, through villages, not on the tracks into the cities that made trains enter through the back door, but loud, self-confident, and often violent. The discourse these motorists provoked were expectations, really, but they were formulated as if the car had already conquered 100 percent of the flow of traffic. We have seen several examples of the power of these expectations. The most revealing example, however, is the quote from a fascist newspaper in 1938 that remarked on the soon-to-be-expected Volkswagen: "Now, the silent passionate desire of the German worker to possess a car, has been realized."[228] Slip of the tongue? No: fantasies are often so powerful that they seem real—indeed realized. Growth of car ownership and car use thereafter looked like an automatism, as if it only had to finish what was already there at the very beginning. An important role in this discourse was played through the reformulation of the automotive adventure and the efforts it took to make sure that car technology allowed the development of this adventure, as the next chapter will show.

Notes

1. E. C. L., "Motoring is more than a summer pastime," *Autocar* 67, no. 1876 (16 October 1931), 666–667, here 667.
2. Archie Chadbourne, "Debt is the Only Adventure a Poor Man Can Count On," *The American Magazine* 104 (December 1927) 44–45, 108–112.
3. Alvan Macauley, "Industry Need Not Ignore Man's Right to Pursuit of Happiness; Automobile contributes materially to pleasure of the American people and manufacturers should not be ashamed of the fact. Enjoying oneself is not a vice," *Automotive Industries* 50, no. 12 (20 March 1924) 652–653, here 652.
4. Ronald Kline, *Consumers in the Country: Technology and Social Change in Rural America* (Baltimore, MD, and London: Johns Hopkins University Press, 2000), 55.
5. Paolo Laconi, "De Summer of Love in vrijstaat Fiume," *Historisch Nieuwsblad* 18, no. 9 (November 2009), 42–50; women: Leif Jerram, *Streetlife: The Untold History of Europe's Twentieth Century* (Oxford and New York: Oxford University Press, 2011), 127; Michael McGerr, *A Fierce Discontent: The Rise and Fall of the Progressive Movement in America, 1870–1920* (Oxford and New York: Oxford University Press, 2003), 305 (Klan), 311 (normalcy).
6. Robert Wohl, *The Generation of 1914* (Cambridge, MA: Harvard University Press, 1979), 54, 113, 224; Alan Kramer, *Dynamic of Destruction: Culture and Mass Killing in the First World War* (Oxford and New York: Oxford University Press, 2007), 268–327, quotes: 305 (de Gaulle), 308 (addicted); real wages: Hasso Spode, "Der Aufstieg des Massentourismus im 20. Jahrhundert," in Heinz-Gerhard Haupt and Claudius Torp (eds.), *Die Konsumgesellschaft in Deutschland 1890–1990: Ein Handbuch* (Frankfurt and New York: Campus Verlag, 2009), 114–128, here 121; Hartmut Kaelble, *Nachbarn am Rhein: Entfremdung und Annäherung der französischen und deutschen Gesellschaft seit 1880* (Munich: Verlag C. H. Beck, 1991), 139–140; radical: Modris Eksteins, *Tanz über Graben: Die Geburt der Moderne und der Erste Weltkrieg* (Reinbek bei Hamburg: Rowohlt Verlag, 1990), 381.
7. David Stevenson, *With Our Backs to the Wall: Victory and Defeat in 1918* (London and New York: Allen Lane/Penguin, 2011).
8. Jean Orselli, "Usages et usagers de la route: Pour une histoire de moyenne durée, 1860–2008" (unpublished doctoral dissertation, Université de Paris I Panthéon Sorbonne, 2009), Deuxième Partie, 20 (table 15); Alfred Sauvy, "L'automobile en France depuis la guerre," *Bulletin de la Statistique générale de la France* 22 (October 1932–September 1933), 565–615, here 591; Alfred Sauvy, "Production et vente d'automobiles en France," *Bulletin de la Statistique générale de la France* 22 (October 1932–September 1933), 57–66, here 59. Unless there is more useful (quantitative or qualitative) information available, I follow mainstream transport economics and traffic engineering in expressing 'use' in passenger-kilometers and (for freight) ton-kilometers. They are the algebraic product of passengers or tons and the number of 'vehicle-kilometers' and are considered to be a measure of a transport mode's 'performance.' Vehicle-kilometers are used when the

average number of passengers or tons are not known or difficult to estimate. In this case it is assumed that the average vehicle occupancy did not change much during the interwar years.
9. Sauvy, "L'automobile en France depuis la guerre," 568–573, 583 (highest per capita); two-thirds: Christophe Studeny, *L'invention de la vitesse: France, XVIIIe–XXe siècle* (n.p. [Paris]: Gallimard, 1995), 328; Frédéric Vieban, "L'Image de l'automobile auprès des Français 1930–1950" (unpublished Mémoire de maîtrise, Histoire Contemporaine, Université F. Rabelais, Tours, France, 1987), 18 (table 5; values for 1930), 17 (second-hand; value for 1936), 52 (credit). I thank Étienne Faugier for providing me with this source.
10. Hermann Gurtner, *Automobil, Tourismus, Hotellerie: Eine Untersuchung über die Bedeutung des Automobils* (Bern: Touring-Club der Schweiz, 1946), 31.
11. Reiner Flik, *Von Ford lernen? Automobilbau und Motorisierung in Deutschland bis 1933* (Cologne, Weimar, and Vienna: Böhlau Verlag, 2001), 39, 54–60, 150; Andrea Wetterauer, *Lust an der Distanz: Die Kunst der Autoreise in der "Frankfurter Zeitung"* (Tübingen: Tübinger Verein für Volkskunde, 2007), 78; right on pleasure: R.Br., "Das Recht auf Freude," *Der Motorfahrer* 16, no. 9 (8 November 1919), 9; "Kraftwagen in der Landwirtschaft," *ADAC Motorwelt* 28, no. 4 (25 January 1931), 32–33; two-thirds: Generalinspektor für das deutsche Strassenwesen, *Der Kraftverkehr auf Reichsautobahnen, Reichs- und Landstrassen im Dritten Reich I: Die dritte allgemeine Verkehrszählung 1936/37 auf Reichsautobahnen, Reichsstrassen und Landstrassen I. Ordnung; Die Verkehrsentwicklung auf den Reichsautobahnen bis Dezember 1938; Die dritte österreichische Verkehrszählung (Winterhalbjahr 1937/38); Die Zusatzverkehrszählung 1936/37 auf Reichsautobahnen, Reichs- und Landstrassen* (Berlin: Volk und Reich Verlag, 1939), 100 (table 28). A. E. Cohen, "A tour through the occupied zone," *Autocar* 47, no. 1455 (8 October 1921), 644–6446, here 645, encountered eight cars during his first two hundred miles into Germany from the Dutch border.
12. R. C. v. Gorrissen, "Einiges Wissenswerte vom deutschen Motorradverkehr," *ADAC Motorwelt* 24, no. 3 (21 January 1927), 10–12; "Eine Seite Statistik," *ADAC Motorwelt* 28, no. 28 (10 July 1931), 34; Maxim Kreissel, "Der Auspufflärm der Motorräder," *ADAC Motorwelt* 25, no. 39 (28 September 1928), 12–13; Alfred Klaiber, "Das Wochenend-Auto, eine Neuerfindung," *ADAC Motorwelt* 28, no. 44/45 (6 November 1931), 18.
13. Josef M. Jurinek, "ADAC und Oeffentlichkeit: ADAC-Presse-Empfang in Berlin," *ADAC Motorwelt* 24, no. 8 (25 February 1927), 2–6, here 2; one-third, AA and AAA: "Presse-Empfang beim ADAC in der Präsidial-Abteilung Berlin," *ADAC Motorwelt* 25, no. 20 (18 May 1928), 2–4, here 2. ADAC represented from 1928 to 1931 about 10 percent of the motorized German population; ANWB represented 40 percent in 1935. Hans-Christoph Graf von Seherr-Thoss, *75 Jahre ADAC 1903–1978: Tagebuch eines Automobilclubs* (Munich: ADAC Verlag, 1978), 23; Gijs Mom and Ruud Filarski, *Van transport naar mobiliteit: De mobiliteitsexplosie (1895–2005)* (Zutphen: Walburg Pers, 2008), 145. Salzburg: Rüdiger Hachtmann, *Tourismus-Geschichte* (Göttingen: Vandenhoeck & Ruprecht, 2007), 130.
14. "Ein Jahr ADAC-Arbeit," *ADAC Motorwelt* 28, no. 3 (16 January 1931), 4–6; "Die nationale Verwaltungsarbeit," *ADAC Motorwelt* 28, no. 3 (16 January 1931), 7–50, here 50.
15. Josef M. Jurinek, "Steuerwacht des ADAC," *ADAC Motorwelt* 28, no. 5 (30 January 1931), 4–7, here 5; "ADAC-Abwehrkampf," *ADAC Motorwelt* 28, no. 29 (17

July 1931), 2; taking out of registration: "Die Kraftfahrzeuge im Deutschen Reich im Jahre 1931," *ADAC Motorwelt* 28, no. 42/43 (23 October 1931), 7–8, here 7.
16. "Kraftfahrer in Not—durch Steuer und Zoll!" *ADAC Motorwelt* 29, no. 17 (22 April 1932), 2–4, here 4; "ADAC-Kampf gegen Steuerdruck: Eindrucksvolle Kundgebungen in allen Gauen Deutschlands," *ADAC Motorwelt* 29, no. 43/44 (28 October 1932), 2; "Letzter Termin: 1. April 1933! Noch drei Monate Not des deutschen Kraftverkehrs?" *ADAC Motorwelt* 30, no. 1/2 (6 January 1933) 2; "Wir machen nicht mehr mit! Die Kraftfahrer wehren sich gegen die Benzinpreis-Politik!" *ADAC Motorwelt* 30, no. 5/6 (3 February 1933), 3–4.
17. Marinebaurat a. W. Mendelssohn, "Der Erfolg der Kraftfahrzeugsteuer-Senkung in Danzig," *ADAC Motorwelt* 30, no. 5/6 (3 February 1933), 10–12; motorcycles to cars: Gustav Plum, "Autobahnen: Die Zuflucht aus der Arbeitslosigkeit," *ADAC Motorwelt* 30, no. 30 (28 July 1933), 8–10; October 1932: Gustav Plum, "Die deutsche Kraftfahrzeug-Industrie 1933," *ADAC Motorwelt* 30, no. 50/52 (22 December 1933), 2–3; foreign cars: Hugo Junker, "Reifenpannen und Fehlzündungen," *ADAC Motorwelt* 30, no. 39 (29 September 1933), 23–24, here 23; March 1933: "Nachrichten aus aller Welt," *ADAC Motorwelt* 30, no. 42/43 (27 October 1933), 37. Class contradictions: Hochstetter, *Motorisierung und "Volksgemeinschaft"*, 163; preceded: Gregor M. Rinn, "Das Automobil als nationales Identifikationssymbol: Zur politischen Bedeutungsprägung des Kraftfahrzeugs in Modernitätskonzeptionen des 'Dritten Reichs' und der Bundesrepublik" (unpublished dissertation, Humboldt-Universität, Berlin, 2008), 31.
18. Walther G. Hoffmann (in cooperation with Franz Grumbach and Helmut Hesse), *Das Wachstum der deutschen Wirtschaft seit der Mitte des 19. Jahrhunderts* (Berlin, Heidelberg, and New York: Springer-Verlag, 1965), 692–693 (table 186).
19. *50 Jahre ADAC im Dienste der Kraftfahrt* (Munich: ADAC, n.d. [1953]), 105, 108–109.
20. Hochstetter, *Motorisierung und "Volksgemeinschaft"*, 192–193, 223, 228, 235–241 (*Motor-HJ*), 244 (membership), 185 (3.4 million), 268 (*Motorsportschulen*); trebling: Wolfgang König, *Geschichte der Konsumgesellschaft* (Stuttgart: Franz Steiner Verlag, 2000), 305; recovery: Anders Ditlev Clausager, "Motorisierung: The German Motorization Program 1933–1939," *Automotive History Review* no. 52 (Summer 2010), 23–35, here 27.
21. Sammy Kent Brooks, "The Motorcycle in American Culture: From Conception to 1935" (unpublished dissertation, George Washington University, Washington, DC, 1975), 126 (cheaper), 132 (comfort), 142 (war), 143 (males), 148 (stability), 207 (production), 235 (social), 242 (registrations), 273 (fourth); registration figures 1930: Malcolm M. Willey and Stuart A. Rice, *Communication Agencies and Social Life* (New York and London: McGraw-Hill Book Company, Inc., 1933), 31 (table 10), 38 (table 13).
22. P. M. S. N., "De dans der getallen," *Het Motorrijwiel en de Lichte Auto* 19 (1931), 1044–1046; "Eine Seite Statistik," *ADAC Motorwelt* 28, no. 28 (10 July 1931), 34; Austin Seven: Andrew Gavin Altman, "Motoring for the Million? Cars and Class in Pre-1952 Britain" (unpublished PhD thesis, Boston College, Graduate School of Arts & Sciences, 3 April 1997), 217–218.
23. Helmut Braun and Christian Panzer, "The Expansion of the Motor-Cycle Industry in Germany and in Great Britain (1918 until 1932)," *The Journal of European Economic History* 32, no 1 (2003), 25–59; Uwe Fraunholz, *Motorphobia: Antiautomobiler Protest in Kaiserreich und Weimarer Republik* (Göttingen: Vandenhoeck & Ruprecht, 2002), 40–41; "Eine Seite Statistik," *ADAC Motorwelt* 28, no.

28 (10 July 1931), 34; "Motorrad : Auto = 1000 : x," *ADAC Motorwelt* 25, no. 46 (16 November 1928), 28.
24. Clausager, "Motorisierung," 28; DKW: Wolf Dieter Lützen, "Radfahren, Motorsport, Autobesitz: Motorisierung zwischen Gebrauchswerten und Statuserwerb," in Wolfgang Ruppert (ed.), *Die Arbeiter: Lebensformen, Alltag und Kultur von der Frühindustrialisierung bis zum "Wirtschaftswunder"* (Munich: Verlag C. H. Beck, 1986), 369–377, here 368; European Volkswagen: *Allgemeine Automobil-Zeitung* (1933), quoted by Wolfgang König, *Volkswagen, Volksempfänger, Volksgemeinschaft; "Volksprodukte" im Dritten Reich; Vom Scheitern einer nationalsozialistischen Konsumgesellschaft* (Paderborn, München, Wien, Zürich: Ferdinand Schöningh, 2004),, 156; 1.2 percent: Heidrun Edelmann, *Vom Luxusgut zum Gebrauchsgegenstand: Die Geschichte der Verbreitung von Personenkraftwagen in Deutschland* (Frankfurt: Verband der Automobilindustrie, 1989), 210.
25. König, *Volkswagen, Volksempfänger, Volksgemeinschaft*, 189; middle class: Sasha Disko, "Men, Motorcycles and Modernity: Motorization in the Weimar Republic" (unpublished dissertation, New York University, May 2008), 89.
26. Walther G. Hoffmann (in cooperation with Franz Grumbach and Helmut Hesse), *Das Wachstum der deutschen Wirtschaft seit der Mitte des 19. Jahrhunderts* (Berlin, Heidelberg, and New York: Springer-Verlag, 1965), 693 (table 186).
27. *How American Buying Habits Change* (Washington, DC: US Department of Labor, n.d. [1959?]), 186; Martha L. Olney, *Buy Now Pay Later: Advertising, Credit, and Consumer Durables in the 1920s* (Chapel Hill and London: The University of North Carolina Press, 1991), 30–31. Olney defines a durable good as "a product whose services can be consumed for three years or more" (ibid., 6); Eurocentric: Koen Koch, "'De laatste dagen der mensheid': De catastrofale Eerste Wereldoorlog als cesuur," *Geschiedenis Magazine* (November 2007), 33–36, here 36. For an analysis concluding that the structural shift to durable goods took already place in the nineteenth century and that in the twentieth only qualitative changes (such as electrification and the car) can be observed, see Harold G. Vatter, "Has There Been a Twentieth-Century Consumer Durables Revolution?" *The Journal of Economic History* 27, no. 1 (March 1967), 1–16.
28. Emil Merkert, *Personenkraftwagen, Kraftomnibus und Lastkraftwagen in den Vereinigten Staaten von Amerika: Mit besonderer Berücksichtigung ihrer Beziehungen zu Eisenbahn und Landstrasse* (Berlin: Springer, 1930), 4.
29. John S. Gilkeson, Jr., *Middle-Class Providence, 1820–1940* (Princeton, NJ: Princeton University Press, 1986), 334–335 (quote on 335).
30. Sandra Coyner quoted in Michael Wildt, *Am Beginn der "Konsumgesellschaft": Mangelerfahrung, Lebenshaltung, Wohlstandshoffnung in Westdeutschland in den fünfziger Jahren* (Hamburg: 1995), 25.
31. Quoted in Catherine Gudis, *Buyways: Billboards, Automobiles, and the American Landscape* (New York and London: Routledge, 2004), 101.
32. Olney, *Buy Now Pay Later*, 40, 47–48; costs: Kevin H. O'Rourke and Jeffrey G. Williamson, *Globalization and History: The Evolution of a Nineteenth-Century Atlantic Economy* (Cambridge, MA, and London: The MIT Press, 2000), 29; Marina Moskowitz, *Standard of Living: The Measure of the Middle Class in Modern America* (Baltimore, MD: Johns Hopkins University Press, 2004), 133 (first a home), 135 (Hoover quote); Willey and Rice, *Communication Agencies and Social Life*, 38. Clothes and shoes: Christine Frederick, "New Wealth, New Standards of Living, and Changed Family Budgets," *Annals* (American Academy of Political and Social Science) 115 (September 1924) 74-82, here: 78.

33. "Who Owns a Motor Car?" *Literary Digest* (11 November 1922), 78.
34. Frederick, "New Wealth," 82.
35. The results of the survey are paraphrased by David Blanke, *Hell on Wheels: The Promise and Peril of America's Car Culture, 1900–1940* (Lawrence: University Press of Kansas, 2007), 180.
36. Olney, *Buy Now Pay Later*, 92; Hazel Kyrk, *Economic Problems of the Family* (New York and London: Harper & Brothers, 1933), 427 (table LXXXV).
37. Olney, *Buy Now Pay Later*, 95, 137; more than half (55.7 percent): John Henry Mueller, "The Automobile: A Sociological Review" (unpublished doctoral dissertation, University of Chicago, August 1928), 8. That year, 57 percent of American families owned a telephone, 46 percent a phonograph, 40 percent a piano, and 24 percent a radio (ibid., 9). Such aggregate overviews should be treated with care: the conclusion, for instance, in Heather B. Barrow, "The Automobile in the Garden: Henry Ford, Suburbanization, and the Detroit Metropolis, 1919–1940" (unpublished dissertation, University of Chicago, Department of History, June 2005), 268, that "each American household [had] its own" car by the end of the 1920s, because the statistics gave an average of "one car for every 4.5 people" is certainly premature.
38. Lendol Calder, *Financing the American Dream: A Cultural History of Consumer Credit* (Princeton, NJ: Princeton University Press, 1999), 203; Clair Brown, *American Standards of Living 1918–1988* (Oxford and Cambridge, MA: Blackwell, 1994), 12–13 (table 1.3), 135.
39. Merkert, *Personenkraftwagen*, 26 (table 13).
40. Carle C. Zimmerman, *Consumption and Standards of Living* (London: Williams and Norgate, 1936), 314.
41. Donald R. G. Cowan, "Regional Consumption and Sales Analysis," *The Journal of Business of the University of Chicago* 8, no. 4 (1935), 345–381, here 348, 356.
42. George Kirkham Jarvis, "The diffusion of the automobile in the United States: 1895–1969" (unpublished dissertation, University of Michigan, Ann Arbor, 1972). For an inversely proportional relationship between population and car density see also: [The Bureau of Public Roads, US Department of Agriculture/The Connecticut State Highway Department], *Report of a Survey of Transportation on the State Highway System of Connecticut* (Washington, DC: Government Printing Office, 1926), 36.
43. William L. Cavert, *Sources of Power on Minnesota Farms* (doctoral dissertation, University of Minnesota, reprinted from *Minnesota Agricultural Experiment Station Bulletin* 262, February 1930), 53; [Bureau of Agricultural Economics], "Horses, Mules, and Motor Vehicles: Year Ended March 31, 1924; With Comparable Data for Earlier Years" (Washington, DC: United States Department of Agriculture, January 1925; Statistical Bulletin, no. 5), 85 (table 98), 87; Hal S. Barron, *Mixed Harvest: The Second Great Transformation in the Rural South, 1870–1930* (Chapel Hill and London: The University of North Carolina Press, 1997), 195; Kline, *Consumers in the Country*, 5 (figure 2); Dutch and Italian farmers: Harold C. Hoffsommer, *Relations of Cities and Larger Villages to Changes in Rural Trade and Social Areas in Wayne County, New York* (Ithaca, NY: Cornell University Agricultural Experiment Station, Bulletin 582, February 1934), 9, 13; no cars mentioned: Marion Fish, *Buying for the Household as Practiced by 368 Farm Families in New York, 1928–29* (Ithaca, NY: Cornell University Agricultural Experiment Station, June 1933; Bulletin 561; Contribution from Studies in Home Economics); Marianne Muse, *The Standard of Living on Specific Owner-Oper-*

ated Vermont Farms (Burlington, VT: Free Press Printing Co., June 1932; University of Vermont and State Agricultural College, Bulletin 340), 10, 22, opposite 24 (table 6), 27, 34. Farmer women: Howard W. Beers, "A Portrait of the Farm Family in Central New York State," *American Sociological Review* 2, no. 5 (October 1937), 591–600, here 598–599.

44. Small family farms: James J. Flink, "Three Stages of American Automobile Consciousness," *American Quarterly* 24 (October 1972), 451–473, here 466; Hoffsommer, *Relations of Cities and Larger Villages*, 29; Malcolm M. Willey and Stuart A. Rice, "The Agencies of Communication," chapter IV in *Recent Social Trends; Report of the President's Research Committee on Social Trends, Volume I* (New York and London: McGraw-Hill Book Company, Inc., 1933), 167–217, here 180.

45. Peter David Norton, *Fighting Traffic: The Dawn of the Motor Age in the American City* (Cambridge, MA, and London: The MIT Press, 2008), 153; Kyrk, *Economic Problems of the Family*, 3 ("urban" meant as having 2500 inhabitants or more; this share was 41 percent in 1900).

46. Harold Katz, *The Decline of Competition in the Automobile Industry, 1920–1940* (New York: Arno Press, 1970), 41 (table 2.1), 54 (table 2.4), 62 (table 2.6); Mark Adams, "The Automobile—A Luxury Becomes a Necessity," in Walton Hamilton et al., *Price and Price Policies* (New York and London: McGraw-Hill, 1938), 27–81, here 42. Zimmerman, *Consumption and Standards of Living*, 333. Roy D. Chapin, "The Motor's Part in Transportation," *Annals of the American Academy of Political and Social Sciences* 116 (The Automobile: Its Province and Problems; November 1924), 1–8, here 6, claimed that "the number of automobile owners in this country is roughly equal to the number of persons receiving incomes in excess of $1,400." 2.5 times: Robert E. Ireland, *Entering the Auto Age: The Early Automobile in North Carolina, 1900–1930* (Raleigh: Division of Archives and History, North Carolina Department of Cultural Resources, 1990), 128. Greg Thompson, "'My Sewer': James J. Flink on His Career Interpreting the Role of the Automobile in Twentieth-century Culture," in Peter Norton et al. (eds.), *Mobility in History: The Yearbook of the International Association for the History of Transport, Traffic and Mobility; Volume 4* (New York and Oxford: Berghahn Journals, 2013), 3–17, here 5.

47. Edmund de S. Brunner and J. H. Kolb, *Rural Social Trends* (New York and London: McGraw-Hill Book Company, 1933), 9, 23–27; portray: Beers, "A Portrait of the Farm Family in Central New York State," 594.

48. I thank Clay McShane for this insight.

49. David J. St. Clair, *The Motorization of American Cities* (New York, Westport, CT, and London: Praeger, 1986), 126–127; 30 million: "A Review of Findings by the President's Research Committee on Social Trends," in *Recent Social Trends*, xxxix.

50. For a critical evaluation of the "transportation thesis" (that suburbanization can be explained by middle-class use of first tramways and then the car), see Fred W. Viehe, "Black Gold Suburbs: The Influence of the Extractive Industry on the Suburbanization of Los Angeles, 1890–1930," *Journal of Urban History* 8 (1981), 3–26. Viehe claims that the oil industry (and the industry in general, building factories) was responsible for the trend.

51. Robert A. Beauregard, *When America Became Suburban* (Minneapolis and London: University of Minnesota Press, 2006), 24, 33, 124 (school), 126 (Mrs. Consumer), 131 (lifestyle).

52. Tim Cresswell, *On the Move: Mobility in the Modern Western World* (New York and London: Routledge, 2006), 38 (nomads); enslaved: T. Arthur Leonard, *Adventures in Holiday Making: Being the story of the Rise and Development of a People's Holiday Movement* (London: Holiday Fellowship, n.d. [1934]), 81; Catherine Simpson, "Volatile Vehicles: When Women Take the Wheel; Domestic Journeying & Vehicular Moments in Contemporary Australian Cinema," in Lisa French (ed.), *Womenvision: Women and the Moving Image in Australia* (Melbourne: Damned Publishing, 2003), 197–210, here 198–199.
53. John Modell, *Into One's Own: From Youth to Adulthood In The United States 1920–1975* (Berkeley, Los Angeles, and London: University of California Press, 1989), 72–89.
54. Daniel J. Walkowitz, *Working with Class: Social Workers and the Politics of Middle-Class Identity* (Chapel Hill and London: The University of North Carolina Press, 1999), 6; John S. Gilkeson, *Anthropologists and the Rediscovery of America, 1886–1965* (Cambridge and New York: Cambridge University Press, 2010), 85–86; Cotten Seiler, "Anxiety and Automobility: Cold War Individualism and the Interstate Highway System" (dissertation, University of Kansas, 2002), 134; C. Wright Mills, *White Collar: The American Middle Classes* (New York: Oxford University Press, 1953). For a critique of the "caricature" Mills makes of the middle class, see Robert D. Johnston, "Conclusion: Historians and the American Middle Class," in Burton J. Bledstein and Robert D. Johnston (eds.), *The Middling Sorts: Explorations in the History of the American Middle Class* (New York and London: Routledge, 2001), 296–306, here 297. See also Daniel Horowitz, *The Morality of Spending: Attitudes toward the Consumer Society in America, 1875–1940* (Baltimore, MD, and London: Johns Hopkins University Press, 1985), 151. On the Lynds' ideological blindness, see Richard Jensen, "The Lynds Revisited," *Indiana Magazine of History* 75 (1979), 303–319.
55. Robert S. Lynd and Helen Merrell Lynd, *Middletown: A Study in American Culture* (New York: Harcourt, Brace and Company, 1929), 23 (business class), 64 (two-thirds), 253 (Fords), 253–255 (quote on 254), 262 (vacation); six editions: Sarah E. Igo, *The Averaged American: Surveys, Citizens, and the Making of a Mass Public* (Cambridge, MA, and London: Harvard University Press, 2007), 69; miscalculations and projection: Horowitz, *The Morality of Spending*, 151.
56. For instance, "Confessions of a Ford Dealer" published in *Harper's Monthly Magazine* in 1927 and reproduced in George E. Mowry (ed.), *The Twenties: Fords, Flappers & Fanatics* (Englewood Cliffs, NJ: Prentice-Hall, 1963), 26–31.
57. Mowry (ed.), *The Twenties*, 47–51.
58. Adams, "The Automobile—A Luxury Becomes a Necessity," 40 (n. 3).
59. Day Monroe et al., *Family Expenditures for Automobile and Other Transportation: Five Regions* (Washington, DC, 1941), 1–3. For motorization among the rural poor, especially in the South, see Ireland, *Entering the Auto Age*, 127–130.
60. Frank Stricker, "Affluence for Whom?—Another Look at Prosperity and the Working Classes in the 1920s," *Labor History* 24 (1983), 5–33; "a large segment of the market consists of travelling salesmen, car rental agencies, and other fleet owners": Katz, *The Decline of Competition in the Automobile Industry*, 355.
61. Richard Sterner (in collaboration with Lenore A. Epstein, Ellen Winston, and others), *The Negro's Share: A Study of Income, Consumption, Housing and Public Assistance* (Westport, CT: Negro University Press, 1971; reprint; first ed. 1943), 144; Lillian Rhoades, "One of the Groups Middletown Left Out," *Opportunity, Journal of Negro Life* (March 1933), 75–77, 93.

62. Stephen L. McIntyre, "'The Repair Man Will Gyp You': Mechanics, Managers, and Customers in the Automobile Repair Industry, 1896–1940" (unpublished dissertation, University of Missouri-Columbia, May 1995), 320–321.
63. Thompson, "'My Sewer,'" 7.
64. Sven Beckert, "Propertied of a Different Kind: Bourgeoisie and Lower Middle Class in the Nineteenth-Century United States," in Bledstein and Johnston (eds.), *The Middling Sorts*, 285–295, here 288–290; decline: Michael Lind, *Land of Promise: An Economic History of the United States* (New York: Harper-Collins, 2012), 262.
65. Angus Campbell and Philip E. Converse (eds.), *The Human Meaning of Social Change* (New York: Russell Sage Foundation, 1972), 74–75 (table A).
66. Adams, "The Automobile—A Luxury Becomes a Necessity," 42–43.
67. Mark Wyman, *Hoboes: Bindlestiffs, Fruit Tramps, and the Harvesting of the West* (New York: Hill and Wang, A Division of Farrar, Straus and Girous, 2010), 257.
68. Hazel Kyrk et al., *Family Expenditures for Household Operation: Five Regions* (Washington, DC: United States Department of Agriculture, The Bureau of Home Economics in cooperation with the Work Projects Administration, 1941), 2–4.
69. Robert S. Lynd and Helen Merrell Lynd, *Middletown in Transition: A Study in Cultural Conflicts* (New York: Harcourt, Brace and Company, 1937), 267–269, 573 (table 43): 3 percent drop was reported for car density (cars per one thousand inhabitants); soul quote in: Steven Mintz and Susan Kellogg, *Domestic Revolutions: A Social History of American Family Life* (New York: The Free Press, 1988), 136; houses: Becky M. Nicolaides, *My Blue Heaven: Life and Politics in the Working-Class Suburbs of Los Angeles, 1920–1965* (Chicago and London: The University of Chicago Press, 2002), 58.
70. Julius Weinberger, "Economic Aspects of Recreation," *Harvard Business Review* 15 (1937), 448–463, here 449.
71. Gilkeson, Jr., *Middle-Class Providence*, 300, 309; Ashleigh Brilliant, "Some Aspects of Mass Motorization in Southern California, 1919–1929," *Southern California Quarterly* 47, no. 2 (July 1965), 191–208, here 191. For an excellent overview of budget studies, see Horowitz, *The Morality of Spending*, 154–165.
72. Gilkeson, Jr., *Middle-Class Providence*, 4; Clark Davis, "The Corporate Reconstruction of Middle-Class Manhood," in Bledstein and Johnston (eds.), *The Middling Sorts*, 201–216, here 204–205 (masculine: 213); Mills, *White Collar*, 63–67, 182, 238.
73. Arthur J. Vidich and Joseph Bensman, *Small Town in Mass Society: Class, Power and Religion in a Rural Community* (Princeton, NJ: Princeton University Press, 1968), 58, 321.
74. Moskowitz, *Standard of Living*, 242 (n. 29); Gilkeson, Jr., *Middle-Class Providence*, 300, 309; meaningless: Jeffrey M. Hornstein, "The Rise of the Realtor®: Professionalism, Gender, and Middle-Class Identity, 1908–1950," in Bledstein and Johnston (eds.), *The Middling Sorts*, 217–233, here 233; Lind, *Land of Promise*, 471.
75. Hans-Jürgen Puhle, "Einleitung," in Puhle (ed.), *Bürger in der Gesellschaft der Neuzeit: Wirtschaft—Politik—Kultur* (Göttingen: Vandenhoeck & Ruprecht, 1991), 7–13, here 8; Ronald Edsforth, *Class Conflict and Cultural Consensus: The Making of a Mass Consumer Society in Flint, Michigan* (New Brunswick, NJ, and London: Rutgers University press, 1987), 32.
76. Gregory Anderson, "Angestellte in England 1850–1914," in Jürgen Kocka (ed.), *Angestellte im europäischen Vergleich: Die Herausbildung angestellter Mit-*

telschichten seit dem späten 19. Jahrhundert (Göttingen: Vandenhoeck & Ruprecht, 1981), 59–73, here 60; Alan A. Jackson, *The Middle Classes 1900–1950* (Nairn: David St. John Thomas Publisher, 1991), 15–16, 26–27, 105–107, 315 (strong belief); Lawrence James, *The Middle Class: A History* (London: Little and Brown, 2006), 232, 256, 359 (birth control); F. D. Klingender, *The Condition of Clerical Labour in Britain* (London: Martin Lawrence, 1935), xiv–xvii, 17–21, 108; prices: E. G. E. Beaumont, "The influence of the automobile user upon the automobile engineer," *The Automobile Engineer* (October 1927), 393–396, here 393 (fig. 1). I thank Clay McShane for providing me with a copy of this source.
77. Philip Massey, "The Expenditure of 1,360 British Middle-Class Households in 1938–39," *Journal of the Royal Statistical Society* 105, no. 3 (1942), 159–196, here 180, 185 (table XXII), 193 (table).
78. Christopher Thomas Potter, "Motorcycle Clubs in Britain During the Interwar Period, 1919–1939: Their Social and Cultural Importance," *International Journal of Motorcycle Studies* 1 (March 2005), http://ijms.nova.edu/March 2005/IJMS_ArtclPotter0305.html (last accessed 16 July 2010). Average car prices in 1936 were nearly half of 1924 prices, but running costs remained high, according to Rudy Koshar, "Cars and Nations: Anglo-German Perspectives on Automobility between the World Wars," *Theory, Culture & Society* 21, no. 4/5 (2004), 121–144, here 135; Owen John, "On the Road," *Autocar* 47, no. 1355 (8 October 1921), 621–623, here 623.
79. Colin Pooley, Jean Turnbull, and Mags Adams, "The impact of new transport technologies on intraurban mobility: a view from the past," *Environment and Planning A* 38 (2006), 253–267, here 255 (table 1); unprecedented: Colin G. Pooley and Jean Turnbull, "Commuting, transport and urban form: Manchester and Glasgow in the mid-twentieth century," *Urban History* 27, no. 3 (2000), 360–383, here 363; A. M. Warnes, "Estimates of Journey-to-Work Distances from Census Statistics," *Regional Studies* 6 (1972), 315–326, here 319 (table 1).
80. Gustav Sjöblom, "The Political Economy of Railway and Road Transport in Britain and Germany, 1918–1933" (unpublished dissertation, Darwin College, University of Cambridge, 16 April 2007), 65 (figure 6).
81. Reinhard Spree, "Angestellte als Modernisierungsagenten: Indikatoren und Thesen zum reproduktiven Verhalten von Angestellten im späten 19. und frühen 20. Jahrhundert," in Kocka (ed.), *Angestellte im europäischen Vergleich*, 279–308; Heinrich August Winkler, *Liberalismus und Antiliberalismus: Studien zur politischen Sozialgeschichte des 19. und 20. Jahrhunderts* (Göttingen: Vandenhoeck & Ruprecht, 1979), especially chap. 13: "Extremismus der Mitte?" The debate was triggered by a study by the American historian Seymour Lipset. For a revision of the easy identification of lower-middle-class culture and fascism, see Richard Hamilton, *Who Voted for Hitler?* (Princeton, NJ: Princeton University Press, 1982), passim, but especially 60–62, 90–91, 121–122, who convincingly shows that urban middle classes did not support the NSDAP (young white collars in towns, for instance, voted social-democratic SDP), while the upper and upper middle classes, especially if Protestant, did. His evidence is partly based on an elegant argumentation, namely that on key election day 31 July 1932, one-tenth of the population in some case study cities like Hamburg were on holidays, and that a large part of the votes given in railway stations or holiday resorts were for NSDAP (ibid., 200–228). For American white collars being less susceptible to right-wing propaganda, see Jürgen Kocka, *White Collar Workers in America 1890–1940: A Social-Political History in International Perspective* (London and Beverly Hills, CA: Sage Publications, 1980), 249, 252–256.

82. René König, "Zur Soziologie der Zwanziger Jahre, oder Ein Epilog zu zwei Revolutionen, die niemals stattgefunden haben, und was daraus für unsere Gegenwart resultiert," in Leonhard Reinisch (ed.), *Die Zeit ohne Eigenschaften: Ein Bilanz der zwanziger Jahre* (Stuttgart: W. Kohlhammer Verlag, 1961), 82–118, here 85; Geiger: Dominik Schrage, "Der Konsum in der deutschen Soziologie," in Haupt and Torp (eds.), *Die Konsumgesellschaft in Deutschland 1890–1990*, 319–334, here 325; surpass: Eric D. Weitz, *Weimar Germany: Promise and Tragedy* (Princeton, NJ, and Oxford: Princeton University Press, 2007), 156.
83. Hamilton, *Who Voted for Hitler?* 24–25.
84. Spree, "Angestellte als Modernisierungsagenten," 287–288, 291 (see for family sizes of different occupational groups: ibid., 295, table 2); Richard Hamilton, "Die soziale Basis des Nationalsozialismus: Eine kritische Betrachtung," in Kocka (ed.), *Angestellte im europäischen Vergleich*, 354–375, here 371–372; 460 families: *Was verbrauchen die Angestellten? Ergebnisse der dreijährigen Haushaltsstatistik des Allgemeinen Freien Angestelltenbundes* (Berlin: Freier Volksverlag, 1931), 24, 53; Sandra Jean Coyner, "Class Patterns of Family Income and Expenditure during the Weimar Republic: German White-Collar Employees as Harbingers of Modern Society" (PhD dissertation, Rutgers University, 1975), 96, 108–109, 121–124, 128, 148, 271, 284, 291, 301–303, 316, 353, 355, 364–366; family oriented: Sandra J. Coyner, "Class consciousness and consumption: The new middle class during the Weimar Republic," *Journal of Social History* 10, no. 3 (Spring 1977), 310–331, here 314; family size: Spode, "Der Aufstieg des Massentourismus im 20. Jahrhundert," 151. I thank Clay McShane for his suggestion about the cost of suburban land in the United States.
85. *Die Lebenshaltung von 2000 Arbeiter-, Angestellten- und Beamtenhaushaltungen; Erhebungen von Wirtschaftsrechnungen im Deutschen Reich vom Jahre 1927/28, Teil I, Gesamtergebnisse (Bearbeitet im Statistischen Reichsamt)* (Berlin: Verlag von Reimar Hobbing, 1932), 20.
86. Robert Dinse, *Das Freizeitleben der Grossstadtjugend: 5000 Jungen und Mädchen berichten; Zusammengestellt und bearbeitet in Verbindung mit dem Deutschen Archiv für Jugendwohlfahrt von ––, Stadtjugendpfleger, Berlin* (Eberswalde-Berlin: Verlagsgesellschaft R. Müller, n.d. [1932?]), 7–8 (many workers), 87 (outings), 91 (tinkering); golden rule: Wolf Dieter Lützen, "Radfahren, Motorsport, Autobesitz: Motorisierung zwischen Gebrauchswerten und Statuserwerb," in Wolfgang Ruppert (ed.), *Die Arbeiter: Lebensformen, Alltag und Kultur von der Frühindustrialisierung bis zum "Wirtschaftswunder"* (Munich: Verlag C. H. Beck, 1986), 369–377, here 370.
87. Gijs Mom, "Transporting mobility: The emergence of two different car societies in a transatlantic perspective," in Hans Krabbendam, Cornelis A. van Minnen, and Giles Scott-Smith (eds.), *Four Centuries of Dutch-American Relations 1609–2009* (Amsterdam: Boom, 2009), 819–830; equaled: Rudolf Hoenicke, *Die amerikanische Automobilindustrie in Europa* (Bottrop: Buchdruckerei Wilhelm Postberg, 1933), 8.
88. Hoenicke, *Die amerikanische Automobilindustrie in Europa*, 61, 14; Hermann Röttger, *Die Struktur der deutschen Automobil-Einzelhandels mit Personenkraftwagen* (dissertation, Universität Köln, 1940), 102; Vincent van der Vinne, *De trage verbreiding van de auto in Nederland 1896–1939: De invloed van ondernemers, gebruikers en overheid* (De Bataafsche Leeuw, 2007), 251; one-quarter and France, and Belgium largest importer: "Der Welthandel in Automobilen," *ADAC Motorwelt* 25, no. 34 (24 August 1928), 26; one-third and one-fifth: E. S. von Oelsen, 'Die Automobil-Ausfuhr der Vereinigten Staaten nach Europa in

den letzten Jahren," *ADAC Motorwelt* 26, no. 25 (21 June 1929), 25–27, here 26.
89. J. W. Drukker and Timo de Rijk, "American Influences on Dutch Material Culture and Product Design," in Hans Krabbendam, Cornelis A. van Minnen and Giles Scott-Smith (eds.), *Four Centuries of Dutch-American Relations 1609–2009* (Amsterdam: Boom, 2009), 442–456.
90. The following pages are partially based on Gijs Mom, "Mobility for Pleasure: A Look at the Underside of Dutch Diffusion Curves (1920–1940)," *TST Revista de Historia: Transportes, Servicios y Telecomunicaciones* no. 12 (June 2007), 30–68.
91. A. F. Napp-Zinn, "Eigenart, Stand und Probleme der Verkehrswirtschaft der Niederlande," *Zeitschrift für Verkehrswissenschaft* 14, no. 3 (1937), 157–172, here 159.
92. US Department of Commerce, Bureau of Foreign and Domestic Commerce, *Installment selling of motor vehicles in Europe* (Compiled in Automotive Division; From Reports by Representatives of the Department of Commerce, Trade Information Bulletin, no. 550, May 1928), 1, 28–29; J. A. G. Pennington, *Advertising Automotive Products in Europe* (US Department of Commerce, Bureau of Foreign and Domestic Commerce, Trade Information Bulletin, no. 462, February 1927), 11.
93. Mom and Filarski, *Van transport naar mobiliteit*.
94. Reinhold Stisser, *Der deutsche Automobilexport unter besonderer Berücksichtigung des niederländischen Kraftfahrzeugmarktes* (Kiel, 1938).
95. Sean O'Connell, *The Car and British Society: Class, Gender and Motoring, 1896–1939* (Manchester and New York: Manchester University Press, 1998), 22, 31–32, 35; Gurtner, *Automobil, Tourismus, Hotellerie*, 31; Sue Bowden and Paul Turner, "The Demand for Consumer Durables in the United Kingdom in the Interwar Period," *The Journal of Economic History* 53, no. 2 (June 1993), 244–258, here 245–248.
96. Mom, "Mobility for Pleasure."
97. Van der Vinne, *De trage verbreiding van de auto in Nederland 1896–1939*, 384, 392; Weber, "Automobilisering en de overheid in België vóór 1940," 331.
98. "Der ADAC und die Selbstfahrer-Union Deutschlands e.V.," *ADAC Motorwelt* 26, no. 14 (5 April 1929), 2.
99. Orselli, "Usages et usagers de la route," Troisième Partie, 19.
100. See, however, Flik, *Von Ford lernen?* who explains the reluctance of the German farmers to motorize during the interwar years by pointing at their much lower income in comparison to their American counterparts.
101. Jarvis, "The diffusion of the automobile in the United States." This could also (at least partly) explain the exceptionally high diffusion rates in Denmark as an 'empty' country. Contemporary sources, however, point at the crucial role of a distribution plant in Copenhagen. "Der Welthandel in Automobilen," *ADAC Motorwelt* 25, no. 34 (24 August 1928), 26. For a discussion of global mobility diffusion patterns, see Hanna Wolf, *Following America? Dutch Geographical Car Diffusion, 1900 to 1980* (Eindhoven: ECIS, 2010).
102. Quoted in Andrea Wetterauer, *Lust an der Distanz: Die Kunst der Autoreise in der "Frankfurter Zeitung"* (Tübingen: Tübinger Verein für Volkskunde, 2007), 18.
103. Russell Bond, "Is the Motor Car a Non-Essential?" *The Club Journal* 10, no. 4 (September 1918), 14–16; "Gasolineless Sundays," *The Club Journal* 10, no. 4 (September 1918), 16; passenger car: F. H. Sweet, "Which Car Shall I Buy—And Why?" *American Motorist* 12, no. 1 (January 1920), 39–40.

104. Kline, *Consumers in the Country,* 85, 104, 122.
105. Monroe et al., *Family Expenditures for Automobile and Other Transportation,* 1–3. Unfortunately, this survey does not distinguish between commuting, shopping, social, and recreational activities within the "family use" category. Cavert, *Sources of Power on Minnesota Farms,* 7, 54–55 (table LI).
106. St. Clair, *The Motorization of American Cities,* 15 (table 1.10); from 1942, however, urban travel dominated again. G. P. St. Clair, "Trends in Motor-Vehicle Travel, 1936 to 1945," *Public Roads* (October–December 1946), 261–267, here 262 (figure 1).
107. Muse, *The Standard of Living on Specific Owner-Operated Vermont Farms,* 18; roadsters: Christopher Wells, "Car Country: Automobiles, Roads, and the Shaping of the Modern American Landscape, 1890–1929" (dissertation, University of Wisconsin-Madison, 2004), 189.
108. Michael L. Berger, *The Devil Wagon in God's Country: The Automobile and Social Change in Rural America, 1893–1929* (Hamden, CT: Archon Books, 1979), 62, 123. Rural ownership: [The Bureau of Public Roads], *Report of a Survey of Traffic on the Federal-Aid Highway Systems of Eleven Western States, 1930, by The Bureau of Public Roads, United States Department of Agriculture and the Highway Departments of Arizona, California, Colorado, Idaho, Nebraska, New Mexico, Nevada, Oregon, Utah, Washington, and Wyoming* (Washington, DC: United States Government Printing Office, 1932), 36; *Woman's World:* Brunner and Kolb, *Rural Social Trends,* 267–268.
109. [The Bureau of Public Roads], *Report of a Survey of Traffic on the Federal-Aid Highway Systems of Eleven Western States, 1930,* 38; salesmen, etc.: Edsforth, *Class Conflict and Cultural Consensus,* 16.
110. Horowitz, *The Morality of Spending,* 110.
111. James W. Follin, "Taxation of Motor Vehicles in the United States," *Annals of the American Academy of Political and Social Sciences* 116 (November 1924), 141–159; George A. Lundberg, Mirra Komarosky, and Mary Alice McInerney, *Leisure: A Suburban Study* (New York: Columbia University Press, 1934), 186; on the approach to the American family by traditionalists and progressives and the "discovery" of the middle-class family: Paula S. Fass, *The Damned and the Beautiful: American Youth in the 1920's* (New York: Oxford university Press, 1977), 37–42 and chap. 2 (53–118).
112. St. Clair, *The Motorization of American Cities,* 83, is one of the few urban historians who explicitly asks the questions: "What were automobiles 'preferred' for? Weekend trips? Daily commuting? Pleasure driving? Daily errands? Buying an automobile cannot be readily construed as an expression of a preference (or equal preference) for all of these different uses." St. Clair does not answer these questions, however, apart from pointing out that the demise of the streetcar was not caused by the popularity of the car, but, conversely, that the car was increasingly used because of the decline of the streetcar.
113. John H. Hepp, IV, *The Middle-Class City: Transforming Space and Time in Philadelphia, 1876–1926* (Philadelphia: University of Pennsylvania Press, 2003), 177, 189.
114. Brunner and Kolb, *Rural Social Trends,* 161; Kline, *Consumers in the Country,* 71; Virginia J. Scharff, "Putting wheels on women's sphere," in Cheris Kramarae (ed.), *Technology and Women's Voices: Keeping in Touch* (New York and London, 1988), 135–146, here 140. In 1927 still, female drivers represented one-fifth of the total in Connecticut. Beth Kraig, "Woman At The Wheel: A History Of Women And the Automobile In America" (unpublished doctoral dissertation,

University of Washington, 1987), 224; 15–20 percent women: Morris S. Viteles and Helen M. Gardner, "Women Taxicab Drivers: Sex Differences in Proneness to Motor Vehicle Accidents," *Personnel Journal* 7 (1929), 349–355, here 350–351; Bernard Fabienne, "Les femmes et l'automobile de 1900 à 1930" (Thèse de Maîtrise d'Histoire Contemporaine, Université Blaise Pascal, 1977), 7–8.

115. Scott L. Bottles, *Los Angeles and the Automobile: The Making of the Modern City* (Berkeley, Los Angeles, and London: University of California Press, n.d. [1987]), 55. In Philadelphia, between 1850 and 1880, commuting by horse streetcar accounted for only 17 percent of the rides, whereas 28 percent of the rides had shopping as purpose, 25 percent recreation, and 10 percent through traffic. Clay McShane and Joel A. Tarr, *The Horse in the City: Living Machines in the Nineteenth Century* (Baltimore, MD: Johns Hopkins University Press, 2007), 75–77. For a similar conclusion (only 15–20 percent of all passenger-miles used for commuting) on the use of steam railroads in the 1920s, see *Recent Social Trends*, 170; Class I: John B. Rae, *The Road and the Car in American Life* (Cambridge, MA, and London: MIT Press, 1971), 89 (table 5.1).

116. Brown, *American Standards of Living*, 133.

117. International Labour Office, *A Contribution to the Study of International Comparisons of Costs of Living: An Enquiry into the Cost of Living of Certain Groups of Workers in Detroit (U.S.A.) and Fourteen European Towns* (New York and London: Garland, 1983; reprint of 1932; The World Economy, A Garland Series, ed. Mira Wilkins), 195; Dynamic City: Barrow, "The Automobile in the Garden," 342; Edsforth, *Class Conflict and Cultural Consensus*, 94; Nicolaides, *My Blue Heaven*, 35 (density), 72–73 (claim). Shopping: Blanke, *Hell on Wheels*, 49. Some downtown business owners in Los Angeles supported the proposed parking ban because they believed that most parked cars were used for commuting. However, most CBD business interests opposed the ban vigorously, but later accepted parking bans during gridlock hours, suggesting that it was the shopping function rather than the commuting function that was first taken from the streetcar. Bottles, *Los Angeles and the Automobile*, 66, 90.

118. Joseph Anthony Interrante, "A Movable Feast: The Automobile and the Spatial Transformation of American Culture 1890–1940" (PhD thesis, Department of History, Harvard University, Cambridge, MA, June 1983), 120.

119. Lundberg, Komarosky, and McInerney, *Leisure: A Suburban Study*, 47–48, 371 (table XV). This study is part of a wider tradition that aims at complexifying the easy connection between suburb and white middle class, and between suburban life and commuting. For another study showing that the suburb with the nuclear middle-class family as its core is an ideological construct based on Progressives' literature, see Mary Corbin Sies, "North American Suburbs, 1880–1950," *Journal of Urban History* 27, no. 3 (March 2001), 313–346.

120. McIntyre, "'The Repair Man Will Gyp You,'" 322–324.

121. Adams, "The Automobile—A Luxury Becomes a Necessity," 37.

122. Bottles, *Los Angeles and the Automobile*, 55, 198–203; Joel A. Tarr, *Transportation Innovation and Changing Patterns in Pittsburgh, 1850–1934* (Chicago: Public Works Historical Society, 1978), 26–28; Howard L. Preston, *Automobile Age Atlanta: The Making of a Southern Metropolis 1900–1935* (Athens: The University of Georgia Press, 1979), 70–71; Reynold M. Wik, *Henry Ford and Grassroots America* (Ann Arbor: The University of Michigan Press, 1972), 22; Joseph Interrante, "The Road to Autopia: The Automobile and the Spatial Transformation of American Culture," in David L. Lewis and Laurence Goldstein (eds.),

The Automobile and American Culture (Ann Arbor: The University of Michigan Press, 1983), 89–104, here 102. Mark S. Foster, *From Streetcar to Superhighway: American City Planners and Urban Transportation, 1900–1940* (Philadelphia, PA: Temple University Press, 1981), 61, quotes a national survey suggesting that 52 percent of car owners used the car to drive to work. This is highly unlikely. Many urban historians, just like the surveys, do not distinguish between commuting and shopping.

123. *Family Expenditures in Selected Cities, 1935–36, Volume VI: Travel and Transportation* (Washington, DC: United States Government Printing Office, 1940), 11.

124. A very raw estimate puts this in perspective. In 1926 twenty million vehicles consumed nine billion gallons of gasoline (*Journal of the Society of Automotive Engineers* 20, no. 2 [February 1927], 312). With an estimated fuel consumption of, say, 15 mpg (18 mpg in 1939, including the several million trucks with a much higher consumption; *Journal of the Society of Automotive Engineers* [*Transactions*] 44, no. 1 [January 1939], 34) they drove an annual average of 6750 miles. Suppose every vehicle was used either to commute a distance of 10 miles per day for 330 days per year or bring produce to the market over a distance of 60 miles once a week, then about half the annual mileage was left to spend on 'pleasure': social visits, shopping, trips during the weekends, and short- and long-range tourism.

125. *American Motorist* (December 1929), 46. On increased winter traffic during the Depression, see Owen D. Gutfreund, *Twentieth-Century Sprawl: Highways and the Reshaping of the American Landscape* (Oxford: Oxford University Press, 2004), 69, 146.

126. Calder, *Financing the American Dream*, 267. Sales of Fords among farmers plummeted from 650,000 in 1929 to 55,000 in 1932. Wik, *Henry Ford and Grass-roots America*, 186.

127. Jesse Frederick Steiner, *Americans at Play; Recent Trends in Recreation and Leisure Time Activities* (New York and London: McGraw-Hill, 1933), 96; Roger B. White, *Home on the Road: The Motor Home in America* (Washington, DC, and London: Smithsonian Institution Press, 2000); Wik, *Henry Ford and Grass-roots America*, 183; Paul S. Sutter, *Driven Wild: How the Fight against Automobiles Launched the Modern Wilderness Movement* (Seattle and London: University of Washington Press, 2002), 33 (400 companies); Thomas M. Heaney, "'The Call of the Open Road': Automobile Travel and Vacations in American Popular Culture, 1935 – 1960" (unpublished dissertation, University of California, Irvine, 2000), 18 (25,000); amazed: Gavin Altman, "Motoring for the Million?" 225.

128. Mildred Strunk (under the editorial direction of Hadley Cantril), *Public Opinion 1935 – 1946* (Westport, CT: Greenwood Press, 1978) (reprint of 1951 edition, Princeton University Press), 31 (half); Carroll D. Clark and Cleo E. Wilcox, "The House Trailer Movement," *Sociology and Social Research* 22, no. 6 (July–August 1938), 503–519, here 510–511; Thomas M. Heaney, "'The Call of the Open Road': Automobile Travel and Vacations in American Popular Culture, 1935 - 1960" (unpublished dissertation, University of California, Irvine, 2000), 21–22 (conservative), 63 (undermined), 16 (alternative), 22 (way of life).

129. Ulrich Kubisch, "Häuser am Haken: Der Wohnwagen: Geschichte, Technik, Urlaubsfreuden," *Kultur & Technik* no. 3 (1998), 11–18, here 12.

130. [The Bureau of Public Roads], *Report of a Survey of Traffic on the Federal-Aid Highway Systems of Eleven Western States, 1930*, 4–5, 8. For a comparable anal-

ysis of Sunday traffic on the road between San Francisco and San Diego in the early 1920s, see Henry R. Trumbower et al., "Recensement de la circulation" (report for the Vth PIARC conference, Milan 1926). I thank Clay McShane for his observation that shops were closed on Sundays during the 1930s (email, 16 February 2010).

131. Merkert, *Personenkraftwagen*, 259 (table 90), 267; John Nolen and Henry V. Hubbard, *Parkways and Land Values* (Cambridge, MA: Harvard University Press, 1937), 83–84 (I thank Clay McShane for making this report available to me); California: Frederic W. Cron, "Highway Design for Motor Vehicles—A Historical Review," *Public Roads* 38, no. 3 (December 1974), 85–95, here 94; [The Bureau of Public Roads, US Department of Agriculture / The Connecticut State Highway Department], *Report of a Survey of Transportation on the State Highway System of Connecticut* (Washington, DC: Government Printing Office, 1926), 72–73; the report does not clarify whether commuting was part of the "business use." It probably was not.

132. Jørgen Burchardt, "User practices, intermediary actors, and the automobile system," paper presented at the International Workshop "Dutch Mobility in a European Context: A Comparison of two Centuries of Mobility Policy in Seven Countries," Utrecht, 5–7 February 2009; Mom and Filarski, *Van transport naar mobiliteit*, 158–159.

133. US Office of the Federal Coordinator of Transportation, Passenger Traffic Report (Washington, DC, 1935), quoted by Gregory Thompson, "From Rails to Rubber; from Public to Private: Passenger Transport Modernization in California from 1911 to 1941," http://www.coss.fsu.edu/durp/research/working-papers (last accessed 31 August 2011).

134. Strunk, *Public Opinion 1935–1946*, 31.

135. David A. Revzan, "Trends in Economic Activity and Transportation in the San Francisco Bay Area," in *Parking as a Factor in Business, containing five papers presented at the thirty-second annual meeting January 13–16, 1953* (Washington, DC, 1953; Highway Research Board, Special Report 11), 161–308, here 264 (table 62); St. Clair, *The Motorization of American Cities*, 121; *How American Buying Habits Change*, 205.

136. Christine McGaffey Frederick, "The Commuter's Wife and the Motor Car," *Suburban Life* no. 13 (July 1912), 13–14, 46.

137. For an observed average car occupancy of 1.8 in Atlanta in 1924, see John A. Beeler, *Report to the City of Atlanta on a Plan for Local Transportation, December, 1924* (New York: The Beeler Organization, 1924), 5; for an estimate of an occupancy of 2.2, see Chapin, "The Motor's Part in Transportation," 1. [The Bureau of Public Roads, U.S. Department of Agriculture/The Connecticut State Highway Department], *Report of a Survey of Transportation on the State Highway System of Connecticut*, 72, gives 1.8 for "business cars" and 2.7–3.5 for "nonbusiness cars." The average for all cars was 2.71–2.77. For an analysis of combined use for shopping and commuting in Seattle in 1947, see Louis C. Wagner, "Economic Relationships of Parking to Business in Seattle Metropolitan Area," in *Parking as a Factor in Business*, 53–90, here 71. *Family Expenditures in Selected Cities, 1935–36, Volume VI*, 3, did not distinguish between car functions "since these diverse uses are not separable." For a critique of the car's "flexibility" as a concept in mobility historians' work, see St. Clair, *The Motorization of American Cities*, 84–86.

138. Follin, "Taxation of Motor Vehicles in the United States," 144; Chapin, "The Mo-

tor's Part in Transportation," 1, gives 5000 miles as an estimate and F. R. Pleasonton, "The Automotive Industry: A Study of the Facts of Automobile Production and Consumption in the United States," *Annals of the American Academy of Political and Social Science* 97 (September 1921), 107–127, here 11, gives 6840 miles; Steiner, *Americans at Play*, 186; St. Clair, "Trends in Motor-Vehicle Travel," 264. Adams, "The Automobile—A Luxury Becomes a Necessity," 34, gives only 2697 miles as the mileage in 1930 of the car (and 218.3 miles for the train). This can be explained by assuming that the other mileages are averages of car drivers, whereas the latter values are per capita averages, also to be found in Willey and Rice, "The Agencies of Communication," 178.
139. On the other hand, the emergence of the closed car (see chapter 5) can be interpreted as evidence for the importance of commuting. On the family car in the 1920s and 1930s, see, for instance, Rae, *The Road and the Car*, 133–137. This analysis does not go into the question whether *every* car trip was a source of psycho-physical pleasure, a point of view for which ample historical evidence seems to exist.
140. Quoted in Norton, *Fighting Traffic*, 220; Ireland, *Entering the Auto Age*, 118.
141. Mueller, "The Automobile: A Sociological Review," 74; "What Do Folks Use Their Cars For?" *The Literary Digest* (17 November 1923), 66, 68–69.
142. Quoted in Claude S. Fischer, *Made in America: A Social History of American Culture and Character* (Chicago and London: The University of Chicago Press, 2010), 67.
143. For instance by Cotten Seiler, *Republic of Drivers: A Cultural History of Automobility in America* (Chicago and London: The University of Chicago Press, 2008), 59, who quotes an issue of *Motor* from 1924 that claimed that 60 percent of car use was for business, or by Bruce Seely, "The Automotive Transportation System: Cars and Highways in Twentieth-Century America," in Carroll Pursell (ed.), *A Companion to American Technology* (Malden, Oxford, and Carlton: Blackwell Publishing, 2005), 233–254, here 241: "During the 1920s, cars became a necessity to many Americans." One historian even took over the concept to indicate the type of use of the period: "Fully 96 percent of the 26 million cars on the road [by 1940] were used for necessity driving on a weekly basis." Blanke, *Hell on Wheels*, 49.
144. Lynd and Lynd, *Middletown in Transition*, 265; Cavert, *Sources of Power on Minnesota Farms*, 56. "Every day we see cars full of people passing us by to go to Norfolk or Richmond to buy," one shop owner observed. "Often as not they could do as well at home, but they enjoy the excitement of travel and the contacts with the larger city life." W. O. Saunders, "Business Is a Pleasure," in *These Are Our Lives: As Told by the People and Written by Members of the Federal Writers' Project of the Works Progress Administration in North Carolina, Tennessee, and Georgia* (Chapel Hill: The University of North Carolina Press, 1939), 277–289, here 285.
145. Lynd and Lynd, *Middletown*, 255. Also Zimmerman, *Consumption and Standards of Living*, 328, distinguished between "'necessity' expenditures and investments for automobiles and travel."
146. John E. Crowley, "The Sensibility of Comfort," *The American Historical Review* 104, no. 3 (June 1999), 749–782, here 760.
147. Horowitz, *The Morality of Spending*, 117–118, 158 (table 19); acceptability: Adams, "The Automobile—A Luxury Becomes a Necessity," 39.
148. Quoted in Edsforth, *Class Conflict and Cultural Consensus*, 88.

149. Quoted in Ted Ownby, "The Snopes Trilogy and the Emergence of Consumer Culture," in Donald M. Kartiganer and Ann J. Abadie (eds.), *Faulkner and Ideology: Faulkner and Yoknapatawpha, 1992* (Jackson: University Press of Mississippi, 1995), 95–128, here 101–102 (quote on 102).
150. Flink, *The Car Culture*, 160, quotes a questionnaire mailing by the National Automobile Chamber of Commerce as early as 1920, resulting in an absolute majority (90 percent) saying that their cars were "used more or less for business" and that 60 percent of the car's mileage was driven "for business purposes."
151. "Necessities, Not Luxuries," *Automotive Industries* 45, no. 7 (18 August 1921), 334.
152. Michael McGerr, *A Fierce Discontent: The Rise and Fall of the Progressive Movement in America, 1870–1920* (Oxford and New York: Oxford University Press, 2003), 272 (culture of pleasure), 301 (War Industries Board, aristocratitis).
153. Norton, *Fighting Traffic*. For the founding of a short-lived pedestrian activist magazine called *The Nemesis* which appeared in January 1925, see Mueller, "The Automobile: A Sociological Review," 123–124; class war: Brilliant, "Some Aspects of Mass Motorization in Southern California," 201. For a revival of resistance in the post-WWI Netherlands, see Mom and Filarski, *Van transport naar mobiliteit*, 147–148; for Swiss resistance, where "la calamité des automobiles" (the car disaster) and their "abus de vitesse" (speed abuse) was prohibited on Sundays from 1919 to 1923, only to be lifted if motorists would behave "reasonable" (*raisonable*), see Jean-Philippe Matthey, "Un aspect de la motorisation dans le canton de Vaud: la question de l'interdiction dominicale de la circulation automobie (1906–1923)" (unpublished Master's thesis, Université de Lausanne, Faculté des Lettres, 1992), 66–91; for the United Kingdom, see Robert Graves and Alan Hodge, *The Long Week-end: A Social History of Great Britain 1918–1939* (London: Faber and Faber, 1940), 181, who point at the "road hog" as a "recurrent newspaper theme throughout the early Twenties"; for Germany where "we motorists are shouted upon by the great masses of the people," see Franz Gissinger, "'Rücksichtlose Autofahrer,'" *Der Motorfahrer* 16, no. 24/25 (21 June 1919), 11 and "Präsidialbekanntmachungen; Autowacht," *Der Motorfahrer* 17, no. 37/38 (16 September 1919), 203; see also Fraunholz, *Motorphobia*.
154. Cavert, *Sources of Power on Minnesota Farms*, 56; Adams, "The Automobile—A Luxury Becomes a Necessity," 37; Vidich and Bensman, *Small Town in Mass Society*, 10
155. *The Automobile and the Village Merchant: The Influence of Automobiles and Paved Roads on the Business of Illinois Village Merchants* (Urbana: University of Illinois, 1928), *University of Illinois Bulletin* 25, no. 41 (12 June 1928) 10 (roads), 16 (before car), 26 (since 1900).
156. Pleasonton, "The Automotive Industry," 107–127, here 116.
157. Kyrk, *Economic Problems of the Family*, 388; Moskowitz, *Standard of Living*, 5.
158. James West, *Plainville, U.S.A.* (New York: Columbia University Press, 1958; first ed. 1945), 10. On the debate on the distinction between luxuries and necessities among contemporary reformers (and the proposal by one of them to see the car as a form of "nonluxurious abundant living," classifying the car as "travel"), see Horowitz, *The Morality of Spending*, 116–123. Wheatland: John S. Gilkeson, *Anthropologists and the Rediscovery of America, 1886–1965* (Cambridge and New York: Cambridge University Press, 2010), 105.
159. Steiner, *Americans at Play*, 60. Remarkably, Steiner estimated the share of recreational car use quite conservatively, as he called it himself, at only one-quarter (ibid., 187); motive: Silas Bent, "Speed on the Highways; Study of the Auto

and Its Safety Problems," *Current History* 45, no 3 (December 1936), 95–99, here 95.
160. Bradley Flamm, "Putting the brakes on 'non-essential' travel: 1940s wartime mobility, prosperity, and the US Office of Defense Transportation," *Journal of Transport History,* Third Series, 27, no. 1 (March 2006), 71–92. Transport experts recently estimated that less than half of British motorists' annual driving is "essential." Susan Handy, Lisa Weston, and Patricia Mokhtarian, "Driving by choice or necessity?" *Transportation Research Part A* 39 (2005), 183–203, here 184.
161. Kyrk, *Economic Problems of the Family,* 375–383 (quotes on 382–382); Martha Wolfenstein, "The Emergence of Fun Morality," *Journal of Social Issues* 7 (1951), 15–25.
162. Otto Suhr, *Die Lebenshaltung der Angestellten: Untersuchungen auf Grund statistischer Erhebungen des Allgemeinen freien Angestelltenbundes* (Berlin: Freier Volksverlag, 1928), 21.
163. König, *Volkswagen, Volksempfänger, Volksgemeinschaft,* 151 (*Geschäftswagen*), 154 (Hitler). König (ibid., 175) quotes a fascist economic newspaper (*Die Deutsche Volkswirtschaft*) that refers to an analysis of the German national statistics office from 1937, but his figure of 80 percent "cars used commercially" is taken from a dissertation from 1936.
164. Flik, *Von Ford lernen?,* 154; functionalities: Edelmann, *Vom Luxusgut zum Gebrauchsgegenstand,* 94–95. Sjöblom, "The Political Economy of Railway and Road Transport in Britain and Germany, 1918–1933," 68, gives 30 to 40 percent of car sales to "business" by the end of the 1930s. Pleasures not mentioned for instance in Statistisches Reichsamt, *Statistisches Jahrbuch für das Deutsche Reich* 48 (1929; Berlin: Verlag Reimar Hobbing, 1929), 160 (table "*Kraftfahrzeuge im Deutschen Reich, b: Verwendungszweck;*" [Motorcars in the German Reich, b: purpose of use]).
165. *Neue-Arbeiter-Zeitung* (communist party newspaper) quoted in Richard Birkefeld and Martina Jung, *Die Stadt, der Lärm und das Licht: Die Veränderung des öffentlichen Raumes durch Motorisierung und Elektrifizierung* (Seelze [Velber]: Kallmeyer'sche Verlagsbuchhandlung GmbH, 1994), 132. The German term is "jeder Werktätige" (which is masculine).
166. Rudy Koshar, *German Travel Cultures* (Oxford and New York: Berg, 2000), 117 (Sundays); Schäfer quoted in Klaus-Dieter Oelze, *Das Feuilleton der Kölnischen Zeitung im Dritten Reich* (Frankfurt, Bern, New York, and Paris: Peter Lang, 1990), 252.
167. Michael Prinz, "Das Ende der Standespolitik: Voraussetzungen, Formen und Konsequenzen mittelständischer Interessenpolitik in der Weimarer Republik am Beispiel des Deutschnationalen Handlungsgehilfenverband," in Jürgen Kocka (ed.), *Angestellte im europäischen Vergleich: Die Herausbildung angestellter Mittelschichten seit dem späten 19. Jahrhundert* (Göttingen: Vandenhoeck & Ruprecht, 1981), 331–353, here 334–337, 343.
168. Josef M. Jurinek, "ADAC-Fünfländer-Tourenfahrt: Sportlich-touristisches Gesamterlebnis und Auswertung," *ADAC Motorwelt* 24, no. 22 (3 June 1927), 10–12 (political act: 12); "Reichshuldigungsfahrt," *ADAC Motorwelt* 24, no. 34 (26 August 1927), 2–3 (map on 2); S. Doerschlag, "ADAC-Reichshuldigungsfahrt," *ADAC Motorwelt* 24, no. 40 (7 October 1927), 2–4; "Reichskanzler Dr. Marx im besetzten Gebiet," *ADAC Motorwelt* 24, no. 42 (21 October 1927), 3.
169. A. Stadie, "Die erste Gebrauchs- und Wirtschaftlichkeitsfahrt," *ADAC Motorwelt* 25, no. 10 (9 March 1928), 3–5.

170. Dorothee Hochstetter, *Motorisierung und "Volksgemeinschaft": Das Nationalsozialistische Kraftfahrkorps (NSKK) 1931–1945* (Munich: R. Oldenbourg Verlag, 2005), 56–57.
171. Quote: E. Rosemann, "2000 Kilometer durch Deutschland: Eine Spitzenleistung von Organisation und Fahrern," *ADAC Motorwelt* 30, no. 30 (28 July 1933), 2–7, here 2; tadellose Ordnung: Siegfried Doerschlag, "Korso 'Einst und jetzt,'" *ADAC Motorwelt* 30, no. 16 (21 April 1933), 6; for photographs of SA members holding flags along the route, see *ADAC Motorwelt* 30, no. 30 (28 July 1933), 16–17; "Auftakt zur Ostland-Treuefahrt," *ADAC Motorwelt* 30, no. 35 (1 September 1933), 2–3; Ernst Rosemann, 'Ostland-Treuefahrt," *ADAC Motorwelt* 30, no. 36 (8 September 1933), 13–18; for ADAC members making the Hitler salute, see "Ostland-Treuefahrt," *ADAC Motorwelt* 30, no. 37 (15 September 1933), 17; on Hühnlein: "Die ersten Richtlinien für den ADAC," *ADAC Motorwelt* 30, no. 44/45 (10 November 1933), 2–7, here 2 (photo on 3); working classes: "Richtlinien für die Organisation des DDAC," *ADAC Motorwelt* 30, no. 44/45 (10 November 1933), 4–5, here 5.
172. *Recent Social Trends.* The star or spoke structure of suburbs ("bourgeois corridors") was already shaped in the horse streetcar years: McShane and Tarr, *The Horse in the City*, 67–70.
173. David Louter, *Windshield Wilderness: Cars, Roads, and Nature in Washington's National Parks* (Seattle and London: University of Washington Press, 2006), 70 (half); occupancy and family size: *Recent Social Trends*, 177, 661; Preston, *Dirt Roads to Dixie*, 126. On the flow of working-class families in search for work, see Heaney, "'The Call of the Open Road,'" 44–47.
174. Interrante, "A Movable Feast," 71–72; Nicolaides, *My Blue Heaven*, 59 (photo of squatter camp).
175. James N. Gregory, *The Southern Diaspora: How the Great Migrations of Black and White Southerners Transformed America* (Chapel Hill: The University of North Caroline Press, 2005), 5–6, 14–15, 25–28; Cindy Hahamovitch, *The Fruits of Their Labor: Atlantic Coast Farmworkers and the Making of Migrant Poverty, 1870–1945* (Chapel Hill and London: The University of North Carolina Press, 1997), 4–5; Corey T. Lesseig, *Automobility: Social Changes in the American South 1909–1939* (New York and London: Routledge, 2001), 198–199; 1.6 million blacks: Lind, *Land of Promise*, 260; drought: Michael W. Schuyler, *The Dread of Plenty: Agricultural Relief Activities of the Federal Government in the Middle West, 1933–1939* (Manhattan, KS: Sunflower University Press, 1989), 174. On blacks in the Great Migration, see Mark S. Foster, "In the Face of 'Jim Crow': Prosperous Blacks and Vacations, Travel and Outdoor Leisure, 1890–1945," *The Journal of Negro History* 84, no. 2 (Spring 1999), 130–149, here 137.
176. Hal K. Rothman, *Devil's Bargains: Tourism in the Twentieth-Century American West* (Lawrence: University Press of Kansas, 1998), chap. 6; *Recent Social Trends*, 141, 179; Julius H. Parmelee and Earl R. Feldman, "The Relation of the Highway to Rail Transportation," in Jean Labatut and Whearon J. Lane (eds.), *Highways in our National Life: A Symposium* (New York: Arno Press, A New York Times Company, 1972), 227–239, here 233; Melvin M. Webber, "Transportation Planning for the Metropolis," in Leo F. Schnore and Henry Fagin (eds.), *Urban Research and Policy Planning* (Beverly Hills, CA: Sage, 1967), 389–407, here 391.
177. Graves and Hodge, *The Long Week-end*, 275; Sutter, *Driven Wild*, 177; for the construction of the Appalachian Trail as a walking trail, see ibid., 153–175; youth hostels and walking: Leonard, *Adventures in Holiday Making*, 130, 217.

178. Cord Pagenstecher, *Der bundesdeutsche Tourismus: Ansätze zu einer Visual History: Urlaubsprospekte, Reiseführer, Fotoalben 1950–1990* (Hamburg: Verlag Dr. Kovač, 2003), 91.
179. Mark Foster, "City Planners and Urban Transportation: The American Response, 1900–1940," *Journal of Urban History* 5, no. 3 (May 1979), 365–396, here 385. However, a five-mile car trip in 1929 was seven times more expensive than a streetcar ride (375).
180. Paula S. Fass, *The Damned and the Beautiful: American Youth in the 1920's* (New York: Oxford university Press, 1977), 41, 55, 61, 65, 91 (working class), 97 (happiness); Edward Shorter, *The Making of the Modern Family* (New York: Basic Books, 1975), 5 (shield), 205 (shelter), 259 (marketplace); Ellis W. Hawley, *The Great War and the Search for a Modern Order: A History of the American People and Their Institutions, 1917–1933* (New York: St. Martin's Press, 1979), 140.
181. *Outing* 70 (April 1917), 130; *Outing* 72 (September 1918), 389; *Outing* 80 (July 1922), 162–164; Warren James Belasco, *Americans on the Road: From Autocamp to Motel, 1910–1945* (Cambridge, MA, and London: MIT Press, 1979), 85; Sutter, *Driven Wild*, 34.
182. Arthur S. Link, "What Happened to the Progressive Movement in the 1920's?" *The American Historical Review* 64, no. 4 (July 1959), 833–851, here 843; Rachel P. Maines, *Hedonizing Technologies: Paths to Pleasure in Hobbies and Leisure* (Baltimore, MD: Johns Hopkins University Press, 2009), 99.
183. Marguerite S. Shaffer, *See America First: Tourism and National Identity, 1880–1940* (Washington, DC, and London: Smithsonian Institution Press, 2001); Michael Berkowitz, "A 'New Deal' for Leisure; Making Mass Tourism during the Great Depression," in: Shelley Baranowski and Ellen Furlough (eds.), *Being Elsewhere; Tourism, Consumer Culture, and Identity in Modern Europe and North America* (Ann Arbor: The University of Michigan Press, 2004), 184–212, here 195; *Annual Report of the Secretary of the Interior for the Fiscal Year Ended June 30, 1918* (Washington, DC: Government Printing Office, 1918), 6.
184. Frank E. Brimmer, "Forty-four Million Awheel: National Motor Tourist Business Aggregates $3,590,400,000 in 1928," *American Motorist* 21, no. 1 (January 1929), 33–36, here 33.
185. Steiner, *Americans at Play*, 59; *How American Buying Habits Change*, 186; Jesse F. Steiner, *Research Memorandum on Recreation in the Depression Prepared under the Direction of the Committee on Studies in Social Aspects of the Depression* (New York: Social Science Research Council, 1937; reprint: n.p. [New York]: Arno Press, 1972), 60, 64; Sutter, *Driven Wild*, 49.
186. Preston, *Dirt Roads to Dixie*, 5
187. Sutter, *Driven Wild*, 108.
188. Willey and Rice, *Communication Agencies and Social Life*, 54 (table 18), 57.
189. *American Motorist* 16, no. 1 (January 1924), 13; *American Motorist* 21, no. 2 (February 1929), 23; Sutter, *Driven Wild*, 108; Belasco, *Americans on the Road*, 93–96; Preston, *Dirt Roads to Dixie*, 154; second half 1920s: Wells, "Car Country," 366 (quoting Bureau of Public Roads chief MacDonald). During the 1920s, one commentator asserts, some 300,000 cars annually made the trip along the Lincoln Highway. James Flagg, *Boulevards All the Way—Maybe* (New York, 1925), quoted in Carey S. Bliss, *Autos Across America: A Bibliography of Transcontinental Travel: 1903–1940* (Los Angeles: Dawson's Book Shop, 1972), 37. For the physical and ideological "construction" of the Lincoln Highway, see Shaffer, *See America First*.

190. Thomas Weiss, "Tourism in America before World War II," *Journal of Economic History* 64, no. 2 (June 2004), 289-327, here 316.
191. Quoted in Jonas Larsen, "Tourism Mobilities and the Travel Glance: Experiences of Being on the Move," *Scandinavian Journal of Hospitality and Tourism* 1, no. 2 (2001), 80-98, here 84.
192. Norman Hayner, "The Auto Camp as a New Type of Hotel," *Journal of Applied Sociology* 15 (1931), 369-372, here 369, claimed that "five years ago [the autocamp] was practically unknown."
193. *American Motorist* 15, no. 5 (May 1923), 2; *American Motorist* 20, no. 7 (July 1928), 8; Belasco, *Americans on the Road*, 83.
194. Belasco, *Americans on the Road*, 11; *American Motorist* 16, no. 6 (June 1924), 16; *Outing* 71 (October 1917), 5; *Outing* 75 (October 1919), 4; last quote: Sutter, *Driven Wild*, photo caption on non-numbered page (ninth photo after p. 48). Sociologists: Glen Jeansonne, "The Automobile and American Morality," *Journal of Popular Culture* 8, no. 1 (Summer 1974), 125-131, here 128.
195. Sutter, *Driven Wild*, 9 (manicure), 27 (experience).
196. *American Motorist* 17, no. 11 (November 1925), 75; *American Motorist* 18, no. 1 (January 1926), 32-34.
197. *American Motorist* 15, no. 5 (May 1923), 8; Preston, *Dirt Roads to Dixie*, 122.
198. *American Motorist* 15, no. 8 (August 1923), 14; *American Motorist* 20, no. 6 (June 1928), 9, 11, 47; "Facilities for European Motor-Touring," *Journal of the Society of Automotive Engineers* (June 1928), 643.
199. *American Motorist* 16, no. 6 (June 1924), 16; citation: *American Motorist* 18, no. 1 (January 1926), 11; *American Motorist*, 20, no. 1 (January 1928) 10; *Outing* 74 (May 1919) 106; Tin-can and 1500: "Car camping in the United States," *Autocar* 47, no. 1365 (17 December 1921) 1246-1249, here 1246, 1249.
200. Aaltje Hessels, *Vakantie en vakantiebesteding sinds de eeuwwisseling: Een sociologische verkenning ten behoeve van de sociale en ruimtelijke planning in Nederland* (Assen: Van Gorcum & Comp., 1973), 82-84, 133-134; J. F. R. Philips, "Recreatie en toerisme, spiegelbeeld of reactie?" in K. P. C. de Leeuw, M. F. A. Linders-Rooijendijk, and P. J. M. Martens (eds.), *Van ontspanning en inspanning: Aspecten van de geschiedenis van de vrije tijd* (Tilburg, 1995), 35-62, here 38-40.
201. "Die im Ausland beheimateten Kraftfahrzeuge im Deutschen Reich vom 1. Juli 1928 bis 30. Juni 1929," *ADAC Motorwelt* 26, no. 45 (8 November 1929), 9; "Die im Ausland beheimateten Kraftfahrzeuge im Deutschen Reich," *ADAC Motorwelt* 28, no. 46/47 (20 November 1931), 16; Karl G. Kühne, "Deutschland und die internationale Auto-Touristik," *ADAC Motorwelt* 29, no. 26 (24 June 1932), 23-25, here 23; 10,000: W. Feilchenfeld, "Kraftwagenverkehr und Fremdenverkehr," *ADAC Motorwelt* 25, no. 39 (1928), 19-21, here 20.
202. P. J. Mijksenaar, "Wie, waar en hoe men in Holland over de bruggen komt," *Het Motorrijwiel en de Populaire Auto* 23 (1935), 898-900; "Over tellingen en tolopbrengsten," *Auto Kampioen* 29, no. 29 (20 July 1935), 984-985; *Verkeerswaarnemingen van den Rijkswaterstaat 1935: Bijlagen* (Den Haag, 1936), annex 13, 16, 17.
203. Weber, "Automobilisering en de overheid in België vóór 1940," 329.
204. Christine Keitz, "Die Anfänge des modernen Massentourismus in der Weimarer Republik," *Archiv für Sozialgeschichte* 33 (1993), 179-209, here 47 (women), 48 (10 percent workers), 187 (necessity), 206; Herbert Jost, "Selbst-Verwirklichung und Seelensuche: Zur Bedeutung des Reiseberichts im Zeitalter des Massentourismus," in Peter J. Brenner (ed.), *Der Reisebericht: Die Entwicklung einer*

Gattung in der deutschen Literatur (Frankfurt: Suhrkamp Verlag, 1989), 490–507, here 498 (ten million); Spode, "Der Aufstieg des Massentourismus im 20. Jahrhundert," 121–122; Hartmut Berghoff, "Träume und Alpträume: Konsumpolitik im Nationalsozialistischen Deutschland," in Haupt and Torp (eds.), *Die Konsumgesellschaft in Deutschland 1890–1990*, 268–288, here 279.

205. Volker Berghahn, *Europa im Zeitalter der Weltkriege: Die Entfesselung und Entgrenzung der Gewalt* (Frankfurt: Fischer Taschenbuch Verlag, 2002), 90–103; boys: Elke Gramespacher, "Sport—Bewegungen—Geschlechter," *Geschlechter—Bewegungen—Sport* (Freiburger GeschlechterStudien 23 [2009]), 13–30, here 20; worried: Wa.O., "Übergangswirtschaft des Kraftfahrwesens: Betrachtungen zum neuen Jahr," *Der Motorfahrer* 15, no 1 (5 January 1918), 2–4, here 4.

206. König, *Geschichte der Konsumgesellschaft*, 289–291; one-third: Wolfgang König, "Nazi Visions of Mass Tourism," in Laurent Tissot, *Development of a Tourist Industry in the 19th and 20th Centuries: International Perspectives* (Neuchâtel: Editions Alphil, 2003), 261–266, here 263; Adelheid von Saldern, "Massenfreizeitkultur im Visier: Ein Beitrag zu den Deutungs- und Einwirkungsversuchen während der Weimarer Republik," *Archiv für Sozialgeschichte* 33 (1993), 21–58, here 21 (senses), 54–55 (German specificities); Christiane Eisenberg, "Massensport in der Weimarer Republik: Ein statistischer Überblick," *Archiv für Sozialgeschichte* 33 (1993), 137–177, here 160 (sports), 166 table 7 (preferences), 171 (women), 186 (12.4 percent), 207 (workers); Christiane Eisenberg, *"English sports" und Deutsche Bürger: Ein Gesellschaftsgeschichte 1800–1939* (Paderborn, Munich, Vienna, and Zürich: Ferdinand Schöningh, 1999), 211 (middle class), 330–331 (rudeness); Hans-Ulrich Wehler, "Die Geburtsstunde des deutschen Kleinbürgertums," in Puhle (ed.), *Bürger in der Gesellschaft der Neuzeit*, 199–209, here 201 (normalcy); Detlev Peukert, "Das Mädchen mit dem 'wahrlich metaphysikfreien Bubikopf': Jugend und Freizeit im Berlin der zwanziger Jahre," in Peter Alter (ed.), *Im Banne der Metropolen: Berlin und London in den Zwanziger Jahren* (Göttingen and Zürich: Vandenhoeck & Ruprecht, 1993), 157–175, here 168–169 (*Wandern*); Peter Gay, *Weimar Culture: The Outsider as Insider* (New York and London: W.W. Norton & Company, 2001), 140 (rightward); Geoffrey Crossick and Heinz-Gerhard Haupt, "Shopkeepers, master artisans and the historian: the petite bourgeoisie in comparative focus," in Crossick and Haupt (eds.), *Shopkeepers and Master Artisans in Nineteenth-Century Europe* (London and New York: Methuen, 1984), 3–31, here 15 (shift to the right); privilege: Kristin Semmens, "'Tourism and Autarky are Conceptually Incompatible': International Tourism Conferences in the Third Reich," in Eric G. E. Zuelow (ed.), *Touring Beyond the Nation: A Transnational Approach to European Tourism History* (Farnham and Burlington: Ashgate, 2011), 195–213, here 197; Kristin Semmens, *Seeing Hitler's Germany: Tourism in the Third Reich* (Houndsmill and New York: Palgrave Macmillan, 2005), 112 (manner), 117 (buses), 120 (Heimat and motor show); apace: Shelley Baranowski, "Radical Nationalism in an International Context: Strength through Joy and the Paradoxes of Nazi Tourism," in John K. Walton (ed.), *Histories of Tourism: Representation, Identity and Conflict* (Clevedon, Buffalo, and Toronto: Channel View Publications, 2005), 125–143, here 127; prototype: Rinn, "Das Automobil als nationales Identifikationssymbol," 59; collective gaze: Rüdiger Hachtmann, "Tourismusgeschichte—ein Mauerblümchen mit Zukunft! Ein Forschungsüberblick," in H-Soz-u-Kult, http://hsozkult.geschichte.hu-berlin.de/forum/2011-10-001 (last accessed 6 October 2011), 14, 16, 32.

207. Clausager, "Motorisierung; The German Motorization Program 1933–1939," 32 (n. 6).
208. *Motorwandern:* König, *Volkswagen, Volksempfänger, Volksgemeinschaft,* 190; Eisenberg, "Massensport in der Weimarer Republik," 297 (75 percent); Hochstetter, *Motorisierung und "Volksgemeinschaft",* 21 (1922), 26 (ten thousand members), 34–35 (SA 1932), 36 (*Motor-SA*), 40 (conquest of the street), 44 (nature), 47–48 (50,000 km), 102–103 (membership NSKK), 153 (monuments); Bernd Weisbrod, "The crisis of bourgeois society in interwar Germany," in Richard Bessel (ed.), *Fascist Italy and Nazi Germany: Comparisons and Contrasts* (Cambridge: Cambridge University Press, 1997), 23–39, here 34 (Sorelian).
209. Jan Thiels, "50 Jaar heilige koe: Een analyse van 50 jaar wisselwerking tussen de auto, het toerisme en het Vlaamsnationalisme binnen het kader van VTB-VAB; Van 1922 tot 1971" (Master's thesis, Vrije Universiteit Brussel, academic year 2008–2009), 113.
210. James, *The Middle Class,* 395–397, 401–404, 417; Ellen Furlough, "Making Mass Vacations: Tourism and Consumer Culture in France, 1930s to 1970s," *Comparative studies in society and history* 40 (1998) 247–286, here 250, 253; Nicole Samuel, "Histoire et sociologie du temps libre en France," *Revue internationale des sciences sociales* 107 (1986), 53–68, here 58; Julian Jackson, "'Le temps des loisirs': Popular tourism and mass leisure in the vision of the Front Populaire," in Martin S. Alexander and Helen Graham (eds.), *The French and Spanish Popular Fronts: Comparative Perspectives* (Cambridge and New York: Cambridge University Press, 1989), 226–239, here 226–228; *syndicats* and *Touring Club*: Patrick Young, "A Place Like Any Other? Publicity, Hotels and the Search for a French Path to Tourism," in Eric G. E. Zuelow (ed.), *Touring Beyond the Nation: A Transnational Approach to European Tourism History* (Farnham and Burlington: Ashgate, 2011), 127–149, here 129, 133; Sunday tickets: André Rauch, *Vacances en France de 1830 à nos jours* (Paris: Hachette Littératures, 2001), 100–101; 40 percent: Christopher Kopper, "Eine komparative Geschichte des Massentourismus im Europa der 1930er bis 1980er Jahre: Deutschland, Frankreich und Grossbritannien im Vergleich," *Archiv für Sozialgeschichte* 49 (2009), 129–148, here 132–133.
211. Hachtmann, *Tourismus-Geschichte,* 106–107 (unions); 318,000: R. G. Pinney, *Britain—destination for tourists?* (London: The Travel and Industrial Development Association of Great Britain and Ireland, November 1944), 21; Harvey Taylor, *A Claim on the Countryside: A History of the British Outdoor Movement* (Edinburgh: Keele University Press, 1997), 226–252 (cycle sales: 232; Blackpool: 252).
212. Jean-Claude Richez and Léon Strauss, "Un temps nouveau pour les ouvriers: les congés payés (1930–1960)," in Alain Corbin, *L'avènement des loisirs 1850–1960* (Paris, Rome, Aubier, and Laterza, 1995), 376–412, here 376–380; Furlough, "Making Mass Vacations," 257 (prestige); sixty million: Berkowitz, "A 'New Deal' for Leisure," 184–212, here 187; Peter J. Bloom, "Trans-Saharan Automotive Cinema: Citroën-, Renault-, and Peugeot-Sponsored Documentary Interwar Crossing Films," in Jeffrey Ruoff (ed.), *Virtual Voyages: Cinema and Travel* (Durham, NC, and London: Duke University Press, 2006), 139–156, here 140 (fantasy), 142 (350,000); Libbie Freed, "Networks of (colonial) power: Roads in French central Africa after World War I," *History and Technology* 26, no. 3 (September 2010), 203–233, here 216 (vagabonds); [Général] Boullaire, *La divi-*

sion légère automobile (Paris: Berger-Levrault Éditeurs, 1924), 5; Orvar Löfgren, "Know Your Country: A Comparative Perspective on Tourism and Nation Building in Sweden," in Baranowski and Furlough (eds.), *Being Elsewhere*, 137–154, here 146–147; Semmens, *Seeing Hitler's Germany,* 133–135; travel relevant vacation: Pagenstecher, *Der bundesdeutsche Tourismus,* 119; infrastructure and middle class: Derek H. Aldcroft (ed.), *Europe's Third World: The European Periphery in the Interwar Years* (Aldershot and Burlington, VT: Ashgate, 2006), 14; Mussolini: Carlo Luna, "One hundred years of Italy on the move," in *Motu proprio: ACI's first hundred years, heading onward* (Rome: Automobile Club d'Italia, October 2005), 21–76, here 35. For a road traffic survey in Germany on Pentecost 1938, in which also quite substantial cross-border traffic into Austria is registered (and an even larger flow is indicated for the *autobahn* in the direction of Arnhem, the Netherlands), see Rudolf Hoffmann, "Der Kraftverkehr auf deutschen Strassen: Pfingsten 1938," *Die Strasse* 5, no. 12 (1938), 388–392; *eigene Heimat*: Wetterauer, *Lust an der Distanz,* 123 (n. 466; the quote is from 1923, during the Inflation); three times as high: Generalinspektor für das deutsche Strassenwesen, *Der Kraftverkehr auf Reichsautobahnen, Reichs- und Landstrassen im Dritten Reich I,* 24 (figure 11).

213. Donna Allen, *Fringe Benefits: Wages or Social Obligation? An Analysis with Historical Perspectives from Paid Vacations* (Ithaca, NY: Cornell University Press, 1964), 75, 95; Heaney, "'The Call of the Open Road'", 41 (quarter); 40 percent: Berkowitz, "A 'New Deal' for Leisure," 188.

214. "A Review of Findings by the President's Research Committee on Social Trends," xi–lxxv, here xxxvi, lxii; *How American Buying Habits Change,* 199; Lizabeth Cohen, *Making a New Deal: Industrial Workers in Chicago, 1919–1939* (Cambridge and New York: Cambridge University Press, 1990), 103; Strunk, *Public Opinion 1935–1946,* 992.

215. Willey and Rice, "The Agencies of Communication," 60 (table 20), 62, 85 (table 28); R. G. Pinney, *Britain—Destination for tourists?* (London: The Travel and Industrial Development Association of Great Britain and Ireland, November 1944; second revised edition; first ed. April 1944), 39, 6, 9, 13, 22; F. W. Ogilvie, *The Tourist Movement: An Economic Study* (London: P. S. King & Son, 1933), 211. The next countries on American tourists' expenditure lists are Mexico, Germany, the United Kingdom, and Italy, in this order. Herbert M. Bratter, *The Promotion of Tourist Travel by Foreign Countries* (Washington, DC: Government Printing Office, 1931; US Department of Commerce, Bureau of Foreign and Domestic Commerce, Trade Promotion Series, no. 113), 2. Armchair: Frank Schipper, *Driving Europe: Building Europe on Roads in the Twentieth Century* (Amsterdam: aksant, 2008), 232. Twelve countries: Harvey Levenstein, *Seductive Journey: American Tourists in France from Jefferson to the Jazz Age* (Chicago and London: The University of Chicago Press, 1998), 253. On American sea travel to Europe during the interwar years, see Lorraine Coons and Alexander Varias, *Tourist Third Cabin: Steamship Travel in the Interwar Years* (New York, Houndsmill, and Basingstoke: Palgrave MacMillan, 2003), 6–23; on American tourism to Paris, see Christopher Endy, *Cold War Holidays: American Tourism in France* (Chapel Hill and London: University of North Carolina Press, 2004; The New Cold War History, ed. John Lewis Gladdis), 16–17.

216. M. C. Isacco, E. Mellini, and A. Mercanti, *Les moyens propres à assurer la sécurité de la circulation...etc."* (report, Association Internationale Permanente des

Congrès de la Route, *VIIe Congrès Munich—1934*) (Paris: Imprimerie Oberthür, n.d. [1934]), 7; seaside resorts: Charles M. Mills, *Vacations for Industrial Workers* (New York: The Ronald Press Company, 1927), 28; Ogilvie, *The Tourist Movement*, 101–102, 163, 181; Jackson, *The Middle Classes*, 304–305; O'Connell, *The Car and British Society*, 88–92 (the 2 percent relate to 1930), 219.

217. Gurtner, *Automobil, Tourismus, Hotellerie*, 26–28 (if for traffic shares a bandwidth was given, the largest percentages have been taken), 66 (playground), 54 (the canton of Tessin did not take part in the census), 71 (4.6 percent), 74–78 (foreign tourist traffic), 229 (more cars); Ogilvie, *The Tourist Movement*, 190–192.
218. *Survey of Tourist Traffic considered as an International Economic Factor* (Geneva: League of Nations, Economic Committee, 22 January 1936), 22.
219. Elke Kölcke, "Die Sommerfrische—vom 'reisenden Mann' zum 'Familienurlaub,'" in Wiebke Kolbe, Christian Noack, and Hasso Spode (eds.), *Voyage: Jahrbuch für Reise- & Tourismusforschung 2009* (Munich and Vienna: Profil Verlag, 2009), 35–45, here 35.
220. Quoted (and translated) in Lewis H. Siegelbaum, "Soviet Car Rallies of the 1920s and 1930s and the Road to Socialism," *Slavic Review* 64, no. 2 (Summer 2005), 247–273, here 258–259 (quote on 259).
221. Nicolaides, *My Blue Heaven*, 95.
222. Nadine Rossol, "Tagber: Gemeinschaftsdenken in Europa 1900–1938. Urspruenge des schwedischen 'Volksheim' im Vergleich," H-SOZ-U-KULT@H-NET.MSU.EDU, 16 January 2012.
223. Weinberger, "Economic Aspects of Recreation," 456. Weinberger refers to the AAA for his estimate of 25 percent "pure pleasure."
224. McIntyre, "'The Repair Man Will Gyp You,'" 319–320; 60 percent: Lawrence B. Goodheart and Richard O. Curry, "A Confusion of Voices: The Crisis of Individualism in Twentieth-Century America," in Richard O. Curry and Lawrence B. Goodheart (eds.), *American Chameleon: Individualism in Trans-National Context* (Kent, OH, and London: The Kent State University Press, 1991), 188–212, here 192–193.
225. Walton Hamilton et al., *Price and Price Policies* (New York and London: McGraw-Hill, 1938), 39.
226. Colin Pooley, Jean Turnbull, and Mags Adams, *A Mobile Century? Changes in Everyday Mobility in Britain in the Twentieth Century* (Aldershot: Ashgate, 2005), 117, 127, 204, 225 (quotation).
227. This approach also receives support lately from transport experts, some of them Dutch. See, for instance, Emmelina Meintje Steg, *Gedragsverandering ter vermindering van het autogebruik: Theoretische analyse en empirische studie over probleembesef, verminderingsbereidheid en beoordeling van beleidsmaatregelen* (dissertation, Rijksuniversiteit Groningen, 1996); Linda Steg, Charles Vlek and Goos Slotegraaf, "Intrumental-reasoned and symbolic-affective motives for using a motor car," *Transportation Research Part F: Traffic Psychology and Behaviour* 4, no. 3 (September 2001), 151–169. For a similar American approach, see Sangho Choo and Patricia L. Mokhtarian, "What type of vehicle do people drive? The role of attitude and lifestyle in influencing vehicle type choice," *Transportation Research Part A* 38 (2004), 201–222.
228. König, *Volkswagen, Volksempfänger, Volksgemeinschaft*, 176.

CHAPTER 5

Translation and Transition
Readjusting the Technology and Culture of Middle-Class Family Adventures

> "The uniforms you wear are called Fiat, Rolls-Royce, Citroën or Ford."[1]

In chapter 4 it is implicitly assumed, just like in classic diffusion studies, that during the spread of the car through the North-Atlantic realm the artifact itself did not change, nor did the people who were supposed to purchase it. In reality, it is exactly the constant adjustment to the perceived 'needs' of the buyers, and, on their turn, their adjustment to the properties of the artifact, that triggers and drives the diffusion process when seen from an evolutionary perspective. This is all the more true in a phase, under consideration here, in which the car evolved into a commodity characterized by rapid style changes. In this chapter we will analyze this mutual adjustment process in two stages.

First we will focus on one crucial aspect of the car's adjusted technology: its development from an open tourer to a closed sedan. We will call this process encapsulation, because the car's form, as a capsule, succeeded the open (clam) shell, as we have characterized the prewar car in chapter 2. Encapsulation is a process that intensifies the cocooning effect and thus directly influences the sensorial experiences while driving the car. We will argue that by dealing with the problem of car sound and noise, the visual properties of the car trip were strengthened, thus enhancing its touring function.

In the next sections we will then address the sensorial experiences themselves as enabled and constrained by the closed-bodied automobile and we will ask ourselves whether the changes in the car's technology were the result of (and in their turn enabled) a change in the car as an 'adventure machine,' first by examining the changes in driving skills the technology of the car enabled, enforced, and constrained, then by trying to characterize the Interbellum version of the automotive adventure resulting from these technological changes, as far as this can be reconstructed on the basis of the traditional sources such as trade and club journals. In other words, we will need, just like for the previous period, unusual, literary sources to describe and analyze the bodily process resulting in an early form of automotive cyborg, an insight we will address in chapter 6.

In the next chapter, which is closely linked to the present one, we will argue that the creation of an 'interior' was not just an issue of increased functionality: it was an answer to a much broader cultural trend of a double desire for intimacy (of the family) and for distance (from the environment, and from the people in this environment). Confirming our main findings in chapter 4, we will see in this and the following chapter that a 'tamed adventure,' a delicate mix of riskiness and comfort, characterized interwar car culture, a counter-argument to the accepted wisdom in the historiography that the success of the car can be explained by pointing at the change in the functionality of the car from a toy into a tool. The car, as commodity, was desired because it fitted in an emerging petit-bourgeois culture of 'domesticated hedonism.' To enable that hedonism, the car itself had to be remodeled.

Orchestrating Car Technology: Constructing the Closed Automobile

The construction of the closed car was one of the most complex and costly operations in the history of automotive technology. It started in the United States, but was soon undertaken as a parallel process in Europe, virtually at the same time.[2] It took American automotive engineers, on whose work we will focus in this section, and the users looking over their shoulders and sometimes outright steering them, about fifteen years to reconcile the new way of traveling with a consensus of what a car was all about. Culturally, it was part of a general process of 'cocooning,' a redefinition of travelers' relationship to the environment and to the Other, a process in which the senses played a crucial role. Technically, it took place in one of the most innovative phases in car production and technology in which the relationship between the engineer and the user took on a new dimension.[3]

Closing the body was a process that linked up with myriad changes in other components. During the adjustment process, car development was crucially shaped by the parallel discourse about sound and noise, especially its 'scientification.' Therefore, in this section, we will first describe how a specific set of properties (dominated by 'comfort') came to shape a specific American car culture necessitating new skills by users and the construction of a new type of 'riding and body engineers' responsible for new ensembles of combined properties (such as 'roadability' and the abatement of 'harshness'). We will then give several examples of how acoustical engineering crucially influenced the eventual shape of the car and how the car developed into a multisensorial 'room on wheels,' which was dominated by vision, but which was crucially supported by a 'sound cocktail' deliberately 'orchestrated' by manufacturers and users alike.[4]

"We Give the Customer What He Wants": How to Include the (Stereotyped) User?

As we have concluded in the previous chapter, the car evolved into a vehicle specifically tuned to long-range tourism. To enable this function, the technical properties had to be geared toward what in the 1930s became known as 'the affordable family car,' a phrase that has retained its powerful attractiveness to this very day, conceived as a not-too-fast, "sober, not to say sedate" type of car.[5] Although Henry Ford promoted his Model T as the quintessential 'universal' car, it was quite uncomfortable on good roads (people got seasick, one historian asserts). It also was considered to be a 'man's car,' and only the Model A in 1927, which copied the suspension from Chevrolet, could explicitly be marketed as the "natural unit" for the urban middle-class family, just as the bathtub and the telephone.[6] The shift of focus to the family was an Atlantic phenomenon: in the United Kingdom, for instance, the family car was considered to "differ ... but little from the largest of the overgrown light cars" and to "drive one in the warmth and shelter of a draughtless, comfortable body almost in the bosom of one's family, so to speak, is pure joy." Whereas the British family car "does not necessarily imply size" (but it did mean a rejection of "'sporty-boyee' two-seaters"), German and French car advertisements still seemed to suggest that family meant size.[7] Crucial in the emergence of this new type of car, enabling a new type of modern touring experience was the co-construction, by users, manufacturers, and engineers alike, of 'comfort.'

How did this process of co-construction come about and evolve? As the story is told in the historiography of car production, it is implicitly understood that car technology followed every whim and desire the users might have felt. This kept the mill of automobile diffusion churning, except perhaps for the first half of the 1920s when Henry Ford proved to be insensitive to the cry for comfort, and many Americans themselves decided to adjust the car to their liking.[8] The discourse on the hundreds of meetings within the Society of Automotive Engineers (SAE) tells a different story, however. Close reading of a quarter-century of *The Journal* of the *Society of Automotive Engineers* and its *Transactions* (proceedings of meetings) reveals that engineers responded to seemingly isolated customer complaints communicated through the dealers and, increasingly, redirected via the ever more powerful sales departments, constantly checking their solutions against a shortlist of very general functions, such as comfort, economy, performance, and speed. Several of the items on this shortlist came from customer surveys such as the questionnaire the National Automotive Chamber of Commerce (NACC) circulated in 1922 among twenty thousand car owners (of which about 10 percent replied), putting "endurance" and "economy of operation" on top of their wish list, immediately followed by "comfort," "price," and "appearance," while, remarkably, "speed" was at the bottom, just before "appointments."[9] Manufacturers and engineers did not follow this wish list slavishly, however: on the contrary, they 'translated' such lists into their own state of the art, for instance when they assured each other (to give just one of dozens of potential quotes) "to give the public comfort and speedy transportation, even if this uses somewhat more gasoline than slower or less comfortable transportation."[10]

Car comfort, many American engineers had witnessed when in Europe during the war, came from Europe: "Our cars come in for a great deal of criticism. They say we sit on our cars while they sit in theirs, and when you ride in their cars you agree with them. We spent 10 days in different makes of European cars. They ride remarkably 'easy.' ... Their cars are most comfortable and they are very low."[11] Soon, American engineers were showing much more self-confidence (and a recognition of emerging differences on both sides of the ocean), for instance after a visit of several months by one of them through England, Belgium, Italy, and France in 1920, although most often 'Europe' was reduced to the Anglophone sister across the ocean. "In Europe ... the automobile is not transportation but adventure. [In America, the motorist] jumps into his four-door sedan as one boards a street car. ... The European motorist demands less convenience than the American." Especially on the European continent, the average motorist is "a courageous speed-hound." They let their "noisy engines" roar at 4000 rpm "without flinching for hours," whereas America is "the paradise of the lazy driver."[12]

Such stereotyping was not an American prerogative, but universal among car engineers all over the North-Atlantic realm.[13] "Because European car-producing countries are small compared with the United States," the British journal *Autocar* opined, "because there is close technical interconnection, because they are developed under the same grinding system of taxation, it might be imagined that there would be little difference between the products of France, England, Germany and Italy. But take the bare chassis, clothe them in foreign coachwork, equip them according to local taste, and the native characteristics will still shine through." The Italian car, from this perspective, "has Art not only written across it but stamped through and through it." And of "the Frenchman" it was said that "his modernism to some extent overcomes his respect for purity of outline," whereas "solidity is the feature of German cars" and "comfort is the characteristic of the British product. ... Calm, ordered, slow to action, dignified in his movements, the Britisher at the wheel rarely allows himself to be carried away. ... Impetuous, nimble-witted, quick of thought, and as quick of action, the Frenchman on the road is constantly finding himself in such situations that he needs all the braking power that human ingenuity can devise to save him from the effect of his *élan*." Hence, so the argument goes, the early introduction in France of the four-wheel braking system.[14]

No doubt it will not be difficult to find similar opinions in the trade journals of other European countries, but what is documented here first and foremost is that from an early date the car leant itself as an excellent object to project 'meaning,' including national prejudices and chauvinist folklore. After all, as soon as the British engineers introduced four-wheel brakes later in the Interbellum, they did not change into 'Frenchmen' overnight.

To American engineers, such British subtleties were lost. For them "French people took their exercise by driving cars" (because of the heavy steering) and in general, European cars were so small and underpowered and lacked so much comfort that no American would ever contemplate buying one.[15] Viewed with a mixture of envy (small companies are easier innovators) and contempt (taxes had led to small engines with raw, 'uncomfortable' characteristics), European car technology was never far away in the United States when it came to making important decisions in technical matters, either as an example to emulate or an option to avoid. "We have been accustomed to look to Europe for the best examples of high-class passenger automobiles," a body maker remarked in 1921, "but I believe that in the near future this will not be true."[16] And indeed, American engineers and users soon followed their own route, although the look over the ocean did not diminish. Slowly, a typical American technology evolved, typical in comparison to the European example. This started to hinder American export to Europe: "The riding-quality problem ... is hardly known abroad. In the operation of American cars in Europe, we find that the conditions are very different, requiring totally different springing and shock-absorbing apparatus."[17] Of course, riding quality was as much a concern in Europe as it was in the United States, but the more 'sporting,' harsher suspension in Europe was morphed in the American engineer's mind into technological primitivism, a process fed by a technicized form of exceptionalism that saw the automobile as an essential American contraption.

Not surprisingly, other lists circulated in Europe as indications of the customers' desires, which, typical for Europe, were ventilated through the organs of the car and touring clubs. The German ADAC's journal saw the gestation of a "typical European car," a combination of the loudly praised American properties like silence, elasticity, and high torque at low engine speed, with a European tradition of low fuel consumption and high speeds from relatively small engines. Later during the decade, endurance was added to this list, on both sides of the ocean, when long-range car trips were increasingly undertaken. In fact, it was quite an accomplishment to give the European small car comparable driving characteristics as a larger car, although the ADAC realized that this could only be implemented by a "rejection of the standard type" and a breach with the "blind imitation of American constructions." The German industry had slavishly followed the American example, the ADAC's journal complained, and had separated itself from the European development. "Let the Americans build their standard," an increasingly self-confident *ADAC-Motorwelt* exclaimed in 1932, feeling the fresh breeze of a new pro-car government, "but give us vehicles that we can use." And it saw with satisfaction a "European" car technology emerge: "Even at an engine volume of about one liter the latest [German] four-seaters are elastic and mountain-friendly."[18]

One wonders how these customer barometers, manufacturers' sales departments on the American side and consumer organizations on the European, could come to such different conclusions in feeling the pulse of their respective constituencies. A more detailed look at two decades of American engineering discourse can illustrate how engineers struggled to adjust the car to the *perceived* wishes of the users.[19] It seems as though a large part of this process was highly ritualized, and characterized by the continuous incantation of 'what the public wants.'

"No one knows better than the automobile manufacturer that the car buyer wants performance," one American commentator remarked in the early 1920s. "The carbureter manufacturer knows that the public is demanding that he produce an instrument capable of throttling to speeds hardly in excess of 2 m.p.h. and upon application of the throttle of picking up rapidly and surely to speeds in excess of 50 m.p.h." It is not very likely that users were able to formulate their desires in these terms, so the engineer in question had already started a process of *translation* of what he perceived to be the user's wishes. "What does the average user demand?" he went on, speculating and translating at the same time:

He wants a car that does not cost him too much, in the first place; he wants one that will be in the service station for the minimum amount of time. It is a matter of pride with every user that his car is dependable, and it is a matter of necessity with most users that the car will be able to perform more than 300 days a year without falter. *If we took an average* of all the users in the country, we would probably find that there is a demand for 15 miles to the gallon of gasoline; there is a demand for quick acceleration, good hill climbing, easy gear shifting, comfort in the seats and springs, accessibility for small adjustments and repairs, a minimum, or rather an absence of rattle, and when it comes to tires, anything short of 8000 miles is always occasion for a fit of peevishness.

The staggering self-confidence of this engineer about his ability to read the users' minds becomes understandable if we read on and learn how he knows all this: "We all know what the public demands because we know exactly what we like ourselves in a car. We like to step into a roomy, comfortable front compartment, into which we do not have to squeeze in order to seat ourselves behind the steering wheel," which was, considering the context of many other articles, a clear rejection of European car culture. "We are always pleased when our first glance tells us that the control members can be readily seen and readily reached. We want the engine to start almost on the first touch of the starting button; we like the starting pedal to engage with little effort, with the engine running smoothly and quietly; we want a clutch pedal which is depressed by very light touch of the foot and which when engaging gives a smooth pick-up without unnecessary slipping and without excessive grabbing." And so the list went on, the engineer meanwhile translating *his own desires* into property specifications, chiseling out a very detailed image of what comfort was all about.[20] In other words, we see a coevolution of a car and its discourse about what a car should be.

And yet, this translating process was also heavily filtered. For instance, the perceived wishes of the users in the realm of an ever more luxurious interior trimming were clearly not translated as such: "It has taken a long time," a Cadillac engineer complained, "to educate the automobile-buying public, and every year the upholstery and trimming of the open and closed bodies have gradually become simpler." Car buyers may have wanted luxury in their car (and they may have been inspired by electric vehicles' interiors), but they shouldn't go too far. Also, the 'user' was not a monolithic entity: "One of the greatest difficulties encountered is to produce a body which will please both Mr. Short and Mr. Long. It is equally difficult to construct a seat cushion so that Mr. Fat will be comfortable as well as Mr. Lean." This engineer deplored that he could not use 'the public' anymore to decrease his uncertainty: "The wise manufacturer is not turning his car over to the public to try out, as in former years." Instead, the public had to be "educated."[21] But educating the public should not be confused with fooling it: "In the automobile business you can kid the public for a week, you can kid it for a month, but nobody has ever yet been smart enough to kid it for a year, ... because the public has become motorwise." Instead, engineers had to "learn how to make [cars] look better than they are.... Then if your advertisement department has been slick enough to pick out those things that the public probably would say about that car and to put the words into the customers' mouth and egg them on and help them on, don't you see how, through the engineering and the production and the advertising in combination, you are beginning to build up that word-of-mouth advertising that is the essence of what sells motor cars?"[22]

Sales, of course, were a clear indication of consumer's choices, but last year's success was by no means a guarantee for the coming year So, with the user at a safe distance, the translation of his wishes (interpreted as 'envisaged functions') into 'properties' proceeded in an atmosphere of constant uncertainty. The general attitude towards 'the public' was respectful and condescending at the same time: most engineers recognized that it was the buyer who delivered the ultimate judgment (government legislation hardly played a role for the passenger car in this period[23]), but the buyer had to be guided in order to make the right choices, especially if these threatened to deviate from what engineers deemed good engineering. In general, the attitude was: "We give the customer what he wants," with the emphasis on *give*.[24]

Despite the self-declared dominance of the engineer in the car's technical evolution, engineers were at a loss, confronted with "the folly of attempting anything in the nature of an exact analysis of such a variable thing as 'customer preference.'" The new user was not like the old, knowledgeable type.

> One would think that, with the great pressure brought by the public upon automobile engineers to develop ease of control and handling of an automobile, a demand would also arise for the elimination of gages and indicators which go in the instrument panels, and that, if these gages are necessary, they should be as inconspicuous as possible. However, such is not the case; vanity comes into play, and almost every man or woman who drives an automobile likes to see an impressive row of gages and dials in front of him and her, probably because it suggests great power and speed under expert control.[25]

Or, as another engineer sighed under general laughter: "Perhaps, after all, a wild guess might be just as good, but it is a lot of fun to analyze these things."[26] It was this "wild guess" that, apparently, had brought the Ford Motor Company to near ruin when it failed "to transform the model T, which had been the world's most popular automobile during the open-car era, into an appealing five-passenger, closed-body sedan."[27]

This trend of generalizing on the basis of ever more detailed (but nonetheless perceived) user expectations and practices and translating these generalizations into a function defined as widely as possible, culminated, by the end of the period, into an outright caricature of the user. Now, the user's wishes were simply redefined as being part of 'human nature' ("man is instinctively lazy" and "human beings are always restless"), reflecting not only a desperate effort to maintain the 'universality' of the car, but also a conviction (if not an arrogance) about the exemplary role of American automotive engineering. For some, simple cynicism was a way out ("The average driver wants to drive his car the way he wants to drive it, even though his way is wrong"), but eventually a consensus seemed to emerge about the ignobility of "John Q. Public," who

> cannot design the new vehicle; he does not know what he wants until he sees it. When he sees it you may tell him it is streamlined or that it has an automatic this or that, or it may be just another automobile; but if his wife looks at it and says, "John, I want one," then the design is a success. The business of the automotive engineer therefore is to lead the purchasing public along the paths in which they should go; pointing the way to betterments, not just trick sales decoys, and to new things which are so obvious to the purchaser that you don't have to write a book for him to read before he knows what it is all about.[28]

If this debate of nearly two decades indicates anything, it is that reading the mind of the user was not a straightforward process, to say the least. The result was an effort to put the user at a distance: "An engineer would be crazy to take an owner's word for how he should design an automobile; but it's just as crazy not to know first hand what that owner wants."[29] At a safe distance, but certainly not out of sight, the user could exercise his or her perceived 'conservative' influence, although even that role was slowly taken over by forces within the industry, such as the sales and production departments: "Where the public used to be the balance wheel to keep us from becoming too radical, it is now the industry itself."[30]

And yet, even at a distance, user culture could be very powerful. The following examples, worked out in more detail, illustrate how these negotiations evolved and the process of coevolution was created. The cases will make clear that 'comfort' functioned as a guiding principle in the struggle between the engineer and the user about who had what to say about the car's technology. These sections will also reveal a remarkable power of the users in getting their habits and skills acknowledged by the industry, illustrating a reversal of the educating process the necessity of which the engineers were so convinced of. And last, the following pages show that the technology itself functioned as a would-be actor, resisting adjustment in the form of 'critical properties,' which we define as analogous to Thomas Hughes's 'reverse salients.'[31] Thus, a critical property is a technical bottleneck, generating tensions between consumers and manufacturers, users and engineers. Such bottlenecks are proof that the car's technology had meanwhile acquired a systems character, thus mimicking the external 'automobile system' emerging during this period.

The User Strikes Back: The Cases of Tires and Bodily Vibrations

Tire manufacturers had noticed during the early 1920s that users deliberately underinflated their tires in order to get a more comfortable ride when car speeds gradually increased, consciously accepting the higher costs of rising tire wear. These suppliers developed 'balloon tires' with a stronger carcass, enabling a lower air pressure, which led to the tire taking up more of the springing work of the total wheel suspension. Engineers from the car manufacturers fiercely opposed this innovation, because they soon found out that such tires would increase the danger of 'shimmy,' a heavy vibration of the wheel around its vertical axis, which endangered steering and driving stability.[32]

Within two years users had massively changed to the new tire type, forcing the car industry to start scientific research into the very elusive phenomenon of 'ride comfort.' In such an atmosphere, the number of alternatives seems to increase. Indeed, every manufacturer developed their own solution, mostly by making the geometry of the wheel-guiding components more exact and increasing the resilience of the leaf springs. This gave American engineers the reputation of being the world's specialists in wheel guidance systems and laid the foundation for the 'typical' American feel of very soft, 'comfortable' springing.[33] This enabled engineers to increase slightly the tire pressure, but a decade later the "terrors" of the user were repeated, when "super balloon tires" appeared on the market, which again were immediately embraced by users.[34]

This case also illustrates that "the engineering community" is not a monolithic entity (nor is the expanding group of users), but consists of factions, schools (often related to car makes), and leading individuals who manage to convince their colleagues of the necessity and the direction of problem solving. The role of the engineering association in this negotiation and translation process is often one of smoothing opposing interests and initiating research in fields that either are too difficult to manage by one single manufacturer or are too far from the production process. Realizing that the American car industry had "no satisfactory yardstick with which to measure riding-comfort," but that at the same time "vibrations ... are by far the greatest annoyance in a car," the SAE itself took the initiative, in 1925, to approach the psychology professor Fred A. Moss at George Washington University in Washington, DC, and set up a Riding-Comfort Research Subcommittee. They developed a "wabble-meter" that could measure "fatigue" as an indicator of the much more elusive concept of comfort. *Comfort*, then, was defined as lack of fatigue. By placing living humans on a specially prepared "vibrating chair," the body was used as a seismograph—in other words, the corporeal senses themselves were used to get a grip on the comfort issue. Later, this research was expanded to Purdue University where Professor Ammon Swope subjected 135 men and women (mostly students) to tests in order to measure groups of "sensory qualities." These included motion, sound, sight, smell, spatial relations, and esthetics, each the result of a myriad of measurements, well distinguished between men and women, such as speed, noise from brakes, desirable amount of visibility, leg room, and type of floor covering. Skidding, for instance, was found to be more objectionable to women, but engine noise appeared to be disagreeable to everybody, while the feeling of acceleration was also contingent upon sex. Not surprisingly, whether man or woman, the customer preferred a closed car above an open one.[35] Comfort, it was now said, was a "state of mind" influenced by "mysterious rattles, squeaks and grunts."[36]

Closing the Body, Opening the Eyes

Just like the adoption of the balloon tire, the change to the closed body (according to some "the last major improvement of the automobile as a personal passenger vehicle"[37]) was triggered by car users in the 1910s who started to buy soft "tops" (often "makeshift winter tops" with "emergency curtains" adorned with celluloid windows) to put on their open cars enabling them an all-season use.[38] It was the Hudson Car Company that saw this trend as one of the first, and who developed, together with the Fisher Body Company, the all-steel closed-bodied Essex in 1919. By 1924 Chrysler and Dodge (the latter's body made by Edward G. Budd Mfg. Co) had followed Hudson's example and when from 1925 a wave of closed models hit the market they were mostly offered at the same price as the open versions, with "astonishing" results. By the end of the decade nearly all passenger-car production was of the closed variety, a classic example of substitution of a body style known (before the war) as phaeton, sedan (in the United States), or saloon (in the United Kingdom; figure 5.1). The substitution is also classic because initially both alternatives experienced a similar increase in popularity, and only in a second phase the sedan developed into the 'winning' option. Once the sedan had won, so-called convertibles appeared on the market (derived, a German observer claimed, from the European 'cabriolet' body type).[39]

This change also led to a decrease of the interior size: whereas most cars traditionally were equipped with seven-passenger bodies, now the norm became five or even four passengers, aimed at the average American family, rather than the extended family of the elite before the war.[40] Of course, the diffusion of the sedan in the entire vehicle park took longer than the production figures suggest: by 1923 about 10 percent of the American "fleet" had a closed body and even in 1926 the open tourer was still in the majority, at 59.1 percent. It illustrates another lesson from a critical appraisal of traditional history of technology: the 'old' technology dominates long after historiography has declared that a certain innovation has taken place.[41] Ford, for instance, only changed to a closed sheet-steel body in 1925, and Cadillac, Chrysler, and Packard followed as late as 1928, while a full-steel roof appeared only in 1935 on most American cars.[42]

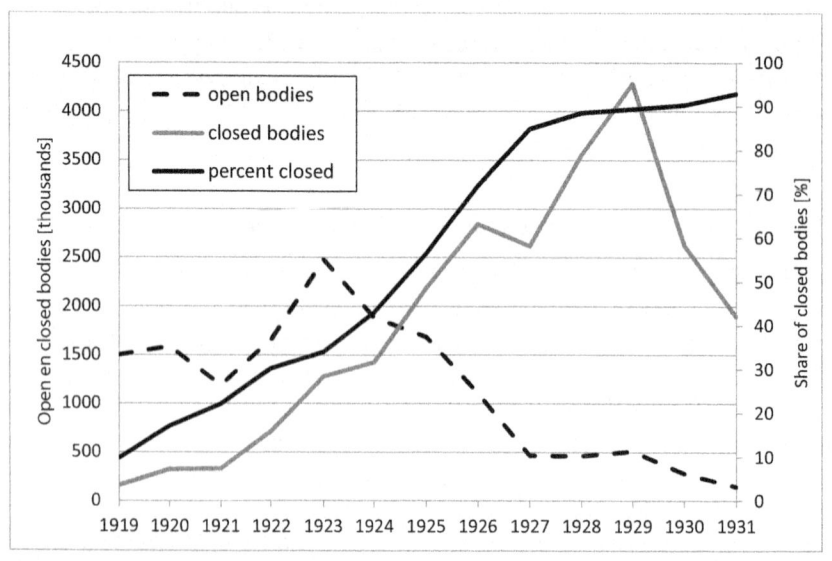

Figure 5.1. Production of open and closed bodies in the United States and Canada (1919–1920: US only), 1919–1931.
Source: Willey and Rice, 39, table 14

The introduction of the closed body had an enormous impact on the car's technology as well as its production and user culture, so much so that a British journal spoke of "almost a new motoring."[43] First, the immediate ("depressing," one designer opined later) effect on the user was a decrease in visibility, because the metal roof and small windows (in cold weather fogged by bad ventilation) were blocking the motorists' view.[44] Second, the nearly 40 percent weight increase of the smaller cars (even bodies with aluminum panels were 10 percent heavier than the previous wooden versions) necessitated a further growth of engine power, an increase that in this decade was nearly used entirely to carry the heavier load, instead of using it for an increase of acceleration and speed.[45]

But the most impact no doubt was in the gradual shift from technology to appearance and aesthetics in both the industry's marketing and the users' buying motives, which according to some contemporary observers (and historians in their footsteps) led to a crucial increase in the role of women in co-constructing the car. Now, women and "younger members of the family" were identified as the *real* customers, as "more precise and definite in knowing what they wanted than the average man," a reflex Virginia Scharf has characterized as a masculine alibi to hide man's own 'feminine' wishes in a largely masculine car culture.[46] By the 1930s in some states (especially California) car dealers reported that one-third to even half of their customers were women.[47]

The closing of the body thus cannot be separated from the other major change in the production and marketing of the car: the annual model change. Introduced during the 1920s by General Motors, this "excuse ... to buy new cars," led to the first massive shakeout of the industry during the Interbellum.[48] The closed body definitively made the car into a commodity, a consumer good influenced by fashion and, in general, a curious mix of engineering rationality and the creativity of the artist-designer.[49] One of the new engineer's tasks was to follow the constant pounding of the sales department to further lower the body, making it appear 'speedy' even as it stood at the curb. Specially designed coloring of the body enhanced the impression of a 'speedy' look.[50]

By the beginning of the 1930s body engineers started to fantasize about unit construction (making a separate chassis superfluous, with the streetcar as an example, and saving 15 percent of steel), while allowing the body to be further lowered and given a streamlined shape (rounded corners, slanted windscreen) for a new type of use: long-range touring. Meanwhile, some body engineers (the famous Joseph Ledwinka among them, working at Budd) criticized the new body style as "a mole-hill on wheels," only accessible to a "contortionist."[51] Indeed, the streamlined body was highly controversial among American engineers of the 1930s (and the users as well). While several engineers ridiculed the concept as ill-conceived (using a scientific argumentation: the arrow-shape was effective to penetrate a solid body, not air), "the public in general, and automotive management in particular" saw the concept as "synonymous with grotesqueness and freakishness."[52] In the end, wind tunnel tests led to the slow acceptance of the principle, although there was a clear tension with conceiving the interior at the same time as "a moving room with homelike atmosphere." The car as "extra room" became so attractive, that an American architectural historian saw the parlor in the home atrophy "as the automobile became the place of choice for intimate socialization." The car's interior got gradually equipped with air conditioning, and all kinds of power-assisted properties such as electrically assisted window lifters and hydraulic power steering.[53]

By the end of the 1930s leading officers of the SAE started to notice that "appearance" was such a hot item that car owners "will accept a certain degree of discomfort."[54] Thinking from the point of view of the interior, engineers had been trying to eliminate vibrations in general and sound in particular to enable vision: "shakes" and "harshness" were smoothed enough to open a vista on the realization of "flight," a fantasy that had accompanied the driving experience for the past forty years: with all bumps engineered away, with all accelerations ironed out, the automotive 'capsule' was now ready to be driven around as a rolling advertisement of itself. In John Dos Passos's words, on "the new highway ... the new-model cars roared and slithered and hissed oilily past (*the new noise of the automobile*)."[55]

It would be misleading to describe this process of turning the car into a 'comfortable' long-range family touring car as a typical American endeavor. For at the same time comparable engineering was undertaken in Europe, where engineers referred to the other side of the ocean, just as had happened at the American side. European makes developed their own close-bodied models, such as Citroën's Traction Avant in 1934, the "Petite Voiture" that also offered front-wheel drive much earlier than American manufacturers were able or willing to do.[56] "Americanization," therefore, was also fueled by counter-movements from Europe, although it is hard to deny that the bulk of the flow of knowledge transfer was a movement over the ocean in the direction of Europe, just as it had been in the opposite direction before the war.

The Process of Prosthetization:
Mutually Adjusting Skills and Technology

In order to enable a tourist gaze for the new, less technically versatile middle-class users a multisensorial repertoire had to be designed allowing drivers and passengers to indulge in touring. It meant, for instance, that more of the driver's tasks should be delegated to the technology. This translation, as we have called it, from software to hardware as partial automation of the driving tasks, took place as what cultural historian Kurt Möser, referring to Norbert Elias, has called "de-sportification" (*Entsportlichung*). It focused mainly on the first of the four tasks he distinguishes of the motorist while using his car: controlling the machine, driving it, taking part in traffic, and navigating.[57] In German novelist Heinrich Hauser's words, in describing a driving lesson, "I didn't see much of traffic. But it was remarkable that [the instructor] even in very crowded streets never had to use the brake and hardly the horn; it looked as if the other vehicles gushed along like a channelled stream. ... I loved this car already. The steering wheel responded to the pressure of my hands through a solid counter-pressure."[58] This was a new experience compared to the initiating rite described in the Williamsons' *The Princess Passes* of 1903 (see chapter 2). Although, as we have seen in chapter 2 and will see again in chapter 6, 'automotive adventure' as a social practice covers more than these four tasks, they form a useful basis for a more systematic investigation of car culture's corporeal substrate. It would fit, for instance, in the recent efforts within anthropology to "model" the "cultural appropriation" of "things" in several respects: purchase, adjusting, naming, giving meaning, incorporation, including in a tradition. *Incorporation* covers the skills we are interested in here.[59]

However this may be, the shift from the first to the second of Möser's tasks (from controlling to driving), which he sees happening during the Interbellum,[60] can also be characterized as 'prosthetization' (where prosthesis is a metaphor for the car as extension of the body rather than as a replacement of an organ, for example, glasses rather than an artificial leg).[61] It can be interpreted as a struggle between driver and engineer, consumer and producer, over the 'behavior' of the commodity, answering the question as to who has the power to decide on what belongs to user competence and what does not, a struggle that took about half a century when it, in the 1970s, ended in a clear victory of the engineers during the 'electronics revolution.'[62] This struggle was first and foremost fought between engineers and highly skilled and trained users such as airplane pilots, who struggled against stabilization efforts to make flying and steering independent from the pilot's skills, but it also extended to the bread-and-butter user culture, as we have seen in the previous section in the case of gear shifting, thus 'deskilling' the car-handling practice to the benefit of new, lay users. "Enskilment" (Ingold) is a process in which learning and doing are inseparable. It is also a process that develops in what Leo Marx has called a "covert culture," as it gets intertwined with driving pleasure (*Fahrlust*) in the terminology of Möser.[63]

Prosthetization started at the component level, such as the car's electric ignition, where levers at the steering wheel for advancing and retarding the ignition timing were replaced by an automatic system on the basis of small centrifugal weights corrected by a pneumatic subsystem.[64] Another example is the introduction of power braking, where the 'feel' of the wheels and the road was replaced by a technical simulation of this feel through pneumatic and hydraulic control systems. Mixture formation in the carburetor and lubrication of the chassis were automated, as was the handling of the gearbox, either by introducing synchromesh gearing (avoiding the need for double clutching) or by automation of the entire gear change process in the case of the automatic transmission. Only when the process of automation was well underway, during the first massive motorization wave of the 1920s, did driving education become institutionalized and sanctioned by the state.[65]

The automation of the transmission in the United States (a development not followed in Europe to this day) illustrates how technology proper and culture had to be mutually and continuously adjusted in order to enable the skills of the users and the properties of the technology to mesh.[66] In the United States, the habit to drive in one gear threatened to ruin the whole concept of 'comfort' as soon as the renovated and expanded highway network started to invite users to speed up. Keeping only three speeds would cause the engines to over-rev on highways, but adding an extra gear set would force users to shift gears more often in city driving. So, the function 'urban use' threatened to interfere with the function 'speeding on the highways,' and engineers had to find a solution. It seemed an insurmountable dilemma and, thus, dozens of alternatives were discussed.[67] In the process special constant-mesh gears were introduced, considerably easing the gear-shifting procedure because it eliminated the necessity to double clutch when gearing down, an innovation that in the midst of the turmoil was hardly noticed and was only later recognized as one of the turning points of gearbox innovation, perhaps because it was soon taken over by the entire engineering community, including the European. In America, the "acceptance by the public was immediate," especially when the spur gears were provided with helical teeth to decrease noise.[68] On top of that, some manufacturers decided to offer a four-speed box with the highest gear designed as an 'overdrive,' enabling the engine to run at a lower speed when speeding along the new highways, a solution first introduced in Germany to allow for high-speed driving on the *Autobahnen*. It is remarkable that as soon as consensus started to be reached about the desirability of adding a fourth gear set, the concept of 'comfort' was redefined to include the four-speed transmission.[69] It shows again that 'comfort' is not an absolute category, but is constantly rethought and redefined to cover changing preferences and engineering constraints and possibilities.

In Europe, the development stopped about here (and continued as incremental improvements in the manual gearbox), but on the other side of the Atlantic the debate became so heated and specialized as in a church on the verge of a schism. Part of the confusion was caused by uncertainty among engineers about the importance of long-range tourism.[70] In the end, General Motors and Chrysler introduced an automatic transmission that used a complicated version of 'internal gears' (so-called epicyclic gearing) that could be automatically shifted as a function of the accelerator pedal's position. Instead of a clutch, a 'fluid flywheel' was built in, invented in the 1920s in Germany for ship propulsion and functioning as a cushion during acceleration and deceleration.[71]

The "Brain Box," as one proud GM engineer called it, was a clever piece of engineering, but it immediately raised concerns whether the comfort paradigm was not stretched too far.[72] In fact, by automating the gear-shifting process, the designer had taken from the user the decision whether and when to perform a gear change; it was "mechanical mind reading."[73]

Less technical analyses of the car have traditionally identified the windscreen as another example of the car becoming an extension of the driver's body. Here, the environment, visible previously as a 360-degree panorama, became condensed as though on a movie screen on which only the forward part of the landscape seemed to be projected. Meanwhile, the side- and rearward gazes were reduced and focused—intensified—into a repeated glance in the rear mirrors. It meant that drivers had to develop a special skill of concentrating on the view in front of them, in the distance, while at the same time keeping a larger part of the landscape, including the foreground, in the corner of their eye.[74]

Thus, closing the body went along with automation, freeing the driver for the other three tasks given by Möser: driving, participating in traffic, and navigating. Many elements of these tasks, however, could not be automated, and had to be learnt during driving lessons, or in practice, as the "dressage" of a horse, where "motion and emotion are linked recursively" and a "personthing" is shaped.[75] Taking a corner, overtaking a fellow motorist, braking, and at the same time using the horn had to be (re)learnt and habitualized, stored in the bodily hardware as it were, in order to free the senses for the ever more complex navigating and traffic-participating tasks governed by the visual sense with the ultimate aim to indulge in automotive adventure (or just to get to work and pick up the dry cleaning on the way home). How long it took to acquire the new skills is not known, but German novelist Arnolt Bronnen describes a scene in which a "mechanic" is forced to act as driver, without much problem. Whatever the duration of the learning curve, the end result was a cyborg-like mix of organic and inorganic bodies. Literature appears to be an excellent locus to express this process very precisely. Erich Maria Remarque quite literally describes the 'becoming cyborg' as the creation of a homosocial (if not homosexual) bonding:

> "Faster." Köster said. The tires began to whistle. Trees and telegraph poles flew along whirring. . . . "More gas," Köster said. "Can I still hold *him* then? The street is wet." "Will see. Before the curve, shift down to the second gear and around it with gas." The engine roared. The air slapped in my face. I ducked behind the windscreen. And suddenly I slipped into the thunder of the machine, vehicle and body became one, a sole tension, a high vibrating, I felt the wheels beneath my feet, I felt the ground, the street, the speed, with a jerk something shifted at its place, the night shrieked and whizzed, *she* hit everything else out of me, the lips clenched together, the hand became vices, I only consisted of driving and racing, senseless and at the same time on high alert.[76]

We witness the transformation into a cyborg, including a gender transformation from masculine to feminine after the bonding succeeded.

The mutual adjustment process of car and driver had deep repercussions for the automotive adventure. A Belgian magazine saw a "new race" of car users emerge, who would be bodily merged with the car, of which "the first [ignition] interruption comes over him as an injury to the heart." Such motorists would learn from the machine "power, honesty, calm and the uselessness of violence."[77] This promise of a nonviolent automotive hybrid would mean a revolution, considering the violent roots of automobilism in the previous period. Möser, a specialist in this crucial, and often forgotten, mini-history of mobile technologies (from the bicycle, to the airplane, to the motorcycle and car), dedicates a long chapter to the development of what he calls the "machine sensibility" of the driver and the process of "simplification of multitasking," but his examples are mostly drawn from the two-wheeler (whether motorized or not), the airplane, and the boat (especially the sailing boat and the canoe). This is clearly because the car, on its four wheels, is less prone to instability. The construction of 'fast' roads though not only made a partial relearning of the steering and braking skills necessary, it also called for more precision and some alternative solutions in wheel suspension.[78] The German *Schnellgang* or *Schongang* (an overdrive to spare the engine while driving on the *Autobahn*) is an example of the latter, a 'technical fix.' As to the former, the relearning of skills: the Dutch touring club warned its members in the 1930s, as soon as the first stretches of German *Autobahnen* became available, not to push the accelerator to the floor, as this would result in overheating and the breakdown of the engine. Also, overtaking at high speeds had to be learnt, as steering systems were designed for low-speed use, with large directional changes of the steered wheels made at a slight turn of the steering wheel. When the Pennsylvania Turnpike was opened in 1940, nearly 28,000 motorists crowded onto it on the first Sunday, and an equal number the following Sunday, to race on the concrete at an average speed of 63 mph (the average top speed of American cars in 1936 was 80 mph), burning their engine bearings, blowing up their radiators, and blowing out their tires. "Velocitization" was what AAA post-WWII analysts called the "inability of drivers to judge speed accurately," while others saw "highway hypnosis" as causing some motorists to race at high speed into highway restaurants.[79]

If the start of the new age of automobility in which the sensorial rearrangement of the car necessitated a redefinition of the automotive adventure, should be pinpointed in time, then 1924 makes a good chance. That was the year that Belgian law (as one of the few defining the shape of the human side of the cyborg) stipulated which minimum requirements turned a human being into a motorist: he or she had to have two eyes, at least one ear, and no epilepsia.[80] By the end of the interwar years some psychologists, defining driver practice as "predominantly a perceptual task," approached the car as "simply a sort of physical extension of the driver's body."[81]

Multiple Adventures: Skills, Thrills, and Risks

The closing of the car body afforded much more than the adjustment of driver skills, however. It enabled the prewar hybrid driver/mechanic to evolve into an 'automobilist.' This process can best be studied through belletristic literature just as for the previous period (in chapter 2), but its complexity suggests an approach in stages. Therefore, in this section, we will first grasp the shift in automotive adventure by contextualizing automobile driving in more general societal trends regarding adventure, violence, and aggression, as far as this can be reconstructed on the basis of secondary literature, following up on the prewar experiences as narrated in chapters 2 (and for the war in chapter 3).

Most contemporary observers and historians agree that the adventurousness of travel and automobility had diminished after the First World

War. Many saw aviation as the true successor of the mobile adventure. We should not forget, however, that most motorists during the Interbellum were novices, and for them the adventurous experience was new, even if they knew that the 'real' adventure was experienced high above their heads, for most of them inaccessible because of its costs.

Attentive commentators had already observed a trend toward a more "petty-bourgeois" adventurousness before the war. The German entertainment magazine *Elegante Welt* opined in 1912 that "the pleasure (*die Lust*) in small adventures, in the coincidence, in the willingness to endure, in high spirit, a defect while on the road, a delay in having a meal, and wind and weather, is the counter-gift of the car for the recent acceleration in our tempo of life."[82] That was quite different from the chicken-killing automotive avant-garde that dominated the public image of earliest automobility, a situation that made us conclude that a pre-war motorist could feel a poet, could experience transcendence. It was even less ambitious than Bierbaum's biedermeier Sentimental Journey (see chapter 2). A comparable 'tamed adventure' was praised in the American journal *Motor Travel* (successor of *The Club Journal* of the Automobile Club) in 1917, where the modern motorist was said to be "no longer satisfied to travel with his eyes glued to a book or card.... He wants to look around." And he had learned this, the journal concluded (with its elite constituency in mind), during his "extensive travel over the magnificently sign-posted roads of Europe."[83] The new automobilism reaped "the joys that come with turning down some attractive side road," as an American guide for "unusual motor trips" promised its readers, "perhaps without definitive knowledge of the place to which it goes, or moving aimlessly along off the beaten path."[84]

The British historian Craig Horner has coined the term 'modest motoring' for this.[85] For some commentators it was quite clear what was happening: marketing consultant Bruce Barton complained in 1922 that middle-class males lacked "the courage to dive off the dock" (and start their own enterprise). The opinionated historian Daniel Boorstin reminded his readers in the 1960s (when a "cheap cafeteria at the corner" offered "adventure in good eating" as the likes of Barton had taught) that travel as long ago as during the Interbellum had become "a bland and riskless commodity."[86] Such commentators described what recent consumption studies characterize as domestication, an integration of consumption in the middle-class family and its home without, however, fully grasping the significance of this process for the changes in user practices. Domestication in mobility not only related to the family home and its 'average' occupants, but also to the home country as the preferred excursion goal. The taming was temporal and spatial, as well as functional, structured just like the tripartite adventure of speed, roaming and tinkering.[87] Motorists, a German automobile journal stated in 1929, enjoyed an "automobile world" presented to them "through the windows of a closed car.... A new, exciting, and desirable world, full of variety, and more entertaining, a world that approaches from the outside, less harsh and imperious, but rather charming and humble."[88] This kind of marketing talk that passed as journalism was constitutive of the domesticated adventure itself: without it the driving experience seemed to be reduced to bread-and-butter transportation from point A to point B. Although older motorists lamented the closing of the car body as a loss of experience, the new breed of automobilists presented the leftovers in adventurousness as crucial for the new automotive experience, if only to prevent bare utilitarianism from creeping into car culture. If transcendence, the feeling of being a poet while driving, was made unattainable, this new breed seemed to think, then other practices had to be developed that could rescue automotive adventure.

And thus, the automotive adventure in its domesticated, 'modest' form presupposed the production of the written word as much as the production of the driving experience.

Indeed, car driving and travel account intensified their intermediality. In the 1930s *ADAC Motorwelt* started to publish warnings that too much adventure was unhealthy, especially because of wrong body posture in the car for members who hitherto had been used to walking to the streetcar stop. The thrill of acceleration, so constitutive of prewar automotive adventure, had to be replaced by the soft satisfaction of being on the move. Traveling quietly was opposed to the habit of the "poor motorists" who undertook travel at high speed, "and were not able to give an account of their trip." For Bertolt Brecht, the war had "interrupted all connections with adventure," so much so that "adventurers in our society are criminals," an attitude which led to the final acceptance of risks and accidents (conceived as 'chance,' as we will see in chapter 7). But Brecht managed to wreck two of his cars, which seems adventurous enough. In the end, even the Nazi regime favored "a modest, disciplined consumption" for financial and ideological reasons (to counter a feared "materialism"), although one should be careful to treat them as a monolithic entity. The predominance of Nazism in German motorization at the same time led to a revival of prewar macho sentiments in the realm of the automotive adventure. Officials of the Nazi motoring organization NSKK downplayed the attraction of (feminine?) touring: "We won't organize beauty contests nor will we undertake tourist outings. We have written the hardest and manliest sport on our banners: motor vehicle off-road sport. Here the man struggles with himself and his machine."[89]

And yet, domestication also meant that more people could participate in a less technicized, less elite adventure. As Georg Simmel said in his essay "Das Abenteuer" (1910; The Adventure), the fact that in the modern world the adventure is just one extreme on a scale where the other extreme is full calculability "makes us all into adventurers." For the adventurer (for which Simmel took the player, the man who likes to play games, as exemplary figure), the process is more important than the result.[90] Simmel characterized adventure as related to chance, and as something with a beginning and an end isolated from the daily and the ordinary. "Life of the adventurer is characterized by restlessness and an ever again renewed search for happiness." Adventure individualizes, and every adventure is unique.[91] From this perspective, the domestication of the adventure, its democratization and the efforts to make it into something within easy reach, is a *contraditio in terminis,* and it would take quite some effort to perform the balancing act of the new automotive adventure. From this point of view it is not surprising that the norms for adventurousness remained high, as a controversy about an *"Alpenfahrt"* (Alpine Trip) organized by ADAC and criticized by the populist *Berliner Tageblatt* illustrates. ADAC maintained that such trips, set up as "sport, in this form involved some danger," but not "irresponsibly" so.[92] When *American Motorist* in 1930 printed the "great adventure" of a "grandmother" who got lost in the mountains, every AAA member knew that the new adventurousness was within everybody's reach. In the same year, AAA announced the publication of its new journal called *Holiday,* subtitled *A-wheel A-wing A-float: A Magazine of Travel—Recreation—Adventure* offering "New Mobility" for new users, "oncoming Americans with modest incomes but with rich desires." Soon, "Budget Auto Trips" were offered to "small groups" to go touring in Europe "in own automobiles."[93] A comparable moderate form had taken hold of the functional (tinkering) aspect of the automotive adventure: some of ADAC's motorcycling members were not so much irritated by large defects, but by the "small mishaps that in a modern motorcycle should not be allowed to occur." On a car trip to Venice two members (clearly belonging to the functional adventurers) had to change a punctured tire "only five times," and as this took only ten minutes each time, it did not get in the way of the enjoyment of the trip, and the occasional tinker-stop was simply accepted as part of the package. As if to emphasize the biedermeier character of their joy they nearly literally quoted Bierbaum (probably without knowing it): "Learn to travel instead of to race."[94] Typical for this type of petty-bourgeois family touring were the "Mystery Hikes" initiated by the British railways, the *"pique-nique surprise"* organized by the Touring Club de France or the trips "M.O.B." (*met onbekende bestemming:* with unknown goal), set up by the Dutch touring club for its members who liked to indulge in an adventure of which they themselves did not know the outcome, even if the organizers did.[95]

The promise of "some danger" and "small mishaps" indicated a search for a new balance in the tripartite automotive adventure. In the following three sections the temporal and spatial side of the car adventure (the functional tinkering adventure will be dealt with in chapter 7) and the role of the family, including its age, gender, and racial aspects will be investigated, in order to get a first grip on this shift.

The Temporal Adventure

As to the aspect of the adventure in time (speed), and seen with hindsight, car driving now entered a transition phase toward a postwar "risk society." Now that taking part in traffic increasingly meant an involvement in a potential accident (as we will see in chapter 7), the danger of automobility had to be reassessed, by every novice car owner. In Norbert Elias's terms, it took several decades for the "*process* of civilization" consisting of "self-regulation" and "internal pacification" to take effect.[96]

This was not a linear process, as it interfered with a counter-current of indulging in a "covert culture" of aggressiveness and violence as part of an emerging culture of personal "sensation seeking" or narcissistic "stress-seeking." In his analysis, Elias (and other analysts in his wake) seems to underestimate this compensating mechanism, in males especially. The insight that leisure and stress are no enemies[97] was a revival of the debate around 1900, when nervousness was seen as a constituent of the automobile movement (see chapter 1). American motorists, for instance, made a new sport of surreptitiously opposing the police control of speeding, hiding in the growing 'swarm' of car traffic and defining themselves as an army against "a dictatorship such as has not existed in this country." As if they had read Brecht they declared to be "proud to belong to the criminal class." Most of them had never had encounters with the law before.[98] Such risky activities, later commentators would assess, "fulfill a need for arousal ... or for stimulation ... as a way to develop capacities for competent control over environmental objects." Formulated by a postwar representative of the "intrinsic motivation approach" of high-risk behavior, this sensational aspect was not included in what Ulrich Beck came to coin as "risk society."[99] According to Beck (who wrote his first book on the risk society under the influence of the Chernobyl nuclear disaster), accidents do not happen as the result of a personal failure, but because systems transform human mistakes into "unintelligible destructive forces." In the risk society, the destructive power of wars is absorbed, generalized, and normalized, while risks are not solved, but cosmetically treated and pushed from sight.[100] This was not a covert operation, as we will see in chapter 7: 'accident-prone' motorists had to be ostracized and eliminated from the ranks of those who obeyed the rules. The problem, though, was that such risky drivers could not be identified, so Beck's insights should be expanded with notions about the ubiquitous culture of risk taking as suggested by Jörg Beckmann in his effort to theorize about an "auto-risk society," in which "old dangers of transportation [have been transformed] into risks of automobilisation." Risks, according to Niklas Luhman and Anthony Giddens, cited by both Beck and Beckmann, are a new type of danger, the result of a choice by an actor between two or more options, for instance to drive to work or to take the bus. Apart from being part of reality, risk is also a social construct and thus differs per actor: what is risk for one actor becomes a danger (in which no decision making is involved) for the Other (as an 'accident') or for the environment (which also can be seen as an Other), as pollution. Risks and benefits of automobilization are thus "unequally distributed." As risk was constructed in the eighteenth and nineteenth centuries in the sphere of maritime insurance, defining a danger as risk implied that "it allowed an activity to be carried on even though it was dangerous," and this is exactly what happened during the interwar years in all motorizing countries.[101]

Nowadays, proponents of extreme sports bask in the illusion that skill can overcome risks, which they have now redefined as "difficulties," but interwar motorists already knew, however hard they tried to forget it, that no matter how skillful one is, a certain statistically determined part of the motoring movement will die each year in a car or motorcycle accident.[102] That was part of the package, too: motorization created the illusion of individuality whereas the phenomenon itself became a more and more collective one, which is more than just observing that the number of motorists increased. The increasing swarm of automobilists saw itself as a societal force, as a group involved in the modernization of a risky society.[103] Just like for aggression and violence, novels are excellent sources to study risk, as we will see in the next chapter.[104]

Not only were risk and skills closely related in the new automotive adventure, the notion of risk also seemed to change the relation between the motorist and the environment. Alongside Beck's structuralist and institutional approach, Beckmann places Mary Douglas's anthropological notion of the "blaming system" as a "modern strategy for dealing with danger," while others stress the function of risks in the establishment of Foucaultian power relations and dominance. Thus, risks are means to establish an "Other." They are defined by insurance companies, who exclude some forms of behavior (such as drunken driving) and sanction others. Risks generate different risk takers, such as Douglas's four ideal types of individualists, hierarchists, egalitarians, and fatalists, or the 'road-hogs' as identified by early twentieth-century car and touring clubs.[105]

Thus, risks reveal the layeredness of automobilism. They have also lately been connected with the playful dealing with risks such as in extreme sports. Called "deep play" by Mihaly Csikszentmihalyi, the "lust for risks" results in intensive sensorial feelings of self-realization, "personal transcendence," authenticity, and happiness. These feelings of "flow" appear in practices such as competition including games, sports, and religious and political contests; in gambling; in mimicry including theater and dance; and in vertigo.[106] In this sense, car driving did not become 'addictive,' as some popular theories claim (the notorious 'love affair,' as coined by Wolfgang Sachs[107]), but its influence went much deeper: it is, to vary on Csikszentmihalyi's phrase, "deep movement" based on "sensation," an insight already expressed by Thorsten Veblen and revived in 1936 in the context of research on consumption and the American standard of living.[108]

To historicize this concept of deep movement we can consult the German writer Ernst Jünger who, in the tradition of the prewar Futurists, did not see a principal difference between the warrior, the criminal, and the racecar driver if it comes to the romance of violence. In the 1920s Jünger was still struggling with the question of "whether the hidden end goal of technology is a space of absolute comfort (*Bequemlichkeit*) or a space of absolute danger," perhaps referring implicitly to the United States for the first possibility, but in any case setting himself apart from the biedermeier petty bourgeois. He also used the cyborg motif (which he called "*Kentaur*") to stress the unity of man and machine and to argue that risk and danger were inherent parts of automotive culture. It is, indeed, remarkable that Jünger used the car as a *pars pro toto* for the entire society. In the cyborg, the motorist delegates agency to the machine, "sensations of feeling are replaced by fascination and movement of the machine."[109] Jünger developed his own theory of perception (in a text tellingly titled "*Das abenteuerliche Herz*," The adventurous heart), which he called "stereoscopy," in which sharp observation is combined with a dreamy openness to impressions. It is this combination of alertness and drowsiness that seemed constitutive for the new form of interwar automotive adventure (while as a kind of manifesto, in two versions of 1929 and 1938, Jünger's text was intimately related to many avant-garde movements between the wars). This combination would have been impossible (because it was potentially deadly) in the previous period with its lust for acceleration and speed sensations.[110]

As Nietzsche had theorized: "the adventurer's fascination with risk—as a warrior, a lover, a gambler—is a way of contending with death; a will to make of death an adversary. Every risk successfully run is a triumph over death." A recent ethnographic study of "voluntary risk taking" distinguishes between three types: one which is "highly safe" (bungee jumping, climbing wall practice), one which is based on "individual choice" (cave diving, cliff jumping), and a "grey area" in between (deep-sea diving, mountaineering). Depending on the application, the car fit into all three areas, although the 'highly safe' type only applies to the driver and his passenger, not to the pedestrians and cyclists, as we will see in chapter 7.[111] Later research connected the violence of racing not only with Jünger's "fraternity of manliness," but also with a new, risky version of the "technological sublime." NASCAR racing, for instance, emerged in the American rural South as a "controlled violence" of "trading paint": bumping and pushing against competing cars in a thrilling effort to avoid catastrophic accidents.[112] It was aggression deployed in order to avoid, rather than inflict death, a solid basis for (masculine) erotic experiences. Tuning kits became available in the United States for Ford's Model T from about 1915, not only to improve the body's touring comfort on long trips (something the industry was not ready for at the time), but also for "reducing friction, vibrations, and unnecessary weight" in order to enhance its performance. The Chevrolet brothers, once loosened from Durant's emerging General Motors conglomerate, developed their own high-performance engine for the Model T. While this was happening in the Midwest, in California a true racing culture around later Ford models (such as the Model A from 1928, and especially the 1932 model with its "flathead V8") and other cars emerged, on dry lake beds, organized by the Southern California Timing Association (notice the name!) founded in 1938.[113] Apart from a "hot rod culture," the 1930s also saw a secret street-racing culture emerge, "the eventual arrival of the police only 'add[ing] to [its] hazards.'" "Playing chicken" was meant to test cold-bloodedness by letting go of the steering wheel until one of the passengers would grab it. "Bumper thumping" was a game to strike other cars or even pedestrians.[114] Thus, skills and thrills formed an explosive cocktail laying the corporeal foundation of the automotive adventure. "The thrill," narrative psychologist Karl Scheibe concluded, "is the expected product of adventure," and he makes an implicit connection with the later work of Csikszentmihalyi by pointing at the "fleeting moment" in literary renderings of athletic performance, at the same time giving us another bodily foundation for the experience of 'flight' we identified in chapter 2. Scheibe distinguishes between three elements of 'thrill': "fear, voluntary entry, and hope for survival."[115]

Such eroticized practices of car culture raise the question about the gender of the car itself. Literary historian Charles Grivel's solution seems too easy when he declares that "the car represents the gender of the other," at least if he means 'the other sex.'[116] His opinion is shared by American literary analyst Deborah Clarke who calls the car both masculine and feminine, and it is this ambivalence that makes that "the symbol of the car unsettles [its] gendering, for the automobile is both phallic symbol and female object."[117] This ambivalence of the car's sexual identity however tends to privilege a heterosexual view upon car culture. However this may be, there is no doubt that the car may stand for desirability itself.

Although one should distinguish between extreme sports and conscious risk-taking as sensation on the one hand, and speeding on the highway, or taking a corner at the edge of one's capability on the other, their roots are the same. In the United States, for instance, there was a clear shift, halfway during the Interbellum, from the previous bohemianism of the touring pleasure without a goal to "making miles" (also during night-driving, done by one-third of the AAA members surveyed in 1929) and "making up lost time." "Touring driving is fast," confirmed *American Motorist* in 1928, when half of the American families owned an automobile. "Go-fever . . . drives you on and on, insists that you reach this or that place before dark, when there is absolutely no reason for reaching it."[118]

Thus, every motorizing nation seemed to be speed-prone. "Speed," an Italian book on tourism declared during the 1930s, "is also preferred the most by our public. If the Italian who wishes to buy a car would be without any need and limitation and if he could fulfill his ideals, he would opt for the fastest, most powerful car, even if he would have to refrain from its transport function, its comfort and even its safety."[119] According to some, speed was a special character of the American culture, as "a pedestal upon which American business and constructive thought are founded." Indeed, even bus tourists "urge the driver to go faster, and it seems not to make much difference how fast he goes; they still want him to go faster."[120] Even nonusers indulged in the speed experience connected to car driving. Ilya Ehrenburg describes a movie theater in his *The Life of the Automobile* (1929) as follows: "A car raced across the screen. The entire audience was racing in that car."[121] This makes also clear, again, that in order for the mobile adventure to become an all-pervasive experience all media had to cooperate: cars, movies, and novels.

The Spatial Adventure

It seems likely, however, that the motoring movement during the Interbellum, when it shifted from the driver/mechanic to the automobilist, also made a shift from the temporal to the spatial aspect of the automotive adventure. Of course, automobilism remained fast (and became faster, as we have seen) but the spatiality of the motoring experience only now could encompass the entire globe, not only because of improved car technology, but also because of an expanding 'automobile system,' as we will see in chapter 7. For Sigfried Gideon (who clearly had the new experience of fast highway driving in mind when he wrote this), "the space-time feeling of our period can seldom be felt so keenly as when driving, the wheel under one's hand, up and down hills, beneath overpasses, up ramps, and over giant bridges."[122] Space was not simply extra-urban landscape; it now included the cityscape as well, and both were loaded with cultural and political connotations.

Space, as a "relational arrangement of bodies," cannot be experienced without movement.[123] For automotive movement, the spatial adventure can be related to the theorists of new urban experiences, such as Walter Benjamin and George Simmel, because "the culture of the 1920s is shaped by and in the town."[124] The former has become famous, among others, because of his theory of the *flâneur* who, mobile but moving against the flow of others and at the same time being part of it, developed a strategy in the crowded city to observe without participating, as a detective, with his sphere of action preferably in the half-public sphere of the 'passages' and the terraces of the cafés, half inside, half outside. "Now landscape, now room," in Benjamin's words. Although his concept may fit better with the previous period for nonmotorized traffic participants and for the peripatetic practices of car pioneers in the immediate vicinity of their towns, his blurring of the inside and outside can be applied to the period under investigation here, especially when this pair is connected to other pairs such as distant/near, and familiar/strange.[125] From this perspective car driving became an extended, mechanized and prosthetic form of *flânerie*.

Paradoxically, car driving allowed closeness and distance at the same time (in Simmel's terms: "the distant becomes near, the closeness becomes far away"): it brought motorists much nearer to remote regions and their people than the train or the horse carriages or even the bicycle, but in doing so it kept them at an 'automotive distance,' again much more than the other vehicles were capable of. Around 1900 Simmel developed, in the context of his philosophy of money, a "theory of distancing" (*Theorie der Distanz*) to analyze the behavior of urbanites who construe around them an "ideal sphere" of "spiritual private ownership" that forces the other to keep a certain "spatial distance." This sphere can also materialize, for instance in the form of an automobile as an extension of the body, and the car trip can thus be interpreted as an exercise of distancing, from classes other than the middle class in one's own country, or from the 'exotic' Other in the 'periphery,' be it domestic or abroad. The act of distancing turns the Other into a 'spectacle' (and, conversely, in order to experience 'exoticism' one needs distance and proximity at the same time: to go there, but keeping a distance) and this act of "objectivation" prepares the other for being under control, the violence of the gun replaced by that of the camera, 'shooting' a picture. Thus, the 'poetics' of prewar automotive adventure morphed into a new transcendental experience: the feeling of power toward everything and everybody outside of the automotive capsule. This also explains why humans (and, it seems, especially men) needed vision instead of the 'feminine' and more 'primitive' 'proximity senses' (touch, taste, smell) to enable this act, although it is the synesthetic combination of touch (of vibrations, 'flight' during touring) and vision that seems to be responsible for the extremely powerful attractiveness of the automotive spatial adventure. The proximity senses seem to benefit the most when moving, unless one enters the closed-car body: then, as we have seen in the previous section, vision regains its primordiality.

What is specifically modern, though, according to anthropologist Tim Ingold, is not so much the privileging of vision over hearing or touching, but the way of seeing, objectivating, and distancing, as well as the insight that in order to perceive, one has to move.[126] According to recent literary theory in the process of making a 'spatial turn,' keeping distance is done to come closer to what is far and strange, which as the same time leads to a perceptual shift: the gaze "plunges in," the other senses do not.[127] Hearing and seeing are focused, synchronized with the car's technology, which offers a space in-between home and destination, work and leisure. This space directs the attention from the world to the flow, the beginning of a second shift (after the one from shell to capsule, see chapter 2) from capsule to corridor, which will fully blossom in the post-WWII years.

Although the new adventurers of spatiality, as an avant-garde movement, preferred the airplane over the car, the two-dimensional car trip was the closest most people could come to this experience. German communist travel writer Egon Kisch in his travelogue on "paradise America" has a grandma explain to her grandchildren what cars were: "airplanes moving on the ground, so they advanced only very slow." However fast motorists drove, they remained slow compared to the 'real' adventure in the air, even if, initially, airplanes did not fly that much faster than cars. The old trope, though, of the car trip as 'flight' became ever more literally formulated: landscape architect Simonson saw in 1932 how the new Mount Vernon Memorial Highway "simulated in its flowing lines, the spatial curves, the horizontal and vertical transitions, and the banked turns of a fast transport aircraft in flight."[128] The connection between car driving and flying was also made through the emphasis on skill: The "cool" German pilot Wolfram von Richthofen (a "prototypical Prussian officer," a fitness freak with one lung, smoking forty cigarettes per day) drove his 3.7-liter Mercedes "like a Messerschmidt machine."[129] On the new 'speedy' roads, car driving came as near to flying as possible, and much more safe, so it seems.

Distancing went along with a certain harshness and lack of empathy.[130] American motorists in the 1930s identified with their fellow motorists rather than with accident victims. They developed a "relatively cold" attitude toward each other resulting in a "shift to callousness," if not outright cynicism—the traffic engineer who opined that the best one could do in case of a lethal car accident was "to send or omit flowers, as the relatives request."[131]

Car driving, in its speedy and tranquil versions (both violent and distanced, and often in a combination of the two) was thus not only a part of modernity, it enhanced it at the same time. Car travel, in other words, is not about combining modernist and anti-modernist elements of contemporary culture. The interpretation of car travel experience as a highly and fully modern practice (and as such not only a primordial expression but also constitutive of the modernization process itself) is consistent with recent theorizing about modernity. This interprets the importance of the past (so ubiquitous in the touring experience as we have seen in chapter 2) not as nostalgia (and anti-modern) but as a part of an unhistorical, future-oriented "alternative modernity," as a rebirth starting with a clean slate, to be forced upon a decadent and doomed society.[132]

Family, Age, Gender, Race, and Aggression

The nuclear family, even (or perhaps especially) now that it started to decline as the core of middle-class daily activity, was emphasized as an important cornerstone of civilized life. In a special issue of the Annals of the American Academy of Political and Social Science in 1924 the extended "family rather than the individual" was heralded as "the unit of pleasure.... Everybody goes to the big picnic." Some traffic experts regarded driving alone a waste of energy and space.[133] When, in a recent analysis, the car as "a centrifugal and atomizing unit" is set against "Eurasian" collectivism and made responsible for enabling an easy assimilation in the system, it seems that the rhetorics are taken too much at face value. In reality it was the increasingly systemic character of the car society and the swarm-like behavior of its drivers (being individual as part of a crowd without a leader) that enabled historian James Flink to boldly claim that "automobility was probably the main barrier to the development of class consciousness among workers in the 1920s and 1930s."[134] We will later, in chapters 6 and 7, investigate the usefulness of the concept of the swarm for the motorists' integration in the emerging "automobile system"; for now it suffices to observe that the swarm, as a moving mass, is nonreflexive, is self-organizing at a subconscious level and as such fully resonating with a cool, unemotional attitude. According to the German publicist Ernst Jünger, quoted by Helmut Lethen, the "automaticism of traffic" reveals itself through the movement of people around "some secret center in accord with 'silent and invisible commands.'"[135]

In France, the family as core of the fabric of daily life was hotly debated. The bachelor was attacked as an "effeminate, pacifist, pleasure-seeking, and sterile *célibataire*, often coded as homosexual." Readers of Georges Ferré's celebration of the bachelor published in 1929 in which he "describe(d) the pleasure of immoderate behavior, the thrill of fusing with one's automobile and becoming one, like a centaur, and most importantly, the excitement of a new morality that was nothing more than 'enjoy yourself' (*jouir*)," were warned against his praise of "mechanization, a taste for profit, cocktails, speed, violent exercise, mass-produced elegance, serialized pleasures, anti-sentimental love, and Americanism." Family promoters (most explicitly the Catholic Church, but also the natalists propagating a higher family fertility, from both the right and the left) had a hard time integrating the car in a family culture that did not seem to accept the father spending his leisure time in the bistro. In fascist Germany, where fatherhood was in higher esteem than motherhood, the man was considered to support the *Volk* more than his family, to be a fighter rather than a father. While the public sphere (the state, the high culture, the army) was the place for homosocial structures, the private sphere was part of the female realm and shaped by the family.[136]

It will probably never be possible to construe a causal connection between the importance of the middle-class family in a national culture and the spread of the car in that culture. Nevertheless, in the United States the shift from the Victorian family to the modern urban family resembled the shift from public transport to the car; in both domains "democracy," "private sphere," and "individualism" emerged as highly appreciated values. For such a motoring family, hitchhikers (or "guest riders," as AAA journal *American Motorist* called them) and other "strangers" should be avoided ("Thumbs down!"). Within that family, the suburban house seemed to turn every American husband into a "domestic man" (a "contented suburban father, who enjoyed the security of a regular salary, a predictable rise through the company hierarchy, and greater leisure") as a result of a major shift from a homosocial to a heterosocial culture. In a situation in which the women's clubs' attention shifted from political to leisure-centered topics, and the car had become "a part of the family," domesticity not only curtailed the way women were liberating themselves, but automotive adventure also had to change drastically.[137]

For the United States detailed sociological surveys enable a glimpse on these changes. The car played either a contributing or a central role in all six "areas of social life" distinguished by the Lynds in Muncie—"getting a living, making a home, raising the young, using leisure, engaging in religious practices, and participating in community activities."[138] In Westchester County, a suburb of New York City, the leisure activities of all income classes (except students) could be divided between seven types: eating, visiting, reading, public entertainment, sports, radio, and motoring, in this order (expressed in the amount of time spent); motoring occupied 16 percent of leisure time spent outside the home by white collars, and 8 percent by professionals and executives. In a county in which nearly every family had a car (and where five motorcycle clubs existed), "motoring as a leisure pursuit" became a crucial activity. When the (mostly female) members of the Parent Teacher Association were asked what form of recreation they especially liked, more than half mentioned "sports," including "golf, other outdoor sports, and motoring." If they had an extra $1000 to spend, they declared they would spend it on travel, followed by "the purchase of a car, [or of] a better car."[139]

Another aspect of the new interwar automotive adventure was its increasing variety. Specific groups created their own subculture around the car. The American urban youth developed a whole new mobile adventure of dating and "petting" ("all forms of erotic behavior short of intercourse") as some boys managed to buy a second-hand "collegiate flivver" for only $25, decorated with whitewash or splotches of paint or graffiti such as "Flappers' delight—Open all night." In Middletown, of the thirty girls charged with "sex crimes" in 1924, nineteen were caught in a car. They saw the car as "a new form of courtship." In California as early as 1917, according to a policewoman, girls liked to "roam ... up and down our crowded streets" When they came back to Middletown a decade later after their original research, the Lynds witnessed how its youth culture, which had started shortly after WWI at college, had expanded to high school. The "Scott Fitzgerald wave" also reached the farm youth, who, according to a survey in 1927, showed a remarkable correlation between "liking farm life" and living in a family with a car, and disliking it if they had no car at their disposal. A recent study concludes that the car had much more impact on the small-town culture in the countryside than in the big towns, simply because cars were less ubiquitous in an urban setting, and most urban kids used public transport, or walked. For such kids, stealing cars and joyriding could be a solution to enhance their adventurous experiences; these practices exploded in Los Angeles between 1914 and 1928, most (male) youth declaring that they were more interested in the ride itself than in the destination. Members of girl gangs also indulged in joyriding. Joyriding was an adventure indulged in mainly by white middle-class youths (black youths, for instance, showed much lower car-theft rates) as if to testify that frustrated desire (mostly among boys) was a very potent driver of what one could call subversive motoring.

Contemporary observers explained this was due to an absence of the father and an urge to "assert ... their masculinity via actions they perceived as being far removed from feminine 'goodness.'" Despite their subversiveness, however, they acted from the same impulse of a longing for "deep movement" as their parents: they saw car driving as a "bid for independence." It was "thrill-seeking" behavior (one of the Los Angeles kids was caught joyriding with two car keys of his father's cars in his pockets), according to a sociologist. The car allowed more privacy and excitement than the other two favorite pastimes, going to the dance hall and to the movies (if possible both in a car, no doubt). It was also seen as less violent, because violence was perceived to be located in street gangs populated mostly by worker youth. Another study found little difference between working-class and middle-class children's behavior in small towns in the 1940s: "Both liked riding long distances, seeking a good time, but working class youth ... seemed to place particular value on possession of a car."[140]

Nonwhite mobility history differs substantially from the mainstream story in this respect. Spanish-American Fabiola Cabeza de Baca's decision to buy a car as a home economist in New Mexico is only one of a potential multitude of stories of nonwhite car culture we know virtually nothing about. For African-Americans the car has mostly been depicted as a "liberating device," a means of "escape and adventure." The black middle class was not only smaller, but also characterized by a more dominant rural background, while education and income differed, too. Such "primitive drivers" (as they were feared by some psychologists) "comprised doctors, ministers, teachers, newspaper editors, and small businessmen, rather than clerical workers, salespeople, and managers," but like women their war service had strengthened their claim to "the freedom to purchase the new consumer goods." For blacks, if they could afford it, car purchase ("a gesture of manliness and independence") meant "escape from the Jim Crow segregation they encountered [in] trains and street cars," and to be "free of discomfort, discrimination, segregation and insult." But in order to enjoy this freedom, blacks had to set up their own system of travel information (such as the *Negro Motorist Greenbook*) to find hotels and resorts open to blacks, and to develop their own resorts and places to stay, as even the national parks discouraged entrance by racial and ethnic minorities. Some of them went even further against the grain, experimenting with car sharing as early as the 1920s.[141] They also started their own resorts (such as Idlewild in Michigan and Elsinore in California) and their racing clubs gathered in the Afro-American Automobile Association (AAAA) founded in 1924 in Chicago.[142] According to a contemporary witness, the issues faced by African American motorists diminished the potential to experience 'adventure': "somehow it takes the joy out of gypsying about, when you have to be at a certain place by a certain time."[143]

Is it, therefore, coincidence that black music, when it became part of the mainstream in the jazz age of the 1920s, showed so much affinity with the train? "Why a train instead of say, a plane or a bicycle?" musician Wynton Marsalis said, tellingly excluding the car from his shortlist, "Jazz music actually is a train." For him, "The train gave voice to feelings 'I could never have found words for.'" The train, Joel Dinerstein in his excellent analysis of interwar black culture asserts, "let [blacks] dream about faraway places" as they sat by the hundreds along the tracks "to watch the Special pull out" of Birmingham, Alabama. "Unable to afford diversions like the circus and excluded from participating in organized sports," for African Americans the train was much more than a "nostalgic symbol of techno-progress," as Dinerstein coins it.[144] Although the story about cars and ethnic minorities in this phase is a story of exclusion, at the same time the *inclusion* as a spectator in the ever-expanding culture of interwar mobility, confirms our thesis that one did not have to be an owner or passenger to be influenced by (auto)mobility. And perhaps, there is also a technical side to this story. Although Dinerstein, not very convincingly, tries to present car sound as potential music, the 'orchestrating' of the 'sound cocktail' of the car, as we have seen in the first section of this chapter, was aimed a getting rid of and retuning the resulting noises, quite different from the attention Count Basie gave to the myriad of train sounds while on tour. Although George Gershwin integrated taxi horns in his "An American in Paris," the "beat-driven . . . clackety-clack of train rhythms" did not have their parallel in automotive acoustic culture, where the haptic and the visual senses dominated, as we have seen. And yet, the identification with the locomotive rather than the entire train by nineteenth-century popular culture (as expressed, for instance in Edward Ellis's 1868 dime novel *The Steam Man of the Prairies*, who could function off the tracks!), the fascination for being at the helm and creating wonders, was transferred to the car once it became more reliable, and once the "swing" of "machine aesthetics" entered mainstream white popular music. While Steam Man was modeled after a black vehicle (prefiguring the "blackface minstrelsy" by white musicians), the 'car face' was white.[145]

The increasing number of self-driving women during the Interbellum coincided with their ascendency to objects of the erotic male gaze, on a par with the car, in advertisements, art, and the earliest forms of car porn. The question arises whether we should not identify a separate 'female adventure' as part of the expanding car culture (just as the separate existence of a 'black adventure' seems defensible), where the construction of the pin-up should be as much an object of investigation as the role of the mother in the 'car family.' Or is the term itself too much connected with male violence and aggression: is there something "intrinsically masculine about travel," as Janet Wolff asked herself (and didn't answer)? Could the 'adventure' concept encompass the taxonomy of relation types between women and cars as given by Julie Wosk?[146] Or are we to believe the efforts of people like Virginia Scharff to Georgine Clarsen to art historian Maud Lavin, who claim, explicitly or implicitly, that "sex difference" is a nonproductive way of approaching women's behavior? Their position is supported by feminist psychologists who maintain that women behave as aggressively as men, as long as one defines *aggressiveness* much more broadly than the exertion of physical violence alone: "sex differences seldom appear when aggression is allowable behavior to women." According to a literature overview of a vast array of recent psychological research, the two sexes differ "in degree rather than pattern," as women "are angered by different things or to different degrees by the same thing."[147] Other feminist scholars tend to reserve some space for a typical male adventure because of "men's greater willingness to risk taking" which is related to "strengthening masculinity," whereas "women's aggression still tends to be aimed primarily inward.... Most women are still expressing their aggressions indirectly, as they were socialized to—through gossip, nonparticipation, interiorizing, hiding, manipulating, excluding and as always fantasy." Also, the "preservation of the self" cannot be realized "without aggression and recognition of the other," and the same applies to "the specter of narcissism ... as one of the saddest and loneliest and most dangerous states of the self," a state which, as we saw in chapter 2, is especially apt for the production of automotive experience, at least among males.[148] Such arguments find support in Georg Simmel, whose definition of the adventure as an eroticized conquest (Casanova being one of his champion adventurers) made him believe that adventurousness "perhaps only could be found on the side of man; ... the love relationship means generally 'adventure' only for the male, for the female the same uses to fall into other categories."[149]

If it is true that women's (accounts of) aggression have hardly been researched so far, the car could be used as an effective lens to do so. After all, as we will see in chapter 7, psychologists invested a lot of effort in identifying and measuring aggressive traits in motorists, and in general, psychotechnic approaches were aimed a singling out (and excluding from the group of motorists) all forms of deviant behavior, thereby threatening the persistence of 'adventurousness' at its very core. Much was at stake here: although Freud only reluctantly acknowledged the existence of an "aggressive instinct" (*Aggressionstrieb*), which he reserved for the outward directed aggression, his reluctance was inspired by his conviction that every instinct was in a sense aggressive and as such related to the bodily "motricity." Later psychological theories broadened the definition of *aggression* to a "tendency to be enterprising, energetic, active" in which "all suggestion of hostility [was lost]." This opened the possibility of a (nowadays obsolete) distinction between Type A ("ambitious, aggressive, business-like, controlling, highly competitive, impatient, preoccupied with his or her status, time-conscious, and tightly-wound") and Type B personalities ("patient, relaxed, easy-going, and at times lacking an overriding sense of urgency"), and through this, between accident-prone, and 'normal' motorists.[150] Psychologists, indeed, love dichotomies: a recent "micro-sociological" theory of violence distinguishes between violent situations rather than individuals and identifies two groups of situations: "a set of pathways around confrontational tension and fear" and fights performed with an audience in mind. While the former contains situations varying between military combat, police violence, and riots, but also domestic violence, the latter can be found in sports, games, and "violence for fun and honour." Despite efforts to rescue certain forms of legitimized, normalized behavior from the extremes, both sets of violent situations seem to apply to car culture, especially so when one realizes that the first set is about "finding weak victims to attack," such as pedestrians or cyclists. This study (which deals with a nonexhaustive array of thirty "types" of violence) rejects Bourdieu's notion of "symbolic violence" as "mere theoretical world play," which denies the importance of the physicality of violence.[151] In the end, experts as late as 1988 claim that "human aggression remains largely a puzzle."[152]

If we are prepared to call the family outing on Sunday a (redefined) adventure, why could we not use the term (as 'adventures,' plural) for the troubles black families had in finding a place to stay for the night during their holidays, or for the chauffeuring of suffragettes by an American women during the 1910s? Amidst the plethora of automotive adventures, however, the male violent and aggressive version clearly dominated: "Violence, rage, and aggression [were] signifiers of masculinity, except when they are used as signifiers of a compromised or defective femininity."[153] And although a recent course book on *The Male Experience* warns against stereotyping male aggressors (and female pacifiers), it mentions five elements of the male role that all can be called automotive traits inducing to aggression: nonfeminine, success, aggressive, sexual, and self-reliant. Stereotyping can be counteracted by recognizing that there are "multiple ideal types" of masculinity, as well as by the assertion of students of violence that "trans-Atlantic comparisons suggest that the United States has not been uniquely violent among Western societies." Men's studies distinguish between at least two forms: "violence to sustain ... dominance" against women and violence as part of gender politics, among men, such as heterosexual violence against gay men, or within street gangs. Likewise, recent research into female aggression distinguishes between "styles," "very different ... from men as [women] oppose, argue, get their way, and intimidate and hurt others." This study speaks of the "mythology of [male] aggression" (and by implication, female nonaggressiveness), but defines aggression as something intentional, a definition that is not very useful to describe the shift from personal automotive aggression toward a hidden, systemic aggression taking place in the period under discussion here. Nonetheless, "a reluctance to focus on female aggression may be a reluctance to consider similarities between women and men."[154]

Perhaps this initial, orientational study of the gendered aspects of motorized violence and aggression can best be closed by a contemporary witness, Sherwood Anderson (whom we met already in a previous phase [see chapter 2]) in a remarkably modernist novel expressing a confused awareness of the gendering of automotive adventure. In *Perhaps Women* (1934) the 'I' asserts: "It's a woman's age. Almost any man will admit that. He admits it rather sadly."[155] The sadness comes from the insight that "the machine has taken from us the work of our hands. Work kept men healthy and strong." Now he drives a car but he did not build it: "How many men are like myself, going restlessly from place to place, seeking something they cannot find? ... The scientists have taken from us old mysteries, and, as yet, no poets have arisen who can give us new ones." Men are no match for machines, while women "are affected less. It must be because every woman has a life within herself that nothing outside her can really touch except maybe a mate." Anderson's confusion is revealed in his ambiguity about this observation: is this to be applauded, or is it "about men being humiliated by their wives"? Against the "impotence" of the men stands the consumerism of the women: "I am trying to proclaim a new American world, a woman's world. The newspapers are all run for women, the magazines, the stores. The cities are all built for women. Whom do you suppose the automobiles are built for?" Anderson's complaint boils down to the insight that the technical sublime is only a poor ersatz for the real, natural sublime, where "your male should be the adventurer. He should be careless of possessions, should throw them aside." Coupled with outright fear that the middle class will be "wiped out" by the Depression, masculine "impotentence" should be countered by a fight against women: "They are the worst of all." On the other hand, women are the hope for America's future as their "living potence," their "hunger," the "potentiality of new strength in them ... may save American civilization in the very face of the machine." In trying to "say the unsayable," the novel ends in decrying the "fake feeling of power" the car allows. "I am sitting in my own car. I step on the gas. It requires but a slight movement of my foot. The car shoots away. It goes at terrific speed. It climbs mountain.... Do I take the power in the car to be my own power?... I am doing it. You are doing it. All Americans are doing it.... I am on the road of impotence." On the other hand, "the power of women is more personal. It is a matter of human relations. It operates directly on others. It is a power the machine cannot touch." On the last page, during a visit to a car factory, the lights went out. "'I want a man,' the girl's [one of the workers] voice said. It was a clear young voice. There was an outburst of laughter from many women—ironic laughter it was, down there in the darkness—and then the light came on again."[156]

Conclusions

This analysis of nearly a quarter-century of coevolution of technical properties and user culture in the Atlantic automotive sphere results in the thesis that the tourism wave did not alter the main impulse behind general automotive engineering, namely to preserve at all costs the golden combination of the car as a city car and a touring car. New functions apparently did not lead to substitution, but to accumulation: at the outbreak of the Second World War the car had acquired a high-speed, continuous touring function without abandoning the city function, keeping threatening alternatives at a safe distance.

Universality implies that specialized functions can only partially be accommodated. That is probably the reason why the American shortlist of functions was so general and abstract. Manufacturers were not interested in developing an exclusive touring car. Instead, by carefully designing their properties, engineers allowed the touring function to creep in, without jeopardizing the 'universal' function of the car. The 'family touring car' that should also function satisfactorily in an urban context was the translation into metal and glass of these contradicting preferences.

A crucial element of this family touring car was its sensorial characteristics. They emerged as a result of a complex 'translation' effort by engineers, constrained by sales managers and fed by car users. The constant emphasis on comfort led to an increasing 'soft ride,' liberating the senses for the visual, crucial in creating the automotive version of the 'tourist gaze.'

This means that in the designing process of technical properties, the user's influence is often rather indirect, but nevertheless very real. Users were increasingly enclosed in a capsule of which the sounds were consciously changed such that the acoustic 'feel' of the exterior was softened, even decoupled: 'comfort' was defined as smoothness, as perfect flight, and, ultimately, as a dampening of the sensorial input from the environment. Only thus could vision be rescued as the dominant sense to engage in touring. Comfort made noise into sound, and this allowed vision to develop into a tourist gaze.[157]

This gaze, we saw, was developed upon and fed by a violent and aggressive substrate and the resulting emotion we can call a 'violent sublime' or perhaps better a 'sublime of violence' as a derivative of David Nye's 'technical sublime.' This sublime goes along with a double feeling of transcendence, a spiritual one (which we will investigate further in the next chapter) and a physical one, expressed in the 'prosthetic' effect of the car as an extension and enhancement of the bodily capacities. In other words, driver and car evolved in this phase in a subliminally aggressive cyborg.

The breakthrough of the car coincided with the emergence of the "new woman" (who was considered to be able to repair her own car), the emancipation of middle-class youth and the increase of interracial violence. Hence, understanding the emergence and persistence of automotive adventure (as a complex of adventures) cannot be realized without including the potentially subversive adventures of women, nonwhites, and children.[158] Sheila Robowtham concludes in her analysis of "women who invented the twentieth century" that several women were longing for a "more hedonist self" although female "adventuring was being recast as a purely inward affair."[159] But traditional sources (archives, club journals) to substantiate this claim historically are scarce. It seems that, just like for the first period, we need to reenter the literary and related artistic realms to get a better grip on the changing emotional and motivational complex that governed interwar automotive adventure. We will do so in the next chapter.

Notes

1. Francis Picabia, *Caravansérail: 1924; présenté et annoté par Luc-Henri Mercié* (Paris: Pierre Belfond, 1974), 63.
2. Stefan Krebs, "The French Quest for the Silent Car Body: Technology, Comfort, and Distinction in the Interwar Period," *Tranfers* 1, no. 3 (Winter 2011), 64–89.
3. Paul Nieuwenhuis and Peter Wells, "The all-steel body as a cornerstone to the foundations of the mass production car industry," *Industrial and Corporate Change* 16, no. 2 (2007), 183–211. There does not exist much scholarship on the history of automotive technology proper, let alone the details of the development of the closed car body. For a basic overview, see Gijs Mom, *The Evolution of Automotive Technology: A Handbook* (forthcoming).
4. The following pages are based upon Gijs Mom, "Orchestrating Automobile Technology: Comfort, Mobility Culture, and the Construction of the 'Family Touring Car', 1917 – 1940," *Technology and Culture* 55 No. 2 (April 2014) 299-325, and Karin Bijsterveld et al., *Sound and Safe: A History of Listening Behind the Wheel* (Oxford: Oxford University Press, 2014).
5. Alan R. Fenn, "The English Light-Car and Why," *Journal of the Society of Automotive Engineers* (hereafter *SAE-J*) 20, no. 2 (February 1927), 203–213, here 212; sedate: "English Comment on the American Made Automobile," *Automotive Industries* 42, no. 24 (10 June 1920), 1359.
6. Leonard P. Ayres, "Saturation-Point for Motor Cars Pushed Ahead to 27,000,000," *SAE-J* 16, no. 2 (February 1925), 195–196, 208, here 196. For a technical composition of the "ideal American family car," see "Improvements in 1927 Cars," *SAE-J* 20, no. 2 (February 1927), 197–200. Poor comfort: Peter J. Hugill, "Technology and Geography in the Emergence of the American Automobile Industry, 1895–1915," in Jan Jennings (ed.), *Roadside America: The Automobile in Design and Culture* (Ames: Iowa State University Press, for the Society for Commercial Archaeology, 1990), 29–39, here 38–39.
7. "Recognition of the owner-driver," *Autocar* 49, no. 13 (17 November 1922), 1054–1056, here 1054 (joy); "On the road," *Autocar* 51, no. 66 (23 November 1923), 1058–1060, here 1059; "The popular family car," *Autocar* 49, no. 11 (3 November 1922), 858–859, here 858 (overgrown).
8. Reynold M. Wik, *Henry Ford and Grass-roots America* (Ann Arbor: The University of Michigan Press, 1972), 64–65; on the personal adjustment of cars in the 1920s, see Kathleen Franz, *Tinkering: Consumers Reinvent the Early Automobile* (Philadelphia: University of Pennsylvania Press, 2005), 74–102.
9. Norman G. Shidle, "Practical Data Gathered for Use in Selling Cars," *Automotive Industries* 45, no. 8 (25 August 1921), 351–354, here 351; see also "Motorists Have Little Use for Speed," *Motor-Travel* 13, no. 4 (July 1921), 18.

10. A. L. Putnam, "Chassis Design for Fuel Economy," *SAE-J* 8, no. 5 (May 1921), 441–448, here 441. On the role of engineers as 'translators' of the user wishes, transforming (technical) properties into (relational) functions, see Gijs Mom, "Translating Properties into Functions (and Vice Versa): Design, User Culture and the Creation of an American and a European Car (1930–1970)," *Journal of Design History* 20, no. 2 (2007), 171–181.
11. David Beecroft, "Conditions in the Automotive Industry Abroad," *SAE-J* 4, no. 6 (June 1919), 521–525, here 523.
12. Maurice Olley, "Comparison of European and American Automobile Practice," *SAE-J* 9, no. 2 (August 1921), 109–117, here 109–111; Andrew F. Johnson, "Passenger-Automobile Body-Designing Problems," *SAE-J* 8, no. 4 (July 1922), 306.
13. The following paragraphs have been taken from Gijs Mom, "Diffusion and technological change: Culture, technology and the emergence of a 'European car,'" *Jahrbuch für Wirtschaftsgeschichte* no. 1 (2007), 67–82.
14. "National Characteristics in Design: Temperamental Attributes Revealed in Mechanism and Coachwork," *Autocar* (27 October 1922), 821–822 (italics in original).
15. "SAE World Automotive Engineering Congress Stresses Thought Freedom Vital to Best World Technical Progress," *SAE-J* 45, no. 1 (July 1939), 13–21, here 19.
16. A. F. Johnson, "Passenger-Automobile Body-Designing Problems," in *SAE-J* 8, no. 4 (April 1921), 306.
17. "Independently Sprung Front Wheels: Discussion of the D. Sensaud de Lavaud Annual Meeting Paper," *SAE-J* 23, no. 4 (October 1928), 385–390, here 385.
18. Josef M. Jurinek, "ADAC und Oeffentlichkeit: ADAC-Presse-Empfang in Berlin," *ADAC Motorwelt* 24, no. 8 (25 February 1927), 2–6, here 4; Piscator, "Die Jahresbilanz der deutschen Kraftfahrzeugindustrie," *ADAC Motorwelt* 25, no. 51/52 (21 December 1928), 4–8, here 4–5; endurance: Siegfried Seher, "Die Zukunft gehört dem Gebrauchssport," *ADAC Motorwelt* 26, no. 25 (21 June 1929), 23; Ernst Schäfer, "Abkehr vom Standardtyp," *ADAC Motorwelt* 29, no. 2/3 (15 January 1932), 11–13 (quotes on 11 and 13); European technology: Joachim Fischer, "Die Charakteristik des deutschen Wagens," *ADAC Motorwelt* 29, no. 32 (5 August 1932), 9–11, here 9.
19. The following pages are taken from Gijs Mom, "'The future is a shifting panorama': The role of expectations in the history of mobility," in Weert Canzler and Gert Schmidt (eds.), *Zukünfte des Automobils: Aussichten und Grenzen der autotechnischen Globalisierung* (Berlin: edition sigma, 2008), 31–58.
20. J. Edward Schipper, "Present Requirements of the Automobile User," *SAE-J* 2, no. 3 (March 1918), 214–217, here 214, 215 (italics added).
21. E. W. Goodwin, "Automobile Body Design and Construction," *SAE-J* 2, no. 4 (April 1918), 271–278, here 273, 271. For a description of the process to push the user out of the design process during the Interbellum, see Franz, *Tinkering*.
22. R. H. Grant, "What Sells Motor Cars?" *SAE-J* 20, no. 2 (February 1927), 270–272, here 271.
23. Pierre Schon, "Effect of Legislation on Motor-Vehicle Design and Operation," *SAE-J (Transactions)* (November 1932), 426–435.
24. Austin M. Wolf, "Striking Engineering Progress Revealed in 1933 Cars," *SAE-J*, 31 no. 1 (January 1933) 1-20, here: 16.
25. J. W. Frazer, "Bodies Considered from Car Buyer's Viewpoint," *SAE-J* 31, no. 1 (July 1932), 294, 298–299, here 294.
26. Alex Taub, "Economics of the Chevrolet Engine," *SAE-J* 25, no. 3 (September 1929), 264–276, here 264; "Factors governing car choice," *SAE-J* 28, no. 5 (May

1931), 592–593, here 592. Reports on early SAE meetings are often verbatim and contain indications of audience response such as laughter and applause.
27. Christopher Wells, "Car Country: Automobiles, Roads, and the Shaping of the Modern American Landscape, 1890–1929" (dissertation, University of Wisconsin-Madison, 2004), 312.
28. *SAE-J* 41, no. 1 (July 1937), 303; *SAE-J* 42, no. 5 (May 1938), 16; *SAE-J* 44, no. 4 (April 1939), 141; William B. Stout, "What the Traveling Public Wants in the Future," *SAE-J (Transactions)* 33, no. 3 (September 1933), 293–297, here 295.
29. Norman G. Shidle, "Looking at the Future of Car Design Through 1937 Trends," *SAE-J* 40, no. 1 (January 1937), 13–16, here 16.
30. Edwin L. Allen, "Body Engineering—Past, Present and Conjecture as to the Future," *SAE-J (Transactions)* 45, no. 3 (September 1939), 365–371, 384, here 368.
31. Thomas P. Hughes, *Networks of Power: Electrification in Western Society, 1880–1930* (Baltimore, MD, and London: Johns Hopkins University Press, 1983); see also Ann Johnson, *Hitting the Brakes: Engineering Design and the Production of Knowledge* (Durham, NC, and London: Duke University Press, 2009).
32. O. M. Burkhardt, "Wheel Shimmying: Its Causes and Cure," *SAE-J* 16, no. 2 (February 1925), 189–191.
33. Tore Franzen, "Suspension Types Will Be Developed for Each Country," *SAE-J (Transactions)* 33, no. 4 (October 1933), 347.
34. B. J. Lemon, "Judging Super-Balloon Tires," *SAE-J (Transactions)* 31, no. 4 (October 1932), 403–411.
35. "Impressions That Are An Insult: Horning Gives Cleveland Section Some Thoughts on Riding-Qualities To Mull Over," *SAE-J* 16, no. 4 (April 1925), 392–394, here 392; S. P. Hess, "Automobile Riding-Comfort," *SAE-J* 15, no. 1 (July 1924), 82–85, here 82; and *SAE-J* 15, no. 6 (December 1924), 543–547, here 543; annoyance: H. L. Horning, "Bearing of Research Department Work on Car Developments," *SAE-J* 17, no. 2 (August 1925), 189–191, here 190; "Riding-Qualities Research: Six Tests of Muscular and Nerve Fatigue Selected as Result of Dr. Moss's Work," *SAE-J* 26, no. 1 (January 1930), 99, 101; "Riding-Comfort Investigations: Valuable Data on Fatigue, Vibration and Shock-Absorbers Presented and Extensively Discussed," *SAE-J* 26, no. 2 (Feburary 1930), 137; F. A. Moss, "Measurement of Comfort in Automobile Riding," *SAE-J* 26, no. 4 (April 1930), 513–521; "Review of S.A.E. Research Activities: Initiation, Progress and Results of Programs That Have Kept Pace with Advancement of the Industry," *SAE-J* 26, no. 6 (June 1930), 712–717, here 716; F. A. Moss, "Bodily Steadiness—A Riding-Comfort Index," *SAE-J* 26, no. 6 (June 1930), 804–808; "Bodily Steadiness—A Riding-Comfort Index; Discussion of Dr. F.A. Moss's Summer Meeting Paper," *SAE-J* 27, no. 1 (July 1930), 111–114; G. C. Brandenburg and Ammon Swope, "Preliminary Study of Riding-Qualities," *SAE-J* 27, no. 3 (September 1930), 355–359; F. A. Moss, "New Riding-Comfort Research Instruments and Wabblemeter Applications," *SAE-J (Transactions)* 30, no. 4 (April 1932), 182–184; Purdue: "Wind Resistance and Comfort; Effects of Body Shapes at Different Speeds—Physical and Mental Effects of Riding," *SAE-J* 30, no. 4 (April 1932), 851–854 (photo of Moss on 853).
36. E. C. Gordon England, "The Body Problem and Its Solution," *SAE-J* 27, no. 1 (July 1930), 69–77, here 70–71.
37. Robert Paul Thomas, "Style Change and the Automobile Industry During the Roaring Twenties," in Louis P. Cain and Paul J. Uselding (eds.), *Business Enterprise and Economic Change: Essays in Honor of Harold F. Williamson* (Kent, OH: Kent State University Press, 1973), 118–138, here 121.

38. George J. Mercer, "Style in Automobile Bodies," *SAE-J* 8, no. 2 (February 1921), 123–126, here 123; winter tops: L. C. Hill, "The Convertible Body—Its Evolution and Possible Future," *SAE-J* 26, no. 2 (February 1930), 171–176, here 172; the costs of the body increased threefold during the 1920s and were one-third of the total costs of the car by the beginning of the decade: Mercer, "Style in Automobile Bodies," 123.
39. "Closed-Car Demand Increasing," *SAE-J* 12, no. 1 (January 1923), 85; Edward G. Budd and J. Ledwinka, "Building of All-Steel Vehicle Bodies," *SAE-J* 16, no. 2 (February 1925), 219–222, 231; astonishing: Thomas, "Style Change and the Automobile Industry During the Roaring Twenties," 129; cabriolet: Rudolf Hoenicke, *Die amerikanische Automobilindustrie in Europa* (Bottrop: Buchdruckerei Wilhelm Postberg, 1933), 56.
40. George J. Mercer, "Cheaper Closed-Body Construction," *SAE-J* 12, no. 2 (February 1923), 213–216, here 213.
41. David Edgerton, *The Shock of the Old: Technology and Global History since 1990* (Oxford: Oxford University Press, 2007).
42. Mercer, "Cheaper Closed-Body Construction," 213; Norman G. Shidle, "Lest We Forget—There are Still More Open Cars Than Closed," *Automotove Industries* 55, no. 10 (2 September 1926), 361–362. Ford and GM makes: James J. Flink, "The Ultimate Status Symbol: The Custom Coachbuilt Car in the Interwar Period," in Martin Wachs and Margaret Crawford (with the assistance of Susan Marie Wirka and Taina Marjatta Rikala), *The Car and the City: The Automobile, The Built Environment, and Daily Urban Life* (Ann Arbor: The University of Michigan Press, 1992), 154–166, here 165.
43. "Where cars are better," *Autocar* 59, no. 1667 (14 October 1927), 735–736, here 735.
44. "The Passenger Car of the Future," *SAE-J* 5, no. 3 (September 1919), 236–241, here 237; depressing: Hermann A. Brunn, "Trends in Body Design," *SAE-J* 22, no. 6 (June 1928), 679–683, here 680.
45. In 1921 the closed-bodied car weighed, on average, 362 lbs more than the open car. "Weights of 1921 Cars on Which Kansas Bases License Fee," *Automotive Industries* 45, no. 7 (18 August 1921), 330–331; wood: William Brewster, "Automobile Body Design," *Automotive Industries* 6, no. 4 (April 1920), 215–218, here 218.
46. O. T. Kreusser, "Automobile Bodies, from the Abstract Customer's Viewpoint," *SAE-J* 23, no. 4 (October 1928), 397–404, here 398; average man: George J. Mercer, "Body Designing Procedure," *SAE-J* 39, no. 3 (September 1936), 358–361, here 359; Virginia Scharff, *Taking the Wheel: Women and the Coming of the Motor Age* (Albuquerque: University of New Mexico Press, 1992), 124. However, Scharff does not substantiate her claim that this delayed the introduction of the closed car.
47. Stephen L. McIntyre, "'The Repair Man Will Gyp You': Mechanics, Managers, and Customers in the Automobile Repair Industry, 1896–1940" (unpublished dissertation, University of Missouri-Columbia, May 1995), 352.
48. Bruce Seely, "The Automotive Transportation System: Cars and Highways in Twentieth-Century America," in Carroll Pursell (ed.), *A Companion to American Technology* (Malden, Oxford, and Carlton: Blackwell Publishing, 2005), 233–254, here 241.
49. Paul Thomas, "The Secret of Fashion and Art Appeal in the Automobile," *SAE-J* 23, no. 6 (December 1928), 595–601.

50. Mercer, "The Trend of Automobile Body Design," 267; coloring: H. Ledyard Towle, "Projecting the Automobile Into the Future," *SAE-J* 29, no. 1 (July 1931), 33–39, 44, here 38.
51. "Session on Front Drives: Participation of Five Past-Presidents in the Discussion Shows Interest in Reversed Design," *SAE-J* 26, no. 2 (February 1930), 142–144, here 143; 15 percent: "Design Trends Clarified at First National Passenger Car Meeting," *SAE-J* 42, no. 5 (May 1938), 13–18, here 18. "There is no question that they [the low cars] are uncomfortable," opined a participant to the discussion at the SAE annual meeting in 1932: H.M. Crane, "How Versatile Engineering Meets Public Demand," *SAE-J (Transactions)* 32 no. 1 (January 1933) 21–32, here: 31.
52. Edwin L. Allen, "Body Engineering—Past, Present and Conjecture as to Future," *SAE-J (Transactions)* 45, no. 3 (September 1939), 365–371, 384, here 371; controversial: *SAE-J* 38, no. 2 (February 1936), 52.
53. L. W. Child, "Air Conditioning of Automobiles and Buses," *SAE-J (Transactions)* 42, no. 6 (June 1938), 263–268, here 263; window lifts: "Bodies," *SAE-J* 47, no. 5 (November 1940), 460D; wind-tunnel tests: *SAE-J* 39, no. 4 (October 1936), 15; extra room: Roger B. White, *Home on the Road: The Motor Home in America* (Washington, DC, and London: Smithsonian Institution Press, 2000), 35. The architectural historian is Folke T. Kihlstedt and the quote is a paraphrase by White.
54. Henry M. Crane, "The Car of the Future," *SAE-J (Transactions)* 44, no. 4 (April 1939), 141–144, here 141
55. John L. Grigsby, "The Automobile as Technological Reality and Central Symbol in John Dos Passos's *The Big Money*," *Kansas Quarterly* 21, no. 4 (1989), 35–41, here 38 (italics in original).
56. Jean-Louis Loubet, "La naissance du modèle automobile français (1934–1973)," *Culture technique* 25 (1992), 73–82, here 74.
57. Kurt Möser, *Fahren und Fliegen in Frieden und Krieg: Kulturen individueller Mobilitätsmaschinen 1880–1930* (Heidelberg, Ubstadt-Weiher, Neustadt a.d.W., and Basel: Verlag Regionalkultur, 2009), 164 (*Entsportlichung*), 178 (four tasks). '*Entsportlichung*' is clearly an effort to escape the dichotomy between pleasure and utilitarianism, as the loss of 'sportiness' does not necessarily mean a gain in uilitarian applications.
58. Heinrich Hauser, *Friede mit Maschinen* (Leipzig: Philipp Reclam jun., 1928), 40–41.
59. Hans Peter Hahn, *Materielle Kultur: Eine Einführung* (Berlin: Dietrich Reimer Verlag, 2005), 103–104; for an overview of car "practices," including skills, see Peter Adey, *Mobility* (London and New York: Routledge, 2010), 133–174.
60. Kurt Möser, "'Der Kampf des Automobilisten mit der Maschine'- Eine Skizze der Vermittlung der Autotechnik und des Fahrenlernens im 20. Jahrhundert," in Lars Bluma, Karl Pichol, and Wolfhard Weber (eds.), *Technikvermittlung und Technikpopularisierung—Historische und didaktische Perspektiven* (Münster, New York, Munich, and Berlin: Waxmann, 2004), 89–102, here 96.
61. Thomas Alkemeyer, "Mensch-Maschinen mit zwei Rädern: Überlegungen zur riskanten Aussöhnung von Körper, Technik und Umgebung," in Gunter Gebauer et al. (eds.), *Kalkuliertes Risiko: Technik, Spiel und Sport an der Grenze* (Frankfurt and New York: Campus Verlag, 2006), 225–246, here 234, unjustifiably rejects the prosthesis metaphor on the basis of the second meaning (limb replacement).
62. For a detailed analysis of the electronic revolution, see Gijs Mom, *The Evolution of Automotive Technology: A Handbook* (forthcoming).

63. Gijs Mom, "Die Prothetisierung des Autos: Kultur und Technik bei der Bedienung des Automobils" (unpublished presentation, Workshop at the Museum für Technik und Arbeit, Mannheim, 26 and 27 January 1995); Christian Kehrt, "'Das Fliegen ist immer noch ein gefährliches Spiel'—Risiko und Kontrolle der Flugzeugtechnik von 1908 bis 1914," in Gebauer et al. (eds.), *Kalkuliertes Risiko*, 199–224, here 215; Tim Ingold, *The Perception of the Environment: Essays on Livelihood, Dwelling and Skill* (London and New York: Routledge, 2000), 416. For a cognitive-psychological description of the "action-plan" of gear changing, see John A. Groeger, *Understanding Driving: Applying Cognitive Psychology to a Complex Everyday Task* (Hove: Psychology Press, 2001), 29–33.
64. Gijs Mom, "The Dual Nature of Technology: Automotive Ignition Systems and the Evolution of the Car," in Christian Huck and Stefan Bauernschmidt (eds.), *Travelling Goods, Travelling Moods: Varieties of Cultural Appropriation (1850–1950)* (Frankfurt and New York: Campus Verlag, 2012), 189–207.
65. On driver education in Germany, see, for instance, Dietmar Fack, *Automobil, Verkehr und Erziehung: Motorisierung und Sozialisation zwischen Beschleunigung und Anpassung 1885–1945* (Opladen: Leske + Budrich, 2000).
66. The following section is taken from Mom, "Translating Properties into Functions (and Vice Versa)."
67. See, among many others, P. M. Heldt, "Some Recent Work on Unconventional Transmissions," *SAE-J* 17, no. 1 (July 1925), 127–141.
68. F. C. Pearson, "Constant Mesh or Sliding-Gear Transmissions," *SAE-J* 27, no. 2 (August 1930), 161–162; citation: A. W. Frehse, "Some Thoughts on Present Day Automobile Transmission," *SAE-J (Transactions)* 38, no. 1 (January 1936), 13–22, here 13.
69. Thomas L. Fawick, "Two Desirable Qiet Driving-Ranges for Automobiles," *SAE-J*, 21 no. 1 (July 1927) 99–106; C.A. Neracher and Harold Nutt, "A Four-Speed Internal-Underdrive Transmission," *SAE-J*, 21 no. 1 (July 1927) 72–76.
70. Herbert Chase, "Comment on American Passenger-Car Gearsets," *SAE-J* 26, no. 6 (June 1930), 727–735, here 729.
71. "Chrysler Acquired Fluid Flywheel Patents in 1933," *SAE-J* 44, no. 4 (April 1939), 26.
72. S. O. White, "Can a Transmission Have a Brain?" *SAE-J* 32, no. 6 (June 1933), 9–11; Brain Box: "Youngren Explains the Hydra-Matic Transmission," *SAE-J* 46, no. 5 (May 1940), 16.
73. John Sneed, "Mechanical Mind Reading (Transmissions)," *SAE-J (Transactions)* (December 1935), 449–458.
74. The following section has been published previously as Gijs Mom, "Encapsulating Culture: European Car Travel, 1900–1940," *Journal of Tourism History* 3, no. 3 (November 2011), 289–307.
75. Rhys Evans and Alexandra Franklin, "Equine Beats: Unique Rhythms (and Floating Harmony) of Horses and Riders," in Tim Edensor (ed.), *Geographies of Rhythm: Nature, Place, Mobilities and Bodies* (Farnham and Burlington: Ashgate, 2010), 173–185, here 179–180. "Personthing" is borrowed from Nigel Thrift (ibid., 179).
76. Rolf Parr, "Tacho. km/h. Kurve. Körper: Erich Maria Remarques journalistische und kunstliterarische Autofahrten," in Thomas F. Schneider (ed.), *Erich Maria Remarque: Leben, Werk und weltweite Wirkung* (Osnabrück: Universitätsverlag Rasch, 1998), 69–90, here 75 (italics added).
77. Donald Weber, "Automobilisering en de overheid in België vóór 1940: Besluitvormingsprocessen bij de ontwikkeling van een conflictbeheersingssysteem" (dissertation, University of Ghent, 2008), 352.

78. Möser, *Fahren und Fliegen in Frieden und Krieg*, 195–196.
79. "Super-Highway Talk Spurs Discussion on Car Design," *SAE-J* 47, no. 6 (December 1940), 18; 80 mph: *SAE-J (Transactions)* (February 1936), 66; "Motorists Make Pennsylvania Turnpike an Amateur Testing Ground," *National Petroleum News* 32 (November 1940), 27–31; Jepson, "How to drive on a superhighway—and how to drive away from it alive," *Changing Times* 10 (June 1956), 13–15.
80. Weber, "Automobilisering en de overheid in België vóór 1940," 394.
81. James J. Gibson and Laurence E. Crooks, "A Theoretical Field-Analysis of Automobile-Driving," *The American Journal of Psychology* 51, no. 3 (July 1938), 453–471, here 453, 470. On the blurring of the distinction between the human and the nonhuman part of the cyborg (including a taxonomy of cyborg types), see Peter-Paul Verbeek, "Cyborg intentionality: Rethinking the phenomenology of human-technology relations," *Phenomenology and the Cognitive Sciences* 7 (2008), 387–395.
82. Quoted in Dorit Müller, *Gefährliche Fahrten: Das Automobil in Literatur und Film um 1900* (Würzburg: Königshausen & Neumann, 2004), 75.
83. "Road Signs and Sign-Posting," *Motor Travel* 8, no. 8 (January 1917), 32.
84. John T. Faris, *Roaming American Playgrounds* (New York: Farrar & Rinehart, 1934), 297.
85. Craig Horner, "'Modest Motoring' and the Emergence of Automobility in the United Kingdom," *Transfers* 2, no. 3 (Winter 2012), 56–75.
86. Daniel J. Boorstin, *The Image: A Guide to Pseudo-Events in America* (New York and Evanston, IL: Harper & Row, 1961), 77, 116; Barton: T. J. Jackson Lears, "From Salvation to Self-Realization: Advertising and the Therapeutic Roots of Consumer Culture, 1880–1930," in Richard Wightman Fox and T. J. Jackson Lears (eds.), *The Culture of Consumption: Critical Essays in American History, 1880–1980* (New York: Pantheon Books, 1983), 1–38, here 34.
87. Caren Kaplan, *Questions of Travel: Postmodern Discourses of Displacement* (Durham, NC, and London: Duke University Press, 1996), 101; domestication: Marilyn Strathern, "Foreword: The Mirror of Technology," in Roger Silverstone and Eric Hirsch (eds.), *Consuming Technologies: Media and Information in Domestic Spaces* (London and New York: Routledge, 1992), vii–xiii, here ix.
88. *Allgemeine Automobil-Zeitung*, quoted in Stefan Krebs, "Closing the Body: Car Technology, Acoustic and Driving Experience" (paper presented at the T^{2}M Conference, Lucerne, Switzerland, 5–7 November 2009), 5.
89. Fritz Strube, "Autofahren ist ungesund! Die ungesunde Haltung; Ausgleich schaffen; Gymnastik des Autofahrers," *ADAC Motorwelt* 29, no. 4/5 (29 January 1929), 17–18; Alfr. Klaiber, "Arme Kraftfahrer!" *ADAC Motorwelt* 29, no. 27 (1 July 1932), 15; modest: Shelley Baranowski, "Radical Nationalism in an International Context: Strength through Joy and the Paradoxes of Nazi Tourism," in John K. Walton (ed.), *Histories of Tourism: Representation, Identity and Conflict* (Clevedon, Buffalo, NY, and Toronto: Channel View Publications, 2005), 125–143, here 131, 134; Brecht quoted in Michael Nerlich, *Abenteuer oder das verlorene Selbstverständnis der Moderne: Von der Unaufhebbarkeit experimentalen Handelns* (Munich: Gerling Akademie Verlag, 1997), 246–247; Nazi machismo: Gregor M. Rinn, "Das Automobil als nationales Identifikationssymbol: Zur politischen Bedeutungsprägung des Kraftfahrzeugs in Modernitätskonzeptionen des 'Dritten Reichs' und der Bundesrepublik" (unpublished dissertation, Humboldt-Universität, Berlin, 2008), 55.
90. Georg Simmel, "Das Abenteuer," in Simmel, *Philosophische Kultur: Über das Abenteuer; die Geschlechter und die Krise der Moderne; Gesammelte Essays,*

Mit ein Vorwort von Jürgen Habermas (Berlin: Verlag Klaus Wagenbach, 1986), 13–26, here 25 (quote), 23 (process).
91. Daniel Fritsch, *Georg Simmel im Kino: Die Soziologie des frühen Films und das Abenteuer der Moderne* (Bielefeld: transcript verlag, 2009), 131–135 (quote on 135).
92. "'Auslandspropaganda'; Auf Kriegsfuss mit dem 'Berliner Tageblatt,'" *ADAC Motorwelt* 25, no. 33 (17 August 1928), 25–27, here 25.
93. Pearl H. Doremus, "I am Lost in the Mountains: Grandmother Has a Great Adventure," *American Motorist* 22, no. 1 (June 1930), 20; "In the Service of the New Mobility," *American Motorist* 22, no. 7 (July 1930), 1–2; "New Kind of Budget Trip Is Popular," *American Motorist* 3, no. 4 (April 1935), 1.
94. "Englische Zahlen!" *ADAC Motorwelt* 26, no. 6 (8 February 1929), 4; Georg Dittmann and Hans Sehring, "Mit 2 PS über sechs Alpenpässe nach Venedig! (Schluss)," *ADAC Motorwelt* 26, no. 8 (8 February 1929), 14–16, here 16. On Bierbaum, see chapter 2, and on tinkering during the Interbellum, see chapter 7.
95. Harvey Taylor, *A Claim on the Countryside: A History of the British Outdoor Movement* (Edinburgh: Keele University Press, 1997), 235; Gijs Mom and Ruud Filarski, *Van transport naar mobiliteit: De mobiliteitsexplosie (1895–2005)* (Zutphen: Walburg Pers, 2008), 242; "Nos excursions automobiles: Pique-nique surprise du dimanche 1er juillet," *Revue du Touring Club de France* 44, no. 475 (June 1934), 187 (the first 'surprise picnic' was organized in the summer of 1933).
96. Norbert Elias, "Technization and Civilization," *Theory, Culture & Society* 12 (1995), 7–42, here 9 (italics in original), 21.
97. Richard G. Mitchell, Jr., *Mountain Experience: The Psychology and Sociology of Adventure* (Chicago and London: The University of Chicago Press, 1983), 222–223.
98. David Blanke, *Hell on Wheels: The Promise and Peril of America's Car Culture, 1900–1940* (Lawrence: University Press of Kansas, 2007), 190.
99. Gabe Mythen, *Ulrich Beck: A Critical Introduction to the Risk Society* (London and Sterling, VA: Pluto Press, 2004), 145–146.
100. Ulrich Beck, *Risikogesellschaft: Auf dem Weg in eine andere Moderne* (Frankfurt: Suhrkamp Verlag, 1986), 8, 74 (war). I thank the members of the discussion group on the risk society at the Rachel Carson Center in the summer of 2010 for sharing their thought with me on this issue: Lawrence Culver, Heike Egner, Stefania Gallini, Sherry Johnson, Patrick Kupper, Agnes Kneitz, Cheryl Lousley, Uwe Lübken, Diana Mincyte, Alexa von Weik, and Gordon Winder.
101. Jörg Beckmann, *Risky Mobility: The Filtering of Automobility's Unintended Consequences* (Copenhagen: Copenhagen University, Sociological Institute: 2001), 62, 66; Stephen Lyng, "Edgework: A Social Psychological Analysis of Voluntary Risk Taking," *The American Journal of Sociology* 95, no. 4 (January 1990), 851–886, here 853. For "nature" as "other," see, for instance, Theodore R. Schatzki, "Nature and Technology in History," *History and Theory, Theme Issue* 42 (December 2003), 82–93, here 82.
102. Mitchell, Jr., *Mountain Experience*, 156.
103. On the distinction between an individual and a "collective self," see Russell Belk, "Possessions and the Extended Self," *Journal of Consumer Research* 15, no. 2 (September 1988), 139–168, here 145; for an overview of "collective movement" including "affective nebula" (Maffesoli) and the "esprit de corps" of traffic participants, see Adey, *Mobility*, 169–172.
104. Willem Schinkel, *De nieuwe democratie: Naar andere vormen van politiek* (Amsterdam: De Bezige Bij, 2012), 304; see also Schinkel, *Aspects of Violence: A*

Critical Theory (Houndsmill and New York: Palgrave Macmillan, 2010; Cultural Criminology, ed. Mike Presdee).

105. Beckmann, *Risky Mobility*, 76–78 (quotes: 77); Mary Douglas, *Risk and blame: Essays in Cultural Theory* (London and New York: Routledge, 1992); Jens O. Zinn and Peter Taylor-Gooby, "Risk as an Interdisciplinary Research Area," in Peter Taylor-Gooby and Jens O. Zinn (eds.), *Risk in Social Science* (New York: Oxford University Press, 2006), 20–53, here 40–42.
106. Mitchell, Jr., *Mountain Experience*, 153; Gebauer et al. (eds.), *Kalkuliertes Risiko*, 8. Csikszentmihalyi quoted in Stefan Kaufmann, "Technik am Berg: Zur technischen Strukturierung von Risiko- und Naturerlebnis," in ibid., 99–124, here 99.
107. Wolfgang Sachs, *Die Liebe zum Automobil: Ein Rückblick in die Geschichte unserer Wünsche* (Reinbeck bei Hamburg: Rowohlt Verlag, 1984); see also Peter Norton, "Americans' Affair of Hate with the Automobile: What the 'Love Affair' Fiction Concealed," in Mathieu Flonneau (ed.), *Automobile: Les cartes des désamour; Généalogies de l'anti-automobilisme* (Paris: Descartes & Cie, 2009; Collection "Cultures Mobiles," eds. Mathieu Flonneau and Arnaud Passalacqua), 93–104.
108. Carle C. Zimmerman, *Consumption and Standards of Living* (London: Williams and Norgate, 1936), 576–577. Veblen did not refer to the car, but to "conspicuous consumption" resulting in "economic sensation."
109. Kurt Möser, "Zwischen Systemopposition und Systemteilnahme: Sicherheit und Risiko im motorisierten Strassenverkehr 1890–1930," in Niemann and Hermann (eds.), *Geschichte der Strassenverkehrssicherheit*, 159–167, here 162–163; sensation: Kurt Lindner, "Mein Auto – mein Selbst? Was passiert, wenn Männer Auto fahren," in: Matthias Bisinger, Ulla Büntjen, Sigrid Haase, Helga Manthey and Eberhard Schäfer (eds.), *Der ganz normale Mann; Frauen und Männer streiten über ein Phantom* (Reinbek bei Hamburg: Rowohlt Taschenbuch Verlag, 1992) 114–134, here: 121.
110. Ernst Jünger, *Das Abenteuerliche Herz (Sämtliche Werke, Zweite Abteilung: Essays, Band 9: Essays III)* (Stuttgart: Klett-Cotta, 1979). This volume contains both versions. For the relation of Jünger's work to the European avant-garde, see Karl Heinz Bohrer, *Die Ästhetik des Schreckens: Die pessimistische Romantik und Ernst Jüngers Frühwerk* (Frankfurt, Berlin, and Vienna: Ullstein Verlag, 1983).
111. Paul Zweig, *The Adventurer* (London: Basic Books, 1974), 218; Patrick Laviolette, *Extreme Landscapes of Leisure: Not a Hap-Hazardous Sport* (Farnham and Burlington: Ashgate, 2011), 14–15.
112. Roger Horowitz, "Introduction," in Horowitz (ed.), *Boys and Their Toys? Masculinity, Technology, and Class in America* (New York and London: Routledge, 2001), 1–10, here 8; Ben A. Shackleford, "Masculinity, the Auto Racing Fraternity, and the Technological Sublime: The Pit Stop As a Celebration of Social Roles," in Horowitz, *Boys and Their Toys?* 229–250, here 243.
113. David N. Lucsko, *The Business of Speed: The Hot Rod Industry in America, 1915–1990* (Baltimore, MD: The Johns Hopkins University Press, 2008), 16–26, 34–39 (Chevrolet), 40–64 (California); not ready: Möser, *Fahren und Fliegen in Frieden und Krieg*, 308.
114. Blanke, *Hell on Wheels*, 108–109.
115. Karl E. Scheibe, "Self-Narratives and Adventure," in Theodore R. Sarbin (ed.), *Narrative Psychology: The Storied Nature of Human Conduct* (New York, Westport, CT, and London: Praeger, 1986), 129–151, here 136–137.
116. Charles Grivel, "Automobil: Zur Genealogie subjektivischer Maschinen," in Martin Stingelin and Wolfgang Scherer (eds.), *HardWar/SoftWar: Krieg und Medien 1914 bis 1945* (Munich: Wilhelm Fink Verlag, 1991), 171–196, here 196.

117. Deborah Clarke, *Driving Women: Fiction and Automobile Culture in Twentieth-Century America* (Baltimore, MD: The Johns Hopkins University Press, 2007), 46. Historian of technology Judy Wacjman "argues how the simple patterns and routines of men and women are really a world apart," according to Adey, *Mobility*, 110.
118. *American Motorist* 20, no. 6 (June 1928), 46; *American Motorist* 21, no. 2 (February 1929), 27; *Outing* 75 (January 1920), 232; *Outing* 80 (June 1922), 99; Belasco, *Americans on the Road,* 86 (making miles), 87 (lost time), 115 (one half), 132 (night driving).
119. Daniela Zenone, "Das Automobil im italienischen Futurismus und Faschismus: Seine ästhetische und politische Bedeutung" (Berlin: Wissenschaftszentrum Berlin für Sozialforschung, n.d. [2002]), 55.
120. *SAE-J (Transactions)* 31, no. 2 (August 1932), 22; *SAE-J* 27, no. 1 (July 1930), 22.
121. Ilja Ehrenburg, *Das Leben der Autos* (Amsterdam: Paco-Press, 1973; first ed. 1929), 278; translation from Enda Duffy, *The Speed Handbook: Velocity, Pleasure, Modernism* (Durham, NC, and London: Duke University Press, 2009), 48.
122. Quoted in Timothy Davis, "The Rise and Decline of the American Parkway," in Christof Mauch and Thomas Zeller (eds.), *The World Beyond the Windshield: Roads and Landscapes in the United States and Europe* (Athens, OH, and Stuttgart: Ohio University Press / Franz Steiner Verlag, 2008), 35–58, here 50.
123. Wolfgang Hallet and Birgit Neumann, "Raum und Bewegung in der Literatur: Zur Einführung," in Hallet and Neumann (eds.), *Raum und Bewegung in der Literatur: Die Literaturwissenschaften und der Spatial Turn* (Bielefeld: transcript Verlag, 2009), 11–32, here 15, 20; Ingold, *The Perception of the Environment,* 191–193.
124. Anke Gleber, "Die Erfahrung der Moderne in der Stadt: Reiseliteratur der Weimarer Republik," in Brenner (ed.), *Der Reisebericht,* 463–489, here 463.
125. Hallet and Neumann, "Raum und Bewegung in der Literatur," 17; Benjamin quoted in Esther Leslie, "Flâneurs in Paris and Berlin," in Rudy Koshar (ed.), *Histories of Leisure* (Oxford and New York: Berg, 2002), 61–77, here 65.
126. Andrea Wetterauer, *Lust an der Distanz: Die Kunst der Autoreise in der "Frankfurter Zeitung"* (Tübingen: Tübinger Verein für Volkskunde, 2007), 74, 85–92; spectacle: David Howes, "Introduction: Commodities and cultural borders," in Howes (ed.), *Cross-Cultural Consumption: Global Markets, Local Realities* (London and New York: Routledge, 1996), 1–16; Roger Célestin, *From Cannibals to Radicals: Figures and Limits of Exoticism* (Minneapolis and London: University of Minnesota Press, 1996), 2; objectivation and camera: Kenneth Little, "On Safari: The Visual Politics of a Tourist Representation," in Howes (ed.), *The Varieties of Sensory Experience: A Sourcebook in the Anthropology of the Senses* (Toronto, Buffalo, NY, and London: University of Toronto Press, 1991), 148–163, here 155–157; proximity senses: Constance Classen, "Foundations for an Anthropology of the Senses," *International Social Science Journal* 49, no. 153 (1997), 401–412, here 405, 407; Constance Classen, "Engendering Perception: Gender Ideologies and Sensory Hierarchies in Western History," *Body & Society* 3, no. 2 (1997), 1–19, here 4; Ingold, *The Perception of the Environment,* 253–254. This is not to say that sight necessarily inividualizes and abstracts from reality; for a plea to see sight as a social practice, see: ibid., 272.
127. Kai Marcel Sicks, "Gattungstheorie nach dem *spatial turn*: Überlegungen am Fall des Reiseromans," in Hallet and Neumann (eds.), *Raum und Bewegung in der Literatur,* 337–354, here 344–345.
128. Davis, "The Rise and Decline of the American Parkway," 52. Egon Erwin Kisch, *Beehrt sich darzubieten: Paradies Amerika* (Berlin: Eich Reiss Verlag, 1930), 218.

129. Volker Berghahn, *Europa im Zeitalter der Weltkriege: Die Entfesselung und Entgrenzung der Gewalt* (Frankfurt: Fischer Taschenbuch Verlag, 2002), 131.
130. In general, distancing has a long tradition, as Norbert Elias has shown, for instance in the use of the fork to replace direct touching; quoted in Jill Steward and Alexander Cowan, "Introduction," in Alexander Cowan and Jill Steward (eds.), *The City and the Senses: Urban Culture Since 1500* (Aldershot: Ashgate, 2007), 1–22, here 13.
131. Blanke, *Hell on Wheels*, 185, 95 (flowers).
132. Roger Griffin, *Modernism and Fascism: The Sense of a Beginning under Mussolini and Hitler* (Houndsmill and New York: Palgrave, 2007), 158.
133. Quoted in Karen Beckman, *Crash: Cinema and the Politics of Speed and Stasis* (Durham, NC, and London: Duke University Press, 2010), 56.
134. Cotten Seiler, "Anxiety and Automobility: Cold War Individualism and the Interstate Highway System" (unpublished dissertation, University of Kansas, 2002), 48–51 (progressivism), 57–58 (Eurasia), 180 (atomizing); James J. Flink, *The Car Culture* (Cambridge, MA, and London: MIT Press, 1975), 155.
135. Eva Horn and Lucas Marco Gisi (eds.), *Schwärme—Kollektive ohne Zentrum: Eine Wissensgeschichte zwischen Leben und Information* (Bielefeld. transcript Verlag, 2009), 14, 79–80; Hemut Lethen, *Cool Conduct; The Culture of Distance in Weimar Germany* (Berkeley/Los Angeles/London: University of California Press, 2002), 155.
136. Kristen Stromberg Childers, *Fathers, Families, and the State in France 1914–1945* (Ithaca, NY, and London: Cornell University Press, 2003), 71 (bistro), 76 (quotes; italics in original), 111 (fighter); Catholic Church: Susan Pedersen, *Family, Dependence, and the Origins of the Welfare State: Britain and France, 1914–1945* (Cambridge: Cambridge University Press, 2006), 394. For the tensed relationship between Nazism and family (and the role of the father as absolute authority and the mother as child-bearer from whom the right to educate was partly taken), see Klaus Theweleit, *Männerphantasien 1 + 2; Band 1: Frauen, Fluten, Körper, Geschichte; Band 2: Männerkörper—zur Psychoanalyse des weissen Terrors* (Munich and Zurich: Piper, 2009), 249; and Bernd Widdig, *Männerbunde und Massen: Zur Krise männlicher Identität in der Literatur der Moderne* (Opladen: Westdeutscher Verlag, 1992), 29–30.
137. Margaret Marsh, *Suburban Lives* (New Brunswick, NJ, and London: Rutgers University Press, 1990), 71 (domesticity), 76, 173; Marina Moskowitz, *Standard of Living: The Measure of the Middle Class in Modern America* (Baltimore, MD: The Johns Hopkins University Press, 2004), 141; hitchhikers: "The A.A.A. Ban on Aid to 'Guest Riders,'" *American Motorist* 6, no. 3 (June 1931), 21; James North, "Thumbs Down!" (cartoon), *American Motorist* 4, no. 6 (June 1936), 4.
138. Quoted in Michael L. Berger, "The Car's Impact on the American Family," in Wachs and Crawford, *The Car and the City*, 57–74, here 73.
139. George A. Lundberg, Mirra Komarosky, and Mary Alice McInerney, *Leisure: A Suburban Study* (New York: Columbia University Press, 1934), 77–79, 83, 99–101, 178.
140. Corey Todd Lesseig, "Automobility and Social Change: Mississippi, 1909–1939" (dissertation, University of Mississippi, May 1997), 174–179; Ellen K. Rothman, *Hands and Hearts: A History of Courtship in America* (New York: Basic Books, Inc., Publishers, 1984), 290–298 (thrill-seeking: 290); John A. Clausen, *American Lives: Looking Back at the Children of the Great Depression* (Berkeley, Los Angeles, and London: University of California Press, 1993), 83, 170–171. In the

United Kingdom, at the same time, 37 percent of the joyriders who got caught had a skilled working-class or middle-class background. Among 210 youth investigated, there was only one girl. Sean O'Connell, "From Toad of Toad Hall to the 'Death Drivers' of Belfast: An Exploratory History of 'Joyriders,'" *British Journal of Criminology* 46 (2006), 455–469, here 458–459, 461 (explanation), 462 (independence); graffiti: Joseph Anthony Interrante, "A Movable Feast: The Automobile and the Spatial Transformation of American Culture 1890–1940" (PhD thesis, Department of History, Harvard University, Cambridge, MA, June 1983), 244; Marlou Belyea, "The Joy Ride and the Silver Screen: Commercial Leisure, Delinquency and Play Reform in Los Angeles, 1900–1980" (unpublished dissertation, Boston University Graduate School, 1983), 207 (car keys), 233 (long distances), 299 (roam), 302 (courtship), 333 (joyrides by girls), 465 (less violent).
141. Mark S. Foster, "In the Face of 'Jim Crow': Prosperous Blacks and Vacations, Travel and Outdoor Leisure, 1890–1945," *The Journal of Negro History* 84, no. 2 (Spring 1999), 130–149, here 141 (liberating device); Kathleen Franz, "'The Open Road': Automobility and Racial Uplift in the Interwar Years," in Bruce Sinclair (ed.), *Technology and the African-American Experience: Needs and Opportunities for Study* (Cambridge, MA, and London: The MIT Press, 2004), 131–153, here 133–137, 142 (car sharing); Babeza de Baca: Virginia Scharff, *Twenty Thousand Roads: Women, Movement, and the West* (Berkeley and London: University of California Press, 2003), 126–128; Daniel M. Albert, "Primitive Drivers: Racial Science and Citizenship in the Motor Age," *Science as Culture* 10, no. 3 (September 2001), 327–351.
142. Franz, "'The Open Road,'" 145.
143. Alfred Edgar Smith, "Through the Windshield," *Opportunity* 11, no. 5 (1933), 142–144, here 144.
144. Joel Dinerstein, *Swinging the Machine: Modernity, Technology, and African American Culture between the World Wars* (Amherst and Boston: University of Massachusetts Press, 2003), 94 (why), 97 (words, the entire quote is from Dinerstein), 98 (faraway), 104 (nostalgic).
145. Dinerstein, *Swinging the Machine*, 70 (Basie), 72 (beat-driven), 100 (Gershwin), 80–81 (Steam Man), 139 (potential).
146. Janet Wolff, "On the Road Again: Metaphors of Travel in Cultural Criticism," *Cultural Studies* 7 (May 1993), 224–239, here 229; Julie Wosk, *Women and the Machine: Representations from the Spinning Wheel to the Electronic Age* (Baltimore, MD, and London: Johns Hopkins University Press, 2001), chapter 5: "Women and the Automobile."
147. Georgine Clarsen, *Eat My Dust: Early Women Motorists* (Baltimore, MD: Johns Hopkins University Press, 2008); Scharff, *Taking the Wheel;* Maud Lavin, *Push Comes To Shove: New Images of Aggressive Women* (Cambridge, MA, and London: The MIT Press, 2010); Kaj Björkqvist, "Sex Differences in Physical, Verbal, and Indirect Aggression: A Review of Recent Research," *Sex Roles* 30, no. 3/4 (February 1994), 177–188; last quotes: Ann Frodi, Jacqueline Macaulay, and Pauline Ropert Thome, "Are Women Always Less Aggressive Than Men? A Review of the Experimental Literature," *Psychological Bulletin* 84, no. 4 (1977), 634–660, here 655, 645, 649. I thank Georgine Clarsen for co-organizing a workshop on "Gendered Mobilities" at the University of Wollongong on 15 and 16 December 2010 to address this question. I thank the participating scholars for their inspiring papers and invaluable comments.
148. Dag Balkmar, "Men, Cars and dangerous driving—Affordances and driver-car

interaction from a gender perspective" (paper presented at Past Present Future, 14–17 June 2007, Umeå, Sweden), citeseerx.ist.psu.edu/viewdoc/download? doi=10.1.1.124.2591... (last accessed 10 November 2010), 5; Lavin, *Push Comes To Shove*, 14 (inward), 119–120 (narcissism), 145 (gossip).
149. Simmel, "Das Abenteuer," 20.
150. "Aggressiveness (or Aggression or Aggressivity)," in J. Laplanche and J.-B. Pontalis, *The Language of Psycho-Analysis* (London: The Hogart Press and the Institute of Psycho-Analysis, 1973), 17–21; Daniel M. Albert, "Psychotechnology and Insanity at the Wheel," *Journal of the History of the Behavioral Sciences* 35, no. 3 (Summer 1999), 291–305; "Type A and Type B personality theory," http://en.wikipedia.ork/wiki/Type_A_and_Type_B_personality_theory (last accessed 22 January 2011).
151. Randall Collins, *Violence: A Micro-sociological Theory* (Princeton, NJ, and Oxford: Princeton University Press, 2008), 1–2 (situations), 8–10 (two sets, pathways, and weak victim), 24–25 (symbolic violence), 463 (thirty types).
152. Stephen Kern, *A Cultural History of Causality: Science, Murder Novels, and Systems of Thought* (Princeton, NJ, and Oxford: Princeton University Press, 2004), 210.
153. Mary Elizabeth Strunk, *Wanted Women: An American Obsession in the Reign of J. Edgar Hoover* (Lawrence: University Press of Kansas, 2010), 9.
154. James A. Doyle, *The Male Experience* (Madison, WI, and Dubuque, IA: Wm. C. Brown Communications, 1995), 73 (stereotype), chapters 7 to 11 for the elements of the male role. The second element (aggressive) is illustrated by "street and motorcycle gangs, fast cars, and fraternity hazing" (ibid., 159); multiple types: Sasha Disko, "Men, Motorcycles and Modernity: Motorization in the Weimar Republic" (unpublished dissertation, New York University, May 2008), 128; two types of violence: R. W. Connell, *Masculinities* (Cambridge and Oxford: Polity Press, 1995), 83; Dana Crowley Jack, *Behind the Mask: Destruction and Creativity in Women's Aggression* (Cambridge, MA, and London: Harvard University Press, 1999), 22 (styles), 35 (intention); comparisons: Ted Robert Gurr (ed.), *Violence in Amerca*, vol. 2: Protest, Rebellion, Reform (Nembury Park, London, and New Delhi: Sage, 1989), 12.
155. Sherwood Anderson, *Perhaps Women* (Mamaroneck, NY: Paul P. Appel, 1970; first ed. 1931), 41.
156. Anderson, *Perhaps Women*, 41–42, 48 (affected less), 50 (humiliated), 55 (newspapers), 57 (possessions), 72 (wiped out), 83 (worst of all), 97 (potence), 106–107 (climb mountains), 126 (unsayabale), 139 (fake feeling), 142 (human relations), 144 (laughter).
157. John Urry, *The Tourist Gaze: Leisure and Travel in Contemporary Societies* (London, Newbury Park, and New Delhi: Sage), 1990.
158. On subversive types of mobility, see Adey, *Mobility*, 117, 122, 125, 127.
159. Sheila Rowbotham, *Dreamers of a New Day: Women Who Invented the Twentieth Century* (London and New York: Verso, 2010), 55–56; repair own car: Bärbel Kuhn, "Die Familie in Norm, Ideal und Wirklichkeit: Der Wandel von Geschlechterrollen und Geschlechterbeziehungen im Spiegel von Leben, Werk und Rezeption Wilhelm Heinrich Riehls," in Werner Plumpe and Jörg Lesczenski (eds.), *Bürgertum und Bürgerlichkeit zwischen Kaiserreich und Nationalsozialismus* (Mainz: Verlag Philipp von Zabern, 2009), 71–79, here 76.

CHAPTER 6

Redefining Adventure
Domesticated Violence and the Coldness of Distance

"Adventure is the gleam that covers the threat."
Ernst Jünger.[1]

This chapter investigates the motives and motifs behind the new automotive middle-class family adventure as presented in the previous two chapters. This 'tamed adventure' in the form of a 'domesticated hedonism' will be analyzed on the basis of a mix of vernacular travel accounts, belletristic literature, and elements of popular culture such as film and songs. The mix is the result of the insight that in this Interbellum period, which is at the same time an intermediate period between 'elite' and 'folk' automobilism, literature does not express anymore to the full the ever more multifaceted culture of those who decided to buy and use a car, let alone of those who did not use or buy them. Whereas in the previous period the literary and automotive avant-garde more or less coincided, now the avant-garde not only walked ahead (or at least: in a different direction) in its reflexive exploration of automotive culture, it also commented on—instead of talked on behalf of—the mass of motorists who increasingly belonged to the middle and lower strata of the middle classes. On top of that, the petty bourgeois preference for the family does not fit comfortably in the literati's inclination to depict the car trip as a highly individualized affair, as a Quest of the Self, as literary analysts have come to call it. Instead, the analysis of the surprisingly rich and variegated and yet so univocal mix of utterances enables us to be more specific about the new type of aggression and violence deployed by the motorized family (the 'sublime of violence,' as we called it in the previous chapter), and allows us to distinguish between several adventures, both male and female, elite and petty bourgeois, white and nonwhite, American and European. We will do so by shifting between bread-and-butter travelogues and 'high' literature, mainstream and avant-garde utterances, movies and songs. Although the borders between vernacular and literary texts are not clear-cut, the analysis of texts of increasing perceived or intended artistry enables a more precise investigation of the motives of the middle-class automotive world in an effort to explain its persistence rather than its emergence, which has been attempted for the previous period in chapter 2.

In this chapter we will first approach the shifts within the automotive adventure through an analysis of more vernacular travel accounts, and then proceed to more literary and other artistic (such as filmed) evocations of the new automobile adventure, not limited to car travel, but related to the possession (or lack) of the car in general. While in the previous chapter we demarcated the realm of the new, 'domestic adventure' as the result of a shift from a dominance of the temporal and functional (speed and tinkering) to the spatial aspect of automotive adventure produced within the framework of the petty bourgeois nuclear family, this chapter will delve much deeper and more extensively into the (psychological) motives and related (literary) motifs in an effort to specify this adventure in the eventual forms ready to conquer the rest of the planet, within the setting of a fully developed 'automobile system.' To this end, we will follow a loosely chronological order but emphasize thematic clustering of the empirical material, such as the lost generation's conquest of the 'periphery,' the family as collective subject in an urban setting, the question of the existence of a female adventure, the dominating spatiality of the new interwar automotive adventure, the urban culture of the car, the cult of 'cool' spun around the car, especially in Germany, and we close this chapter, just as we did in chapter 2, with an analysis of the car symbols and affinities in both the highbrow utterances and those of the popular culture. We will then wrap up our argumentation and form a bridge to the final empirical chapter 7.

An Avant-Garde in Autopoetic Travel Experience: The Conquest of the 'Periphery'

In terms of autopoetic adventure, the immediate postwar years offered an abundance of evidence that the war had functioned as a catalyst in a fundamentally continuous evolution. That was first of all visible in a veritable explosion of travel accounts during the interwar years. Where the rail tracks stood for predictability and loss of vision, first the dirt road for carriages and then the paved highways for cars promised an "adventurous 'left and right.'"[2] For the French novelist Blaise Cendrars travel as action was a continuation of waging war (he lost an arm during that war), enabling both a longing for renewal and an escape from the "now."[3] The continuity between war experiences and postwar travel was also established through the delight of recounting the suffering after the fact, just like experienced mishaps heightened the storytelling to friends after the holiday: "Yes, it was hell, but it made men of us, and you who didn't serve should be filled with awe and envy," said Eddie Rickenbacker, who was a pilot during the war, and after the war wondered how he could enjoy life "without this highly spiced sauce of danger?" Continuity was also expressed in the juxtaposition of war, road accidents, and mountain climbing: "Most men like adventures. Anyone who has ever been through a street accident, anyone who has climbed a mountain, knows that. It is one of the strange attributes of the mind that we enjoy what makes our flesh creep."[4] One is tempted to call this masculine fascination by death, danger, and violence (so intense that it equals street accidents with mountain climbing) a universal European phenomenon: when the Portuguese writer Fernando Pessoa created his alter ego Álvaro de Campos, engineer and member of the European avant-garde, he had him drive his Chevrolet "almost smoothly," but he could not resist his "terrible desire" to accelerate "suddenly, violently and difficult to understand." For him, driving is dreaming, while the car, "obedient to my unconscious movements at the steering wheel, jumps under me, with me."[5] Continuity (in this case with prewar adventurousness) went even so far as outright pastiche, when Belgian writer Didier de Roulx (pseudonym of André Janssens) published, in 1923, *La 629-E9,* a polemical "souvenir" of Octave Mirbeau's *La 628-E8.*[6]

As a genre the travelogue was "open, critical, non-dogmatic, anti-authoritarian, radically practical, and not geared towards the confirmation of the well-known, but toward the new." And most of all, for a generation that could not stop being shocked by the war, it was "a source of irrational happiness, the celebration of the simple joy to be alive." Although the travel account as a narrative of liberation was "passé," at least for the heterosexual and masculine mainstream, it could function as a means for the "reinvention of adventure" conceived as a "risky amplification of the self (*riskante Selbstverstärkung*)."[7] Indeed, the car fitted seamlessly in this program, even more so: it helped to (re)shape it.

In literary (and general) history, the 'lost generation,' an epithet coined by Gertrude Stein, among others,[8] has been identified by historians as a preoccupation of the male elite whose members had been killed proportionally more than other classes, and hence indulged in self-pity. In the war "an older Europe, an agrarian Europe, with its warrior class, committed suicide."[9] For mobility history, this concept also overemphasized the role of the car out of a similar mechanism: just like the prewar avant-garde of Futurists and Expressionists they were at the forefront of the motorizing middle classes, and as such certainly not representative of the average citizen, and they enjoyed it.[10]

And yet the sheer size of the group of Anglophone writers that directly got involved in the war as "ambulance or camion drivers" (Malcolm Cowley, one of them, mentions twelve in his contemporary analysis of the mood of this generation, published in 1934) and the example they formed for fellow writers suggest that we may have discovered the Interbellum equivalent of the Belgian/French group of writer-motorists from the previous period. These writers, Cowley asserts, "taught us courage, extravagance, fatalism, these being the virtues of men at war; they taught us to regard as vices the civilian virtues of thrift, caution and sobriety; they made us fear boredom more than death." They taught us a "spectatorial attitude," he says; "we ourselves were watchers."[11]

The image of the traveler, of wanderers and vagabonds adventuring without tour guides, was ubiquitous among intellectuals of the "war generation," making "a religion out of speed and movement. They loved the feeling of domination and power that came with driving a car."[12] For Aldous Huxley, who published a collection of car adventures in 1925 as "a Tourist," "the temptation of talking about cars, when one has a car, is quite irresistible."[13] Even more so, the visit to the colonies, by train and boat, complemented locally by car and truck trips, was characterized by a "sense of dominance and superiority of Europeans and North Americans" that was crucially supported by technology in general, and automotive technology in particular. The same was true for the European and North American peripheries as a host of travelogues to the south on both continents testifies.

It was the war, French writer Paul Morand argued in his *Conseils pour voyager sans argent* (1930; Advice for Traveling Without Money), which had made traveling into a massive pastime, as in 1914 "entire peoples departed without a ticket."[14] The presence of the war is equally undeniable for the British travelogue. J. B. Priestley, in his famous *English Journey* into the Depression-stricken countryside of mining and textile industries, observed that he would have met many more "younger folk" along his journey "if the finest members of my generation had not been slaughtered in the war." His journey by motor coach and train was the trip of a member of "the new urban mob" (Priestley started his career as a clerk in the wool trade), protesting against the "uglification" by the industrial revolution in the countryside and against "cheap mass production and standardized living." Priestley, in this "later age of bread and circuses," defended a "liberal democracy" as an alternative to the Soviet Union, and his praise of "little England" was aimed at defending modernity against a greedy monopoly capitalism-in-the-making. The modern England he saw emerging was the land "of arterial and by-pass roads, of filling stations and factories that looked like exhibition buildings, of giant cinemas and dance halls and cafés, bungalows with tiny garages, cocktail bars, Woolworths, motor coaches, wireless, hiking, factory girls looking like actresses." In short, he saw the car society emerging, in which the middle class dominated, including the "commercial travellers of the more morose kind." Nevertheless, he preferred "a dirty diseased eccentric to a clean healthy but rather dull citizen," and the motor bus to the passenger car. While the former "offer(ed) luxury to all but the most poverty-stricken," the latter had effectively "cut oneself off from the same reality of the region one passes through, perhaps from any sane reality at all." At home he possessed a Daimler, driven by a chauffeur "because I am a very bad driver."[15]

Priestley was certainly not the only Briton who discovered, in the wake of the war, the domestic roundtrip ("travel writing about *home*") as a lucrative genre welcomed by many fellow countrymen. In 1927 the journalist Henry Vollam Morton became famous with his *In Search of England,* a trip undertaken in a "little blue motor car" (a Morris), celebrating British beauties with the war still in the back of the mind, in a mildly critical attitude toward mass tourism. Written in an "easy reading" style, Morton "seemed to speak with the very voice of middle England itself, as it spoke in the 1920s before disillusionment crept in." *In Search of England* was the start of an impressive array of Morton's travelogues, first on British regions, then on the Middle East, undertaken by a typical representative of the new motorist: the technical layman. When he is stopped by "a girl" who asks for "half a pint of petrol," he is much relieved: "I had seen myself plunged in the entrails of that small stubborn car, pretending, for such, under a woman's eyes, is the terrible vanity of man, to be a mechanic, blundering round its interiors with a careless and ignorant spanner; and probably cutting myself, losing my temper, swearing, and doing something quite serious and expensive to that small stubborn machine." Not only is the functional adventure reduced to nil, the car itself disappears from view during the remainder of the trip, thus continuing a tradition which we already observed in the travelogues of Edith Wharton (see chapter 2).[16] Morton's "motoring pastorals" were a literary as well as motoring construction: he "'deliberately shirked realities' by making 'wide and inconvenient circles to avoid modern towns and cities,'" even if he, now and then, did make some concessions in describing the 'middle landscape,' such as the coal fields of Wales.[17]

George Orwell, however, directly inspired by Priestley's account and, like him, traveling mostly by train, went on a quest for a socialist England. He opined that the motor car, like "the machine" in general, could best be accepted "rather as one accepts a drug—that is, grudgingly and suspiciously. Like a drug, the machine is useful, dangerous and habit-forming." In the future, Orwell predicted, even the Priestleys among the British would drive, as "in twenty years' time [the car] may need no nerve or skill at all" and "the transition from horses to cars has been an increase in human softness." Orwell saw this softness emerge in the new suburbs, with its "Tories, yes-men and bum suckers" who would be willing to die on the battlefield against Communism.[18]

Whether on the Left or on the Right, the autotechnical layman crossed his home country revealing his amazement before its uglification and massification, including the emergence of suburbs and their new middle-class mediocrity. The poet John Betjeman, for instance, would publish a collection of poems in 1937 in which the clerks in the suburbs were castigated as follows: "And talk of sport and makes of cars / In various bogus-Tudor bars / And daren't look up and see the stars / But belch instead."[19]

Several members of the 'lost generation' ventured farther away, where they tried to integrate the meeting with the Other in their Quest of the Self, witnessing the spread of modernity toward the periphery. D. H. Lawrence, for instance, traveled with his wife through Sardinia mostly by motor bus. Fascinated by the violent character of the automobile (for instance when a hare at night races away from the road "with its ears back"), he saw in Italian car drivers the ultimate cyborgs: "It all seems so easy, as if the man were part of the car. There is none of that beastly grinding, uneasy feeling one has in the north. A car behaves like a smooth, live thing, sensibly."[20] Lawrence was still caught in the fascination for the skilled daredevil, while the automotive world around him was already moving in another direction. The most extreme example of the avant-gardist for whom "the whole world becomes a mirror of his mind" was Raymond Roussel, born in the same posh neighborhood in Paris as Marcel Proust and like him looking for "ways of insulating [himself]," the latter "in his cork-lined bedroom," the former "when travelling by boat, in his cabin." During his trips around the world he maintained a strict separation between the practice of traveling and the practice of writing, insisting "that none of his experiences of these countries had any influence on his writings." "In me, imagination is everything," he confessed in his *Comment j'ai écrit certains de mes livres* (1935; How I wrote some of my books). In Europe, he traveled in his "*automobile roulotte.*" Exhibited at the 1925 auto show in Paris, this thirty-foot caravan consisted of "a sitting room, a bedroom, a study, a bathroom and a small dormitory for the three domestics (two chauffeurs and a valet)" and was equipped with "electric heating, a paraffin stove and a paraffin hot water system"—and a radio. Both Mussolini and the Pope admired his "revolutionary vehicle." Thus having built his perfect automotive capsule, "the curtains closed, immersed in his daily ration of Loti or Verne, indifferent to the landscapes through which he was passing," one wonders why he traveled at all. Through experimentation with all kinds of affinities (such as that between writing and chess), and addiction to barbiturates, the enigma becomes only bigger, until one realizes that it was the movement itself, and its aimlessness, which he needed to set the creative process in motion. When Roussel could not afford his roulotte anymore, he asked his driver to take him "at random through the streets of Palermo," in the evening swallowing "nine kinds of barbiturate," the same procedure he would soon use to commit suicide. And although his fellow writers (such as the American John Ashbery, but also, after the war, Michel Foucault) admired him because he taught them that one "did not have to write out of [one's] 'experience,'" we know better: the experience was only shifted, from the visual to the haptic and, perhaps (the radio!), the acoustic.[21] In Roussel, the loneliness within the mobile swarm became apparent.

There were, however, many other modalities among travel writers when it comes to their relationship to 'experience.' Priestley, Morton, Lawrence, and Orwell are only the most famous of a much larger group. Literary historian Paul Fussell even speaks of a "British Literary Diaspora" characterized by "masculine discourses, adventure and exploration . . . instrumental in the construction of rationales for imperialism," but also (quoting Freud) used "to escape the family and especially the father."[22] Writers from other countries joined them: Ernest Hemingway, Paul Morand, T. E. Lawrence, and Louis-Ferdinand Céline all made well-read literary works conceived as travelogues. Several of them had been involved in the "private ambulance brigades" of the war, such as Ernest Hemingway, E. E. Cummings, and John Dos Passos, the latter basing a large part of his literary career on this experience.[23] These highbrow authors form the tip of a gigantic iceberg of middlebrow and lowbrow publications (such as the Biggles series by "Captain W. E. Johns" or thrillers and adventure novels) that help relativize the "myth" of the senseless war atrocities by adding alternative narratives on "the war as adventure and as masculine self-test." They were often written by "the middle classes and addressed a middle-class audience," with "balance" as their trademark.[24] This is a first clear sign of an elite that no longer represented the experiences of the masses.

Klaus and Erika Mann clearly belonged to the highbrow part of this generation. They published a travelogue (*Rundherum*, 1930) on their trip around the world undertaken in 1927 and 1928. Only twenty-one and twenty-two years old, respectively, they used ships on the oceans, trains to cross the continents, cars in the cities and for excursions, a multimodality one can observe in most transcontinental and global travelogues. Fantasizing about future world travel, they see rocket planes that enable one to dine in Paris and have tea in Tokyo a couple of hours later. Although this will no doubt result in a loss of "truly wondrous and mysterious adventures," they wonder "which new, unimaginable adventures these generations will experience."[25]

In a little-known travelogue not meant for publication, American women Rose Wilder Lane and Helen Dore Boylston reported on their trip from Paris to Albania in 1926 in a Ford Model T bought (and produced) in France. Lane had written biographies of Henry Ford and Herbert Hoover, divorced in 1918 and began a freelance career of traveling in Europe and the Middle East, doing publicity work for the Red Cross.[26] While traveling by train in Europe she met Boylston, who had been a nurse in a field hospital during the war and would later write the Sue Barton nurses' book series for girls, and both appeared to be driven by "a kind of enthusiasm for unconventional experience that in those times precluded them from careers as wives and mothers," as the editor of their journal formulates. The car, as a persona nicknamed Zenobia (perhaps a reference to the "androgynous potential" of the female protagonist with the same name in Nathaniel Hawthorne's 1852 novel *The Blithedale Romance*?), enables this. Lane had made big trips earlier, such as through the Syrian desert, and she had once, back in the United States, taught her father how to drive.[27]

The trip to Albania (accompanied by a servant who shares the interior of their car but not their hotel rooms), where Lane intends to build a house with money earned on the stock market, is a personal colonization project. Written years after the events, the journal is quite critical of Europe, which they (and the journal's editor as well) see as "hopelessly corrupt at heart—in its government most obviously, but ultimately in a citizenry yet to discover radical human freedom."[28] The Ford, as symbol of that freedom, thus enabled a trip to a double periphery: Europe as peripheral to the United States, and the south and Albania as peripheral within Europe. Apart from the tinkering part (undertaken by "Trouble," Boylston's nickname) their adventure seems to be geared toward the symbolic 'conquering' of southern Europe, and the adventure's female flavor stems from the fact that men are seen as obstructions to that goal, to be overcome by irony and cunning mixed with some condescendence, or by making fun of their lack of mechanical skills cloaked in macho helpfulness.

Other women found their adventure in a purely Whartonesque indulgence in the past, such as in the case of the British "Mrs. Rodolph

Stawell," who in her account of some tours through Yorkshire immediately shifts into a poetic mood as soon as the car starts moving,[29] or the American Katharine Hooker during her 1927 trip, *Through the Heel of Italy*, in which the car (and the chauffeur) is nearly hidden from sight and the road takes over as an actor. Seventy-eight years old when she undertook this trip with her daughter-photographer, Hooker's adventure is devoid of any violence, except some mild racism and distancing from the "cheerful but unwashed poor." But the account is about adventure nonetheless, such as what she calls "book adventures," the collecting and purchasing of books. This is not to say that Hooker doesn't have "the spirit of the discoverer," as her biographer remarks, nor that she did not experience any corporeal adventures; on the contrary: she went through two shipwrecks in her childhood, and was seen by her biographer as "always adventurous and athletic" with a fascination for "unconventional, strange, even outré, bizarre characters."[30] But her territorial adventure was devoid of aggression: "We soon discovered we had fallen upon a market day for our excursion, ... They and their various animals encumbered the road. Encumbered did I say? I ask pardon; we, with our interloping automobiles were the cumberers. To whom do the country roads belong, if not to the *contadini* whose houses and fields border them?"[31] This form of empathic spatial adventure was very rare indeed among writing travelers and traveling writers, and reminds us of Vernon Lee's utterances in a previous phase.

Like Erika and Klaus Mann, the German leftist professional travel writer Arthur Holitscher also practiced worldwide multimodality: in *Der Narrenführer* (1925; The Fools Guide, with four editions in two years), one of his more than thirty travel books, he uses taxis, trains (to reach Paris), tourist and city buses, and his own feet, while he flies from Paris to London. Remarkably, he finds the flow of urban cars extremely slow; during his stay in Paris, he describes a half-hour trip between the Madeleine church and the end of Boulevard des Italiens that would have cost him three minutes on foot. "The large town is one single traffic barrier." While in former times in the Bois de Boulogne people greeted each other, "now the cinematographically blurred silhouet of sedans shoot[ed] past, eye forward, gaze at the steering wheel and at the back of the car rolling in front."[32]

Whereas Paul Morand may have been the French champion of travel writing, Holitscher had to share a similar role in Germany with giants such as the communist and *neusachlich* "racing reporter" (*rasender Reporter*) Egon Erwin Kisch (who, like the Austrian Robert Musil, had worked at the army press quarters during the war), and the less well-known Alfons Paquet. Paquet (who wrote the screenplay of Ruttmann's movie *Melodie der Welt* (1929) and in 1935 criss-crossed Europe in a plane) was a clear proponent of the train, which made him define the concept of "adventure" in a narrow sense as spatial adventure, that is, as a "break from everyday life and a set off into new cultures and unknown regions." In his multimodal *Weltreise eines Deutschen* (World Journey by a German, 1934), Paquet uses the Siberian Express to reach Manchuria and a steamboat to reach Japan, and locally uses the motor bus (equipped with a "long-stroke engine and good springs"), the rickshaw, a carriage, a streetcar, a sled, an airplane, and, in Europe, an occasional car from a friend.[33] Likewise, another German travelogue writer, the actor and theatre director Wolfgang Hoffmann-Harnisch, endulged in multimodality, taking the plane from Berlin to Frankfurt, and then the zeppelin Hindenburg to Rio de Janeiro, enjoying the "sensorial confusion" (*sinnverwirrend*) emanating from the combination of "the frog and bird perspective."[34] Egon Kisch also traveled mostly by train in "Paradise America" (1928–1929), a "free America where everyone may starve according to his own manner."[35]

Holitscher, as one of the few, described what happened inside the mobile capsule, where "every passenger in the speeding car transforms automatically into the driver. The landscape disappears, is not there as soon as a chicken crosses the road. Every curve in the road is a problem, more important, life-determining than the sudden vision flash of a fairy town on a steep rock." Holitscher's trips are not a "quest for the self" but a "quest for community (*Gemeinschaft*)," one long-failed search for "belonging" to a societal middle position he himself characterizes as "neither boss, nor servant."[36] Holitscher's *sachlich* account documents the change of perception we have described earlier in chapter 2, with an immediate effect on the narrative style itself: he is only capable of quoting half a sentence from a monument they speed by. "A town pops up: what's her name? We're in the open field again, between grain, wine, olives."[37] The emphasis on congested city traffic can serve as a metaphor here: for Holitscher, traveling not only meant writing, but also a political act of personal liberation, against a background of collective behavior, of the flow of 'others.'

In the spectrum of travel types, 'colonial' trips further away were often more open in showing a dose of aggression and a feeling of superiority toward the 'periphery' and the Other. Although such long-range journeys were mostly focused on speed, their spatial or territorial character dominated. Such stories produced adventures in which the violence was moderated by women, as Georgine Clarsen explains, stressing "the diversity of women's colonial travel writing" and their "desire for pleasure and new experiences." But in her analysis of several accounts of long-range car trips through Africa, Clarsen also gives examples of women mimicking male aggression, such as in the case of British Diana Strickland who on her trip from Dakar to Maswa from West to East in 1927–1928 destroys a native hut when its inhabitants refuse to give her water. On such long trips the functional adventure (the challenge of technical reliability) becomes primordial, indeed, although in the case of Stella Court Treatt's 1927 trip from *Cape to Cairo* hundreds of Africans are recruited to make the expedition possible.[38]

On the other hand, eccentric and well-traveled (including a voyage to the West Indies) Lady Dorothy Mills is traveling alone (well, with a 'boy') through the French and Portuguese colonies of West Africa, reaching her destination by steamboat and train, then locally traveling by the occasional Ford of a fellow tourist, hitchhiking in trucks between cities, or in some cases even "walking" through the jungle, as she calls it, carried in a hammock. All of the classic adventures are experienced, including the usual tongue-in-cheek vandalism such as the killing of chickens, the taking of hairpins at 80 km/h "scattering broadcast men, women, and children, cattle, goats, chickens, and other lorries," the (nearly ecstatic, and at the same time racist) devotion to the "barbaric beauty" of the "exotic other," and probably many more 'adventures', as her reference to an attempted assault by her 'boy' in a hotel suggests. Her defense of Portuguese colonialism as a civilizing mission is outspoken, as is her ambiguous eulogy of the truck as a better vehicle than the car in these remote circumstances: "One of these days I shall write a book on *camions*; I seem to know so much about them, from their infancy to their grave, which is generally nearly my grave also. I have driven them and slept in them; I have pushed them; I have sworn at them; I have known them ill and well, going and not going, unable to go and refusing to go; I have taken them to pieces and put them together again. One nearly ran over my precipice; another blew up with me inside it; many have fallen to pieces beneath me. In fact, the genus *camion* and I have lived much life together, and never have I learnt to like one."[39]

For Germany, Karolina Fell has analyzed a surprisingly large set of travelogues written by women. She concludes that a separate genre of female travelogue cannot be distinguished, in sharp contrast with other analysts of female travel accounts, who conclude that "women's accounts follow a different 'descriptive agenda' to that of men," sometimes specific to an "interactive" rather than an "objectivist" rhetoric, resulting in "'centripetal' accounts of their journeys, which moved outwards from a personal, domestic space and back again."[40] Although these assessments are generally a lot more sophisticated than the stereotypes about "the reckless male and the impulsive female," the hope that the "Woman

Driver myth" would be substituted by the "Androgynous Driver" (as an amalgam of the Woman Driver and the Macho Motorist, "as an 'intermediate sex' that existed between and thus outside of the biological and social order") did not materialize during the interwar years: "By the 1930s, women and men alike had disowned the New Woman's brave vision."[41] Nor did Willa Cather's hope materialize, who explained her (what some feminist critics have called misogynist) "project to take over a male tradition of writing" as follows: "When a woman writes a story of adventure, a stout sea tale, a manly battle yarn, anything without wine, women and love, then I will begin to hope for something great from them, not before."[42]

From a mobility perspective male colonial travelogues, indeed, do not seem to differ substantially from female accounts as long as one limits one's judgments to the general level of moving around. Graham Greene's *Journey Without Maps* (1936) through Liberia, one of his four travel books, starts, like Dorothy Mills's journey, in a train, followed by taking a ride, with his cousin Barbara, in a truck: "The old engine boiled, and the metal of the footboard burnt through my shoes; the driver was bare-footed. We drove wildly up- and down-hill for an hour on a road like a farm track, but the impression of reckless speed was deceptive, formed by the bumps, the reeling landscape, the smell of petrol and the heat; the lorry couldn't have gone more than twenty miles an hour."[43] But by far most of his two hundred–mile voyage was made on foot. Greene represents the younger generation who did not experience the war, a "generation brought up on adventure stories," as he himself remembered, "who had missed the enormous disillusionment of the First War; so we went looking for adventure."[44]

Evelyn Waugh, satirist of British aristocracy, also belonged to that generation and he too related the "rough journeying of all these men" to the missed war. Between 1930 and 1935 Waugh, already established as a writer and "a large car at my disposal because I am *nouveau riche*," discovered the travel account. His trip to the coronation of Haile Selassi in Abyssinia, financed by the *Times*, not only depicted the usual colonial distancing toward the local people, it also, for the first time as far as it is known, testifies of the normalcy of the car in this type of travel account, even to Africa, as the narrator falls asleep every now and then during the car trip.[45] Like Greene, Waugh took part in what James Wolf has called "imperial integration," undertaken by middle-class "tourists and casual travelers" of "the second and third generations of British 'adventurers,'" only a few of whom "found imperial rule unfair." Their visits were clearly gendered as they contrasted the "effeminacy" of the farmers with the "masculinity" of the hillmen (as Kipling did in India), the community with individualism, the vegetarian with the carnivorous, the desert with the town, and the nomad with the hunter.[46]

Even André Gide, critical toward the car as we saw in chapter 2, is not convinced he is "perhaps a bit old to throw myself into the bush and into the adventure" during his *Voyage au Congo* (1927). In his *Retour de Tchad* (published a year later) he also admits that "a trip too well prepared" would not allow enough room for adventure, although it is not clear that he considered the sixty porters and the thirty-five cases sent ahead in two vans as overpreparation. From a mobility point of view Gide's trip is a classic: the physical movement is accompanied by a lot of spiritual journeying (such as reading Joseph Conrad), and the travelogue here and there slips into a sociological treatise. At the same time, "one does not travel to Congo for one's pleasure" and even if one does, the use of a car would stand in the way because of its speed "which does not allow me the leisure to examine [the environment] better." The "hurried" stages done by car are "a kind of parenthesis, [as it] could only give us a provisional glimpse" of the landscape: the car trip is used to get through monotonous and ugly parts and therefore often "seems to be longer." The "too hasty flight confuses the senses, renders all gray." To really enjoy the environment one has to go on foot, along paths that have been cleared prior to the arrival of his party. During these foot trips Gide finds out about the atrocities against the indigenous people, on which he will report in his *Voyage,* causing a stir back home. The only existing road Gide used was expressly made for one car trip a month by the French rubber company. A gang of indigenous women had worked an entire night to get the road ready for Gide's party. A director of the company held responsible for the atrocities by Gide, answered in a long letter to a French newspaper suggesting that Gide's accusations were the result of an "adventurous imagination."[47] Adventure, in Gide's context, had not much to do with the thrill of speed or the pleasure of spatial enjoyment (let alone the tinkering on the cars): it was unpleasant, and in the eyes of the company director even subversive, dangerous in a negative way, if not communist.

Quite the opposite was Gustav Stratil-Sauer's highly adventurous motorcycle trip from Leipzig to Kabul, which had him land in an Afghan prison cell accused of murder.[48] With a PhD in geography after an aborted pursuit of music studies he describes his exploration of the East as a piece of music: sections of the trip are named after music parts (such as Furioso), while his engine sings in "discants" (a treble voice). Tellingly, his imprisonment was the result of a race (as he had done several times before) with a native whose horse bolted when hearing Stratil's car. Such travelogues impressed upon their readers that adventure was impossible without the motorized vehicle, despite (or as we meanwhile know, also because of) the many mishaps, accidents, and defects. To master the latter, Stratil had himself educated as a motorcycle mechanic, able to take the engine and the rest of the vehicle fully apart. A recent study by Sasha Disko places this and similar German colonial travelogues in a context of Edward Said's "Orientalism," in which "the mobility afforded by a motor vehicle was central to claiming modernity that the other was purported to lack."[49]

Most American would-be travelers did not need colonies to indulge in similar adventures: they had their own frontier.[50] Winifred Dixon, for instance, in 1921 embarked with her friend Toby on a westward trip in an open Cadillac, narrated in a literary style full of frontier symbolism and bearing all of the ingredients of the automotive adventure: the speeding and hurrying ("each [sheep] herd we met meant a wasted half hour"), the joys of the driving skills as a form of art, but most of all the functional adventure the two women managed to handle with cunning and adroitness. "I had taken no course in mechanics, and had, and still have, a way of confusing the differential with the transmission. But I love to tinker!" And thus the female functional adventure started:

> Our jack was the kind whose advertisements show an immaculate young lady in white daintily propelling a handle at arm's length, while the car rises easily in the air. Admitting she has the patience of Job, the strength of Samson, and the ingenuity of the devil, I should like to meet her just long enough to ask her if she stood off at arm's length while she put the jack in place, rescued it as it toppled over, searched vainly for a solid spot in which the jack would not sink, pulled it out of the mud again, pushed the car off as it rolled back on her, hunted for stones to prop it up, and a place in the axles where the arm would fit, and then had the latch give way and be obliged to do it all over again. And, with no reflections on the veracity of the lady or her inspired advertiser, I should demand the address of her pastor and her laundress.

At last Dixon stops a passing Mexican "and ask(s) him to lend us his brute strength, which he smilingly (does), pleased as a child at being initiated into the sacred mysteries of motoring." Just like the self-qualification as *Westward Hoboes*, the tinkering is often narrated in a tone of self-mockery, as in the case of the "important looking pipe beneath the car broken in two," which appears to be the exhaust pipe, or the moment when their brakes failed, and "neither Toby nor I knew how to tighten brakes." When they succeed, she confesses: "Never had we known a prouder moment. The incident gave us courage to meet new contingencies." It also strengthens them in their independence: "As for doing without a man, we found Providence always sent what we needed, in any crisis we could not meet ourselves. In Tucson we found two old friends, Miss

Susan and Miss Martha, who shared our brash confidence in ourselves enough to consent to go with us as far as Phoenix."[51]

For Dixon there is no doubt about the existence of female adventure. She distinguishes between several types, including "misadventures" (such as "driving fourteen miles with the emergency brake on") and "little adventures," such as the gendered fear of being followed at a distance by a "half-breed Mephisto."[52] The journalist Kathryn Hulme, however, who made a cross-country trip with a female friend in 1928, could only indulge in "the pleasure of self-sufficiency" after driving their defective car far off the road, which meant they could not be bothered by "helping" men, who knew not much more about car technology than they did.[53] In some cases of women's travel such a Mephisto is quite near, for instance in Marieluise Fleisser's *Andorranische Abenteuer*, a trip of three months undertaken with a dominating, if not abusive male friend, the Swede Hellmut Draws-Tychsens. Fleisser's adventures are mostly of a tamed, lower-middle-class bohemian character, driven by a constant lack of money. In other words, the spectrum of the female adventure is as broad and variegated as men's, including a new, pecuniary adventure of the pseudo-nomad: the uncertainty of having enough "capital" (including mobility capital) to reach one's goal. The relationship between traveling and writing is made very direct: Fleisser partly financed their trip from publications on her adventures in the *Vossische Zeitung*.[54] Such female adventurers varied from Bertha Eckstein-Diener, a proponent of matriarchy and recently celebrated as one of the German "women warriors," to the communist Maria Leitner who, despite the title of her travelogue (*Eine Frau reist durch die Welt;* 1932, A Woman Travels through the World), only has an eye for the labor-related adventures at her multiple destinations and does not waste much space on the question how she got there in the first place.[55] Other examples of this expanding group of female adventurers were Sophia Skorphil, who married a motorcycle repair shop owner in Vienna, made a weeklong, 2000-km trip to the Adriatic coast in 1930 with her sister, enjoying an independent "marriage vacation."[56] In Germany Hanni Köhler (who undertook a 20,000-km nine-month motorcycle trip to India in 1931), Susanne Körner, Ilse Lundberg, and Trüdell Hoffmann became well-known adventurers who celebrated their 'independence' through the creation of a cyborg-style unity with their engines.[57]

Domesticating Adventure: The Family as Collective Subject, and Speed

American literature, as "a literature of movement, of motion," has been analyzed as having a "journey structure," the Western being the iconic example. The linear itinerary structures the narrative, in which space and time become interchangeable, and the "outward voyage" also represents the "inward voyage." "The American poet, indeed American poetry itself," a recent overview claims, "has been on the move." Distinguishing among several types of journeys (the heroic quest, the wandering journey, the home-seeking journey), literary historian Janis Stout acknowledges that all literatures use such structures, but the American literature knows them as the dominant form.[58] Many literary analysts have used this state of affairs as a basis for their thesis that travel literature is a Quest of the Self (which makes their products into "identity papers" [*Papiers d'identité*] as globetrotter Paul Morand called one of his travelogues). Although travel literature, and its belletristic versions in particular, privilege the individual experience, the fusion of the body and the machine in many of the travelogues can not only be called a "collective symbol" transcending the individual. In exceptional cases travelogues also present the family as a "collective subject" (as a sociological group between the single individual and the collective, in Simmel's definition) such as in John Steinbeck's "family journey novel" *The Grapes of Wrath*.[59] Depicting autopoetics as a celebration of individualism would be as much off the mark as mistaking this image of the car-trip-as-self-quest for car culture in general.

Steinbeck's fictional Joad family was certainly not of the average kind, but there were thousands of them. *The Grapes of Wrath*, his controversial and subversive novel on the "Dust Bowl Migration," sold 400,000 copies during its first year. It was bought by five million since its publication in 1939, brought him the Nobel Prize after the war, and deals with the road, not the car, as central symbol. Perhaps one can even claim that it is the flow, the collectivity of cars and trucks of migrants that forms the central actor and (literal) driver of the narrative.

Steinbeck was not alone in emphasizing the collective character of migration. In Marie De L. Welch's poem "The Nomad Harvesters" migrants are depicted as "a great band, they move in thousands; / Move and pause and move on." They are described as a swarm of grasshoppers: "Ours is a land of nomad harvesters, / Men of no root, no ground, no house, no rest; / They follow the ripening, gather the ripeness, / Rest never, ripen never, / Move and pause and move on."[60]

In earlier journalistic texts Steinbeck had made a distinction between "migrants by nature" and "gypsies by force of circumstance," placing the Joads in the same category as the autocamp tourists. For him, the fifteen government camps (in 1940) enabled a transcendence of the family: "Use'ta be the fambly was fust," Ma Joad concluded at the end of the novel. "It ain't so now. It's anybody."[61]

The Joad family of sharecroppers, "tractored out" from their land, form a protected, cocooned monad in their battered Hudson vehicle (half car, half truck) as part of a giant swarm of thousands of migrants to the West along highway 66: "The cars of the migrant people crawled out of the side roads onto the great cross-country highway." They come "from the wagon tracks and the rutted country roads. 66 is the mother road, the road of flight." Only when on the move (their houses shattered or left behind) do they first become a "unity" as a family, then the group expands with newcomers who make the family concept obsolete, and then these expanded families combine into groups of twenty 'families' that occupy squatter camps—camps, incidentally, that would later be used for the internment of Japanese, Italians, and Germans during the war. This is not a typical travelogue: a more-than-family described as a collective subject ("and the family called: 'Good-by, Muley'" and "the twenty families became one family, the children were the children of all") ventures on a scary one-way, noncircular trip that can hardly be called an "automotive adventure" in the multifaceted sense, as we have been using this term so far. The purchase of the Hudson Super-Six itself is done out of pure necessity: "What you want is transportation, ain't it?" the salesman asks, and a trade against their wagon and mules fails: "Didn't nobody tell you this is the machine age?" This promises to become an adventure on the very edge of existence, perhaps because the trip is not rooted in sedentarism, but is an expression of real nomadism.

Indeed, here we are at the extreme end of the spectrum of automotive adventures, a world where the term itself threatens to lose its meaning as the types of mobility start resembling those by non-Western, circular, and noncircular migration, a world where the middle class does not dominate and where the rickshaw puller's experiences in Beijing or the impoverished Chinese farmer bringing his produce to the market by illegally using the train must be described and analyzed in different metaphors.[62]

Likewise, in Steinbeck's novel the elements of the adventure start to shift. The novel does not contain much sightseeing, and the functional 'adventure' is stretched beyond the limit. Al Joad is listening with his entire body to what may go wrong: "The brakes squealed when they stopped, and the sound printed in Al's head—no lining left.... Al was one with his engine, every nerve listening for weaknesses, for the thumps or squeals, hums and chattering that indicate a change that may cause a breakdown. He had become the soul of the car." With no house to protect them, the truck is now the central place. Technical reliability becomes an issue of life and death: "Two hundred and fifty thousand people over the road. Fifty thousand old cars—wounded, steaming. Wrecks along the road, abandoned." This reminds us of the description of the *Voie Sacrée* in Verdun (see chapter 3). The nightmarish character of the functional adventure becomes clear when Al and a newcomer, former convict Casy who learned to maintain cars in prison, decide to change a connecting rod in the engine of a couple that joined them on their pilgrimage. In addition to what literary historiography reveals about how well Steinbeck prepared himself in documenting the situation of the camps, he must have done at least as thorough a study of car technology (not mentioned in literary historiography), considering the use of technical jargon and the description of the trick to reinsert the piston into the cylinder without using a ring clamp.[63]

Steinbeck's novel does offer a glimpse on the middle-class automotive adventure, though, but now seen from the point of view of those who are normally the observed objects. The highway appears to carry two distinct flows: the Joads of America (the "Others," but now on the move themselves) and the "folks (in) them big new cars" who are used to coming and gazing upon the Others. The former are engaged in physical mobility with only one purpose: social mobility. "They [the white poor] ain't a hell of a lot better than gorillas." Such others have different priorities: before they will buy a new car in California, young Connie and Rose of Sharon fantasize that they will first get a house. Steinbeck's "proletarian novel" subverts the middle-class notion of mobility as freeing, but at the same time the novel is an eloquent reminder of the power of mobility's promise, its expectation, its illusion, of a new American frontier.[64]

This mobility is subversive, because it intends to help another world emerge: "At first the families were timid in the building and tumbling worlds, but gradually the technique of building became their technique [in the squatter camps]. Then leaders emerged, then laws were made, then codes came into being."[65]

In a similar vein Boxcar Bertha also provides a narrative that tends to explode the family structure.[66] Called "Boxie," because her mother found her often in box cars when she was looking for her daughter, she tells a revealing story about women hoboes who (earlier than men) started to use automobiles rather than train wagons to enable their nomadic existence: they'll "take longer, but it's cleaner and safer." Most women hitchhiked, but some possessed "dilapidated Fords" or they traveled in "rattling side-cars of men hoboes' motorcycles. A few of them even had bicycles." In one of their 'jungles' (Camp Busted near Jacksonville) she counted "thirty men, fifteen women, and eighteen children, all white. At the edge of the camp were four or five dilapidated cars and one truck. There were also motorcycles and a couple of bicycles. . . . The talk was largely about the road. The car owners had long tales. Gas was high, the cars they had were pieces of junk, needing everything they couldn't afford. 'When the charities know you've got a car they won't help you,' one of the men wailed. 'They always ask, "Why don't you sell it? You'd get fifty or twenty-five, or at least ten dollars for it." Well, I've got four kids and all the junk we own in the world is in that car. I couldn't get more'n ten dollars for it ... and it's our home. You know they wouldn't ask a man to sell his home for ten dollars.'"

Like the Joad family moving on the very edge of automotive 'adventure,' Boxie wanted "freedom and adventure" when she, as a child, saw older women take the road: "the rich can become globe-trotters, but those who have no money become hoboes." One of her female friends confirmed that hitchhiking "most of the time ... was simple and good fun, though often she found men difficult. More than once she had been dumped out of a car in the country because she wouldn't give a man what he wanted." All parts of automotive adventure, including the erotic, are turned upside down in this form of subversive mobility. Another woman hobo later told her: "When it comes to hitch-hiking, there's nothing to it. Anybody with a skirt on can hitch-hike. I go down between New York and Chicago just like a business man. It never takes me more than three days and I always end up with more money than when I started. I go down to Palm Beach and Muscle Shoals in automobiles just as if I owned a fleet of them." Lesbian hoboes traveled in small groups and "had little difficulty in getting rides or obtaining food. The majority of automobilists ... senses that they were queer and made little effort to become familiar.... Many of the lesbians in Chicago hunted in packs and traveled in automobiles." When she decides to hike east, she "found that a great army of women had taken the road, young women mostly, gay, gallant, sure that their sex would win them a way about, far too discontented to settle down in any one place." Bertha opted for the roughed version of hoboism: for a while she was part of a gang of thieves ("who had their own cars"), and she prostituted herself to keep alive, in an anti-bourgeois atmosphere: "'What have you been reading lately ... *Good Housekeeping?*' he asked scornfully." Hilariously enough, and just like in Steinbeck's novel in "them big cars," the paths of the hoboes and the "rich" sometimes cross, as when one of her hitchhikes ends in a tourist camp.[67]

From other sources we know that more than half of the "vagabonds" in a Los Angeles mission at the end of the 1920s hitchhiked. Hoboes, used to traveling by train, complained that the car ruined their culture because of the emergence of "gasoline tramps" and "automobile floaters," who substituted individualism for the collectivity of the hobo "jungles." Already in 1914 a disgruntled Frederick Mills complained that it was "impossible to hobo in peace thru this country. Willy-nilly one is picked up and carried along by kind-souled auto-owners. The country just swarms with machines." After the war, hoboes were among the "auto campers" who played a major role in the Californian harvests, at their peak in the 1920s forming a ten thousand–strong membership of the Industrial Workers of the World (IWW) opposed violently by the police and the Ku Klux Klan. The 'democratic,' dilapidated car seems to have substituted a hierarchy of train-travel cultures: riders on the passenger trains were at the top, the "speed-obsessed youths who rode express train" next, and the majority riding the freights, gondolas, or boxcars on the bottom.[68]

In New York Bertha meets Andrew Nelson who gives her a job helping him get his "federal survey of transients" done. From this research she cites that around 1934 (when she was around thirty years old) between 300,000 and 400,000 'transients' were roaming the country, 8000 of them (2 percent) women. Most were looking for work, but 4 percent were motivated by a "desire for adventure," while an equal share were considered "habitual hoboes." Another survey gave a much higher share of women, and elsewhere in her autobiography she gives a figure of 1.5–2 million hoboes; the majority (83 percent) of which was white. In the end, after she had become involved in relief work for her fellow hoboes, she finds their situation hopeless now that the car is displacing the train as preferred tramp vehicle: "Outside of feeding and clothing them, nothing can be done worthwhile without changing our economic system. And now the great dear public wants the capitalistic system."[69]

Although this conclusion may well have flowed from the pen of the Chicago anarchist Dr. Ben Reitman (whose biographer suspects him to be the inventor of Bertha's story, "an absorbing if implausible tale," a story that was filmed in the early 1960s by Martin Scorsese), it was shared by the "traveling journalist" James Rorty who, as an American Orwell, traveled around the country during a seven-month, fifteen thousand–mile trip in a car at a speed of about 100 km/h, toward California, *Where Life is Better,* at about the same time that Boxcar Bertha told her story. There he was put in jail for a short while, while his companion, "a near-sighted Russian Jew," tried to appease the "mob of Mack Sennett cops and stools." He is one of the car drivers who picks up hitchhikers, "although I was repeatedly warned against (it)." His left-wing anti–New Deal pamphlet written for the *New York Post* starts with a new frontier perspective reminiscent of the swarm idea: "Whereas in the earlier time the prevailing motion was a slow, irresistible drift from east to west, now the movement is rapid, accelerative, and circular, almost centrifugal."

They move like "bewildered ants." There is even a suggestion of continuity with war: "Is it possible that before long the hawks may see that milling movement of men in automobiles... straighten out into marching lines, military convoys, a new surge of conquest destined this time to hurdle the ocean barrier and hurl itself upon the Asiatic mainland?" Like Boxcar Bertha's account, Rorty's report contains a diatribe against the lower middle classes, but also the transients are beyond hope as "the flop-house contingent ... are almost impossible to organize."[70] "Grown with the railroads" (in the nineteenth century hoboes seem to have been the dominant casualty of the railroads, the number of deaths and injured running as high as five thousand per year), it was the car (as a vehicle but also as a symbol of nationwide mechanization of labor) that heralded the end of traditional hoboism. The "flivver bums" were the harbingers of the Okies and Arkies of a decade later.[71]

Such travel stories tended to question the family as traveling subject. But they were exceptions: in most interwar travelogues, the family, if present at all, functions as a narcissistic subject, replacing the narcissistic individual (or couple) of the previous phase, and was just as "outer-directed" as that individual.[72] In the United Kingdom in 1932 W. R. Calvert published his *Family Holiday: A Little Tour in a Secondhand Car* with all the tongue-in-cheek, witty platitudes of the nuclear family's closed culture, complete with nicknames for family members and father's naughtiness about women. Situated at the opposite extreme from the Toads and the Boxcar Berthas on the spectrum of automotive adventure, this travelogue was published in a phase identified by a student of Scottish tourism as "the end of family holidays."[73] During the holiday the husband is "much more faithful and devoted to [the Old Mole] than to the various beautiful young ladies," the Old Mole being the secondhand open-bodied Humber in which the group travels from London to Wales and back, in sixteen days, following a route guide provided by the autoclub.[74]

Highbrow literature could not refrain from satirizing this kind of domesticated adventure. In the United States playwright Thornton Wilder, master of Brechtian alienation, dedicated an entire one-act play to a family outing in which the adventure is reduced to the "interior monologue" in the automotive cocoon, after the risk of a technical breakdown has been more or less eliminated by a careful checkup at the garage before the trip. "The best little Chevrolet in the world" is represented, on stage, by a platform on which four chairs are placed, for the parents and the two kids, and departure is accompanied by a chorus of hellos from every neighbor within sight. "The air's getting better already. Take deep breaths, children," the domineering mother orders, and "they all inhale noisily." They have to slow down for a funeral car column and mother reminds her children that "we'll all hold up traffic for a few minutes some day." Interrupting the traffic flow is only permitted for a brief moment in the middle-class universe, and only if no other option is available. The terror of the family culture during this three-day "happy journey" to visit a daughter in nearby Camden, NJ (during which "no scenery is required" on stage, as Wilder instructs the play's director), appears at the halfway mark in their journey, when the crying children apologize to the mother for being rude to her, a scene which is soon followed by singing in the car accompanied by the following conversation, started by the son:

> ARTHUR: Ma, what a lotta people there are in the world, ma. There must be thousands and thousands in the United States. Ma, how many are there?
>
> MA: I don't know. Ask your father
>
> ARTHUR: Pa, how many are there?
>
> ELMER: There are a hundred and twenty-six million, Kate.
>
> MA [*giving a pressure about ARTHUR's shoulder*]: And they all like to drive out in the evening with their children beside'm.

Shortly before the curtain falls, while the group is singing another song, Ma opines: "There's nothing like being liked by your family," the tautology of this phrase mirroring the claustrophobia of the group, despite their trip 'into the open': the capsule here serves as a prison.[75]

From this perspective it is no wonder the literati had trouble combining the family ideal with the mobile adventure, however redefined. Many of them simply wrote as if the car was still used by and exclusively tailored to the lonely white male. Literary historians in their footsteps tend to see the family as a "collective subject," "alone together."[76] The most extreme case in the denial of the collective character of the car trip is presented by German travel writer Colin Ross, who took his family on many of his trips, but described these trips as a fully individual affair.[77] Even William Carlos Williams's account of a trip to Europe (*A Voyage to Pagany*) in 1924 does not mention his wife, who accompanied him during the entire journey. Instead, he split his wife up, so to speak, into several women (including an incestuous relation with a sister; even Paris is depicted as a woman to be conquered), thereby reshaping this trip into a true Quest of the Self. It enables Williams to have his protagonist, Dev Evans, spend a large part of his trip conquering a "virgin continent" at the same time as he tries to conquer a number of women. Evans mostly travels by train, but the local taxi trips sometimes set in motion a form of surrealistic, automatic, and associative writing.[78]

In the end, whether individual at the personal level or at the level of the small, situational party or family, in the minds of both the practitioners themselves and the onlookers, the car stood for subjective freedom. As such it differed substantially from the joys of collective freedom and the social "fellowship" (if not socialism) of the bicycle and walking 'movement,' for instance in Britain. There, motives such as "stoical self-sufficiency," which found satisfaction in "adverse natural conditions," were still celebrated as much as the delights of "improvised roadside repairs."[79] And although there, too, the 'pleasurable hardships' tended to become domesticated, this domestication was especially observable in the culture of the car.

Domesticating Car and Motorcycle, and the Cyborg Motif

The domestication of the car went along a multitude of paths, but the taming of the automotive adventure took on a special flavor when it was undertaken by the family in the 'affordable family car,' even if travelogues and highbrow autopoetics did not favor this type of collective subject. Domestication was portrayed as an individual project and it used all the well-known elements from an earlier period (see chapter 2): the flight metaphor, the violence (including racist characterizations of nonusers as bush niggers and heathens[80]), the imperialist instrumentalization of the car, the car as a moving camera and, in general, as a machine for "sensory travel," and the eulogy of repair as a therapeutic practice. It was as if a new layer of automotive novices, belonging to the lower middle classes, mimicked the elitist project of the prewar years, but then in a manner adjusted to the new circumstances. Indeed, the autopoetic literature of this period can also be characterized by its ambivalent depiction of the car trip: euphoria (happy transport) on the one hand, ride to hell (accidents, suicide in the car, war and mass slaughter) on the other. Often, the accounts seemed nearly a literal reproduction of prewar narratives, with only slight variations. Perhaps Interbellum travel was a bit more focused on the "real, living, everyday," whereas the prewar bourgeois cultural overload of an Edith Wharton or a Marcel Proust became increasingly rare, although we should not forget that Proust's major work was published after the war.[81] What, then, were the salient differences (if any) from the previous period?

According Siegfried Kracauer, a German analyst of the middle classes, traveling resembled dancing, because the destination had lost its importance, and the aim of the traveler was to be in "simply a new place." Indeed, since the 1880s professional dancers had begun a process of self-reflection on the relationship between motion and modernity such that female dancers, for instance, "did not call upon sexuality for effective self-display. Rather they enhanced their popularity by being shapes in motion."[82] But while Kracauer connected dancing to a temporal adventure, he saw traveling as being "reduced to a pure experience of space."[83] Indeed, this was not the beat-driven dancing inspired by the train (as described in the previous chapter), but a swarm-like rhythmic movement enforced by the curves in the roads and the inverted, wave-like passing of the landscape, rather than the sounds of the machine. In Kracauer's words, "Dancing gives man, violated by intellect, the possibility to grasp the *eternal*." Man "steps out of time, out of the profane business (*Getriebe*) into another, rhythm as such."[84] Recent research into collective dancing ("clubbing") emphasizes the "oceanic experience" generated by both drug use and crowd-based movement: "through dancing and the embodiment of ... music, through self-mastery and the use of body techniques in the expression of a dancer's understanding, the dancer is able to transcend or escape the self and strive for a realm beyond the confines of the body."[85] The expanding automotive swarm offered similar experiences to the motorist.

Kracauer was a regular contributor to the 'feuilleton' section of the *Frankfurter Zeitung* (where he also published his study on German white-collar workers, *Die Angestellten*), a body of texts that has been studied in detail recently by Andrea Wetterauer. "Feuilletonism," the writing and reading of newspaper contributions on a selection of topics, including travel, was a massive phenomenon indeed: the *Kölnische Zeitung* alone between 1933 and 1945 counted thirty thousand contributions written by four thousand authors, of which the section *Die Reise* (until 1939) formed an important part.[86] According to Wetterauer's analysis of the *Frankfurter Zeitung*, travel accounts between 1923 and 1929 (711 in number) formed half of all feuilleton contributions, written in a "flaneurish writing style," but only 5 percent of them mentioned cars. In these journalistic-essayistic pieces, the car is portrayed as a new "experience tool" (*Wahrnehmungsgerät*) of an insecure middle class that used the travelogue as an exercise in renewed self-definition and appreciation of the rapidly changing environment. The "automotive world of the senses" rather than the landscape was the object of perception, emphasizing the cocooning effect of car travel in this phase. Traveling was conceived as narrating, both of these practices "self-reflexive projects" according to Anthony Giddens. In both, adventure functioned as a way to investigate and assess identity. This self-definition was largely masculine (the only female driver identified traveled in a group), and closely connected to risk taking and domination. "If you cannot dominate a woman," a father in New Objectivist Marieluise Fleisser's play "Pioniere in Ingolstadt" (1929) reproaches his son, "how can you master a car?"[87] The 'mastering' was apparently necessary in a postwar culture in which women—in industrializing countries, at least—started to join men as motorists in increasing numbers, according to several contemporary observers, which decisively influenced car purchases and car design alike. They were attracted to the traveling adventure no less than men, as we saw in chapter 4.

'Mastering,' in the sense of 'domesticating,' was especially welcome in Germany where the travel account functioned as a means to regain a world lost during and through World War I and to create "a newly stabilized (male) subjectivity." By the end of the 1920s, 'sachlichkeit' was considered to be insufficient to convey such a personal quest, and the "impressionistic," "poetisizing" style became the norm, satirized in Kurt Tuckolsky's *Pyrenäenbuch* in which he dealt "more with my own world than with the Pyrenees." Domestication, then, was a common project of writers and motorists alike, and could thus also be called feuilletonization (*Feuilletonisierung*). In other words, the domestication of the car trip coincided with the domestication of the travel account.[88]

The role of the car (or the motorcycle) is remarkable here: mediating between the two states distinguished by Friedrich Nietzsche (the domestic, sedentary way of living and the adventure, see chapter 2), the machine was now put not so much in the garden but in front of the home, as an ostentatious token of suppressing the most dangerous adventures and enabling the ones that suited the lower-middle-class family.[89] This is as far as the "taming of technology" went: Werner Sombart published a book with that title (*Die Zähmung der Technik*, 1935) in which he pleaded for the prohibition of cars and motorcycles in protected natural landscapes, but it was ignored by the Nazis. According to the "conservative revolution" of Ernst Jünger and his peers, technology should be "bravely endured."[90] Domestication is a very concrete practice, and requires skills, most of them noncognitive.[91]

The prototype of at least the European version of a domesticated adventurer was Heinrich Hauser, born in 1901 and thus belonging to the "young generation" whose adolescence took place during the war. "Restless," Hauser tried to relive his missed war experience in the right-wing *Freikorps* immediately after the war, and when that did not work he made a trip around the world as a sailor, got domesticated himself in 1925 when he married, but broke out again through a multitude of shorter and longer journeys, in the process becoming very popular as a travel writer.[92] His fascination for machines and technology in general not only made him write a photo documentary about the Ruhr area,[93] he also dedicated several publications to the Opel factories in

Rüsselsheim and he used Opels for his trips through the Balkans, and to take part in rallies or perform test drives. We have met him already, in chapter 4, as the owner of a self-built trailer in which he traveled through Nazi Germany with his wife and daughter. Hauser's construction (and selection) of his personal adventure can be reconstructed quite precisely in his prolific writings. Many of the family experiences, for instance, "are in fact not at all worth telling," but tinkering certainly was, revealed by a remarkable slip of the pen in his trailer book: there he wanted to recount "one of the most beautiful and work-rich weeks of my life. The first 400 km of our trip brought a plethora (*Unmenge*) of experiences, which are worth to be remembered." This promising sentence is immediately followed by: "It appeared that the brake lever in the trailer that actuates its brakes requires an extraordinarily large force. The muscles of my left arm still hurt."[94] Other experiences worth telling are the observation of large flows of cyclists, motorcyclists, and cars as soon as they approach a big town; another is the encampment of the *Hitlerjugend* in the woods, with their "hardness against themselves and (their) stoicism."[95]

Hauser can be considered as the personification of a new form of adventure, characterized by "non-committence and inability of any (social) connection." His adventure was aimed against the family (to which he returned, of course, time and again, in his three marriages) and perhaps even against the woman. Rejecting the big city and its "inhuman traffic," he was continuously looking for a regressive form of harmony, which he found, for instance, in the village-like male community of the ship crew.[96] On his voyage to Chile in a sailing ship he not only showed his fascination for large technical objects (recounting transcendent experiences vis-à-vis sea storms, large sails, high speed, and flying), he also celebrated the "order" of "our village" of independent men (in fact, most sailors were recruited from the German countryside) who did not allow any "cosiness" (*Gemütlichkeit*) to creep in. Hauser's adventure is a constant search for order that he only can find in the premodern village of a (totally masculine) sailing ship crew. In one of his many daydreams on board he confessed not to like chickens "because they are disagreeably feminine." Technology (and, paradoxically, especially 'old' and large technology such as sailing ships, but also car factories) could transport him into a "rush" (*Rausch*) and thus made him into one of the German representatives of a form of technological sublime, in which Leo Marx's 'middle landscape' played a crucial role. Like Sherwood Anderson (who wondered how he could say the 'unsayable'; see chapter 5) he used other media such as film and photographs whenever the "beauty" and the "unbelievable" largesse of technology or the (urbanized) landscape showed itself as "undescribable."[97] Landscapes, he declared in his Ruhr-area documentary, "should be studied through a microscope." The "old landscape" is "unreal, as if one enters a painting of another century." Farms have water piping, electricity; farm hands wear the same clothes as industrial workers. Prefiguring Leo Marx, Hauser can find in the Ruhr area "hardly any landscape... which is not cut through by a white track" of the railroad. Now, car swarms and their roads ("ant roads," *Ameisenstrassen*) have been added to these landscapes.[98]

While practicing the new, domesticated adventure, Hauser invented a new version of flight. In the liberal *Frankfurter Zeitung* (of which he was briefly editor, until he felt imprisoned by this fixity) he coined the term *Autowandern* (car tramping) in a description of "car driving in the Fall": the "real adventure" was not "the race against time and against other subalterns of the machine," but consisted in a new slowness enabled by improved roads. We know that this new slowness was fast. On the German *Autobahnen* "the man behind the steering wheel slides into a tranquility, a recuperation (*Erholung*) that he did not know up to now: the car seems to roll without any driving assistance, the grip on the steering wheel relaxes" and the driver experiences "an incredibly blessed feeling of safety, a floating without gravity, fully like flying." The car trip becomes a "happy journey guided by the landscape" and the motorist experiences a "resonance of the tramp (*Wanderer*) not only with the landscape, but also with its people." In other words, Hauser celebrated the experience of being an ideal swarm member, enjoying a nearly railway-like traveling, guided by the road rather than by the driver.[99] In this particular case Hauser's flight also took on a different connotation: as flight from fixity represented by the woman, if not flight from the woman as such. The car becomes the woman's more reliable substitute, a "prude lover" (*keusche Geliebte*), the true protagonist of his travelogues, for instance in his Balkan journey in an open, expensive Opel (borrowed from the factory). In this journey, 10,000 km long, undertaken during ten weeks in 1937 through seven countries, Hauser declares his "love of the car," which "corresponds mostly with an explicit rejection of the railroad." The car begets agency (one analyst speaks of "zoomorphing" as opposed to anthropomorphing) so much so that by the end of the trip, it is the car (driven on a short stretch of road without Hauser's hands on the steering wheel) that indicates the direction to follow: back home.[100] Hauser is the prototypical observer, writing in a *neusachlich* style, leaving out all psychological interpretation of people's behavior, registering like a camera, categorizing the 'periphery' (of Europe in the Balkan, of the West in Chile) along a measure of 'modernity,' often calibrated by the size of the 'car system': "Austria (is) in the range of motorisation somewhere between an African colony and a South American small nation." His observations, like his filmmaking, are steered by the rhythms of the landscape or the cityscape. It is an "image-rich gaze" (*bildhaftes Staunen*), the glance of the hurried tourist.[101] Hauser's freedom of mobility is a regulated freedom: "traffic regulation is modern, it is western, it is a symbol of the metropolis." In the East, however, "everyone drives as he likes," resulting in chaos. It is this paradox, Hauser's search for order while away from home, an order he hopes to find in the premodern male community, which characterizes his "travel technique," which he rejects to call adventurous: "My journey is simple like a pre-war trip. It would be ridiculous to call it in any way adventurous—I could have put my tent as easily in the middle of Germany." What Hauser desires is a home away from home, a mobile home, heavily gendered.[102]

There is no doubt that the prosthetic properties of the car formed an important part of the automotive adventure. Perhaps as a reaction to this process, the motorcycle culture in several countries seemed to continue the more risky aspects of prewar automotive adventure, including its territorial aggression. "The road belongs to us," a German motorcyclist opined in 1928 (echoing the arrogance of prewar motorists), and not to the "vermin" of the dogs and other animals.[103] Indeed, the relationship between bodily skills and the motorcycle's stability resembled early aviation practices at a moment in time when car driving seemed to become more temperate. But one had always to be ready for short circuits between body and machine, as a hilarious text from Kurt Schwitters testified, in which the 'I' races along on a motorcycle, unable to stop the machine until it runs out of petrol, a metaphor for his country.[104]

In Walter Julius Bloem's novel *Motorherz* (1927; Motor Heart) falls are skillfully avoided by subtle counter-steering through the body and the cyborg motif, unity with the machine, is ubiquitous: the "iron comrade" is the main character's "best friend"; he becomes a "human automaton" while the motorcycle has a "rare soul, difficult to understand."[105] This was apparently so important that the German handbook series *Autotechnische Bibliothek* (containing dozens of volumes full of practical advice) dedicated a separate volume to motorcycle accidents, in which it was claimed, just like in the case of later extreme sports, that many accidents could be prevented through the proper handling of the machine. "The number of real misfits (*Stümper*) is remarkably small. For, many unfit climb in the driver's seat [of an automobile], but the motorcycle is only mounted by those who feel up to it."[106] In F. Stoll's 1940 book *Jungens, Männer und Motoren* (Boys, Men and Motors) a father refuses that his son become a member of the *Motor-Hitlerjugend*, but he changes his mind when he realizes that even agricultural youths should acquire the skill of driving for their future role as a soldier. Nonetheless, the British C. K. Shepherd shortly after the war traveled across the United States on a motorcycle, and fell 150 times, not counting his many breakdowns.[107]

Bloem, son of a well-known family of writers (using Kilian Koll as a pseudonym), had been an officer in the war, fought in a *Freikorps* against the revolution, and greeted Hitler's coming to power with great enthusiasm; he studied philosophy and wrote an influential book on film theory. Later during the Second World War, he would become a pilot. Although not strictly a travelogue, the scarce sections in *Motorherz* that have a touring flavor (for instance when the two protagonists flee to Italy, in the train, their bikes nearby) only heighten the militaristic and aggressive high-speed character of the fictional Berlin motorcycle culture. This culture is collective, and governed by a strict discipline in which a constant discourse is going on about the relationship between men and women (the latter forming about one-tenth of the membership), and in particular the fitness of women for this form of sport. The motor club's emblem is a skull, but its members are middle-class citizens (old and new *Mittelstand*, including some older officers supervising the motoring youth) with "cold and muscle-hard face[s]" and longing for heroic lives, their engines "cracking satans" when they take part in their collective outings that exhibit the behavior of a swarm. Like in Steinbeck's *Grapes of Wrath*, motorized travel is primarily a group practice in which the male adventure is embedded. This adventure is a constant "struggle between his heart and his head—and all great victors are chain-destructors (*Kettensprenger*)." Thora Moebius, a divorced but beautiful 'new woman,' with her "magnificent contempt for danger," irritates the men with her "great self-confidence" and (consequently, one is inclined to conclude) will die in a violent accident in the end, but until then she functions as a controversial part of a racing culture presented as a Darwinian struggle, where the weak (the spoilt women, the automobilist in their closed cocoons) have no place. The "soldiers of speed" celebrate their sport as a religion (literally, in the Cathedral of Cologne!) and motor sport is presented as an *ersatz* army where every member has a ceremonial but meritocratic function, their clothing functioning as cocoons, their races described in a filmic style. Within this culture speed is very carefully defined: club members are "no kings of speed— no record chasers, but gray-clad, impersonal, serving soldiers." When a woman spectator remarks that she does not see sport, which should be "a drive into the open" instead of "hellish obsessivity," an older club member confesses that "something secret makes him feel jubilant—as jubilant as during the war." Motorcycle sport should be fast, not speedy, the difference being between control and cool rationality on the one hand, and rush and emotion on the other, cold versus warm, manly versus feminine, hard versus soft. This racing culture does not have time for the landscape, which is "without exciting beauty, only the road before the eyes lures and lures."[108]

Domestication provoked parts of the mobility culture to escape into motorcycle adventure, a practice providing the basis for an adventurous counter-narrative. Samuel Beckett, in Ireland in 1925, dreamt about participating in a motorcycle race, when his professor, in a discussion of Beckett's essay on Valery Larbaud, promised him he could become the best in French language if he only would stop using his motorcycle. Beckett, who loved going to the pub and the movie theatre as much as tinkering on his bike, answered that driving "sometimes took him far away, not simply some miles, far away, until he came to himself—but sometimes that could also be accomplished by a verse or an intelligent comment."[109] Indeed, motorists who wrote highbrow literature had two paths to escape: through racing or writing and reading.

In Julius Donny's *Garage 13* this culture is depicted as a world of swindle, with members interested in adventures of an anti-bourgeois, unconventional flavor, combined with lots of alcohol and meetings in a sleazy bar. In the end, the two protagonist-lovers will get permission to marry through another swindle, which boils down to a deal in which a motorcycle is exchanged against a woman. Presented as a "humoristic novel," business ethics are depicted as fraudulent and unreliable. The narrative shows Berlin's youth, anti-American, and prone to experience an aggressive unity with their machines, who enjoy the "machine-gun-like" pinging of the engine.[110]

Violence and eroticism were close neighbors in these novels, triggered and enhanced by the cyborg experience. A scene in *Garage 13* with Wolf steering and Karl (the heroine adorned with a male name) on the buddy seat demonstrates:

> "So, now you please cling to me for sixty seconds."—Obediently two young girl's arms embraced him. . . . With a jerk the machine moved into racing speed. . . . A fierce air current groomed her bob (*Bubikopf*) backwards like steel wire. A group of agricultural workers . . . observed with satisfaction that even the white-silky underwear of the fast-passing slender girl on the backseat had been shifted upward, fully against the rules. After a kilometer, well before the highway crossing, Wolf killed the engine and let the machine roll out. With a jubilant sigh Karl released him from her embrace and adjusted her hair and clothes. Wolf turned and observed with satisfaction the same scene as the workers. "Well, did you like it?" "It was nice, but

one needs the right clothes." "But it is also possible in a different way." And Wolf switched a small lever at the petrol tank. "Half throttle!" he explained. "Now we move so slow and silent like cosy cyclists moving through the landscape. Isn't that at least as attractive? I will now lead you in this tempo through the nicest routes of the Mark."[111]

Apparently, the taming of the car adventure eroticizes the new violent unity of man and two-wheeled machine. A French novel on motorcycles titled *Ma Kimbell* (My Kimbell, the title a reference to the bike's make) depicts the long-range trip of a white-collar salesman of agricultural tractors as both dangerous and autoerotic: he once nearly dozes off, and on another occasion just manages to avoid an accident in a sharp curve because he is dreaming away. Truck drivers are his "enemy." In this "adventure of love" written by Luc Durtain, in which the motorcycle is "the major heroine," cars are "boxes for legless cripples." Fleeing from another "adventure" with a lover in Paris, the 'I' (Claude), not only enters an "Empire of Speed" that knows no national boundaries, he also expresses his love of the machine with its "war-like exhaust explosions," enjoying the cocooning effect that the noise provides: "The picturesque? Nature? I don't give a damn. *Tap, tap, tap:* the piston, with big blows such as the blows of an axe, sets me apart from everything around me (*me détache de toutes choses*)." A petrol tank functions as "third buttock." This indulgence in technology aimed at "conquering the world," as its subtitle reads, cannot prevent the author, reputed to be one of the most passionate French landscape poets of his era, from delivering examples of "pure" landscape descriptions. In *Ma Kimbell* the tension between literary individualism and the collectivity of the trip reaches a new zenith: the theme of the lonely motorcycle rider can only be reconciled through Durtain's membership in the literary school of Unanimism, which stresses the social role of the individual within a community. Indeed, Claude is not alone on his trip. Sometimes other members of the traveling party are mentioned, one of them, a secondary school teacher, driving in a car. Pacifist (against World War I), socialist (having visited the Soviet Union, by train), and author Luc Durtain (pseudonym of André Nepveu) saw the collective as the real actor, in which the individual plays its indispensable role, but the novel does not succeed in depicting the true collectivity of the narrative's subject. Instead, Durtain's reputed fascination for the architecture of his body includes the technology too. "How beautiful she is, this mechanism! How long she is! And low! It crawls to jump. It resembles a tiger." At the same time, the protagonist wishes to merge with his object of desire: "(a) motorcyclist experiences a strange solidarity with that force he holds between his knees." More than that: it is self-love, narcissism. Apart from the temporal and spatial adventure, the cyborg motif necessitates inclusion of the functional adventure as well: tires are checked, carburetors have to be adjusted, and cylinders are taken apart to remove carbon deposits. "The thicker the dirt, the more pleasurable one experiences their removal." Remarkably, the novel ends with a recalcitrant machine and the narrator's desperate but ultimately successful attempts to reconquer its "perverse obstination" during its repeated refusals to work. The result is respect for "this eternal miracle of force and matter, where I find the same nodes as in my own body. I listen to it just like I listen to my heartbeat. Acceptance. Order. Necessity. When I at last lift my head and look around, I have a moment of perfect certainty."[112]

A similar ecstatic feeling of purification must have taken hold of the Jewish German philologist Victor Klemperer, whose diaries between 1936 and 1938, when the Nazis took away his driving permit, express the angst-lust of his "desperado-adventure," even if he had to confess that "the car absorbs me (*frisst mich auf*) . . . the car is never all-right, something is always failing . . . perhaps the joy will come later." This sounds like the complaint of an addict. He indulges in high speeds on the *Autobahnen*. Varying on the German national anthem, Klemperer sums up his "devouring passion" with "*Auto, Auto über alles*" despite his two accidents in this period: a collision with a pedestrian and an accident in which he is thrown out of his car.[113]

Speed and Slowness

During the Interbellum, with the increasing importance of the systems character of the automotive experience, the further spread of the car among the lower middle classes led to a reappraisal of the balance between the temporal, spatial, and functional adventures. This was also caused by automobile culture entering the urban multimodality. As we have seen in chapter 4, Atlantic automobilism had always been an 'urbanizing' culture, but the city streets were mostly used to speed into 'nature.' Now, car culture became truly urban as it started to push away competing modalities (as we will see in the next chapter) and a separate 'urban autopoetics' emerged. In town, speed took on a different character, according to Robert Musil. The urban automotive adventure's temporal aspect shifted toward the only form of speed the city still had in stock: the speed of connection, "the hurry of the transfer and the uncertainty of progress on time."[114] Instead, high speed could now be enjoyed on smooth, straight highways with the entire family. The German *Autobahnen*, for instance, initially were littered with "fellow citizens . . .

who had expected too much of their decent, faithful vehicle," as a contemporary observer remarked, a situation necessitating the adjustment of car technology able to endure long-distance high-speed usage, as we saw in chapter 3. With the new cars the new high-speed adventure seemed at last to make the old dream of indulging in 'flight' (see chapter 2) into a reality, so much so that Hitler himself declared that "the motor vehicle belongs in its full essence more to the airplane than to the railroad." The "anti-gravity movement" (*schwerelose Bewegung*) on the "rivers of German fascism" (as Klaus Theweleit has called them) was praised in brochures in which "flying carpets" appear and the observation is made that "on the autobahnen, one flies." Now, Proust's inversion seemed to have been democratized over the entire car-driving community, and not only the landscape but the entire infrastructure itself started moving: "On the road," another observer remarked, "we seem not to be active anymore.... We are so passive and the dynamism of the road has so much impact upon our experiences that in the end the relation seems to turn upside down: it is the road which is active: it moves fast and smooth, without friction and violence towards us and irresistibly absorbs the car." Remarkably, while speed increased on the freeways, the temporal aspect of adventure was further domesticated by the slowly increasing regularity and 'democratic' character of automotive flows, a practice called *Autowandern* by the novelist Heinrich Hauser, as we have seen. High speed under these conditions increased passivity in the automotive capsule. The increased noise of higher speeds, generated at the same time, did not prevent a German poet of the *Autobahnen* to hear how "the street sings through the clatter of the engines" and "a trush sings a song in the woods."[115]

Speed now became an all-pervasive trait of autopoetics, including that segment of travelogues and novels that addressed the subversive part of mobility. In Edward Anderson's *Thieves Like Us* (1937), for instance, a story about an outlaw subculture written in a New Objectivity style, cars (two Ford V-8s) are used to perform bank robberies. The cars have a "gun-metal" color, race at 80 mph (which of course leads to a collision) and function as fragile cocoons of intimacy, for sex and violence (and violent sex: "I can snap her little body in my hands.") From the perspective of the cool 'underside' of American society, the car is the alpha and omega of their culture, even if one of the protagonists has to concede that "cars don't fly."[116] In B. Traven's *Die Weisse Rose* (1929; The White Rose) cronies of capitalists use the car as a weapon, to deliberately run "at full speed" over the body of the protagonist.[117] Subversive use of the car is especially pronounced in the interwar movies, for instance those that are inspired by real-life criminals, such as *You Only Live Once* (1937) referring to Bonnie and Clyde (Clyde preferred the Ford V8 "because of its speed and its steel body"). Likewise, John Dillinger (shot in 1934 by the FBI after watching *Manhattan Melodrama*) was admired by many because of his "superhuman driving skill," which made him avoid dangerous shootouts, and was perhaps because of that "excluded from direct portrayal in contemporary films." Anecdotal evidence tells us that his robberies were made easier because people thought they were shooting a movie.[118]

The fascination for the subversiveness of the dandy, the *flâneur*, and the criminal was based on the admiration—not hampered by morality—for their skills, for the criminal act as much as the technical dexterity as an "aesthetic act."[119] The same relation between speed and subversion can be found in books written by the avant-garde. For Francis Picabia in his surrealist novel *Caravansérail* (1924, but published half a century later), the car is the embodiment of speed and fear: in a taxi the protagonists (Rosine and the 'I') are driven "at an absolutely vertiginous speed, risking a collision at any moment.... Yes, I am scared, I said. Deep inside me I hid carefully that I longed for an accident." Speed and danger unite man and woman, but only those that are part of the Dada elite. "Petrol, that's Dada, the engine is the public: but don't worry, you are not Dada, you are like those well-regulated carburetors that absorb the gas, not aware of the energy they bring into the combustion chamber."[120]

Picabia, sitting at the steering wheel of one of his *bolides*, was photographed ("en grande vitesse," at high speed) by his friend and photographer Man Ray in 1924. In his novel he depicts the car-speeding culture as being close to the use of other drugs, such as opium.[121] Klaus Mann's short story "Speed" (1940) also plays with the 'rush' aspect of speed. "'They call me Speed,'" he said, and he added—half out of pride, half as an apology: 'Because I am mostly in a hurry.'" Speed had "an insatiable desire for adventures, deeds, continuous fast movement." He is adored by young women, who take him along in their "giant, elegant cars," and he consumes marijuana, thus chasing away the "dull comfort of the petty bourgeois home" of the protagonist, whenever he appears. Marijuana causes flight ("like a bird," *vogelgleich*) and needs to be procured at breakneck speed when the supply runs out.[122] Curiously, the modern traveling writer, on the quest of the self, is interested in experiences that create "a feeling of identity loss," an exaltation taken literally.[123]

The French diplomat, novelist, and world traveler Paul Morand, although originally a speed lover (he boasted that on one of his car journeys he raced from Paris to the Mediterranean at 100 km/h; he found speed to be "truly the only new vice"), wrote already at the end of the

1920s, in an essay titled "On Speed" (1929; *De la vitesse*) that art and speed are irreconcilable, despite the fact that his six novels of the cycle *Ouvert la Nuit* were "based... on the notion of speed." This ambiguity, if not contradiction, was lost on Morand (although literary historians have discovered in his work some passages against speed *excesses*), and like Gide, he declared: "The artist is an aristocrat, he travels slowly." Like Gide he criticizes the "flying along" of the landscape, which he observes especially from a train. His change of mind came, according to one literary scholar, after his trip to the United States, where he saw the passion for speed functioning as a "flight offence" (*délit de fuite*), an alibi for not having to think, but in his essay Morand also likens "very high speed" with Communism, as it "kills the individual." In any case, one gets used to high speed, and driving fast becomes monotonous. For the German literary historian Birgit Winterberg, Morand's attitude toward speed is, contrary to Marinetti's speed celebration in the pre-WWI phase, an expression of a "deep-rooted pessimist world view."[124]

No wonder that the democratization of the automotive adventure, and especially its seemingly quintessential speed experience, led to a revival of an anti-speed sentiment among the more well-to-do, reminiscent of André Gide's position before the war (see chapter 2). The idea of a benevolent slowness emerged first and foremost in France, but in Berlin, which considered itself the real hub of modern Europe, Franz Hessel had reinvented *flânerie:* "going slow through crowded streets is a particular joy," and he likened it to "a kind of reading of the street, in which human faces, display booths, shop windows, café terasses, tramways, cars, trees become mere letters with equal rights." Most of his urban travelogue, however, was dedicated to a sightseeing trip in a "giant car" (*ein Riesenauto*), a tourist bus that sometimes went "brutally fast" and granted him a filmic panorama of his own city.[125] The Russian writer and literary theorist Viktor Shklovsky, who also lived in Berlin for a number of years, is skeptical about the benefits of speed as well. "But to what purpose does one need speed?" he asks the reader. "Only he who flees needs it, or his perpetrator. The engine pushes people to what is aptly called crime."[126]

There were also racist aspects involved in the anti-speed attitude among the literary elites: the "racy tabloid" *Tempo*, published by the (Jewish) Ullstein Verlag in Berlin, was renamed *Die jüdische Hast* (The Jewish Hurry). Morand's ambiguity toward speed (directly after his essay "De la Vitesse" in his *Papiers d'identité* follows a eulogy of racing titled "Comme le vent" [Like the Wind] on several circuits all over the world) ends in a half-hearted acceptance: "Let's love speed, the modern wonder, but let's always check our brakes." Only some months after he wrote his unconvincing anti-speed pamphet, Morand was joined (and explicitly mentioned) by his compatriot Valery Larbaud in a piece ostentatiously called "Slowness" (*La lenteur*), although the latter acknowledged how difficult it is to struggle against something which is stronger than oneself: "the demon of Speed, which is in the engine, keeps on trying to dominate."[127] Speed was inside the man, as a drug, and the car-as-catalyst enhanced its effects as soon as the desire was unleashed.

Even Dadaist Tristan Tzara (pseudonym of Samuel Rosenstock), still busy trying to *épater les bourgeois,* in the 1930s described a future society "in which haste has been abandoned in favor of oblivion" and "symbols of the past" are discarded, such as literature, currency, and public transport. But violence is retained "as a natural means of human expression."

To sum up: whereas the nuclear family became the true subject of interwar (increasingly urban) automobilism, speed emphasized the individual experience of the human part of the cyborg, feeding the literary bias in the same ways it had before the war, but now sanctioned through the emerging high-speed road networks.

Flows and Violence: Urban Culture and the Middle-Class Family

If one wishes to analyze the shift toward and within the automotive middle-class family adventure, Sinclair Lewis's *Babbitt* seems an obvious starting point. Domestication, such is clear, took place in the city first. Lewis's literary career ran parallel to the career of the middle-class car: before *Main Street* and especially *Babbitt,* which made him world famous (he received the Nobel Prize for Literature in 1930) and rich, his coming-of-age as a writer took place through the car trip, just as it did for thousands of fellow middle-class Americans. Considering himself to be an author with a "Midwestern theme," his membership in the Socialist Party was less middle ground (but much more easily accepted than it would have been half a century later). Belonging to the Wells and Shaw rather than the Marx persuasion, he also acted as a "suffragent," giving speeches from a car. At the end of 1919 Lewis wrote (often in the train, commuting to his job at a New York City publishing house) a three-part series for the *Saturday Evening Post* on a transcontinental trip he undertook with his wife. Its title addressed the core of the automotive adventure during the Interbellum: "Adventures in Autobumming: Gasoline Gypsies." Around the same time he finished a fictional rendering of his experiences called *Free Air,* which also first appeared in the same newspaper (an earlier version was called "Danger—Run Slow").

Described as the novel that stands in the middle of the American transition from a paternalistic to a companionate family culture, but at the same time as a new woman's tale of losing independence during the gradual construction of the "companionate couple," Lewis's fictional trip contained all the ingredients we have discussed so far as constitutive of the automotive adventure. As a matter of fact, he had studied the cyborg motif as a crucial element in this adventure in an earlier novel on aviation that, according to a contemporary reviewer, gave the reader "a kindred thrill of breathless flight, of danger that is a fearful joy, and of confident omnipotence that is superhuman."[128] His novel has also been called an incorpation of "freedom and democracy, pioneering spirit of the west and a technology expanding human possibilities"—in short: a quintessential 'frontier novel.' The cars in this story "are problematic, but not malevolent. Their animate characteristics are 'friendly.' But the incautious, unskilled or inexperienced driver is flirting with disaster."[129]

In *Free Air* a young, wealthy East Coast woman (Claire Boltwood) decides to go on an automotive adventure from Minneapolis to Seattle in an expensive car. Her father, who accompanies her and is involved in railroad building, initially is quite skeptical about her desire, but gradually starts to enjoy the adventure. Claire soon becomes fascinated by the skills of a mechanic (Milt Daggett) who starts to follow her at a distance in a cheap car. Meanwhile, she enjoys both the temporal and functional aspects of adventure, including the flight and the cyborg experience. Her "pleasure of being defiantly dirty" (when she is not in the company of men), competes with her "feeling more like an aviator than like an automobilist." The cyborg motif is applied in a quite special way as the car functions as an artificial limb shared by both: when her new lover is driving, "every time he touched the foot-brake, she could feel the strain in the tendons of her own ankle." Despite the double entendre of the title (the "free air" to fill the tires also promises to provide the freedom of the open skies while at the same time setting her free from the constraints of her social background), Lewis's is a bread-and-butter romance with a happy ending: in the end it is her lover who gives her the stamp of approval ("You drive like a man") while she slips back into a passive role. Her adventure then undergoes a twist: "'Do you really care for things like that, all those awfully expensive luxuries?' begged Milt. 'Of course I do. Especially after small hotels.' 'Then you don't really like adventuring?' 'Oh yes in its place! For one thing, it makes a clever dinner seem so good by contrast!'" Meandering between irony and romanticism, Lewis seems to struggle with his heroine's position vis-à-vis adventure. Claire's and Milt's difference of opinion later develops into a row. Milt: "'All these things are kind of softening,' And he meant that she was still soft.... 'They're absolutely trivial. They shut off.'" Claire: "'They shut off rain and snow and dirt, and I still fail to see the picturesqueness of dirt! Good-by!'" Claire likes to be dirty but not as an ostentatious masculine way of showing off some kind of adventurousness, which at the same time is domesticated to such an extent that the couple can quarrel about who is soft and who is not. Despite their different expectations about the automotive adventure, though, they share a contempt for "limousinvalids," who, "insolated from life by plate glass, preserved from their steady forty an hour from the commonness of seeing anything along the road, looked out at the campers for a second, sniffed, rolled on, wearily wondering whether they would find a good hotel that night and why the deuce they hadn't come by train." Her admiration for the "Sagebrush Tourists" leads Claire at the end of the novel to sigh: "Didn't know I could like roughing-it so well," immediately adding that this kind of adventure was "not dangerous at all, but rather vigorous." *Free Air* stands for a society in which car *ownership* is less important than the skill to deal with the car, car *use*. It is written early enough to vent skepticism about the usefulness of a closed body.[130]

Whereas *Free Air* was still a part of the travelogue tradition analyzed earlier in this chapter, Lewis's next novel, *Babbitt* (1922), left the nomadic thrill of mobility behind and focused on the sedentary part in which the automotive adventure was further marginalized, reduced to the daily commute of the protagonist with the same name in the fictional town of Zenith. Remarkably, the train was used for long-range, adventurous trips away from the family, reminding us that the roles of the various means of mobility where not yet firmly defined. In a letter to a fellow novelist, Lewis showed himself "very tired of hearing about motor cars."[131] Although reputed, among recent literary critics, to be "old-fashioned" ("a provincial writer of materialist romances, apparently left behind by Modernism" and "a clumsy and over-productive factionaliser of obvious social problems"), his critical treatment of both the suffocating aspects of middle classness and the conformist crisis of American manliness is still very much worthy of analysis, because it does so in the language and on the terms of the petty-bourgeois business community itself, so much so that the novel made Lewis popular overnight, initially praised by the very people he seemed to chastise.[132] A possible explanation has been given by Simone Weil Davis, who analyzed American advertizing culture from the perspective of belletristic literature, and who concluded that Lewis's diatribe against conformism was aimed at the then much-debated issue of "other-directedness," thus implicitly advocating a "true," "self-directed" individualism, excluding any sense of solidarity or community allegiance. Thus the car, and not the train, became a vehicle of American conformism.[133] Another analyst sees Babbitt being taken up in "a brotherhood of consumption," because he "experiences consumption not as social competition, in the Veblenian account, but as social cohesion." The car is crucial to this, too: "The out-of-town business trip is supplanted by the circular journey of the domesticated consumer." The car, from this perspective, is used to reconcile men with (feminine) consumption. "Emotionally," American historian David Blanke concludes, "the car comforted those drawn to consumption." And comforting was necessary, as the blurring of the borders between work and play had to be accepted against a puritan ethic; to this end, the adventure had to be literally domesticated, taken into the home, the family, where, to give only one example, school books in mathematics were suddenly called "Adventures in Numbers."[134] "The car as symbolic house" was, indeed, a powerful metaphor in this phase.[135] Other observers compared *Babbitt* with the Lynds' studies in *Middletown*, most of them considering the "scientific" approach of the Lynds superior, but the homology between the two struck many as revealing: "when I am reading about Middletown I seem to be revisiting Zenith, where I first met Babbitt," a journalist opined.[136]

And yet, to compare Babbitt and their neighbors (whose life, according to Lewis, was "dominated by suburban bacchanalia of alcohol, nicotine, petrol, and kisses") with the Everyman may be a bit too easy: abroad, Babbitt's far-from-average income and his related lifestyle was observed more sharply as a means to distinguish himself from average America, such as in the case of the Dutch travel-writer Ellen Forest, who called some parvenues in Monte Carlo "fat Babbiths [sic]."[137] In the United Kingdom the petit bourgeois was also seen with more contempt than it was for Lewis: George Orwell's traveling salesman George Bowling, in a novel with the Lewisian title *Coming Up For Air* (1939), is not only disgusted by the "Americanized" milk bar ("shiny and stream-lined") in his hometown, he also uses his 1927 model company car with "a biggish mileage, ... tied together with bits of string" to escape from his suffocating family situation, even if he and his wife attend a lecture of the Left Book Club on "The Menace of Fascism." For him, a fascist England would "probably [not] make the slightest difference." Orwell (pseudonym of Eric Blair) leaves the reader in the dark as to whether Bowling would tell his wife what he "really [had] been doing" or "let her go on thinking it was a woman, and take my medicine." Bowling's adventure is certainly not high speed, but purely spatial. Such authors remind us of J. B. Priestley's "rural-centered thinking" of the prewar period. But Bowling knows his escape is in vain before he starts: "There's a chap who thinks he's going to escape! There's a chap who says he won't be stream-lined! ...

I trod on the gas and the old car rattled into the thirties." Later, after he had "let the car run down the hill slowly," "the cows and elm trees and fields of wheat rushed past till the engine was pretty nearly red-hot."[138] Despite Babbitt's and Bowling's differences in income and status, both protagonists share a small-town, philistine culture in which the car has literally been domesticated (made into a home away from home).

Babbitt is the story of the failed rebellion of a conformist, a small-town real estate businessman who tries to compensate for the American male's loss of power within the family, in "his transition from macho to domesticated man," by an "unemotional," hard and cold manhood experienced within a "fratriarchal" conformism of the "self-made man." After his brief escapade toward a soft liberalism and even bohemianism (including an extramarital flirt: "Don't you love to sit on the floor? It's so Bohemian!"), his eventual reaffirmation of his manliness entails a loss of individuality: his "achievement of the standard [becomes his] measure of individuality," a condition the German literary historian Jürgen Link calls "normalization."[139] Babbitt's adventures are along "the well-worn routes of male society: poker games and noisy lunches with the clan of good fellows, out-of-town conventions, infidelities, boozing." Remarkably, the car does not seem to have a place in this "yearning for adventure": the visits to the conventions as well as a fishing holiday are done by train (where the macho stories fill the smoky Pullman car), while the automobile is used as a pure city vehicle to get him to the train. Babbitt's "middle-class discontent . . . is so thoroughly domestic," his rebellion so "unrevolutionary," that parking his car, in the words of Lewis the author, becomes a token of "a virile adventure masterfully executed. With satisfaction he locked a thief-proof steel wedge on the front wheel." Such is Babbitt's territorial macho-adventure.[140] Indeed, Lewis, as early as 1922, had done with the adventurousness of the car. Instead came the hierarchy of possession.

Babbitt's parking adventure is undertaken from the safe haven of the closed body (as a cocoon for the "spiritually weakened self"): "Oh, Dad," his older daughter Verona begs, "why don't you get a sedan? That would be perfectly slick! A closed car is so much more comfy than an open one." And although Babbitt retorts that an open version would give "more fresh air," the closed car of the neighbors wins: "They went, with ardour and some thoroughness, into the matters of streamline bodies, hill-climbing power, wire wheels, chrome steel, ignition systems, and body colours. It was much more than a study of transportation. It was an aspiration for knightly rank." It puts Babbitt's apparent utilitarianism ("poetry and French . . . subjects . . . never brought in anybody a cent") into question. His adventures have to be confessed in church: "Been going on joy-rides? Squeezing girls in cars?' The reverend eyes glistened." Tellingly, in his later novel *Dodsworth* (1929) an engineer-entrepreneur intending "to supplant the horseless carriage with the sculptured automotive streamline" is portrayed, in the words of a later literary critic, as a man of "unwasteful and extremely unadventurous precision." When such "folks own their homes," it is said in *Babbitt*, "they ain't starting labour-troubles, and they're raising kids instead of raising hell!" No wonder that Lewis, the author, later also spoke against the American culture of "the new divinity, the God of Speed." Playing with his own fame, Lewis characterized Dodsworth as "not a Babbitt," but the man to whom Dodsworth sells his company is a philistine nonetheless, albeit an educated one: "Europe? Rats! Dead's a doornail! Place for women and long-haired artists.... More art in a good shiny spark-plug than in all the fat Venus de Mylos they ever turned out." At the same time, it is a reference to Francis Picabia's painting of a spark plug representing a nude female. This car tycoon is quite outspoken about his civilizing mission: "By making autos we're enabling half the civilized world to run into town from their pig-sties and see the movies, and the other half to get out of town and give Nature a once-over. Twenty million cars in America! And in twenty more years we'll have the bloomin' Tibetans and Abyssinians riding on cement roads in [our] cars! Talk about Napoleon! Talk about Shakespeare! Why, we're pulling off the greatest miracle since the Lord created the world!"[141]

That Lewis focuses on the commute as the quintessential function of suburban car driving may not be representative of the actual use of the car (as we have seen in chapter 4), though it is all the more revealing: this most standardized and ritualized of all automotive practices puts the little adventures in a special light, such as the pride Babbitt feels about his electric cigar lighter, "a priceless time-saver." Lewis had earlier written a short story on a rebellion by suburban commuters, but Babbitt's rebellion was special: "He noted how quickly his car picked up. He felt superior and powerful, like a shuttle of polished steel darting in a vast machine." Even being a part of the system while commuting (or perhaps rather because of it) does not prevent the protagonist from indulging in his "three and a half blocks of romantic adventure."[142] Lewis's 'car society' was already present in his earlier novel that made him famous in the United States, *Main Street,* in which a car appears on every third page. In *Babbitt*, the name of the protagonist is a clever pun on efficiency, as Babbitt was the name of the inventor of a friction-bearing material (applied in the piston rods encircling the fast-turning crankshaft in the combustion engine), without which the 'engine of society' cannot run efficiently.[143]

Lewis's satire is so effective because he seems to identify partially with his protagonist, as has often been observed by literary critics: he satirizes himself, including his remembrance of his own indulgence in a petty-bourgeois automotive adventure across the United States. We should also be aware of the fact that his readers, in the myriad American 'Middletowns' where his fame was greatest, perhaps did not see the satire in the 'three and a half blocks of romantic adventure': he was considered "not a teacher but a brother."[144] His own automotive (and social) mobility took off in a Model T that he took on a camping tour (and which he provided with a makeshift starter to prevent him from injuring his writing hand; he wept when he had to sell it), then he bought a new Hupmobile when he started to earn money (and left the suburbs), and after two accidents (in one of which his son was injured) he "upgraded" to a beige Cadillac. Babbitt himself made it from a Model T to a Willys, but even this did not help: "Gosh, I'd like Some day I'm going to take a long motor trip," Babbitt confesses to his wife. "'Yes, we'd enjoy that,' she yawned." That was the way forward: "We've got a lot to do in the way of extending the paving of motor boulevards, for, believe me, it's the fellow with four to ten thousand a year, say, and an automobile and a nice little family in a bungalow on the edge of town, that makes the wheels of progress go round!" Meanwhile, "Babbitt went to church regularly, except on spring Sunday mornings which were obviously meant for motoring." To this end he needs a minimum of technical knowledge ("he kicked about the garage and swept the snow off the running-board and examined a cracked hose-connection"), but if need be, there is always his tinkering and "motor-mad" son, who becomes a car mechanic able to fill a "differential with grease, out of pure love of mechanics and filthiness."[145]

Lewis's books, literary analyst Alfred Kazin notes, "have really given back to Americans a perfect symbolic myth, the central image of what they believed themselves to be."[146] His ambiguity about his protagonist (whom he depicts as "'two Babbitts,' a boosting conformist and a rebel wanabe") and his hesitancy in choosing "between satire and sociology, or between romance and the novel," seems to point at the very essence of American middle-class car driving during the Interbellum. At the same time, Lewis's merging of his indulgence in the joys of motoring and its mild 'auto-critique' may be seen as typical for the middleclass writer: "For George F. Babbitt, as to most prosperous citizens of Zenith, his motor car was poetry and tragedy, love and heroism. The office was his pirate ship but the car his perilous excursion ashore."[147] The attractiveness of driving a car, we have seen this before, is exactly this ambiguity between sociology and poetry, business and pleasure, even if some contemporary commentators were "appalled by the prospect of 'millions of Ford-driving, silk-shirted Babbitts.'"[148]

Although Lewis made several attempts to write a novel on workers, he failed because they "have no special speech which I can hear," even if they were "Babbitts in overalls." "He was middle-class," his biographer asserts, "his ear was attuned to Babbitt's speech," which Lewis himself once compared to Italian fascists and the Ku Klux Klan. He was one of the few who had written on white-collar female office workers (with a feminist twist, as one of his biographers has pointed out) and "America the mediocre" as depicted in *Main Street* ("100 percent Americanism & God's countriness") as well as *Babbitt* shaped his career as a writer. "He really *was* Babbitt," his contemporary colleagues gossiped about him: "He was so *square*." And yet, he felt deeply insulted when he, during one of his several trips across the Atlantic, was identified by his European colleagues as a voice for their anti-Americanism. Lewis didn't seem to bother, for instance, about Virginia Woolf's annoyance about his "vigorous, buoyant, but philistine air," or when he was seen as second-rate author by the "Joyce and Stein brigade" in Paris. Instead, he dedicated his novel to Edith Wharton, and wept when he saw the immense poverty of the London slums. "Damn it. You can't realize how much of Babbitt there is in me—in all of us." In fact, his novel *Dodsworth*, written in Weimar Berlin and during his honeymoon in a luxurious mobile home in Britain (a publicity stunt concocted by Lewis, as his new wife, Dorothy Thompson wrote to her friends), ended, in a first version, in the protagonist finding his redemption in European culture. But eventually Lewis did not favor European culture over the American, and depicted Dodsworth's wife Fran as "corrupted by the false dream of Europe." In a way, Lewis was a nomad of sorts himself: he never settled down, and every house he bought or occupied, was temporary.[149] Was there a relationship between his foot-loose homeliness and his inclination to depict the car as a central vehicle driving his early novels? The car as an ersatz home, a safe haven of intimacy?

On a less satirical level, in Helen V. Tooker's *The 5:35*, announced as *A Novel of Suburban Life*, the car is not even used for commuting (to New York City), but only as a "feeder" to the daily commute by train, and for the occasional errand. The protagonist, a young white-collared woman (Neil Fleming) is picked up from the station by her mother in the family car, and when she and her fiancé, at the end of the novel, decide to take a look at their future home, they take the car "as a pleasanter means of travelling" through the "rolling country." The "glow of family unity" in the suburb appears to be rather oppressive for Neil: she is forbidden to follow her boss to Chicago and has to give up her job "to stay home and keep house." Obstinate Neil is literally educated by her lover about the "very holy thing" of the family, something not in opposition to "the individual," but a "practical social unit formed precisely for the purpose of protecting and developing your individual." The family is the means through which societal rules are appropriated and internalized, as Foucault's theory of governmentality teaches us. Meanwhile Neil, who her lover thinks would "hate to turn out to be a flat tire," tries to adjust to the white-collar family culture, criticizing "the extreme women on the covers of *Vogue*." When her father slips into a depression and they have to sell their car, the limit is reached, for Neil's brother at least: "'But, gosh, Mother, have a heart! We can't get along without a car. You haven't any idea how much we use it—you'd soon find out how much if you sold it. Fact!' His tone tightened, 'You can't really mean it. Everybody knows a car is indispensable nowadays.' . . . Neil listened with a detached lack of interest as they argued. What did cars matter? She and [her new lover] Anthony would not be able to afford a car—not for years and years—except a Ford, maybe. She wondered how she would like driving a Ford. Some people couldn't get the knack of it." For the middle-class suburban family in 1928 (right at the time Ford sales were eclipsed by General Motors) a Ford is not a real car, nor is it a fine car for a woman as we saw in chapter 5. Whatever the brand, however, mother is not convinced: "'I'm sure I don't take any pleasure riding nowadays, what with thinking that a tire may go at any moment." We are far away from the functional adventure of a previous phase. In the end the children get their way, especially when the son offers to pay the upkeep. "'We've gotta have a car, Mums. By hook or by crook.' 'Well,' she hesitated, 'if you think you can—Of course, I shall be glad to have something to get about it, and I do hate to think of you having to walk all the way back and forth from the station. But I don't quite see how you are going to manage.' He lit a cigarette."[150]

Although the novel suffers under a lack of irony or satire, the loneliness of the suburban crowd is convincingly depicted:

> An automobile in the next driveway began to racket. It was almost the only sound to be heard, but it burst upon the quietness of the afternoon and shattered it. The engine spluttered and choked and coughed, died away as though it had finished, and began again. It roared and banged and exploded violently, died down, and exploded again. It kept this up for an incredible length of time—for seventeen minutes after Neil thought to look at her watch. She thought she would fly out of her skin, and she could see her father's forehead corrugated with the agony of it. She was on the point of dashing out and begging the owner to stop it somehow,

when it gave a terrific bang and rattled off down the street, coughing from a greater and greater distance.

It may be hell outside the car, but inside there is always the comforting cocooning, for instance when Neil sits next to her first boyfriend, although there is always the uncertainty of the car's functioning, as if it has its own agency: "The engine whirred and stopped, whirred and stopped, whirred again, and settled into a regular vibration. They moved slowly forward. The automatic cleaner swung back and forth fanlike, *tick, tick,* across the windshield. The little car shot out into the maelstrom of the night, following its headlights down a street that glittered with ice and rushing water and crackled under the wheels." The limited reliability of the car is a metaphor for the uncertainties of a beginning companionship: one never knows when it stalls. Once running, she enjoys his skills: "She watched his hands on the wheel and drew contentment out of their sure, light grip. There is, she decided, something essentially thrilling about sitting beside a good driver, especially for a woman riding with a man." She even enjoys a true narcissistic inversion, when she and her new boyfriend go out on a "picnic lunch" against a "green, racing background of trees and fields."[151]

While the attractiveness for extra-urban use seems to be inspired by the skills of driving, in an urban setting it is the possession of the car that makes the difference, as *Babbitt* so convincingly reveals. Likewise, in F. Scott Fitzgerald's *The Great Gatsby*, as well as Michael Arlen's *The Green Hat*, it is the place of the car in the hierarchy of cars on the market that counts: in both novels the protagonist drives a yellow super-car, as if their authors had spies in each other's fictional car fleets. Fitzgerald's *Great Gatsby*, a more "glossy version" of American life than *Babbitt* but the book that "really created for the public the new generation," dealing with America's "decadent rich" and the promiscuous college youth, is a famous, and perhaps overanalyzed case of literary criticism. But literary critics are apparently not aware of thirty years of mobile prehistory: in a literary critics' feast of multi-interpretability the yellow Rolls-Royce stands for "restlessness," for the "destruction of moral and societal norms," for "phallic extension of Gatsby, the man," and even for "richness, society and fate."[152] For other critics, the novel is about the inseparable connection between "the American dream and American wealth," organized around a series of parties where the car functions both "as showfront and mobility—the same thing, in essence." Through this double motive, Fitzgerald "weaves" another idea, "that of combined excitement and irresponsibility," the "irresponsibility of rotten driving" by a party of drunks, for instance.[153] Similarly, for Leo Marx

Fitzgerald's novel stands for the end of the American pastoral dream.[154] And yet, most analysts fully miss the point made by German literary historian Wolfgang Munzinger about the "untypical" character of Fitzgerald's novel, which, together with Nabokov and Miller and a poet such as Cummings represents a "Europeanized" view of the car as a luxurious token of conspicuous consumption.[155] We should, however, be careful with such stereotyping: there is nothing much 'European' to be found in Cummings' poem "She being Brand," where he plays with the affinities between the skills of handling the car and handling the woman ("I touched the accelerator and gave / her the juice"); instead, Cummings is positioning himself in a tradition that started already in the bicycle period (see chapter 2).[156]

Apart from *Gatsby*'s exceptional status, the emergence of a 'car society' indeed implies that its central symbols get to carry a heavy burden of metaphoric energy. Fitzgerald manages to distribute this burden as he individualizes and personalizes the cars, identifies them by brand name, describes their accessories sometimes more extensively than he does human actors: he creates, in the words of one analyst, a world of cars "among themselves" (*unter sich*), as a parallel, allegorical metaworld adjacent to the world of the humans, as ambivalent as these humans.[157] According to Munzinger, Fitzgerald's novel is inspired by (if not partially copied from) *The Green Hat: A Romance for a Few People*, published in 1924 and written by the British Michael Arlen (pseudonym of Dikran Kouyoumdjian of Bulgarian descent) in which a woman owns a yellow Hispano Suiza, which she uses to drive herself into death. Made into a movie (starring Greta Garbo), Arlen's car universe reveals the same variety and luxuriousness as Fitzgerald's, as Munzinger shows by a meticulous comparison between the 'car parks' in both novels. The car is celebrated (and criticized) as a cocoon: "a great yellow beast with shining scales" and Iris, the protagonist, has "enslaved the beast" in which she "lay aslant in the driving seat." The narrator is not impressed: "One hundred and twenty horses drew us. . . . 'Let him pass, Iris, cried I, a little scared. A woman driving, you never know, might lose her head, boy's head, curly head, white and tiger-rawny, but too white, too intent, too infernally reckless. . . . 'Can do seventy-five if you like,' cried the lips of the dancing hair. 'Let him pass, Iris!'" The car is a weapon of cyborg quality: it makes a sound "like the roar of a thousand rifles" and when, at the very end of the novel, it races out of control and the crash is inevitable the difference between (wo)man and machine disappears: "Suddenly the moaning of the wrecked car ceased."[158] Like Evelyn Waugh's *Vile Bodies* (1930), Arlen's novel deals with the Bright Young People, a group of wealthy, pleasure-seeking young urbanites, "led by a number of upper-class and upper-middle-class young women," who stirred the media landscape of interwar London during the second half of the 1920s. Remarkably, nearly all novels dealing with this group (by Brenda Dean Paul, Jean Cocteau, Scott Fitzgerald, Michael Arlen, and Evelyn Waugh) "include disastrous car crashes."[159]

Waugh's satire also indulges in the modernity of both "wireless" and all varieties of mobility one can think of, including an airship used to organize a party. One of the actors, Eddy Littlejohn, buys a racing car, of course. Father Rothschild owns a motorcycle. Mrs. Ape, an evangelist, owns a Packard "bearing the dust of three continents." A middle-aged gentleman is adorned with a Morris Oxford saloon and others alight from their Rolls Royce or from an "electric brougham," while the father of the 'I''s lover is taken to the station in the neighbor's motor car. An accident occurs in a runaway motor car "which would not stop." A motor race forms part of the bored urbanites' adventure, and like so many novels of this period, the narrative ends with a lethal crash, partly observed from an airplane, from where the road below "unrolled like a length of cinema film."[160]

The accident, which the interwar novelists appear so fascinated by, came to stand for the breaking open of the car cocoon, as Sinclair Lewis already in 1916 suggested in his *Saturday Evening Post* series: suddenly the occupants in the "mobile private space" (open or closed, that did not seem to matter much) were "vulnerable to intrusion by strangers." Indeed, the closed car provided intimacy for the middle-class nuclear family against the masses.[161] And the accident became connected to violence: According to Paul Virilio, the arts kept being caught in violence during this period. When Pablo Picasso saw the camouflage on the first tanks during World War I he seems to have said: "*We* [avant-garde artists] are the ones who did that." But there is even a more radical way of constructing the continuity between war and peaceful violence, war and art, than giving art the task of camouflaging aggression. That more radical way is declaring art the victim of war: "Now is the time," Virilio explained in an interview in a book about *The Accident of Art*, "to recognize that we are the products of major accidents, and war is one of them. But today, accident and war are just one and the same thing. You just have to look at the World Trade Center." From that perspective, German Expressionism and French Surrealism were making "war by other means." "You can't understand the 20th century," Virilio contends, "without the death drive."[162]

During this phase, the system character of the car enters into the fictional world of autopoetics. The system character (everything is related to everything) is first experienced in an urban context. It is brought to life in the case of the false fire alarm in George Milburn's *Catalogue,* when "all over town" (the original title of the novel) car engines are started to bring volunteers to the fire. A similar car "network" in this small Arizona town supports the distribution of Sears' and other mail-order houses' catalogues as well as the orders sent back to the headquarters in Kansas City, including spare parts for Model Ts. While the car meanwhile (the novel is from 1936) forms the backbone of the Rural Free Delivery system of mail, mule carts are often used as feeders to this network (between the mailbox and the farm, for instance). The railroad still functions as deus ex machina, which brings the catalogues to the local post office, the last part done by truck. In this nested, hierarchical mobility system even a jitney functions, but it has hardly any clientele and most people use the bus. Amid many familiar car clichés (the car "sped away" or "leaped away," etc.) the use of the car is only worth mentioning when in the hand of "niggers," or in the case of an accident where the black driver's nose is smashed by the jitney driver, or a black man is lynched. Here, the (motorized) mob is depicted as dangerous.[163] The closed car protects from this danger, while the mob's 'adventure' is found elsewhere: in the ordering of commodities through the mail. In this consumption adventure the car plays a constituting role in supporting the creation of a network that enables the emergence of three communities: those who consume by catalogue, those who are part of the fire-fighting community, and those who want to lynch a fellow citizen, falsely accused of killing a white with whom he collided in his Model T.

This was no different in Europe. Erich Maria Remarque (pseudonym of Erich Remark), who came from a lower-middle-class background (he worked as a primary school teacher, his father was a bookbinder), not only became world famous overnight with his anti-war novel *Im Westen nichts Neues* (1929; translated as *All Quiet on the Western Front*), he also wrote about cars, boats, and cocktail recipes and he made publicity poems for a tire company. A notorious drinker, his car-related work (especially a series of short articles in *Sport im Bild, Das Blatt der guten Gesellschaft* (Sports Illustrated, The Good Society Magazine) testifies of the narcotic but at the same time systemic character of participating in urban traffic. In Germany at least, it was Remarque who brought automotive adventure into the city (proceeded, as we saw in chapter 2, by the Expressionists from 1910), in the shape of a war against nonusers, and of users among themselves. Car driving and drinking both created an "intensity of life." Like Lewis and Faulkner (as we will see later), Remarque longed for such intensity at an early age, as his inclination in his youth to (illegally) wear officer uniforms testified. Unlike Brecht, who gladly accepted his second Steyer car as a present from his publisher when he ruined his first in an accident (and who wrote in his diary that the first thing he wanted to learn was "driving," only then followed by the writing of "modern iambic verse"), Remarque rejected a comparable offer from his publisher on the ground that it did not have any travel gear on its luggage carrier.[164]

Remarque's *Sport im Bild* contributions, published during the second half of the 1920s, right in the middle of the boom years of the Weimar Republic, luxuriously illustrated by Bernd Reuters, can be characterized as one long, repetitive eulogy of the skills and cyborg pleasures of the automotive adventure, alternated with evocations of the urban traffic flow in which the car seamlessly seemed to fit. The "courage to live at full throttle" resulted in a plethora of experiences constituting the Weimar middle-class automotive adventure: from a near erotic description of the protagonist's love for the car's technology and comfort (in *Kleiner Auto-Roman* [Little Car Novel], less than two pages long), via the song of the car trip in an even shorter *Hinter den Scheinwerfern* (Behind the Headlamps) and the Kiplingesque (see chapter 2) alibi of racing at 130 km/h to save a diseased relative (*Josefs Moment*) to the cyborg-racing adventure in which a stone "smashed against the car. He ducked, as if he himself was hit." Car culture, these short stories declare, was a separate world: the cocooning started to cover the entire system. A man who married a "fast" woman he met in a car regretted this after a couple of weeks, "for what works excellently in the car, can be unsupportable in daily life. Gregor felt it bitterly. Courage in the curve turned into imperiousness in the *salon*, security while driving into unbearable dogmatism in conversation, energy into intolerability, sporting comradeship into marital boredom." Remarque's more system-oriented pieces confirm that the car system really emerged in an urban context: "frictionless smooth traffic asks for a certain minimum speed," and when that is achieved "magnificently ordered car colums slide along each other, like endless *paternoster* elevators." Here, Remarque was nothing more than an extra intermediary between the classical intermediary (the clubs, especially the touring clubs) and the car users: "the entire traffic problem is largely a pedestrian problem" and "tramways are the worst traffic barriers," while measures against sloppy car drivers "have to be [taken] with ruthless strictness!" What the pedestrian lacks, he asserts, is discipline enforced upon her by technology. Traffic, Remarque concludes, is a war on the street, where a culture of "raw robber barons" rules: "Knights and gentlemen behind the steering wheel exist, but the journeymen and farmhands dominate, and in the end the law of the jungle reigns: the strongest engine." Like in Milburn's *Catalogue,* the mob started to invade the automotive world. No wonder, then, that Remarque used words like pack (*Meute*) and herd (*Rudel*) to characterize the swarm-like movements on the urban streets.[165] His autopoetic work is the opposite of that of the "futuristic auto fanatics," as one literary analyst asserts: "who, although they also make a connection with the technological vehicle, do not do so to generate models for regulation back to normality, but more to transgress normalcy's border, to destroy these borders." In Remarque's universe the adventure is literally tamed through the cyborg mechanism. Coolness, from this perspective, is self-regulation.[166] No wonder that his anti-war novel was criticized, by his fellow-writer Stefan Zweig for instance, as depicting the "old pleasure of war as a non-bourgeois way of life, as an invitation to adventure."[167]

In 1936, while in exile, his New Objectivist *Three Comrades* appeared (only a year later followed by the German version, *Drei Kameraden*), partly inspired by a very popular film operette *Die Drei von der Tankstelle* (1930; The Three from the Gas Station) and prepared during several years under the provisional title *Pat*. The novel, depicting the early 1930s in Germany, tells the struggle of three friends (a homosocial model of a "struggling community of men" as an alternative to the family) who in the crisis have to sell one of their cars in order to purchase a workshop where cars are restored. One of the few Interbellum literary works on the functional adventure of tinkering, repair, and maintenance, *Three Comrades* was, like so many books in this period, written with the Great War in mind. Unlike the *Sport im Bild* pieces, however, the narrative combines the intoxication of booze and car, praises the distancing effect of the car cocoon and acknowledges the racing fever that takes hold of every driver, even when driving with a family. Inversion, in a previous period the privilege of the few (such as Proust), is now triggered by drunkenness: "The steering wheel shivered in my hand. The streets fluctuated past me, the houses wavered, and the lamp posts stood aslant in the rain." The story is embedded in a traffic context: "Outside, the muffled roar of the street poured on, punctuated with the vulture cries of motorcars. They screamed whenever anyone opened the door. They screamed like a nagging, jealous woman." The novel became so popular within Germany and abroad (in the Netherlands, alone, eleven printings have been produced since the 1930s) that it was developed into an Oscar-winning Hollywood movie, its scenario written, among others, by F. Scott Fitzgerald.[168]

Ilya Ehrenburg's *Leben der Autos* (1929; translated as *The Life of the Automobile*, "perhaps the only true automobile novel" according to literary historian Jens Peter Becker) enumerates six exemplary ways to die by car and ends with a loud crash resulting in "iron splinters, glass slivers, a chunk of warm flesh." Ehrenburg's semi-fictional montage of histories also tells a story of consumption, in which Charles Bernard, formerly dealing in cigarette paper but now, on his "soft carpet slippers," living on a small pension, experiences the car chase in the cinema we have quoted previously. Then he finds an old guide through the Pyrenees and eventually not only buys a radio but also a car, on credit and equipped (a wink at Lewis?) with an electric cigar lighter and even a "beautiful flower vase." He knows he has to drive slowly for the first 500 km, but already a speed of 30 km/h appears to him as a "racing flight. He could not clearly distinguish hill or trees or people. All around him flickered like once in the cinema." But then the speed fever takes hold of him: "He hadn't bought the car to crawl. ... But stop, Charles! Why are you in a hurry?" A "second Charles" has the answer: "flickering and wind.... He turns right. Perhaps it is not he who turns. Some people cry: Hey! ... He doesn't hear them. Then—a thought: what happened to the car? ... Well, it's all very clear! Bernard even opens his eyes for a short while: the car has gone crazy."[169]

Ehrenburg's history is one of the few of the period that deals with the car's dehumanizing mass and series production (privileging Citroën's story over Ford's) and the colonial violence involved in rubber and oil production. In other words, he adds the production system as foundation to the system of traffic flows. The novel stresses the character of the car system as a global network without a fatherland, in which the automobiles have gained agency, as the title already indicates. For Karl Lang, chauffeur, "the engine consists of many hundreds of components. The city consists of many hundreds of streets. Karl is part of the engine and part of the city. He presses the pedal. He turns the steering wheel. ... In town there are only cars. They hurry, overtake each other, they fall ill, curse eachother with their shameless honking. ... Karl thinks about nothing." Tired and on the brink of dozing off, he "often hears the word 'Faster!' He doesn't know who pushes him to hurry: the people or the engine. ... Karl doesn't think about this. He is an ordinary chauffeur—and there are many chauffeurs, they all push on the pedal, throw blood-red turning signals out and keep silent. For them the horns speak." "All [motorists] drive because they own cars. They don't drive, the cars drive, and they drive because they are cars. Suddenly one of the cars stops, in the middle of the desolation of the suburb. ... Around there are pillars and pumps. The car wants to feed." Car and driver are one. "The engine of the Mercedes car and the heart of the gentleman ... beat in the same rhythm. Both are strong, and both are beautiful. Their kinship establishes itself in every curve."[170]

In this system, what previously were called catastrophes are now called accidents. "Soon, they will stop talking at all. Silently they will drag the knocked-down to the side and carry statistics silently.... Long before its birth, when it only consisted of metal layers and a pile of drawings, it carefully kills Malayan coolies and Mexican workers. Its labour contractions are painful. It smashes flesh, blinds eyes, eats lungs, takes sanity.... The lilac withers, chickens and dreamers run to and fro in agony. The car laconically overruns the pedestrian.... It is not to blame for anything. Its conscience is as pure as mister Citroën's. It is just fulfilling its fate: it is destined to annihilate the people."[171]

The only other novel from this period with an extensive interest in the production side of the fledgling 'automobile system' is Sherwood Anderson's *Perhaps Women* (1931).[172] Anderson (whom we met earlier in chapters 2 and 5) also, like Ehrenburg, deals extensively with the automotive cyborg motif, but as none other bases this upon acoustics, written in a *sachlich* style: "I went in the machine from Chicago to Miami. ... It ran gaily. There was a soft murmuring sound. Something within the machine sang and something within me sang. Something within me beat with the steady rhythmic beating of the machine." By now it should not come as a surprise anymore that this is immediately followed by a fantasy of violence: "The machine gave its life to me, into my keeping. My hands guided it. With one turn of my wrists I could have destroyed the machine and myself. There were crowd of people in the streets of some of the towns and cities through which I passed. I could have destroyed fifty people and myself in the machine." The cyborg is part of a swarm-like motorized crowd producing "new music, not heard, felt in the nerves," music that has to be "orchestrate(d)."[173]

The fusion of man and automobile is also a motif in Robert Musil's short story "Der Riese Agoag" (The Giant Agoag, the latter term a wordplay on Berlin's major public transport company), where an entire bus functions as armor. It is a violent fantasy played out against

> the dwarves... who swarm on the street.... As they crossed the street, he shot towards them like a big mongrel does towards sparrows. He looked down on the roofs of the smart passenger cars which previously always intimidated him because of their distinction, in about the same manner as a man, knife in hand, looks down on the fluffy chickens on a poultry farm. ... if it was true what they say, that clothes make the man, why would that not also be possible with a motor bus? One is loaded or surrounded by an enormous force, just like someone puts on an armour or a gun.[174]

In a sense, Musil continues the elitist prewar tongue-in-cheek, laconic attitude toward others (which explains why this short story has been labeled a "small 'psychology' of fascism"). His major work, the monumental *Der Mann ohne Eigenschaften* (1930–1942; The Man Without

Qualities), starts with urban congestion and a deadly accident ("something" shoots from the boardwalk under a truck; the German term ["*aus der Reihe*"] can be interpreted as a military term) observed by a bystander and his companion. The story takes place in 1913 in Vienna. The bystander not only enormously exaggerates (by a factor of ten, as one analyst later calculated) American road accident statistics, but the exaggeration at the same time functions as an introduction to a cool and distanced assessment of the case: "this ghastly incident [could be] inserted in some order [and could be made] into a technical problem, which didn't concern [the companion] directly anymore." Like Kracauer, Musil couples car driving to dancing, leading to a conscious switching off of thinking, a "dethronement of the ideocracy of the brain," and an unreflexive, cyborg-like conduct as a driver.[175] Is that (the fragile and fearful unity of man and machine) perhaps the reason why the "phantom car" or the "fantasy car" (that has its own agency) pops up so regularly in autopoetics?[176] However this may be, Musil's contemporary, Heinrich Hauser, mused in his *Friede mit Maschinen* (1928; Peace with Machines) that cars meanwhile functioned so well "that they seemed to alienate the driver from their mechanical nature."[177] Beware of the perfect cyborg! Musil's novel reveals that the 'car society' is first identified and experienced in the city. From this perspective, it is not surprising that the bee metaphor (standing for the swarm of people, of motorists?) appears throughout the entire novel.[178]

Alfred Döblin's "traffic novel," *Berlin Alexanderplatz*, also contains an accident and depicts the development of the protagonist in relation to the car. Döblin, too, sees cars and machine guns as equal agents of violence in his universe. The novel starts with "transport worker" Biberkopf, who takes the tram to go to the inner city. Later he loses an arm in a car accident. Like Hermann Hesse (as we will see later), Döblin depicts traffic as a "hunt" and compares it with the military.[179]

In the Netherlands the most famous New Objectivity writer, F. Bordewijk, also depicted cars as personae in a depersonalized context. In his controversial *Knorrende beesten* (1933; Purring Beasts) hardly any humans are present.[180] Written in a filmic and *neusachlich* style in brief sentences with few adjectives, the central place of action is a mundane, drab seaside resort, more particularly a parking lot, where two classes of cars are parked, expensive ones and cheap ones. This "novel of a parking season," as Bordewijk called his novella of barely fifty pages, does not describe feelings but registers movements. The novel has been interpreted alternately as an attack on fascism or a critique on communism, but what remains as an attitudinal bottom line is the distancing from other people's feelings.

With or Without a Car: A Women's Adventure?

How then does women's adventure fit into this context? To them, just like for the Jewish Klemperer or the blacks in George Milburn's universe, the car seems first and foremost to have signaled a road to liberation. Nonusers are equally involved in some way (mostly through the expectation and hope of being a motorist one day). Especially in Germany around twelve million women worked outside the house, more (in percentage) than elsewhere in Europe.[181] For German novelist Irmgard Keun's lower-middle-class heroines, the car is a distant but actively dreamed-about conveyance. Sometimes it allows for a new, filmic experience in the urban culture, such as in *Das kunstseidene Mädchen* (1932; The Artificial-Silk Girl) where a taxi trip is described that shows the city streets as if on a movie screen. Writing, in the "cold order" of the Weimar Republic as well as in the novel, was like filming: "I want to write as film, for such is my life," the artificial-silk girl explains. *Das kunstseidene Mädchen* is a bitter, stream-of-consciousness novel about eighteen year old Dorit, *Angestellte,* who tries to survive in early 1930s Berlin, unemployed, while having a short opportunistic affair with a married man, and with him, at last, enjoys chauffeured drives in his Mercedes and taxi rides without looking at the meter. Her experiences with cars are ambiguous. The car stands for inaccessible wealth when it comes to ownership, and being in debt when ambitions in the own ranks are translated into ownership of a second-hand, dilapidated version. It also stands for intimacy and protection against a harsh world, and for comfort, 'flight,' "soft, as a Mercedes wheel rolling on bumpy pavement."[182]

For Keun's most famous heroine, Gilgi, a stenotypist who becomes unemployed in her highly successful New Objectivist urban novel from 1931 of that name (30,000 copies in six editions during the first year alone), only the tram and the train are accessible.[183] She once kissed in a car, and when she falls in love her boyfriend rents a Cadillac, but when she dreams of individual mobility she figures on buying a perambulator for her illegitimate baby. Gilgi's adventures are torn between her love for Martin and her wish to stay independent, rent her own room, read novels, and design her own clothes. While he and her friend Olga represent adventurousness and bohemianism (Olga is truly independent and hence she travels), Gilgi does not have time for this: she must invest all her power to keep straight, and that means working. Not much leisure for Gilgi. Nor do Keun's heroines seem to be adventurous, at least not in terms of their mobility. They have to be careful, hardly move. In sport, her adventure is being part of the crowd on the tribune. Her cyborg ex-

Redefining Adventure   **479**

perience is with her typewriter: "her brown, small hands with the neat, short-nailed typing fingers belong to the machine, and the machine belongs to them." Her mobile adventure is virtual, for instance when she wanders on the radio waves: "Radio—Rrroma—Napoli—delicious! The whole world is with me in my room." Her visiting aunt boasts that her daughter "is engaged with a Mercedes-Benz." She is "in her own way an ultramodern mother: *Auto, Auto über ahalles....*' If only a man would have a good character,' says aunt Hetty. Character, character! If he has a firstclass car, it would be character enough, one would think. ... She hopes that her little [daughter] Gerda soon will be lucky, even if it would be a motorcycle with sidecar." While the automotive adventure is related to man and marriage, and mobility is horizontal as much as vertical, Gilgi cannot afford adventure as she has to work to maintain independence, nor can she share the "planless" *flâneries* of her boyfriend. When unemployment comes her inner world disintegrates, even though she tries to develop a hardness that should prevent her feelings from "eating inwards." Sometimes she dreams of being aggressive to her adopting parents, and that is why being "cool" is so important: "when I now cry only half a tear, I will beat everything to pieces." The whole novel stresses the importance of being cool, cold, and hard, toward Nazis and Communists fighting against each other in the streets, toward her former friends who commit suicide, toward her mother who gave her away when she was a baby. Coldness is seen as something negative, as equivalent to "being egotistic," as not being prepared to kiss a man, as something sad and as the end situation of her disintegration process, but it is also necessary, protection, for instance when she is interrogated by a doctor at her request for an abortion. At the end of the novel she chooses to flee to Berlin, following the illusion of her dream, in a train.[184]

For female lower-middle-class Germans during the Depression, automotive adventure was part of a dream, and the specific female adventures they produced were aimed at *social* mobility. Adventure is existential rather than (physical) mobility related. This is also the case for another famous Weimar "Girl" (the capital C indicating that the word had meanwhile entered normal German parlance), Marieluise Fleisser's heroine, the traveler in flour, Frieda Geier.[185] Subtitled *Novel of Smoking, Sporting, Loving and Selling,* Fleisser's novel also deals with the tensions between men and women, and the danger the former represent for a woman who has not yet managed to become "cold": "man is a dangerous principle." In her novel, published in 1931, which, together with the work of Brecht, Döblin, and Arnolt Bronnen (see later in this chapter) belongs to the icons of literary New Objectivity, the sporting man, a small shopowner in a countryside town, fails to live up to Frieda's expectations, and falls back into a patriarchal attitude, ready, it is suggested, to be integrated into the fascist mob.[186]

Fleisser's earlier work provides a fascinating, but ultimately tragic play (and more than that!) on male violence and female fear of this violence. Her female protagonists are not socially integrated, but speechless, uncertain, poor. Life is a "battlefield." Fleisser's tragic life, starting in the bohemian and semi-criminal circles of Munich's district of Schwabing (which helped her to distance herself from a small-town petty-bourgeois background), formed a source of irritation as well as empathy of generations of feminist scholars since her rediscovery in the 1970s. In Munich she joined the Luxemburgian writer Alexander Weicker's "adventure." Then, Bertolt Brecht (whom she characterized as a bicyclist, "an inconspicuous man of the street") took her into his instrumentalized universe, at which point she buried herself in a petit-bourgeois marriage in her Bavarian hometown without any energy left for writing. Tellingly, her only story written through the eyes of a man is the short story with the title "Abenteuer aus dem Englischen Garten" (Adventures from the English Garden [located in Munich]).[187]

In her work the "sane world" of the nuclear family is conspicuously absent and there seems to be a relationship with the absence of the car. Like Keun's Gilgi and Dorit, Frieda is not "wild... certainly not." Although she owns an Opel *Laubfrosch* for her work as a traveling saleswoman, the car disappears from sight once it has been mentioned, as if to stress that it cannot play a role in the leisure time of a single, not very wealthy white-collar girl in the Weimar Republic. The car (the German equivalent of a Ford, known to be conceptually derived from the French small car Citroën 5CV) radiates (in the words of Frieda herself) "something quietly philistine" while "it looks ridiculous if the light vehicle is pushed to the border of her performance." Like the Ford in Helen Tooker's novel, Frieda's small car is not a 'real' car. She takes the train to visit her sister who is raised in a monastery. In Frieda's world it is not the car that is the vehicle of adventure, but the man. And the man (Gustl Gillich) is a typical representative of the petty-bourgeois patriarchal philistine, owner of a tobacco shop (tobacco, as we know, is the poor man's equivalent of the car, together with cinema), but very mobile in his corporeal behavior.[188]

Fleisser's adventures included a hidden admiration for "border crossers," men who despised bourgeois morality, such as the French writer Jean Genet, the American slapstick film actor Buster Keaton with his "stoneface," the British racecar driver Ashby with his "cold-blooded composure" (*kaltblütige Ruhe*) whom she observed during a race during his "20 seconds of refueling" or her 'own' bohemian adventurer Alexander Weicker, her first love, with whom lust and violence were seductively mixed and who wrote a novel subtitled *Adventurous Chronicle of a Superfluous*. In the end of Fleisser's novel, leaving the bourgeois order only seems to work for men: Frieda can defeat her man only by withdrawal.[189]

Gilgi's male counterpart, Hans Fallada's "Little Man" Johannes Pinneberg, largely confirms the relationship of the German urban lower middle classes, whether man or woman, to the car in this period. He commutes by train (the outing with his partner, Emma Möschel or Lämmchen [Little Lamb], also takes place via train), while his boss owns a car and, in general, the car forms a part of urban background noise. This is what makes such utterances into a true 'urban' novel. It is the car as part of flowing traffic that makes the city into an actor in this type of novel: the noise of the metropolis is "compounded from thousand single sounds into one big sound. It grows and shrinks, it becomes very loud and is nearly gone, as if the wind had swallowed it." In Pinneberg's urban universe it is the wealthy who own "fat cars." Remarkably, Fallada (pseudonym of Rudolf Ditzen, 1893–1947), whose novel was already made into a movie one year after it was published (in 1932), a year later followed by an English version, integrated a household budget overview in his narrative, which confirms our findings of chapter 4: the couple can spend 5 to 6 percent on transport (*Fahrgeld*), cannot afford a bicycle, let alone a car, and does not wish to risk the taking of credit. The hierarchy of vehicle ownership reflects social hierarchy, just as the hierarchy of household commodities were a function of income. And it is the woman who is strict and hard when it comes to surviving: "You know, little lamb, I had thought you to be very different. Much softer." The novel ends in misery and unemployment, their newborn baby sleeping.[190]

The adventure of Karoline in Ödön von Horváth's play *Kasimir und Karoline* (a "ballad on an unemployed chauffeur," first staged in 1932, written in 1931), about the Munich Octoberfest, consists in the intimacy she seeks with Kasimir, a truck driver who owns a convertible. Her "pettybourgeois opportunism" makes her desire the car more than the man. Intimacy takes place in the car, partially offstage. The action happens mostly on a parking lot for "*hochkapitalistische*" sedans. Horváth's "folk play" is about order that has to be established after the disorder of the Octoberfest, during which Karoline wants to have fun and forget her husband for a while. The car functions as a means to seduce, to make short erotic trips, or it plays the role of (according to some commentators, "bizarre") metaphor, to characterize a woman, for instance, "on whom nothing really works properly." Nonetheless, Karoline knows how to apply the brakes and thus prevents a man from killing himself (and her).[191]

Living in the Bavarian village of Murnau (hometown of several members of the expressionist painter group *Der blaue Reiter*) and himself a son of the petty bourgeoisie (*Kleinbürgersprössling*), Horváth studied the philistines in the flesh. Like Sinclair Lewis he related the philistine atmosphere to "the age of the business men." Characterized as "the most analytical dramatist in the German language, next to Brecht," he was accused by his colleagues of being a "parasite of the Weimar Republic," because of his efforts to compromise with the Nazi culture. Started as a playwright who posed as a worker to get his first play staged in Berlin, he was marked and harassed as a communist when the local fascists started to emerge in his village. Although he tried to accommodate, his authorship of the novel *Der ewige Spiesser* (1930; The Eternal Philistine) and other utterances about "degenerated love of the fatherland and narrow-minded nationalism" pushed him to flee to Paris, where he died from a fallen tree branch. As has been emphasized by literary and other historians time and again, Horváth's analysis of *Kleinbürgertum* was not Marxist, but indicated the blurring of the borders between a proletariat becoming petty bourgeois and an ever more impoverished *Mittelstand*, "that financially hardly distinguishes itself from the proletariat, but clings to its *Mittelstand* mentality." For Horváth by far the largest part of (German) society was petit bourgeois, a conclusion based on the indeed "enormously wide layer of declassed *Mittelstand*." In the universe of Horváth, "office girls, waiters, car salesmen, those residing in half-silk artistic circles, failed people, small swindlers and imposters" represent the "philistines of the future" with their "false and kitchy opinions." Horváth's entire work, some contemporaries opined, was one large "demonology of *Kleinbürgertum*." Like Lewis, Horváth not only wrote ironically *on* this very heterogeneous group, he also wrote *for* them. Initially hoping for a "proletarian romanticism" based on "collectivism," he changed his mind when collectivism and the radicalism of the Right got intertwined, just like Remarque felt he had to give up his collectivist fantasies and fled in homosocial comradeship. According to Walter Benjamin, "fascism appealed to the collective in its *unconscious*, dreaming state." This collective was not a class, but a heterogeneous mass consisting of "the audience of the theater, an army, the inhabitants of a city." In Horváth's sketch of a novel called *Der Mittelstand* he emphasizes the "other-directedness" of this societal group, devoid of any liberal ambitions and governed by the coldness of the *Aufsteiger* (the parvenu), which also "infected" the proletariat. The "motif of coldness" dominated his later work.[192]

Horváth's "eternal philistine" Kobler, a sacked car salesman, defined as "the most deformed version of the petty bourgeois," uses the train to travel to the World Exhibition in Barcelona (just like the author himself did in 1929), which affords him the chance to extensive, cliché conversations about the unification of Europe, fascism and women, Jews, and the coming world war. Kobler is the "hypochondrical egotist," the interwar version of what we characterized as narcissist for the prewar phase, the German Babbitt whose conformist "life is determined by utilitarian impulses and profit interests," and who is willing "to raise the arm for the fascist salute" if that would ease crossing the border. The car is used only at the beginning and in the much shorter second part, respectively to swindle, and by a sporting protagonist to get repaid, sexually, for a dinner he treated a woman to. "Harry was a splendid wealthy driver-owner [Herrenfahrer]. He overtook simply everything and followed the curves as they came. Exceptionally he didn't talk about ice hockey, but elucidated traffic problems. Thus he explained to her that of every car accident certainly some pedestrian was guilty. ... In general, the German state, he opined, should deal with the question of how to get people to work more, so that we at last will emerge again! Pedestrians would be overrun anyway, and now certainly! In this respect our former enemies were very right, when they slandered Germany." In Horváth's universe the woman's "adventure" is to prostitute herself.[193]

Remarkably, and contrary to the novels discussed so far, women's different adventures are, for Horváth, related to the family structure ("the Christian as well as the Jewish"), on which the bourgeois *Mittelstand* crucially depends. According to one literary historian, in Horváth's work the "double aspects of consolidation and disintegration" of the *Mittelstand* are played out within the family, which "on the one hand appears intact as patriarchal structure, in which the man in sovereign dominance uses the woman before marriage as a sexual object, during marriage oppresses her and as a father educates his daughter for this form of marriage. On the other hand Horváth also shows the *Mittelstand* and *Kleinbürger* mentality, which through compensation of normal satisfaction of impulses in the form of abnormal satisfaction (sadism and masochism) and preparation to authoritarian subjugation was made ready for fascism." Indeed, Horváth's work lends support to the idea of the "extremism of the center," coined by American sociologist Seymour Lipset and meanwhile rather controversial among historians. Horváth's work, from this perspective, can be read as a literary illustration of the theses of Sigmund Freud and Wilhelm Reich, the latter known for inventing the term "enforced family" (*Zwangsfamilie*). Time and again Horváth emphasized the role of the youth, for whom marriage has become "impossible" because the woman has to earn money as well. In Horváth's play *Glaube Liebe Hoffnung* (Faith Love Hope, published in 1932, staged in 1936) family morality has become public morality represented by the power of the state, when the vice police rebukes one of the female protagonists. In the same play a car scene shows love making that is as mechanical as the car itself. In Horváth's urbanized world, instead of the public spaces in trams, urban public spaces like squares, parking lots, and sport arenas become places of action (just like in Dutch author Bordewijk's New Objectivist novel), enabling mobility to take on a new role, that of intimacy in the closed car. The car, and technology in general, functions as an optimistic antidote against a culture of irrationality. By the end of the interwar years the car as capsule truly seems to have been firmly embedded in the middle-class universe. Kasimir, one analyst observes, is even proud of the (German) car he does not possess. The car becomes an object of chauvinistic pride irrespective of ownership.[194] It reintroduced a certain aura, seemingly lost in the era of the technical reproduction of the work of art, in the same way as a cloud of smoke around the head of the 'flapper' does.[195]

To sum up: if we are to believe the novels discussed here, (lower-) middle-class women's adventures did not take place in the car, although the urban car culture had meanwhile so far developed that the car nonetheless played a backstage role of modernity, as taxi, or as the man's car.

Whereas Dorit, Gilgi, Frieda, and Emma (heroines from the novels by Keun, Fleisser, and Fallada) represent a dependent, female adventure not firmly related to the car, their lives are quite distant from the self-conscious appropriation of the car by travelers like Rose Lane and Helen Boylston and their Zenobia, or by Gertrude Stein, who, even if she did not know how to get the car in reverse, "by integrating the female car into a lesbian household ... challenged the hetero-normativity of the family and of automobile culture." Like Lane and Boylston, Stein and her partner Alice Toklas gave their cars names: Auntie for their first Ford, Godiva for their second. They bought their third, a Ford V8, after the success of her autobiography in the early 1930s. Stein's functional adventure was adequate: although she confessed to be "very good" at repairing the cars' defects, "she never did anything for herself, neither changing a tyre, cranking the car or repairing it." She explained this out of a "democratic" attitude, a deep feeling of "equality," a principle she also tried to apply to her art, as she avoided separating her protagonists from their "background." Stein's adventure was, according to her own diary, twofold: it consisted of paintings and cars. Like William Carlos Williams (who applied her literary techniques) she wrote in her cars, "much influenced by the sound of the streets and the movement of automobiles. She also liked them to set a sentence for herself as a sort of tuning fork and metronome and then write to that time and tune." Her style, as later critics have observed (some of whom saw in her "hermetic" poems "a feminist reworking of patriarchal language") was rhythmic and tuned to the sound of her sentences rather than their meaning. A wealthy private art collector (especially of Picasso's work), she established stylistic connections between writing and painting and in general was a master of a synesthetics in which also the car played a role. Stein was a real "car writer," her controversial theory of grammar containing ideas about "landscape plays," the construction of poetical forms that followed the structure of play rather than that of "logic." Sitting on the steps of a car in a repair shop, "watching her own [car] being taken to pieces and put together again, she began to write" and finished her entire talk on "Composition As Explanation" (1926).[196] After the war Stein took part in motorized *flâneries* in the Bois de Boulogne, and "many expatriates," as her biographer relates in a rather misogynist passage, recalled "the sight of Getrude Stein, perched high atop the driver's seat of her Ford runabout, gripping the steering wheel, peering intently ahead and proceeding down Parisian streets, like some comically large goddess of the machine." Their first long-range car outing after the war took place in 1922, when they went to Provence for a stay of several months. Stein's American "tour" when she was already famous was also a car tour, in a "Drive Yourself-Car" in and around Chicago, consisting of trips that included "a few minor adventures" in the form of flat tires.[197]

Even Virginia Woolf, like E. M. Forster a prominent member of the Bloomsbury group of British intellectuals, who never wrote a travel book, had the car enter her oeuvre as—according to the literary canon—an outright threat ("a pistol shot," a reminiscence of the war) to her characters who often (but not always, as we will see) moved on foot. The role of the car in Woolf's work has been debated inconclusively, but it cannot be denied that the thrill and excitement of her continental car trips shaped her thinking on modernity, a nice example of what we called affinity between the motorized world and the world of fiction (see chapter 2). For American cultural studies scholar Gillian Beer, Woolf clearly preferred the airplane to the car as a metaphor of modernity. "Instead of the muffled superplus of attributed meaning represented by the car," Beer observes reading *Mrs. Dalloway* (1925), "the aeroplane is playful, open, though first received as ominous." British literary historian Andrew Thacker could not disagree more: he depicts her as making a "*volte-face*" by considering, before the war, the use of photographs in a book of fiction "an unnecessary act of violence," and after that war using maps to write her novels, realizing that "the streets of London have their map; but our passions are uncharted."[198]

Thacker is right. In her novel *Jacob's Room* (1922) Woolf describes the vehicles in a traffic jam (the first in belletristic literature?) as "red and blue beads . . . run together on the string."[199] Woolf, indeed, is an interesting example of the immediate postwar generation who—as a woman—is 'converted' by the car, and subsequently experimented with affinity between writing and motoring. "When Woolf searches for an analogy in the material world to the creation of modern literature, in the 1925 version [of her essay "Modern Fiction"] she substitutes the motor-car for the bicycle of the 1919 version." In *Orlando* (1928), in which her friend Vita Sackville-West is portrayed as an "expert driver," aggression toward fellow-motorists, but especially toward pedestrians, pops up, remarkable for a novelist whose reputation still is based on her peripatetic pedestrianism: "'Why don't you look where you're going' she snapped out ... while the motor car shot, swung, squeezed, and slid" through urban traffic.[200] *Orlando* is the novel (like *Mrs. Dalloway*, it is one of the first "shopping novels," but now by car) that makes Woolf into a true autopoetic writer, a fact ignored by many of her commentators. Research by some of them has indeed revealed that the "motorist point of view" in this novel was the result of a biographical shift accompanied by "a heated argument with [her husband] Leonard about spending money on the garden (his intention) as opposed to travel (her desire)."[201]

Woolf indeed became fascinated by the "nice light little shut up car" (a Singer) she and Leonard bought in 1927: "This is the way to live," she wrote to a friend, after one of her motoring trips through France. And on another occasion she wrote that "nothing ever changed so profoundly my material existence ... as the possession of a motor-car." It was an "absorbing subject ... that filled our thoughts to the exclusion of Clive & Mary & literature & death & life—motor cars ... we talk of nothing but cars." Her Singer (bought second hand and nicknamed the Umbrella, followed by another Singer with an open body two years later and a posh Lanchester in 1932, all paid for by Leonard) was not male, nor female, but a hermaphrodite. The car allowed her "to see the heart of the world uncovered for a moment," because motoring shines "accidentally ... upon scenes which would have gone on, have always gone on, will go on, unrecorded, save for this chance glimpse."[202]

In his published diary Leonard Woolf confirms Virginia's excitement about "holiday 'touring' on the Continent—I do not think that anything gave Virginia more pleasure than this. She had a passion for travelling, and travel had a curious and deep effect upon her. When she was abroad, she fell into a strange state of passive alertness [we seem to hear Heinrich Hauser's "rush," quoted previously, here!]. She allowed all these foreign sounds and sights to stream through her mind." The car "opened to one a new way of life," gave a "wonderful feeling of liberation" especially at low speed: "The slower the better—it is the journey that matters." But Leonard does not tell his readers that he forbade Virginia to drive because she backed into a wall during her first driving lesson he gave her. According to her, she "only knocked one boy very gently off his bicycle." Leonard himself, however, also "knew very little about either [the car's] inside or its outside." With worn tires during a trip to France in 1929, they had punctures every twenty-five miles. Bereft of the hoped-for driving adventure, Virginia fantasized about buying a motorcycle, but she never did. For her, "increased mobility" was an antidote to depression. The trip they undertook in 1935 through Holland (with its "highest manifestation of the complacent civilization of the middle classes," its "featherbed civilization," in Leonard's words) to Germany must have had such a therapeutic effect. In Germany they were greeted by "an unending procession of enthusiastic Nazis" waiting for Goering and shouting "Heil Hitler!" while Jewish Leonard, a marmoset pet on his shoulder, waved back.[203]

The way Woolf introduced the car in her following work (in *The Years* [1937] there is again a car in urban traffic, this time stuck in a traffic jam) suggests that there was a connection (and perhaps a shift between) driving their car and driving her pen. From the moment she discovered the car, one of her commentators rightly remarks, Woolf's writing as well as motoring practices can be characterized as "scene making." If Sinclair Lewis stands for the American middle class discovering the motor car, Virginia Woolf's conversion in 1927 represents the volte-face of the British upper middle class, who just had enough money to buy a second-hand Singer. As Woolf herself declared self-consciously: "I am middle class but I never buy my own beef."[204] There is also a parallel with the German iconic car convert (Remarque), as Woolf contributed to *Vogue* and *Vanity Fair,* also on topics related to her newly conquered mobility. In her short story "Evening over Sussex: Reflections in a Motor Car" (1927) she reveals how even a well-versed writer can be "unable to master the stimuli" generated by the car trip. Beauty was "escaping all the time" and perception boiled down to "sit and soak; to be passive" and she observes how the subject's "self splits up." In *Orlando,* too, her motoring protagonist "was ... changing her selves as quickly as she drove—there was a new one at every corner." Woolf is one of the first who observes the fragmentation of perception engendered by the car; reading signs and panels along the roads can only be done in passing: "Undert—Nothing could be seen whole or read start to finish."[205] In the end Leonard appears to be no less absorbed by their car: on the day, in 1941, that Virginia drowns herself he "recorded the mileage for the day" in his diary.[206]

Ascribing a certain type of adventure to women will not work, that much is clear. "Dada Queen" Elsa von Freytag-Loringhoven frightened the urban bourgeois by walking around in her own outfit adorned with a backlight on her butt, while Erika Mann participated in a car rally through North Africa with her brother Klaus, enjoying a heavy marijuana trip. Like Remarque in *Sport im Bild,* she published her car-related *feuilletons* in Ullstein's tabloid *Tempo* and followed a mechanic's course as preparation for her world tour. Her car trips often were like trance, triggered by exhaustion. Erika Mann's adventure did not include the killing of chickens: she avoided them gracefully, the radio at full volume. An overview of German "extra-ordinary women" contains several adventurers, including "the wildest woman of the Weimar Republic," the dancer Anita Berber who, painted by Otto Dix in a garishly colored red dress and red lips, according to legend wanted to found a "Club of Icy Women." But it also contains Marieluise Fleisser and her lust for violent men, as well as the pilot Elly Beinhorn, who in 1931 flew to Africa alone (and wrote a book on it: *Ein Mädchen fliegt um die Welt* [1932; *A Girl Flies Around the World*]), married car racer Bernd Rosemeyer, and is depicted as the typical representative of the Weimar 'fast lane.' Her adventure was also experienced at night, in the car capsule: "Whoever drove at night on hopeless country roads in a car with beautiful, very soft springs and a radio, which makes it hard to establish whether perhaps there is really an English dancing band present, will experience somehow this unreal feeling which overcomes me whenever nature and technology overlap." Beinhorn's supermobility is exceptional, but just like several women in autopoetic novels we have encountered, she too will give up her independence: "Nowadays I am of course very well aware that I resisted [his] strong personality from the first day—until I in the end succumbed, with drums and trumpets." Her husband keeps racing in his Horch convertible, which he calls Manuela.[207]

When one is looking for them, a surprisingly large group of women travelers can be unearthed. Karolina Fell, in her dissertation on "calculated adventure," identified dozens, from Annemarie von Nathusius, who drove to Persia, searching for "the Middle Age 'Aventiure,'" to the world traveler Clärenore Stinnes, and Margret Boveri, who toured through Africa and the Middle East—all of whom wrote extensive travelogues. Nathusius, Fell observed, used a chauffeur and a mechanic and developed a "deeply romantic relationship" to her Mercedes with its "strong iron heart," whereas the less aristocratic Stinnes drove a high-up middle-class Adler Standard 6, in which she broke records in repairing and driving speed. But Fell's conclusion is that no "uniformly 'female' representation pattern" could be identified, only "common European patterns." The adventure of a trip, she contends, may exist only "in the heads of the writers and the readers."[208] Women's subversiveness in mobility, some feminists claim, is often embodied in their immobility.[209]

This is not to say that everybody would agree with Fell: Mary Schriber's overview of female travel writers looks, on the contrary, for an "*écriture feminine* [sic]" and finds it: "To an attentive reader, a woman's 'world' is visible in 'the chinks,' the sites of difference, where the gender of the 'world' according to men becomes visible as well." There are no large differences, she concludes, but gender is written in the "fissures" of the texts, in four domains, one of them being "modes of transportation." According to Schriber, it is "the vehicle convention" that triggers differences, a conclusion largely confirmed by the present analysis. For instance, "Not danger to life and limb, but danger to sexual integrity is the likely issue in the encounter of a woman with indigenous males."[210] Likewise, Shirley Foster in her analysis of nineteenth-century women's travel writing, claims that women "focused more on social experiences and family life, more easily accepted foreign cultures and were more willing to embrace 'difference.'" Tim Cresswell theorized that "even when women are moving, their movements may be experienced very differently from those of men." And Martin Green, who "want(s) to overcome the mutual hostility between the adventure tale and literature as a body of standards," is convinced that "adventure belongs to men (and vice versa) for the profoundest of reasons."[211] On the other hand, Karen Beckman, in a recent analysis of autopoetic movies, warns us to "avoid aligning feminism with a moralistic rejection of thrills, speed, and humor." She suggests that the "crash," as a 'shocking' interruption of the flow, might be a productive entrance to such a nonmoralistic analysis: "the crashed car opens up potentially productive temporal and spatial uncertainties, in spite of the motorcar's privileged position in patriarchal sexual structures."[212] If it is true, as Margaret Walsh states in an overview of women's car and bus travel in the United States, that their "influence on automobility was cultural and indirect rather than technological or economic," then women's adventures deserve to be studied in detail since it is more than likely that their discourse, like men's, differs by continent, and by period.[213] What they have in common, though, is thriving against a background of a "destabilization of traditional views on masculinity."[214]

Women's adventures as depicted by men also take place far from the nuclear family. Vivian Norwood, protagonist in E. J. Rath's novel *Gas-Drive In,* knows her car's technology and uses this knowledge to put arrogant men in their place. In one scene Vivian is instructing a man to repair a car: "I've heard men say they knew something about mechanics; they seem to think it's their special field—The other way with that wrench, I said! Are we going to be here until morning?—I never drove a car until I knew something about one. Easy there! If you shear the head off that bolt we're finished—That's better, but don't be so *slow*!" But the heterosexual order is soon redressed: "'Another word like that and I'll kiss you.' She fell back as though he had struck her." By the end of the novel she is weakened (deiced, one could say) as a woman in love: "Where was all the old crisp decision that had seemed to be an inseparable part of her nature?"[215]

Part of a rare novel dedicated to a large extent to the functional adventure of repairing and maintenance (she even buys a garage to try to recover her stolen car, in which a mysterious letter is hidden), Vivian's 'temporal adventure' is as vivid as that of men's. But high speed is not necessarily coupled to physical aggression: "if she felt the impulse to violence, she managed to curb it." Instead, of course, she experiences inversion: "the tree trunks were merging themselves" when she drove along them, and "the road began twisting." The author stresses Vivian's femininity: she hates boyish women, but causes confusion by sometimes dressing like a man, especially when repairing cars. Skilled, outperforming men in technotalk, but initially owning a purple car, her 'spatial adventure' is eroticized in heterosexual play. "Driving soothed her mood. It was good to feel the rush of air against her cheek—good just to be moving, going somewhere, anywhere," in an open car of course. As an adventurous woman, she can give "cold look(s)," but mostly the men are "chilly, self-contained and competent." The most remarkable passage in this novel from the point of view of automotive adventure is the decision by a male protagonist to sell his car because "nobody but a wild man has a right to own a car like that." If he wants to conquer the heroine, he should not present himself as a risk taker. Just like in Bloem's *Motorherz,* speed must be 'rational.'[216]

The Ubiquitous Car: A Spectrum of Adventures, Adjusted to Middle-Class Taste

The ubiquity of the car, as a physical object as well as within discourses of mobility, resulted in a variation of automotive adventures. Willa Cather, whose androgynous protagonists and their rebellious mobility (on the bicycle) we already met in a previous phase (see chapter 2), added another perspective to this kaleidoscope of possible American attitudes toward the car, that of the small-towners and farmers who gave the movement such a decisive boost before the war. To Cather, echoing Lewis, "the 1920's seemed ... the period not of great creative men of business, but of their lackeys, their secretaries, managers, lawyers, vice-presidents." But when she conceived her famous (Pulitzer

Prize-winning) novel *One of Ours* (1922), her optimism about the catalytic effects of technology had faded. Subject of "highly polarized readings," especially among feminist literary scholars, Cather's (according to some, misogynist) "novel of development" not only testifies to the continuity of tourism and war experience, but also depicts machinery as the epitome of disaster. In her short story "Coming Aphrodite!" (1920) she already discerned in prewar Washington an occasional automobile, "misshapen and sullen, like an ugly threat in a stream of things that were bright and beautiful and alive." In an essay written a year after the publication of *One of Ours* ("Nebraska: the End of the First Cycle," 1923) the car has become a metaphor for all that is rotten in the current state of affairs: "The generation now in the driver's seat hates to make anything, wants to live and die in an automobile, scudding past those acres where the old men used to follow the long corn-rows up and down. They want to buy everything ready-made: clothes, food, education, music, pleasure." As a prelude to that point of view, in *One of Ours* Cather describes a seemingly peaceful agricultural world in Nebraska, cruel in its dullness and killing gentle people who lack the guts to fight against it. That happens to the protagonist Claude Wheeler, member of a rather wealthy farmer family that uses their cars (plural) not so much for far-off outings. The emphasis within the microculture of the car adventure is on pleasurable use around the town, for shopping, courting, social visits, going to the movies, the car wash, the trip itself giving as much pleasure as the activities at the destination. Adorned with a rather onerous family name, Claude is secretly in love with a girl with an even more onerous name (Enid Royce), "cool and sure of herself under any circumstances, and that was one reason why she drove a car so well,- much better than Claude, indeed." Surprised by a storm, Enid not only adjusts the wheel chains around the tires, she also ("calm and motionless . . . amiable, but inflexible") takes over the steering wheel when Claude runs the heavy car into a ditch. Some analysts consider Enid, "[a] prohibitionist, a rigid Christian, and a car-oriented woman," as "one of the most unpleasant of Cather's female characters," an example of "the ideological work done by Cather's pejorative picture of the New Woman." Indeed, Enid uses her car to evangelize and to campaign against drinking. The car, from this perspective, is the vehicle of hypocrisy: an instrument of pleasure under a smokescreen of work. Claude eventually chooses a less-determined partner, while Enid becomes a religious prohibition fanatic and his farmer friend comments: "Having a wife with a car of her own is next thing to having no wife at all. How they do like to roll around! I've been mighty blamed careful to see that [his wife] Susie never learned to drive a car."

Literary critics have called the novel "a study of erotic war motivation unequaled" during the interwar years. While Claude dies in the French trenches (where he had at last found the real adventure of intense and empathic living), his comrades that survive either commit suicide or start garages and repair shops where they could "look at the logical and beautiful inwards of automobiles for the rest of [their] life." Women, such as Claude's mother and her maid Mrs. Mahailey, dominate the end of the novel as the sole survivors, refusing to take part in male adventure. For the rest of her creative life taking refuge in premodern themes and considering industrial society as "a lost lady," Cather continued to depict "new men" rather than "new women," as "destroyers of civilization."[217]

The car as a multifaceted adventure machine for the white-collared lower middle classes crept into the literary universe alongside a car culture that, at first sight, was nothing more than a continuation of the prewar elite adventure. Often quoted as characteristic in histories of automotive literature (unjustifiably so, as we will see!), such novels kept the idea of the car as a luxurious statement of distinction alive. This trope had meanwhile become so ubiquitous that historians easily make questionable claims, such as those about a "borrowing" relationship between William Faulkner and Scott Fitzgerald, whose descriptions of their respective protagonists' cars as voluptuous and protective cocoons look so much alike. "Sitting down behind many layers of glass in a sort of green leather conservatory, we started to town," Fitzgerald writes.[218]

Yet, their cars' similarity is not so amazing and not exclusively American: Pierre Frondaie's successful novel in France, L'Homme à l'Hispano (The Man in the Hispano Suiza; a novel very probably unknown to both Faulkner and Fitzgerald), starts in a very similar way: "White, magnificent as a royal barge, but terrestrial and put on powerful wheels, the Hispano received the last glimmers of the day on its ivory and silvery body."[219] The borrowing is not literary but social, through the ubiquity of car ownership in literary circles. A pseudonym of René Fraudet, Frondaie uses the luxurious Spanish car as a driver of his successful novel, which was filmed twice (in 1926 and 1933). He tells the adventures of the young imposter Georges Dewalter who conquers a married woman (Stéphane, herself owner of several cars) in the French posh sea resort of Biarritz using an extremely expensive car borrowed from a friend. Described as a projectile, the car allows Proustian inversion ("the trees seemed to flee while keeping their distance") as well as the flight experience (the car "went over the main road with the softness of a barge"), but also an unusual comparison with railways (Heinrich Hauser springs to mind), as the "song of the road" is enjoyed as soon as "the speed (allure) became regular and rhythmic like a train." The quote confirms our earlier thesis that true acoustic pleasure can only be produced in a train. Whereas Stéphane's rich husband, the Englishman Sir Oswill, is called "a man of the train" (and thus non-adventurous), Dewalter is depicted as a petty bourgeois who confesses to being "an adventurer, a crook." Why does one travel, he later asks his conquest. She answers: "Life is a journey. When one is rich, and sensitive to visions, one travels to fill one's head."[220] Frondaie's novel hints at what was different from the prewar elitist car novel: in the next chapter we will see how highway construction allowed car swarms to behave like trains. Also, the systemic character of car traffic had made the defect, and especially the accident, into a powerful destroyer of life and interrupter of the narrative.

In interwar car novels the traffic accident is increasingly portrayed as part of (and disrupting) an emerging car system. In *The Great Gatsby* "all protagonists . . . are directly or indirectly involved in an accident."[221] Dreiser used the overrunning of a child during a joyride in a heavy Packard in *An American Tragedy* (1925) and a subsequent collision resulting in a lot of injuries as a token of the futility of wealth, as a deus ex machina, and as a way of showing the tension between lust and reality.[222] Lewis, in his *Free Air,* used the accident as a symbol for "the conflict between man and technology" and as a sign that the desire for adventure can never fully be satisfied and 'adventurers' therefore are constantly looking for more, for repetition of the feeling of flow.[223] For Franz Kafka, employee at the insurance agency *Arbeiter-Unfall-Versicherungs-Gesellschaft* for the kingdom of Bohemia in Prague, as for so many other novelists of this phase like Döblin and Musil, the accident is a "topos of modernity" (*Leitfigur der Moderne*), as German literary historian Claudia Lieb, who dedicated an entire analysis to the road traffic accident in twentieth-century literature, asserts.[224] German poet Erich Kästner, the interwar belletristic attorney for the nuclear family par excellence, seems to quote the title of a prewar movie nearly literally in his poem "Gedanken beim Überfahrenwerden" (1930; Thoughts While Being Run Over):

> Hey, my hat! Is this the end?
> Gosh, these motor buses are enormous.
> And where are my hands?
> Why did this have to happen to me? . . .
> If my Dorothee could see me like this!
> Good that I am on my own.[225]

And Brecht ("probably the first important poet who has something to say about urban man") opines that, once the car has been integrated in daily life, the only way we can understand its role is through alienation, "understanding it as something strange, new, result of a construction, as such something artificial." For Brecht, the street was a theater (the accident as street theater), with the mob as audience. "In the year 1936 the car has become so commonplace that its observation assumes the de-automatisation of perception."[226]

When the German literary historian Jens Peter Becker, in one of the earliest and most extensive analyses of car literature, emphasizes both the symbolic and the structural (narrative-driving) role of the automobile in Interbellum literature, the structural role of the car should be further qualified in order to prevent it from losing its analytical potential.[227] If the car in the novels of this period is still called, in literary criticism, an actor (a German literary historian even called it a "hidden protagonist," *eine heimliche Romanfigur*), it plays a similar role as the landscape, or the home. It is always there, only occasionally mentioned, but it does not seem to drive the narrative anymore, except perhaps in a very literal sense: just as the landscape functioned as hardly mentioned background, and the house as a similar location, the car moves the protagonists around, and these protagonists are more and more of the common stock. The car, in other words, increasingly becomes part of a system of society, the latter meanwhile turning into a 'car society.' Interwar novels were increasingly dealing with what transport sociologist Vincent Kaufmann has coined 'motility,' the promise, the expectation of being mobile. Indeed, several literary analyses suffer from calling the car a "universal stereotypical point of view on our culture."[228]

Numerous, if not nearly infinite, are the examples of the car's ubiquitousness in interwar literary production. In fact, the car is perceived to be so all-pervasive that Theodore Dreiser in his *An American Tragedy* (published in 1925 but dealing with the turn of the century) indulges himself in anachronism by depicting his protagonist as fully embedded in a rich car culture. Dreiser had made a habit of anachronistic exaggeration, as in his *The Titan* (1914) in which his protagonist owns an 80 horsepower car.[229] Now, the car even penetrates into the 'periphery' as a matter of course. In William Faulkner's world, women moderate men's aggression and prevent them from driving dangerously fast, reminding us of a recent study of American literature that concluded that boys are on the move, while women "prevent moving." When, in Faulkner's novel *The Flags in the Dust* (first published in 1929 as *Sartoris*), Miss Jenny, who as an "indestructible woman" is not at all intimidated by men's gadgets, and Old Bayard discuss the daredevilry of his grandson, she asks: "When he finds that car wont go fast enough for him?" She answers herself—"He'll buy an aeroplane"—but immediately gives the solution to her problem: "He ought to have a wife."[230] When old Bayard buys a car and becomes converted to a speed maniac himself, his doctor asks:

"Why don't you stay out of that damn thing?" he replies: "What business is it of yours? ... By God, can't I break my neck in peace if I want to?"[231] Car speeding is described as a mechanical form of rage, often ending in an accident: "He jerked the throttle all the way down its ratchet and she clutched him and tried to scream.... The small car swayed on the curve, lost its footing and went into the ditch, bounded out and hurtled across the road. 'I didn't mean—' he began awkwardly. 'I just wanted to see if I could do it,' ... and she was sobbing wildly against his mouth."[232] In Faulkner's work, and especially in *The Sound and the Fury*, his stylistic experiments with "stream of consciousness" and "interior monologue" techniques reflect and support the cocooning effect of the closed-bodied car, in which violence and erotics intertwine. In general, his view on writing as "a mode of action rather than substitution" made the writing process itself into "a form of adventure rather than evasion."[233] In that sense, Faulkner achieved for the Interbellum period what the prewar avant-garde had practiced for the pioneering phase: to create an affinity (see chapter 2) between writing and driving. And he really was qualified to do so: like Sinclair Lewis, Faulkner's literary career started in a car—or better, *on* a car—when he climbed on the hood while his friend was driving, mimicking with his arms the flight of a plane that he had been so eagerly (and vainly) hoping to fly during his military service in Canada during World War I. Like Lewis, too, he wrote a travelogue; unlike Lewis, his was never published. Inclined to embellish (and especially make more adventurous) his past, his first car was a Ford Model T with a modified race-car body.[234]

Drinking heavily and spending much of his time in the company of bootleggers and criminals in Memphis, Faulkner spiced his car descriptions with machismo and misogyny: "My friend the bootlegger's motor car," his short story *Country Mice* starts, written during his first trip to Europe in 1925, "is as long as a steamboat and has the color of a chocolate ice cream soda. It is trimmed with silver from stem to stern like an expensive lavatory. It is upholstered in maroon leather and attached to it, for emergencies and convenience, is every object which the ingenuity of its maker could imagine my friend ever having any possible desire for or need of. Except a coffin. It is my firm belief that on the first opportunity his motor car is going to retaliate by quite viciously obliterating him." There is always violence hidden somewhere, including the tongue-in-cheek attitude we know so well from the previous elite period of car culture: "My friend the bootlegger turned a corner viciously. The pedestrian, however, escaped."[235] With all his bravado, Faulkner, like Remarque, made his literary oeuvre into a witness of the last transition to a 'car society,' the society in which the 'automobile system' had expanded beyond the urban realm. Faulkner "equate[d] life with motion." After his breakthrough with *The Sound and the Fury*, one of his biographers relates, he bought a "touring car" in which he sometimes, during the evenings, drove "forty miles along the silvery beach, past Biloxi, to Gulfport for dinner."[236]

Like Faulkner, John Dos Passos experimented with language in his *U.S.A.* trilogy, but he also, more so than Faulkner, stressed the systems character of the car. Although most protagonists in his chronicle of "a society of stenographers, interior decorators, and advertising men" travel by rail and steamship, the car functions as a cocoon for intimacy, courtship, and sex; however, owning one is reserved for the wealthy. More important still is Dos Passos's "pluralised method of narration" and his "harder [than in his earlier work], more expressionistic, and collective style" which at the same time 'rhymes' with the swarm-like character of the automotive flow. Just as the structure of his book "oscillates between two possibilities: the collectivist and expressionist epic, and the modern anti-epic—where the sum of the parts represents not a collective meaning but a loss of meaning," modern traffic oscillates between flow and anti-flow without meaning congestion. Like Lewis and Faulkner, Dos Passos started his career with a travelogue, to anarchistic Spain in the early 1920s, where he learned about the mere "destructive" influence of "a man of the middle classes . . . in the reorganization of society." Nonetheless, several critics call the car in his work "the central symbol of industrial society's, of the 'big money,' power run amok, technological and automotive power which is satirically presented as corrupting the characters' values, controlling their individual choices, and often violently and prematurely ending their lives."[237]

But how central can the car become? There comes a moment, in every national literary culture, that the killing becomes so systemic and "normalized" that the car cannot maintain its central steering role, and becomes a covert driver of the novels' universe. 'Car' then becomes a kind of background noise, always there but never to blame, just like 'family' or 'marriage.' This normalization, as literary historian Jürgen Link has argued, is much more than striving toward routine, and is different from succumbing to normativity. For Link, "normalism" can substitute for a set of terms used to characterize Western society, such as "modernity," "capitalism," and "technocracy." In modern society, what is "anormal" is crucial, as well as the answer to the question as to where the border lies between "normality" and "anormality." There are two "artistic-literary strategies in the game of transgressing the borders of normality," a "realistic one" through the increasing intensity of the "thrill," and a "surrealistic" one, by following other "intensities of a transnormal type." Link gives episodes from Döblin's *Berlin Alexanderplatz* and Kafka's *Der Verschollene (Amerika)* as examples of the two strategies, respectively. In its role of regulating the passenger back to normalcy the car trip (as a self-therapeutic practice) is a very crucial element indeed: it is a "model of normalistic living." Whereas earlier in this chapter American literature has been characterized as a "literature of the journey," now all literature can be analyzed as a "(not) normal ride."[238]

The Cult of Cool: Becoming Cyborg

Despite its ubiquity, the car played at least a catalytic, enhancing, and consolidating role in several cultural trends and traits of the period. This always seems to be the case when new acolytes join the growing mass of car owners. Germany is seen as the epicenter of this trend. The cultural sociologist Helmut Lethen has pointed at the German intelligentsia as a new group of would-be motorists who developed a "cool conduct" (*kaltes Verhalten*) as a behavioral response to the modernization process. Reacting to the emergence of an other-directed "shame culture" (as opposed to an inner-directed "guilt culture") New Objectivity (New Sobriety) artists depicted individuals as "no more than motion-machines," while feelings were "mere motor reflexes, and character is a matter of what mask is put on."[239]

New Objectivity (*Neue Sachlichkeit*, according to some analysts "peculiarly German," just like "avant-garde" was a French invention, the former claiming primacy in applied arts, the latter in fine arts) took an "egalitarian" delight in urban circulation and saw society not as a community but as an "open system of traffic among unconnected individuals." Literary history has identified six characteristics of a *neusachlich* attitude: coolness, urban, objective thinking, the subject as type, the new postwar generation, and new social groups such as the intellectuals and the *Angestellte*. In Germany the *sachlich* attitude was part of an "ideology of life" (*Lebensideologie*) going back into the last two decades of the nineteenth century, and prevalent until the early 1960s.[240] By keeping distance, *neusachlich* artists gained "freedom of movement," observing their motto "Not expression—but signals; not substance—but motion!" The literary equivalent of automotive distancing was the stylistic figure of irony, especially powerful in the German 'petty-bourgeois novels,' such as Fleisser's. As the New Objectivist writer Arnolt Bronnen, echoing Remarque, remarked in 1930: "The German, whose warrior nature embraces all its mutations, such as ambition ... and contempt for death, apprehends traffic, in the first instance, as a warlike state."

Famous and controversial New Objectivist painter Otto Dix painted a "stone-cold" self-portrait of himself and his wife, while his depictions of prostitutes as the personification of unemotional sex fascinated him. His interest in murder and rape resonated with the obsession by French surrealists in aggressive sex and violence. André Breton saw the firing of a pistol in a crowd as the "simplest surrealist act" and one can indeed wonder about the parallels between such an act and the act of car driving in a city. If surrealism is a way to "call forth the unknown through chance and play," then the automotive adventure certainly qualifies to be part of the surrealist universum.[241]

In a period governed by "cultures of hatred" in which proposals for "codes of conduct" were booming, Max Weber developed the "scientific type of the cool persona" in the form of the Pole explorer, while others were fascinated by the dandy-soldier (dandy because one was recommended to be "as artificial as possible in life"). Jünger, Brecht, and others played with the concept of the man with the "*Kältepanzer*" (armour of coldness), a "mobile subject without inner depth, whose movements are neither constraint through moral interventions nor through the voice of consciousness." This man lives in a "chronic state of alarm" and is constantly afraid to be shamed or hurt, and his personality is expressed in the 'ideal types' of the *Freikorps* soldier, the communist cadre, the disciplined worker, the intellectual nomad, and the boastful dandy.[242] The cold persona was neither a metaphor nor a mere word play: it was consciously, if not scientifically constructed through the development of military psychology in the 1920s, which intended to enhance the "strength of will" and make "steel hard personalities," such as was the case in the German "Psychological Laboratory" founded in 1927 for the selection of officers, but also influencing the armies in other European countries. If the will failed to develop such "aggressive subjectivity" (such as in France, according to some observers) there was always the possibility of the armored car, as a protective capsule. Volker Berghahn, who analyzed this development of increasing societal violence, speaks of several types of "violent man" in the making, such as the party and ministry bureaucrats, and the (often academically educated) "cool-calculating power technicians."[243] The rejection of the "tyranny of intimacy" may partially explain why the idea of the family as the core of Interbellum society was less easily spread in Germany and Europe than in the United States. In Germany family values were projected on "communities of comradeship." Coolness rather than intimacy was asked for: "Give [a man] a mask, and he will tell the truth," Oscar Wilde concluded; and we could add: give such a man a car, and he will act upon it, for instance by killing others in an increasingly unsafe traffic situation during the Interbellum.[244]

When Lethen constructed two opposing shortlists of terms that either represented the "humanist warm" or the "*neusachlich* cold" attitude, "mobility" came first on his "cold list," followed by "separation" and a bit further down "transparency," "planning," and "apparatus." Other observers have identified two aspects of *sachlichkeit*, both equally effective: on the one hand, *sachlich* smacked of alienation (*Verdinglichung*, thingness, fetishism), on the other it referred to (American) democracy. New Objectivity used traffic, Lethen claims, as a "perception model," and many New Objectivists novels were "traffic novels" with their "egalitarian joys" and "delight in urban circulation" between "unconnected individuals," their keeping of distance. "Traffic does not oblige its participants to a heroic pose," while "the solitary figure moving against the flow is granted the special status of the flaneur."[245] In view of the parallel shift toward an "urban poetics" (as we have observed earlier) Lethen's term should be expanded to the concept of the "*urban* traffic novel," and his *flaneur*, certainly if motorized, should be seen as an "accelerated flaneur," in the words of Patrick Laviolette.[246]

The "cool gaze" (*kühle Blick*), apparently necessary to negotiate traffic, was not a German privilege, though: the French author Frondaie also had a woman characterize his protagonist approvingly as "nice, cold" (*net, froid*), and there were also "traffic novels" in France, such as the one written by Valery Larbaud (in the previous period), or Blaise Cendrars, or especially Paul Morand. World-traveler Morand, skeptical about American "inhuman" culture, as he explained in the preface of his French translation of *Babbitt*, was "cool," too, and also advocated the "distanced, detached and nonchalant attitude" and the "amused tone of the dandy" as well as his "cold reserve."[247] Several European realistic painters show it in their portraits, including Picasso's "Olga" (1928) and Tamara de Lempicka's woman portrait "Das Telephone II" from the same year. In her famous "AutoPortrait" from 1925 (published in 1929 on the cover of *Die Dame*, an upper-middle-class magazine for the 'new woman') she reveals her hidden desires for mobility, as she depicts herself in a green Bugatti, while she owned a small, yellow Renault. Not only does the Bugatti emphasize the importance of the exterior, both body and car are masks, both spotless, metallic, and geometrically built up from spheres and cylinders. New Objectivity artists excelled in expressing this gaze in their preferred portrait of the "small man." These portraits not only are "emotionless, unsentimental," but each painter or writer depicted them in their coldness as objects: people thus become like things, commodities.[248]

"Ice-cold" and "fast as lightning" (*eiskalt* and *blitzschnell*) were used as "fashionable words" (*Modewörter*) in Germany to describe car races: the public whistled in disgust when a German car racer had to give up because of a broken finger. Hitler pointed at hardness of body and soul as a weapon in the "struggle for survival" of German youth.[249] And as far away as the United States, John Dos Passos's "paratactic style" (in which fragments of texts are put in juxtaposition without any subordination, just like the impressions from a speeding car) is especially impressive in his sections titled "Camera Eye," in which the 'I' of the spectator-author, as a tourist, and in a stream-of-consciousness style, observes the world through a nonhuman, cold camera, "draining off the subjective," as the author stated in an interview.[250] It reminds one of Walter Benjamin's loss of "aura" whenever a camera is used, and fits in a tradition of the "thanatourism" of the Great War, as we saw in chapter 3. Babbitt, too, wants "to flee out to a hard, sure, unemotional man-world." And William Carlos Williams wanted to write poetry "sculpting the words themselves," as his biographer states, "words with hard, concrete, denotative jagged edges."[251]

Others have observed how in this phase sexuality and emotionality switched their public roles: whereas the former was more easily accepted compared to the previous Victorian period, the latter suffered from "a desire to conceal anger," especially in the United States. The American equivalent of the European "cool persona" was Superman, conceived in 1938 as hero of a comic strip by a young male author who confessed to have been "inspired" by his failures in conquering girls at high school. Superman was immune to fear, impervious to emotion, and hostile to women.[252]

Although critics ridiculed the "worship of elevators" among proponents of the *Sachlichkeit* movement, and rejected the "engineers' romanticism which does not understand how a carburettor works and hence hears the breath of time in the pounding of a six-cylinder," driving (in) a car became the equivalent of seeking protection while moving around, of wearing a mask or mounting a filter. Jünger saw this also expressed in the masks and uniforms motorists were carrying in the previous period, a protection they now could enjoy in the closed body of the car.[253]

Lethen's seminal (and very influential) analysis now seems somewhat one-sided. Cold behavior is often accompanied by the heat of hidden passion, just like the car as cool cocoon is propelled by the explosions of a throbbing combustion engine. Ernst Jünger's dream of a new race of "modern" humans and his plea for a "total mobilization" (*totale Mobilmachung*) was governed by this combination of "an icy brain and a glowing heart." For the purpose of our inquiry into the history of mobility it is good to remember that, for Nietzsche, cold equaled fast, and heat represented dirtiness, illness, and slowness. For Jünger, who approved of this development and who conceptualized his *soldatischer Mann* (soldier man) as an alternative to the *Bürger* (the petty bourgeois) who "avoids danger," even the landscape "is becoming more technological and more dangerous, colder and more glowing; the last little bits of the old *Gemütlichkeit* are disappearing." The mobilization does not only cover the "armies that meet on the battle fields," but also "the new kinds of armies of traffic, food, re-armament—the armies of work." It is Klaus Theweleit who has shown how this new type of man could perform a "bodily transformation" between sexuality and violence.[254] The "imperial praxis" of controlling the world can only be achieved by controlling the self as well. The resulting inner poverty can best be compensated through a "derived liveliness in relation to and through an identification with machines—and what machine would be better suited for that purpose than the car?"[255] This is the skill Kurt Möser forgot in his listing of automotive skills (see chapter 5): the ability to compensate for a loss of feeling through the controlling of the car. In order to control the car, however, the driver has to obey it. And in order to obey it, he or she has to forget his or her origin of a pedestrian. "Every pedestrian who is installed behind a wheel," Adolf Hitler unbosomed on the attendants of his "Table Talk," "at once loses his sense of the consideration to which he is convinced he is entitled whilst he is a pedestrian." Hitler apparently had studied his drivers carefully: "Now, [my chauffeur] Müller never stopped thinking of the people on the road. He drove very carefully through built-up areas. He believed that anyone who runs over a child should be put in prison at once. He didn't skirt the edge of the road, as many people do, but instead he stuck rather to the top of the camber, always mindful of the child who might unexpectedly emerge." Now, Schreck, one of his later drivers, had developed another skill: "He used his car as a weapon for charging at Communists." And like a Babbitt (who was so proud of his electric cigar lighter) with a fascist twist he added: "I can rely on the clock [in Schreck's car] on the instrument-panel. All the instruments are in perfect working order." Indeed, Hitler was no adventurer in our definition, which does not mean that he was not involved in 'adventures.' Schreck was, however: he once "crushed the bicycles" of a group of "Reds" under his car, and "we often had very painful incidents of this kind." Göring, for instance, Hitler continues his Table Talk tongue-in-cheek, "made a point of always driving on the left-hand side of the road. In a moment of danger, he used to blow his horn. His confidence was unfailing, but it was of a somewhat mystic nature."[256] By the time the car became available to the lower middle classes, some of its drivers had morphed into vandalizing killers of communists, indulging in intimidating macho-driving.

Coldness migrated even further down into the social stratification. A more cynical writer like Erskine Caldwell, with his "harsh and derogatory portrayal of poor, white southerners," in his *Tobacco Road* in a sometimes *sachlich* style described how even the marginal figures in the countryside during the Depression indulge in car driving once they manage to get hold of one. Sister Bessie and Dude not only marry and buy a car with Bessie's saved money in one blow, they also immediately take a drive and make love in one blow, in a gluttonous and fully untamed form of car consumption: while Dude can hardly hide his desire to marry the car rather than Bessie, he not only ruins it within a couple of days, he also kills a black man on a horse wagon: "The nigger driving it ought to have had enough sense to get out of our way when he heard us coming." The car for them is like the factory where they hate to work: "they're all right to fool around in and have a good time in, but they don't offer no love like the ground does." Meanwhile, grandmother is slowly dying and receives the fatal blow when they run her over: "Looks like she's dead," one of the protagonists observes. "'Is she dead, Jeeter?' Jeeter looked down and moved one of her arms with his foot. 'She ain't stiff yet, but I don't reckon she'll live. You help me tote her out in the field and I'll dig a ditch to put her in.'"[257]

Only very rarely do interwar novelists turn the structural violence around into a conscious murder. This happens in *Die Weisse Rose,* by B. Traven, a mysterious, anarchist German novelist living in Mexico, whose protagonist Jacinto is lured from Mexico to San Francisco: "And there they kill him.... [literally: they beat him to death, *dort schlagen sie ihm tot*] Sorry: the man had a car accident. He is found dead on the street." Traven, author of *Der Schatz der Sierra Madre* (1927; The Treasure of the Sierra Madre), was one of the most popular German writers (also read, it seems, by working-class people), perhaps because he connected adventurousness with anti-capitalism.[258] If aimed at (or experienced by) layers 'beneath' the middle class, we have seen that earlier in Steinbeck's *Grapes of Wrath* and Caldwell's *Tobacco Road,* automotive adventure acquires a grim flavor.

The car as *panzer,* as a tank-like capsule but with soft interior upholstery, is a powerful symbol in a period of increasing 'democratized' street violence. Spanish 'ultraist' (avant-garde) writer Xavier Bóveda created a poem "Un automóvil passa" (1919; A Car Passes) in which the passengers of a car remain fully untouched by the car's speed, creating a cleft between an inner and an outer perspective, and confirming our earlier conclusions about the insolating function of the closed car.[259] The fascist worker-poet Heinrich Lersch, reviving Italian futurism, experienced the unity with the car: "I am enclosed in metal and steel."[260] This symbolic role of the car as a token of modernity takes on a crucial role in this phase, enough to add the 'experience of being modern' and the practice of distinction to the adventurous use profile of the car. Such symbolic adventures provoked a response by nonusers, of course. In Germany Hermann Hesse described such an aggressive adventure from the perspective of the pedestrian. In a hallucinating and surrealistic passage in *Der Steppenwolf* (1929) protagonist Harry Haller, an intellectual torn between disgust for the petty bourgeois and the lure of middle-class comfort, does not see any solution other than to shoot down motorists during a "merry hunt on automobiles," a "narcotic vision" (of anti-Americanism, as some literary historians suggest) as if to counteract Jünger's cyborg dreams: "On the streets cars, partially ironclad, were hunting down pedestrians, crushed them to pulp, pushed them dead against the walls of the houses. I understood immediately: this was the struggle between man and machine, long prepared, long awaited, now at last released. Everywhere dead and maimed people were lying, everywhere, too, destroyed, bent, half-burnt automobiles, and above the wild chaos planes were circling, and from many roofs and windows they were shot at with rifles and machine guns.... 'Aim at the driver!' [his fellow-pedestrian and former school mate] Gustav ordered. 'Done!' Gustav laughed. 'The next one is mine.'" Hesse himself traveled by train and does not seem to have been particularly fond of technological innovation, let alone "crazy cars," as he called them. To him, crossing a street was risking "mortal danger." He was right, of course, but for Hesse his surrealist inclination to "magic" may have helped him to support modernist reality, including its cars. A recent analysis also connects this scene to the ubiquitous chase scenes in movies of the 1920s, and thus sees Haller jump into a movie: "The lust of movies is at the same time a lust of driving and shooting"; 'shooting' is of course used here for both senses of the word: shooting a movie and shooting a gun.[261]

The car as capsule enforced a fusion of machine and driver and this, in its turn, enhanced the cocoon-like structure, sometimes up to the point of pure magic: the car's cocoon transcends, indeed, a process that at this stage is not yet felt as "de-humanization" causing "road rage."[262] The most extreme case (reminiscent of Marie Corelli from a previous period, see chapter 2) of this magical unity with the car is the novel *Das Zauberauto* (1928; The Magic Car) by the journalist and poet, and former *angestellte*, Friedrich Schnack, who five years later, with eighty-eight colleagues, would sign a creed of allegiance to Hitler. Schnack's cars are not yet ubiquitous like in the novels and poems of many of his colleagues: his car euphoria is based on its property of compressing time and space and, hence, its potential of connecting people. *Das*

Zauberauto is a story about confused senses (*sinnverwirrt,* as the author calls it) written by a champion of "expressionist nature lyricism." In it I found the only 'olfactorial inversion'—to my knowledge—in interwar literature: "the swelling, blossoming meadows ruddered along, through the lowered windows swirled fresh odors." The plot is quickly told: a young, small-town shop owner, Valentin Langguth, buys a new car on credit and enjoys the functional car adventure in his study of the "car manual" which had to make him into an "experienced owner-driver" (*Selbstfahrer*). Although he motivates his purchases as utilitarian ("how easy and fast one could execute any distant deliveries, urgent orders") and distances himself from motorcyclists and speed "bandits," his real motives reveal themselves in his initiation into the technical and driving skills, when "the steering wheel conveys, from below, a hardly felt, secret tremble, the shivering engine forces. Seductive! He had the steel racer firmly in his hands." A mysterious double, seated next to him, incites him to speed up, and thus he drives into a magical dream world, a "boy's forest" enabling him to relive experiences from his youth in which the spirit of a woman appears. This novel confirms that the car still can trigger man's boyish fantasies, even in a Germany where the utilitarian car seems dominant. These are experienced in a bacchanal of speed and flight, his trips described as films. "After the closing of the performance [of a movie he just saw] he went out on the street, still in a spell, filled with the rush of the distance and the afterglow of foreign images." It is the smells of cinnamon and other "colonial" spices stored in his shop that trigger his adventurous dreaming, but it is the car that functions as a catalyst in this transcendence into the sublime, each time his speedy drives get him lost. As a metaphor for the rush of fast but aimless meandering through the magic woods, he starts a filling station that offers both petrol and wine. The woman tempers his speed desires: "One does not drive to race past everything as quickly as possible, doesn't one?" The novel also contains dreamt trips, expressions of his "100 km life," trips without a trip, when Valerian, intoxicated by his love for the woman, enjoys the intimacy of the virtual car, which subsequently vanishes in thin air. The basis for these fantasies, the novel tries to convey, is the cyborg experience: "What is a driver without a car? A sad figure, like a groom from whom one has stolen the bride."[263] The magic unification with the technology seems to replace (or rather add to) the functional adventure of the previous phase.

The German author par excellence of the emerging cyborg as a mobile observer is Arnolt Bronnen (pseudonym of Arnold Bronner), who in one of his short stories ("Triumph des Motors") describes the emergence of the open roadster with the engine in the front, in which "man

[is] only nerve, and only a nerve among many." Man is dominated by the machine in this cyborg bond: "His eye is a telescope, his brains are already unable to determine direction and goal of the trip; he climbs in the car and dashes off."[264] In another short story, "Der Autodieb: Eine Geschichte" (1931; The Car Thief: A History), the thief destroys his "beloved" but stolen car at the end of an intoxicating race.[265] As usual, cocooning leads to distancing and aggression: "Grass is beautiful if visible at a distance from a car; the closer the vehicle comes to the grass, the more it loses its beauty." In Bronnen's stories tourists have only a limited willingness to indulge in nature: "We found [the Austrian province of] Styria as green as it is mentioned in the song, we would even have been prepared to find it blue, if it only had better streets," and they take revenge by killing a dog.[266]

Bronnen's mobility was social as well as physical: having started as a controversial playwright, and publishing his short stories, like Remarque, in *Sport im Bild*, after the prepublication of his first novel in the glossy magazine *Die Dame* (for which he received 12,000 Goldmarks) he wrote: "one goal only: a car at last" and he immediately bought an expensive Wanderer 6/30 hp. "Being a car owner," an actor in one of his short stories opines, "meant belonging to the caste of modern knighthood, it means to be sovereign within the subjugation of these colossal cities, it means to be a tiger in the asphalt jungle." In his novel *O.S.*, one of the German *Freikorpsromane* on the immediate postwar counter-revolutionary movements, the car is a "wonderous initiating rite for the hero, functioning as an entrance key to the initially strange world of the adventure." One of the "traffic novels" identified by Helmut Lethen, *O.S.* shows how mobility is impossible without communication technologies such as the telephone, resulting in a description of the antirevolutionary counter-movement as a "nervous system." The train is the main vehicle used, but in an emergency the car is handed over to a taxicab mechanic for a 500 km trip to *Oberschlesien* (O.S., Upper Silezia) who immediately learns to drive: "I don't know at all to drive, he thought, but he swallowed it, out of pride." The skills of using both a car and a gun are signs of his manhood.[267] We have to take this very literally, as Bronnen confessed that "my entire being, in the act of literary production, made me into one sole, giant genital organ." Bronnen's enigmatic, sexually driven anti-patriarchal and anti-bourgeois attitude made him not only portray himself in an earlier novel as a woman (and made him famous as a provocateur), his protagonist in *O.S.* has been compared to Gramsci's *"intellettuale organico,"* a new man "with new organs." Bronnen was fascinated by the new media, but his novel is also replete with ubiquitous, empty violence, sometimes depicted in the style of the slapstick movie. Criticized as "sexual smut" (*Schweinerei*) and "outright pornography," the intention of his "literature of adventure" was to convey "the high feelings and cerebral weaknesses (*Denkschwachheiten*) of primitive people, in trivial form." Brecht and several others (most notably Kurt Tucholsky, who called him a "fascist piccolo") distanced themselves from him, especially when he further drifted to the political right and to "national bolshevism," thus making the controversy about *O.S.* into the "largest of the numerous literary feuds of the Weimar Republic."[268] For Bronnen's friend Ernst Jünger, Bronnen's car "fitted [him] like a well-dressed suit."[269]

But the Austrian Bronnen was by far not the only "cyborg artist": Brecht's famous poem "Singender Steyerwagen" (Singing Steyer Car) declares: "We have: / Six cylinders and thirty horsepower / ... We lay in the curve like sellotape (*Klebestreifen*)," and the poet Melchior Ernst observes in *Sport im Bild*:

> Where at the steering wheel end the levers
> I end too with flesh and nerves.
> Force and blood from my veins
> flow into the mechanics....
> Whether I am lord, or whether I serve:
> only together we are rhythm.

The poem ends with a flight metaphor: "Like a star in empty ether / flies the unbound car."[270]

There were subcultural pockets and 'peripheries' that deviated from this remarkably uniform appropriation of the car by interwar petit bourgeois: not everywhere in Europe was the car seen, and criticized, as a token of *sachlichkeit* and harshness. The Russian social revolutionary Viktor Shklovsky, during his Berlin exile in the early 1920s, wrote lonely letters about his unrequited love; he followed his beloved's advice not to write about love, and the first topic that came to his mind instead was the car, whose "speed separates the driver from mankind. Start the engine, open the throttle—and you already move outside space, and time seems to move only through the speed dial." The Russian Formalist and "dynamite thrower" Shklovsky, who wrote a *Sentimental Journey* (1922) describing his flight from Russia (but in his poor Berlin exile was forced to write publicity for motorcycle firms), is one of the few who attempted a "marxist description of reality" in which he depicted cars as fetishes meant to be used by philistines with their fast spreading banality. Whereas in the first phase (see chapter 2) Julius Bierbaum could still entertain the illusion that he could combine sentimentality with the machine, Shklovsky's reference to Sterne is purely ironic, as his *Journey* "describes the travels of a bewildered intellectual through Russia, Persia, the Ukraine, and the Caucasus—all convulsed by violence and cruelty of every variety." A Hispano-Suiza is not a seduction tool (such as in Frondaie's novel), but a parvenue vehicle, with too much hood length, which "would have earrings if he were human." But deployed in the context of revolutionary struggle, the car becomes a catalyst of collectivity, the communist equivalent of what Ernst Jünger had described for the protofascist 'soldier,' with his "rather special, detached intelligence, this double consciousness, which at the same time participates directly in an action and remotely but meticulously observes its own participation." For Jünger, this self-reflective car driver becomes part of a swarm: "Depersonalized, they became cogs in a vast machine, yet cogs which, in order to function, required a high degree of technique, imagination, intelligence, courage—all, human qualities that depend precisely on a high degree of personal development." In Russia, to the contrary, the car was a metaphor of happiness: "You have poured the revolution into the city like foam, oh you automobiles. The revolution shifted gears and drove off. Springs bent, mudguards flexed, the cars raced through town and where there appeared to be two, it seemed to be eight. I love the automobiles. In those days they rocked the entire country awake. The revolution went through its foam phase and then went to the front on foot." Shklovsky thus formed a part of a Russian tradition of benign appropriation of the car, "with headlights by Marx and Company," where the car was "a metaphor of revolutionary transformation."[271]

In his *Sentimental Journey* Shklovsky, as a literary scholar (who worked on a "theory of prose," especially in *Don Quixote* and Sterne) and head of a car unit in Petrograd at the same time, writes about writing ("I begin to write again," his chapter on Persia begins. "Now I tell you what kind of country it was that I found myself in") just as he writes about driving. No wonder, then, that he uses the car as a reservoir of metaphors to describe his world. The Bolsheviks, for instance, when they took over Russia ran a "mechanism [that] was so imperfect that it could run even when improperly serviced. On oil instead of gasoline. The Bolsheviks held out, are holding out and will hold out, thanks to the imperfections of the mechanism which they control. . . . If I could just relate the things that happened to automobiles alone!" For Shklovsky, the world is like a car or a motorcycle: gulls in Stettin have "the voice of a motorcycle."[272] Shklovsky had synesthetic experiences in the car, as a Russian Proust: its pull resembled an "increasing voice," and that is probably also the reason why he rejected the electric car as having no heart (one of the few passages in Western literature that deals with this forbidden vehicle) and why he likens his lonely fellow Russian exile members to "a dead accumulator, without sound and without hope."[273]

Symbolisms and Affinities: Avant-Garde and Popular Culture

As we have seen in chapter 2, apart from providing content as narrative, belletristic literature also functions as a reservoir of symbols and metaphors that, in the long run, may even be more influential within the automotive culture and beyond. Also, in the previous pages we have given several examples of affinity, of structural homology between the practices of reading and writing on the one hand, and driving on the other. In this section we will wrap up the symbolic and homological aspects of interwar textual and filmic utterances, of both highbrow and popular culture, claiming that the differences between them were smaller than the differences between both these cultures on the one hand and tamed middle-class culture on the other.

Symbols

Both types of metaphors identified in chapter 2 are abundantly present: the metaphors that adorn the driving experience ("the mileage counter beat like a heart run crazy," Paul Morand; "the burning eyes of the Renault," Michael Arlen) and, more and more once the car became ubiquitous, the reverse: the car as reservoir of images to describe society ("my heart was full as a just-filled tank," and "his eyes sparkled like a fresh spark plug," both Friedrich Schnack). For Alfred Döblin, his "short circuit" between movement and perception is "the petrol on which my engine runs." Victor Klemperer observes in his diaries how "a multitude of mechanical terms" entered the German language during the Third Reich. Especially in the language of Goebbels men were compared to "engines running at high speed," whereas the totalitarian "*Gleichschaltung*" (synchronization) as a term was derived from automotive practice (at the end of the 1920s the gear box got 'synchronized,' allowing easier gear shifting).[274] When Hans Fallada's poor Small Man uses the railway as a metaphor instead of the car we know enough about his economic condition: "and now it is high railway that we go to bed," Pinneberg jokes to his little son.[275]

In Lewis's *Babbitt* a woman is compared to "a Ford going into high" or a man is criticized because "he never will learn to step on the gas."[276] Horváth's Kobler compares international politics with the car trade.[277] Kurt Tucholsky uses an entire poem as car symbol (and, tellingly, as car *system* symbol) for the right-wing Kapp Putsch of 1920:

> Stop, chauffeur, you drive into death!
> Listen to the bleak singing!

The wrong signal warns: Red!
And you drive in fourth gear.[278]

And the American poet Edward Estlin Cummings described "she being Brand" (which is the title of his poem from 1926), having:

thoroughly oiled the universal
joint tested my gas felt of
her radiator made sure her springs were O.

K.) I went right to it flooded-the-carburetor
cranked her
up, slipped the
clutch (and then somehow got into reverse she
kicked what
the hell)[279]

Jens Peter Becker, in his analysis of American autopoetic literature found surprisingly few poems in which the car played a carrying role, a discovery that is replicated in this author's analysis of Dutch lyricism.[280] In Dutch poetry of this period only the dark sides of the car are exposed, if at all. J. A. dèr Mouw, for instance, observes "the stinking car" as a "gob [fluim] on wheels," and vitalist poet H. Marsman sees the car as an enemy: "The commercial nitwits drive in their Buick / with expensive dolls on God's roads / whereas poets sing in their corners / between the mob in the despicable streets."[281]

In Germany, however, Bertolt Brecht not only became "the car poet par excellence," but he also created affinities between the production of cars and of poems: his montage techniques not only characterize his texts, they are even applied to the people he describes. In his poem-as-play "Mann ist Mann" (1924–1926) a man is "re-modelled (ummontiert) like a car / ... One can, if we do not control him properly / Transform him in the wink of an eye (über Nacht) into a butcher." In a discussion with Marieluise Fleisser, Brecht compared her play *Pioniere in Ingolstadt* with "certain cars as you see them drive around in Paris, cars built by their owners from separate parts assembled by the tinkerer himself, but it runs, after all, it runs!"[282] For Brecht, as car lover, this was not a compliment. Fleisser's protagonists are shown while in motion, physically and spiritually: "Move, so that I know you," the German editor of her collected works observed.[283]

Erich Kästner used the car symbol in a critical way in his poem "Die Zeit fährt Auto" (1928; The Times Drive a Car): "But no man can steer."[284] Kurt Tucholsky's pun to give his bundle of poems the title *Mit 5 PS* (At 5 hp) already shows how the car had become an accepted (and stabilized) symbol. A German book-review journal asked five workers what they thought of Tucholsky's publication, and one opined: "As a mechanic I thought... the book is a travel book which depicts the car trip of a rich man through the world. But then it struck me as odd, that it was 5 and not 6 or 12 hp," referring to the tax levels according to which cars were categorized. All in all, European poetry seems to have been the refuge of car criticism and resistance, more so than in the United States. An exception to this conclusion is the *Autobahnen* poetry, which praised the new freeways of the Third Reich.[285]

This selection of German car poetry already shows that it is not easy to create structure in the abundance of symbols. Frédéric Monneyron and Joël Thomas, in a study of the car as an *"imaginaire contemporain,"* distinguish, based upon a Jungian analysis of the entire imaginary corpus, between three "regimes" and "structures": a "diurnal regime" with "heroic structures" (keywords: separate, split up), a "nocturnal regime" with "mythic structures" (keywords: fusion), and a mediating regime with "synthetic structures" (keywords: reconciliate, travel). Applied to the car, they conclude that the automotive symbol is "totalizing" (*totalisant*), expressing the concept of liberty as well as intimacy, covering the social and the sexual, representing a weapon and a cocoon (and a cocoon as weapon, such as the tank, and the 'safe car' in general, one is inclined to add) and "contributes to a definition of a gender identity of masculinity as well as femininity." Indeed, the car's imagery during the interwar years became so ubiquitous that it covered all three regimes and all three structures.[286]

The German literary historian Rolf Parr, focusing on the car, distinguishes between five types of metaphors (the anthropomorphing of the car; the car as instrument to avoid danger; as the coupling of man and machine; as a border and as a way to lose normalcy), but he misses the car and motorcycle as weapon and, in general, the violent symbolism of the car. According to Hanno Ehrlicher, metaphors of violence were ubiquitous, from Walter Benjamin's "shock aesthetics" of the movies to André Breton's "shot in the masses" as a form of surrealist art. Especially the violence of and in avant-garde art functioned as an "aesthetic procedure," as an "intransitive act of aggression." This not only applied to the artistic avant-garde, but also to the mobile avant-garde: in every motorizing country the motorcycle's exhaust sounded like a machine gun (in Morton's England trip it even resembles "a machine-gun corps showing off to a general"). Indeed, the car's and motorcycle's potential as a symbolic source seems to be without limit, and as contradictory and ambiguous as the real world: for Remarque the car is male (called "Karl"), for Steinbeck and William Carlos Williams it is female, for Virginia

Woolf it is a hermaphrodite; for some, as we saw extensively in this chapter, the car is fast, for others (such as Holitscher on Paris traffic, or Robert Musil) it is slow.[287]

Dadaist and Surrealist André Breton had the "high ambition to possess the world in one metaphor," and for this the car was a perfect candidate.[288] German literary research has brought to light that the car symbol operated at least in six versions or series of sub-symbols: the car as vehicle, as machine, as a plant, a body, a commodity, and a host of not very much interrelated sub-symbols, such as a prison, or a raft. This taxonomy however, apart from again neglecting the role of violence and aggression, misses the central point of Interbellum car imagery, which is the metaphor of 'traffic,' and which since the end of the nineteenth century was evolving into a "primary emblem of modern life," and was reaching its first zenith during the interwar phase, as Lethen has shown.[289] Arnolt Bronnen, for instance, sees the dangers of traffic as a "new school of morality" with courses on "elasticity and resoluteness."[290]

When Brecht talks about two strategies of behavior in traffic, he talks about society like this: "I know a driver who knows the traffic rules well, keeps to them and knows how to benefit from them. He understands adroitly how to press forward, then again to keep a regular speed, spare his engine, and thus he finds carefully and unflinchingly his way between the other vehicles. Another driver whom I know behaves differently. More than in his route he is interested in the traffic as a whole, and he feels just like a little part of it.... He drives with the car in front of him and the car behind him in his head, enjoying constantly the progress of all cars, and the pedestrians as well."[291] The shift from car metaphors to traffic metaphors (or better: the inclusion of the latter in the metaphorical reservoir) is the ultimate act of synchronization between art and mobility in this phase.

Affinities

It was not only literature that appeared sensitive to the car as symbolic reservoir. Gerd Arntz, for instance, used the car in his graphic art as a token of wealth: in his *Oben und Unten* (1931; Above and Below) the car is at the upper part of the image providing opportunity for a couple to make love; at the same time, as part of the picture, the symbol is democratized as a spectacle.[292] Painter and photographer László Moholy-Nagy, for whom art was "the grindstone of the senses," dedicated an entire study to *Vision in Motion* (1947) in which he explained the "simultaneous grasp" of modern plastic arts and in which he told the anecdote of an experience with a Toulouse-Lautrec poster from 1900 and "a contemporary poster." Moved at car speed, Toulouse-Lautrec became a blur, while the modern poster could be "read."[293]

While the plastic and literary arts were struggling to deal convincingly with motion (reinvestigating the role of the senses, as we saw), the photograph managed, first, to get "motorized," and then started to talk—or better: make noise—because it was the noise rather than the speech that, according to Jean Renoir, made the 'talkie' special. It was this development, invented at the same time as the car, that inspired numerous literary artists. Sometimes this inspiration was fed through the content, like in Artaud's "theater of cruelty" derived directly from silent cinema.[294] Often, however, the inspiration took a transmedial turn, and occurred along the path of what we have called 'affinities,' the third level in autopoetics (see chapter 2). We have already highlighted Gertrude Stein's practice of having her texts be quasi dictated by the car's rhythms. In his surrealist novel *En joue!* . . . (1925), which perhaps best can be translated as 'Aiming' (of a gun alongside the cheek, *la joue*), Philippe Soupault also plays with the affinity between car driving and writing. The protagonist Julien, when waking up, asks for the newspapers, and his maid gives him both *Auto* and *Comœdia*. "If Julien starts his day by reading *Auto*, he will be sporting; if it is *Comœdia* which he reads first, he will be a poet." The 'situational' character of car driving and poetry making, and, implicitly, the transcendent function of the former, cannot be better expressed. Julien is partly modeled after Soupault's fellow-surrealist Jacques Rigaut, but while the latter committed suicide with a gun, Soupault's novel contains a suicide by car. Indeed, the roles of poet and driver are connected through violence: before the war a cousin committed suicide in a car, and years later Julien follows the road his cousin used, driving faster and faster.[295] Rigault, a member of surrealism's predecessor movement Dada (which started in 1918, its sister urban movement of surrealism started in 1924, both "firmly rooted in the city"), and thus part of France's 'lost generation,' struggled against "mediocrity" ("Every Rolls-Royce that I encounter prolongs my life by fifteen minutes") and wrote one of the shortest novels ever, *Roman d'un jeune homme pauvre* (Novel of a Poor Young Man): "Poor young man, mediocre, 21 years of age, clean hands, would marry woman, 24 cylinders, health, erotomaniac or fluent in Annamite."[296]

The *"écriture automatique"* (automatic writing) of the Dadaists and the Surrealists, just like the *"monologue intérieur"* (stream of consciousness) of the New Objectivists were the answers of the avant-garde to the same cultural tendencies that also enabled new automotive experiences in a closed car. Soupault proposed, in the words of historian Peter Conrad, to use a "literary speedometer to calculate how fast they were travelling across the page."

The Parisian expatriate Gertrude Stein studied psychology at Harvard before she went to live in Europe, and her public controversy with the behaviorist Skinner on her interpretation of 'automatic writing' is very relevant indeed for our search for the motifs behind and the skills necessary for middle-class car driving. Like the surrealists Breton and Soupault, automatic writing represented "pure performance, with the method itself more important than its 'effects.'" Aimed at "a radical de-hierarchization of literature," this form of writing was a means "to link inner and outer worlds" and was criticized by Louis Aragon as an effort "to drown all personal responsibility ... by attributing [Breton's and Soupault's] words to the dictation of some shared parasitical organ." It is as if we see the fledgling middle-class automobilization described, complete with a transfer of individual responsibility to the 'swarm' of the traffic flow. For William Carlos Williams, who "experimented with ways of controlling motion: speeding up a line and breaking it at unexpected places, letting syntax expand and contract as the feeling or subject dictates, and varying rhythm to defeat monotony," movement "must always be considered aimless, without progress." Movement, as the literary historian Charles Grivel has argued, is intrinsically connected to writing: "Traveling means placing the body into a *state of writing*.... I move and excite my body, and it writes." Probably without knowing this, Grivel is applying Schatzki's and Gibson's practice and ecological theories. Car driving can indeed be compared to how Stein defined writing, as a "distracted production" of texts: she wrote her "Literature Without Meaning" (as Skinner called it) "in the presence of distracting noises."[297] In that double sense, the avant-garde (literary and mobile) won: they "establish[ed] common ground for the majority," although this could not yet have been foreseen when super-Dadaist Tristan Tzara wrote his poem "La Revue Dada 2": "Five Negro women in a car / exploded following the 5 directions of my fingers." Tzara saw Dada conquer Paris "in a snow-white or lilac-colored car, passing down Boulevard Raspail through a triumphal arch made from his own pamphlets, being greeted by cheering crowds and a fireworks display," but his extra-poetic reality, just like Tamara de Lempicka's in her Renault, was much more prosaic: he arrived at Gare de l'Est and walked to his colleague Picabia's home, "without anyone expecting him to arrive."[298]

However they tried, the very existence of an avant-garde presupposes a hierarchy of speed, a fast and a slow lane. In the fast lane, even when surrealist writers exploded the syntax, which then did not convey meaning anymore, the "aggression, sadism, hatred" remained observable: "Dogs gorged on gasoline and set afire will be turned loose in packs against naked women, just the most beautiful among them, of course. Old people will be pressed and dried between the leaves of great wooden books and then stretched out in the carpets of middle-class salons."[299] For others, it sufficed if the speed of car culture could be expressed in 'fast' verbs, such as Friedrich Schnack did throughout his entire novel, using verbs such as fly (*fliegen*), chase (*jagen*), shoot (*schiessen*), and dash (*stäuben*; causing dust at the same time) to describe the car's movement.[300] Even Theodor Adorno, in his conservative criticism of jazz and entertainment music from 1939 (written in the United States), constructs a connection with "car religion (which) makes, in the sacramental moment of the phrase 'That is a Rolls Royce' all people into brothern (*alle Menschen zu Bruder*)." Adorno compares car driving with infantile "regressive listening," which desires the same acoustic experience, over and over again, and a revolt against it sends such listeners only deeper into their state of alienation. For Adorno, the driver (of his "old second-hand car") is the model of the "cool" listener. His personal skill is nothing more than the mindless submission to the powers that be. His control of the machine is the perfection of his submission, a thesis we see, many years later, resonate with Cotten Seiler's analysis of the United States as a 'Republic of Drivers.' If Adorno had chosen the train as a symbol (as Joel Dinerstein did in his seminal analysis of jazz), he might have concluded otherwise.[301]

While the content of travel writing became decoupled from actual travel experience (see Roussel), avant-garde artists put mobility on the throne of modernity, not so much through their content, but through affinity. Apart from this by now quite obvious homology between car driving and experiencing something bigger than oneself, there are many more parallels between avant-garde writing and car driving, albeit on a more abstract level, and especially when it concerns violence. A recent German literary study emphasized the analogies between car, film, and weapon, identifying terms like "shooting," and "projectile" (*Geschoss*) as topoi.[302] This was not the prerogative solely of men: German journalist and motorist Friedel Spada wrote a travelogue titled *With a Shotgun and Lipstick* (1927) about her journey to the East.[303]

Recent studies of the ubiquity of the car accident in interwar literature result in another powerful example of affinity: the words themselves are becoming "automobile," they "roll" as if following their own logic, texts are "assembled" (*Montage*), so much so that they sometimes "crash" and the analyst herself becomes captured by the power of the metaphors:

"Just as the car runs over Franz," Claudia Lieb writes in an analysis of *Berlin Alexanderplatz*, "the text thunders over the traditional forms of the novel." The accident is as much mechanical as it is linguistic.[304] The crash can also be seen, according to Karen Beckman, as a way of restoring the distance between those inside and the 'others,' as a "leaning toward the other without fusion." The focus on collisions "bring[s] difference to the fore within a framework of uncomfortable, sometimes painful, and even fatal, proximity."[305] British member of the lost generation, Christopher Wren, to give only one example from many, has a 'bad' woman in his very popular masculinist first novel, *Beau Geste*, die in a car crash, a litteral deus ex machina that soon would be used by many novelists as a fate one didn't have to discuss any further.[306]

A literary history of French modernist literature emphasizes that in modern novels, "in the end, the genesis appears to be more interesting than the product," just as, one is inclined to think, the trip in the adventurous realm of interwar car culture is preferred to the destination. In a parallel argument, the "subjective realism" of the "deconcentrated work" (*œuvre déconcentrée*) of French surrealism, which nonetheless is equipped with a "central consciousness," affinities are revealed with the swarm-like behavior of the increasing traffic flows. Travelogues by Paul Morand are considered to be part of the genre of the adventure novel (*roman d'aventure*).[307] Dada's French journal *Aventure* also indicates that there existed a true 'adventure movement' in Europe, against which Bertolt Brecht stood opposed, as seen in his assertion that "Adventurers in our society are criminals."[308] Adventure and anarchism were "the conspiratory code words" of young French surrealist writers Aragon and Breton, while they also literally used the term *"coeurs aventureux"* (adventurous hearts) as perhaps a direct reference to Ernst Jünger's theory of adventure. For Jünger, the adventure is a means of knowledge acquisition and the "adventurous heart" is an "organ of perception" to enable just that. One of Jünger's analysts calls this a "neoromantic reflex": "The task is life, but adventure is poetry. Duty makes the task bearable, but the lust of danger makes it light. That is why we don't feel ashamed to call ourselves adventurers."[309]

The French *nouveau romancier* Alain Robbe-Grillet, after the Second World War, would characterize the modern novel as "less *the writing of an adventure than the adventure of writing*," echoing Rivière's theory of the *"roman d'aventure"* we dealt with in chapter 2.[310] In other words, the affinity between car driving and car writing is not shaped through transcendence. Neither practice meets in a realm beyond social reality. Instead, the content of this transcendence, its motor, is 'adventure,' which means that one is prepared to expect the unexpected. Simmel, too, connected transcendence to adventure. Other thinkers have done so as well: "Adventure," Paul Zweig concluded in his analysis of *The Adventurer*, "is clearly a form of 'experiential transcendence.'"[311] This transcendence had a multisensorial 'flavor,' especially in the hands of Surrealists, who, as successors of the multisensorial Symbolists, developed their own "multisensory aesthetics."[312]

This, now, could be felt and experienced by ever more middle-class people in industrialized countries. For some of them, the transcendence was toward a group feeling, for others it was more abstract, a feeling of "communal identity" of sorts.[313] Multisensorial transcendence, then, seems to be the core of avant-garde experience, whether in writing or through driving. This transcendence, the previous pages have made abundantly clear, carries overt, but most of all covert, aggressive and violent overtones. Technological sublime (David Nye) has shifted in the automotive realm to the 'sublime of violence.'

The most radical juxtaposition of writing (if not literature as such) and the car is to be found in William Carlos Williams's anti-novel *The Great American Novel* (1923), "a satire on the novel form" in his own words,[314] a text without much structure but that exclaims repeatedly that "the words" must be set free, especially from their European (English) influence on American literature. In this novel Williams, who "freed American poetry from the irons of rhyme and meter," has nearly every incantation on the liberation of language been followed by a passage on cars. On one occasion the car even acquires agency as it seems to fall in love with a truck:

> And the little dusty car: There drawn up at the gutter was a great truck painted green and red. Close to it passed the little runabout while conscious desire surged in its breast. Yes there he was the great powerful mechanism, all in its new paint against the gutter while she rolled by and saw—The Polish woman in the clinic, yellow hair slicked back. Neck, arms, breast, waist, hips, etc. This is THE thing—The small mechanism went swiftly by the great truck with fluttering heart in the hope, the secret hope that perhaps, somehow he would notice—HE, the great truck in his massiveness and paint, that somehow he would come to her. Oh I wouldn't like to live up there![315]

The machine and the words are interdependent:

> On the side of the great machine it read: Standard Motor Gasoline, in capital letters. A great green tank was built upon the red chasiss, FULL of gas. The little car looked and her heart leaped with shy wonder.
>
> Save the words. Save the words from themselves. . . .

> Puh, puh, puh, puh! Said the little car going up hill. But the great green and red truck said nothing but continued to discharge its gasoline into a tank buried in the ground near the gutter. . . .
>
> What then is a novel?[316]

Elsewhere in the text, the narrator grants James Joyce that he "has in some measure liberated words, freed them from their proper uses. . . . For me as an American it is his only service."

> It would be a pity if the French failed to discover him for a decade or so. Now wouldn't it? Think how literature would suffer. Yes think—think how LITERATURE would suffer.

> At that the car jumped forward like a living thing. Up the steep board incline into the garage it leapt—as well as a thing on four wheels could leap—But with great dexterity he threw out the clutch with a slight pressure of his left foot, just as the fore end of the car was about to careen against a mass of old window screens at the garage end. Then pressing with his right foot and grasping the handbrake he brought the machine to a halt—just in time—though it was no trick to him, he having done it so often for the past ten years.[317]

In the end, when there is no story to tell anymore, an anonymous male plays tricks with a car in a garage.

What is crucial for this phase, too, and another expression of the shift from car artifact to car system, is that one does not have to *own* a car anymore in order to *use* it, as a provider of metaphors, or outright fantasies. The Dutch New Objectivist Constant van Wessem uses the car image as a vehicle for his love to *Adelaïde* (1924):

> I have no car. I have no riding horse. But you, Adelaïde, I love. I want to travel to Monte Carlo, visit casinos and gambling houses to put the 100,000 *francs* on one figure and to win, which would enable me to get a car, a riding horse, a servant, as you wish. And then, you the terrible, you will sit in the car and I will drive. We will travel, we will make trips, we will fly up on mountain slopes and shoot over ravines and race through chute brooks until we have reached the top in a spiral move. And when we have reached the top I will put the steering wheel over and we will at full speed dash against the earth. Ha, Adelaïde, then our souls will meet in Heaven, to dance a menuet of sweet understanding! Hahaha![318]

Popular Culture

When we read the last fantasy we realize that in this second phase of automobilism, avant-garde and average driver only overlapped very partially: the former was no protagonist in car driving anymore (like Mirbeau or Dreiser in a previous phase), and the latter was the subject of constant criticism by the former, especially in what literary historian Elisabeth Tworek-Müller has called "petit-bourgeois novels" (*Kleinbürgerromane*). She analyzes how in the "*Trivialliteratur*" (dime novels) an appeal was made to the German "resentment" (*Ressentiment*) whereas "those that could have immunized them, such as Thomas Mann and Hermann Hesse and Alfred Döblin and Kasimir Edschmid, were not read in the circles we are talking about."[319] According to her dissertation, the "petty-bourgeois mentality" can only be analyzed by combining the individual elements of this mentality. Literature, as the "most sensitive ... organ of the public sphere," is able to do so par excellence.[320] In this literature, against the critical, self-reflexive attitude of the 'small man' stands the flight forward into boastful noisiness of Babbitt. Both are desperately looking for some anchor in life. Consumption can provide such an anchor, and consumption of (automotive) mobility in particular. The latter was Babbitt's privilege, and out of reach for Fleisser's Gilgi, Fallada's Pinneberg, and Horváth's Kobler, although they managed to approach Babbitt's example, in the bus, in the taxi, or through swindling. Although the role of the (lower) middle classes in the emergence of fascism should be relativized by pointing to the role of the industrial high-bourgeoisie and the collaboration of the existing bureaucracy and power structure, Tworek-Müller refers to Adorno's famous F-scale, which should be able to predict fascist tendencies. This F-scale measured "subsumption under conventional values of the *Mittelstand*, ... uncritical submission under the authorities of the own group, ... the tendency to judge, reject and punish people who don't respect conventional values, ... repulsion of the subjective, the fantastic, the sensitive, ... superstition and the disposition to think in strict categories, (and) thinking in dimensions such as dominance—subjugation, strong—weak, leader—followers." According to Tworek-Müller, "determining factors of the petit-bourgeois socioculture such as family life, leisure, forms of sociability (*Geselligkeit*), political attitudes and partisanship" are much easier described by literary writers than by social scientists. Especially pronounced is the petit bourgeoisie's inclination to compensate for "social declassification" by "ideal superiority," and either to radicalize toward right-wing attitudes or to "flee into the idyll," or both, and preferably in the context of the family. For our automotive adventure the aggressive behavior toward strangers is important, toward Jews and "reds" in the first place, but also toward the (foreign) other. At the same time, the car can be dreamt about, even if (or perhaps especially because) one does not own one, as in the case of Horváth's protagonist Agnes Pollinger, who produces an entire idyllic, touristic dream.[321]

Redefining Adventure

In the slow lane, however, the average motorist became silent, if measured in terms of literary production. In the *Revue du TCF* of the Touring Club de France hardly any members not belonging to the institutional avant-garde took up the pen to put their automotive adventure on paper. Similarly, when in 1938 the Dutch journal *Toeristenkampioen* (Tourist Champion) of the touring club ANWB initiated a contest for "holiday memories," only a hundred of the one hundred thousand members responded, and most of their products were rejected because of their "factual" character. "Moods" had to be depicted, a desperate editor emphasized, and this was apparently not given to many. Gone were the times of the sentimental utterances of the car club members of the prewar period.[322] This was confirmed by a novelist like E. M. Forster, who observed in 1920 that the middle class's "solidity, caution, integrity [and] efficiency" was now the "dominant force" in Britain, at the price of "smothering ... imagination and creativity."[323] Another reason for the motorists' silence, however, may be found in the fact that 'moods' were no currency anymore among the New Objectivist or Surrealist avant-garde. Instead, the avant-garde started to discover the 'common man' as 'stuff' for novels and poems. Babbitt was, of course, the quintessential icon of this. For him, as well as for the majority of drivers, apparently, writing was no longer the intellectual equivalent of motor driving. And although highbrow literature tried to cope with the new experiences in its New Objectivist coldness (mixed with a shot of nostalgia and sentimentality), it was the cinema and popular song that took over that role. The 'dance of the car' (Kracauer) was expressed in the consecution of images and the rhythm of songs.

Becker concludes that, for this period, it is popular culture where the impact of the automobile first and foremost has to be sought. American dime novels, British "penny dreadfuls," German *Groschenhefte*, and Dutch *stuiverromans* were replete with car chases and all other characteristics that seemed to be borrowed, at least in America, from the Western and its "ideology of manhood," and in Europe from the urban crime novels, as well as from America. Youth novels, especially in the United States, received a new injection of automotive adventurousness when they started to emphasize the skills necessary to drive and repair a car, and greatly diversified the typology of adventures that the car offered, meanwhile hardly making a difference between boys' and girls' experiences. Even in the conservative Netherlands girls' books presented car driving as a rite of passage for well-to-do girls. In Nellie Weesling's *Een tante die chauffeert* (1934; An Aunt Who Drives) two girls undertake a spatial (and functional: three punctures, engine trouble, a leaking radiator) adventure with their 23-year-old aunt who drives her old Ford (bought upon advice of the girls) on an adventurous odyssey through the southern part of the country where men seem to be either stupid or dangerous.[324] In the United Kingdom, however, if we are to believe the most authoritative study on the subject, girls' novels seem to have largely neglected the car and jumped immediately to the next-higher form of aviation adventure, either as hostesses or as pilots. Like in other cultures (see the early novels by the American Willa Cather), middle-class girls initially enjoyed the bicycle, such as in Isabel Orman's "The Strike of the Sisters" (1899, published in the journal *Girl's Realm*) or H. G. Wells's *Ann Veronica* (1909). But with the emergence of "violent feminism" around 1910, and especially when woman icons appeared such as Amelia Earhart and Amy Johnson (who flew from England to Australia in 1930), girls' literature became "airborne," as can be seen in the serial *Mistress Of The Air* (1937–1939) by Marise Duncan. When, in 1933, the long series on the girl detective Valerie Drew started, one of the (male) members of the team of authors opined: "In a car race—if one is injured the other boy goes on to win. When it comes to girls you've got to have that girl stopping and losing the race and you show what a fine girl she is not to care about winning the race, but to stop for the male rival who's crashed. Different angles—that's how you have to do it. If you don't do it that way, you know, she drives heedlessly on although the dog's injured at the roadside, the readers wouldn't take it." The stereotyping even goes further: many "girl criminals" in this series are French "adventuresses" who did not enjoy "British upbringing."[325]

Typical for the shift in meaning of the car in this phase is the role of illustrations in this form of popular literature. The Dutch illustrator Hans Borrebach later remembered how he was not very charmed by the existing habit during the Roaring Twenties to draw women "without much of a bust." By the 1930s he was glad the gender confusion was over and he could contribute to a new lifestyle magazine in which the eroticized female body and the car could be more openly connected, such as women in overalls, a monkey wrench in hand, and he could illustrate a piquant story about a car trip through Germany in which the author picked up a fifteen-year-old school girl. Borrebach's drawings were the first Dutch car-related pin-ups of the 1930s.[326]

In a way, by focusing on popular culture, the crass masculinity of highbrow avant-garde driving and writing somehow got moderated, just like low- and middlebrow war novels brought 'balance' into the memory of war, as we saw in chapter 3. In the case of the British interwar years, literary historian Alison Light found a more feminine "redefinition of Englishness, ... at once less imperial and more inward-looking, more domestic and more private." She sees this expressed, among others, in the detective stories of Agatha Christie, who herself lived an "itinerant" life, but who let most of her adventures develop in trains, no doubt also because they provide such a claustrophobic setting and allow extensive collective conversations. Christie's stories convey the domesticated middle-class adventurousness, always "staying firmly on the rails," predictable but—paradoxically—unexpected, where the good and the bad in the end are clearly separated and the latter can be put at a distance by the former.[327]

In another segment of interwar popular culture, songs in the United States expressed the "Truck Driver's Blues" (1936), or contained sexual innuendoes such as in "Sport Model Mama" by Bertha Chippie Hill, who reported to "receive punctures every day," or Virginia Liston who in 1926 accused her "Rolls Royce Papa" of having "a bent piston rod." Every national culture got its own vocal culture. In the Netherlands the most popular song in the 1930s was from the revue artist Louis Davids about a lower-middle-class neighborhood salesman who pawns his radio and buys a shaky Model T at a scrap yard that does not start when he wants to take his family on a ride. Every Dutchman knew the refrain: "And the oil man, the oil man, he's bought himself a Ford. / And he drives it through the city, acting like a lord." In Germany, in 1929, Kurt Gerron and Friedrich Hollaender proposed an anti-car "war song" (*Kriegsgesang*) from the perspective of pedestrians in their *Die Grossstadt-Infanterie* (The Metropolis Infantry).[328]

Indeed, nonliterary popular culture seemed to be able to express the automotive adventure better, or at least easier, than the written word, at least as far as the immediate experience, near to the everyday practices, is concerned. Like the car itself, film as a medium accommodated the "restlessness of the urbanite" and focused on the "everyday experience" as a basis to extol the unusual, the adventure. Film, as a "language of movement" (*Sprache der Bewegung*) showed the "sensations" and the "gestures" rather than the inner life and the reflectivity of the actors. The "optic manner of narrating," German film theorists of the period commented, produced "something sketchy, abrupt, fragmentary," while actors' behavior seemed to be outer-directed. The privileging of sight in early silent movies until the end of the 1920s heightened this impression of witnessing a "hyper-reality" (which is, after all, nothing less than an English translation of "*surréalisme*").[329] For Walter Benjamin, movies rescued "us" from the prison of our taverns, offices, and railroad stations by allowing that "we calmly and adventurously go traveling."[330]

Film and photography also seemed more prone to not only express, but at the same time produce the automotive experience. This was literally the case with the new hobby of amateur filmmaking that, according to a recent analysis, not only produced "a sense of family . . . *because* it was badly made," it also showed a typically modern characteristic in that it was highly reflexive, as it was used "to film the act of filming, or, in fact, to make seeing visible." Remarkably, although hardly analyzed, such films often did not show the means by which the family had traveled, and as such stood in a tradition started with Edith Wharton (see chapter 2).[331]

The car was certainly a central actor in the 16-mm thirty-minute film Kiyooka Eiichi (a Japanese student in America) shot during his trip across the United States in a Model T in 1927 from the back seat while his two American friends were driving. Eastman Kodak had introduced the equipment in 1923, and he paid as much for it as for the car: $100. Eiichi had graduated with a degree in English literature from Cornell University, but he had just started an engineering study, and in preparation for his trip he took the engine of his Ford fully apart and put it back together again. He also modified the interior of the car. The Japano-American excursion's filmic account is characterized by "the thrill of movement and the thrill of reproducing movement through the cinema." Eiichi's diaries tell a tale of intense functional adventures, not only around the car's breakdowns (such as a wheel coming off), but also in negotiating the roads: "The best way was to run into the mud hole and get stuck and wait until another car comes in the other direction. They would stop before running into the mud hole. . . . Then this man who helped us goes into the mud hole and waits for another car to come along to pull him out. When you get used to it, it isn't too bad if you take in the right spirit." Part of Eiichi's adventure, however, was also the shocking racism encountered during their trip.[332]

Much better organized were the famous motorized adventures into the French colonies set up by competing French car manufacturers and embedded in a military tradition of conquest and adventurism. Citroën's *La traversée du Sahara* (1923; Crossing the Sahara), and the black and yellow "cruises" of 1926 and 1932 (to Africa and China respectively) were done on half-track cars, and documented by film company Pathé Nathan. The 'yellow cruise' to China (the *Croisière jaune*, 1931–1932) was supported by British intelligence services and the French geologist and philosopher Pierre Teilhard de Chardin. The Black Cruise took, as a French popular book on cars and films observed, eleven thousand kilometers of road and fifty-two kilometers of celluloid, but thereafter the desert lay open for winter tourism by car. Such cars took their own rails with them, so to speak, and they surpassed earlier failed efforts to construct a railway line through the desert, not coincidentally the same type of space where in the United States the *Grapes of Wrath* could ripen. Renault (*Les mystères du continent noir*, 1926, Mysteries of the

Black Continent; and *La première traversée rapide du désert (329 heures)*, 1931, The First Rapid Crossing of the Sahara [329 hours]) and Peugeot (*Images d'Afrique*, 1926) produced their own versions of what after the Second World War would become institutionalized as the colonial car adventure par excellence, the Paris–Dakar car rally. The "depiction of African women in various states of undress" and the use of the term "penetration" in one of the films left no doubt about the message of masculine taming of both the car and the continent: before this kind of travel could be commodified as tourism, its technology first had to be domesticated.[333] When, in 1931, French archaeologist Marthe Oulié published her account of a Sahara car rally commemorating the taking of this part of Africa by France a century ago (but justified in terms of the 'conquest': it should prove the usability of the road from Algiers to the river Niger), the adventure of the desert had been domesticated into a thrilling contest between ten groups of four cars in which sixteen women took part. For those who found even this too cumbersome, there was always the possibility to cross the desert in a luxurious motor bus.[334]

The cyborg as part of the system illustrates the paradox of dynamism and stability (the mobile sessility, as it will be called later, or "the dialectics of restlessness and coming to rest") in the car: while the latter moves, the driver and passengers are static, whereas in a train, on a ship, in a plane, even in a bus, passengers can be mobile. In the car the movements of the driver's limbs are prescribed and well defined, like a pilot, or ship's captain, or a bus driver, or, for that matter, a visitor to the movies, as we will see later.[335] Thus, car drivers and passengers are doubly nested, as the car itself is embedded in the traffic flow. Film, a flow of pictures itself, appears to be a preferred medium to show this, such as in Walther Ruttmann's famous documentary movie *Berlin: Die Sinfonie der Grossstadt* (1927; Berlin; Symphony of a Metropolis) in which not only the title emphasizes the composedness and controlledness of urban culture, but in which traffic flows suggest that the world "constantly reproduces itself," like a turning machine, or a river, always the same and constantly changing. Heinrich Hauser's new adventure and technological sublime became common knowledge: "The colossal dynamism (*Schwung*) ... puts the onlooker in a hypnotic dusky state" and "produces slowness," just like car flows become monotonous and congested.[336]

Because of the collapse of the European film industry during WWI and other more domestic reasons Hollywood witnessed its breakthrough during the 1920s and a decade later only five big studios were left, mirroring the concentration of car manufacturers in Detroit. There were more similarities between both industries, such as their "parallel processes of self-regulation" and the way their respective products were advertised, nicely illustrated in a movie produced in 1936 called "How You See It: How Persistence of Vision Makes Motion Pictures Possible," in which the car was used as example.[337]

Surveys indicated that females between ages eight and nineteen went to the movies nearly once a week. They were, as the Lynds jotted down from the mouth of one of the Middletowners, "more aggressive today" in making dates and even if they did not possess cars, they could produce car-related fantasies by enjoying standardized formulas such as "the adventure story, western, comedy, and love story." For the Lynds, who were drawing comparisons with an earlier phase of the extended family, movie culture was "an individual, family, or small group affair," and while both industries emphasized their products as "sanctuaries of individualism" and later historians concluded that both "reinforced personal isolation," it seems that the collective as true subject of both subcultures has largely been ignored, also by film historians. Hence, the function of the automotive cocoon to enable "shared journeys" has been vastly under researched.

In Europe, Berlin functioned as the hub through which American movies spread over the continent. German film theorists seemed to be more insightful in this respect, as they emphasized that individuals in the audience represented subgroups of businessmen, workers, elderly women, salesgirls, etc., although they connected this with hope (just as earlier commenters had done in connection with the car) for a "democratic impulse" leading to a "classless audience." The basis for this hope, again, was the observation that film (like cars) "makes people alight from their own selves."[338] Transcendence, again, in both cases.

Film also influenced car culture through its star system, the most instructive example for this period perhaps being Gloria Swanson, who, at seventeen, played in *The Danger Girl* (1916), in which she raced a car in her normal dress (but switched to overalls to change a tire). Used to racing her car on screen as much as off screen, she "disguised herself as a man complete with top hat, tuxedo and cane, smoking a cigarette and even flirting with another girl."[339]

Film covered all three aspects of the automotive adventure. The functional adventure seems a motif mostly occurring in the silent movies during the 1910s and 1920s, such as *Tire Trouble* (1924), in the comedy series of Our Gang (problems with a Model T bus), or *'Twas Henry's Fault* (1919), a movie about the reliability of the Model T passenger car.[340] In a period when the car's reliability was still a hot issue it was not a coincidence that early cars in movies were either taxicabs (James Cagney in *Taxi*, 1932) or luxury items well attended by (mostly invisible) mechanics, such as in the first filmed version of *The Great Gatsby* (1926, dir. Herbert Brenon, now considered lost).[341] Race movies, celebrating the orgy of speed for millions just as car novels in the previous period had done for a much smaller audience, abounded during this phase: *The Roaring Road* (1919), with footage from a Santa Monica road race; *Speed Along* (1927), about a farmer getting a race car for his anniversary; or *Getting Gertie's Goat* (1927), a motorcycle comedy. Howard Hawks's *The Crowd Roars* (1932), with a "cool" James Cagney involved in an accident, was "the first true racing movie," in which stunt men pulled the car to a speed of 180 km/h and then have it lose a wheel. Close-up shots of the occupants were taken from a camera on the hood, shooting through the windscreen. The list can be extended nearly endlessly: the twelve episodes of *Burn 'Em Up Barnes*, from 1934, focused on Indianapolis racing but also on going down the stairs, in a car, of the Sacré-Coeur in Paris; *Speed,* 1936, with James Stewart as a test-racing driver at the Indy 500; *California Straight Ahead,* 1937, with John Wayne in a racing truck; *Daredevil Drivers,* 1937, depicting outlaw race tracks and a collision with a motor bus; *Ten Laps to Go,* 1938, with a crash during a race.[342]

Speed coupled with armed violence was one of the preferred topics of early cinema (just as the movie also appeared to be an excellent medium to convey the "timeless adventure" of the Great War[343]), although the first full-length gangster movie appeared only in 1927 (*Underworld,* by Joseph von Sternberg). To give only one example of the popular combination: the drive-by shooting in *Scarface: The Shame of the Nation* (1932), a movie that deals with Chicago's gangster culture. Much more modest was film production in Europe, but the popularity of *Die Drei von der Tankstelle,* to mention only one, testifies that the car had meanwhile conquered the cinematic universe. It is also less well known that German filmmaker Fritz Lang made *Spies* (1927), with a motorcycle chase. A special genre in the high-speed car movie is made up of *Autobahn* propaganda movies, especially because they seem to break with the modernist obsession for individuality: they depict the car as part of a system, a flow along a corridor of gigantic dimensions. *Strassen ohne Hindernis* (1935; Roads Without Obstacles) is the first of this set, stressing the *Autobahn*'s role in unemployment abatement, while in the feature film *Mann für Mann* (1939) the *Autobahn* functions as "the site for resolving male subjectivity in crisis."[344]

Through the movie series of the Keystone Kops (in the previous period, between 1912 and 1917), Harold Lloyd (1916–1963), Laurel and Hardy (1921–1944), the Three Stooges (1930–1934), and Charlie Chaplin (for instance in *The Car Cheap*), millions of viewers all over the world could participate in the fun of destroying cars, of laughing away the tensions generated by the dangers these cars produced, especially in the cities, and could learn to internalize the idea that modernity and the individual possession and use of the car were intimately coupled.[345] Lloyd, a collector of cars and (later) hi-fi systems, started to make slapsticks for Keystone, and then made autopoetic movies such as *Get Out and Get Under* (1920) starring a Model T (the title inspired by a car song), and *Girl Shy* (1924) and *Speedy* (1928). Keaton, whose hobby was model trains, seems to have developed his cool, indifferent face because he saw in his vaudeville years that the audience did not like the victim of violence to smile. Chaplin started his film career with autopoetic themes, such as *Kid Auto Races at Venice, Mabel at the Wheel,* and *Mabel's Busy Day,* all made in 1914 and shown to the soldiers in the French trenches.[346] The German travel reporter Egon Erwin Kisch relates a fascinating anecdote during his visit (together with Upton Sinclair) to Chaplin in Hollywood, who, working on *City Lights,* was struggling in take after take to get a car as an actor into one of his scenes. When asked by Chaplin, both visitors totally overlooked the presence of the car in the scene. "And what does the car do?" Chaplin asks them. "'I don't know,' I say. And Upton Sinclair opines: 'I think he drives away.'" 'Devil, devil,' Chaplin murmurs, 'all to nothing.'" In another scene, when Charlie offers a flower to a blind flower salesgirl, both again fail to see the car that is parked during the entire scene at the curb. The normalcy of the car, even for the European visitor, had increased so much that a film director needed to rethink its role in order to keep attracting attention. This reminds us of Brecht who found he needed alienation to deal with the car's ubiquity.

When it comes to violence and cars in popular culture Laurel and Hardy knew where the anxieties of their audience were situated: their comedy *Big Business* (1929) paired "car demolition (with) house smashing" and thus enacted, as a laugh, "the subject [of] 'war' itself." And Groucho Marx (in *Monkey Business,* 1931) knew how to approach a woman in the language of the time: "'You're a woman who's getting nothing but dirty breaks. Well, we can clean and tighten your brakes, but you'll have to stay in the garage all night.'"[347]

From this period, in which the car's fragile *assemblage* was still in peoples' minds, stems the hilarious Laurel and Hardy movie *Two Tars,* in which in a typical tit-for-tat slapstick a traffic jam is converted into a scrapheap of cars by the motorists themselves, pulling components from each other's cars. The film also defuses the tensions around the need to have a minimum amount of driving skill; in an earlier scene in *Two Tars* Ollie takes over the steering wheel from Stan because of his bad driving, and he then drives the car right into a lamp post.[348] The humoristic play about who masters who should be interpreted against the background of increasing automotive domestication. The message was this: technology (the "mischievous machine") can fail, and so can we, as drivers. Remarkably, automotive domestication went hand in hand with the taming of the "anarchistic" slapstick, with "bad drivers and ridiculous traffic cops." While initially comedy may have protected directors from increasing censorship more than was the case with film in the "dramatic realism" genre, in the 1930s "the studios ... brought their vehicles into greater conformity with classical storytelling conventions and established social standards."[349]

In general, slapstick was an excellent tool to incorporate the violence and (tongue-in-cheek) aggression of the early automotive culture, building upon earlier European examples as we have seen in chapter 2, but soon developing its own grammar of senseless falling, beating, kicking, and other forms of bodily harm. Certainly, Hollywood movies, including their fascination for the (often motorized) adventures of gangsters, were thriving on a violent cultural substrate. If the car did not provide adventure any more, film did: "All the adventure," a *Saturday Evening Post* advertisement ran in the 1920s, "all the romance, all the excitement you lack in your daily life are in Pictures. They take you completely out of yourself into a wonderful new world. ... Out of the cage of everyday existence! If only for an afternoon or an evening—escape." The parallels between film viewing and car-driving experiences were striking: if they did not offer transcendence ("out of yourself") they at least enabled "escape," out of your social self.[350]

The classic road movie, which according to the consensus among film historians started in the 1960s in the United States and should be connected to "the nation's frontier ethos," was prefigured during the Interbellum in movies such as W. C. Fields's parodic *It's a Gift* (1934) and its serious counterpart *The Grapes of Wrath* (1939, with Henry Fonda). In Frank Capra's Oscar-winner *It Happened One Night* (1934) Clark Gable and Claudette Colbert hitchhike and flee by motor bus, living out a "predominantly male fantasy about the possibilities of automobile romance."[351] Westerns and "Depression-era social conscience films" formed the main inspiration for prewar road movies, such as *I am a Fugitive from a Chain Gang* (1932) in which a truck is stolen to get away from prison. Humphrey Bogart plays one of the brothers who start a wildcat trucking business to the West Coast in Raoul Walsh's *They Drive By Night* (1940); his character loses an arm in an accident. The trip itself, as physical endeavor, is the main theme of these early road movies.[352]

Contrary to the consensus in film studies that films convey the individuality of the protagonists, film seems particularly able to visualize urban scenes of flow and other system characteristics. The semi-documentary movie *Menschen am Sonntag* (1930), for instance, made by a collective of amateur directors, shows a motor bus drive through Berlin. For the professional movies, special techniques were developed, such as 'traveling' shots or the 'phantom rides,' where the camera is mounted on the vehicles and directly documents the acceleration of urban life. This affinity between film and life is literally performed in the accelerating walks of Buster Keaton. It is even more celebrated in the experimental documentary *Man with a Movie Camera* by the Russian Dziga Vertov (pseudonym of Denis Kaufman) from 1929, in which not only images are shown four times faster than usual, but the lack of human actors also strengthens the machinic rhythm of the city's traffic in which the flow of water is juxtaposed to the flow of traffic. We see the city through the camera, but we also see the man with the camera, placed on a car, filming a bus, an ambulance. The movie is a rapid montage of urban perceptions, and perceptions of perceptions (reminiscent of the amateur tourist shots we saw earlier), not of speed or movement (as that in and of itself cannot be observed) but of fast people, fast vehicles, fast water, fast products in a factory. We see mobility as part of a system, as a system in and of itself, in its modal split (tramways, cars, pedestrians). Vertov opposed the approach of directors such as Sergei Eisenstein, who tried to influence their audience through the emotion and psychology of the movie's content. Instead, he believed that people would evolve into a "perfect electric man," whose kinship to the machine would improve his perception.[353] In the end, the cyborg motif haunted both popular and highbrow autopoetic culture.

Conclusions

What does an analysis of literary and popular culture add to our findings of chapter 4, where we concluded that new users (in terms of class, gender, and ethnicity) joined the prewar elite motorists and in the process reshaped automotive adventure? Indeed, the reading of more than two hundred and fifty novels and poems and the viewing of a dozen movies confirm the remarkable continuity of what we have seen before World War I. Like the car, its culture democratized, and this was accompanied by domestication—that is, by taming and literally making 'familiar,' both the car and the man.[354] In other words, the further taming of the car was a form of appropriation within a petit-bourgeois family culture that incorporated it into a societal system. Kracauer and Musil, who likened the driving of the car to dancing, may not have included this perspective, but the domestication of "hot jazz" into ballroom dancing was a parallel process, taking place through the standardization of steps and the elimination of movements that were considered too wild, such as the lifting of the partner and the raising of a leg. Tim Cresswell, who analyzed this process, also shows how American Supreme Court rulings redefined mobility in such a way that a citizen's freedom of mobility was recognized, while "shadow-citizens" (such as hoboes and tramps) were considered not to have a "right" to free movement.[355] From this general perspective, European and American novels convey the same basic attitude: they experience the car as present as a matter of course, especially toward the end of the Interbellum. This, of course, has to do with the background of the authors, who for the most part came from the same societal groups that started to motorize in this phase.

In terms of content, all aspects of the prewar car adventure (temporal, spatial, functional, erotic, and colonial) continued to appear, but their specific mix, per subculture (Europe, United States), per country, per group (including gender and ethnicity), and period (Roaring Twenties, Depression era, 1930s) differed and an impressive variation of car-related behavior was developed and displayed. They also mixed in a myriad of ways, such as in the case of the eroticized maintenance culture in Steinbeck's *The Grapes of Wrath*: "'Jesus!' he said. 'They ain't nothin' I love like in the guts of a engine.' 'How 'bout girls?' 'Yeah, girls too! Wish I could tear down a Rolls an' put her back. I looked under the hood of a Cad' 16 one time an,' God Awmighty, you never seen nothing so sweet in your life!'"[356]

The new elements of interwar Atlantic automotive culture made it distinctive on its own terms. The democratization of the automotive adventure led to its domestication in the context of the family, while new groups (women, blacks) now started to instrumentalize the car as a (still contested) tool of liberation and emancipation. "Black automobile owners felt a sense of escape and adventure," Mark Foster opines when he analyzes prosperous blacks' leisure. And in a magazine article titled "Why Take a Man Along?" Laura McClintock asked her readers in 1923 "What woman having tasted the joys of entire freedom, would want to go back to man-chaperoned obscurity in the back seat?"[357] And although not a few of the adventures of black motorists turned out nasty, one of them concluded in 1930 that "all Negroes who can do so purchase an automobile as soon as possible in order to be free of discomfort, discrimination, segregation and insult."[358] It seems as though, within this spectrum of cultural utterances, the spatial adventure started to dominate. Perhaps because of that the celebration of violence and aggression (the temporal adventure) witnessed a new apex, especially in popular culture, where the spatial adventure had been made unthinkable unless at high, reckless speed.

The distinction between highbrow and popular culture points to one of the two major differences with the previous period. The new users did not write to the same extent as their elite counterparts of the prewar phase; they were discovered, so to speak, by an avant-garde that described their lives critically in a context of automotive ubiquity of which even the nonusers were a part. The car was so pervasive in this culture that it often was hardly worth mentioning, and if so, only as a prop to the narrative, just like the house, or food. And yet (as we know from other sources—see chapters 4 and 7), many members of the lower middle class could not afford to buy a car and thus other vehicles helped to make mobile subjects streetwise, such as the motorcycle or the moped, or the motor bus. These vehicles were much less covered in literature, let alone celebrated. For those lower-middle-class members, however, the car functioned as an important, personal objective in life, an instrument that allowed passage into adulthood and citizenship. In doing so they co-constructed, together with the elite and an all-too-eager car industry, a hierarchy of car models, and of automotive distinction.

The second main difference with the prewar period is the systemic character of the car culture, and its related, albeit hidden collectiveness. The car was depicted as part of traffic, of a flow, and the driver as part of a swarm, although typically, many novels and poems continued to spread the image of a purely individualistic relationship between owner/user and vehicle. This, however, turned into an illusion. The systems character of the car culture was first observed to emerge in an urban setting. It also led to a shift in the violent aspects of the automotive adventure. Now, the traffic accident became the deadly interruptor of flows, in traffic and in life, and the stealthy way this shift took place necessitates a separate analysis, which we will undertake in the next chapter.

As to the second, symbolic level, the situation was of course similar to that of the content: the symbolic arsenal became enormously variegated. The car's ubiquity, however, led to a new symbolic realm, in which the metaphorical direction was inverted. Now, the car stood for all kinds of societal situations and processes. This was done by the literary avant-garde, who thus could spread the impression (if not the illusion) that everyone benefitted from the car, users and nonusers alike. One did not need to own a car to be part of its adventure. Even more: the car as symbol thus supported the illusion of its ubiquity before real mass motorization took place in the 1950s and 1960s: in the 'car society,' the emphasis shifted from car ownership to car use, and use (at least metaphorically, but often also in 'would-be cars' such as motorcycles and buses) became ubiquitous.

The third level, of affinity between cultural behavior and its utterances, also underwent an important shift. Prefigured by the prewar elite, new users learned to perceive the environment in a filmic way, like a movie, aided by the development of the closed body and other technical changes, such as the shift in sound engineering. Literature included the movie's technical means, such as simultaneity and change of perspective, means that appealed to the feelings rather than the knowledge interests of the audience.[359]

The new car users were not writers. They did not need to be, as the avant-garde and their epigones had made films, dime novels, Hollywood movies, and songs that reflected, for better and for worse, the new, encapsulated family experience. The production of this experience rhymed, so to speak, with the consumption of mass media, and had a distinctive collective, social flavor. It was, as we have seen, a form of multi-sensorial transcendence. It was this tense relationship between collective consumption (of movies, of the car as commodity) and individuality as a mainstream discourse that explains the enormous persistence of car culture: although not every driver could feel like a poet (as was the case in the previous period), he or she certainly experienced a form of transcendence of his or her own body, a feeling of being part of a wider phenomenon, a group in movement, a swarm, modernity *tout-court*, a distinctive middle-class modernity. Part of this modernity was being 'cool' and distant (ironic, as we called it as well) toward the Other, both domestic and exotic. Transcendence then had two sides: an individual, bodily aspect and a social one—*esprit de corps* in a double sense.

The car became, in this phase, the carapace, the armor of an insecure but basically (although covertly) aggressive petty bourgeois male and his family. Women and blacks, and new users hitherto less touched by the car (in the colonies, for instance, or in Eastern Europe), had no choice but to liberate themselves in a car that killed. Women, for instance, used the closedness of the car as protection and although they may have wished to be less violent (and, as we will see in the next chapter, often were praised for their 'defensive' accident-avoiding behavior), they could not escape the inherent lethality of a one-ton projectile in ever more dense traffic.

Literary analysis of the two consecutive periods provides insight into the modulation of the automobile adventure: in the first period, until well into the 1920s, increasing car casualties provoked a flight forward, culminating in a loud futuristic celebration of maiming, death, and destruction. During the interwar years, though, a gradual process of internalization set in, a potent combination of suppression at the individual level and delegation to the experts who started a hunt for the black sheep among the motorists (as we will see in the next chapter), while at the same time providing the scientific justification for the normalization of the mobile risk society. From a weapon or projectile the car evolved into a part of the (middle-class) human body. From a toy, through a tool, to a prosthesis, just like books, one is inclined to say, quoting the French philosopher Bernard Stiegler.[360] Indeed, Sigmund Freud characterized writing as prosthetic, as a compensation for a loss and thus emphasizing the 'negative prosthesis.' But for Freud writing can also be seen as 'positive prosthesis,' "producing in a range of modernist writers a fascination with organ-extension, organ-replacement, sensory-extension; with the interface between the body and the machine which Gerald Heard, in 1939, labeled *mechanomorphism*."[361] In this literal and literary sense, and only in this sense, the car did evolve from a toy to a tool: the car as an extension conquering the environment, the surrounding territory, was realized to the full in the second period of the 'closed body,' when the car became both a fashionable form of clothing and a *panzer*.

But what is tool, who actor? Karl Marx saw workers as a "living appendage of the machine." The motorist also had to become such an appendage before he or she could develop into a manager of his or her own mobility. The appropriation of modernity, in less than half a century, had resulted in the unification of driver and machine, of adventure and comfort (as extensively displayed in what literary historian Roberto Cagliero has called "middle-class fiction"[362]) and modernity could thus be experienced in the very practice of driving, of motorized movement itself.

When German historian Heinrich Winkler, in an effort to repudiate Seymour Lipset's thesis of the "extremism of the center," characterized the "1932 idealtypical voter on the national socialist party (as) a non-wage earning Protestant member of the *Mittelstand*, who either lives on a farm or a small town and who previously voted for a party of the political center or on a regional party, which resisted power and influence of large industry and union," he could have been describing the American middle class as well. The ubiquity of the car, its social versatility and ideological flexibility, is testified by the fact that what in the United States and many Western European countries was a liberal project, turned into a fascist project in Germany and Italy, and a revolutionary project in the Soviet Union.[363]

The thrill of mastering the 'purring beast' is since then firmly incorporated in automobility, and every novice has to go through an initiating rite of domesticating this thrilling experience during the first driving lessons. For the period under consideration here, it cannot be denied that American literary production associated car use with less violence than did the Europeans, but this seems to be compensated for by a more outspoken, destructive mood in popular culture and films. In the end, despite a difference in 'car aggressiveness' on both sides of the Atlantic, the lethal accident statistics prove that for the car embedded in a system, this difference did not play much of a role in the number of people killed per capita. The question, then, is how was this remarkable convergence achieved? And what was the role of the car in this process? We will address this question from several systemic perspectives in the next chapter.

Notes

1. Ernst Jünger, quoted in Karl Heinz Bohrer, *Die Ästhetik des Schreckens: Die pessimistische Romantik und Ernst Jüngers Frühwerk* (Frankfurt, Berlin, and Vienna: Ullstein Verlag, 1983), 165.
2. Jürgen Link and Rolf Parr, "Schiene—Maschine—Körper: Eine kleine Archäologie der Symbolik des Vehikel-Körpers," *Uni-Report: Berichte aus der Forschung der Universität Dortmund* 19 (1995), 15–21, here 17.
3. Robert Wohl, *The Generation of 1914* (Cambridge, MA: Harvard University Press, 1979), 227–228; lost an arm: Kimberley J. Healey, *The Modernist Traveler: French Detours, 1900–1930* (Lincoln and London: University of Nebraska Press, 2003), 55.
4. Michael C. C. Adams, *The Great Adventure: Male Desire and the Coming of World War I* (Bloomington and Indianapolis: Indiana University Press, 1990), 113 (hell), 114 (Rickenbacker), 126 (accident).
5. Álvaro de Campos, "Au volant de la Chevrolet," describes an event in 1928 and is quoted in Antonio Tabucchi, *La Nostalgie, l'Automobile et l'Infini: Lectures de Pessoa* (n.p. [Paris]: Éditions du Seuil, 1998), 52–53.
6. Paul Aron, "De *La 628-E8* d'Octave Mirbeau à *La 629-E9* de Didier de Roulx," in Éléonore Reverzy and Guy Ducrey (eds.), *L'Europe en automobile: Octave Mirbeau écrivain voyageur* (Strasbourg: Presses Universitaires de Strasbourg, 2009), 231–238.
7. Alexa Geisthövel, "Das Auto," in Alexa Geisthövel and Habbo Knoch (eds.), *Orte der Moderne: Erfahrungswelten des 19. und 20. Jahrhunderts* (Frankfurt and New York: Campus Verlag, 2005), 37–46, here 42; reinvention: Anne Beezer, "Women and 'Adventure Travel' Tourism," *New Formations* 21 (1993), 119–130, here 119; Gerrit Walther, "Auf der Suche nach der 'Gattung': Interdisziplinäre Reiseliteraturforschung," *Archiv für Sozialgeschichte* 32 (1992), 523–533, here 524 (amplification), 528 (passé); happiness: Auke Hulst, "Van loopgraaf naar bergtop," *NRC Handelsblad* (16 December 2011), "Boeken," 1–2.
8. Jay Winter, "The Lost Generation of the First World War; Paul Fussell: *The Great War and Modern Memory* (1975)," in Uffa Jensen et al. (eds.), *Gewalt und Gesellschaft: Klassiker modernen Denkens neu gelesen* (Göttingen: Wallstein Verlag, 2011), 337–350, here 337–338; lost generation: Mellow, *Charmed Circle*, 273; self-pity: Wohl, *The Generation of 1914*, 120; Adams, *The Great Adventure*, 122–123; for an example of American self-pity, see ibid., 129, on Faulkner.
9. Winter, "The Lost Generation," 340.
10. Ellis W. Hawley, *The Great War and the Search for a Modern Order: A History of the American People and Their Institutions, 1917–1933* (New York: St. Martin's Press, 1979), 161.

11. Malcolm Cowley, *Exile's Return: A Literary Odyssey of the 1920s*, edited and with an introduction by Donald W. Faulkner (New York and London: Penguin Books, 1994; first ed. 1934), 38–39.
12. Wohl, *The Generation of 1914*, 226 (quotation: 227).
13. Klaus Plonien, "Re-mapping the World: Travel Literature of Weimar Germany" (unpublished dissertation, University of Minnesota, 1995), 195; for an overview of American car travelogues and novels, see Marguerite S. Shaffer, "Seeing America First: The Search for Identity in the Tourist Landscape," in David M. Wrobel and Patrick T. Long (eds.), *Seeing and Being Seen: Tourism in the American West* (Lawrence: University Press of Kansas, 2001), 165–193; Aldous Huxley, *Along the Road: Notes and Essays of a Tourist* (London: Chatto & Windus, 1948; first ed. 1925), 16.
14. Paul Morand, *Conseils pour voyager sans argent* (Paris: Émile Hazan & Cie., 1930), 10–11.
15. J. B. Priestley, *English Journey: Being a rambling but truthful account of what one man saw and heard and felt and thought during a journey through England during the autumn of the year 1933* (Harmondsworth and New York: Penguin Books, 1977; first ed. 1934), 379 (war), 66 (mob), 386 (liberal democracy), 45 (morose), 186 (circuses), 54 (uglification), 67 (mass production), 389 (little England), 386 (dull), 9 (luxury), 146 (closed car), 145 (bad driver); factory girls, etc., quoted in Lawrence James, *The Middle Class: A History* (London: Little and Brown, 2006), 423.
16. H. V. Morton, *In Search of England* (Cambridge, MA: Da Capo Press, 2002; first ed. 1927), 6 (war), 23 (petrol), 248 (mass tourism); Jan Morris, "Introduction to the Da Capo edition," in ibid., ix–xi, here xi (middle England).
17. David Matless, *Landscapes and Englishness* (London: Reaktion Books, 1998), 66; shirked: Michael Bartholomew, "H.V. Morton's English Utopia," in Christopher Lawrence and Anna-K. Mayer (eds.), *Regenerating England: Science, Medicine and Culture in Inter-War Britain* (Amsterdam and Atlanta, GA: Rodopi, 2000), 25–44, here 32.
18. George Orwell, *The Road to Wigan Pier* (Milton Keynes: Lightning Source UK Ltd., 2009; Oxford reprints; first ed. 1937), 145, 139. Inspired: Dame Margaret Drabble, "Dark felicity: Following Priestley to the Potteries," in J. B. Priestley, *English Journey* (Ilkley: Great Northern Books, 2009), 29–34, here 29; the battlefield, etc., quote is from Orwell's *Coming Up for Air* (1939), as quoted in James, *The Middle Class*, 418.
19. John Betjeman, "Slough," http://www.cdr.stanford.edu/intuition/Slough.html (last accessed 5 December 2011).
20. D. H. Lawrence, *Sea and Sardinia*, ed. Mara Kalnins (London and New York: Penguin Books, 1999; first ed. 1921), 118 (hare), 114 (peasant), 115 (grinding).
21. Mark Ford, *Raymond Roussel and the Republic of Dreams* (Ithaca, NY: Cornell University Press, 2000), 27 (Proust), 30 (insulating), 170 (roulotte), 171 (curtains), 182 (barbiturates), 207 (chess), 211 (Palermo), 232 (experience); Raymond Roussel, *Comment j'ai écrit certains de mes livres*, série "Fins de siècles," ed. Hubert Juin (n.p.: Jean-Jacques Pauvert, 1977), 27.
22. Caren Kaplan, *Questions of Travel: Postmodern Discourses of Displacement* (Durham, NC, and London: Duke University Press, 1996), 51, 53 (the quote on imperialism is a paraphrase by Kaplan); Paul Fussell, *Abroad: British Literary Traveling Between the Wars* (Oxford and New York: Oxford University Press, 1980), 16 (family).

23. Malcolm Bradbury, "The Denuded Place: War and Form in *Parade's End* and *U.S.A.*," in Holger Klein (ed.), *The First World War in Fiction: A Collection of Critical Essays* (London and Basingstoke: The MacMillan Press, 1976), 193–209, here 202.
24. Barbara Korte, "Erfahrungsgeschichte und die 'Quelle' Literatur: Zur Relevanz genretheoretischer Reflexion am Beispiel der britischen Literatur des Ersten Weltkriegs," in Jan Kusber et al. (eds.), *Historische Kulturwissenschaften: Positionen, Praktiken und Perspektiven* (Bielefeld: transcript Verlag, 2010), 143–159, here 148–150; Brooke L. Blower, *Becoming Americans in Paris: Transatlantic Politics and Culture between the World Wars* (Oxford and New York: Oxford University Press, 2011), 221–224.
25. Erika and Klaus Mann, *Rundherum: Abenteuer einer Weltreise; Mit Originalfotos; Nachwort von Uwe Naumann* (Reinbek bei Hamburg: Rowohlt Taschenbuch Verlag, 2001; first ed. 1929), 24.
26. Marion Sanders, *Dorothy Thompson: A Legend in Her Time* (Boston: Houghton Mifflin, 1973), 59 (Rose Lane), 101 (Albania).
27. "Prologue," in William Holtz (ed.), *Travels with Zenobia: Paris to Albania by Model T Ford; A Journal by Rose Wilder Lane and Helen Dore Boylston* (Columbia and London: University of Missouri Press, 1983), 1–23, here 8. Hawthorne depicts Zenobia as an androgynous woman characterized by "independence, assertiveness, self-reliance, courage, and strength of personality." She ends her life through suicide. Ellen Walker Glenn, "The Androgynous Woman Character in the American Novel" (unpublished dissertation, University of Colorado, 1980), 28 (independence), 44 (androgynous), 47 (suicide).
28. "Prologue," 21. On Lane's and Boylston's trip, see also Deborah Clarke, *Driving Women: Fiction and Automobile Culture in Twentieth-Century America* (Baltimore: Johns Hopkins University Press, 2007), 62–65.
29. Mrs. Rodolph Stawell, *Motor Tours in Yorkshire* (London: Hodder and Stoughton, 1919), 3.
30. Samuel Marshall Ilsley, *Katharine Hooker: A Memoir* (Santa Barbara, CA: The Schauer Printing Studio, 1935), 41 (discoverer); 47 (*outré*).
31. Katharine Hooker, *Through the Heel of Italy* (New York: Rae D. Henkle Co., 1927), 15–16 (distancing); 97 (daughter as photographer); 157 (book adventures), 161 (quote; italics in original). Road as actor: "The road led first down a shrub-covered slope" (ibid., 223). Hooker has been advised not to touch the visited people because of their possible skin disease (ibid., 51).
32. Gert Mattenklott, "Der Reiseführer Arthur Holitscher," in: Arthur Holitscher, *Der Narrenführer durch Paris und London; Mit Holzschnitten von Frans Masereel und einem Nachwort von Gert Mattenklott* (Berlin: S. Fischer Verlag, 1986) (first ed.: 1925) 156-174, here: 160; Holitscher, *Der Narrenführer durch Paris und London*, 7 (four editions), 10 (half an hour by taxi), 12 (barier);
33. Sabine Brenner, Gertrude Cepl-Kaufmann, and Martina Thöne, *Ich liebe nichts so sehr wie die Städte . . . Alfons Paquet als Schriftsteller, Europäer, Weltreisender* (Frankfurt: Vittorio Klostermann, 2001), 127 (train), 129 (adventure); Alfons Paquet, *Weltreise eines Deutschen: Landschaften, Inseln, Menchen, Städte* (Berlin: Büchergilde Gutenberg, 1934), 52, 82, 174, 125, 181, 198 (long-stroke), 247, 262; Kisch and Paquet: Gert Mattenklott, "Der Reiseführer Arthur Holitscher," 156.
34. Wolfgang Hoffmann-Harnisch, *Wunderland Brasilien: Eine Fahrt mit Auto, Bahn und Flugzeug* (Hamburg: Deutsche Hausbücherei, 1938), 8.

35. Bodo Uhse, "Egon Erwin Kisch—Werk und Leben," in Egon Erwin Kisch, *Der Mädchenhirt: Schreib das auf Kisch! Komödien; Gesammelte Werke 1* (Berlin and Weimar: Aufbau Verlag, 1992), 5–28, here 15 (starve).
36. Mattenklott, "Der Reiseführer Arthur Holitscher," 158 (*Gemeinschaft*), 162 (servant).
37. Manfred Chobot, "Arthur Holitscher (1869–1941)," *Literatur und Kritik* (1 November 2004), 99–110; Holitscher, "5000 Kilometer durch Südwesteuropa mit 120 PS," 191 (monument), 200 (quote). For a comparison with Theodore Dreiser, see Alfons Goldschmidt, "Holitscher und Dreiser," *Die Weltbühne: Wochenschrift für Politik, Kunst, Wirtschaft* 25 (1929), 282–284.
38. Georgine Clarsen, "Machines as the measure of women: Colonial irony in a Cape to Cairo automobile journey, 1930," *Journal of Transport History* 29, no. 1, Third Series (March 2008), 44–63, here 48; Georgine Clarsen, *Eat My Dust: Early Women Motorists* (Baltimore, MD: Johns Hopkins University Press, 2008), 147 (hut), 150 (new experiences). Stella Court Treatt, *Cape to Cairo: The Record of a Historic Motor Journey* (London, Calcutta, and Sydney: George G. Harrap, 1927). On British novelists and colonialism, see, for instance, Raymond Williams, *The Country and the City* (London: Chatto & Windus, 1973), 279–288. On (mostly Anglophone) travelogues crossing Africa, see Gordon Pirie, "Non-urban Motoring in Colonial Africa in the 1920s and 1930s," *South African Historical Journal* 63, no. 1 (March 2011), 38–60.
39. Lady Dorothy Mills, *The Golden Land: A Record of Travel in West Africa* (London: Duckworth, 1929), 34 (scattering), 40 (exotic other: "The sculptural beauty of the human body is lost as it crouches, monkey-wise, over the wheel of a Ford car"), 55 (hammock), 99 (speedy travel), 106 (West Indies), 118 (a boy with "the charm of a ... gorilla"), 120 (barbaric beauty), 130 (defense), 184–185 (camion quote; italics in original), 203 ("But the full story of [her boy] Amara is not one that I can relate to a polite public").
40. Anne Beezer, "Women and 'Adventure Travel' Tourism," *New Formations* 21 (1993), 119–130, here 122.
41. Beth Kraig, "Woman At The Wheel: A History Of Women And the Automobile In America" (unpublished doctoral dissertation, University of Washington, 1987), 215 (reckless), 296 (androgynous); vision: Carroll Smith-Rosenberg, "Discourse of Sexuality and Subjectivity: The New Woman, 1870–1936," in Martin Bauml Duberman, Martha Vicinus, and George Chauncey, Jr. (eds.), *Hidden from History: Reclaiming the Gay and Lesbian Past* (New York and Ontario: New American Library, 1989), 264–280, here 265.
42. Hermione Lee, *Willa Cather: Double Lives* (New York: Pantheon Books, 1989), 12–13.
43. Graham Greene, *Journey Without Maps* (London and New York: Penguin Books, 1978), 57.
44. Fussell, *Abroad*, 70.
45. Douglas Lane Patey, *The Life of Evelyn Waugh: A Critical Biography* (Oxford and Cambridge, MA: Blackwell, 2001), 86–87; falling asleep: Evelyn Waugh, *Remote People: A Report from Ethiopia and British Africa 1930–1931* (New York: The Ecco Press, 1990; reprint of 1931 edition), 58.
46. James B. Wolf, "Imperial Integration on Wheels: The Car, the British and the Cape-to-Cairo Route," in Robert Giddings (ed), *Literature and Imperialism* (Houndsmill and London: MacMillan, 1991), 112–127, here 112; gendered: John M. MacKenzie, "T.E. Lawrence: The Myth and the Message," in Giddings (ed.), *Literature and Imperialism*, 150–181, here 170.

47. André Gide, *Voyage au Congo, suivi de Le retour du Tchad: Carnets de Routes* (n.p., n.d [Paris: Gallimard, 1985]; first ed. 1927), 97 (bush), 100 (35 cases), 105 (once a month), 180 (path is cleared), 219 (gray); André Gide, "Le retour du Tchad," in Gide, *Voyage au Congo*, 287–554, here 291 (preparation), 518 (imagination), 533 (no pleasure).
48. Gustav Stratil-Sauer, *Fahrt und Fessel: Mit dem Motorrad von Leipzig nach Afghanistan* (Münster: Verlagshaus Monsenstein und Vannerdat, n.d.; reprint of 1927 edition).
49. Sasha Disko, "'The World is My Domain'—Technology, Gender and Orientalism in German Interwar Motorized Adventure Literature," *Transfers* 1, no. 3 (Fall 2011), 44–63, here 47.
50. For an analysis of Australian continent-crossing travelogues, see Peter Bishop, "Driving Around: The Unsettling of Australia," *Studies in Travel Writing* 2 (1998), 144–162. For the difference vis-à-vis American and European travelogues, see Georgine Clarsen, "Automobility 'South of the West': Toward a Global Conversation," in Gijs Mom et al. (eds.), *Mobility in History: Themes in Transport (T²M Yearbook 2011)* (Neuchâtel: Alphil, 2010), 25–41.
51. Winifred Hawkridge Dixon, *Westward Hoboes: Ups and Downs of Frontier Motoring* (New York: Charles Scribner's Sons, 1928; first ed. 1921), 2 (Cadillac; love to tinker), 36–37 (jack quote); 39 (herd), 60 (driving as art), 87 (prouder moment).
52. Ibid., 204–205
53. Kathleen Franz, *Tinkering: Consumers Reinvent the Early Automobile* (Philadelphia: University of Pennsylvania Press, 2005), 56. When Dixon and her friend encounter such a helpful man, the women ask him how the ignition system works. "The man muttered something ... and drove off." Quoted in ibid., 55.
54. Marieluise Fleisser, *Andorranische Abenteuer* (Berlin: Gustav Kiepenheuer Verlag, 1932), 76 (lack of money). *Vossische Zeitung*: Sissi Tax, *Marieluise Fleisser, Schreiben, Überleben: Ein biographischer Versuch* (Frankfurt: Stroemfeld/Roter Stern, 1984), 100–101.
55. Helen Watanabe-O'Kelly, *Beauty or Beast? The Woman Warrior in the German Imagination from the Renaissance to the Present* (Oxford and New York: Oxford University Press, 2010), 247–248; Maria Leitner, *Eine Frau reist durch die Welt: Mit einem Nachwort von Hartmut Kahn* (Berlin: Dietz Verlag, 1986); Maria Leitner, *Hotel Amerika* (1930), http://nemesis.marxists.org (last accessed 4 March 2011), 1, 5. Special travelogues: Karolina Dorothea Fell, *Kalkuliertes Abenteuer: Reiseberichte deutschsprachiger Frauen (1920–1945)* (Stuttgart and Weimar: J. B. Metzler, 1998), 115–116.
56. Sasha Disko, "Men, Motorcycles and Modernity: Motorization in the Weimar Republic" (unpublished dissertation, New York University, May 2008), 250.
57. On Köhler, Körner, Lundberg, and Hofmann, see Disko, "Men, Motorcycles and Modernity," chap. 5.
58. Janis P. Stout, *The Journey Narrative in American Literature: Patterns and Departures* (Westport, CT, and London: Greenwood Press, 1983), 3, 6, 14–15, 18; poetry: Kurt Brown (ed.), *Drive, They Said: Poems about Americans and Their Cars*, preface by Edward Hirsch (Minneapolis: Milkweed Editions, 1994), xv.
59. Stout, *The Journey Narrative in American Literature*, 53, 161; Link and Parr, "Schiene—Maschine—Körper," 16, 21; Jürgen Link and Rolf Parr, "Das Kollektivsymbol 'Auto': Normalisierende 'Körper/Vehikel'-Katachresen in Medien und Literatur," *Uni-Report: Berichte aus der Forschung der Universität Dortmund* 19 (1995), 22–29; Georg Simmel, *Soziologie: Untersuchungen über die Formen der*

Vergesellschaftung (Berlin: Duncker & Humblot, 1958), 538. Simmel also deals with "a kind of spiritual communism" of the traveling group (*die Reisebekanntschaft*) as a situational collective that "often develops an intimacy and cordiality (*Offenherzigkeit*), for which in fact no inner reason can be found" (ibid., 500).

60. Richard Nate, *Amerikanische Träume: Die Kultur der Vereinigten Staaten in der Zeit des New Deal* (Würzburg: Königshausen & Neumann, 2003), 254.
61. Nate, *Amerikanische Träume*, 247 (quote Ma Joad), 261 (gypsies). Tim Cresswell, *The Tramp in America* (London: Reaktion Books, 2001), 46–47, like Steinbeck, distinguishes between the hoboes and the Okies, as the latter traveled mostly in family groups, with their "paraphernalia of domesticity" in their "Jalopies."
62. John Steinbeck, *The Grapes of Wrath* (London and New York: Penguin Books, 2000; first ed. 1939), 11 (tractored out), 71 (transportation), 74 (machine age), 133 (Muley), 116 (half truck), 117 (Super Six), 137 (road of flight), 227 (crawled), 141 (fifty thousand), 143 (soul), 144 (scary); David Strand, *Rickshaw Beijing: City People and Politics in the 1920s* (Berkeley: University of California Press, 1989); for a taxonomy of forms of nomadism and migration, and the differences between nomadism and sedentism in (mostly) preindustrial societies, see Tim Ingold, *The Appropriation of Nature: Essays on Human Ecology and Social Relations* (Manchester: Manchester University Press, 1986), 165–197.
63. Steinbeck, *The Grapes of Wrath*, 114 (brakes), 141 (fifty thousand), 143 (soul), 144 (scary), 227 (one family), 192–217 (conrod repair). By the end of the interwar years, there were twenty-six "permanent migrant camps" in the United States. See Michael W. Schuyler, *The Dread of Plenty: Agricultural Relief Activities of the Federal Government in the Middle West, 1933–1939* (Manhattan, KS: Sunflower University Press, 1989), 212; for a comparison with other camps, such as German concentration camps, see Tim Cresswell, *On the Move: Mobility in the Modern Western World* (New York and London: Routledge, 2006).
64. Steinbeck, *The Grapes of Wrath*, 147 (new cars), 148 (social mobility), 151 (house). On the road as a "religious image and as a rhetorical structure," see Christine Marie Hilger, "Sacred Path: Considering 'Road' as Religious Image in John Steinbeck's *The Grapes of Wrath*," in Michael J. Meyer (ed.), *The Grapes of Wrath: A Re-Consideration*, vol. 1 (Amsterdam and New York: Rodopi, 2009), 73–98; Japanese: Henry Veggian, "Displacements and Encampments: John Steinbeck's *The Grapes of Wrath*," in Meyer (ed.), *The Grapes of Wrath*, vol. 1, 351–374, here 364; Kristen Marie Haven, "'An Away Thing and a Stay Thing': Alfredo Véa, Jr.'s Refiguring of Steinbeck's Temporary Structures," in Meyer (ed.), *The Grapes of Wrath: A Re-Consideration*, vol. 2 (Amsterdam and New York: Rodopi, 2009), 831–854, here 838 (freeing), 842 (gorillas); proletarian: Lisa A. Kirby, "A Radical Revisioning: Understanding and Repositioning *The Grapes of Wrath* as Political 'Propaganda,'" in Meyer (ed.), *The Grapes of Wrath*, vol. 1, 243–260, here 245.
65. Florian Schwieger, "The Joad Collective: Class Consciousness and Social Reorganization in *The Grapes of Wrath*," in Meyer (ed.), *The Grapes of Wrath*, vol. 1, 183–217, here 199 (building technique).
66. For the following, see Roger A. Bruns, "Afterword: Dr. Ben Reitman's Paean to the Road," in *Boxcar Bertha: An Autobiography, As Told to Dr. Ben L. Reitman*, introduction by Kathy Acker, afterword by Roger A. Bruns (New York: Amok Press, 1988), 281–285.
67. Burns, *Boxcar Bertha*, 125 (Boxie), 33 (cleaner), 68 (sidecars), 80 (Camp Busted), 85–86 (charities), 16–17 (adventure and the rich), 52 (dumped), 61–62 (Palm

Beach), 66 (lesbians), 106 (thiefs), 251 (gallant), 110 (*Good Housekeeping*), 253 (tourist camp), 276 (capitalistic system). For a similar experience of five men buying an old Ford to look for work in 1930 and selling it for $5 after "she knocked a few loud whacks, then threw off a connecting rod and busted the block," see John H. Abner and George L. Andrews, "Grease Monkey to Knitter," in *These Are Our Lives: As Told by the People and Written by Members of the Federal Writers' Project of the Works Progress Administration in North Carolina, Tennessee, and Georgia* (Chapel Hill: University of North Carolina Press, 1939), 165–179, here 169–170. "That summer I met whole families wandering homeless and broke, even women with babies in their arms." Once this subject of the Federal Writers' Project gets a job as a knitter one of the first things he does is buying a car (ibid., 171, 176).

68. Marlou Belyea, "The Joy Ride and the Silver Screen: Commercial Leisure, Delinquency and Play Reform in Los Angeles, 1900–1980" (unpublished dissertation, Boston University Graduate School, 1983), 172–174; Mark Wyman, *Hoboes: Bindlestiffs, Fruit Tramps, and the Harvesting of the West* (New York: Hill and Wang, A Division of Farrar, Straus and Girous, 2010), 257–258, 264; hierarchy, Mills, and IWW: Jon Savage, *Teenage: The Creation of Youth Culture* (London: Pimlico, 2007), 281, 258, 263.

69. Burns, *Boxcar Bertha*, 254–255 (survey), 48 (two million), 249 (1934); Tim Cresswell, "Embodiment, power and the politics of mobility: the case of female tramps and hobos," *Transactions of the Institute of British Geographers*, NS 24 (1999), 175–192, here 184; Tim Cresswell, *The Tramp in America* (London: Reaktion Books, 2001), 39, gives 500,000 in 1906. For an overview of tramping around 1900, see Eric H. Monkkonen, "Regional Dimensions of Tramping, North and South, 1880–1910," in Monkkonen (ed.), *Walking to Work: Tramps in America, 1790–1935* (Lincoln and London: University of Nebraska Press, 1984), 189–211; and John C. Schneider, "Tramping Workers, 1890–1920: A Subcultural View," Monkkonen (ed.), *Walking to Work*, 212–234. Most tramps were "unmarried white men in the prime of life" (Monkkonen [ed.], *Walking to Work*, 226).

70. Roger A. Bruns, *The Damndest Radical: The Life and Work of Ben Reitman, Chicago's Celebrated Social Reformer, Hobo King, and Whorehouse Physician* (Urbana and Chicago: University of Illinois Press, 1987), 262–263; the authenticity of Bertha Thompson's story is not questioned by Lynn Weiner, "Sisters of the Road: Women Transients and Tramps," in Monkkonen (ed.), *Walking to Work*, 171–188; James Rorty, *Where Life is Better: An Unsentimental American Journey* (New York: Reynal & Hitchcock / John Day, 1936), 9 (trip), 15 (centrifugal), 17 (ants), 32 (Asiatic), 54 (warned), 70 (flop-house), 83 (traveling journalist), 90–97 (an entire chapter 7 against a lower-middle-class hitchhiker).

71. Bruns, *The Damndest Radical*, 266; casualties: Cresswell, *The Tramp in America*, 31.

72. David Riesman (in collaboration with Reuel Denney and Nathan Glazer), *The Lonely Crowd; A Study of the Changing American Character* (London: Geoffrey Cumberlege / Oxford University Press, 1953).

73. Alastair Durie, "No holiday this year? The depression of the 1930s and tourism in Scotland," *Journal of Tourism History* 2, no. 2 (August 2010), 67–82, here 70.

74. W. R. Calvert, *Family Holiday: A Little Tour in a Second-Hand Car* (London and New York: Putnam, 1932), 10–11.

75. Thornton Niven Wilder, "The Happy Journey to Trenton and Camden," in Wilder, *The Long Christmas Dinner and Other Plays* (London, New York, and

Toronto: Longmans, Green and Co., 1931), 111–135, here 113 (no scenery), 117 (garage), 118 (Chevrolet), 119 (inhale), 120 (hold up traffic), 126–127 (crying children), 128 (long quote; italics in original), 134 (liked by family).
76. Priscilla Lee Denby, "The Self Discovered: The Car in American Folklore and Literature" (unpublished dissertation, Indiana University, May 1981), 120–121.
77. Plonien, "Re-mapping the World," 98.
78. William Carlos Williams, *A Voyage to Pagany* (New York: The Macaulay Company, 1972; first ed. 1927), 24 (Paris as woman), 38–39 (example of 'automatic writing'), 253 (virgin continent); Paul Mariani, *William Carlos Williams: A New World Naked* (New York: McGraw-Hill Book Company, 1981), 230 (in Vienna with his wife Flossie), 266 (his wife split up), 291 (Pagany as Europe).
79. Harvey Taylor, *A Claim on the Countryside: A History of the British Outdoor Movement* (Edinburgh: Keele University Press, 1997), 252 (fellowship), 266 (motives).
80. Andrea Wetterauer, *Lust an der Distanz: Die Kunst der Autoreise in der "Frankfurter Zeitung"* (Tübingen: Tübinger Verein für Volkskunde, 2007), 189.
81. Koshar, *German Travel Cultures*, 79.
82. Elizabeth Coffman, "Women in Motion: Loie Fuller and the 'Interpenetration' of Art and Science," *Camera Obscura* 17, no. 1 (2002), 72–105, here 73, 77.
83. Siegfried Kracauer, "Die Reise und der Tanz," in Siegfried Kracauer, *Das ornament der Masse: Essays* (Frankfurt: Suhrkamp Verlag, 1977), 40–49; Chris Rojek and John Urry, "Transformations of Travel and Theory," in Rojek and Urry (eds.), *Touring Cultures: Transformations of Travel and Theory* (London and New York: Routledge, 1997), 1–19, here 6.
84. Kracauer, "Die Reise und der Tanz," 47 (italics in original).
85. Ben Malbon, *Clubbing: Dancing, Ecstasy and Vitality* (London and New York: Routledge, 1999; Critical Geographies, eds. Tracey Skelton and Gill Valentine), 108–109.
86. Klaus-Dieter Oelze, *Das Feuilleton der Kölnischen Zeitung im Dritten Reich* (Frankfurt, Bern, New York, Paris: Peter Lang, 1990), 4, 252–254.
87. Bernhard Weyergraf (ed.), *Literatur der Weimarer Republik 1918 – 1933* (Munich and Vienna: Carl Hanser Verlag, 1995), 569; Wetterauer, *Lust an der Distanz*, 64 (flaneurish: "*flanierende Schreibweise*"), 119, 121, 127 (5 percent), 136 (senseworld: "*automobiler Sinnenwelt*"), 144 (female), 163, 169, 193. For a similar analysis of American 'car literature' as an exercise in "self-discovery," see Ronald Primeau, *Romance of the Road: The Literature of the American Highway* (Bowling Green, OH: Bowling Green State University Popular Press, 1996), 84. Fleisser: Siegfried Reinecke, *Mobile Zeiten: Eine Geschichte der Auto-Dichtung* (Bochum, 1986), 105. Giddens quoted in Torun Elsrud, "Risk Creation in Traveling: Backpacker Adventure Narration," *Annals of Tourism Research* 28, no. 3 (2001), 597–617, here 597, 599.
88. Peter J. Brenner, "Schwierige Reisen: Wandlungen des Reiseberichts in Deutschland 1918–1945," in Brenner, *Reisekultur in Deutschland: Von der Weimarer Republik zum "Dritten Reich"* (Tübingen: Max Niemeyer Verlag, 1997), 127–176, here 134–137; subjectivity: Plonien, "Re-mapping the World," 217.
89. Paul Zweig, *The Adventurer* (London: Basic Books, 1974), 209.
90. Dorothee Hochstetter, *Motorisierung und "Volksgemeinschaft": Das Nationalsozialistische Kraftfahrkorps (NSKK), 1931–1945* (Munich: R. Oldenbourg Verlag, 2005), 171.
91. Eric Laurier, "Driving: Pre-Cognition and Driving," in Tim Cresswell and Peter

Merriman (eds.), *Geographies of Mobilities: Practices, Spaces, Subjects* (Farnham and Burlington: Ashgate, 2011), 69–81, here 77.
92. Grith Graebner, *"Dem Leben unter die Haut kriechen . . ." Heinrich Hauser—Leben und Werk; Eine kritisch-biographische Werk-Bibliographie* (Aachen: Shaker Verlag, 2001), 13, 24–31; young generation: Gregor Streim, "Flucht nach vorn zurück: Heinrich Hauser—Portrait eines Schriftstellers zwischen Neuer Sachlichkeit und 'reaktionärem Modernismus,'" *Jahrbuch der Deutschen Schillergesellschaft* 43 (1999), 377–402, here 380.
93. Heinrich Hauser, *Schwarzes Revier, Herausgegeben von Barbara Weidle: Mit einem Vorwort von Andreas Rossmann* (Bonn: Weidle Verlag, 2010; first ed. 1930).
94. Heinrich Hauser, *Fahrten und Abenteuer im Wohnwagen: Herausgegeben, kommentiert und mit einem Nachwort versehen von Robert Hilgers* (Stuttgart: DoldeMedien Verlag, 2004; reprint of original 1935 edition), 31 (not worth telling), 37 (brake lever).
95. Hauser, *Fahrten und Abenteuer im Wohnwagen*, 71.
96. Graebner, *"Dem Leben unter die Haut kriechen . . ."* 237, 275, 277.
97. Heinrich Hauser, *Die letzten Segelschiffe: Schiff, Mannschaft, Meer und Horizont* (Berlin: S. Fischer Verlag, 1932, Elfte bis vierzehnte Auflage; first ed. 1930), 25 (*Rausch*), 63 (village), 69 (order), 99 (*Gemütlichkeit*), 123 (chickens), 25 (undescribable).
98. Hauser, *Schwarzes Revier*, 12–13.
99. Heinrich Hauser, "Herbstfahren im Auto," *Frankfurter Zeitung, Reiseblatt* 79 (21 October 1934); Heinrich Hauser, "Autowandern, eine wachsende Bewegung," *Die Strasse* (1936), Heft 14, 455–457, here 455.
100. Heinrich Hauser, *Süd-Ost-Europa ist erwacht: Im Auto durch acht Balkanländer* (Berlin: Rowohlt, 1938), 8 (love), 103 (anthropomorphing), 282 (end of trip); prude lover: Gregor Streim, "Junge Völker und neue Technik: Zur Reisereportage im 'Dritten Reich,' am Beispiel von Friedrich Sieburg, Heinrich Hauser und Margaret Boveri," *Zeitschrift für Germanistik* 9, no. 2 (1999), 344–359, here 353.
101. Periphery: Streim, "Junge Völker und neue Technik," 347; image-rich: Volker Pantengen, "Weltstadt in Flegeljahren: Ein Bericht über Chicago" (film review, 30 November 2003) (http://newfilmkritik.de/archiv/2003-11/weltstadt-in-flegeljahren-ein-bericht-uber-chicago/; last accessed 26 September 2014), 1; Austria: Hauser, *Süd-Ost-Europa ist erwacht*, 19.
102. Hauser, *Süd-Ost-Europa ist erwacht*, 92 (nonadventurous), 267–268 (regulated); mobile home: Streim, "Flucht nach vorn zurück," 397.
103. Disko, "Men, Motorcycles and Modernity," 180.
104. Kurt Schwitters, "Mein neues Motorrad" (1926) in Schwitters, *Das literarische Werk, Band 2: Prosa 1918–1930, Herausgegeben von Friedhelm Lach* (Cologne: Verlag M. DuMont Schauberg, 1974), 269–270.
105. Walter Julius Bloem, *Motorherz* (Berlin: August Scherl, 1927), 149 (best friend), 98 (automaton); 98 (soul); 159 (iron comrade); on Bloem: Jürgen Hillesheim and Elisabeth Michael, *Lexikon nationalsozialistischer Dichter: Biographien—Analysen—Bibliographien* (Würzburg: Königshausen & Neumann, 1993), 301–303.
106. Albert Sachs, *Motorradunfälle* (Berlin: Carl Schmidt & Co., 1929), 20, 59, 119 (misfits).
107. Tom Goodman, "The Road Worst Traveled: Book Review," *International Journal of Motorcycle Studies* 5, no. 2 (Fall 2009). The book reviewed by Goodman is Tim Fransen's *An Anthology of Early British Motorcycling Literature* (2009). Stoll quoted in Hochstetter, *Motorisierung und "Volksgemeinschaft"*, 246–247.

108. Bloem, *Motorherz*, 64 (*Kettensprenger*), 76 (one-tenth), 17 (cold face), 23 (soldiers of speed), 44 (satans), 49 (swarm), 66 (*ersatz*), 68 (kings of speed), 76 (cocoon), 80 (without beauty), 89 (rush), 102 (Cathedral of Cologne), 151 (contempt), 159 (iron comrad), 175 (selfconfidence: *grosse Sicherheit*), 176 (weak), 185 (weak), 143 (control); 190 (jubilant), 207 (example of filmic style), 210 (death of Thora Moebius), 212 (Thora's contempt for "morals").
109. Quoted in Walter Kappacher, "Vor dem Rennen in den Wicklow Mountains," *Frankfurter Allgemeine Zeitung* (9 July 2011), "Bild und Zeiten," Z1–Z2 (my translation).
110. Julius Donny, *Garage 13: Humoristischer Roman* (Berlin: Verlag Georg Koenig, n.d. [1930]), 29, 44, 93, 108 (fraudulent), 113, 123, 143 (machine gun), 145 (canon).
111. Donny, *Garage 13*, 17–19.
112. Luc Durtain, *Ma Kimbell: Conquêtes du monde* (Paris: Librairie Gallimard, 1925), 9 (warlike), 10 (Empire of Speed), 10 (picturesque; italics in original), 12 (tractor), 13 (fleeing from a lover), 22 (adventure), 38 (dozing off), 52–53 (traveling party), 57 (beautiful), 80 (near accident), 95 (enemy), 102 (better than man), 111 (cylinders), 134 (dirt), 144 (cripples; landscape description), 177 (empty tank), 192 (solidarity), 198 (perverse), 202 (certainty); Yves Chatelain, *Luc Durtain et son œuvre* (Paris: Les Œuvres representatives, 1933), 53 (adventure of love), 66 (train); Marcel Thiébaut, "Luc Durtain," *La revue de Paris* 41 (1934), 919–939, here 920–921 (architecture), 926 (buttock).
113. Quoted in Andreas Kelletat, "Philologie, Textwissenschaft und Kulturwissenschaft: Konkurrenz oder friedliche Koexistenz?" lecture in the series 'Wege der Kulturwissenschaft,' Johannes Gutenberg Universität, Mainz, http://www.fask.uni-mainz.de/inst/romanistik/ringvorlesung/vortrag_AFK/vortrag-A... (last accessed 10 February 2008); desperado quote from Gernot Böhme, "Das grosse Ereignis: das Grammophon; Victor Klemperers Tagebücher als Quelle der Technikgeschichte," in Harm-Peer Zimmermann (ed.), *Was in der Geschichte nicht aufgeht: Interdisziplinäre Aspekte und Grenzüberschreitungen in der Kulturwissenschaft Volkskunde* (Marburg: Jonas Verlag, 2003), 71–85, here 78–79; for a partial English translation of the diaries, see *I Shall Bear Witness: The Diaries of Victor Klemperer 1933–41, abridged and translated from the German edition by Martin Chalmers* (London: Weidenfeld & Nicholson, 1998).
114. Musil's essay "Geschwindigkeit is eine Hexerei" (Speed is witchcraft, 1927) quoted in Matthias Bickenbach, "Robert Musil und die neue Gesetze des Autounfalls," in Christian Kassung (ed.), *Die Unordnung der Dinge: Eine Wissens- und Mediengeschichte des Unfalls* (Bielefeld: transcript Verlag, 2009), 89–116, here 108.
115. Erhard Schütz, "'. . . Eine glückliche Zeitlosigkeit . . .': Zeitreise zu den 'Strassen des Führers,'" in Brenner, *Reisekultur in Deutschland*, 73–99, here 83 (new car technology), 87 (flight), 91 (Hitler), 92 (inversion); flying carpet: Benjamin Steininger, *Raum-Maschine Reichsautobahn* (Berlin: Kulturverlag Kadmos, 2005), 107. Hauser quoted in Rudi Koshar, "Germans at the Wheel: Cars and Leisure Travel in Interwar Germany," in Koshar (ed.), *Histories of Leisure* (Oxford and New York: Berg, 2002), 215–230, here 219–221. Ferdinand Oppenberg, "Reichsautobahn," quoted in Erhard Schütz, "'Jene blassgraue Bänder'; Die Reichsautobahn in Literatur und anderen medien des 'Dritten Reiches,'" *Internationales Archiv für Sozialgeschichte der deutschen Literatur* 18, no. 2 (1993), 76–120, here 114; Theweleit quoted in Jürgen Link and Siegfried Reinecke, "'Autofahren ist wie das Leben': Metamorphosen des Autosymbols in der deutschen Literatur," in

Harro Segeberg (ed.), *Technik in der Literatur: Ein Forschungsüberblick und zwölf Aufsätze* (Frankfurt: Suhrkamp Verlag, 1987), 436–482, here 470.
116. Edward Anderson, *Thieves Like Us* (New York City: The American Mercury, 1937), 38 (gun), 43 (80 mph), 52 (collision), 62 (sex), 68 (fly). For an example of New Objectivist style in this novel, see 83.
117. B. Traven, *Die Weisse Rose: Roman* (Frankfurt: Diogenes Verlag / Büchergilde Gutenberg, 1983), 199.
118. Anedith Jo Bond Nash, "Death on the Highway: The Automobile Wreck in American Culture, 1920–40" (unpublished dissertation, University of Minnesota, June 1983), 83 (Bonnie and Clyde), 85 (Dillinger); shooting a movie: Maurice Girard, *L'automobile fait son cinema* (n.p. [Paris]: Du May, n.d.), 81.
119. Bohrer, *Die Ästhetik des Schreckens*, 33.
120. Francis Picabia, *Caravansérail: 1924; présenté et annoté par Luc-Henri Mercié* (Paris: Pierre Belfond, 1974), 45 (taxi), 49 (unites), 63 (Dada).
121. Picabia, *Caravansérail*, 8–9.
122. Klaus Mann, "Speed," in Mann, *Speed: Die Erzählungen aus dem Exil* (Reinbek bei Hamburg: Rowohlt, 2003), 105–150, here 114 (hurry), 115 (insatiable), 122 (cars), 124 (marihuana), 128 (comfort), 143 (break-neck), 146 (flight).
123. Árpád von Klimó and Malte Rolf, "Rausch und Diktatur: Emotionen, Erfahrungen und Inszenierungen totalitärer Herrschaft," in von Klimó and Rolf (eds.), *Rausch und Diktatur: Inszenierung, Mobilisierung und Kontrolle in totalitären Systemen* (Frankfurt and New York: Campus Verlag, 2006), 11–43, here 25.
124. Morand quoted in Stéphane Sarkany, *Paul Morand et le cosmopolitisme littéraire, suivi de trois entretiens inédits avec l'écrivain* (Paris: Éditions Klincksieck, 1968), 129 (artist); and in Michel Collomb, *La littérature Art Déco; Sur le style d'époque* (Paris: Méridiens Klincksieck, 1987), 94 (offense), 157. Kills the individual: Paul Morand, *Voyages: Édition établie et présentée par Bernard Raffalli* (Paris: Robert Laffont, 2001), 896. Paul Morand, "De la vitesse: Un essay," in Paul Morand, *Papiers d'identité* (Paris: Berard Grasset, 1931), 271–296, here 274–275 (monotonous), 296 (brakes); *Ouvert la Nuit* and speed excesses: Bruno Thibault, *L'Allure de Morand: Du Modernism au Pétainisme* (Birmingham, AL: Summa Publications, Inc., 1992), 15–16, 50; Birgit Winterberg, *Literatur und Technik: Aspekte des technisch-industriellen Fortschritts im Werk Paul Morands* (Frankfurt, Bern, New York, and Paris: Peter Lang, 1991), 18 (pessimism), 38–40 (train), 145–147 (example of Morand being a car lover); for an overview of Morand's travelogues, see ibid., section 3.3 (185–197); vice: Healey, *The Modernist Traveler*, 58; "De la vitesse" has been translated into English as "On Speed," in Jeffrey T. Schnapp, *Speed Limits* (Miami Beach, FL, Montréal, and Milan: The Wolfsonian—Florida International University / Canadian Centre for Architecture / Skira, 2009), 268–271. The car trip is described in his *La route de la Méditerranée: Documents commentés par Paul Morand* (Winterberg, *Literatur und Technik*, 147, 70) and was accompanied by photographs of Germaine Krull and Kertesz, among others.
125. Franz Hessel, *Ein Flaneur in Berlin: Mit Fotografien von Friedrich Seidenstücker, Walter Benjamin's Skizze "Die Wiederkehr des Flaneurs" und einem "Waschzettel" von Heinz Knobloch* (Berlin: Das Arsenal, 1984; Neuausgabe von *Spazieren in Berlin* [1929]), 7 (slow), 51 (giant car), 128 (fast); reading quote on back cover.
126. Viktor Shklovsky (transliterated as "Schklowskij"), *Zoo: oder Briefe nicht über die Liebe* (Frankfurt: Suhrkamp Verlag, 1965).

127. Paul Morand, "Comme le vent," in Morand, *Papiers d'identité*, 325–335, here 328–329; Valery Larbaud, *Aux couleurs de Rome* (Paris: Gallimard, 1938), 147. Translated as "Slowness" in Schnapp, *Speed Limits*, 274–278; racism: Peter Gay, *Weimar Culture: The Outsider as Insider* (New York and London: W. W. Norton & Company, 2001), 134; speed as "élément israélite," also in "De la vitesse," 291. For an elegant analysis of car speed and slowness (which according to him can better be studied in popular culture than in the official discourses), see Rudy Koshar, "Organic Machines: Cars, Drivers, and Nature from Imperial to Nazi Germany," in Thomas Lekan and Thomas Zeller (eds.), *Germany's Nature: Cultural Landscapes and Environmental History* (New Brunswick, NJ, and London: Rutgers University Press, 2005), 111–139. For a comparison between a prewar (Mirbeau) and postwar (Morand) attitude toward speed, see Jelena Novaković, "La vitesse dans *La 628-E8* de Mirbeau et *L'homme pressé* de Paul Morand," in Reverzy and Ducrey (eds.), *L'Europe en automobile*, 255–267.
128. Shaffer, *See America First*, 221–225; Richard Lingeman, *Sinclair Lewis: Rebel from Main Street* (St. Paul, MN: Borealis Books, 2002), 40–41 (Wells), 67 (writing in train), 69 (superhuman), 73 (suffragent), 124, 132 (Run Slow); midwestern theme and Nobel Prize: William Holtz, "Sinclair Lewis, Rose Wilder Lane, and the Midwestern Short Novel," *Studies in Short Fiction* 24, no. 1 (1987), 41–48, here 45, 46; pitfalls and companionate: Franz, *Tinkering*, 58–59. On Lewis's "radical politics" and his fascination for the IWW, see also Paul S. Sutter, *Driven Wild: How the Fight against Automobiles Launched the Modern Wilderness Movement* (Seattle and London: University of Washington Press, 2002), 151–153.
129. Wolfgang Munzinger, *Das Automobil als heimliche Romanfigur: Das Bild des Autos und der Technik in der nordamerikanischen Literatur von der Jahrhundertwende bis nach dem 2. Weltkrieg* (Hamburg: LIT Verlag, 1997), 100 (freedom), 102 (frontier); Nash, "Death on the Highway," 18 (disaster).
130. Sinclair Lewis, *Free Air* (n.p.: General Books, 2009; reprint; first ed. 1919), 7 (dirty), 13 (skills), 105 (aviator), 125 (ankle), 24 (tires), 76 (drive like a man), 102 (adventuring), 103 (softness), 62 (limousinvalids, Sagebrush), 164 (dangerous). Skills: Munzinger, *Das Automobil als heimliche Romanfigur*, 105.
131. Lingeman, *Sinclair Lewis*, 132.
132. Joel Fisher, "Sinclair Lewis and the Diagnostic Novel: *Main Street* and *Babbitt*," *Journal of American Studies* 20 (1986), 421–433, here 421 (provincial), 426 (own terms). About the conviction that Americans should "thank that entrepising realtor, Babbitt, for the suburb," see Margaret Marsh, *Suburban Lives* (New Brunswick, NJ, and London: Rutgers University Press, 1990), 150.
133. Simone Weil Davis, *Living Up to the Ads: Gender Fictions of the 1920s* (Durham, NC, and London: Duke University Press, 2000), 77–79.
134. Catherine Jurca, *White Diaspora: The Suburb and the Twentieth-Century American Novel* (Princeton, NJ, and Oxford: Princeton University Press, 2001), 63; feminine: Disko, "Men, Motorcycles and Modernity," 142; numbers: Martha Wolfenstein, "The Emergence of Fun Morality," *Journal of Social Issues* 7 (1951), 15–25, here 15, 23.
135. Denby, "The Self Discovered," 106.
136. Quoted in John S. Gilkeson, *Anthropologists and the Rediscovery of America, 1886–1965* (Cambridge and New York: Cambridge University Press, 2010), 81. The study *Small Town* by sociologist Joseph Bensman (pseudonym: Springdale) "read like a Sinclair Lewis novel." Charlotte Allen, "Spies Like Us: When

Sociologists deceive their Subjects," *Lingua Franca* 7 (November 1997), 31–39, here 35.
137. Clare Virginia Eby, "*Babbitt* as Veblenian Critique of Manliness," *American Studies* 33–34 (1992–1993), 5–23, here 20 (old-fashioned); Ellen Forest, *Een vacantiereis van 23.000 k.m. per Hudson door Europa* (Amersfoort: S.W. Melchior, n.d.), 91, 116. Babbitt's income of $8,000 was well above the average: Frank Stricker, "Affluence for Whom?—Another Look at Prosperity and the Working Classes in the 1920s," *Labor History* 24 (1983), 5–33, here 32; bacchanalia: Sinclair Lewis, *Babbitt* (Harmondsworth and New York: Penguin Books, 1985; first ed. 1923), 259; emotionally: David Blanke, *Hell on Wheels: The Promise and Peril of America's Car Culture, 1900–1940* (Lawrence: University Press of Kansas, 2007), 40.
138. George Orwell, *Coming Up for Air* (London and New York: Penguin Books, 2000), 117 (Red Book Club), 122 (fascism), 132 (mileage), 142 (chap), 146 (hill), 183 (cows); rural centered: Andrew Gavin Altman, "Motoring for the Million? Cars and Class in Pre-1952 Britain" (unpublished PhD thesis, Boston College, Graduate School of Arts & Sciences, 3 April 1997), 173. On the use of terms such as streamline (including by Orwell) as "a kind of shorthand for all manners of imagined decadence," see Dick Hebdige, "Towards a cartography of taste 1935–1962," *Block* 4 (1981), 39–56, here 45. On other work by Orwell with similar themes ("English middle class life ... permeated by disabling boredom, restlessness, meanness and callousness"), see Alan A. Jackson, *The Middle Classes 1900–1950* (Nairn: David St. John Thomas Publisher, 1991), 316–317.
139. Eby, "*Babbitt* as Veblenian Critique of Manliness," 8 (unemotional), 13 (fratriarchal), 20 (achievement). Jürgen Link, *Versuch über den Normalismus: Wie Normalität produziert wird* (Göttingen: Vandenhoeck & Ruprecht, 2009); Bohemian: Lewis, *Babbitt*, 257.
140. Denby, "The Self Discovered," 88, 178; Lewis, *Babbitt*, 110–111.
141. Lewis, *Babbitt*, 62 (Verona), 63 (knightly rank), 70 (French), 130 (raising hell), 298 (confession); Sinclair Lewis, *Dodsworth* (New York: Harcourt, Brace and Company, 1929), 20 (doornail), 10 (unadventurous), 20–21 (Tibetans); Cecilia Tichi, *Shifting Gears: Technology, Literature, Culture in Modernist America* (Chapel Hill and London: University of North Carolina Press, 1987), 234 (God of Speed). Tichi claims that Lewis's prose in its "verbal economy" is the literary equivalent of the "efficient ideals of streamlining" (ibid., 90); weakened self: Munzinger, *Das Automobil als heimliche Romanfigur*, 108.
142. Lewis, *Babbitt*, 46–47; Lingeman, *Sinclair Lewis*, 75 (rebellion).
143. Munzinger, *Das Automobil als heimliche Romanfigur*, 106
144. Mark Schorer, *Sinclair Lewis: An American Life* (New York, Toronto, and London: McGraw-Hill, 1961), 344.
145. Lewis, *Babbitt*, 77 (long motor trip), 142 (progress), 172 (church), 274 (hose connection), 174 (motor-mad), 178 (filthiness). Ted is experimenting with a "wireless telephone set" (ibid., 301). Lingeman, *Sinclair Lewis*, 87 (Model T), 95 (wept), 102 (Hupmobile), 109 (accident), 145 (son), 147 (sell Hupmobile), 177 (Willys), 200 (Cadillac).
146. Nash, "Death on the Highway," 12.
147. Jurca, *White Diaspora*, 47 (unrevolutionary, discontent); Lingeman, *Sinclair Lewis*, 193 (boozing); last quote: Lewis, *Babbitt*, 25.
148. Jurca, *White Diaspora*, 55.
149. Lingeman, *Sinclair Lewis*, 97 (white collar), 99 (feminist), 110 (mediocre), 155 (*Main Street* success), 159 (100 percent), 167 (Woolf), 184 (Wharton), 198 (slum),

199 (Damn it), 205 (overalls), 206 (ear), 232 (Joyce), 237 (Klan), 262 (permanent home), 320 (Weimar), 327 (honeymoon), 332 (false dream); Sanders, *Dorothy Thompson,* 112 (square; italics in original), 137–138 (trailer).
150. Helen V. Tooker, *The 5:35: A Novel of Suburban Life* (Garden City, NY: Doubleday, Doran & Company, 1928), 1 (commute to New York), 12 (feeder), 23 (family unity), 319 (keep house), 54 (very holy), 57 (flat tire), 46 (*Vogue*), 239–242 (row about selling the car), 289 (rolling country).
151. Tooker, *The 5:35,* 222–223 (tick, tick; italics in original), 268 (picnic).
152. Jens Peter Becker, *Das Automobil und die amerikanische Kultur* (Trier: Wissenschaftliche Verlag Trier, 1989), 118 (fate), 120 (restless), 123 (norms), 129 (phallic); new generation: *The Autobiography of Alice B. Toklas* (New York: Harcourt, Brace and Company, 1933), 268 (Stein's remark is first and foremost directed to Fitzgerald's *This Side of Paradise,* but "she thinks this equally true of The Great Gatsby"); decadent rich: Paula S. Fass, *The Damned and the Beautiful: American Youth in the 1920's* (New York: Oxford University Press, 1977), 27; glossy: Lingeman, *Sinclair Lewis,* 203.
153. Milton R. Stern, *The Golden Moment: The Novels of F. Scott Fitzgerald* (Urbana, Chicago, and London: University of Illinois Press, 1970), 163 (dream), 188 (parties), 232 (motif), 235 (drunks).
154. Leo Marx, *The Machine in the Garden: Technology and the Pastoral Idea in America* (Oxford and New York: Oxford University Press, 2000; first ed. 1964).
155. Munzinger, *Das Automobil als heimliche Romanfigur,* 20–21.
156. Karen Alkalay-Gut, "Sex and the Single Engine: E.E. Cummings' Experiment in Metaphoric Equation," *Journal of Modern Literature* 20, no. 2 (Winter 1996), 254–258, here 258.
157. Munzinger, *Das Automobil als heimliche Romanfigur,* 120.
158. Ibid., 128–150; Michael Arlen, *The Green Hat: A Romance for a Few People* (Woodbridge, Suffolk: The Boydell Press, 1983; first ed. 1924), 101 (beast), 194–195 (let him pass), 241 (roar), 244 (moaning).
159. Savage, *Teenage,* 245 (young women), 250 (crashes).
160. Evelyn Waugh, *Vile Bodies* (New York, Boston, and London: Back Bay Books, 1999), 2 (dust), 124–125 (Rolls Royce and electric brougham), 163 (Littlejohn), 168 (airship), 186 (motorcycle), 194 (wireless), 195 (Morris Oxford), 258 (would not stop), 284 (film). I thank Rhonda Tripp for bringing this novel to my attention.
161. Nash, "Death on the Highway," 3; intimacy: Frédéric Monneyron and Joël Thomas (eds.), *Automobile et Littérature* (n.p.: Presses Universitaires de Perpignan, 2005), 23–24.
162. Sylvère Lotringer and Paul Virilio, *The Accident of Art* (New York: Semiotext(e), 2005), 14–17 (World Trade Center), 19 (death drive).
163. Gorge Milburn, *Catalogue* (New York: Avon Books, 1977; first ed. 1936, as *All Over Town*), 7 (jitney), 30 (sped away), 122–124 (black man in car), 131 (mule cart), 151 (smashed nose), 156 leaped away), 167–168 (fire), 188 (lynching).
164. On Remarque's role in Interbellum Germany, see Modris Eksteins, *Tanz über Graben: Die Geburt der Moderne und der Erste Weltkrieg* (Reinbek bei Hamburg: Rowohlt Verlag, 1990), 412–443 (middleclass and cocktails: 413, 415; reject car: 423); on his "alcohol and car motifs," see Rolf Parr, "Tacho. km/h. Kurve. Körper: Erich Maria Remarques journalistische und kunstliterarische Autofahrten," in Thomas F. Schneider (ed.), *Erich Maria Remarque: Leben, Werk und weltweite Wirkung* (Osnabrück: Universitätsverlag Rasch, 1998), 69–90, here 69 (intensity

of life: 79); on wearing officer uniforms and on his father as bookbinder, see Bernhard Stegemann, "Autobiographisches aus der Seminar- und Lehrerzeit von Erich Maria Remarque im Roman *Der Weg zurück*," in Schneider (ed.), *Erich Maria Remarque*, 57–67, here 62, 58. Iambic verse: Link and Reinecke, "'Autofahren ist wie das Leben,'" 466. For a reconstruction of Brecht's accident, see Matthias Bickenbach and Michael Stolzke, "Ein lehrreicher Unfall des Dichters Brecht," in Bickenbach and Stolzke, "Schrott: Bilder aus der Geschwindigkeitsfabrik; Eine fragmentarische Kulturgeschichte des Autounfalls," http://www.textur.com/schrott/ (last accessed 26 April 2011).

165. Near-erotic: Erich Maria Remarque, "Kleiner Auto-Roman," *Sport im Bild* 32, no. 11 (1926), 470–471 (full throttle: 471); skill: Remarque, "Das Gehirn des Kraftwagens," *Sport im Bild* 32, no. 8 (1927), 330–332; Remarque, "Josefs Moment," *Sport im Bild* 33, no. 23 (1927), 1366–1368; stone: Remarque, "Station am Horizont," *Sport im Bild* 34, no. 4 (1928), 215–219; Gregor: Remarque, "Die verhängnisvolle Kurve," *Sport im Bild* 34, no. 9 (1928), 1368–1370, here 1370 (italics added); frictionless and *paternoster:* Remarque, "Automobil-Unfälle," *Sport im Bild* 31, no. 7 (1925), 412–413, here 413 (italics added); pedestrian problem: Remarque, "Auto und Weltstadt," *Sport im Bild* 31, no. 24 (1925), 1490–1491, 1530–1531, here 1490, 1530; robber barons: Remarque, "Film auf dem Asphalt," *Sport im Bild* 32, no. 20 (1926), 880–881; pedestrian: Parr, "Tacho. km/h. Kurve," 79 (n. 27).

166. Parr, "Tacho. km/h. Kurve," 84, 86.

167. Quoted in Matthias Schöning, *Versprengte Gemeinschaft: Kriegsroman und intellektuelle Mobilmachung in Deutschland 1914–33* (Göttingen: Vandenhoeck & Ruprecht, 2009), 213.

168. Community of men (*kämpferische Männergemeinschaft*): Rainer Jeglin and Irmgard Pickerodt, "Weiche Kerle in harter Schale: Zu *Drei Kameraden,*" in Schneider (ed.), *Erich Maria Remarque*, 217–234, here 224; Erich Maria Remarque, *Three Comrades*, transl. A. W. Wheen (New York: The Ballantine Publishing Group, n.d.; reprint; first ed. 1936), 6 (war), 13 (racing fever), 23 (distancing), 29 (tinkering), 46 (roar of street); Karen M. Beukers, "Die Remarque-Rezeption in den Niederlanden," *Erich Maria Remarque Jahrbuch/Yearbook* VII (1997), 73–92. The novel *Pat* was ready in 1933: Thomas F. Schneider, "Von *Pat* zu *Drei Kameraden*: Zur Entstehung des ersten Romans der Exil-Zeit Remarques," *Erich-Maria-Remarque-Jahrbuch* 2 (1992), 67–77, here 67; Oscar: Eksteins, *Tanz über Graben*, 443; scenario: Parr, "Tacho. km/h. Kurve," 83 (inversion quote on 73).

169. Ehrenburg, *Das Leben der Autos*, 277–284; Becker, *Das Automobil und die amerikanische Kultur*, 99; for an extensive analysis, see Siegfried Reinecke, *Autosymbolik in Journalismus, Literatur und Film: Struktural-funktionale Analysen vom Beginn der Motorisierung bis zur Gegenwart* (Bochum: Universitätsverlag Dr. N. Brockmeyer, 1992), 169–173.

170. Ehrenburg, *Das Leben der Autos*, 145 (feed), 225–227 (Karl Lang), 241 (kinship).

171. Ibid., 235–236.

172. Sherwood Anderson, *Perhaps Women* (Mamaroneck, NY: Paul P. Appel, 1970; first ed. 1931).

173. Anderson, *Perhaps Women*, 12–13, 16.

174. Robert Musil, "Der Riese Agoag," in Musil, *Nachlass zu Lebzeiten* (Reinbek bei Hamburg: Rowohlt Taschenbuch Verlag, 2010; first ed. 1957), 103–107, here 105. For another European novel in which a heavy motorized vehicle is one of the protagonists, see Bontempelli's *18 BL* (1934), an "elaboratedly choreographed

mass spectacle centred on the life-cycle of a Fiat truck used in the First World War and later by *squadristi* [fascist fighting groups] in their anti-communist expeditions." Roger Griffin, *Modernism and Fascism: The Sense of a Beginning under Mussolini and Hitler* (Houndsmill and New York: Palgrave, 2007), 238.

175. Gastly incident quoted in Jochen Hörisch, "Beschleunigen und Bremsen: Die Entdeckung der Zeit in der Moderne," in Jörg Huber and Alois Martin Müller (eds.), *"Kultur" und "Gemeinsinn"* (Basel and Frankfurt: Stroemfeld Verlag / Roter Stern / Museum für Gestaltung Zürich, 1994), 195–215, here 209, 211; Robert Musil, *Der Mann ohne Eigenschaften: Roman I Erstes und Zweites Buch*, ed. Adolf Frisé (Reinbek bei Hamburg: Rowohlt Taschenbuch Verlag, 2007), 10–11; Musil seems to have based his accident episode on two accidents that happened in front of his house in 1911: Matthias Bickenbach and Michael Stolzke, "Die Logik des Unfalls: Zufall und Notwendigkeit," in Bickenbach and Stolzke, "Schrott," 2; Carl Wege, "Gleisdreieck, Tank und Motor: Figuren und Denkfiguren aus der Technosphäre der Neuen Sachlichkeit," *Deutsche Vierteljahresschrift für Literaturwissenschaft und Geistesgeschichte* 68 (1994), 307–332; ideocracy: Möser, *Fahren und Fliegen in Frieden und Krieg*, 324. *Aus der Reihe* and military: Claudia Lieb, *Crash: Der Unfall der Moderne* (Bielefeld: Aisthesis Verlag, 2009), 239, 242.

176. For an overview of phantom and fantasy cars in early autopoetics, from E. F. Benson's "The Dust Cloud" (1906) through Ken Batten's "The Ghosty Car" (1928) to Betsy Emmons's "The Ghost of the Model T" (1942), see Jean Marigny, "Voitures fantastiques," in Monneyron and Thomas (eds.), *Automobile et Littérature*, 173–187, here 174–175.

177. Koshar, "Organic Machines," 124 (the quote is Koshar's paraphrase). On Hauser as teacher of cyborgs-to-be, see Paul A. Youngman, *We Are the Machine: The Computer, the Internet, and Information in Contemporary German Literature* (Rochester and New York: Camden House, 2009), 35.

178. Lieb, *Crash*, 246. Lieb associates the metaphor to processes of thinking.

179. Quoted in ibid., 210 (transport worker), 211 (*Verkehrsroman*), 212 (military), 214 (hunt); for an analysis of Musil's novel from the perspective of the accident, see ibid., 237–252.

180. F. Bordewijk, *Knorrende beesten* (Utrecht, 1933).

181. Barbara Kosta, "Die Kunst des Rauchens: Die Zigarette und die neue Frau," in Julia Freytag and Alexandra Tacke (eds.), *City Girls; Bubiköpfe & Blaustrümpfe in den 1920er Jahren* (Köln, Weimar and Wien: Böhlau Verlag, 2011), 143–158, here 143.

182. Irmgard Keun, *Das kunstseidene Mädchen: Roman, mit Materialien, ausgewählt von Jörg Ulrich Meyer-Bothling* (Leipzig, Stuttgart, and Düsseldorf: Erbst Klett Schulbuchverlag Leipzig), 65 (soft), 16 (debt); taxi trip quoted in Nina Sylvester, "*Das Girl*: Crossing Spaces and Spheres. The Function of the Girl in the Weimar Republic" (unpubl. diss., University of California, Los Angeles, 2006), 196; Ursula Krechel, "Irmgard Keun: die Zerstörung der kalten Ordnung; Auch ein Versuch über das Vergessen weiblicher Kulturleistungen," *Literaturmagazin* 10 (1979), 103–128, here 112 (film), 113 (cold order).

183. Irmgard Keun, *Gilgi—eine von uns: Roman* (Berlin: List Verlag, 2008; first ed. 1931); Krechel, "Irmgard Keun: die Zerstörung der kalten Ordnung," 106 (thirty thousand).

184. Keun, *Gilgi*, 2008, 16 (typewriter), 21–22 (independence), 24 (Olga), 33 (beat to pieces), 68 (radio), 87–88 (Mercedes), 140 (planless), 142 (sport), 202 (Cadillac), 206 (eating inwards: *hineingefressen*); references to coldness on 72, 89, 104, 175,

228–229, 248, 250; illusion: Carme Bescansa Leirós, *Gender- und Machttransgression im Romanwerk Irmgard Keuns: Eine Untersuchung aus der Perspektive der Gender Studies* (St. Ingbert: Röhrig Universitätsverlag, 2007), 299. For an analysis of Keun's novels as the narrative of a *flâneuse*, see Anke Gleber, *The Art of Taking a Walk: Flanerie, Literature, and Film in Weimar Culture* (Princeton, NJ: Princeton University Press, 1999), 191–213. In her desire "to search for the traces of an alternative form of flanerie" resulting in "a gendered definition of 'modernity,'" Gleber overlooks the fact that Gilgi herself opposes to the *flâneries* of her boyfriend, because it is "planless." Gleber discovered the *flâneuse* ("the active gaze" of "the urban woman as spectator") for the first time in Walter Ruttmann 1927 documentary film on Berlin, in which a woman "focuses her gaze on a man through a shopping window." Remarkably, this woman, mistaken to be a prostitute by Siegfried Kracauer, has only recently been identified as a "window shopper" (ibid., 172, 180). On the *flâneur* as "an exclusively masculine type," see Griselda Pollock, *Vison and Difference: Femininity, Feminism and Histories of Art* (London and New York: Routledge, 1988), 67.

185. On Fleisser's Frieda and several other young Weimar women as Girl, see Sylvester, "*Das Girl*: Crossing Spaces and Spheres."
186. Dabina Becker, "Marieluise Fleisser (1901–1974)," in Britta Jürgs, in cooperation with Ingrid Herrmann (ed.), *Leider hab ich's Fliegen ganz verlernt: Portraits von Künstlerinnen und Schriftstellerinnen der neuen Sachlichkeit* (Berlin and Grambin: AvivA Verlag, 2000), 68–86; for an analysis of Fleisser's *Mehlreisende Frieda Geier* as a New Objectivity novel, see Hemut Lethen, *Cool Conduct: The Culture of Distance in Weimar Germany* (Berkeley, Los Angeles, and London: University of California Press, 2002), 141–145; Dangerous principle: Michael Töteberg, "Spiegelung einer Bohemien-Existenz und Sportroman: Zeitliterarische Bezüge zum Prosawerk Marieluise Fleissers," *Text + Kritik: Zeitschrift für Literatur* 64 (1979), 54–60, here 55.
187. English Garden: Tax, *Marieluise Fleisser, Schreiben, Überleben*, 67; bicyclist: Marieluise Fleisser, "Frühe Begegnung," (1964) in Fleisser, *Gesammelte Werke; Zweiter Band: Roman, Erzählende Prosa, Aufsätze* (Frankfurt: Suhrkamp, 1994), 297–308, here 300; battlefield quoted in Theo Buck, "Dem Kleinbürger aufs Maul geschaut: Zur gestischen Sprache der Marieluise Fleisser," *Text + Kritik* 64 (1979), 35–54, here 42; speechless: Julia Freytag, "'Lebenmüssen ist eine einzige Blamage': Marieluise Fleissers Blick auf stumme Provinzheldinnen und Buster Keaton," in Freytag and Tacke (eds.), *City Girls*, 95.
188. Marieluise Fleisser, *Mehlreisende Frieda Geier: Roman vom Rauchen, Sporteln, Lieben und Verkaufen* (Berlin: Gustav Kiepenheuer, 1931), 28 (wild), 41–42 (philistine: *spiessig*), 54 (train); no nuclear family: Susan L. Cocalis, "Weib ohne Wirklichkeit, Welt ohne Weiblichkeit: Zum Selbst-, Frauen- und Gesellschaftsbild im Frühwerk Marieluise Fleissers," in Irmela von der Lühe (ed.), *Entwürfe von Frauen in der Literatur des 20. Jahrhunderts* (Berlin: Argument-Verlag, 1982), 64–85; sane world and man as adventure: Peter Beicken, "Weiblicher Pionier: Marieluise Fleisse—oder Zur Situation schreibender Frauen in der Weimarer Zeit," *Die Horen* 132 (1983), 45–61, here 60.
189. Gast Mannes, "Der Jappes und die Fleisserin," *Galerie: Revue culturelle et pédagogique* 13, no. 3 (1995), 403–432; and no. 4, 525–551, here 409–412, 424 (self-consciousness), 429 (border crossers), 548 (only men); Alexander Weicker, *Fetzen: Aus der abenteuerlichen Chronika eines Überflüssigen; Roman; Studienausgabe, Vorgestellt und kommentiert von Gast Mannes* (Mersch: Editions

du Centre national de la littérature, 1998), 355; Ashby: Marieluise Fleisser, "Der verschollene Verbrecher X," in Fleisser, *Die List: Frühe Erzählungen; Herausgegeben und mit einem Nachwort von Bernhard Echte* (Frankfurt: Suhrkamp Verlag, 1995), 69-74, here 73; Keaton: Freytag, "'Lebenmüssen ist eine einzige Blamage,'" 103; refuelling: Kosta, "Die Kunst des Rauchens," 164; withdrawal: Beicken, "Weiblicher Pionier."

190. Hans Fallada, *Kleiner Mann—was nun?* (Hamburg: Rowohlt Taschenbuch Verlag, 2009; first ed. 1932), 27 (commute), 50 (softer), 97 (outing), 105 (boss), 211 (fat cars), 224 (budget overview), 254 (thousand sounds); urban noise on 34, 81, 275; for the hierarchy of household commodities (and the cleft between American and German society in this respect), see Wolfgang König, *Geschichte der Konsumgesellschaft* (Stuttgart: Franz Steiner Verlag, 2000), 228.

191. Ödön von Horváth, "Kasimir und Karoline: Volksstück," in *Gesammelte Werke, Band I: Volksstücke, Schauspiele* (Frankfurt: Suhrkamp Verlag, 1970), 253-324, here 289 (truck driver), 306 (*hochkapitalistisch*), 307 (convertible), 313 (brake), 318 (metaphor); order: on p. 5* at end of the book; Reinecke, *Autosymbolik in Journalismus, Literatur und Film*, 161 (opportunism); 162 (desires the car); Benno von Wiese, "Ödön von Horváth," in Traugott Krischke (ed.), *Ödön von Horváth* (Frankfurt: Suhrkamp, 1981), 7-45, here 25 (ballad), 27 (bizar).

192. Elisabeth Tworek-Müller, *Kleinbürgertum und Literatur: Zum Bild des Kleinbürgers im bayerischen Roman der Weimarer Republik* (Munich: tuduv-Verlagsgesellschaft, 1985), 63 (office girls), 64 (kitchy), 66 (on and for them); Axel Fritz, *Ödön von Horváth als Kritiker seiner Zeit: Studien zum Werk in seinem Verhältnis zum politischen, sozialen und kulturellen Zeitgeschehen* (Munich: List Verlag, 1973), 119-120 (Mittelstand), 121 (demonology); Traugott Krischke, "Materialien zur Horváth-Forschung: Aspekte und Möglichkeiten," in Krischke (ed.), *Ödön von Horváth*, 203-218, here 203 (analytical); Benjamin paraphrased by Susan Buck-Morss, "The Flaneur, the Sandwichman and the Whore: The Politics of Loitering," *New German critique* 39 (1986), 99-141, here 116. Meinrad Pichler, "Von Aufsteigern und Deklassierten: Ödön von Horváths literarische Analyse des Kleinbürgertums und ihr Verständnis zu den Aussagen der historischen Sozialwissenschaften," in Krischke (ed.), *Ödön von Horváth*, 67-86, here 71 (*Aufsteiger*), 78 (*Der Mittelstand* and other-directedness); *Kleinbürgersprössling*: Remo Hug, *Gedichte zum Gebrauch: Die Lyrik Erich Kästners; Besichtigung, Beschreibung, Bewertung* (Würzburg: Königshausen & Neumann, 2006), 18; parasite and motif of coldness: Jürgen Schröder, "Das Spätwerk Ödön von Horváths," in Krischke (ed.), *Ödön von Horváth*, 125-155, here 149; Remarque: Jeglin and Pickerodt, "Weiche Kerle in harter Schale," 219.

193. Ödön von Horváth, "Der ewige Spiesser: Erbaulicher Roman in drei Teilen," in Horváth, *Gesammelte Werke, Band III: Lyrik, Prosa, Romane* (Frankfurt: Suhrkamp Verlag,1978), 145-178; Horváth's trip to Barcelona on p. 4*-5* at the end of the book; Elisabeth Tworek, "Der unbestechliche Blick eines Heimatlosen," in Tworek and Brigitte Salmen, *Ödön von Hováth: Ein Kulturführer des Schlossmuseums Murnau* (Murnau: Schlossmuseum des Marktes Murnau, 2003), 25-73, here 28 (*Kleinbürgerstudien*), 37 (nationalism), 48 (proletarian). Death by tree branch: "Zeittafel," in Tworek and Salmen, *Ödön von Hováth*, 74-76, here 76; Tworek-Müller, *Kleinbürgertum und Literatur*, 188 (deformed and hypochondrical), 164 (profit), 171 (fascist salute), 227 (cliché).

194. Fritz, *Ödön von Horváth als Kritiker seiner Zeit*, 174 (Jewish), 174 (marriage), 175 (Reich), 177 (impossible), 184 (car scene), 221 (squares), 227 (ruthless-

ness), 228 (pride); Lipset quoted in Heinrich August Winkler, "Extremismus der Mitte? Sozialgeschichtliche Aspekte der nationalsozialistischen Machtergreifung," *Vierteljahreshefte für Zeitgeschichte* 20, no. 2 (1972), 175–191, here 175. For a four-minute fragment of a 2006 reenactment of Glaube Liebe Hoffnung by *Münchner Kammerspiele*, see http://www.youtube.com/watch?v=DnjWq-un5g8 (last accessed 23 March 2013).

195. Kosta, "Die Kunst des Rauchens," 147.
196. Clarke, *Driving Women*, 59–62 (Stein). *The Autobiography of Alice B. Toklas*, 213–214 (exciting), 218 (very good), 235 (Auntie succeeded by Godiva), 253–254 (influenced), 256 (landscape plays), 286 (repair shop); on Stein's synesthetic practices and her grammar theory, see "Gertrude Stein," http://en.wikipedia.org/wiki/Gertrude_Stein (last accessed 10 April 2011): 10 (feminist reworking); Mellow, *Charmed Circle*, 354 (third Ford).
197. Mellow, *Charmed Circle*, 241(Bois de Boulogne), 255 (Provence), 405 (Chicago).
198. Gillian Beer, "The island and the aeroplane: The case of Virginia Woolf," in Homi K. Bhabha (ed.), *Nation and Narration* (London and New York: Routledge, 1990), 265–290, here 274; Andrew Thacker, *Moving Through Modernity: Space and Geography in Modernism* (Manchester and New York: Manchester University Press, 2009), 152–153, 172 (*volte-face*). "Pistol shot" quoted in Micéala Symington, "La marche et la conduite: Intériorité, extériorité et esthétique du roman (Virginia Woolf et Octave Mirbeau)," in Monneyron and Thomas (eds.), *Automobile et Littérature*, 49–56, here 51.
199. Thacker, *Moving Through Modernity*, 166.
200. Ibid., 176–177.
201. Melba Cuddy-Keane et al., "The Heteroglossia of History, Part One: The Car," in Beth Rigel Daugherty and Eileen Barrett (eds.), *Virgina Woolf: Texts and Contexts; Selected Papers from the Fifth Annual Conference on Virginia Woolf; Otterbein College, Westerville, Ohio, June 15–18, 1995* (New York: Pace University Press, 1996), 71–80, here 77; shopping novel: Jane Garrity, "Virginia Woolf, Intellectual Harlotry, and 1920s British *Vogue*," in Pamela L. Caughie (ed.), *Virginia Woolf in the Age of Mechanical Reproduction* (New York and London: Garland, 2000; Border Crossings, ed. Daniel Albright), 185–218, here 198.
202. Cuddy-Keane et al., "The Heteroglossia of History," 79; second Singer and Lanchester: Hermione Lee, *Virginia Woolf* (New York: Vintage Books, 1999; first ed. 1997), 554; Andrew Thacker, "Traffic, gender, modernism," *The Sociological Review* 54, no. s1 (October 2006), 175–189, here 182 (hermaphrodite), 183 (quote on "Modern Fiction").
203. Leonard Woolf, *Downhill All the Way, An Autobiography of the Years 1919–1939* (New York: Harcourt, Brace & World, Inc.: 1967), 178 (alertness), 181 (way of life), 182 (liberation and slower), 184 (punctures), 188–193 (Holland and Nazis, quotes on 188, 189 and 191); Lee, *Virginia Woolf*, 624 (motorcycle)
204. Garrity, "Virginia Woolf, Intellectual Harlotry," 199; scene making: Makiko Minow-Pinkney, "Virginia Woolf and the Age of Motor Cars," in Caughie (ed.), *Virginia Woolf in the Age of Mechanical Reproduction*, 159–182, here 170.
205. Minow-Pinkney, "Virginia Woolf and the Age of Motor Cars," 163.
206. Lee, *Virginia Woolf*, 748.
207. Dieter Wunderlich, *AusserOrdentliche Frauen: 18 Porträts* (Munich and Zürich: Piper, 2010), 58–67 (Freytag; backlight: 64), 106–115 (Berber), 116–130 (Fleisser), 131–149 (Mann; rallye: 137), 150–163 (Beinhorn); Irmela van der Lühe, *Erika Mann: Eine Biographie* (Frankfurt: Fischer Taschenbuch Verlag, 2001), 57, 70;

dream: Anke Hertling, "Angriff auf eine Männerdomäne: Autosportlerinnen in den zwanziger und dreissiger Jahren," *Feministische Studien* 30, no. 1 (2012), 12–29, here 23–24. On Beinhorn, see also Uwe Day, "Mythos ex machina; Medienkonstrukt 'Silberpfeil' als massenkulturelle Ikone der NS-Modernisierung" (diss., Universität Bremen, November 2004), 235–239; radio and independence quotes: ibid., 240–241. In Lisa St. Aubin de Terán (ed.), *Indiscreet Journeys: Stories of Women on the Road* (London: Sceptre, 1991), however, the selected women do not seem to have much relationshop with mobility or the road.

208. Fell, *Kalkuliertes Abenteuer*, 47–54, 247 (pattern), 251 (heads).
209. Linda McDowell, "Off the road: Alternative views of rebellion, resistance and 'the beats,'" *Transactions of the Institute of British Geographers* 21 (1996), 412–419; this was a response to Tim Cresswell's effort to define subversive mobility as a "counter-hegemonic struggle" based on Jack Kerouac's novel *On the Road*. Tim Cresswell, "Mobility as resistance: A geographical reading of Kerouac's 'On the road,'" *Transactions of the Institute of British Geographers* 18 (1993), 249–263.
210. Mary Suzanne Schriber, *Writing Home: American Women Abroad 1830–1920* (Charlottesville and London: University Press of Virginia, 1997), 11 (écriture), 61 (modes), 75 (convention), 79 (integrity). For a similar conclusion on gender coming out at the "crack at the concept of Self," after an analysis of one female and two male travel accounts through Africa, see Susan Blake, "A Woman's Trek: What Difference Does Gender Make?" in Nupur Chaudhuri and Margaret Strobel (eds.), *Western Women and Imperialism: Complicity and Resistance* (Bloomington and Indianapolis: Indiana University Press, 1992), 19–34, here 32.
211. Foster quoted in Patricia D. Netzley, *The Encyclopedia of Women's Travel and Exploration* (Westport, CT: Oryx Press, 2001), 2 (the quotes are paraphrases from Netzley); Cresswell, *On the Move*, 198; Martin Green, *Seven Types of Adventure Tale: An Etiology of a Major Genre* (University Park: The Pennsylvania State University Press, 1991), 3–4.
212. Karen Beckman, *Crash: Cinema and the Politics of Speed and Stasis* (Durham, NC, and London: Duke University Press, 2010), 18–19.
213. Margaret Walsh, "Gender on the road in the United States: By motor car or motor coach?" *Journal of Transport History*, Third Series, 31, no. 2 (December 2010), 210–230, here 211.
214. Barbara Korte, Ralf Schneider, and Claudia Sternberg, "Einleitung und Grundlegung," in Korte, Schneider, and Sternberg (eds.), *Der Erste Weltkrieg und die Mediendiskurse der Erinnerung in Grossbritannien: Autobiographie—Roman—Film (1919–1999)* (Würzburg: Königshausen & Neumann, 2005), 1–32, here 25.
215. E. J. Rath, *Gas-Drive In: A High-Powered Comedy-Romance that Hits on Every Cylinder* (New York: Grosset & Dunlap Publishers, 1925), 249 (italics in original), 251 (kiss), 288 (crisp).
216. Rath, *Gas-Drive In*, 18 (purple), 44 (technotalk), 50 (excited), 55 (buys a garage), 61 (gender confusion), 86 (soother her mood), 88 (feminine and hated), 162 (competent), 172 (inversion), 175 (wild man), 232 (cold look), 267 (violence).
217. Willa Cather, *One of Ours* (New York: Vintage Books, 1991; first ed. 1922), 107 (cool), 112–112 (ditch), 166 (Susie), 369 (garages), 370 (suicide); pejorative: Pearl James, "The 'Enid Problem': Dangerous Modernity in *One of Ours*," *Cather Studies* 6 (2006), 1–33, here 3, http://cather.unl.edu/cs006_James.html (last accessed 18 September 2011). E. K. Brown, *Willa Cather: A Critical Biography, Completed by Leon Edel* (New York: Alfred A. Knopf, 1953), xiv (1920s), 217

(novel of development), 219 (ugly), 220–221 (diver's seat), 228 (Pulitzer). Nanette Hope Graf, "The Evolution of Willa Cather's Judgment of the Machine and the Machine Age in her Fiction" (unpublished doctoral dissertation; Lincoln, University of Nebraska, 1991), 6–7 ("Nebraska"), 167 (survivors), 181 (new men), 184 (destroyers), 200 (lost lady). Polarized readings: Steven Trout, "Willa Cather's *One of Ours* and the Iconography of Remembrance," *Cather Studies* 4 (2004), 1–15, here 3, http://cather.unl.edu/cs004_trout.html (last accessed 18 September 2011); on the continuity of tourism and war, see Debra Rae Cohen, "Culture and the 'Cathedral': Tourism as Potlatch in *One of Ours*," *Cather Studies* 6 (2006), 1–18, http://cather.unl.edu/cs006_cohen.html (last accessed 23 September 2014). Erotic war motivation: Stanley Cooperman, *World War I and the American Novel* (Baltimore, MD: Johns Hopkins University Press, 1967), 129–137.

218. Randall Waldron, "Faulkner's First Fictional Car—Borrowed from Scott Fitzgerald?" *American Literature* 60, no. 2 (May 1988), 281–285, here 283.
219. Pierre Frondaie, *L'Homme à l'Hispano: Roman* (Paris: Éditions Émile-Paul Frères, 1925; soixante-neuvième édition; first ed. 1924), 1.
220. Ibid., 2 (projectile), 3 (allure; inversion), 12 (owns more cars), 50 (flight), 197 (man of train), 199 (crook), 238 (fill one's head). The 1926 movie edition was directed by Julien Duvivier (Girard, *L'automobile fait son cinema*, 99); see also http://cinema.theiapolis.com/movie-2ODX/l-homme-a-l-hispano/cast.html and http://fr.wikipedia.org/wiki/L'Homme_ percentC3 percentA0_l'Hispano_(film, _1926) (both last accessed 30 March 2011).
221. Reto Sorg and Michael Angele, "'Oh, my Ga-od! Oh, my Ga-od! Oh, Ga-od! Oh, my Ga-od!' Automobil-Unfall und Apokalypse in der Literatur der Zwischenkriegszeit," *Compar(a)ison* 2 (1996), 137–173, here 164.
222. Munzinger, *Das Automobil als heimliche Romanfigur*, 88–91.
223. Ibid., 101.
224. Lieb, *Crash*, 315; on Kafka and the car accident, see also Benno Wagner, "Kafkas Poetik des Unfalls," in Kassung (ed.), *Die Unordnung der Dinge*, 421–454.
225. Quotes in: Lieb, *Crash*, 225; Walter Delabar, "Linke Melancholie? Erich Kästners *Fabian*," in Jörg Döring, Christian Jäger, and Thomas Wegmann (eds.), *Verkehrsformen und Schreibverhältnisse: Mediale Wandel als Gegenstand und Bedingung von Literatur im 20. Jahrhundert* (Opladen: Westdeutscher Verlag, 1996), 15–34, here 16.
226. Bertold Brecht, "[Verfremdung des Autos]," quoted in Sorg and Angele, "Oh, my Ga-od!" 173; urban man: Erdmut Wizisla, *Walter Benjamin and Bertolt Brecht: The Story of a Friendhip* (2009), quoted in Edward Timms, "Beyond the Borders," *Times Literary Supplement* (19 March 2010), 13; street as theatre: Lieb, *Crash*, 253–254, 264 (Brecht quote on 262, n. 183).
227. Becker, *Das Automobil und die amerikanische Kultur*, 112.
228. Munzinger, *Das Automobil als heimliche Romanfigur*, 18.
229. Ibid., 77 (*Titan*), 88 (*American Tragedy*).
230. William Faulkner, *Flags in the Dust: Edited and with an Introduction by Douglas Day* (New York: Vintage Books, 1974; first ed. 1929), 92; boys: Mary Gordon, *Good Boys and Dead Girls and Other Essays* (London: Bloomsbury, 1991), 20–21; Mimi Reisel Gladstein, *The Indestructible Woman in Faulkner, Hemingway, and Steinbeck* (Ann Arbor, MI: UMI Research Press, 1986; first ed. 1974), 38.
231. Faulkner, *Flags in the Dust*, 108.
232. Ibid., 291–292.

233. David Minter, *William Faulkner: His Life and Work* (Baltimore, MD, and London: Johns Hopkins University Press, 1997; first ed. 1980), 72; on his stream of consciousness, see also André Bleikasten, "Faulkner and the New Ideologues," in Donald M. Kartiganer and Ann J. Abadie (eds.), *Faulkner and Ideology: Faulkner and Yoknapatawpha, 1992* (Jackson: University Press of Mississippi, 1995), 3–21, here 12.

234. Carvel Collins, "About the Sketches," in William Faulkner, *New Orleans Sketches: Introduction by Carvel Collins* (New Brunswick, NJ: Rutgers University Press, 1958), 7–34, here 32. On Faulkner's invented heroism in aviation, see Minter, *William Faulkner*, 29–39; flying on the hood of a car: Joseph Blotner, *Faulkner: A Biography; One-Volume Edition* (New York: Vintage Books, a division of Random House, 1991; first ed. 1974), 73, 143 (posing as a war hero); race-car body: Ted Ownby, "The Snopes Trilogy and the Emergence of Consumer Culture," in Kartiganer and Abadie (eds.), *Faulkner and Ideology*, 95–128, here 107.

235. William Faulkner, "Country Mice," in Faulkner, *New Orleans Sketches*, 191, 207 (pedestrian).

236. Blotner, *Faulkner*, 144–145; motion: Linda Welshimer Wagner, *Hemingway and Faulkner: Inventors/Masters* (Metuchen, NJ: The Scarecrow Press, 1975), 134.

237. David Sanders, "The 'Anarchism' of John Dos Passos," *The South Atlantic Quarterly* 60 (1961), 44–55, here 48 (destructive); travelogue: John Dos Passos, *Rosinante to the Road Again* (Breinigsville, PA: Dodo Press, 2010; reprint of edition from 1922); pluralized: Bradbury, "The Denuded Place," 204–205; John Dos Passos, *U.S.A.* (London and New York: Penguin Books, 2001; first eds. 1930–1937), 141 (intimacy), 177 (wealthy), 219 (eroticism); systems view: Tichi, *Shifting Gears*, 202; big money: John L. Grigsby, "The Automobile as Technological Reality and Central Symbol in John Dos Passos's *The Big Money*," *Kansas Quarterly* 21, no. 4 (1989), 35–41, here 36; stenographers: Charles Marz, "'U.S.A.': Chronicle and Performance," *Modern Fiction Studies* 26, no. 3 (Autumn 1980), 398–415, here 412. Dos Passos's last part of the *U.S.A.* trilogy, *The Big Money*, became especially famous because of the few pages dedicated to characterize Henry Ford and his "Tin Lizzie," which can be read as "a critical social and ideological documentation of the Ford era." Munzinger, *Das Automobil als heimliche Romanfigur*, 24.

238. Jürgen Link, "Maintaining Normality: On the Strategic Function of the Media in Wars of Extermination," *Cultural Critique* 19 (Fall 1991), 55–65, here 62; Link, *Versuch über den Normalismus*, 33–35 (normativity; routine), 46–48 (*realistisch*), 461 (model).

239. Lethen, *Cool Conduct*, 12–13 (shame); Helmut Lethen, "Von Geheimagenten und Virtuosen: Peter Sloterdijks Schulbeispiele des Zynismus aus der Literatur der Weimarer Republik," in *Peter Sloterdijks "Kritik der zynischen Vernunft"* (Frankfurt: Suhrkamp, 1987), 324–355, here 336.

240. Plonien, "Re-mapping the World," 4; Lethen, *Cool Conduct*, 26 (unconnected), 27 (egalitarian); six characteristics: Hug, *Gedichte zum Gebrauch*, 41; Frank Trommler, "The Creation of a Culture of Sachlichkeit," in Geoff Eley (ed.), *Society, Culture, and the Sate in Germany, 1870–1930* (Ann Arbor: The University of Michigan Press, 1996), 465–485, here 478–479 (avant-garde), 479 ("peculiar Germanness").

241. Jonathan P. Eburne, *Surrealism and the Art of Crime* (Ithaca, NY, and London: Cornell University Press, 2008), 5–6; irony as distancing tool: Tworek-Müller, *Kleinbürgertum und Literatur*, 64; Lethen, *Cool Conduct*, 26 (freedom), 28 (signals).

242. Jeglin and Pickerodt, "Weiche Kerle in harter Schale," 221; Lethen, *Cool Conduct,* 44 (codes), 62 (dandy).
243. Volker Berghahn, *Europa im Zeitalter der Weltkriege: Die Entfesselung und Entgrenzung der Gewalt* (Frankfurt: Fischer Taschenbuch Verlag, 2002), 118–129; aggressive subjectivity: Bohrer, *Die Ästhetik des Schreckens,* 17; Lethen, *Cool Conduct,* 42–43 (Weber).
244. Lethen, *Cool Conduct,* 62 (Wilde); hatred: Lethen, "Von Geheimagenten und Virtuosen," 332; Gay, *Weimar Culture,* 120–121. On the "cult of coolness" around the dandy in world literature, see Hiltrud Gnüg, *Kult der Kälte: Der klassische Dandy im Spegel der Weltliteratur* (Stuttgart: J. B. Metzlersche Verlagsbuchhandlung, 1988); on the destruction of the family in WWI Germany, see Elisabeth Domansky, "Militarization and Reproduction in World War I Germany," in Eley (ed.), *Society, Culture, and the Sate in Germany,* 427–463; on "communities of comradeship" as replacement for the family, see Thomas Kühne, "Comradeship: Gender Confusion and Gender Order in the German Military, 1918–1945," in Karen Hagemann and Stefanie Schüler-Springorum (eds.), *Home/Front: The Military, War and Gender in Twentieth-Century Germany* (Oxford and New York: Berg, 2002), 233–254, here 245–246.
245. List: Wege, "Gleisdreieck, Tank und Motor," 309; Lethen, *Cool Conduct,* 26–29; Trommler, "The Creation of a Culture of Sachlichkeit," 484–485 (alienation).
246. Sabina Becker, *Urbanität und Moderne: Studien zur Grossstadtwahrnehmung in der deutschen Literatur 1900–1930* (St. Ingbert: Werner J. Röhrig Verlag, 1993), 34; Patrick Laviolette, *Extreme Landscapes of Leisure: Not a Hap-Hazardous Sport* (Farnham and Burlington: Ashgate, 2011), 163–165.
247. Frondaie, *L'Homme à l'Hispano,* 276; Sarkany, *Paul Morand,* 57 (Larbaud and Cendrars), 69 (distanced), 71 (dandy), 84 (reserve), 122 (inhuman).
248. Wieland Schmied, "Der kühle Blick: Der Realismus der Zwanzigerjahre," in Schmied (ed.), *Der kühle Blick: Realismus der zwanziger Jahre, 1. Juni–2. September 2001* (Munich, London, and New York: Prestel Verlag / Kunsthalle der Hypo-Kulturstiftung, 2001), 9–36, here 12, 17; examples of the cool gaze are from Christian Schad, Giorgio de Chirico, and René Magritte; Picasso's "Olga" (1923) is on 8 and 339, *Das Telephon* on the cover. On "AutoPortrait": Julie Wosk, *Women and the Machine: Representations from the Spinning Wheel to the Electronic Age* (Baltimore, MD, and London: Johns Hopkins University Press, 2001), 115. On *Die Dame*: Sylvester, "*Das Girl*: Crossing Spaces and Spheres," 21–22, 36, 42–52.
249. Day, "Mythos ex machina," 188–189 (ice cold), 193 (Hitler).
250. Marz, "'U.S.A.': Chronicle and Performance," 400–401; interview (with Frank Gado) quoted in Townsend Ludington, "The Ordering of the Camera Eye in U.S.A.," *American Literature* 49, no. 3 (November 1977), 443–446, here 443.
251. Mariani, *William Carlos Williams,* 250.
252. Peter N. Stearns, *American Cool: Constructing a Twentieth-Century Emotional Style* (New York and London: New York University Press, 1994), 198, 239 (quote), 279 (Superman).
253. Möser, *Fahren und Fliegen in Frieden und Krieg,* 237; elevator: Peter Borscheid, *Das Tempo-Virus: Eine Kulturgeschichte der Beschleunigung* (Frankfurt and New York: Campus Verlag, 2004), 330 (the quote is from cultural critic Friedrich Sieburg in 1926).
254. William H. Rollins, "Whose Landscape? Technology, Fascism, and Environmentalism on the National Socialist *Autobahn*," *Annals of the Association of American Geographers* 85, no. 3 (1995), 494–520, here 496; icy brain quoted in Möser,

Fahren und Fliegen in Frieden und Krieg, 194; Klaus Theweleit, *Männerphantasien 1 + 2; Band 1: Frauen, Fluten, Körper, Geschichte; Band 2: Männerkörper— zur Psychoanalyse des weissen Terrors* (Munich and Zürich: Piper, 2009; first ed. 1977 and 1978), 487; avoids danger: Burkhardt Wolf, "Schiffbruch mit Beobachter: Zur Geschichte des nautischen Gefahrenwissens," in Kassung (ed.), *Die Unordnung der Dinge*, 19–47, here 19. On heat metaphors in the 1920s, see Jeglin and Pickerodt, "Weiche Kerle in harter Schale," 221. Rearmament: Berghahn, *Europa im Zeitalter der Weltkriege*, 110; Nietzsche: Thomas Fitzel, "Schmutz und Geschwindigkeit oder Warum das *Tempo* Tempo heisst," in Döring, Jäger, and Wegmann (eds.), *Verkehrsformen und Schreibverhältnisse*, 74–98, here 85.

255. Kurt Lindner, "Mein Auto—mein Selbst? Was passiert, wenn Männer Auto fahren," in Matthias Bisinger et al. (eds.), *Der ganz normale Mann: Frauen und Männer streiten über ein Phantom* (Reinbek bei Hamburg: Rowohlt Taschenbuch Verlag, 1992), 114–134, here 120–121.
256. *Hitler's Table Talk 1941–1944*, with an introductory essay on the mind of Adolf Hitler by H. R. Trevor-Roper (London: Weidenfeld and Nicolson, 1953), 310–312.
257. Erskine Caldwell, *Tobacco Road* (Athens, GA, and London: University of Georgia Press, 1995; first ed. 1932), 82–83 (new car), 103 (marrying), 103 (cannot hide his desire), 122 (black man), 123 (nigger), 134 (example of *sachlich* style), 172 (grandmother dead), 182 (factory); Lewis, *Babbitt*, 279; derogatory: Lisa A. Kirby, "A Radical Revisioning: Understanding and Repositioning *The Grapes of Wrath* as Political 'Propaganda,'" in Meyer (ed.), *The Grapes of Wrath: A Re-Consideration*, vol. 1, 243–260, here 249.
258. Quoted in Reinecke, *Autosymbolik in Journalismus, Literatur und Film*, 168; adventurousness: "B. Traven," at http://de.wikipedia.org/wiki/B_Traven (last accessed 23 April 2011); Traven, *Die Weisse Rose*.
259. Quoted in Hanno Ehrlicher, *Die Kunst der Zerstörung: Gewaltphantasien und Manifestationspraktiken europäischer Avantgarden* (Berlin: Akademie Verlag, 2001), 345–346.
260. Erhard Schütz, "'. . . Eine glückliche Zeitlosigkeit . . .': Zeitreise zu den 'Strassen des Führers,'" in Brenner, *Reisekultur in Deutschland*, 73–99, here 92.
261. Hermann Hesse, "Der Steppenwolf," in Hesse, *Gesammelte Werke; Siebter Band: Kurgast; Die Nürnberger Reise; Steppenwolf* (Frankfurt: Suhrkamp Verlag, 1970), 183–415, here 372–383, especially 372, 375; train, crazy cars, and mortal danger: Hermann Hesse, "Die Nürnberger Reise" (1927) in Hesse, *Gesammelte Werke*, 117–181, here 139–140; Hesse, "Kurzgefasster Lebenslauf," in Hesse, *Gesammelte Werke*, 469–489, here 489; narcotic vision (*Rauschgiftvision*) and anti-Americanism: Bernhard Weyergraf and Helmut Lethen, "Der Einzelne in der Massengesellschaft," in Weyergraf (ed.), *Literatur der Weimarer Republik 1918–1933*, 636–672, here 654; jump in movie and lust in movies: Lieb, *Crash*, 220–221.
262. Mike Michael, "Co(a)gency and the Car: Attributing Agency in the Case of 'Road Rage,'" in Brita Brenna, John Law, and Ingunn Moser (eds.), *Machines, Agency and Desire* (n.p.: Centre for Technology and Culture, 1998), 125–141, here 126.
263. "Friedrich Schnack," http://de.wikipedia.org/wiki/Friedrich_Schnack (last accessed 29 October 2011); Friedrich Schnack, *Das Zauberauto: Ein Roman* (Leipzig: Insel-Verlag, 1928), 17 (urgent), 19 (tremble), 26–27 (speed up), 35 (olfactorial inversion), 37 (*Selbstfahrer*), 39 (flight and film), 41 (afterglow), 76 (dream world), 95 (woman), 97 (credit), 96 (cinnamon), 101 (petrol and wine),

135 (tempering woman), 145 (*sinnverwirrt*), 151 (100-km life), 168–169 (trip without a trip), 170 (groom).
264. Arnolt Bronnen, "Triumph des Motors: Kurzgeschichte einer Form," in Bronnen, *Sabotage der Jugend: Kleine Arbeiten 1922–1934* (Innsbruck: Institut für Germanistik, 1989), 134–137, here 136; the story dates from 1929.
265. Arnolt Bronnen, "Der Autodieb: Eine Geschichte," in Bronnen, *Sabotage der Jugend*, 286–289.
266. Arnolt Bronnen, "Autos auf Reisen," in Bronnen, *Sabotage der Jugend*, 137–139, here 138; the story dates from 1928.
267. Friedbert Aspetsberger, "Einige Zitate zu Bronnen in den zwanziger Jahren," in Bronnen, *Sabotage der Jugend*, 343–377, here 344 (*Sport im Bild*), 364 (caste), 370 (Goldmark); Arnolt Bronnen, *O.S.: Nach dem Text der Erstausgabe von 1929; Mit einem Vorwort von Wojciech Kunicki und einem Nachwort von Friedbert Aspetsberger* (Klagenfurt and Vienna: Ritter, 1995), 37; Lethen, *Cool Conduct*, 212; for an analysis of *O.S.* as a New Objectivism novel see ibid., 210–214. On the monopolistic role of *Ullstein* publishers during the Weimar Republic, which could make writers rich overnight and brought others to "near-starvation," see Gay, *Weimar Culture*, 134–135.
268. Friedbert Aspetsberger, "*O.S.*—ein so infames wie gelungenes Werk, vielleicht ein k.u.k-Skandal in der Weimarer Republik," in Bronnen, *O.S*, 369–414, here 369 (feuds), 370 (piccolo), 373 (organ), 382 (Gramsci), 403 (*Schweinerei*), 404 (pornography), 410 (high feelings); new media and national bolshevism: Ulrike Baureithel, "Motorisierung der Seelen: Anmerkungen zu Arnolt Bronnens Konzeption der Mensch-Maschine-Symbiose," in Wolfgang Bialas and Burkhard Stenzel (eds.), *Die Weimarer Republik zwischen Metropole und Provinz: Intellektuellendiskurse zur politischen Kultur* (Weimar, Cologne, and Vienna: Böhlau Verlag, 1996), 131–142, here 132, 135; violence: Uwe-K. Ketelsen, "Die Sucht nach dem 'resistenten Zeichen': Zur Ästhetik der Gewalt in Arnolt Bronnens Roman 'O.S.,'" in Frauke Meyer-Gosau and Wolfgang Emmerich (eds.), *Gewalt: Faszination und Furcht; Jahrbuch für Literatur und Politik in Deutschland 1* (Leipzig: Reclam, 1994), 96–118, here 105. For a comparable effort to see "the entire body (as) a genital," see (Freud's rival) Victor Tausk's 1919 article "On the Origin of the 'Influencing Machine' in Schizophrenia," in which this often-evoked machine type of the 'Influencing Machine' is analyzed as "a representation of the patient's genitalia projected to the outer world," a regression to "a stage of diffuse narcissistic organ libido." Quoted in Tim Armstrong, *Modernism, Technology, and the Body: A Cultural Study* (Cambridge: Cambridge University Press, 1998), 102.
269. Aspetsberger, "Einige Zitate zu Bronnen in den zwanziger Jahren," 370–371.
270. Brecht and Ernst quoted in Aspetsberger, "Einige Zitate zu Bronnen in den zwanziger Jahren," 366.
271. Viktor Shklovsky (transliterated as "Schklowskij"), *Zoo: oder Briefe nicht über die Liebe* (Zoo or Letters Not About Love) (Frankfurt: Suhrkamp Verlag, 1965), 19–20 (foam), 67 (voice), 105–106 ("Ispana Suize" [sic]) Alexander Kaempfe, "Viktor Schklowskij und sein Zoo," in ibid., 121–144, here 123 (Formalist), 132 (marxist), 134 (fetish), 135 (philistines), 139 (happiness; educator), 141 (motorcycle); in the text I follow the English transcription of the author's name Shklovsky; Richard Sheldon, "Making Armored Cars and Novels: A Literary Introduction," in Viktor Shklovsky, *A Sentimental Journey: Memoirs, 1917–1922*, translation and literary introduction by Richard Sheldon; historical introduction by Sidney Monas (n.p.:

Dalkey Archive Press, 2004), ix–xxv, here xvii (cruelty); Sidney Monas, "Driving Nails with a Samovar: A Historical Introduction," in Shlovsky, *A Sentimental Journey*, xxvii–xlvii, here xxxii–xxxiii (intelligence and detached).

272. Shklovsky, *A Sentimental Journey*, 72 (begin), 80 (Persia), 183 (Bolsheviks), 186 (prose), 272 (motorcycle).
273. In Scott Fitzgerald's *This Side of Paradise* (1920) the mother of the protagonist drives an electric car. Munzinger, *Das Automobil als heimliche Romanfigur*, 121–122. Another Russian avant-gardist, Vladimir Mayakovsky, served in the Petrograd Military Training School, and wrote a travelogue on his visit to Detroit. He could not resist buying a Renault himself, describing the love for his wife in car terms. On Mayakowsky, see Lewis H. Siegelbaum, *Cars for Comrades: The Life of the Soviet Automobile* (Ithaca, NY, and London: Cornell University Press, 2008), 177–178. For "the association of cars with fearfulness and death" during the Stalin era (including the rumour that "black cars" appeared on Moscow streets to kidnap strangers whose bodies were subsequently taken apart), see ibid.; Schklowskij, *Zoo*, 103–104.
274. Gijs Mom, *The Evolution of Automotive Technology: A Handbook* (forthcoming).
275. Thibault, *L'Allure de Morand*, 16; Arlen, *The Green Hat*, 128; Schnack, *Das Zauberauto*, 140; Fallada, *Kleiner Mann—was nun?* 349; Döblin quoted in Lieb, *Crash*, 209; Klemperer quoted in Hochstetter, *Motorisierung und "Volksgemeinschaft"*, 168; Schklowskij, *Zoo*, 103–104.
276. Lewis, *Babbitt*, 105 (Ford), 288 (gas).
277. Tworek-Müller, *Kleinbürgertum und Literatur*, 214.
278. Quoted in Reinecke, *Autosymbolik in Journalismus, Literatur und Film*, 177.
279. Quoted from e. e. cummings, *100 Selected Poems* (New York, 1926) in Munzinger, *Das Automobil als heimliche Romanfigur*, 159.
280. Becker, *Das Automobil und die amerikanische Kultur*, 41; Gijs Mom and Ruud Filarski, *Van transport naar mobiliteit: De mobiliteitsexplosie (1895–2005)* (Zutphen: Walburg Pers, 2008), 149.
281. Quoted in Fons Alkemade, "De auto in de Nederlandse literatuur en speelfilm" (report for Foundation for the History of Technology, Eindhoven, The Netherlands; Eindhoven: Nederlands Centrum voor Autohistorische Documentatie NCAD, August/September 2000), 55.
282. Marieluise Fleisser, *Gesammelte Werke, Erster Band: Dramen*, ed. Günther Rühle (Frankfurt: Suhrkamp Verlag, 1983), 442. I thank Walter Delabar (email 4 September 2011) for bringing this passage to my attention.
283. Günther Rühle, "Leben und Schreiben der Marieluise Fleisser aus Ingolstadt," in Fleisser, *Gesammelte Werke, Erster Band*, 5–60, here 48.
284. Erich Kästner, *Gedichte* (Cologne: Verlag Kiepenheuer & Witsch, 1959; Gesammelte Schriften, Band 1), 72.
285. Reinecke, *Mobile Zeiten*, 174–175 (Tucholsky), 197–206 (Autobahnen lyricism).
286. Monneyron and Thomas (eds.), *Automobile et Littérature*, 78–82 (quote on 81).
287. Parr, "Tacho. km/h. Kurve," 71–73 (Karl: 73); Ehrlicher, *Die Kunst der Zerstörung*, 1–2; Morton, *In Search of England*, 121; W. S. [= Webster Schott], "Introduction: The Great American Novel," in William Carlos Williams, *Imaginations: Kora in Hell, Spring and All, The Great American Novel, The Descent of Winter, A Novelette and Other Prose*, edited with introductions by Webster Schott (New York: New Directions Books, 1970), 155–157, here 155.
288. Jean-Yves Tadié (ed), *La littérature française: dynamique & histoire, II* (Paris: Gallimard, 2007), 663.

289. Link and Reinecke, "Autofahren ist wie das Leben," 445–447; emblem: Kristen Whissel, *Picturing American Modernity: Traffic, Technology, and the Silent Cinema* (Durham, NC, and London: Duke University Press, 2008), 215.
290. Arnolt Bronnen, "Moral und Verkehr," in Bronnen, *Sabotage der Jugend*, quoted in Bickenbach and Stolzke, "Die Logik des Unfalls: Zufall und Notwendigkeit," 1–2.
291. Quoted in Reinecke, *Autosymbolik in Journalismus, Literatur und Film*, 149.
292. DE, "Gerd Arntz," in Johannes Bilstein and Matthias Winzen (eds.), *Ich bin mein Auto: Die maschinalen Ebenbilder des Menschen* (Cologne and Baden-Baden: Walther König / Staatliche Kunsthalle, n.d. [2001]), 98–100. For a similar use of the car as a symbol of wealth, see George Scholz's painting *Zeitungsträger* (newspaper carriers): DE, "Georg Scholz," in Bilstein and Winzen, *Ich bin mein Auto*, 92–93.
293. L. Moholy-Nagy, *Vision in Motion* (Chicago: Paul Theobald and Company, 1956), 153 (grasp), 245–246 (anecdote); grindstone quoted by Rosan Hollak, "De logica van de lijn," *Nieuwe Rotterdamse Courant* (4 February 2011), Cultureel Supplement, 11.
294. Lotringer and Virilio, *The Accident of Art*, 38 (Artaud), 58 (motorized); Renoir: Girard, *L'automobile fait son cinema*, 47.
295. Philippe Soupault, *En joue! ... Roman* (Paris: Bernard Grasset, 1925), 23 (journals), 91–92 (suicide).
296. [Editions du Chemin de Fer], "Jacques Rigaut, 'Je serais und grand mort': biographie," http://fr-fr.facebook.com/note.php?note_id=191307637567395 (last accessed 28 March 2011); Dada and surrealism as "firmly rooted in the city": Lynne Kirby, *Parallel Tracks: The Railroad and Silent Cinema* (Exeter: University of Exeter Press / Duke University Press, 1997), 174.
297. Armstrong, *Modernism, Technology, and the Body*, 197–204 (the quotes from Aragon and Williams are theirs, the other quotes are paraphrases by Armstrong); Charles Grivel, "Travel Writing," in Hans Ulrich Gumbrecht and K. Ludwig Pfeiffer (eds.), *Materialities of Communication* (Stanford, CA: Stanford University Press, 1994), 242–257, here 243 (italics in original); Peter Conrad, *Modern Times, Modern Places* (New York: Alfred A. Knopf, 1999), 92 (Soupault); Herbert Leibowitz, "Something Urgent I Have to Say to You": The Life and Works of William Carlos Williams* (New York: Farrar, Straus and Giroux, 2011), 247.
298. Plonien, "Re-mapping the World," 38; Hans Richter, *Dada: Art and Anti-art* (London and New York, 2004) quoted in "Tristan Tzara," http://en.wikipedia.org/wiki/Tristan_Tzare (last accessed 10 April 2011); Eksteins, *Tanz über Graben*, 435 (avant-garde had won).
299. Tristan Tzara, quoted in Philip Beitchman, *I Am a Process with No Subject* (Gainesville: University of Florida Press, 1988), 49.
300. Schnack, *Das Zauberauto*. *Stäuben* on 136, 144; *jagen* on 91, 115, 126; *schiessen* on 74, *fliegen* on 66—the first three in one paragraph on 71.
301. Theodor W. Adorno, "Über den Fetischcharakter in der Musik und die Regression des Hörens," in: *Dissonanzen: Einleitung in die Musiksoziologie* (Frankfurt am Main: Suhrkamp Verlag, 1973) (Gesammelte Schriften Band 14), 14–50, here 20 (religion), 34 (infantile), 39 (over and over), 44–45 (driver as model), 50 (second hand); Cotten Seiler, *Republic of Drivers: A Cultural History of Automobility in America* (Chicago and London: The University of Chicago Press, 2008); Joel Dinerstein, *Swinging the Machine: Modernity, Technology, and African American Culture between the World Wars* (Amherst and Boston: University of Massachusetts Press, 2003).

302. Lieb, *Crash*, 221.
303. Quoted in Disko, "'The World is My Domain,'" 50.
304. Lieb, *Crash*, 215 (thunder), 225 (automobile), 315 (linguistic).
305. Beckman, *Crash*, 24.
306. Adams, *The Great Adventure*, 122-123.
307. Tadié (ed.), *La littérature française*, 637 (genesis), 640 (deconcentrated), 649 (roman d'aventure).
308. Michael Nerlich, *Abenteuer oder das verlorene Selbstverständnis der Moderne: Von der Unaufhebbarkeit experimentalen Handelns* (Munich: Gerling Akademie Verlag, 1997), 246.
309. Bohrer, *Die Ästhetik des Schreckens*, 114-115 (reflex and poetry), 163-167, 362.
310. Quoted in Nerlich, *Abenteuer*, 241 (italics in original).
311. Simmel, "Das Abenteuer," 17; Paul Zweig, *The Adventurer* (London: Basic Books, 1974), 249.
312. Constance Classen, *The Color of Angels: Cosmology, Gender and the Aesthetic Imagination* (London and New York: Routledge, 1998), 8.
313. Schwieger, "The Joad Collective," 197.
314. [William Carlos Williams], *The Autobiography of William Carlos Williams* (New York: New Directions Books, 1967), 237.
315. William Carlos Williams, "The Great American Novel," in Williams, *Imaginations*, 158-227, here 171; freed poetry: W. S. [= Webster Schott], "Introduction: Beautiful Blood, Beautiful Brain," in Williams, *Imaginations*, ix-xviii, here x; anti-novel: W. S. [= Webster Schott], "Introduction: The Great American Novel," in Williams, *Imaginations*, 155-157, here 155.
316. Williams, "The Great American Novel," 172.
317. Ibid., 169.
318. Hans Anten, *Van realisme naar zakelijkheid: Proza-opvattingen tussen 1916 en 1932* (Utrecht: Reflex, 1982), 39-40.
319. Tworek-Müller, *Kleinbürgertum und Literatur*, 14 (*Kleinbürgerromane*), 19 (resentment).
320. Ibid., 84 (the quote is from Kurt Sontheimer).
321. Ibid., 97-99 (F-scale), 102 (factors), 185 (declassification), 231 (idyll), 234 (dream), 250-251 (Jews and reds), 308 (relativize). For an analysis of "petit-bourgeois mentality," see ibid., 180-267.
322. "Is die vacantie nu reeds vergeten?" *Toeristenkampioen* 2, no. 41 (9 October 1937), 1534; "De uitslag van den wedstrijd in vacantieherinneringen," *Toeristenkampioen* 2, no. 45 (6 November 1937), 1639-1642; "Weer een prijsvraag om vacantie-herinneringen," *Toeristenkampioen* 3, no. 38 (17 September 1938), 1505-1506, here 1505; "Prijsvraag vacantieherinneringen," *Toeristenkampioen* 3, no. 45 (5 November 1938), 1729-1732.
323. Quoted in James, *The Middle Class*, 404. The last quote is a paraphrase by James.
324. Nellie Wesseling, *Een tante die chauffeert: Een amusant verhaal voor jonge meisjes* (Helmond, 1934); Orwell, *Coming Up for Air*, 72 (penny dreadfuls).
325. Mary Cadogan and Patricia Craig, *You're a Brick, Angela! The Girls' Story 1839-1985* (London: Victor Gollancz, 1986), 76-77 (Orman and Wells), 81 (violent feminism), 264-266 (Johnson), 316-317 (girl stopping), 319 (French).
326. Gabriel Bos, "Hans en haar zusjes: Het jongsachtige meisje in de meisjesroman," in Aafke Boerma, Erna Staal, and Murk Salverda (eds.), *Bab's bootje krijgt een stuurman: De meisjesroman en illustrator Hans Borrebach (1903-1991)*

(Amsterdam, 1995), 53–72, here 64; Henk van Gelder, "'Zo iets als een artistiek geweten ken ik niet': Over het leven en werk van Hans Borrebach," in Boerma, Staal, and Salverda, *Bab's bootje krijgt een stuurman*, 73–103. Jhr. De Marees van Swinderen, "Autorijden wil ik en . . . autorijden zal ik, oftwel de vrije vertaling van het schone Duitsche begrip: 'AUTO-FIMMEL,'" *Auto-Leven* 19, no. 36 (10 September 1931).

327. Becker, *Das Automobil und die amerikanische Kultur*, 96; David K. Vaughan, "On the Road to Adventure: The Automobile and American Juvenile Series Fiction, 1900–40," in Jan Jennings (ed.), *Roadside America: The Automobile in Design and Culture* (Ames: Iowa State University Press, for the Society for Commercial Archaeology, 1990), 74–81; Alison Light, *Forever England: Feminity, Literature and Conservatism Between the Wars* (London and New York: Routledge, 1991), 8 (redefinition), 99 (itinerant), 100 (on the rails), 101 (good and bad).

328. American blues: John A. Heitmann, *The Automobile and American Life* (Jefferson, NC, and London: McFarland, 2009), 115 (quotations are by Heitmann); Graeme Ewens and Michael Ellis, *The Cult of the Big Rigs and the Life of the Long Haul Trucker* (Northolt: Quarto, 1977), 150; Ute Rosenfeld, "'Von Auto-Hymnen und -Klageliedern': Eind Kulturgeschichte der Automobilität im Widerhall populärer Musik," in Marianne Bröcker (ed.), *Das 20. Jahrhundert im Spiegel seiner Lieder: Tagungsbericht Erlbach/Vogtland 2002 der Kommission für Lied-, Musik- und Tanzforschung in der Deutschen Gesellschaft für Volkskunde e.V.* (Bamberg: Universitätsbibliothek Bamberg, 2004), 203–231, here 207; Louis Davids quoted in Ruud Filarski (in cooperation with Gijs Mom), *Shaping Transport Policy: Two Centuries of Struggle Between the Public and Private Sector—A Comparative Perspective* (The Hague: Sdu Uitgevers, 2011), 92 (my translation).

329. Restlessness: Thomas Koebner, *Lehrjahre im Kino: Schriften zum Film* (St. Augustin: gardez! Verlag, 2000), 23–27 (Sprache der Bewegung: 27); Harro Segeberg, "Literarische Kino-Ästhetik: Ansichten der Kino-Debatte," in Corinna Müller and Harro Segeberg (eds.), *Die Modellierung des Kinofilms: Zur Geschichte des Kinoprogramms zwischen Kurzfilm und Langfilm 1905/06–1918* (Munich: Fink, 1998), 193–219, here 198–201 (quotes on 198, 199, and 201).

330. Tim Cresswell and Deborah Dixon, "Introduction: Engaging Film," in Cresswell and Dixon (eds.), *Engaging Film: Geographies of Mobility and Identity* (Lanham, MD: Rowman & Littlefield, 2002) 1–10, here 4–5.

331. Alexandra Schneider, "Homemade Travelogues: *Autosonntag*—A Film Safari in the Swiss Alps," in Jeffrey Ruoff (ed.), *Virtual Voyages: Cinema and Travel* (Durham, NC, and London: Duke University Press, 2006), 157–173, here 159 (tourist experience), 159 (family), 160 (seeing), 163 (no car). For an American example of 1920s footage of urban traffic scenes taken with a "family 16 mm camera," see "Old Cars," http://www.youtube.com/watch?v=8jo_Iuqw7XU (last accessed 20 May 2010).

332. Jeffrey K. Ruoff, "Forty Days Across America: Kiyooka Eiichi's 1927 Travelogues," *Film History* 4, no. 3 (1990), 237–256, here 246 (thrill), 250 (mud hole), 252 (racism). "In those days," Eiichi remembered later when interviewed, "overhauling a car was part of the fun, part of the fun of having a car" (ibid., 241).

333. Peter J. Bloom, "Trans-Saharan Automotive Cinema: Citroën-, Renault-, and Peugeot-Sponsored Documentary Interwar Crossing Films," in Ruoff (ed.), *Virtual Voyages*, 139–156, here 139 (Citroën), 140 (desert), 150 (Renault and Peugeot); on the American desert as theme in early road movies, see Steven Cohan and Ina Rae Hark (eds.), *The Road Movie Book* (London and New

York: Routledge, 1997), 20; secret service and Teilhard de Chardin: "Croisière jaune," http://fr.wikipedidia.org/wiki/Croisi percentC3 percentA8re_jaune (last accessed 8 January 2011); celluloid: Girard, *L'automobile fait son cinema,* 14; winter tourism: Pirie, "Non-urban Motoring in Colonial Africa in the 1920s and 1930s," 41.

334. Marthe Oulié, *Bidon 5: En rallye à travers le Sahara* (n.p. [Paris]: Ernest Flammarion, n.d. [1931; cinquième mille]), 8 (rallye), 17–18 (conquest; there also an overview of previous Sahara crossings), 33 (sixteen women), 119 (motor bus). In 1925 Oulié also sailed the Mediterranian in a nonmotorized sailing ship with an all-female crew; see Marthe Oulié and Hermine de Saussure, *La croisière de "Perlette": 1700 milles dans la Mer Égée* (n.p. [Paris]: Librairie Hachette, 1926).
335. Bennet Schaber, "'Hitler can't keep 'em that long': The road, the people," in Cohan and Hark (eds.), *The Road Movie Book,* 17–44, here 41, n. 2
336. Wege, "Gleisdreieck, Tank und Motor," 317; Bryony Dixon, *100 Silent Films* (Houndsmill and New York: Palgrave Macmillan, 2011), 27–28 (Symphony); on the "anesthecizing that occurs in [an] overwhelmingly passive viewing experience," see Maud Lavin, *Push Comes To Shove: New Images of Aggressive Women* (Cambridge, MA, and London: MIT Press, 2010), 125.
337. Beckman, *Crash,* 58.
338. Whissel, *Picturing American Modernity,* 223; Mary P. Ryan, "The Projection of a New Womanhood: The Movie Moderns in the 1920's," in Jean E. Friedman and William G. Shade, *Our American Sisters: Women in American Life and Thought* (Lexington, MA, and Toronto: D. C. Heath and Company, 1982), 500–518, here 501 (comedy); Anton Kaes, "Schreiben und Lesen in der Weimarer Republik," in Weyergraf (ed.), *Literatur der Weimarer Republik 1918–1933,* 38–64, here 53 (Berlin); Kenneth Hey, "Cars and Films in American Culture, 1929–1959," *Michigan Quarterly Review* 19, no. 4 (1980), 588–600, here 587 (Lynds), 589 (five left), 592 (visits), 594 (sanctuaries); Koebner, *Lehrjahre im Kino,* 42–43 (democratic). For the "enduring appeal of self-suficient individualism" of the "1920s film themes," see also Philip Davies and Brian Neve, "Introduction," in Davies and Neve (eds.), *Cinema, Politics and Society in America* (Manchester: Manchester University Press, 1981), 1–18, here 7. Laderman, *Driving Visions,* 284, n. 15, for instance, speaks of "the road movie's sense of *individual* adventure and rebellion" (italics in original). For an extensive analysis of individualism as part of American exceptionalism, see Richard O. Curry and Lawrence B. Goodheart, "Individualism in Trans-National Context," in Curry and Goodheart (eds.), *American Chameleon: Individualism in Trans-National Context* (Kent, OH, and London: Kent State University Press, 1991), 1–19. Shared journeys: Denby, "The Self Discovered," 118.
339. William M. Drew, "The Speeding Sweethearts of the Silent Screen," http://www.welcometosilentmovies.com/features/sweethearts/sweethearts.htm (last accessed 30 April 2010), 2.
340. Dave Mann and Ron Main, *Races, Chases & Crashes: A Complete Guide to Car Movies and Biker Flicks* (Osceola, WI: Motorbooks International, 1994), 128–129.
341. Eric Mottram, "Blood on the Nash Ambassador: Cars in American Films," in Peter Wollen and Joe Kerr (eds.), *Autopia: Cars and Culture* (London: Reaktion Books, 2002), 95–114, here 228 (Cagney).
342. For a short description of the movies mentioned, see Mann and Main, *Races, Chases & Crashes;* stunts and Sacré-Coeur: Girard, *L'automobile fait son cinema,* 74.

343. Claudia Sternberg, "Der Erinnerungsdiskurs im Spielfilm und Fernsehspiel," in Korte, Schneider, and Sternberg (eds.), *Der Erste Weltkrieg und die Mediendiskurse der Erinnerung in Grossbritannien*, 243–342, here 253.
344. Gertrud Koch, "In Bewegung: Das Auto des Films," in Bilstein and Winzen (eds.), *Ich bin mein Auto*, 155–165, here 156–157; *Underworld*: Girard, *L'automobile fait son cinema*, 78; Patton, *Open Road*, 248; Edward Dimendberg, "The Will to Motorization—Cinema and the Autobahn," in Jeremy Millar and Michiel Schwarz (eds.), *Speed—Visions of an Accelerated Age* (London, Guelph, and Amsterdam: The Photographers' Gallery / Trustees of the Whitechapel Art Gallery / Macdonald Stewart Art Centre/Netherlands Design Institute, n.d. [1998]), 62–71, here 62, 65; *Scarface* quoted from Sarah S. Marcus in a contribution to a discussion on "Urban Transportation in Literature, Theater, Film & TV" on 11 February 2007 on H-NET Urban History Discussion List (H-URBAN@H-NET.MSU.EDU). For more interwar European films on road and cars as a theme, see Ezio Alberione, "Nastri d'asfalto e pellicole di strada," in Mirko Zardini (ed., with Giovanna Borasi et al.), *Asfalto: Il carattere della città* (Milano: La Triennale di Milano/Mondadori Electa, 2003), 153–157.
345. For a clip of the Keystone Kops, see http://www.youtube.com/watch?v=Sx9Tovevx6w&feature=related (last accessed 23 March 2013).
346. Trenches: Girard, *L'automobile fait son cinema*, 44; http://nl.wikipedia.org/wiki/Buster_Keaton and http://nl.wikipedia.org/wiki/Harold_Lloyd (both last accessed 6 May 2011).
347. Kisch, *Paradies Amerika*, 265–266; Mottram, "Blood on the Nash Ambassador, 224 (the quote on war is from Stan Brakhage); dirty breaks quoted in Joseph Anthony Interrante, "A Movable Feast: The Automobile and the Spatial Transformation of American Culture 1890 – 1940" (PhD thesis, Department of History, Harvard University, Cambridge, MA, June 1983), 254.
348. Stan Laurel and Oliver Hardy, *Two Tars* (1928; Bounty Classics, BF65). For the demolition of cars in *It's the Old Army Game* (1926), see Mottram, "Blood on the Nash Ambassador," 227; for a comparable demolition movie *Hot Water* (1924) starring Harold Lloyd, see Beckman, *Crash*, 69–91; there, too, an extensive analysis of *Two Tars* (ibid., 91–104).
349. Beckman, *Crash*, 60; last quote is from Henry Jenkins, *What Made Pistachio Nuts? Early Sound Comedy and the Vaudeville Aesthetic* (New York, 1992).
350. On the 1930s gangster films, see, for instance, Melling, "The mind of the mob," 31–33; escape: ibid., 23–24.
351. Interrante, "A Movable Feast," 253.
352. For a list of thirty-six silent-movie titles on speed and cars from the 1910s and 1920s, see ibid., 127; Patton, *Open Road*, 249 (Gable); Laderman, *Driving Visions*, 24 (Chain Gang); Cohan and Hark (eds.), *The Road Movie Book*, 1 (frontier ethos); Mark Williams, *Road Movies* (New York: Proteus Publishing Co., Inc., 1982), 106–107 (Bogart); for photos of the set of this "first, and some say best, trucking film ever made," see Ewens and Michael Ellis, *The Cult of the Big Rigs*, 138–139.
353. Koebner, *Lehrjahre im Kino*, 60 (traveling and phantom rides); http://de.wikipedia.org/wiki/Menschen_am_Sonntag (last accessed 2 July 2011); Dixon, *100 Silent Films*, 129–130 (Vertov); "Dziga Vertov," http://en.wikipedia.org/wiki/Dziga_Vertov (last accessed 5 April 2010); movie seen on 4 April 2010 in Van Abbemuseum, Eindhoven, exhibition: "Lissitsky +: Overwinning op de zon, 19 september 2009–5 september 2010"; for *Man With a Movie Camera*, see http://

video.google.com/videoplay?docid=-2809965914189244913# (last accessed 5 April 2010). Remarkably, Kirby, *Parallel Tracks*, 171, analyzes Vertov's film as a film about trains and about women.

354. For an analysis by demographers of the relationship between family formation and the 'taming' of males (resulting in the conclusion that there is no causal relation, but a strong correlation), see Joseph Henrich, Robert Boyd, and Peter J. Richerson, "The puzzle of monogamous marriage," *Philosophical Transactions of the Royal Society B* 367 (2012), 657–669.

355. Cresswell, *On the Move*, chap. 5 (on dancing), chap. 6 (Supreme Court rulings).

356. Quoted in Munzinger, *Das Automobil als heimliche Romanfigur*, 180.

357. Mark S. Foster, "In the Face of 'Jim Crow': Prosperous Blacks and Vacations, Travel and Outdoor Leisure, 1890–1945," *The Journal of Negro History* 84, no. 2 (Spring 1999), 130–149, here 141; McClintock quoted in Wosk, *Women and the Machine*, 143.

358. George S. Schuyler, "Traveling Jim Crow," *The American Mercury* 20, no. 80 (August 1930), 423–432, here 422. Here also more examples of 'nasty adventures.'

359. Wolfgang Müller-Funk, *Literatur als geschichtliches Argument: Zur ästhetischen Konzeption und Geschichtsverarbeitung in Lion Feutchwangers Romantrilogie Der Wartesaal* (Frankfurt and Bern: Peter D. Lang, 1981), 26.

360. David Barison and Daniel Ross, "The Ister: An Excerpt; From Novi Sad to Vukovar," http://www.rouge.com.au/3/ister.html (last accessed 22 November 2010), 4.

361. Armstrong, *Modernism, Technology, and the Body*, 77–79 (see in general his chap. 3, "Prosthetic Modernism").

362. Roberto Cagliero, "The Literature of Exhaust," in Francesca Bisutti De Riz and Rosella Mamoli Zorzi, in cooperation with Marina Coslovi (eds.), *Technology and the American Imagination: An Ungoing Challenge; A.I.S.N.A. Associazione Italiana di Studi Nord-Americani, Atti del Dodicesimo Convegno Biennale, Università di Venezia, 28–30 ottobre 1993*, vol. VIII, n. 10—1994, Rivista di Studi Anglo-Americani, Annuario dell' A.I.S.N.A. (Venezia: Edizioni Supernova, 1994), 575–583, here 576.

363. Winkler, "Extremismus der Mitte?" 175.

CHAPTER 7

~~~

Swarms Into Flows
The Contested Emergence of the Automobile System

> "In the past the man has been first; in the future the system must be first."[1]

"In Southern New Jersey you may see a farm," the contribution on "Business" in Harold Stearn's *Civilization in the United States* (1922) observed,

> now prosperously devoted to berry and fruit crops, on which, still in good repair, are the cedar rail fences built by a farmer whose contacts with business were six or eight trips per year over a sand road to Trenton with surplus food to exchange for some new tools, tea, coffee, and store luxuries. That old sand road has become a cement pavement—a motor highway. Each morning a New York baking corporation's motor stops at the farmhouse and the driver hands in some fresh loaves. Presently a butcher's motor stops with fresh meat, then another one with dry groceries, and yet another from a New York department store with parcels containing ready made garments, stockings and shoes.[2]

In the following lines the author explains what infrastructural and systemic support was necessary to enable this producing farm to evolve into a consuming unit. The car, in short, was not only portrayed as the harbinger of progress and personal pleasure (as we have seen in the previous chapters), its systemic aspects promised (or threatened, depending on the perspective) to change the very way society functioned.

In this chapter we will analyze the emergence of this systemic aspect along four different trajectories, constructing ever more layers around the individual vehicles and cars and their individual users, and witnessing how these vehicles were invited, if not forced into swarms and flows along corridors of 'traffic.' These four trajectories are the changing 'functional adventure' (repair, maintenance, tinkering), the building of a car-friendly road network to enable individual vehicles to move in 'flows,' the efforts to guarantee that this network, and the flows upon it, functioned in a safe and efficient way, and the positioning of the fledgling car system in the existing structure of mobility systems, most particularly the railways. In developing this system the vehicle diversified too, toward the motor bus, the truck, and, illustrated by the quote at the start of this chapter, the delivery van. Their emergence followed patterns different from those of the car and motorcycle, and seemed to threaten the existing mobility modes much more directly.

We will argue in this chapter, which closes the group of chapters on the interwar period, that on the verge of the Second World War an 'automobile system' was fully in place in every motorizing country in the North-Atlantic realm, irrespective of the penetration rate of the car (its density, in cars per capita). The successful emergence of the car, as we have seen in chapter 1, took place on the basis of a collective, swarm-like behavior, well organized (mostly by the clubs) but not centrally steered.[3] The swarm, as part of a comparison between human and insect societies, was often invoked in the raging debates about the collective and individualizing elements of modern society. Entomologist William Wheeler saw human society develop through "a kind of interstitial swarming" that sets it apart from a bee colony where all bees are born from a single mother.[4] As the French sociologist Gabriel Tarde already observed in 1890 when he formulated his concept of 'mass psychology,' the organized mass (la foule) forms the basis of every societal system: "Incoherence becomes cohesion, noise becomes voice, and the thousands of closely-packed people form only one single, incomparable beast, a terrific and monstrous predator persuing its goals with irresistible determination." The swarm functions through communication between its members: mobility and communication, fatefully separated by transport experts around the turn of the century, merge. Yet at the same time within the swarm, individual courage and the willingness to take risks multiplies, and the risk becomes a commodity.[5]

In the next sections we will first focus upon the maintenance and repair substrate built underneath the fledgling automobile system, followed by a reconstruction of the transnational discourse about the necessity of a car-friendly road infrastructure. We will then deal with the controversy between the engineer-builders and the spatial planners, and their differing projects to bring about the 'automobile system.' Both groups were very much concerned about road safety, but while the 'builders' sought the solution in a technical fix (in the form of the freeways), the 'planners' (often emerging from the local, urban level) emphasized the educational and attitudinal aspects. This controversy should be seen against the background of a much larger and fundamental change, from a 'train society' to a 'car society': the 'struggle between road and rail' will be the subject of the following section, after which we wrap up this chapter in the conclusions.

Coping With the Car's Unreliability:
Maintenance, Repair, and the Functional Adventure

The first systemic (material, infrastructural) layer built around the sets of atomized small communities of car users was the loose, noncoercive network of hotels and repair facilities, started by many national touring clubs during the emergence of the bicycle in the 1890s. This was followed by a soft (compared to the hardware of the roads) network of taxation and licensing legislation spread out by the national state, constraining (but sanctioning at the same time) a taming and civilizing process started well before the First World War and in most countries, for the first phase of basic tax and license legislation completed by 1910. The result was that the elite car culture, just starting to expand toward middle-class layers, was recognized as a societal fait accompli in Europe, whereas in the United States (in the middle of the European war) the first expansion to farmers took place, as we have seen in chapter 4. In 1916 American legislation set the scene for the building of a highway network. The Interbellum added a new level of transnational governance, leading (through the Commission for Communication and Transit of the League of Nations) to the standardization of road signs and transborder traffic.[6]

The networks were not only soft, they also had coarse mazes, and holes in the system had to be compensated by the users' own efforts to keep the mobile part of the system, the vehicle, functioning, either by hiring a chauffeur, or by tinkering and delegating insurmountable 'troubles' to a fledgling network of repair facilities. It was, in other words, a system that facilitated rather than coerced, in which the motorist could feel enabled to follow his (and increasingly her) path: the swarm character was largely left in place, despite the construction of a more constraining system around it. This system was as much geared toward use as it was toward purchase: while in 1925 in the United States $3 billion dollars were spent on the latter, $5 billion went to the former as payment for repair and maintenance.[7]

As new groups of lay users, not versed in automotive technology, joined the veterans of the first phase, the functional aspect of the automotive adventure which had been so instrumental in the breakthrough of the car in the previous period, started to come under pressure. Like the other aspects of the automotive adventure (touring and racing, including the erotic and 'colonial' aspects; see chapters 2, 5, and 6) the functional adventure had to be adjusted to the new users. As the car's functional spectrum now became dominated by a mixed form of use in which the 'serious' side of mobility (business use, commuting, a part of the daily routine such as shopping) played a non-negligible role (see chapter 4), the mechanical defect acquired a different meaning as an unpleasant interruption of an adventure that was increasingly geared to moving around, and less to the pleasurable sides of male bonding in clubs and other groups, whether they were on the move or bent over the engine along the road. Reliability became a catchword in a culture that was increasingly colored by discourses on waste and efficiency, and the saving of time and money.[8] This was especially true in the United States, but the efficiency and waste discourse quickly migrated to Europe as well. As a result, the functional adventure became tainted negatively, and the pleasure of unreliability vanished from memory, including academic history writing.

It was Henry Ford who was one of the first to combine mass production with after-sales service. The keyword was "preventive service," the result of detailed rules regarding inventories of relatively cheap spare parts to be kept by officially recognized Ford dealers and repair shops: they spread toward the smallest villages, whereas the sales agencies were to be found in larger towns. In the United States Ford introduced "operation sheets" for mechanics in an effort to rationalize repair work as much as the work in his production facilities, but he had to abandon its flat rate system in the early 1930s due to mechanics' resistance.[9] This idea that the responsibility of the manufacturer did not stop at the factory gates was first developed for the electric vehicle and its maintenance-prone battery, while during the First World War it was perfected into a rudimentary system of logistics for trucks under military conditions, as we have seen in chapter 3. Now, it became extended to the gasoline car, which refused to be made reliable enough for a lay public. It took a 300-page manual called *Ford Service: Detailed Instructions for Servicing Ford Cars,* finished in 1925, to cover the entire technology of the Model T from a user point of view.[10]

From the perspective of the receiving countries in Europe there was a lot of work involved as the new practice often interfered with a tradition of repair, emerging around the turn of the century and based on handicraft. In the Netherlands, for instance, the earliest garages for automotive repair developed from three roots: the horse trade and carriage rent, the bicycle dealer and repairman, and the individual technician who had started his job as an enthusiast and member of the new automobilism movement.[11] As a parallel system a network of petrol stations was created, according to a Dutch study initiated in Europe by the American army during the war.[12] This fuel-supply network emerged at an amazing speed during the 1920s in all motorizing countries, changing the landscape into one openly geared toward the automobile. Billboards along the new highways in the countryside were the most outspoken expressions of this, but the cities, too, made their 'scapes' more easily readable from an automobile.[13]

Generally, automotive technology evolution during the Interbellum was governed by three tendencies, two of them provoked by the 'American invasion,' the third by the European response. The first tendency was the "elimination of the human element," as a Dutch trade journal called it somewhat elliptically: the increasing automation of the handling of the car. In chapter 5 we have given a sketchy overview of these trends, as it intertwined with the closing of the body and the desire for 'comfort.'

The second tendency was a lowering of fuel consumption, which became dominant in the second half of the 1920s in Europe when it became clear that the American makes would not bring the hoped-for mass motorization and European manufacturers had to develop their own solutions. But like in the United States, cars for a mass market had to be geared to a touristic function. They should have, as an expert of the Dutch touring club recommended, "a closed body with enough room for five persons, reach a maximum speed of 95 to 100 km/h on a straight road and an average of 60 to 70 on long trips, due to a rapid acceleration, a valuable pick-up [re-acceleration after a decrease in speed], a good brake system and a high stability on the road."[14]

The third tendency was a general development toward a higher reliability. New types of steel, new mixtures of rubber and stronger tire casings, and roller bearings rather than friction bearings, all combined with more sophisticated testing methods, made the number of defects decrease substantially, but the condition of the road surfaces as well as the gradual increase in trips meant at the same time that the defect remained a commonplace and a recurrent phenomenon in the daily user culture. In the technical sections of the car journals in all motorizing countries this ambiguous struggle against technical unreliability (as a source of irritation and delight, depending on the personal skills, and perhaps on class, too) can be followed, for instance in the letters to the editor from all those new users who wished to become streetwise and versatile in basic car mechanics. Gone were the prewar days of deeply technical treatises of many pages, complete with mathematical formulas and professional technical drawings, prewar days when there was not much difference in technical competence between producers and users. Now, those treatises were to be found in engineering journals, while the 'consumer' journals started sections for the layman who had lost his equal position (in terms of knowledge, and skills) vis-à-vis the editor, and had taken on a role of asking questions to a separate specialist on the editorial team. The latter answered, for all to see and learn from, basing his knowledge on test drives of cars borrowed from the manufacturers. These and other technical sections aimed at the transfer of basic knowledge, handy tips and admonitions not to touch specific components or to execute certain repairs and leave them to the specialist. This was part of a general tendency initiated by the manufacturers and their engineers to push away the knowledgeable user from under the bonnet, as has been argued in chapter 5.

At the same time, clubs started to set up technical support services for motorists who got stuck on the roadside. In Europe, this started in the United Kingdom and was taken over by the German ADAC in 1927 and the Dutch ANWB around the same time, in both cases without much success.[15] One of these "car doctors" in the Netherlands remarked in 1936 that some motorists did not even want to change a flat tire.[16] No doubt, car technology, as manufacturers realized in the interwar years, had to be 'fool-proof,' or '*Narrensicher.*'

In the United States the early proliferation of a service infrastructure (and a less well-developed car and touring-club life) made a mobile support system less necessary. However different the organizational characteristics of the maintenance infrastructure, similar trends of 'co-construction' by users and their struggle against reluctant engineers at the manufacturers have been unearthed for the United States, for instance in a study that analyzed a hundred patents after 1915 and two hundred letters to Henry Ford in the 1920s. Sent in the middle of the car-tourism frenzy, the proposals to change the car's technical structure were often focused on enhancing comfort and adding accessories (such as extra luggage trunks, but also beds and cooking utensils), in short: "Individualizing [their] Automobile(s)," beyond what manufacturers considered as normal. The latter's efforts, according to this analysis, were aimed at pushing the user back from car technology in proportion to the professionalization and scientification of car engineering, to introduce technologies (such as the luggage trunk at the rear) and present them as their own developments, and make the automobile 'tinker-proof.' While 72 percent of automotive patents in the United States were filed by individuals in 1921, by the end of the 1930s they were mostly filed by corporations.[17]

Between simplification of the technical advice and increasing reliability, a gap remained that was gradually filled by the automotive garage. In the United States extensive educational programs were set up to teach (nearly exclusively male) youngsters how to fix (instead of design) cars. Remarkably, while the car became the epitome of modernity, behind the scenes, in and around the greasy pits of the repair shop, a counter-image prevailed of low esteem of mechanics' work. Also, automotive repair became a realm of black employment, the basis of a segregated car culture with separate black garages, filling stations, car-wash facilities, and even training institutes. In 1937 one witness claimed that one-third of this kind of work was done by blacks.[18] Maintenance and repair could also be female, however. Georgine Clarsen has found several examples in the United Kingdom and Australia of garages run by women celebrating "female independence" with their "ethic of experimentation" in gender bending and exhibiting the "noncollective kind of feminism."[19] Whether male or female, the "mechanics problem" and its related issue of trustworthiness, was a "class-based conflict."[20]

In the Netherlands the repair sector was a vivid example of interwar European corporatism, practiced in a sector with an overwhelming majority of very small enterprises whose trade associations were backed by the state with legislation on education for technical personnel, and licenses for recognized members.[21] A recent study of the German maintenance and repair infrastructure, however, claims that the German garage sector (consisting in 1929 of 20,000 repair shops, on 421,000 cars) managed to "solve ... the crisis of trust" vis-à-vis the motorists. This could be done because the repair culture was embedded from the beginning in a wider artisanal culture, and although the sector became regulated rather late (in 1932, followed in 1935 by a licensing system on the basis of a mechanic's diploma) it was integrated in the national socialist motor corps (NSKK). Ideologically near the Nazi worship of small-scale labor, 80 percent of the chief mechanics in the sector were a member of the Nazi party or its *Sturmabteilung* SA. Similar affections between garage personnel and the national socialist movement can be found in the Netherlands for this period.[22]

And yet, some layers of the users refused to adhere to this division of responsibilities. The "S.O.S" section of the Dutch touring club journal *Autokampioen* illustrates nicely what kind of problems car users were confronted with and which of these problems they wanted to deal with themselves. From the more than four thousand questions sent in between 1928 and 1938 it becomes quite clear that there were still quite a lot of tinkerers who experimented with mixing petrol types or the tuning of the engine in order to get more power out of it. The desire to beautify and to enhance comfort was also strong. Among the technical questions those on the electrical system dominated, especially the ignition.[23] A comparable analysis of the do-it-yourself section (*Fragenkasten*) in *ADAC-Motorwelt* gives similar results.

In short: the extent of the technical sections in the journals suggests that the functional element as late as the 1930s was still an important part of the automotive adventure. These same journals also testify, however, that if one did not wish to develop such tinkering skills, it was possible to own and use a car, but at the price of a regular garage bill.

Transnationalizing the Local:
Planning and Building National Road Networks

The same ambivalence governed another element of the fledgling automobile system: its road infrastructure. Good roads invited less adventurous citizens to opt for the car, while they also imposed more discipline upon an 'anarchistic' movement. For the pioneering period before WWI it is difficult to establish a correlation between the existence of good roads and the size of the auto movement, but the later conclusion that the car is unthinkable without the (paved) road is an unfounded projection of 'common sense' into history. In the United States especially cars with large ground clearance indicated that manufacturers saw them used in places where roads were bad or absent. Indeed, except for trucks, a correlation between car ownership and road-network development does not exist for the early period in the United States.[24] But during the interwar years this started to change—in a transnational manner, that is—by diffusing local practices through a transcontinental structure of experts.[25]

In 1908 the French government, acting through its diplomatic channels, called upon its fellow governments of other countries to convene in Paris for a road conference, motivated by two acute problems. The first was the rapid 'degradation' of the condition of roads by the recently introduced automobile, and the second was a direct consequence of the first: the problem of dust formation caused by the suction of rotating tires. A year later the *Association Internationale Permanente des Congrès de la Route* (AIPCR, internationally better known under its English acronym, PIARC, or Permanent International Association of Road Congresses) was founded in Paris.

The French initiative can be read as an answer to a movement led by several truly European (in the sense of 'transnational') actors, such as LIAT and AIACR (see chapter 1). The international character of the very early bicycle and automobile movements placed the issue of transborder travel on the agenda of these clubs (resulting in a struggle between automobile and touring clubs over assumption of the—lucrative!—responsibility for issuing 'triptyques' and 'carnets de douane'). In addition, the touring clubs even managed to agree (as early as 1900 and 1902) upon standardized designs for road warning signs used to mark corners, obstacles, railroad grade crossings, and dangerous crossroads. The national clubs themselves subsequently started to place these warning signs along the roads, using them at the same time as a means of advertising their existence by printing their club names on the signs.[26] Another initiative, Ernest Guglielminetti's founding in Monaco of a League against Road Dust (*Ligue contre la poussière sur les routes*), gained a surprisingly rapid following in most Western-European countries. Although his role was downplayed at the first Paris conference, it cannot be denied that his initiative, like that of the clubs, formed a powerful 'movement from below,' despite (or perhaps even because of) the unmistakably elite character of this movement.

From 1908, PIARC's members (mostly government officials and engineers) started a process of unofficial but very effective consensus on many matters. These related to roads, such as pavement and sub-soil technology, road alignment, bridge and tunnel construction, traffic safety statistics, up-to-date planning and financing, an international context that, remarkably, is fully overlooked in many national histories of road construction.[27] On the massively attended PIARC conferences, held every three to four years during the interwar years, the discourse around the emergence of 'automobile systems,' especially their infrastructural substrates, in the member nations (including the United States) can be followed in detail. In all these countries road building went along with increasing centralization of the planning and funding practices, but each country placed its own emphasis depending on the geological situation, the strength of the organized car movement, the tradition of central planning, and the culture of internal democracy. The motor behind the nonetheless remarkably uniform road-building frenzy in the Atlantic world was formed by a golden formula, briefly characterized as the 'earmarking' of car tax income for road building, a spiral of mutually supporting financial and building practices of enormous proportions.[28] Although national historiographies of automobilism and road building do not yet allow the pinpointing of concrete channels and persons of knowledge transfer from the transnational to the national level (and vice versa), a short description of the themes and problems of the first conferences make it plausible that such transfers were very important indeed. It is an impressive illustration of the futility of the concept of 'Americanization' (see chapter 5) as it reveals the enormously complex ways transnational knowledge was built up.[29]

The Paris conference in 1908, opened by initiator and Minister of Public Works Louis Barthou in the amphitheater of the Sorbonne, was a clear success, measured by the number of attendees, the number of papers submitted, and, especially, the support from governments all over the Western world. Table 7.1 shows that French reports and attending engineers dominated, at least in numbers, the entire period under

Table 7.1. PIARC Conference Characteristics, 1908–1938

Year	Location	A	B	C	D	E	F	G	H
1908	Paris	28	2411	1600	UK (279), Germany (191), Belgium (121), Austria (74), Italy (58), USA (50), Netherlands (48), Switzerland (43), Russia (40), Spain (32), Denmark (25)	46	107	France (40), UK (19), Belgium (13), Germany (12), US (9)	The present road; Maintenance; Dust and Wear; The future road; Effects of vehicles upon roads; Effect of roads on vehicles; Road signs; Road transport services
1910	Brussels	32	2118	1200	France (511), Germany (276), UK (160), Austria (113), Italy (56), USA (55), Switzerland (53), NL (51), Spain (39), Russia (35)	71	125	France (23), Belgium (18), UK (13), Italy (11), Germany (12), US (12), Hungary (8), Germany (6), NL (6)	Paving technology (incl. dust); Soil foundation; Tramways; Cleansing; Paving type choice; Tracing and lighting; Influence of vehicle weight and speed on roads; Specifications of vehicles; Exploitation of public transport.
1913	London	44	3793	2000	France (610), Germany (286), Belgium (173), USA (139), Italy (136), Austria (107), Russia (80), NL (63), Switzerland (51), Spain (32)	52	149	UK (30); France (19); US (19); Germany (17); Austria (14); Russia (13), Hungary (10), Italy (9), Bulgaria (6); NL (4)	New roads; Pavement of bridges; Stone roads with bituminous binders; Wood paving; Lighting of roads and vehicles; Wear detection methods; Traffic regulation; Centralisation and decentralisation of building and maintenance; Financing
1923	Sevilla		1891	600–700	France (495); Spain (375); UK (225); USA (166); Belgium (135); Italy (62); NL (54); Switzerland (42); Sweden (35); Czechoslovakia (32)		59	USA, France, UK, Italy (all 6); Belgium (5); NL, Switzerland (both 4)	Concrete roads; Asphalt roads; Tramway tracks in road surfaces; Development of motorised traffic; Traffic regulation; Congestion.

Swarms Into Flows

		A	B	C	D	E	F	G	H
1926	Milan	55		3429	France (562); UK (410); Poland (174); USA (168); Belgium (142); NL (93); Hungary (92); Switzerland (91); Czechoslovakia (75); Rumania (67)		75	Italy (12); France (6); UK (6); US (6); NL (5); Sweden (5); Belgium (4); Switzerland (4)	48	Concrete roads; Asphalt roads; Testing of asphalt; Road censuses; Town planning and traffic; Special automobile roads.
1930	Washington	64		3380	1000	France (512); UK (431); Poland (165); Italy (123); Belgium (105); Spain (88); NL (81); Sweden (80); Germany (76); Portugal (73)	74	USA, France, UK, Germany, Italy (all 6); NL, Denmark, Switzerland, Siam (all 4)	69	Concrete and bricks; Asphalt; Roads in colonies; Financing; Road transport (coordination); Urban traffic regulation.
1934	Munich	52			2100			France, Austria, Japan, Italy, Sweden, Germany, UK (all 6); Hungary, China (both 5)	86	Concrete and bituminous materials; Economic pavement construction; Road safety; Mutual influence of vehicles and road; Standardization and regulation of vehicle weight and dimensions
1938	The Hague	53		3938	2200	UK (648), France (516); Germany (507), USA (186); Poland (158); Belgium (141), Italy (128), Rumania (105), Switzerland (100), Spain (99)	89	NL (16); UK (7); Germany, Australia, USA, France, Sweden (all 6); Hungary, Japan, Poland, Czechoslovakia (all 5)	93	Concrete, bricks and bituminous materials; Road accidents; Flow separation; Road surface slipperiness and glare; Sub-soils;

Source: *Comptes rendus* of PIARC conferences; Munich: *Verslag van het zevende internationale wegencongres gehouden te München in 1934* (The Hague: Algemeene Landsdrukkerij, 1937), ANWB archives

A = number of officially represented governments at the conference
B = number of PIARC members
C = number of conference attendants
D = main foreign countries represented (indicative of conference attendance)
E = share of foreign members (percentage)
F = number of reports
G = countries dominating as source of official reports (number of reports)
H = official questions discussed at the conference (reformulated and abbreviated by this author)

study, although the Anglophone countries were not far behind: French attendees wrote 15 percent of the 736 reports for the eight conferences under consideration, followed by the United Kingdom, with 13 percent, and the United States, with 9 percent. The table thus testifies that the French were successful in the construction of a truly international organization in which all relevant actors were gathered and all relevant topics pertaining to the construction, financing, and management of automobile-friendly roads were discussed extensively and sometimes even quite fiercely. It cannot be denied though that the distance to the United States, and the special infrastructural situation there, necessitated a partly separate development.

During the 1920s several national PIARC committees were founded; the Dutch *Vereeniging Het Nederlandsche Wegen-Congres* (Road Conference Association) probably was one of the first, in 1920. Other countries with national PIARC structures before the Second World War were Austria, Czechoslovakia, Denmark, Germany, Great Britain, Italy, Poland, Spain, Switzerland, the United States, and Yugoslavia.[30]

At the first conference the manner of treating the issue of road signs highlighted several tendencies that played a role throughout the entire first period until the Second World War. The issue caused quite a discussion between representatives of the clubs and of the governments, because LIAT (representing the national clubs as initiators of sign design) had changed its opinion during its 1908 conference in Stockholm. It now favored one single warning sign in the form of a diagonally placed red bar. LIAT representatives pleaded in vain for such a change during the Paris conference, stressing that bicyclists would also benefit from the simplicity of a single sign. They were opposed by representatives from AIACR (and the Italian Touring Club, which was dominated by motorists) who rejected the single sign as dangerous. In the end, most touring clubs (except the Swiss, Belgian, and Dutch clubs) joined the automobile clubs and voted in favor of the four-sign compromise, although the Dutch finally decided pragmatically to join the ranks of the four-sign proponents "without renouncing its opinion." As a result, the road-sign issue was delegated to a special diplomatic conference held in October of the following year in Paris, where the signs were included in a convention on traffic, a nice example of 'officializing' private initiative at a transnational level. LIAT representatives complained that this convention focused exclusively on car traffic, and they also deplored the fact that responsibility for placing these signs was now centrally controlled by the states. This controversy led to a row, during the second PIARC conference in Brussels in 1910, between auto- and touring-club representatives, but the dice were already thrown.[31] This episode clearly documents the official (state-supported) character of PIARC and the dominance of automobile interests over the interests of other road users. By the time of the second PIARC conference it was clear that the 'road problem' was defined as a 'car problem.' First the horse owners (in Paris, 1908) and then the bicyclists (in Brussels, 1910) were relegated to their own 'paths.'[32] It is, in the light of the previous chapters (especially chapter 1) not a coincidence that this happened at exactly the moment (1910) of a first breakthrough toward a middle-class type of motorists.

Once established as primarily a matter of concern for motorists, the paved road became, during subsequent PIARC conferences, firmly embedded in a rudimentary automobile system of which the highway formed the material spine. During the first conferences before the war (Paris, 1908; Brussels, 1910), the special automobile road was rejected. Instead the attendees emphasized the importance of the improvement of the existing road system, including pavement modernization, broadening and straightening of dangerous curves. Also, from a very early date it was stipulated that *new* main roads should avoid routes through towns (London, 1913). Furthermore, the asphalt pavement option, as an engineering compromise, was rescued from the French, who were in favor of the much more expensive, but—from a maintenance point of view—technically superior solution of pavement by setts or cobblestones (London, 1913).[33]

During these conferences participants formulated a rudimentary state of the art through a careful process of consensus building during the discussions on the 'conclusions.' These included a maximum speed (25 km/h); a maximum axle load (4 tons; in case of 5 tons the maximum speed was to be reduced to 15 km/h); a maximum wheel pressure; a minimum road width (6 m) and curve radius (50 m, later increased to 200–300 m[34]); the construction of only "moderate gradients" and "parabolic tangents" for main-road access and exit; the avoidance, if possible, of level crossings; suppression of dust (first by tarring, then by asphalt paving); the opinion that street cars and trams were road obstacles that preferably should operate on separate tracks (and if that was not possible, their rails should be constructed inside the pavement instead of upon it); the placement of distance markers between large towns; and, of course, the necessity of separate horse and bicycle lanes. A fourth conference, meant to be held in Munich in 1916, did not take place because of the war, which also ended this early program of structuring the conception of roads.

During the first postwar conference, in Sevilla, Spain, in 1923, many attendees agreed that the war had played a decisive role as the real starting point of "automobilism."[35] When most Western nations started massive road-improvement projects financed by new automobile taxes during the second half of the 1920s, the Great War was again invoked to stress the exemplary role of the European leadership in this domain. An American representative, for instance, referred to the 2.5 million American soldiers "who during two years (1918 and 1919) had experienced the vast road networks in good shape, which gave birth to the desire to make the American network similar, if not superior, to the European network."[36]

Later during this phase, the transfer of ideas, both in technology and in management, was reversed, and the United States became the example as soon as it started its own impressive road-improvement campaign. From the painting of white stripes on the road to the struggle against automobile parking in cities by simply putting written notes on the windscreen (with the request to pay the fine at the police station), the European debate became more and more colored by American examples. In this sense PIARC slowly evolved into an open road lobby. For instance, in 1923 in Sevilla, the conference called upon national governments to "encourage the development of motor traffic" and even subsidize motor buses, where it had defined the motor bus before the war solely as a feeder for the railways.[37] No doubt, the rehabilitation of concrete as an alternative to asphalt also can be attributed to a large extent to the preference of many American road engineers for this technology. Concrete was considered the ideal pavement choice, but its application was dependent on the increasing scientification of the road-building profession. This was because the composition of the mixture and the manner in which concrete should be applied during actual road building were much more critical than was the case with asphalt. Concrete was mostly applied in the United States, Italy, Belgium, and Germany (in the latter case also because of autarchic motives), whereas other countries opted for asphalt, which they adjusted to the better properties of concrete during the early 1930s.[38]

Thus, if we are to characterize the Interbellum period from the perspective of a transnational organization like PIARC, two major tendencies emerge. The first is an increasing scientification (and quantification) of the road problem, and the second an accompanying acceptance and promotion of the centralization of planning and financing of the ever-growing national road-improvement projects. Initially, during the third PIARC conference in London (1913), the idea of centralizing road management was rejected. British county engineers had mobilized their colleagues at the London conference against the more centrally oriented engineers and they rejected in massive numbers any plan for centralization. Later conferences, however, clearly embraced the French concept of centralized management of *national* roads, whereas the lower-order roads could be delegated to the counties, provinces, *Länder,* or *départements.*[39] During the fifth PIARC conference in Milan (1926) this tendency revealed itself in the insight that central control had to be supported by 'soft' control of centralized statistics, both of road censuses and accidents.

The conference held in Washington in 1930 came at exactly the right moment to impress most visiting European road engineers. The participants witnessed the management of 'traffic' (defined as a flow of automobiles), and with this the definitive shift in the unit of analysis from the single vehicle to a systems approach. This shift was reinforced by the founding of road research laboratories in most countries (and most American states) during the second half of the 1920s and the early 1930s. These agencies then took up the issue of road-construction technology and materials, as well as the testing of alternative solutions. This division of labor allowed the PIARC conference free time to concentrate on management and control. The reports on road financing written for the Washington conference clearly reveal the differences between national approaches to highway programs, which ranged from the totally decentralized British tradition to, on the other extreme, French centralization. In fact, before the Nazis took power in 1933, the decentralized German financing tradition was even more extreme than the British, because the highest authorities on road building in 1930 were the individual German states. Despite these differences all Western European nations (except Germany) agreed that the national state at least paid for the construction of a network of national roads, and that the construction of most other, lower-order roads could only receive state support if their design complied with a centrally formulated and imposed standard. To enable this approach, most countries worked according to some sort of Road Plan, identifying a list of road-improvement projects that were qualified according to their function within the system: national roads, secondary roads at a regional level, tertiary (local) roads, and, in some countries such as France, also quarternary (agricultural) roads.

Many countries, too, completed the development of these networks well before the Second World War, a fact often neglected in road historiography because of its fascination for the more spectacular freeway projects.[40]

In many countries these huge improvement projects of paving, straightening, and widening existing roads were enabled by new taxes on fuel, a very efficient method because of the 'spiraling effect' of mutually supportive increasing car registrations and increasing road building, use, and wear. At the Washington PIARC conference, congestion and the struggle against it (the priority to keep the flow of vehicles moving) joined the other factors to further enhance this spiraling effect. In fact, had this effect not occurred, the history of Interbellum road building would have looked quite different in Europe, because, as a German road engineer remarked in Washington, "we can hardly think of increasing the tax rate without causing protests by the auto industry. The increase of the revenues from that source can thus only be reached through the increasing number of automobiles." The result of the spiraling effect can hardly be underestimated, although it is not given its proper place in national motorization histories. Once established as a 'growth mechanism,' a surprisingly (also for the contemporaries) rapid process of national and secondary road improvement all over Western Europe and North America was set in motion. For the Netherlands, it is now well proven that national road-building engineers looked with great concern at the diminishing car-registration figures during the depression of the 1930s. They were all too happy when they discovered unemployment relief could compensate for this loss of momentum of the spiraling growth of cars and fuel-tax revenues, which meanwhile represented one-quarter of all state-tax revenues, illustrating at the same time how the state became a major actor in the construction of the 'car society.'[41] Nonetheless, at the Washington conference, French representatives concluded that "the national network could be considered to be more or less finalised," but maintenance required ever-larger sums. Washington was also the place where a consensus was reached on the civilizing role of roads in the opening up of new territory (including the colonies). The French delegation did not succeed, however, in preventing the Washington conference from accepting a conclusion that the principle of a "dedicated tax" (implying the allocation of all car tax revenues to road building and improvement) should be regarded as "inviolable."[42]

At the seventh PIARC conference in 1934 in Munich, although the Americans had refused to submit reports, the centralization ideal was further promoted. Concrete as an alternative received a second boost, without, however, convincing proponents of asphalt (such as France, the United Kingdom, and, for national roads, the Netherlands) to give up their alternative. For the second time in PIARC's history, the asphalt road was rescued, this time from its competition by concrete, because one of asphalt's greatest drawbacks, its slipperiness, appeared to be to a large extent resolved by improvements in automobile construction, especially in suspension and tires. Four years later, at the conference in The Hague, asphalt's slipperiness was unanimously regarded as being largely solved, mainly by carefully mixing the right amounts of broken stones and bituminous substances.[43] The solution to the slipperiness problem clearly showed a systems approach in which the vehicles and the infrastructure were mutually adapted in a process lasting two decades.

The last conference before the war, then, was held in The Hague in 1938 (the conference announced for 1942 in Budapest did not take place). This meeting can be seen as the occasion where, under certain circumstances, special automobile roads came to be considered a safe solution to the "road problem," mostly because of the radical separation of flows (between slow and fast traffic, and between traffic in opposing directions) this road type allowed.[44] Only on the brink of the Second World War, too late for most countries to change their road-construction planning, was the freeway accepted by the engineering communities with close ties to the national governments. How could this have happened?

PIARC did not have a monopoly on the road problem. Recent scholarship on European freeway-system planning has revealed the existence of a separate community, partly overlapping with the PIARC constituency but dominated by a group of road-building contractors and promoters from countries with a tradition of central and even authoritarian state control. These, such as Puricelli from Italy, Kaftan from Germany, and Lucien Lainé from France, promoted the concept of a transnational European freeway network as a means of bringing international peace and unemployment relief during the depression.[45] This group managed to get the support of Albert Thomas of the International Labour Office (ILO) in Geneva, who was close to the main actors within the League of Nations. Most of these plans came to nothing, despite the organization of two dedicated conferences in 1931 and 1932 on the topic and the formation of a *Bureau Internationale des Autoroutes* (BIAR, 1931) in Geneva (in 1932 renamed *Office International des Autoroutes,* or OIAR). However, a plan for the construction of a single road corridor between London and Istanbul was promoted by the British Touring Club AA and adopted by the international tourism association AIT (successor of LIAT), and received some national support in Eastern Europe where parts of this corridor were actually realized.[46] But the most important influence of the BIAR and OIAR initiatives was indirect: it convinced certain engineering factions in the national ministries of the rationality of the freeway concept. The Dutch case can illustrate this.[47]

In the Netherlands it was the unique intermediary role of the touring club ANWB that appeared to be crucial in bringing about national centralization in road building. The Netherlands had a long tradition of land-use planning and engineering executed by a central departmental service, *Rijkswaterstaat,* as part of the Ministry of Waterstaat (Water Management). Founded during the French occupation at the turn of the nineteenth century after the model of the *Corps des Ponts et Chaussées,* Rijkswaterstaat had been responsible for the construction of the railway (after 1860) and waterway network, for which, just like in France at the departmental level, a system of provincial sub-departments existed alongside the independent Provincial Waterstaat departments.[48]

In 1925 ANWB organized a special national meeting right before the elections to put pressure both on parliament to prioritize the road problem and on regional and local officials not to oppose centralization. Backed by a minister who previously had been a member of ANWB's road commission, a Road Plan (1927) was developed based on a scientification (mostly a 'mathematization') of the road census of 1923 (which from then on was to be repeated every three years). This was supported by a Road Tax Law (1926), largely written by a commission from the Society Dutch Road Congress (founded in 1920 after the model of PIARC), that earmarked the revenues from a new fuel tax (as well as the existing bicycle tax) for road construction, to be financed by a special Road Fund.

Although the next national road plan according to the new law was due in ten years, pressure from parliament incited the drawing of the second plan one year after the finishing of the first. This plan, presented to (and approved by) parliament in 1930, was not only more ambitious and more costly than the first, it also shortened the 25-year building and planning period by 5 years and was now fully based on a scientific calculation of the *actual* (not the estimated future) traffic present on the roads to be improved. Because of this, the plan emphasized the importance of local and regional traffic and abstained from promoting freeway building. As we have seen, this was fully in line with international road engineering as discussed (and shaped) within PIARC.

By the time of parliamentary acceptance of the second road plan in 1930, Dutch road building was part of a fully centralized Directory General next to the existing departments within the transport ministry for the other mobility modes. In 1933 an extra fuel tax would create new revenues to start a crash construction program of bridges over the main rivers that had hitherto separated road traffic in a northern and southern part. Although in 1934 the Road Fund was reorganized into a Traffic Fund to allocate a part of the tax revenues to the national railways, the road reconstruction program was hardly influenced by the Depression. Between 1934 and 1937 (the year the crisis had still not receded because of the government's clinging to the gold standard for its national currency) the bulk of the unemployment relief funds were invested in road and bridge construction (figure 7.1).

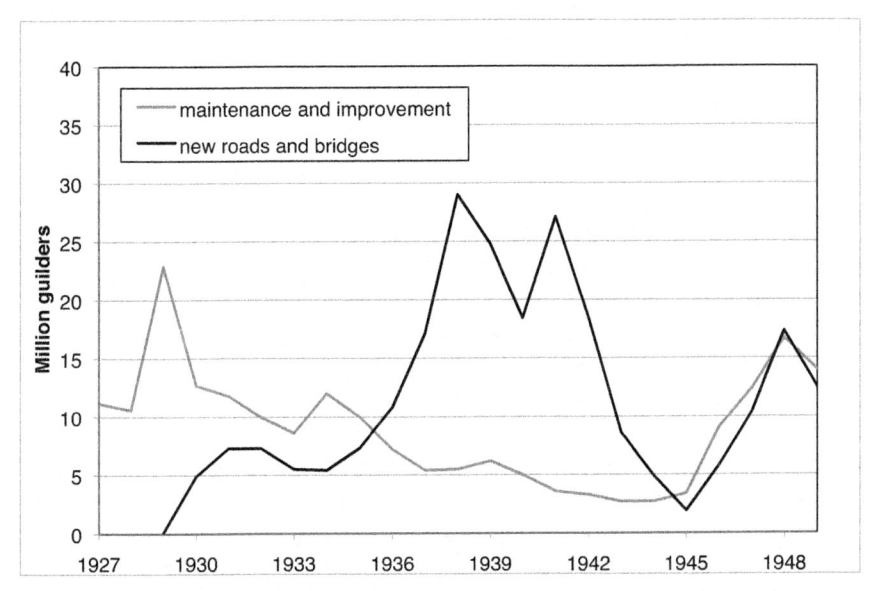

Figure 7.1. Improving and building the Dutch road system during the interwar years: expenditures by the national government on maintenance and improvement versus new roads and bridges.
Source: Wegen, 1950, jubilee issue, XXXI.

It would, however, be a serious mistake to limit this analysis to the national road-building practices, as the secondary roads and (after 1937, when they were included in the national funding) the tertiary roads were improved at a faster rate than the national network. As a result of this, by the beginning of the 1930s the network was fully adjusted to automobile use in most of the eleven Dutch provinces. The constant criticism, however, from parliament and motoring interests, about the Ministry not doing its utmost to resolve the 'road problem' can be explained by a particularity of the Dutch situation: although the distribution formula made the western, most urbanized provinces the main beneficiaries of the road funds, it nonetheless was advantageous for some less densely populated provinces if they had a large existing road network or surface area. The even spread of the road network over the entire national territory was an international phenomenon directly related to the centralization of its planning *and* to the political clout of agrarian interests. It followed railroad network building practice, which also tried to cover the entire country including less densely populated areas. The Dutch road network was also quite evenly spread over the nation's surface, except the western provinces. This was caused by the instability of the soil in these 'wet' provinces, with their large waterway networks and meandering roads on the dikes alongside these waterways. The need for these provinces to construct *new* roads was therefore much bigger, roads that also were much costlier. Therefore, it took at least five years of intensive scientific research (for which a special soil laboratory was founded by *Rijkswaterstaat* in 1934) before road building in the western provinces gained momentum, but by the eve of the war the Netherlands was covered by a dense road network, fully adjusted to car, truck, and bus, and even with some 200 kms of freeways. The latter was the direct result of the same geological conditions: constructing freeways in the West was less costly than building 'mixed roads' for all types of traffic flow, including slow traffic. Such mixed roads, for instance, were favored by Belgian road planners who managed to integrate them (built from 1934 and sometimes up to 60 m wide) in national 'corridors' of canals, railways, and roads along which a conscious policy of industrial concentration was developed. While Belgium (which constructed 28 km of freeways before the war in a far more decentralized institutional setting) was set up as a giant urban transport network, planners in the Netherlands were less inclined to see their task as a part of a more general mobility planning including all other modes.[49]

Although the historiography in other countries does not allow for a full comparison, some parallels, especially with the North-Western European countries, may testify to the fundamentally universal readiness of the road network before the construction of the freeways. In several of these countries the immediate post-WWI years witnessed the founding of a separate Ministry of Transport, such as in Great Britain, Germany (both 1919), and Sweden (1920), or the earmarking of tax revenues for road building, such as in Switzerland from 1928 (half of the extra fuel tax). In France, however, neither was the case (no new transport ministry, nor earmarking): the centralization of the *Corps des Ponts et Chaussées* was surprisingly weak and largely dependent on its rural branches. This meant that, equally surprising for a country with a 'centralized' reputation, rural resistance against the car was more or less institutionalized in the Public Works structure, leading to an extreme contrast between Paris and the rest of the country. The *Corps*, for instance, could not impose any standards on lower-level services and a special road fund was only created after the Second World War. With hindsight, this makes sense in a country that, because of its concentration of urbanization on one single spot, largely remained agricultural. Although France had to invest large sums after the First World War to bring its heavily damaged road network back to its prewar international reputation (with 600,000 km still the densest in the world), leading transport experts in the *Adminstration des Travaux Publics* (Public Works

Department) fiercely (and consistently during the Interbellum) rejected the building of special automobile roads because they were seen as *antipopulaire* and *antidémocratique*.[50] Nonetheless, there were many parallels with the Dutch story: with the United Kingdom as an example, the Touring Club (which had been instrumental in getting touristic Alpine roads built before the war) requested, just as in the Netherlands, that an autonomous "Office national de la route" would be set up; the club also proposed to initiate a road tax paid by all motorists proportional to their wearing effect on the pavement. A TCF petition (signed by fifty thousand) demanded a "road policy," which at least enabled the road-construction and maintenance budget to be increased gradually from 1920. But like the Automobile Club, the TCF was against the construction of freeways, arguing that the national roads were good enough and no "luxury roads" were necessary: the network indeed grew from 340 km in 1920 to 10,500 km six years later. Instead, one member of the car lobby opined, we should "de-iron" (*déferrer*) the railroads.[51]

The French rejection of a 'flight forward' into freeway network building was part of a mainstream opinion within international road-building circles (including the United States), but its argument was no doubt inspired by the pioneering role of Italy and Germany. In the historiography of early freeway planning this role has been related to the authoritarian, heavily centralized structure of transport planning in these countries (enabling the effective silencing of railway interests and the forced unification of the road-building community around the spectacular mega-project, as happened in Germany after the Nazis took over), but the picture is a bit more nuanced and also explains the 'anomaly' of the French situation better than was attempted in an earlier publication.[52] Recent German research also helps to nuance this further. As we will see later, the power of the German railroads was so large that even the car-friendly Nazi leaders could not refrain from investing the larger part of national funds in railroads. Not being able to 'silence the railway interests' did not prevent the Nazis from building an *Autobahnen* network, however. On the other extreme of this comparative yardstick, the situation in France clearly shows that a country can be highly centralized and yet does not opt for the freeway escape. A comparison with other countries can help explain this.

In Germany, which has been researched best in this respect, the founding in 1919 of a special transport ministry (*Reichsverkehrsministerium RVM*) functioned as a compensation for the loss of central planning power by the military during the lost war.[53] Historians so far have failed to systematically investigate the extent of road building in this period, but anecdotal evidence suggests that in Germany as well by the end of the Interbellum most through roads were at least paved, and that, like in the Netherlands, regional networks were ready earlier (by the beginning of the 1930s) than the national network. Nonetheless, here too complaints about the bad state of the roads were heard during the 1920s and early 1930s, probably prompted by the 'light' repavement practices (surface treatment with tar or tar macadam rather than entirely new asphalt layers), causing high maintenance costs, and also because road building stagnated during the Depression.[54]

In the United Kingdom a newly founded Ministry of Transport managed a special Road Improvement Fund, initiated in 1910 when by law a Road Board was founded that channeled the bulk of the tax revenues to the counties. This Road Board functioned as an example for several European countries, for instance in the Netherlands where a proposal by the local PIARC branch was not taken over by the ministry, and similarly in Italy.[55] Centralization of the road-reconstruction program happened faster than in Germany, but also in the United Kingdom the counties remained responsible for road improvement until 1936, when the Trunk Road Act made the ministry responsible for 7242 km of national roads and the Road Fund lost its independence. The minister delegated his authority back to the county councils, one level above the previous centers of gravity, the county boroughs.[56] Until the second half of the 1950s (when the central government started its motorway construction program) local investment in highways was higher than those of the government.[57]

Winston Churchill was one of the eloquent opponents of the "nonsense" and "absurdity" of earmarking car-tax revenues for road construction, and he set an example for subsequent Treasury ministers—so much so, that in the late 1930s even the Labour opposition accused the government of "steadily reducing the possibilities of our lower paid workers from indulging in motoring." Churchill was the one who in 1926 started a tradition of what contemporaries called a "raid on the Road Fund," and rather than the motorists, it was the local county officials (who benefited so much from the earmarking principle) who were able to maintain the fund's independence until 1936. Nonetheless, during the 1920s the expenditure on roads increased from £27 to £66 million. Several consecutive transport ministers soothed the rural authorities by siphoning a larger share of road funding to local and regional road-improvement schemes, but central planning was weak. For the same reason the consecutive central governments, often loudly headed by Churchill, sided with the counties in their fierce opposition to building special automobile roads, "these great new race tracks" for private motorists that, as had been found out, did not really relieve unemployment.[58] That must have been the reason why road improvement took the form of repaving existing, meandering roads. However, 505 miles of bypasses around towns had been constructed by 1934 and many bends in the roads were now banked, which invited British motorists to speed up, causing casualty statistics to skyrocket.[59]

In fact, even for the United States, where the national state engaged in a massive highway-construction program during the 1920s and 1930s, recent research emphasizes a 'rural bias' in federal road funding, but this did not result in the Good Roads movement having its way (of farm-to-market roads). Instead, some older regional studies show how local (urban) business interests managed to focus road construction on one or two levels higher (regional and national roads). Several propaganda committees that pleaded for the construction of continental through roads (such as the Lincoln Highway from east to west or the Dixie Highway to the south) supported this turnaround in general American public opinion. Although such proposals were megalomaniacal if judged on the basis of road use, "the reason for advocating a transcontinental system to serve local traffic was simply due to the fact that federal participation could not be justified on the grounds of local traffic."[60] The initiatives were clearly grassroots based, and not the result of a top-down planning rationale.[61]

At the institutional level this shift of attention from rural post roads to the main roads (backed by Congress in 1921 through its Federal Highway Act) was accompanied by the creation of State Highway Departments: the first was founded in New Jersey in 1891, and in 1917 all states had such an institution, at the same time that the Office of Road Inquiry of the Department of Agriculture was changed into the Bureau of Public Roads (BPR), after the acceptance of the Federal Aid Road Act in 1916. Under this law, and especially under its extension in 1921, the federal government spent $840 million before 1929 on road building and maintenance (which represented up to half of the total costs spent on these roads), plus $47 million on the 13,500 miles in parks and forests. It was the first effort to conceive a national network, the result of a powerful coalition of interests led in Congress by Republican senator Charles Townsend. The state departments, backed by twelve regional offices of the BPR, became the actual level of building activities, while the BPR distributed the funds and formulated the standards and procedures. All this was financed up to 94 percent (1928) from fuel-tax receipts—tiny amounts per fuel purchase so that motorists hardly noticed it, as a German witness observed. As most of this fuel was consumed in urban areas, the road funding was at the same time a giant redistribution machine, because 70 percent of the funds were spent on through roads ($211 million in 1928) and only $8 million in towns. As we will see later, the New Deal added extra funding to this gigantic project. As a result, in the United States before WWII about a quarter of the entire national road network consisted of newly created roads (mostly created before the 1920s; between 1904 and 1914 alone 300,000 miles of rural roads were constructed), while during the Interbellum county and local roads were largely neglected by federal institutions.[62] In fact, in the course of the Interbellum two systems were gradually built: the so-called 7 percent system of roads, "interstate in character," mandated by the 1921 Highway Act and resulting in a network of 200,000 miles of national highways, and a Secondary Road System built up under the 1934 Federal-Aid Highway Act. Both acts were related, because the former stipulated that states should spend three-sevenths of the 7 percent on secondary roads. The funds for the network of national highways were distributed on the basis of a formula containing population and four production factor indices, resulting in a bias toward wealthy counties.[63] Thus, the "sacrosanct" Highway Trust Fund functioned as "a perpetual-motion machine for building roads," so much so that by the beginning of the 1930s, just like in most North-Western European countries, a "truly comprehensive national road system began to emerge."[64]

Despite the recent boom of road building history in the United States, however, our knowledge of actual road improvement at the aggregated local and regional level is far too sketchy to allow a reliable comparison with European road-building practices.[65] The situation will probably appear to differ considerably per state, but it seems that in states with high car density, first the primary network was constructed (in other words, the Good Roads movement was largely bypassed, or redirected toward long-distance traffic), while farm-to-market roads were only dealt with during and after the Depression, and especially during the New Deal. Another American study characterized the Interbellum as a period of "improving and coordinating existing systems," with the exception of the American forests, where between 1916 and 1935 nearly ninety thousand miles of new roads were created, one-fifth of them forest highways.[66] The result, however, seems to have been similar to Europe: an automobile-friendly network was in place by the eve of the Second World War, perhaps earlier, as can be deduced from the extensive report by a German expert observer who in 1930 concluded that "almost the entire union can be traversed on sturdy roads (*auf starken Fahrbahnen*)." Despite the large share of unpaved secondary and tertiary roads, road censuses revealed that most of the vehicle miles were produced on paved roads. Even in Mississippi, where the construction of a statewide road system was not started before 1935, 40 percent of the existing roads were surfaced as early as 1923. In 1928 an "eight-year cycle of road building came to a close in North Carolina," resulting in a highway network covering the entire state.[67]

While the consensus among the road-building community was in favor of the improvement of the existing road network, in a limited number of European countries enthusiasm about the car as a modern vehicle took hold of national governments who saw the car as a means of forced modernization of society. Not coincidentally such governments were authoritarian in nature: first in Italy (1922), then in Germany (1933), the fascist governments appropriated the car movement to fuel hopes for national revival through individual car ownership based on the conviction that special roads (*autostrade, Autobahnen*) would allow the movement to flourish beyond their middle-class constituency.[68] It was no coincidence, therefore, that the Dutch proponent of such plans was the future leader of the national-socialist league (and a provincial Waterstaat engineer).[69]

But even in countries that were skeptical about freeway projects a fascination for the German *Autobahn* project in particular among traffic engineers cannot be denied and thus the freeway concept influenced the design of lower-order roads even if a country did not choose for a separate network. Delegations from the United Kingdom and the United States visited Germany, and PIARC held its conference in 1934 in Munich (just as it had held its 1928 conference in Milan, to admire the Italian freeways). In an effort to exploit the hypermodernist appeal of its *Autobahn* system the German government invited in 1937 the two main British car and touring clubs AA and RAC, who in the same year, at the time of the *Oktoberfest* in Munich, brought a 224-strong delegation for ten days to Germany (a young Colin Buchanan among them who was "enormously impressed"), including road lobby groups and members of parliament, but representatives of the Ministry of Transport were "noticeably absent."[70] The United Kingdom resisted the inclusion of the freeway into its road planning until the 1950s, and most other countries followed the British example. France, for instance, built and planned some freeway trunks to ease the congestion Parisian suburbanites were facing, but its main journal, the *Revue générale des routes et de la circulation routière*, repeatedly printed comments from high-ranking road-building officials against such plans. Nevertheless, within the experts' discourse a turnaround about the usefulness of such networks took place during the 1930s, largely because the freeway lobby managed to integrate the safety issue into its propaganda.[71]

The United States is a special case, because the interstate network of highways could to a large extent not be built upon an existing road network. However, a system of nonconnected regional parkways (with separation of vehicular and pedestrian traffic) has lately been proposed by scholars specialized in road history as a precursor that also influenced the German *Autobahnen* design. Indeed, the restaurant and service buildings along the 160 miles of the I-76 (the Pennsylvania Turnpike) between Harrisburg and Pittsburgh were inspired by the traditional, regional architecture of the *Autobahnen*. As if to emphasize the freeway's railway heritage, the turnpike financed from New Deal funds was built upon an abandoned railway route, and knew no speed limit.[72]

Developed from Hausmannian and earlier European concepts of boulevards and avenues, parkways stemmed from an urban planning tradition. This was more particularly drawn from the practices of landscape architects, and was, according to Frederick Olmsted (who coined the term) a symbol of "the abandonment of the old-fashioned compact way of building towns."[73] Just like the urban traffic safety proponents who we will encounter in the next section, landscape architects and city planners lost the struggle with the centralized (federal or state) road-building engineering community, despite the fact that some of these parkways, most particularly the Blue Ridge Parkway that swept along the Appalachian Mountains for hundreds of miles, could hardly be called urban. But the maximum allowed speed was low (about 30–40 mph [48–64 km/h]) and they were often characterized by a mixed form of use that came to be rejected by the international PIARC community. Indeed, they were perhaps an all-too-open expression of the pleasurable side of mobility in their invitation to consume the surrounding countryside at a leisurely undulating pace, all characteristics that made them difficult to compare with the cross-country arteries such as the *autostrade* and *Autobahnen*, as recently has been done.[74] "Commercial highways" on the other hand, largely disregarded the scenic consciousness of the parkways, until in 1936 when BPR started a landscape section and five thousand miles of roadsides became 'improved' under the New Deal scheme. Instead of presenting the parkways as the American equivalent of European freeways, they can better be seen as the predecessors of urban freeways, which were celebrated in the 1939 Futurama exhibition in New York, designed by Norman Bel Geddes and sponsored by General Motors.[75]

Contested Order: Spatial Planners Versus Engineers

While the national mobility discourses during the interwar years tended toward railroad and inland navigation interests, the car system developed in an asystemic, fragmented way represented at state level by specially founded departments or sub-departments within Ministries of Transport or Public Works. The movement's roots were local, and regional at most, and only in some countries were they strong enough to support and feed a national center able to negotiate effectively with state representatives, especially in those countries where touring clubs had managed to overtake automobile clubs as the true representatives of motoring and tourism interests. In other words, despite its transnational structure the movement was firmly rooted locally, definitely nationally oriented.

This representation was not like a labor union: middle-class motorists kept behaving as if in a swarm, without an acknowledged leadership, and their very ostentatious, noisy, and dangerous presence and the tax revenue they meanwhile generated, was enough to capture the imagination of influential parts of the governing institutions. Swarms, or "collectivit[ies] defined by relationality" provoke the fascination of planners as they have enough discipline built in to promise further streamlining while at the same time (as military planners have recently observed) they represent a planning challenge as "swarming requires autonomous or semi-autonomous operating agents." Especially the military, who cannot seem to give up the idea of central command (nor can traffic engineers), is drawn to the swarm concept because it consists of "large numbers of small units of maneuver that are tightly internetted." The uneasy aspect is the "command element" that only "'knows' a great deal but intervenes only sparingly, when necessary." Applied to road traffic, this 'command' knowledge is twofold: it is 'hardened' in the road network, its curves, width, nature of pavement; it also consists of a set of rules, some of them also 'hardened' in signs and traffic lights, but also internalized as rules of priority. From this perspective, the flow metaphor of transport experts is adequate: it represents the ideal, the end state of a set of measures and infrastructural installations that forces the swarm into a well-ordered battalion. But there is one important difference: water has to be channeled in order to be able to flow as a stream, but individual cars follow their own routes, if their drivers so wish.[76]

The urge to enforce order on an anarchic, erratic traffic fit with a tradition of technical and social engineering and, more generally, orderly thinking (*Ordnungsdenken*), "a transnational discourse formation, carried by experts who experienced their heydays in North-West European countries between the 1920s and the 1950s." Whereas the 'technical fix' of the engineers soon reigned supreme in the national road-building frenzy, social engineering and education seem to have been the paths originally favored by urban traffic engineers.[77] Even the Nazis, who initially favored a "rather anti-authoritarian, truly anarchic position" when it came to issues related to traffic behavior, started to enforce regulations on car traffic from 1938, especially because of the increasing number of traffic accidents.[78] Playwright Hanns Johst had his actors exclaim, in 1933, in a play dedicated to Hitler: "Germany! One last word! One wish! Order!!" a cry that somewhat paradoxically (but all the more adequately) ended in a call for destruction: "Germany!!! Arise! In flames!! Blaze up! Burn monstrously!!!"[79]

Autobahn architect Fritz Todt compared the national socialist state with traffic on his freeways: "We want to pursue our goals in a straightforward and rapid manner; we overcome crossings, unnecessary ties are alien to us. We don't want to swerve, we'll make enough space to go ahead."[80] Destruction, as we see, is never far away if an orderly, uninterrupted movement has to be created.

Whereas pedestrians and cyclists were much more difficult to discipline, the car "medializes its passenger; it ties him in its machine network, the car embodies and internalizes him at the same time."[81] The expertise to plan and build the automobile system that had to perform this task of converting a swarm into a flow, most particularly the road network, was drawn from two distinctive fields of knowledge and practice: municipal traffic engineering and rural planning. Historiography so far has emphasized the former, whereas the emergence of a fully fledged 'car system' right after the Second World War cannot be explained without including the role of the latter. How else can it be explained that before the Second World War urban traffic problems were hardly taken into account at the national, and certainly not at the international, level, especially in Europe, but also in the United States?[82] In the end engineers eager to build corridors of speedy 'flows' (or "auto rivers," as American archaeologist Arthur Krim coined them) beat the more integrally thinking spatial planners.[83] Such corridors became the real 'cocoons' in the post-WWII Age of Abundance, flows of capsules speeding through (but insulated from) the countryside.

Centralization of road planning was not undertaken out of statist altruism, because local governments couldn't finance their infrastructures: leading national figures saw road traffic take on some long-distance characteristics, although their conclusion that for this a new road system was necessary was largely based on wishful thinking, as most traffic experts repeatedly emphasized the largely local and regional character of road traffic.[84] In other words, the conception of such megalomaniacal structures as the planning and building of national freeway networks was a direct consequence of national centralization under a sectoral spell, and was inspired by railway thinking. It is a classic example of social construction, forced upon a mobility practice largely regional in character.[85] Differently put: freeway and modern highway construction ran ahead of traffic developments, including in the United States. Such networks stimulated, rather than answered to, long-range flows, threatening to destroy the very mobility modes most central governments were eager to protect.

While centralization enabled the collection and distribution of tax revenues, the standardization of road design, the decision on the location of roads as well as the scientification of research in central laboratories, the intermediate level of provinces, *départements,* states, cantons, and *länder* saw their executive role increase as well. This double gubernational structure of conceptual and financial centralization and a spread of executive responsibilities to a lower level (while at the same time keeping local interests out of the loop) and the highly successful spiral of earmarked revenues explain to a large extent why in most countries by the beginning of the 1930s nationwide networks for automobile use were in place, independent of the question whether road lobbies could convince national governments to engage in freeway network building. This, much more than the question of whether a country decided to introduce freeways, constituted the real systems breakthrough of a new mobility paradigm based on the 'automobile' (car, motorcycle and moped, bus, truck).

A factor that may have played a role in retarding the application of the freeway concept at a European level is the fact that the early freeway plans were based on another image of 'Europe' than was the case within PIARC with its North-Western European dominance. For instance, ILO director Albert Thomas was an admirer of Francis Delaisi's *Les deux Europes* (1929) in which a plea was made to connect wealthy western Europe with agricultural eastern Europe through the construction of infrastructures. The League of Nations' Communications and Transit Committee however followed PIARC's hesitant policy toward transnational road network building and tried, instead, to stimulate international mobility through regulation of driver's licenses and customs formalities—engineering over politics. Comparable plans advanced by French and Italian railroad engineers favoring railway corridors connecting southern European countries were also turned down during this period.[86]

But the main reason for the (albeit limited) success of the freeway lobby was no doubt the dominance, within this lobby, of entrepreneurs and contractors and their indifference to the axiom of most PIARC members about the predominantly local (or at most regional) character of road traffic, mostly around the larger cities. The historian who looks for prewar predecessors of the postwar International Road Federation

(IRF) and comparable vocal and assertive propagandists for the freeway cause will find them not within PIARC but in the neighborhood of the freeway lobby: the German private lobby and research group HAFRABA, the group around the Italian contractor Puricelli, the French *Compagnie des Autoroutes* of road builder Lainé, the British Road Federation (1932) founded by car manufacturers and road transport companies, and the Dutch *Algemeene Nederlandsche Verkeersfederatie* ANVF (General Dutch Traffic Federation) founded by car manufacturers and importers, oil companies, and the touring club ANWB. Indeed, the personal and ideological bridge between the two communities was formed by the touring and automobile clubs.[87] Although this should be analyzed more closely in several European countries, it was these lobby groups, intent on rescuing the automotive adventure in a phase of Depression, who managed to shift public opinion by mobilizing the safety argument.

Rescuing Automotive Adventure: The Construction of Road Safety

At the first PIARC conference in 1908, road safety problems seemed to result from an unresolved conflict of interest between old and new road users. Initially a form of coexistence was advocated (hence, too, the proposal of 'mixed roads,' combining fast and slow traffic on one very wide corridor), but soon it became clear that motorists' desires of unrestricted flow and, as a counter-movement, experts' concerns about the safety of pedestrians, cyclists, and domestic animals made a more radical solution necessary. To the majority of conference attendants there was no doubt that the solution was to be found in disciplining horses, pedestrians, and bicyclists according to rules formulated by the newcomers and their self-declared custodians, the road builders and planners.[88]

Why did these specialists accept so easily this shift in road-use culture? Apart from a general belief in the modernity of the vehicles (representing both technical and economic 'progress'), the experience of many conference attendants with centrally controlled railways and inland navigation systems earlier in their careers may have provided an inspiration for their automatic claim for monopoly. Nevertheless, because many official representatives had previously served the interests of other transport modes, it is remarkable that the reflex to put the automobile at the front was so easily adopted and shared. No doubt the prominent place of the touring and automobile clubs, as well as the close cooperation between these clubs and the highest echelons of the European nation-states (and especially the French state), played a supporting role in this respect. And we should also not forget the basic attractiveness of the automobile, as we have analyzed in chapter 2. And perhaps it also formed a challenge to many engineers to find ways to domesticate a mass, if not a swarm, of highly atomized, even anarchistic, motorists. In fact, the only actors who initially resisted the claim of dominance were horse owners and riders who (rightfully) feared that paving with smooth surfaces (especially asphalt) would force horses to their own 'paths.'[89] In other words, the substitution of the horse was not the result of a 'natural' process of increasing cars and decreasing horses, driven by new consumer preferences, but was actively brought about by designing the infrastructure such that one mode was accommodated, another willfully pushed into oblivion.

The three prewar PIARC conferences clearly reached an unambiguous conclusion: although special roads for cars were rejected, the automobile, and especially the passenger car, had been accepted as the norm on the brink of the Great War. But while horses could be dispensed with in the ordered universe of road builders and planners, pedestrians and cyclists had to be addressed differently. As they were the primary victims of the first car accidents in big cities, they represented "the majority of the public," and were seen as "the cause [of these accidents] because they are unaccustomed . . . to this new type of traffic." It was assumed that they needed to be educated and as soon as they had gotten used to the novelty of the automobile, "the number of accidents will diminish automatically."[90]

Early debates on this issue reveal how the main actors struggled to define the 'road problem.' Inclined to find technical fixes to this problem, they were not sure whether these fixes had to be applied at the level of the single vehicle or the infrastructure. To the extent that vehicle speed was recognized as a constituent factor in road safety, the choice was clear: the speed of automobiles must be curtailed, with speeds adjusted to "la sécurité publique et la commodité générale" (public safety and general standards), a clear reference to the image of the road as a public good. Most attendants at the London conference of 1913 agreed with a German representative that fixing a maximum speed was necessary, because "one cannot ask from Engineers [sic] that they must construct roads that would be absolutely adapted to the demands of automobilism." On the issue of lighting, however, the opinions were still mixed, as the conclusions of the same conference show. It was an open question whether automobiles should be equipped with lighting systems or the entire road system should be illuminated. This tension between vehicle solutions and system solutions would be present during the entire period until the Second World War. For instance, during the first postwar conference in Sevilla, a discussion took place about the question of whether mirrors for better and safer vision should be constructed inside automobiles only, or also along the roads, and especially at crossings.[91]

In this sense road safety had and has always played a constitutive role at the PIARC conferences, but it was in Munich (1934) and in The Hague (1938) that special sets of questions (to be addressed in the papers of the participants) and conclusions were dedicated to the safety issue in an atmosphere of increasing anxiety about the 'slaughter' on the roads on the basis of the first sets of statistics that became available. These statistics show a general trend of steep increase for the 1920s to an extremely high level compared to post-WWII values. In the United Kingdom, for instance, the 7300 deaths on the roads in 1934 were only slightly less than the postwar peak value of 7700 in 1972, when ten times more cars were on the road.[92] For the 1930s a more diffuse picture results, as some countries increased their road fatalities further, while others stabilized, or even saw numbers of traffic deaths decrease, as was the case in Depression-ridden Germany (figure 7.2). Despite this decrease Germany belonged to the extremely unsafe countries because of the large number of vulnerable motorcycles.[93] At the same time the American and British fatality rates (per capita) during the 1930s were about 25 percent higher than during the postwar period until the beginning of the 1960s.[94] How can these disparities be explained?

The unsurpassed long-term statistics assembled by the French demographer Claude Chesnais (figure 7.3) reveal (either expressed in ab-

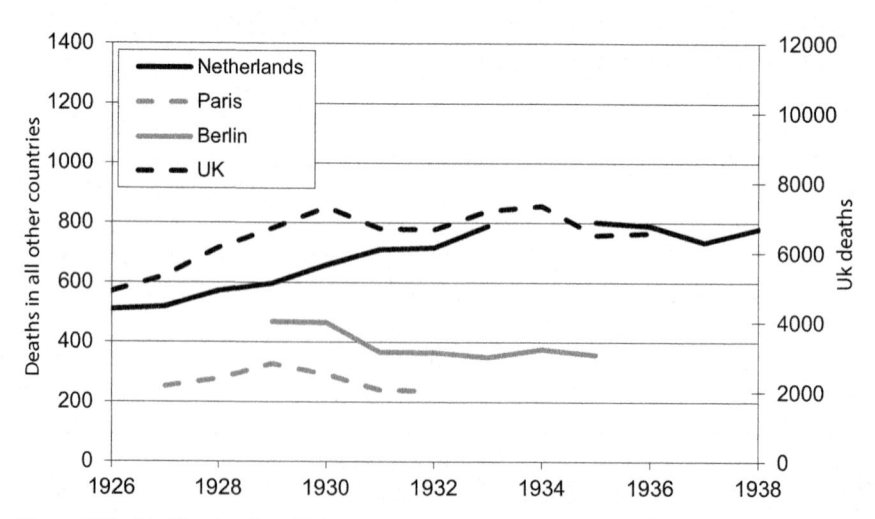

Figure 7.2. Traffic deaths during the interwar years in two European countries and two European capitals.

Source: several reports for the PIARC conference in Munich, 1934, and The Hague, 1938.

Swarms Into Flows 597

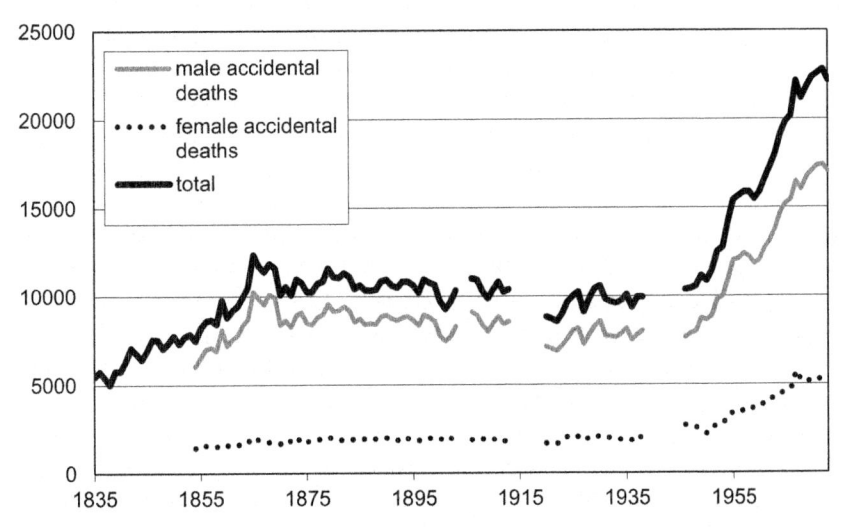

Figure 7.3. Long-term statistics (1835–1973) of French accidental deaths, including gendering.
Source: Chesnais, "La mortalité par accidents en France depuis 1826."

solute numbers or normalized on the basis of population density) how societal risks increased during the nineteenth-century 'horse economy' due to increased construction work and related traffic, resulting in a heavily gendered increase in male deaths. Female deaths seem to have been hardly affected by this: women kept being the victim of accidental falls or burnings in the home. British long-term statistics also reveal that women were the victims half as often from accidental death as men, but for both men and women the car became the main cause of accidental death during the 1920s, overtaking drowning (outside of the house) in the case of men, and burns (inside the house) in the case of women. Of both sexes together, from 1927 road fatalities overtook suicide as main cause of accidental death. In the United States, too, general accident statistics in 1919 were still dominated by domestic accidents, but in 1926 car accidents per capita were more frequent than falls. In France these statistics declined slightly during the first two decades of the twentieth century, whereby the First World War seems to have resulted in a breach in this trend to much lower levels, soon to be compensated by a largely male-dominated explosion of automobile-caused deaths during the Interbellum, prefiguring the second, and much more powerful, explosion during the postwar mass motorization.[95] These examples again show how misleading it is to talk about a cultural gap between the United States and Europe on the basis of car-density figures alone: the threat of the car was as real (and as much an issue in societal discourses) on both sides of the ocean, at exactly the same time.

Post-WWII analyses of national safety statistics (and especially those based on the unrivaled French statistics) have meanwhile made abundantly clear that it was the car that disturbed an evolutionary, incremental development of a constant number of horse-carriage deaths and a slowly rising number of railroad deaths (figure 7.4). The very rapid increase in traffic casualties in most motorizing countries may have been caused by the fact that the secondary road network was often ready earlier than the primary road network: because of their car-friendly surfaces and alignment taken over from the primary network, such roads invited motorists to speed up, increasing the chance of violent encounters with fellow motorists (on crossings, from the opposite direction) and slower traffic. Initially, more automobile roads meant fewer (fatal) accidents, simply because the chance of an encounter with potential victims was reduced as automobile-friendly road length increased more than automobile registrations, while congestion in the larger towns meant more accidents, but fewer lethal accidents. From this perspective, an accident was nothing less than "the spectacular manifestation of inefficiency," as an insurance official declared in the American magazine *Fortune*.[96] From the same perspective it becomes clear why the freeway proponents' safety argument was so powerful.

International comparisons on the basis of fatality statistics are delicate, and give different results dependent on the question asked. If traffic deaths related to the population can be used as a proxy for the violent threat of automobilism toward the population as such (users and non-

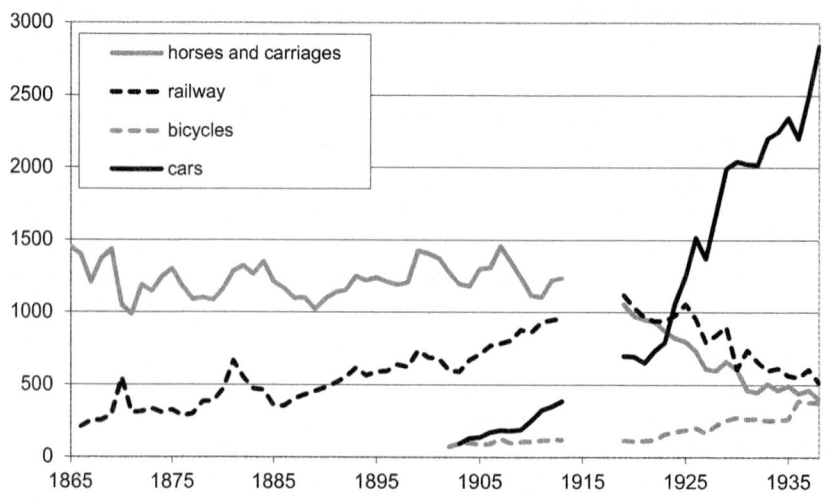

Figure 7.4. Traffic deaths in France, 1865–1939.
Source: Chesnais, "La mortalité par accidents en France depuis 1826."

users together, the latter much more numerous than the former), then, by the end of the 1920s, the United States led this sinister statistical contest with nearly 80 deaths per 100,000, followed by Canada and Switzerland, while Denmark, Belgium, the Netherlands, and France were the least affected (24–30 deaths), and England plus Wales somewhere in the middle (41).[97] There is, of course, a clear relation with car density here, but not fully: Denmark had more cars per capita than the Netherlands, but fewer deaths per capita. Recalculated in deaths per vehicle, however, the United States suddenly appeared the safest (in 1934), while "the fascist record," as an attentive and angry British observer concluded, appeared to be the least "safe." Here, too, there was not a one-to-one relationship with car density, as British roads were nearly twice as "unsafe" along this yardstick as the French.[98]

What most worried many PIARC delegates was the alarmingly large share of pedestrians and cyclists among the deaths on the roads. In the United Kingdom, for instance, cyclists were involved in one-third of all accidents, and they formed one-quarter of all road deaths. British pedestrians fared even worse: they formed 41–54 percent of all road deaths in 1933–1936. In 1936 about 60 percent of the French road casualties were nonmotorists. In cities these figures were even more alarming. In Paris in 1932 cyclists and pedestrians accounted for three-quarters of all road deaths. In and around the towns of Mannheim and Heidelberg in the German state of Baden nearly one-third of the victims killed between 1929 and 1934 were pedestrian, and one-tenth children. In the Netherlands a constant annual cohort of five hundred children younger than the age of five had died or been injured in the forty thousand annual accidents, and nearly a thousand children in the ages of six to eleven. For the United Kingdom the death toll of children since 1927 had reached fourteen thousand by 1937. In that year still half of the victims killed were pedestrians (and 40 percent of the wounded), while the remainder of the victims were motorcyclists (18 percent), bicyclists (18 percent), and nondriving car passengers (10 percent).[99]

Most observers agreed that the carnage on the roads was the result of "human errors," and they pleaded in favor of obligatory education in schools. The optimism at the PIARC conferences was remarkable. American initiatives to organize special safety campaigns in selected cities suggested that something could be done against unsafe roads. Most American cities that started safety campaigns reported drastically diminishing fatality figures.[100] Road safety, at last, seemed 'doable,' in a double sense: by adjusting the infrastructure such that flows caused fewer injuries and deaths, and second (because the figures remained high, despite these measures) by adjusting the way people talked about the onslaught on the road. The observer in the *Atlantic Monthly* was certainly not the only one spicing his comments with a dose of sarcasm on this shaky combination of technical and discursive fixes, which since then became customary: "Eventually we may arrive at the comfortable stage in which the normal rise in casualties will not particularly annoy us, while a temporary drop would be hailed with genuine satisfaction."[101]

But this drop did not come, and so, one is inclined to say, it was created: in every country around this annual onslaught a careful statistical myth began to be constructed of a constantly declining death rate (figures 7.5 and 7.6). This is not to say, of course, that road-building engineers conspired to play down the lethal aspects of their profession, although one study on the British interwar years claims that statistics may have been deliberately faked by the police to acquire more personnel.[102] But it is remarkable how eager any signs (whether real or imagined) of a decrease of traffic risks were received by engineers, both inside and outside PIARC. It is no coincidence, either, that this desire for an optimistic reading of traffic fatality data occurred within the context of the same process of scientification that had driven general road-building culture. Such an optimistic 'reading' was achieved by 'normalising' the resulting graphs, dividing absolute accident, injury, and death numbers

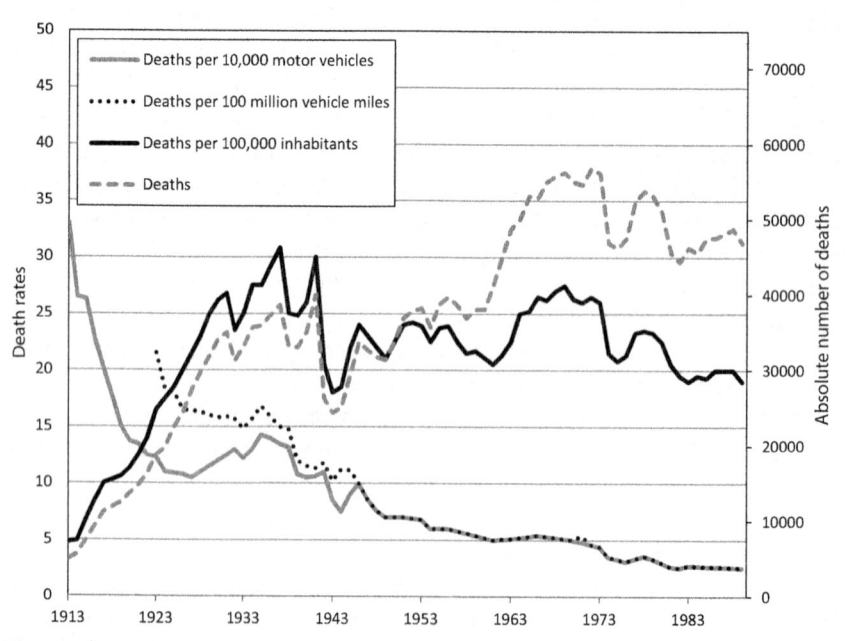

Figure 7.5. Mystification of road-death statistics by constructing death rates; example: United States, 1913–1989.
Source: McShane and Mom.

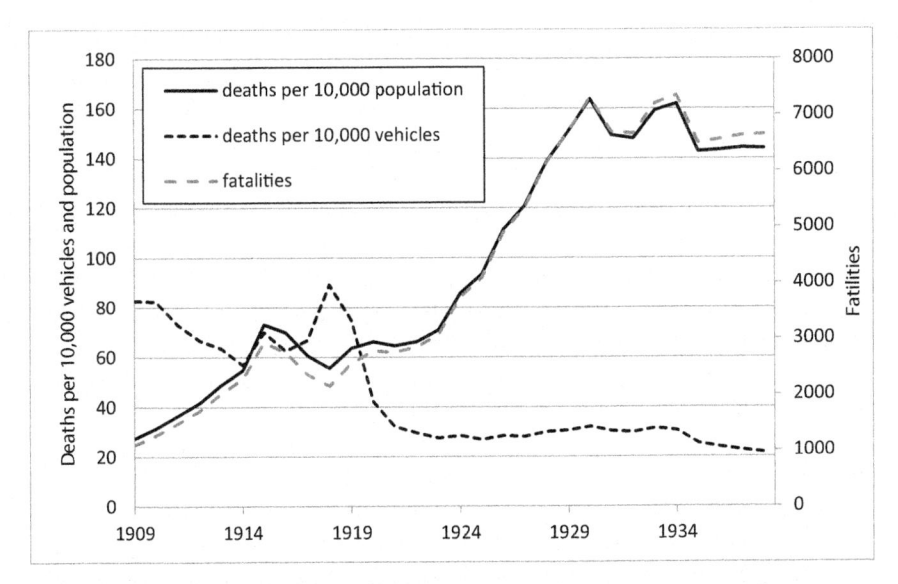

Figure 7.6. Car-related fatality rates in the United Kingdom, 1909–1938.
Source: Jacobs, 15, table 3.

by some constantly increasing factor, mostly the number of cars or the number of vehicle kilometers. In other words and most ironically, the factor that drove the spiraling effect of road-building financing upward (the increasing car use) also fueled the fears and concerns about the car as a novel contribution to what later would become known as the 'risk society.'

Traffic engineers were not alone in feeling embarrassed by the rising road deaths. Criminal statisticians to this day struggle with the question how to deal with them. Whereas murder rates (per capita) in Boston, to give only one well-researched example, decreased since the 1850s (contrary to general opinion), manslaughter and rape increased, both related to the emergence of the car as some commentators observed: whereas American judges initially dealt with lethal accidents as "murder" or "homicide," and only later categorized them as "vehicular manslaughter," these commentators saw, like the Lynds before them, the car incite "young couples ... to seclude themselves from the gaze of society," and thus "the incidence of every type of illicit sexual activity increase ... , including those based on force." As a result crime statisticians 'purified' their data by eliminating "the 'contamination' with auto accidents."[103]

Although further research into this matter is necessary, it is quite probable that planners and engineers at the last PIARC conference before the Second World War found each other in a technocratic compromise. Urban planner and director of the Urban Development department of the city of The Hague ir. P. Bakker Schut proposed a conclusion in which the new freeway concept was presented as a way out of the safety dilemma. He borrowed this argument from the Dutch freeway lobby, which was unsuccessful until it started to use the dual arguments of economy (the freeway is cheaper to build than a 'mixed road' for all types of traffic) and safety, arguments that were also brought forward by freeway proponents in other countries.[104] Fully according to Bakker Schut's functionalist point of view (also expressed in the famous CIAM declaration of Athens of 1933 by functionalist architects), the proposed conclusions stated that "freeways, i.e. roads only accessible to motorised traffic, provide the most satisfying solution for long-distance traffic, both from a safety point of view and from the perspective of traffic speed and of capacity (of the road)."[105] This conclusion, based upon the concept of the separation of flows already introduced at PIARC conferences before the First World War, suddenly put the technical fix of the 'freeway' in a new light, despite the reservations expressed by both the French and the British delegation in The Hague against (what the French still considered as) an extremely expensive and undemocratic solution.[106]

This process of general acceptance and redefinition of traffic danger as an anonymized, statistical risk has been analyzed most extensively for the United States. A recent American study convincingly shows how the legitimation of the car as a universal vehicle was seriously challenged during the 1910s and the early 1920s, because of exploding traffic deaths (especially for children in city streets) and fear of urban congestion. Anecdotal evidence from other countries, such as the Netherlands and Germany, suggests that in these same countries immediately after the war a revival of anti-car sentiment can be observed. Road safety, then, was one of the major controversies that prevented the car from becoming a fully accepted paradigm for future mobility.[107]

An interesting aspect of the massive campaigns (including civic parades, and the distribution of flyers and brochures) in the United States by local business communities, supported by a new type of traffic engineer and the automobile and touring clubs, was the debate about the usefulness of mechanical speed governors in passenger cars, a repetition of the uproar during the prewar years where speed was at stake also, as we have seen in chapter 1. Fiercely opposed by officials (including self-appointed safety expert and car manufacturer Paul Hoffman), a majority of car owners (69 percent) were not against such a device: "the 'motor-minded' American both wanted and feared speed," as a historian of American traffic safety concludes.[108] The campaigns, however, managed by the end of the 1920s to shift the burden of accusation from the motorist to the pedestrian, or, more cynically, to "blame the victim" in case of an accident.[109] Similar efforts to accommodate enraged urbanites were undertaken in other North-Atlantic countries, such as the United Kingdom where some parents refused to send their children to school after so many of them had been killed by cars.[110]

A second strategy was to find the culprit in a minority of motorists themselves, and not in the roads or the cars. The latter strategy became canonized, so to speak, during the 1924 National Safety Council presided by then secretary of commerce Herbert Hoover.[111] It was followed, a decade later, by a true "national crusade" culminating in a famous *Reader's Digest* brochure (which made it into a film in 1936) titled *... And Sudden Death,* meant to shock the nation by depicting the car accident in all its gruesome details, and which was bought by tens of thousands of Americans.[112] It was the "carelessness" of the "incompetent" driver that was to blame for the majority of the accidents, the argument ran, and it was the "roadhogs" (or the "road Mutt," after "a crude and unsocialized personality" depicted in an American comic strip) who had to be eliminated from the ranks of the pioneer motorists in order to make the movement ready for societal acceptance. The search for 'order' thus went along with a tendency of exclusion of the 'bad motorist.' How did this take place?

During the 1920s, when the traffic experts started to dominate the safety debate, this carelessness was redefined and limited to so-called accident-prone drivers. The hunt for this type of driver, undertaken by psychologists, was fueled by the conviction that people who deviated from the norm in society could not possibly be safe drivers. Such drivers were characterized by "physiological deficiencies, lack of knowledge or skill, and wrong attitude." Statistics now proved that women, despite their reputation, were three times more dangerous drivers than men (if statistically corrected for their lower mileages per year). Speeding was identified as by far the main cause of accidents.[113] Adventure, the secret behind the success of the *emergence* of automobility, threatened to collide with order, the secret behind the successful *persistence* of automobility during the Interbellum. The solution to this paradox was a technical fix, as so often when engineers dominated the discourse. Psychology related to road safety made its inroads into road expertise through the discipline of "psychotechnics," a technicized form of personality taxonomy derived from well-developed industrial safety research. One of the first investigators was the German-American H. Münsterberg (1913). The research was directed at munitions factories and in traffic matters at streetcar, locomotive, and truck drivers during the First World War.[114]

But the hunt for the 'misfits' started in earnest around the mid-1920s, when the indignation about the 'carnage on the road' began to build up: the *Unfallneigung* (proclivity to cause accidents) and accident-proneness were simultaneously 'discovered' in Germany and the United Kingdom in 1926 (by the psychologists Karl Marble and Eric Farmer respectively).[115] This hunt had a clear racial flavor. In a sample of 1387 cases between 1936 and 1965 (taken by historian Daniel Albert from a much larger set of 15,000 cases between 1921 and 1987) diagnosed at a "Traffic Court Clinic" in Detroit, blacks were overrepresented as were those with a criminal record (or both). Whites were diagnosed as "neurotic, unstable, and immature" and women (who formed only a 7.3 percent share of the sample) were called "unstable, impulsive, infantile, anxious and irritable" (and "feebleminded," "simple," and "unreliable"). Homosexuality, too, appeared to be a good predictor of dangerous driving.[116] Such socially "maladjusted" had to be kept far from the steering wheel. In the United States the safety discourse aimed at splitting the motorists in a small group of "speed hogs" and the majority of fast but 'rational' drivers who could be portrayed as the true successors of the Progressives.[117] The dichotomy even entered the official documents of American Congress, which in 1938 discussed a study in which "high-accident-rate drivers" were statistically identified.[118]

Remarkably, however, at the individual level persons with an accident-prone personality could not be identified: "Despite a great deal of effort and the invention of a multitude of ingenious tests, the condition known as 'accident-proneness' cannot be diagnosed at the present," one study concluded as late as 1956. And if they could have been identified, some clear-minded psychologists argued, the enormous number of casualties and fatalities would make it much more plausible that the carnage was caused by "a common human characteristic." This is not to say that people who more than proportionally cause accidents do not exists, but research soon found out that their influence was irrelevant (except, perhaps, in large fleets of public vehicles such as trains and planes and trams). It was the 'normal driver' who caused by far the majority of the accidents. After the Second World War the "Percentage Fallacy" was also dethroned as a statistical failure.[119] The hunt for the accident prone was, in other words (and in hindsight), nothing more than an effort to exclude those who were not really able to play the game with death, because in the end, of course, the final (subjective) goal of car use could not be death, but a continuous near-death, or risk. The 'othering' of potential drivers (black, women, homosexuals) was based in quite a different cultural substrate than statistics. It was a form of ostracizing to others without being deprived of the fascination for the atrocities of the carnage, as became clear when in 1935 more than 3.5 million reprints were requested from *Reader Digest*'s article "And Sudden Death."[120] It is also apparent in the highly popular memoirs of racecar driver Eddie Rickenbacker, who describes his youth as a sequence of "brushes with deaths" and the rest of his life as a constant play with fear, wreckages, near-deaths, and similar 'adventures.'[121]

The police, who had to cope with automotive unsafety firsthand, saw in any violator of the traffic and speed laws a potential roadhog. As a result, American motorists waged a virtual war against an increasing number of traffic police, as a careful reading of newspapers in several states by a recent student of this phenomenon testifies: acting like the members of a swarm, they had to be disciplined into 'civilized' behavior such that the 'accident-prone' motorists could be identified as unmodern, and the bulk of the motorists could be seen as the true successors of the progressive movement. Swarm behavior could be observed, for instance, in the systematic exceeding of the speed limits by large groups of drivers. For inhabitants of "Plainville" (a rural counterpart to Middletown) in 1939 the highest number of offenses during two consecutive sessions of the court were "driving while intoxicated." That could be the result of a personal crusade of the local police, and as such would not form a proper base for generalization, but the comments by the town's inhabitants can be seen as typical: for them, in former days "disturbing the peace" was used to "punish people of chronically undesirable conduct. ... Nowadays they are arrested when they start up their car."[122] Even Adolf Hitler felt compelled to denounce aggressive drivers, whom he, in a speech of 1939, called "parasites on the Volk" and who in the large German towns were reprimanded in public by loudspeaker cars stationed at crossings. The louder one shouted against the roadhogs, the more civilized the bulk of the motorizing middle class became, but only if this middle class accepted the 'slaughter' as 'inevitable' or the price or even a "regular sacrifice" for the Western way of life.[123] In other words, the openly and increasingly violent character of road traffic led to a redefinition of the road-safety problem, from one which questioned the very basis of a future 'car society' into a management problem of separated traffic flows at different speeds, drastically lowering the statistical chance of encountering a 'collision partner.'[124]

Why the violence of road traffic became a more or less accepted by-product of modern risk-aversive society has not been much discussed in the historiography, but after our excursion into the violent aspects of mobility literature in chapters 2, 3, 5, and 6, it does not seem too far-fetched to assume that while the central state increasingly claimed the monopoly on the execution of open violence (in war, in the repression of internal uprisings and in capital punishment), at the same time it redistributed part of this violence among its motorized citizens, in a statistical, normalized way, so much so that already at the end of the 1920s an American dissertation on the automobile spoke of "the prosaic nature of the automobile accident." The counterpoint, then, of Elias's self-regulation is this accommodation of "normal accidents": it needed an extra layer of Elias's *"zivilisatorische Rindenbildung"* (civilizational cortex formation) to make the horror acceptable. This was done by making the accident statistically invisible (redefine it as a risk) and at the same time use the thrill of risk taking to enhance the angst-lust of the automotive adventure.[125] The accident became an unalienable part of "the machine" itself and the statistically defined violence of causing or being the victim of an accident became a part of Zygmunt Bauman's "redistribution of [the state's] legitimacy," or, in the words of Paul Virilio, "a profane miracle," as it reveals something (the systemic character of the individual drive) "that we would not otherwise be able to perceive."[126] One step further and figures such as Rickenbacker acquire a god-like status, as he near-died, over and over again, on our behalf. The miracle does not seriously question the single accident, nor the accidents as a mass phenomenon: apart from some sporadic upheavals of indignation, the increasing unsafety did not lead to an increase in a feeling of terror, but in "normalizing."[127] Another factor that may explain the easy acceptance of the annual death toll is the statistical fact that during the 1930s, the share of nonusers among the accident victims diminished: motorists now started to kill each other instead of only women and children along the road.[128]

The Battle of the Systems: Road Versus Rail and the Coordination Crisis

At a moment that an ever more powerful auto lobby managed to convince certain elements of the national governments to support a shift from one mobility mode to another, more powerful parts of central governments tried to protect the state's financial and sociopolitical interests in the railroad system. This was done without preventing the rural tramway system from collapsing under its own weight of deficit and mismanagement, pushed over the edge by public criticism. With the exception of the United States, national governments were clearly dominated by a concern to keep the railways alive, and even in the United States the railroads dominated freight transport until the 1960s. The Dutch mobility infrastructure may serve as an illustration: major investments in basic infrastructures for railroads were not necessary any more, but electrification of railway tracks, trains, and safety systems and the purchase of streamlined diesel trains still absorbed considerable sums. The waterway network, however, enjoyed a lively investment climate and nearly doubled in size during the interwar years. Figure 7.7 nicely shows how despite the political and scientific bias toward the older systems, road building during the 1920s gained a foothold in infrastructural matters, as a gradually expanding niche, a Trojan Horse.

Business interests were divided: in an era in which laissez-faire liberalism was under severe criticism, even the *Reichsverband der Deutschen Industrie* "did not advocate free enterprise in transportation."[129] That is why it took so long for the International Chamber of Commerce (ICC), located in Paris but having its 1931 annual congress in Washington, to adopt a resolution for the setting up of an investigation "to determine the effects of the development of motor traffic upon railroad traffic in its various classes, and particularly in those classes of goods that are most affected" and "to consider the nature of the resulting modifications which may be desirable in the fundamental principles governing the railroad rate schedules." Two years later a Committee of Independent Experts finished an Introductory Report, followed, in 1935, by an overview of *Road and Rail in Forty Countries*.[130]

This is not the first time that the 'coordination debate' has been analyzed by historians,[131] but so far most histories are national in scope (in great contrast with the debate itself!) and are dominated by an eco-

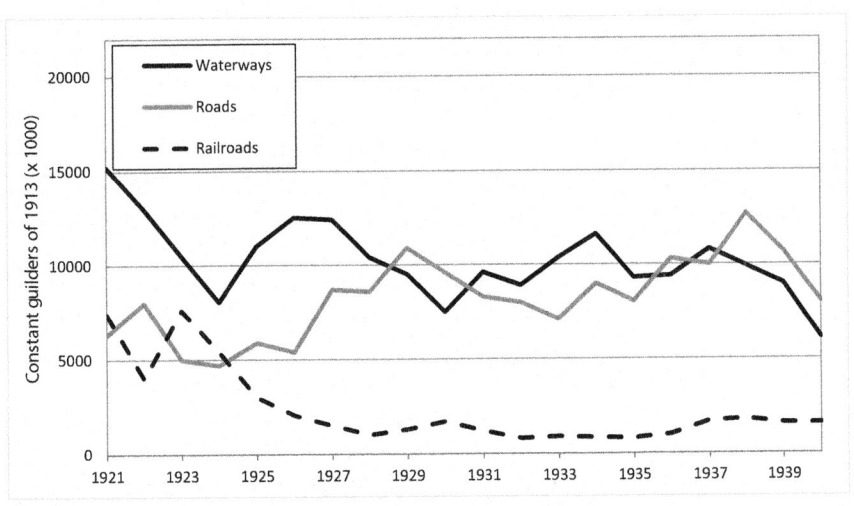

Figure 7.7. Gross capital creation in Dutch mobility infrastructure, 1921–1940 in constant guilders (×1000) of 1913.
Source: Mom and Filarski, 122, figure 6.2.

nomic discourse. This section, which takes the (Dutch) motor bus and especially the truck as a case study, is an effort to overcome both drawbacks of the historiography. As the following analysis will show, the coordination crisis was nothing less than a polarization between two different societies, with two different mobility systems, and two different mobility cultures. This was not primarily a class issue (although the emergence of the middleclass had certainly to do with it) but an issue of two types of citizens, so it seems: those who saw mobility as transport, as a serious but mundane utilitarian function that had to be managed as a service, just as education or mail, and those who saw mobility as an individual adventure, an experience that had as much to do with pleasure as with spatial displacement. Within this tension, the central (and centralizing) state saw this struggle as an argument to increase its interventionism, invited by some of the actors themselves.

The Motor Bus

With the general lack of solid academic work on the emergence of the motor bus it would be pretentious to claim to be able to give an overview of road-passenger transport in the Atlantic world. And yet, it is quite clear that if we want to make an inventory of the spread of motorization the motor bus (of which a German source in 1928 estimated the respective fleets in Europe and the United States to be 112,000 and 80,000) should be included.[132]

Take the Netherlands: a significant part of the large reservoir of potential car users were those Dutch who could not even afford a motorcycle, but started to become 'streetwise' through early bus use.[133] Motor buses in the Netherlands experienced an unprecedented boom during the 1920s. TT Fords (the mechanically strengthened version of the Model T) and, after 1927, AA Fords were re-equipped with a bus body by a growing national industry. Until 1926 these 'wild' buses were not hindered by any regulation. Road censuses during the 1920s indicate that the increase in 'intensity of use' (expressed in number of vehicles passing counting points of the road census) was by far the highest for the bus, higher even than that for bicycles. In the 1930s, if looked upon from an aggregate diffusion point of view, the growth of the Dutch bus fleet seemed to stagnate, but from a user point of view something quite different happened (a reason why focusing upon the artifacts instead of their 'performance' can easily be misleading). Whereas the number of units grew only modestly, not only did the seating capacity steadily increase, but the effective schedule speed kept increasing as well, through the introduction of more powerful engines and four-wheel brakes, even

Swarms Into Flows

during the depression.[134] This has apparently not been taken into account in figure 7.8, which gives the best overview available of the 'modal split' of passenger mobility during the first half of the previous century, drawn shortly after WWII. And yet, in terms of performance at its zenith in 1939 the bus contributed only 9 percent to general passenger road mobility. If compared to the United States (with only 3 percent of total passenger-kilometers traveled in the motor bus), this was an impressive performance.[135]

When the government tried to protect its interests in the railways (and those of the regional and local governments in their tramway systems), it set up a regulation in 1926 to prohibit outright competition with tramway companies by buses, forcing bus operators to acquire a license if they wanted to start a scheduled service. But because the licenses had to be requested at the provincial level, often the interests of the national railways and the municipal tramways were bypassed, so much so that hundreds of bus routes mushroomed, especially in the northern provinces.[136]

The regulation of 1926 could not stem the tide of the 'wild buses' either, for two reasons. First, illegal regular services kept being offered, because bus users were willing to support operators and drivers in unexpected ways. For instance, passengers of illegal line buses were

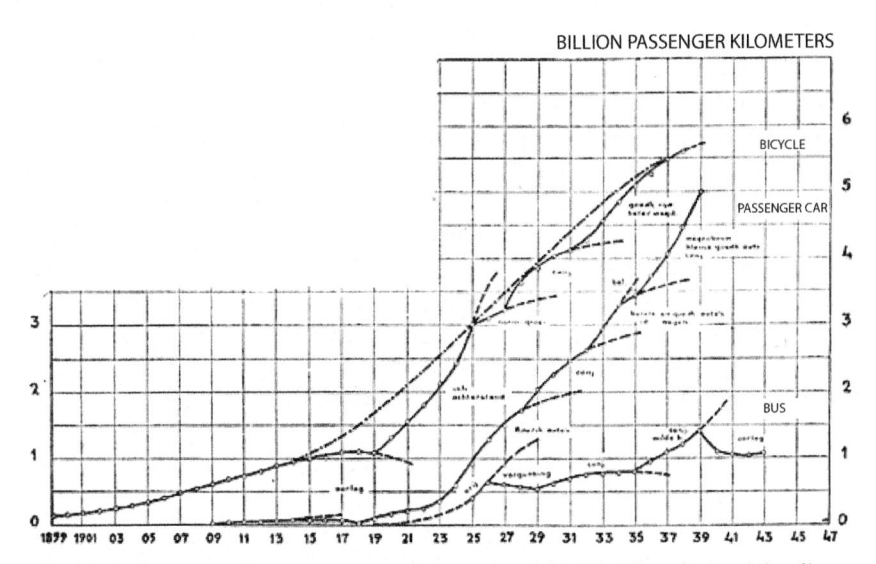

Figure 7.8. Performance (in passenger-kilometers) of the Dutch modal split.
Source: de Graaff, "Groei van het verkeer en zijn problemen." The interrupted lines are predictions extrapolated on the basic of logistic curves; the English terms are translations superimposed upon the Dutch on the original graph.

prepared to sign a document declaring that they were enjoying an 'excursion' (which was excepted from the 1926 regulation). And if that did not help, passengers were, according to an eyewitness, willing "to 'duck' for a while, either by lying down flat-belly between the seats or, if possible, even to climb in the luggage nets, because the driver had spotted a police officer along the road." Most of this line bus travel was local; only in 1935 was the first long-range regular bus service introduced in the Netherlands between Amsterdam and Rotterdam, a distance of 60 km. Nevertheless, whereas the total number of trips in the Dutch public transport sector remained more or less constant during these years, the share of motor bus trips during the 1930s increased from 5 to 31 percent, indicating a direct substitution for tramway use. Because of the capacity increase in the bus system, the growth of the number of bus users was even more spectacular, from 10 million in 1924 (which was about equal to the number of the country's inhabitants) to 154 in 1939.[137] In other words, buses helped lure tramway users toward 'road experience,' expanding the societal base of the road lobby considerably. And although tramways were seen as road users as well, the motor bus road experience took on a special, subversive flavor. One of the bus inspectors remembered that "we were then living in a time when sportiness was a more powerful inspiration for several bus operators than their wallet," no doubt referring to the many unemployed who had bought a second-hand makeshift contraption offering rates far below normal.[138]

Second, and ironically, the 1926 regulation forced outlaw bus operators into long-range irregular tourism trips. To this end a Dutch bus (body) manufacturing industry emerged, first copying the American touring car concept, and then developing designs of their own. One of the larger operators, VIOS in the small town of Wateringen, had a fleet of seventy touring cars that were only used during the summer months, and then 'produced' annual mileages of 15,000 to 19,000 km per vehicle.[139] From the beginning of the 1930s an international bus tourism developed, laying the basis for a special relationship between Dutch travelers and the motor bus after WWII, when the Netherlands, until the early 1950s (when it was overtaken by France) had one of the largest fleets of motor buses per capita in the world, after the United Kingdom and the United States.[140]

This started in 1935, the year of the World Fair in Brussels, when organized touring car trips were scheduled with destinations in Luxemburg, Germany, France, Spain, Switzerland, Austria, Hungary, Italy, and "even" (because of the obligation to use the ferry) England. At that moment 2500 of the 4000 buses in use in the Netherlands were outlaws or for tourists. Shortly before the outbreak of the war, international "folk tourism by bus" (as the touring club ANWB called it in a condescending way) reached its zenith; at that time regular services existed to Sweden, Germany, France, and Switzerland. This type of bus use was dominated by inhabitants of Amsterdam, and, in general, tourist bus companies were located in the big cities (whereas the companies of regular short-range use were located in the smaller cities and villages). Women were often in the majority. To counteract competition among themselves, these operators founded a Central Office for the Promotion of Touring Car Travel (CEBUTO) in 1933, which organized "merry evenings" during the winter months, where "songs at the piano" were played and "a chorus of cooperating drivers" sang, and movies were shown about distant countries.[141]

Other countries no doubt have similar stories to tell (although a recent German study claims that the Dutch bus fleet was, relatively speaking, extremely large), but like the truck, the motor bus does not enjoy much favor among mobility historians.[142] Belgium, for instance, saw its densely populated northern provinces (badly covered by the steam tram) invaded by "dozens of improvised buses" from the Netherlands, provoking legislation in 1923 necessitating a license, followed by a second law a year later that protected the interests of the national railroads much more effectively than was the case in the Netherlands. As in the Netherlands, however, only during the second half of the 1930s was a true 'coordination regulation' reached, which put about 40 percent of all licensed bus lines in the hands of the regional railway company.[143]

The United Kingdom was probably the only country where bus regulation was working effectively, as figure 7.9 shows: whereas the British truck fleet expanded uninterruptedly during the entire period, the number of buses and coaches was drastically reduced in 1925 and soon stabilized during the remainder of the period under consideration. This reduction took place well before the British Road Traffic Act of 1930, which introduced a rigorous licensing system resulting in "a controlled monopoly" that lasted until the 1980s.[144]

By contrast, the German passenger transport coordination did not develop as abruptly as the United Kingdom: the Road Transport Services Act of 1928 exempted the *Reichspost* from licensing. But here, as in the United Kingdom, the trend went clearly toward a road mobility "more similar to railway traffic." The *Autobahnen* were given regular bus services as soon as a stretch was ready, but the Nazi holiday organization *Kraft durch Freude* mainly used the train for its cheap collective excursion program.[145] All in all, the role of the motor bus in the German modal split of semi-public passenger transport was so modest (in pas-

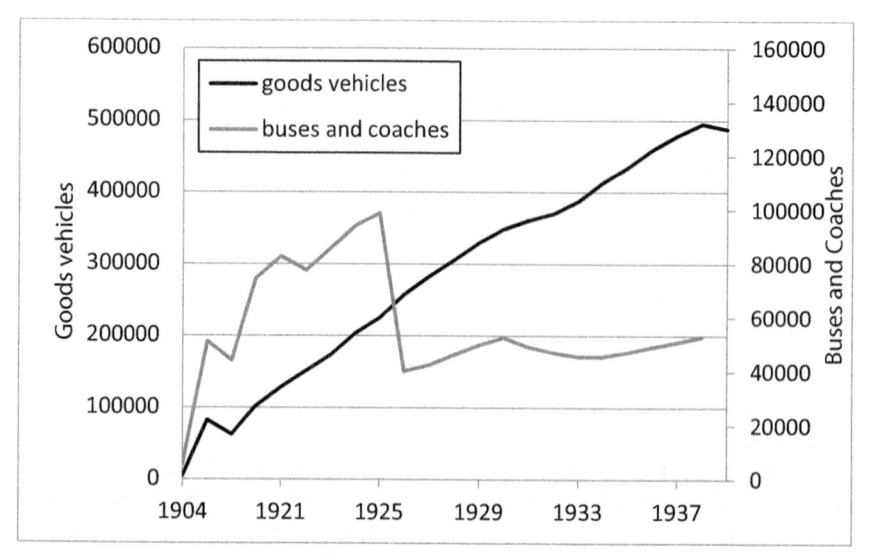

[Figure: line chart showing goods vehicles and buses and coaches in the UK, x-axis years 1904, 1921, 1925, 1929, 1933, 1937; left y-axis Goods vehicles 0–600000; right y-axis Buses and Coaches 0–160000]

Figure 7.9. Buses and trucks in the United Kingdom, 1904–1939. Note: the beginning of the scale on the x-axis is not linear.
Source: Savage, 144–145, tables V and VI.

senger-kilometers), that we need a logarithmic scale to illustrate its role compared to the rail alternatives as it was an order of magnitude smaller (figure 7.10).

In the United States little is known about the number of bus users for touristic purposes, although incidental evidence suggests that there, too, transcontinental bus tours became an important part of the road tourism movement. One third of the 25,000 buses in 1934 (owned by 3500 operators and carrying about half the number of railway passengers over a network of 350,000 miles) was deployed on interstate routes, which were hardly affected by the Motor Bus Industry Code of self-regulation introduced during the New Deal by the Interstate Commerce Commission. By the end of the 1920s it took 5.5 days to cross the country by bus. Trip lengths (at an average of 71 miles) were in-between those of the Class I railroads (about 50 miles) and the Pullman trains (340 miles).[146]

In the United States the 'wild bus' also knew a subaltern and openly subversive version: between 1910 and the early 1920s the 'jitney' became the vehicle for many unemployed (including blacks in southern states) to shape the 'underside of the diffusion curve.' Estimates of 62,000 in the United States (and 3500 to 2800 in Canada) reveal the extent of this phenomenon, which in an official document of the US Coordinator of Transport was held responsible for demonstrating "the

Swarms Into Flows 613

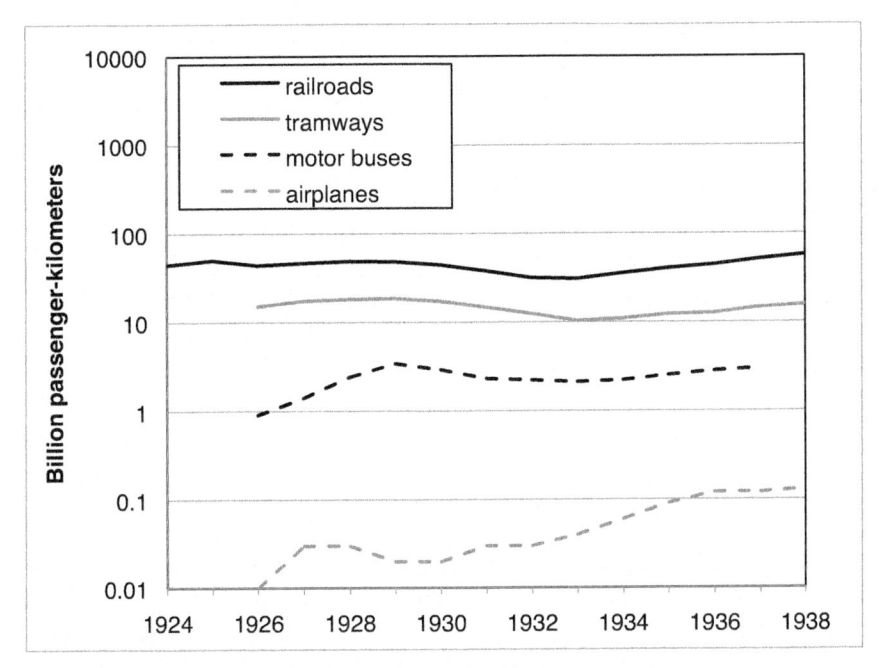

Figure 7.10. Modal split in (semi-)public transport in Germany.
Source: Hoffmann, 400, table 77.

popular appeal of motor-carrier transportation." Thus, many members of the lower middle class and workers gained their first 'road experience' in a more collective automobile culture.[147]

The motor bus added a crucial mode to the 'modal split' of road mobility, strengthening the case of the road against the rail considerably, although the real threat to the railways was not so much the bus, but rather the passenger car.[148] This became clear when a new type of car user joined the increasing group of road users: the businessman, with a clear trip destination, who decisively started to reject the train. If there ever existed a correlation between the number of cars and the quality and spread of the road network (which clearly was *not* the case before, say, the 1930s[149]), then it was in this period that business interests discovered the straight, unhampered road as an alternative to the rail. This cannot only be substantiated by the loss, by national (and often nationalized) railroad companies, of many first class and subscription clients, but also by sporadic road censuses indicating a dramatic increase in annual mileages per car, as we have seen in chapter 4.

For the Netherlands, then, the resulting "automobile system" was fully developed on the eve of the war, not in the least through a careful strategy by the national government to include the automobile and touring clubs as well as other "traffic associations" into a practice of cooperation. Rather, what emerged was a system that left the bulk of the railroad system largely intact. This system was mostly geared toward passenger transport, as the extensive Dutch inland-navigation network, despite being constantly in crisis, prevented the railroads from becoming a major transporter of bulk freight. On the other side of its mobility spectrum, however, the Dutch mobile culture was largely characterized by being slow: as late as 1939, in terms of passenger-kilometers, bicycling still dominated the nation's mobility culture, as figure 7.10 shows. In France, too, the bicycle's performance was higher than that of the motor bus, but there the car had by far overtaken the bicycle (figure 7.11); comparable high figures of cycle use are reported for the United Kingdom on the basis of an analysis of road censuses.[150]

The Emergence of the Truck

Just like its counterparts in the passenger mobility sector, the highly atomized fledgling postwar truck sector was considered to be 'wild' and 'anarchistic' by the existing transport interests, mostly represented by the railways. Among the state's civil servants trucks had a bad reputation as destroyers of roads, with their solid-rubber tires.[151] There was

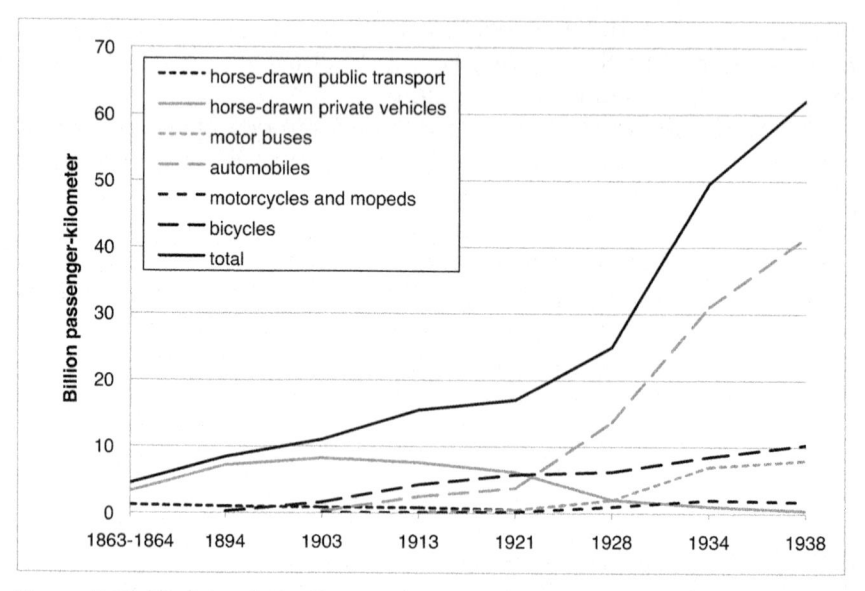

Figure 7.11. Modal split in France in billion passenger-kilometers. Note: the x-axis scale is not linear.
Source: Orselli, 76, table 16.

much continuity between the entrepreneurial anarchism and the revolutionary and counter-revolutionary deployment of the truck in the immediate postwar years in Germany and Italy.[152] Looking for work after the armistice, many former soldiers who had learned to drive during the war became drivers (and often owner-drivers) of the thousands of surplus trucks that flooded the market in the early 1920s. The American army alone had 39,000 very cheap surplus vehicles on offer, which they agreed not to take back to the home country after the war; and the British War Disposals Committee sold 60,000 vans and trucks to the domestic market. Driving trucks was much more than a simple job. It was a 'patriotic duty' when the 'red menace' of striking railroad workers in the United Kingdom threatened to paralyze the country. As such, it also continued the adventurous war experience of many drivers, who had been involved in moving ammunition or road stone in France without much consideration for costs or authority. In Germany surplus trucks formed the backbone of a revived road-transport association controlled by the *länder* and the municipalities.[153]

Elsewhere I have extensively dealt with the freight part of the coordination crisis.[154] Basically, like the passenger side (the bus versus the train), it was a struggle between two concepts of societies. In all industrializing countries, the emergence of the truck took place as an explosion, fueled initially, as we saw, by the surplus trucks of the war, but soon supported by considerably expanded manufacturing facilities and road improvements. In many of these countries the number of trucks doubled during the first two to three years after the war. They introduced new functions, such as door-to-door transport (without much transshipment) and farm-to-city transport of perishable commodities.[155] It seems that the latter transport function has been especially important in the emergence of American trucking, where next to perishable goods cotton also formed an important truck load. In Ohio in 1926 more than a quarter of all hauls were for agricultural purposes, the largest share after "products of manufacture," with 56 percent. With nearly one million trucks in 1930 American farmers formed the largest group of truck owners, followed by owners of wholesale distribution trucks and retailers. By 1932 two-thirds of the American truck fleet belonged to one-man firms: at that moment 2.2 million Americans drove their *single* truck, most of them as "private carriers."[156]

Originally, the 'old' transport systems were not alarmed by the rapid expansion of the trucking sector. They saw the truck as a potential extension to their geographical coverage (as feeders and distributors) and counted on the national governments to protect them from any competition that might jeopardize their monopoly. This self-confidence seemed well founded: most national railroad companies came out of the war with a lot of prestige. Often belonging to a select group of the largest companies in the country (if not *the* largest, such as in Germany, with one million employees), their special tariff structure had enabled the construction of a railroad network covering the entire national surface such that the remotest customer could count on 'fair' transport costs that depended upon the value of the freight. An elaborate system of exceptional (discount) tariffs, developed during the previous decades, softened this so-called *ad valorem* system and supported a social and economic state policy of industrialization and spatial development. As a herald of a new mobility system, if not an entirely different society, the truck represented the decentralization of industries, the increasing need to transport small amounts of commodities rather than large carloads of bulk ('less-than-carloads,' or LTC in railway managers' jargon) and a productivity increase of agriculture rather than the heavy industry of the railroads.[157] In short, the truck represented a society in which liberalism and marginal-costs calculations were more important than central state regulation and spatial policy based on an *ad valorem* transport system.[158] In all countries that tried to set up some sort of regulation of freight haulage, however, the 'small man' was excluded from the deliberations, and the same was true for the 'user.'

The railroads' appeal to restriction of the competition fit with a general societal trend toward regulation, anti-liberalism, and corporatism. Raoul Dautry, France's central coordination personality for many years, looked in envy to the British "public trust" as an instrument of coordination of London transport. Historically, large projects such as the highway networks have been prone to this kind of management, implying that Western parliaments often granted power to public authorities they could not control.[159]

Apart from the politico-organizational aspect of this type of planning, it also has theoretical and ideological repercussions that lately have been discussed under the banner of social engineering. This type of 'planned progress' and its accompanying desire for 'order' and fear for 'chaos,' 'fragmentation,' and 'congestion' saw its first heyday occur in the systemic gigantomania of New Deal and Fascist and Stalinist planning in which the individual tended to disappear and the community of the 'people' (the *Volk*) had replaced the nation. Not coincidentally, this type of planning has its roots in the First World War with its unprecedented state intervention and central, military planning (including logistics, in the sense of 'materials handling' which blossomed for the first time in this war, too, as explained in chapter 3), while the desire for 'order' was crucially strengthened by the postwar middle-class fear for revolution. In the United States it was 'republicanism,' as a fragile amalgam of individualism and community, that provided the middle-class values for a vision of a friction-free, efficient society in which the topos of the flow was already a crucial element as far back as the early 1900s. In the United States, too, after the war and the social unrest in its wake, there was not only an eagerness to centralize, but also a certain willingness among the middle classes to be centralized. Planners had, as (male-dominated) experts with a "cool head," to create "harmony" and spatial "balance," seeing space from a bird's-eye perspective.[160] The entire discourse is very akin to the *sachlich* 'cool conduct' analyzed in the previous chapter. A technocratic version, the civil engineers' practices to create unhampered traffic flows, were just as much an effort "to win the wilderness over to order," practices for which American planners and engineers alike favored the metaphor of "taming."[161]

If several national studies conclude that the coordination did not bring the railroads any economic advantage, it cannot be denied that the coordination crisis can be seen as a giant operation against the single owner-driver and his 'anarchistic' behavior.[162] Apart from a clash between two industrial sectors within capitalism (the heavy industry and its railroads; the light industry and its trucks), it was also a struggle of ideas about the necessity of societal order. Instead of a topic hesitantly taken up by transport historians, the coordination debate should be approached as a highly relevant cultural phenomenon worthy of more study.

Conclusions

As indicated earlier in this chapter, the reports from the International Chamber of Commerce in 1931 and 1935 indeed formed a watershed toward a society in which road transport dominated and which only after the Second World War was to grow into one of the characteristics of Western, industrialized societies: less-than-carload transport from door to door, on the basis of marginal costs and embedded in an ideology of liberalism that had the entire project of the European transport integration, as well as the American road-freight expansion, in its grip. At the same time, inland navigation was revived in many countries during the Interbellum, and it would be worthwhile to analyze the motives behind the decisions to do so. The same applies to the massive road-building projects, which do not seem to have been built upon actual demand, but were as much an expression of the expectation of, and hope for, a new order based on the automobile. Sources from engineering circles call this 'planning,' but the extrapolation of an observed trend into the future remains a fantasy, however sophisticated its undertaking. Why did all societies under investigation invest in alternative transport modes that seemed to directly undermine the very protection they wanted to provide to the railways? It belongs to the salient ironies of mobility history that the state itself provided a train-like road system that allowed both the business man in his passenger car and the small trucker to gradually increase trip length.

To analyze this process further, a transnational approach is necessary. At the same time, the focus on economic aspects of costs should be widened toward a discourse analysis of the coordination debate, as there is ample evidence that much of this debate was rhetoric.[163] Part of that rhetoric was (and is, among historians) the sharp dichotomy between a liberal and a regulatory approach: even in the sanctuaries of liberalism (the United Kingdom and the United States) freight mobility was regulated, if not cartelized.[164] Given the common start of the problem (the explosion of trucking after the First World War) and the different solutions proposed, this analysis confirms that every country seems to devise measures such that in the end the Western societies do not differ too much from each other. This is even true for the United States: despite the large differences in modal split, the overall development is remarkably parallel to Europe, assuming at the same time that both 'European' and 'American' mobilities are not static, but in constant flux, and consisting of multiple mobilities. This dynamism is all the more true for the differences in urban mobilities, which are even larger than the general differences referred to above.[165] What no government could change, however, was the remarkable resilience of the trucking sector against concentration: at the end of the Interbellum (and in fact to this day) it still was a highly atomized sector and as such it still reflected in its heritage, the individually motorized civilian.

When John Merritt was interviewed as part of the federal writers' project of the New Deal, he explained how he and his three brothers-in-law bought four light Ford trucks to haul vegetables and fruit as wildcatters. He declared that, although he had got "tired of it," he "like[d] the uncertainty and chance of trucking."[166] In the end, many trucks do not differ that much from the passenger cars in their affordance of adventure.

In this chapter we have seen the (Atlantic) automobile system emerge as an interconnected ensemble of supporting structures both material and immaterial, such as: infrastructures (roads, including elements less dealt with such as filling stations and parking lots), legislation and regulation, maintenance system (garages, do-it-yourself networks, scrap yards), insurance, and the cultural and ideological mechanisms that demarcate the system vis-à-vis the other transport modes, including cycling and pedestrians. Technically, the closed car had been made into a much more reliable vehicle in the hand of the nuclear family as collective subject, and a service system guaranteed the car's usability even if the reliability was far from perfect. Also, the unreliability of the system as a whole was engineered away to a large extent through the construction of an automobile-friendly road system, including the beginning of a high-speed freeway system. The system's remaining unreliability was subsequently carefully engineered from sight through the construction of safety statistics that not only promised an ever safer future, but also assured each individual user that her exposure to the risks of the car society was manageable and under control. It is the building of this system, and the shift of focus from automobility to mobility in general, that turned the different penetration rates of the automobile (presented at the beginning of chapter 4) into a secondary matter as a descriptor of Atlantic automobilism, making the entire North-Atlantic realm into one transnational, multimodal 'car society.'

'Mass motorization," as we have learnt to characterize post-WWII car society, emphasizing a breach with the prewar society, is inconceivable without the preparatory developments during the Interbellum as presented in this chapter. For the contemporaries of that period, Atlantic automobilism had meanwhile become part and parcel of daily life. Whether one was able to drive a car did not matter that much anymore: the all-pervasive automotive adventure, however tamed and domesticated, now affected every citizen.

Notes

1. F. W. Taylor, *The Principles of Scientific Management* (1911), quoted in Richard Hornsey, "'He Who Thinks, in Modern Traffic, is Lost': Automation and the Pedestrian Rhythms of Interwar London," in Tim Edensor (ed.), *Geographies of Rhythm: Nature, Place, Mobilities and Bodies* (Farnham and Burlington: Ashgate, 2010), 99–112, here 104.
2. Garet Garrett, "Business," in Harold E. Stearns, *Civilization in the United States: An Inquiry by Thirty Americans* (New York: Harcourt, Brace and Company, 1922), 397–415, here 400.
3. For a characterization of modern traffic as a combination of "herds of interacting, patterned and intensely mobile hybrids that roam the *globe*" and the recognition that "there is regulation of automobility but not the ability to determine where the herds of cars might travel to or when," see John Urry, *Sociology beyond Societies: Mobilities for the Twenty-first Century* (London and New York: Routledge, 2000), 189 (italics in original).
4. William Morton Wheeler, *Emergent Evolution and the Development of Societies* (New York: W.W. Norton & Company, 1928), 37.
5. Eva Horn and Lucas Marco Gisi (eds.), *Schwärme—Kollektive ohne Zentrum: Eine Wissensgeschichte zwischen Leben und Information* (Bielefeld: transcript

Verlag, 2009), 75–81; Arwen P. Mohun, "Designed for Thrills and Safety: Amusement Parks and the Commodification of Risk, 1880–1929," *Journal of Design History* 14, no. 4 (2001), 291–306.
6. For a transnational history of traffic control systems, see Clay McShane, "The Origins and Globalization of Traffic Control Signals," *Journal of Urban History* 25, no. 3 (March 1999), 379–404; and for a critique on McShane from a European angle, see Hans Buiter and Peter Staal, "City lights: Regulated streets and the evolution of traffic lights in the Netherlands, 1920–1940," *The Journal of Transport History*, Third Series, 27, no. 2 (September 2006), 1–20.
7. Joseph J. Corn, "Work and Vehicles: A Comment and Note," in Martin Wachs and Margaret Crawford (with the assistance of Susan Marie Wirka and Taina Marjatta Rikala), *The Car and the City: The Automobile, The Built Environment, and Daily Urban Life* (Ann Arbor: University of Michigan Press, 1992), 25–38, here 30; see also his more recent *user unfriendly: Consumer Struggles with Personal Technologies, from Clocks and Sewing Machines to Cars and Computers* (Baltimore, MD: Johns Hopkins University Press, 2011).
8. See, for instance, J. K., '"Service",' *Het Motorrijwiel en de Lichte Auto* 211 (1933), 38–39, where the defect (the "panne") is characterized as "sometimes a nightmare." *De Autokampioen* (3 January 1931), 2, looks back on the previous period of frequent defects as "misery of bygone days" (my translation).
9. Kevin L. Borg, *Auto Mechanics: Technology and Expertise in Twentieth-Century America* (Baltimore, MD: Johns Hopkins University Press, 2007), 104–106; villages: Harold C. Hoffsommer, *Relations of Cities and Larger Villages to Changes in Rural Trade and Social Areas in Wayne County, New York* (Ithaca, NY: Cornell University Agricultural Experiment Station, Bulletin 582, February 1934), 42; on the flat rate system, see especially Stephen L. McIntyre, "The Failure of Fordism: Reform of the Automobile Repair Industry, 1913–1940," *Technology and Culture* 41, no. 2 (April 2000), 269–299.
10. Stephen L. McIntyre, "'The Repair Man Will Gyp You': Mechanics, Managers, and Customers in the Automobile Repair Industry, 1896–1940" (unpublished dissertation, University of Missouri-Columbia, May 1995), 222.
11. J. C. E. S., "Het Ford-systeem te Amsterdam," *Auto-Leven* 11 (1923), 275–278; v. V., "Ford Service," *Bedrijfsauto* 12, no. 42 (20 October 1932), 944–945.
12. Joop Segers, "Benzinestations: Een geschiedenis van de benzinedistributie in Nederland," *Industriële Archeologie* no. 13 (1984), 164–180, here 170, 172.
13. Catherine Gudis, *Buyways: Billboards, Automobiles, and the American Landscape* (New York and London: Routledge, 2004).
14. H. Neumann, "Uit het Fransche autoleven," *Autokampioen* 31 (1931), 743–744, here 744.
15. David Keir and Bryan Morgan, *Golden Milestone: 50 Years of the AA* (London: The Automobile Association, 1955), 137.
16. K. R., "Technische hulp langs den weg," *Motorkampioen* 12 (1924), 150; "Zes jaar autodokter," *Autokampioen* 29 (1936), 1004–1006.
17. Franz, *Tinkering*, 17 (individualizing), 84 (patents), 94 (letters), 126 (control), 128 (72 percent), 137–138 (professionalizing and trunk).
18. Borg, *Auto Mechanics*, 81–89 (fix: 87), 93 (male), 95–96 (segregation); one-third and training: Kathleen Franz, "'The Open Road': Automobility and Racial Uplift in the Interwar Years," in Bruce Sinclair (ed.), *Technology and the African-American Experience: Needs and Opportunities for Study* (Cambridge, MA, and London: MIT Press, 2004), 131–153, here 140.

19. Georgine Clarsen, *Eat My Dust: Early Women Motorists* (Baltimore, MD: Johns Hopkins University Press, 2008), 114 (ethic), 115 (noncollective).
20. John A. Jakle and Keith A. Sculle, *Motoring: The Highway Experience in America* (Athens and London: University of Georgia Press, 2008), 98; McIntyre, "'The Repair Man Will Gyp You,'" 20.
21. "Het leven in de Bovag-families: de afdeelingen!" *Het Automobielbedrijf* 4, no. 5 (28 November 1934), 94–96; "Jaarverslag over 1940," *Bovag* 4, no. 14 (12 april 1941), 189–194, here 194.
22. Stefan Krebs, "'Notschrei eines Automobilisten' oder die Herausbildung des deutschen Kfz-Handwerks in der Zwischenkriegszeit," *Technikgeschichte* 79, no. 3 (2012), 185–206, here 9, 11, 15–16, 20; Gijs Mom and Ruud Filarski, *Van transport naar mobiliteit: De mobiliteitsexplosie (1895–2005)* (Zutphen: Walburg Pers, 2008), 156.
23. This analysis is the result of a class on Long-term Automotive Trends, undertaken by this author at TU-Eindhoven in June 2004. I thank the students for taking part in the analysis and in the subsequent discussions.
24. Gijs Mom, *The Electric Vehicle: Technology and Expectations in the Automobile Age* (Baltimore, MD: Johns Hopkins University Press, 2004), 291–293. This is confirmed by Cecil Kenneth Brown, *The State Highway System of North Carolina: Its Evolution and Present Status* (Chapel Hill: The University of North Carolina Press, 1931), 190: in Virginia the number of passenger cars doubled while in South Carolina it increased only by 124 percent, despite a more rapid highway construction in the latter state.
25. The following pages have earlier been published (unless otherwise indicated) as Gijs Mom, "Building an Automobile System: PIARC and Road Safety (1908–1938)" (proceedings of the PIARC conference, Paris, 17–23 September 2007; on CD-ROM).
26. *Compte rendu Paris 1908*, 293, 364–365; *Compte rendu Bruxelles 1910*, 425. Three of these signs were accepted during the LIAT conference of 1900 in Paris, the fourth during its conference in Geneva in 1902. In the United States another "Uniform Sign System" was developed and implemented. Jakle and Sculle, *Motoring*, 80. Titles of the PIARC proceedings (all published in Paris in French, German, and English versions) differ per conference; they have been replaced by a much shortened formula consisting of *Compte rendu* plus the location and year of the conference, as for this section most of the proceedings have been read in French (as present in the extensive archives of ANWB, The Hague). ANWB also produced a Dutch translation that has been consulted if the French was not available. Also, conference reports have very lengthy titles, including the question and sub-question they are supposed to answer; they have been replaced by author name(s), a shortened version of the title followed by the phrase "PIARC conference" plus location and year of conference.
27. Gijs Mom, "Constructing Multifunctional Networks: Road Building in the Netherlands, 1810–1980," in Mom and Laurent Tissot (eds.), *Road History: Planning, Building and Use* (Lausanne: Alphil, 2007), 33–62.
28. For the following section, see Gijs Mom, "Roads without Rails: European Highway-Network Building and the Desire for Long-Range Motorized Mobility," *Technology and Culture* 46, no. 4 (October 2005), 745–772. This in its turn is largely based upon a study in Dutch: Mom and Filarski, *De mobiliteitsexplosie*, 173–201.

29. Daniel T. Rodgers, *Atlantic Crossings: Social Politics in a Progressive Age* (Cambridge, MA, and London: The Belknap Press of Harvard University Press, 1998).
30. *AIPCR—PIARC 1909–1969*, 20.
31. *Compte rendu Bruxelles 1910*, 424–434 (quote on 431).
32. *Compte rendu Bruxelles 1910*, 451.
33. Mom, "Roads without Rails"; Mom, "Inter-artefactual Technology Transfer: Road Building Technology in the Netherlands and the Competition between Bricks, Macadam, Asphalt and Concrete," *History and Technology* 20, no. 1 (April, 2004), 3–23. For the avoidance of towns, see *Compte rendu Londres 1913*, 315.
34. *Verslag van het vierde internationale wegencongres, gehouden te Sevilla, 1923* (The Hague: Vereeniging het Nederlandsche Wegen-Congres, 1924), 34.
35. *Compte rendu Seville 1923*, 126.
36. H. L. Bowlby, "Transports militaires par routes," quoted in D. Blas Sorribas Bastaran, "Le développement des Transports Automobiles" (*Rapport Générale* IV, PIARC conference Sevilla 1923), 15.
37. *Verslag van het vierde internationale wegencongres, gehouden te Sevilla*, 59 (parking tickets and stripes), 42 (motor buses).
38. Mom, "Inter-artefactual Technology Transfer;" *Verslag van het zesde internationale wegencongres, gehouden te Washington (D.C.) in 1930* (The Hague: Algemeene Landsdrukkerij, 1933), 12–13; Donald Weber, "Automobilisering en de overheid in België vóór 1940; Besluitvormingsprocessen bij de ontwikkeling van een conflictbeheersingssysteem" (dissertation, University of Ghent, 2008), 436–437; for the efforts of the American Asphalt Association to counteract the enthusiasm for concrete see: I. B. Holley, Jr., *The Highway Revolution, 1895–1925: How the United States Got Out of the Mud* (Durham, NC: Carolina Academic Press, 2008), esp. chapter 10; for a reliable but brief overview of pavement history, see Jakle and Sculle, *Motoring*, 55–70.
39. Edo J. Bergsma and L. C. Steffelaar, *Het derde Wegen-Congres te Londen, 23–28 Juni 1913* (The Hague: ANWB, Wegen-Commissie, 1913), 26; French centralization proposal in E. Marion, "Autorités chargées de la construction et de l'entretien des Routes, etc." (Report 55, PIARC Conference London, 1913), 22–23.
40. In this chapter I use 'freeway' as a generic term for all those limited access roads specifically designed for separated high-speed flows of automobiles such as the German *autobahnen*, the Italian *autostrade*, the Dutch *autosnelwegen*, the French *autoroutes*, and the American post-WWII interstate highways. The pre-WWII American highways, superhighways, and parkways do not belong to this category.
41. Mom and Filarski, *De mobiliteitsexplosie*, 195; one quarter: Finla Goff Crawford, *Motor Fuel Taxation in the United States* (Syracuse, NY: published by author, 1939), 11.
42. Fuchs Hellich et al., "Budget des Routes" (report 47 of PIARC conference Washington, 1930), 15 (quote). On the 'civilizing' role of the road network, see *Compte rendu Washington 1930* (Paris, 1931), 120ff.
43. Ibid., 18–19. Nevertheless, 90 percent of the German *autobahnen* in 1938 were paved with concrete, mainly because it, according to a German engineer, was "particularly suited to fast traffic" (Grossjohan, Mallison, and Temme, "Construction and maintenance," report 22 of PIARC conference, The Hague 1938, 3).
44. *Compte rendu The Hague 1938*, 293.
45. Frank Schipper, *Driving Europe: Building Europe on Roads in the Twentieth Century* (Amsterdam: aksant, 2008).

46. Alec Badenoch, "Touring between War and Peace: Imagining the Transcontinental Motorway 1930–1950," *Journal of Transport History*, Third Series, 28, no. 2 (September 2007); Ingrid Strohkark, "Die Wahrnemung von 'Landschaft' und der Bau von Autobahnen in Deutschland, Frankreich und Italien vor 1933" (unpublished dissertation, Hochschule der Künste Berlin, 2001).
47. For the following pages, see Mom and Filarski, *De mobiliteitsexplosie*, 173–201. See also Gijs Mom, "Decentering highways: European national road network planning from a transnational perspective," in Hans-Liudger Dienel and Hans-Ulrich Schiedt (eds.), *Die moderne Strasse: Planung, Bau und Verkehr vom 18. bis zum 20. Jahrhundert* (Frankfurt: Campus Verlag, 2010), 77–100.
48. For the departmental base of the *Corps des Ponts et Chaussées*, see Jean-Claude Thoenig, *L'Ère des technocrates: Le cas des Ponts et Chaussées* (Paris: Les Éditions d'Organisation, 1973), 45 (scheme of institutional structure), 65.
49. Weber, "Automobilisering en de overheid in België," 456, 459–464. See also Greet de Block, "Engineering the territory: Technology, space and society in 19th and 20th century Belgium" (unpublished dissertation, Katholieke Universiteit Leuven; manuscript for preliminary defence, October 2011). For an American proposal of mixed roads from 1906, see Jakle and Sculle, *Motoring*, 66–67.
50. Thoenig, *L'Ère des technocrates*, 60; Stefan Sandmeier, "Vom Eisenbahn zur Verkehrsplanung: Die Institutionalisierungsgeschichte des Verkehrswesens an der ETH Zürich," in Stefan Sandmeier, Andreas Frei, and Kay W. Axhausen, *125 Jahre Verkehrswesen an der ETH Zürich* (Zürich: Institut für Verkehrsplanung und Transportsysteme, ETH Zürich, 2008), 5–33, here 11; Louis Delanney, *Strassenverwaltung und –finanzierung in fünfzehn Ländern* (Paris: Internationale Handelskammer, June 1937), 61. For the French prewar reluctance to build freeways, see J. Nicod, "Les autoroutes de l'Europe Occidentale et la formation d'un réseau de grandes routes européennes," *L'information géographique* 19, no. 1 (1955), 3–19, here 11.
51. Georges Reverdy, *Les routes de France du XXe siècle 1900–1951* (Paris: Presses de l'école nationale des ponts et chaussées, 2007), 62 (Alpine roads), 72–76 (road policy), 140–144 (freeway; "de-iron" on 143).
52. Mom, "Roads without Rails." In this publication France was identified as an exception still to be explained.
53. In this section the German term *Verkehr* is translated as 'transport' or 'mobility' as it embraces much more than a literal translation of 'traffic' would suggest.
54. As late as 1933 only 2 percent of German state and provincial roads were covered with a heavy asphalt layer. Gustav Sjöblom, "The Political Economy of Railway and Road Transport in Britain and Germany, 1918–1933" (unpublished dissertation, Darwin College, University of Cambridge, 16 April 2007).
55. Mom and Filarski, *De mobiliteitsexplosie*, 182; Massimo Moraglio, "The Renewal of the Italian Road Network in the 1920s," *Transfers* 2, no. 1 (Spring 2012), 87–105.
56. P. H. Levin, "Highway Inquiries: A Study in Governmental Responsiveness," *Public Administration* 57, no. 1 (1979), 21–49, here 22; C. D. Buchanan, *Mixed Blessing: The Motor in Britain* (London: Leonard Hill, 1958), 130; Delanney, *Strassenverwaltung und –finanzierung in fünfzehn Ländern*, 47; faster centralization: Sjöblom, "The Political Economy of Railway and Road Transport in Britain and Germany, 1918–1933," 195.
57. James A. Dunn, Jr., "The Importance of Being Earmarked: Transport Policy and Highway Finance in Great Britain and the United States," *Comparative Studies in Society and History* 20, no. 1 (January 1978), 29–53, here 39.

58. William Plowden, *The Motor Car and Politics in Britain* (Harmondsworth and Ringwood, Australia: Penguin, 1971), 303 (Labour quote), 187, 202.
59. Buchanan, *Mixed Blessing*, 125–126; Peter Scott, "Public-Sector Investment and Britain's Post-War Economic Performance: A Case Study of Roads Policy," *Journal of European Economic History* 34, no. 2 (2005), 391–418, here 396. From 1930 there was no speed limit in the United Kingdom outside the towns.
60. Wayne E. Fuller, "Good Roads and Rural Free Delivery of Mail," *Mississippi Valley Historical Review* 42 (June 1955), 67–83; Peter J. Hugill, "Good roads and the automobile in the United States 1880–1929," *Geographical Review* 72 (July 1982), 327–349; Howard Lawrence Preston, *Dirt Roads to Dixie: Accessibility and Modernization in the South, 1885–1935* (Knoxville: University of Tennessee Press, 1991); last quote: David J. St. Clair, *The Motorization of American Cities* (New York, Westport, CT, and London: Praeger, 1986), 123.
61. Jakle and Sculle, *Motoring*, 43.
62. Owen D. Gutfreund, *Twentieth-Century Sprawl: Highways and the Reshaping of the American Landscape* (Oxford and New York: Oxford University Press, 2004), 93, 157, 167; Christopher Wells, "Car Country: Automobiles, Roads, and the Shaping of the Modern American Landscape, 1890–1929" (dissertation, University of Wisconsin-Madison, 2004), 340–343 (Townsend); shift in 1921: Bruce Seely, "The Automotive Transportation System: Cars and Highways in Twentieth-Century America," in Carroll Pursell (ed.), *A Companion to American Technology* (Malden, Oxford, and Carlton: Blackwell Publishing, 2005), 233–254, here 242; *America's Highways 1776–1976: A History of the Federal-Aid Program* (Washington, DC: US Department of Transportation, Federal Highway Administration, 1976), 109, 89 (300,000), 415 (before 1920s); fuel tax: Emil Merkert, *Personenkraftwagen, Kraftomnibus und Lastkraftwagen in den Vereinigten Staaten von Amerika: Mit besonderer Berücksichtigung ihrer Beziehungen zu Eisenbahn und Landstrasse* (Berlin: Springer, 1930), 236–241, 288–289, 290 (German).
63. *America's Highways 1776–1976*, 108–109, 269; Wells, "Car Country," 343.
64. Jakle and Sculle, *Motoring*, 4; comprehensive: Wells, "Car Country," 345.
65. See, for instance, Michael R. Fein, *Paving the Way: New York Road Building and the American State, 1880–1956* (Lawrence: University Press of Kansas, 2008), 103, 126–127.
66. Fuller, "Good Roads and Rural Free Delivery of Mail"; last quote: Mark Adams, "The Automobile—A Luxury Becomes a Necessity," in Walton Hamilton et al., *Price and Price Policies* (New York and London: McGraw-Hill, 1938), 27–81, here 38 (n. 1); forests: Paul S. Sutter, *Driven Wild: How the Fight against Automobiles Launched the Modern Wilderness Movement* (Seattle and London: University of Washington Press, 2002), 62.
67. Charles L. Dearing, *American Highway Policy* (Washington, DC: The Brookings Institution, 1941), 121; Merkert, *Personenkraftwagen*, 235; Corey T. Lesseig, *Automobility: Social Changes in the American South 1909–1939* (New York and London: Routledge, 2001), 60–61, 69; Brown, *The State Highway System of North Carolina*, 252 (see also the enclosed map, "State Highway System in North Carolina").
68. For both countries, see, respectively, Thomas Zeller, *Strasse, Bahn, Panorama: Verkehrswege und Landschaftsveränderung in Deutschland von 1930 bis 1990* (Frankfurt and New York: Campus Verlag, 2002); and Michele Bonino and Massimo Moraglio, *Inventare gli spostamenti: Storia e immagini dell'autostrada To-*

rino—Savona/Inventing Movement; History and Images of the A6 Motorway (Torino, London, Venice, and New York: Umberto Allemandi, 2006).
69. Mom and Filarski, De mobiliteitsexplosie, 119.
70. Peter Merriman, Driving Spaces: A Cultural-Historical Geography of England's M1 Motorway (Malden, MA, Oxford, and Carlton, Australia: Blackwell, 2007), 32; Stephen V. Ward, Planning the Twentieth-Century City: The Advanced Capitalist World (Chichester: John Wiley and Sons, 2002), 89; noticeably absent: Stephen V. Ward, "What did the Germans ever do for us? A century of British learning about and imagining modern town planning," Planning Perspectives 25, no. 2 (April 2010), 117–140, here 124.
71. For an overview of European freeway initiatives during the Interbellum, see Storia Urbana 26, no. 100 (July–September 2002; special issue "La formazione della rete autostradale europea: Italia, Spagna, Francia, Germania"). For some other European countries and the United States, see Mom and Tissot (eds.), Road History. For the founding of a Main Highway Administration (Gushodor, 1936) in the Soviet Union, see Lewis H. Siegelbaum, "Soviet Car Rallies of the 1920s and 1930s and the Road to Socialism," Slavic Review 64, no. 2 (Summer 2005), 247–273; and Lewis H. Siegelbaum, Cars for Comrades: The Life of the Soviet Automobile (Ithaca, NY, and London: Cornell University Press, 2008), 150–157.
72. Christof Mauch and Thomas Zeller, "Introduction," in Mauch and Zeller (eds.), The World Beyond the Windshield: Roads and Landscapes in the United States and Europe (Athens and Stuttgart: Ohio University Press / Franz Steiner Verlag, 2008), 1–13, here 8; Phil Patton, Open Road: A Celebration of the American Highway (New York: Simon and Schuster, 1986), 80; America's Highways 1776–1976, 136; E. Willard Miller, "The Pennsylvania Turnpike," in Donald G. Janelle (ed.), Geographical Snapshots of North America; Commemorating the 27th Congress of the International Geographical Union and Assembly (New York and London: Guilford Press, 1990), 428–431; New Deal: Jakle and Sculle, Motoring, 152.
73. Wells, "Car Country," 137. Thomas Zeller, "Der automobile Blick: Berg- und Alpenstrassen und die Herstellung von Landschaft in Deutschland und den USA im 20. Jahrhundert," in Dienel and Schiedt (eds.), Die moderne Strasse, 265–283; Jakle and Sculle, Motoring, 136–145.
74. They were, however, also compared by contemporaries: the PIARC conference in Washington (1930), in between the ones in Milan and Munich (where delegates could admire the autostrade and the Autobahnen, respectively), would showcase the Mount Vernon Memorial Highway as a "tourist-oriented roadway." Timothy Davis, "The Rise and Decline of the American Parkway," in Mauch and Zeller (eds.), The World Beyond the Windshield, 35–58, here 42; see also his "Mount Vernon Memorial Highway and the Evolution of the American Parkway" (unpublished dissertation, University of Texas at Austin, December 1997). Fritz Todt found the American parkway concept so impressive that he had one of the American Department of Agriculture's brochures on this topic translated. Thomas Zeller argues that this does not necessarily mean that there was "a widespread technology transfer" between both countries, concluding that it would be "a mistake to assume that the autobahn was simply an adaptation of the American parkway." Zeller, "Building and Rebuilding the Landscape of the Autobahn, 1930–70," in Mauch and Zeller (eds.), The World Beyond the Windshield, 125–142, here 130–131. Although Davis does not claim Zeller's assumption, it is remarkable that both do not go into the fundamental difference

in conceptual background (urban versus cross-country) of the two concepts. Zeller also points at the trucks that were allowed on the European freeways.
75. Ward, *Planning the Twentieth-Century City*, 120; *America's Highways 1776–1976*, 133–134; Cliff Ellis, "Lewis Mumford and Norman Bel Geddes: The highway, the city and the future," *Planning Perspectives* 20 (January 2005), 51–68. On Bel Geddes, see also Christina Cogdell, "The Futurama Recontextualized: Normal Bel Geddes's Eugenic 'World of Tomorrow,'" *American Quarterly* 52, no. 2 (June 2000), 193–245. For parkways as predecessors as urban freeways, see Clay McShane, "The Carriage and the Origin of Urban Highways" (paper presented at the T²M Conference, Dearborn, MI, 5 November 2004).
76. Collectivity: Eugene Thacker, "Networks, Swarms, Multitudes," www.ctheory.net/articles.aspx?id=423 (last accessed 16 September 2014), 3; Internetted: John Arquilla and David Ronfeldt, *Swarming & the Future of Conflict* (Rand National Defence Research Institute, n.d.), 22; autonomous: "Swarming (military)," http://en.wikipedia.org/wiki/Swarming_(military) (last accessed 13 August 2011), 10. Flow models assume constant flow rates of the particles, whereas swarms move on the assumption that the 'particles' "take account of what the other agents around them are doing." In other words, human swarms modeling needs to predict human behavior. "Crowd Modeling: Model Behaviour," *The Economist* (26 August 2011), http://www.economist.com/model/13174313 (last accessed 26 August 2011). Recent car-related research tries to integrate these military insights into "Multimodal Driver Assistance" models. Darya Popiv, "Integration of a Component Based Driving Simulator and Design of Experiments on Multimodal Driver Assistance" (TU Munich), http://www.bgsce.de/about/popiv/index (last accessed 4 November 2012).
77. Anette Schlimm, *Ordnungen des Verkehrs: Arbeit an der Moderne—deutsche und britische Verkehrsexpertise im 20. Jahrhundert* (Bielefeld: transcript, 2011), 33; urban traffic engineers: David Blanke, *Hell on Wheels: The Promise and Peril of America's Car Culture, 1900–1940* (Lawrence: University Press of Kansas, 2007), 242.
78. Kurt Möser, *Fahren und Fliegen in Frieden und Krieg: Kulturen individueller Mobilitätsmaschinen 1880–1930* (Heidelberg, Ubstadt-Weiher, Neustadt a.d.W., and Basel: Verlag Regionalkultur, 2009), 343.
79. Quoted in Alan Kramer, *Dynamic of Destruction: Culture and Mass Killing in the First World War* (Oxford and New York: Oxford University Press, 2007), 318.
80. Todt quoted in Wolfgang Munzinger, *Das Automobil als heimliche Romanfigur: Das Bild des Autos und der Technik in der nordamerikanischen Literatur von der Jahrhundertwende bis nach dem 2. Weltkrieg* (Hamburg: LIT Verlag, 1997), 5 (n. 10).
81. Charles Grivel, "Automobil; Zur Genealogie subjektivischer Maschinen," in Martin Stingelin and Wolfgang Scherer (eds.), *HardWar/SoftWar: Krieg und Medien 1914 bis 1945* (Munich: Wilhelm Fink Verlag, 1991), 171–196, here 172.
82. "Not until fairly recently have the nation's urban transportation problems been seriously considered by the rurally dominated federal and state legislatures." Melvin R. Levin and Norman A. Abend, *Bureaucrats in Collision: Case Studies in Area Transportation Planning* (Cambridge, MA, and London: MIT Press, 1971), 27.
83. Arthur Krim, "Route 66: Auto River of the American West," in Janelle (ed.), *Geographical Snapshots of North America*, 30–33.
84. Delanney, *Strassenverwaltung und –finanzierung in fünfzehn Ländern*, 6.

85. For the freeways as railway-inspired arteries, see Mom, "Roads without Rails." See also Zeller, "Building and Rebuilding the Landscape of the Autobahn," 131, who points at the presence of 1959 railway engineers in 1934 and 1935 alone, assigned to *autobahn* work.
86. Irene Anastasiadou, "Networks of Powers: Railway Visions in Interwar Europe," *Journal of Transport History*, Third Series, 28, no. 2 (September 2007); Erik van der Vleuten et al., "Europe's system builders: The contested shaping of transnational road, rail, and electricity networks," *Contemporary European History* 16, no. 3 (2007), 321–347, here 16, 20. Later Delaisi made a plea for the construction of rural roads in Eastern Europe.
87. On Puricelli, see, for instance, Massimo Moraglio, "Real ambition or just coincidence? The Italian fascist motorway projects in inter-war Europe," *Journal of Transport History*, Third Series, 30, no. 2 (December 2009), 168–182.
88. For a history of early automotive safety in Germany, France, and Switzerland, see Christoph Maria Merki, "Die 'Auto-Wildlinge' und das Recht; Verkehrs(un)sicherheit in der Frühzeit des Automobilismus," in Harry Niemann and Armin Hermann (eds.), *Geschichte der Strassenverkehrssicherheit im Wechselspiel zwischen Fahrzeug, Fahrbahn und Mensch* (Bielefeld: Delius & Klasing, 1999), 51–73.
89. *Compte rendu Paris 1908*, 132, 135. Horse traction proponents also criticized the decision to make the inner part of road curves lower than the outer part, meant to support easy cornering of fast automobiles.
90. *Compte rendu Londres 1913*, 504.
91. Ibid., 506, 512, 635–637; *Compte rendu Seville 1923*, 179.
92. Harvey Taylor, *A Claim on the Countryside; A History of the British Outdoor Movement* (Edinburgh: Keele University Press, 1997), 249.
93. Dorothee Hochstetter, *Motorisierung und "Volksgemeinschaft": Das Nationalsozialistische Kraftfahrkorps (NSKK) 1931–1945* (Munich: R. Oldenbourg Verlag, 2005), 374; Dietmar Fack, *Automobil, Verkehr und Erziehung; Motorisierung und Sozialisation zwischen Beschleunigung und Anpassung 1885 - 1945* (Opladen: Leske + Budrich, 2000) 241. The number of road traffic deaths in the German Reich decreased from 10,014 in 1935 to 7,635 in 1937. Hellmuth Wolff, *Kampf dem Verkehrsunfall* (Halle: Akademischer Verlag, 1938), 10. Extremely high: Wolfgang König, "Autocrash und Kernkraft-GAU: Zum Umgang mit technischen Risiken," in Herfried Münkler, Matthias Bohlender, and Sabine Meurer (eds.), *Sicherheit und Risiko: Über den Umgang mit Gefahr im 21. Jahrhundert* (Bielefeld: transcript Verlag, 2010), 207–222, here 218.
94. F. A. Whitlock, *Death on the Road: A Study in Social Violence* (London: Tavistock, in association with Hicks Smith and Sons, 1971), appendix II, figures 2c and 5. See also Bill Luckin, "War on the Roads: Traffic Accidents and Social Tension in Britain, 1939–45," in Roger Cooter and Bill Luckin (eds.), *Accidents in History: Injuries, Fatalities and Social Relations* (Amsterdam and Atlanta, GA: Rodopi, 1997), 234–254.
95. Jean-Claude Chesnais, "La mortalité par accidents en France depuis 1826," *Population* (French edition) 29, no. 6 (November–December 1974), 1097–1136; M. Greenwood, W. J. Martin, and W. T. Russell, "Death by Violence, 1837–1937," *Journal of the Royal Statistical Society* 104, no. 2 (1941), 146–171, here 148 (table 2), 150 (table 4), 153 (table 5); E. George Payne, *A Program of Education in Accident Prevention, with Methods and Results* (Washington, DC: Government Printing Office, 1922; Department of the Interior, Bureau of Education, Bulletin,

1922, no. 32), 6; John Henry Mueller, "The Automobile: A Sociological Review" (unpublished doctoral dissertation, University of Chicago, August 1928), 90.
96. Quoted in Anedith Jo Bond Nash, "Death on the Highway: The Automobile Wreck in American Culture, 1920–40" (unpublished dissertation, University of Minnesota, June 1983), 33.
97. Albert P. Weiss and Alvhh R. Lauer, *Psychological Principles in Automotive Driving: Under the Auspices of the National Research Council 1927–1929* (Columbus: Ohio State University, 1930), 5.
98. J. S. Dean, *Murder Most Foul: A Study of the Road Deaths Problem* (London: George Allen & Unwin, 1947), 9.
99. Rolf Schäfer, *Die tödliche Verkehrsunfälle in Nordbaden in den Jahren 1929–1934* (Heidelberg: Brausdruck, 1935), 4; Bedaux et al., "Die Unfälle auf den Strassen" (report 46, PIARC Conference, The Hague, 1938), 5; W. C. Clemens et al., "Massnahmen zur Trennung der Verkehrsarten auf der Stasse" (report 61, PIARC Conference, The Hague, 1938), 13–15; Delemer et al., "Les moyens propres à assurer la sécurité de la circulation" (report 60, PIARC Conference, Munich, 1934), 5; *Compte rendu The Hague 1938*, 166–167; Joe Moran, "Crossing the Road in Britain, 1931–1976," *Historical Journal* 49, no. 2 (2006), 477–496, here 479; motorcyclists: Dean, *Murder Most Foul*, 12. Care should be taken in interpreting these early road safety statistics, especially because the reporting methods by the police were not standardized, and the question of who was guilty in case of an accident was often dependent on the personal views of the police officer. These were most often critical toward the automobile.
100. Williams, "Die Unfälle auf den Strassen," 17–18.
101. Seth K. Humphrey, "Our Delightful Man-Killer," *The Atlantic Monthly* 148 (1931), 724–730, here 730.
102. Howard Taylor, "Forging the Job: A Crisis of 'Modernization' or Redundancy for the Police in England and Wales, 1900–39," *British Journal of Criminology* 39, no. 1 (Special Issue 1999), 113–135, here 128.
103. Ted Robert Gurr (ed.), *Violence in Amerca: Volume 1: The History of Crime* (Nembury Park, London, and New Delhi: Sage, 1989), 38 (homicide; vehicular manslaughter), 67 (contamination); Theodore N. Ferdinand, "The Criminal Patterns of Boston since 1849," *American Journal of Sociology* 73, no. 1 (July 1967), 84–99, here 92 (illicit).
104. Mom, "Roads without Rails."
105. *Compte rendu The Hague 1938*, 293; on CIAM, see Hans van der Cammen and Len de Klerk, *Ruimtelijke ordening: Van grachtengordel tot Vinex-wijk* (Utrecht: Het Spectrum, 2003), 137.
106. Clemens et al., "Massnahmen," 22; Bressot, Giguet, and Delaigue, "Massnahmen zur Trennung der Verkehrsarten auf der Strasse" (report 60, PIARC Conference, The Hague, 1938), 22.
107. Norton, *Fighting Traffic;* Mom and Filarski, *De mobiliteitsexplosie*, 147–148.
108. Nash, "Death on the Highway," 52.
109. Brian Ladd, *Autophobia: Love and Hate in the Automotive Age* (Chicago and London: University of Chicago Press, 2008), 75.
110. Andrew Gavin Altman, "Motoring for the Million? Cars and Class in Pre-1952 Britain" (unpublished PhD thesis, Boston College, Graduate School of Arts and Sciences, 3 April 1997), 229.
111. Paul M. Leonardi, "From Road to Lab to Math: The Co-evolution of Technological, Regulatory, and Organizational Innovations for Automotive Crash Testing,"

Social Studies of Science 40, no. 2 (April 2010), 243–274, here 248. For an extensive treatment of the 1924 conference, see Daniel M. Albert, "Order out of Chaos: Automobile Safety, Technology and Society 1925 to 1965" (dissertation, University of Michigan, 1997), 68–79.
112. Nash, "Death on the Highway," 38–42.
113. C. H. Lawshe, Jr., *A Review of the Literature Related to the Various Psychological Aspects of Highway Safety: A Progress Report of an Investigation Conducted by The Engineering Experiment Station of Purdue University, In cooperation with State Highway Commision of Indiana* (Lafayette, IN: Purdue University, April 1939; Research Series no. 66, Highway Research Bulletin, no. 2), *Engineering Bulletin* 23 (1939), pt. 2A, 9 (speeding), 31 (quote), 33 (women); a similar conclusion (three to three-and-a-half times as much, expressed per thousand miles and per thousand dollars of revenue, for taxi drivers) in Morris S. Viteles and Helen M. Gardner, "Women Taxicab Drivers: Sex Differences in Proneness to Motor Vehicle Accidents," *Personnel Journal* 7 (1929), 349–355, here 354–355. For an example of the many popular publications praising women for their safe driving, see Robert Clair, "Again the Woman Driver!" *Journal of American Insurance* 8, no. 8 (1931), 13–15. For a book-length outcry against speeding in the context of the American road safety campaign, see Katherine Allen, *Bleeding Hearts: A Solution to the Automobile Tragedy* (San Antonio, TX: The Naylor Company, 1941); road-Mutt: Nash, "Death on the Highway," 77.
114. Morris S. Schulzinger, *The Accident Syndrome: The Genesis of Accidental Injury* (Springfield, IL: Charles C. Thomas, 1956), 34. Industrial roots: Regina Weber, "'Der rechte Mann am rechten Platz': Psychotechnische Eignungsprüfungen und Rationalisierung der Arbeit bei Osram in der 20er-Jahren," *Technikgeschichte* 68, no. 1 (2001), 21–51. See also Wilfried Echterhoff, "Geschichte der Verkehrspsychologie, Teil 1," *Zeitschrift für Verkehrssicherheit* 36, no. 2 (1990), 50–70; and Echterhoff, "Teil 2" in *Zeitschrift für Verkehrssicherheit* 36, no. 3 (1990), 98–112. Industrial roots in the United States: Norton, *Fighting Traffic*, 32–39.
115. John C. Burnham, *Accident Prone: A History of Technology, Psychology, and the Misfits of the Machine Age* (Chicago and London: University of Chicago Press, 2009), 2–4.
116. Albert, "Order out of Chaos," 157–169 (quotes: 168); homosexuality: Daniel M. Albert, "Psychotechnology and Insanity at the Wheel," *Journal of the History of the Behavioral Sciences* 35, no. 3 (Summer 1999), 291–305, here 302.
117. Blanke, *Hell on Wheels*, 114.
118. *Motor-Vehicle Traffic Conditions in the United States, Part 6: The Accident-Prone Driver*, House Documents, 75th Congress, 3rd Session, 3 January–16 June 1938 (Washington, DC: United States Government Printing Office, 1938), 1.
119. Whitlock, *Death on the Road*, 26 (not identified); David Shinar, *Psychology on the Road: The Human Factor in Traffic Safety* (New York and Toronto: John Wiley & Sons, 1978), 31 (maladjusted); Schulzinger, *The Accident Syndrome*, 13, 40 (not diagnosed), 178 (human characteristic); T. W. Forbes, "The Normal Automobile Driver as a Traffic Problem," *The Journal of General Psychology* 20 (1939), 471–474, here 473; dethroned: Frank A. Haight, "Accident Proneness: The History of an Idea" (Irvine: Institute of Transportation Studies, August 2001), UCI-ITS-WP-01-2.
120. Blanke, *Hell on Wheels*, 100–101.
121. Edward V. Rickenbacker, *Rickenbacker* (Englewood Cliffs, NJ: Prentice-Hall, 1967), 17.

122. Blanke, *Hell on Wheels*, 111–119, 144, 161–182 (speed limit: 175); James West, *Plainville, U.S.A.* (New York: Columbia University Press, 1958; first ed. 1945), 93. The quotes are from West, who paraphrases the attitude of the inhabitants.
123. Ladd, *Autophobia*, 37, 55 (sacrifice); parasites: Rudi Koshar, "Germans at the Wheel: Cars and Leisure Travel in Interwar Germany," in Koshar (ed.), *Histories of Leisure* (Oxford and New York: Berg, 2002), 215–230, here 223; loudspeakers: Hochstetter, *Motorisierung und "Volksgemeinschaft"*, 400.
124. Gutfreund, *Twentieth-Century Sprawl*, 31.
125. Mueller, "The Automobile: A Sociological Review," 132; see also Charles Perrow, *Normal Accidents: Living with High-Risk Technologies* (New York: Basis Books, Inc., Publishers, 1984), 5: a "normal accident" is a "system accident." On Elias's self-regulation in traffic, see also Stephan Rammler, *Mobilität in der Moderne: Geschichte und Theorie der Verkehrssoziologie* (Berlin: edition sigma, 2001), 173–174.
126. Zygmunt Bauman, *The Individualized Society* (Cambridge: Polity Press, 2004; reprint of first ed. 2001), 211; machine: David Bissell, "Vibrating materialities: Mobility—body—technology relations," *Area* 42, no. 4 (2010), 479–486, here 483; miracle: Sylvère Lotringer and Paul Virilio, *The Accident of Art* (New York: Semiotext(e), 2005), 63.
127. Reto Sorg and Michael Angele, "'Oh, my Ga-od! Oh, my Ga-od! Oh, Ga-od! Oh, my Ga-od!' Automobil-Unfall und Apokalypse in der Literatur der Zwischenkriegszeit," *Compar(a)ison* 2 (1996), 137–173, here 146.
128. Nash, "Death on the Highway," 67: death rates of children and passengers had declined by the early 1930s.
129. Christopher Kopper, "Passenger Transportation in Inter War Germany," *Transfers* 3 no. 2 (Summer 2013) 89–107.
130. Paul Wohl and A. Albitreccia, *Road and Rail in Forty Countries: Report Prepared for the International Chamber of Commerce* (London: Oxford University Press / Humprey Milford, 1935).
131. The most recent publication for the United States: Mark H. Rose, Bruce Seely, and Paul F. Barrett, *The Best Transportation System in the World: Railroads, Trucks, Airlines, and American Public Policy in the Twentieth Century* (Columbus: Ohio State University Press, 2006); for the United Kingdom: Peter Scott, "British Railways and the Challenge from Road Haulage: 1919–39," *Twentieth Century British History* 13, no. 2 (2002), 101–120; for France: Nicolas Neiertz, *La coordination des transports en France: De 1918 à nos jours* (Paris: Ministère de l'Économie, des Finances et de l'Industrie, Comité pour l'histoire économique et financière, 1999); for Germany: Richard Vahrenkamp, "Lastkraftwagen und Logistik in Deutschland 1900 bis 1955: Neue Geschäftsfelder, neue Kooperationsformen und neue Konfliktlinien," *Vierteljahresschrift für Sozial- und Wirtschaftsgeschichte* 95, no. 4 (2008), 430–455, lately followed by his *Die logistische Revolution: Der Aufstieg der Logistik in der Massenkonsumgesellschaft* (Frankfurt and New York: Campus Verlag, 2011); for the Netherlands: Jac. Verheij, *Wetten Voor Weg en Water (1923–1998): Het experiment van de Wet Autovervoer Goederen en de Wet Goederenvervoer Binnenscheepvaart en de jaren erna* (Delft: Eburon, 2001), and Mom and Filarski, *Van transport naar mobiliteit*, ch. 9 (203–235). Gustav Sjöblom ("The Political Economy of Railway and Road Transport in Britain and Germany, 1918–1933") ventured a comparison between Germany and the United Kingdom, as did Schlimm, *Ordnungen des Verkehrs*.

132. W. Feilchenfeld, "Kraftwagenverkehr und Fremdenverkehr," *ADAC-Motorwelt* 25, no. 39 (1928), 19–21, here 19.
133. The following section has been published previously as Gijs Mom, "Mobility for Pleasure: A Look at the Underside of Dutch Diffusion Curves (1920–1940)," *TST Revista de Historia: Transportes, Servicios y Telecomunicaciones* no. 12 (June 2007), 30–68; parts have also been published in Gijs Mom, "Clashes of Cultures: Road vs. Rail in the North-Atlantic World during the Inter-War Coordination Crisis," in Christopher Kopper and Massimo Moraglio (eds.), *The Organization of Transport: A History of Users, Industry, and Public Policy* (London and New York: Routledge, forthcoming 2015)
134. Jan Erik Grunveld, *Per ATO en per spoor: 20 jaar omstreden autobushistorie* (Utrecht, 1987), 76–93; Maili Blauw, "Met of zonder vergunning? Wetgeving, regulering en toezicht op het vervoerbedrijf in Nederland in de negentiende en twintigste eeuw," in *150 jaar toezicht Verkeer en Waterstaat: Rampen, wetten en inspectiediensten* (Den Haag, 2004), 181–242.
135. W. J. de Graaff, "Groei van het verkeer en zijn problemen," *De Ingenieur* 60, no. 10 (5 March 1948), V.21–V.36; 3 percent: Margaret Walsh, *Making Connections: The Long-Distance Bus Industry in the USA* (Aldershot and Burlington, VT: Ashgate, 2000), 27 (table 2.1).
136. Johan W. D. Jongma, *Geschiedenis van het Nederlandse wegvervoer* (Drachten and Leeuwarden: FPB Uitgevers, 1992), 113; J. J. Stieltjes, "III. Het verkeer te land: C. De exploitatie van het railloos verkeer," in "Het verkeer in Nederland in de XXe eeuw," *Tijdschrift van het Koninklijk Nederlandsch Aardrijkskundig Genootschap*, Tweede serie, 50 (1933), 331–662.
137. P. H., 'Autobus-reminiscenties,' *Bedrijfsauto* 20, no. 29 (18 July 1940), 592–594, here 593; Jongma, *Geschiedenis van het Nederlandse wegvervoer*, 113; *Wegen* (1936), 341–349, 365–375.
138. "Een vergeten jubileum der autobuskeuring," *Autokampioen* 35 (1941), 38–39, here 38.
139. A. C. Q., "Een praatje met den toerwagenondernemer Lipman in Wateringen," *Bedrijfsauto* 19, no. 20 (18 May 1939), 12–13.
140. Peter Rocholl, *Vergleichende Analyse der Entwicklung des Personenkraftverkehrs im westeuropäischen Wirtschaftsraum* (Düsseldorf: Droste Verlag, 1962), 129 (graph 12).
141. H. T., "Autobussen, Touringcars, en het Wetsontwerp Vergunningstelsel," *Bedrijfsauto* 13 (1933), 454–456; J. W. Kwak, "Is het einde van het touringcarwezen nabij?" *Bedrijfsauto* 16 (1936), 143–144; Autobusondernemer, "Feest in de schaduwen van morgen," *Bedrijfsauto* 16, no. 13 (26 March 1936), 192, 194; "Cebuto," *Autobusdienst* 6, no. 1 (1 July 1933), 9; H. T., "Ontwikkeling van het touringcar-wezen," *Bedrijfsauto* 15, no. 44 (31 October 1935), 687, 691; quotation: M., 'Groot-toerisme van den kleinen man,' *Autokampioen* 36 (1942), 617–618, here 618; P. Kuin and H. J. Keuning, *Het vervoerswezen* (De Nederlandse volkshuishouding tussen twee wereldoorlogen, III) (Utrecht and Brussels: Uitgeverij Het Spectrum, 1952), 133. Margaret Walsh also found a female dominance of long-range bus use in the United States: Walsh, *Making Connections*, 11.
142. Kopper, "Passenger Transportation in Inter War Germany," 12. A comparison between Germany and the United Kingdom suggests that the relatively small national motor bus fleet in Germany forms the exception, and the Dutch fleet seems to be more in line with the United Kingdom and France, relatively speak-

ing. Sjöblom, "The Political Economy of Railway and Road Transport in Britain and Germany, 1918-1933," 61 (n. 110).
143. Weber, "Automobilisering en de overheid in België," 467–469.
144. Sjöblom, "The Political Economy of Railway and Road Transport in Britain and Germany, 1918-1933," 232–236 (quote: 234). The classic source on British bus coordination is still John Hibbs, *The History of British Bus Services* (Newton Abbot: David & Charles, 1968), but see also Corinne Mulley, "The background to bus regulation in the 1930 Road Traffic Act: Economic, political and personal influences in the 1920s," *Journal of Transport History*, Third Series, 4, no. 2 (September 1983), 1–19.
145. More like railways: Sjöblom, "The Political Economy of Railway and Road Transport in Britain and Germany, 1918-1933," 237; autobahnen: Richard Vahrenkamp, *Autobahnbau in Hessen bis 1943* (n.p. [Darmstadt]: Hessisches Wirtschaftsarchiv, 2007), 86–88.
146. *American Motorist* 21, no. 2(February 1929), 35; *SAE Journal* 24, no. 1 (January 1929), 64; R. E. Plimpton, "Long-Distance Passenger Services," *SAE Journal* 25, no. 3 (September 1929), 285–297; *Hours, Wages, and Working Conditions in the Intercity Motor Transport Industries, Part I: Motor-Bus Transportation* (Washington, DC: Federal Coordinator of Transportation, Section of Research, Section of Labor Relations, 1936), xv, 8 (n. 1); Code: Walsh, *Making Connections*, 25. On motor bus use in the United States (including the school bus, dominant in numbers), see also Jakle and Sculle, *Motoring*, 183–203.
147. Carlos A. Schwantes, "The West Adapts the Automobile: Technology, Unemployment, and the Jitney Phenomenon of 1914-1917," *The Western Historical Quarterly* 16, no. 3 (July 1985), 307–326; Blaine A. Brownell, "The Notorious Jitney and the Urban Transportation Crisis in Birmingham in the 1920's," *The Alabama Review* 25 (April 1972), 105–118; estimates: Donald F. Davis, "Competition's Moment: The Jitney-Bus and Corporate Capitalism in the Canadian City, 1914-29," *Urban History Review/Revue d'histoire urbaine* 18, no. 2 (October 1989), 101–122, here 101; last quote: *Hours, Wages, and Working Conditions in the Intercity Motor Transport Industries*, part I, 2. For illegal taxis in Paris (socalled cammionage by suburbians), see Mathieu Flonneau, "Une histoire des taxis parisiens au cours de la première moitié du XXe siècle; entre innovation, occasions manquées et blocages," in *Paris et Ile-de-France: Mémoires publiés par la Fédération des Sociétés Historiques et Archéologiques de Paris et de L'Ile-de-France, Tome 62* (Paris, 2011), 37–53.
148. Mom and Filarski, *De mobiliteitsexplosie*, 134–135; a similar conclusion for the United Kingdom in Sjöblom, "The Political Economy of Railway and Road Transport in Britain and Germany, 1918-1933," 80.
149. Mom, *The Electric Vehicle*, 291–293.
150. Sjöblom, "The Political Economy of Railway and Road Transport in Britain and Germany, 1918-1933," 48.
151. Helmuth Trischler, "Der epische Konflikt zwischen Schiene und Strasse: Der Güterverkehr der USA seit dem Ende des 19. Jahrhunderts," in Harry Niemann and Armin Hermann (eds.), *100 Jahre LKW: Geschichte und Zukunft des Nutzfahrzeuges* (Stuttgart: Franz Steiner Verlag, 1997), 243–262, here 251.
152. Kurt Möser, *Geschichte des Autos* (Frankfurt and New York: Campus Verlag, 2002), 138.
153. Thomas Gibson, *Road Haulage by Motor in Britain: The First Forty Years* (Aldershot and Burlington, VT: Ashgate, 2001), 126–127, 138–139, 142–145; Philip

Bagwell, *The Transport Revolution from 1770* (London: B. T. Batsford, 1974), 224; Plowden, *The Motor Car and Politics in Britain,* 366, gives twenty thousand as the number of trucks sold cheaply.

154. Gijs Mom, "Clashes of Cultures: Road vs. Rail in the North-Atlantic World during the Inter-War Coordination Crisis," in Christopher Kopper and Massimo Moraglio (eds.), *The Organization of Transport: A History of Users, Industry, and Public Policy* (London and New York: Routledge, forthcoming 2015); Gijs Mom, "Struggle of the Systems: Freight mobility from a Transatlantic Perspective, 1920 - 2000," in Gérard Duc et al. (eds.), *Histoire des transports et de la mobilité. Entre concurrence modale et coordination (de 1918 à nos jours)/ Transport and mobility history. Between modal competition and coordination (from 1918 to the present)* (Neuchâtel: Alphil, 2014).

155. Plowden, *The Motor Car and Politics in Britain,* 366, 372–373; [The Bureau of Public Roads, US Department of Agriculture / The Ohio Department of Highways and Public Works], *Report of a Survey of Transportation on the State Highway System of Ohio* (n.p. [Washington, DC], 1927), 81.

156. William R. Childs, *Trucking and the Public Interest: The Emergence of Federal Regulation 1914–1940* (Knoxville: University of Tennessee Press, 1985), 12, 20, 33–34; Harold G. Moulton, *The American Transportation Problem (Prepared for the National Transportation Committee)* (Washington, DC: The Brookings Institution, 1933), 575–578.

157. Heidi Rohde, *Transportmodernisierung contra Verkehrsbewirtschaftung: Zur staatlichen Verkehrspolitik gegenüber dem Lkw in den 30er Jahren* (Frankfurt: Peter Lang, 1999), 40; Wolfram Fischer, "Wirtschaft, Gesellschaft und Staat in Europa 1914–1980," in Wolfram Fischer (ed.), *Europäische Wirtschafts- und Sozialgeschichte vom Ersten Weltkrieg bis zur Gegenwart* (Stuttgart: Klett-Cotta, 1987), 1–221, here 94–95. A comparable conclusion also in Sjöblom, "The Political Economy of Railway and Road Transport in Britain and Germany, 1918–1933," 294.

158. See Vahrenkamp, "Lastkraftwagen und Logistik," 438, who shows that the number of employees in the car-friendly economic sector in 1925 Germany hardly formed half of the railway-friendly sector.

159. Susan Tenenbaum, "The Progressive Legacy and the Public Corporation: Entrepreneurship and Public Virtue," *Journal of Policy History* 3, no. 3 (1991), 309–330, here 321. For France, see, for instance, Richard F. Kuisel, "Vichy et les origines de la planification économique (1940–1946)," *Le mouvement social* 98 (1977), 77–101.

160. Anselm Doering-Manteuffel, "Konturen von 'Ordnung' in den Zeitschichten des 20. Jahrhunderts," in Thomas Etzemüller (ed.), *Die Ordnung der Moderne. Social Engineering im 20. Jahrhundert* (Bielefeld: transcript Verlag, 2009), 41–64, here 20 (cool); Michael Hochgeschwender, "*The Nobles Philosophy and Its Most Efficient Use*: Zur Geschichte des *social engineering* in den USA, 1910–1965," in *Die Ordnung der Moderne,* 171–197, here 183 (WWI), 186 (fear), 190 (male); Nadine Klopfer, "'Clean Up'; Stadtplanung und Stadtvisionen in New Orleans, 1880er–1920er Jahre," in *Die Ordnung der Moderne,* 153–169, here 161 (flow), 167 (republicanism); Ariane Leendertz, "Ordnung, Ausgleich. Harmonie; Koordinaten raumplanerischen Denkens in Deutschland, 1920 bis 1970," in *Die Ordnung der Moderne,* 131 (birdseye); John S. Gilkeson, Jr., *Middle-Class Providence, 1820–1940* (Princeton, NJ: Princeton University Press, 1986), 263 (fragmentation), 270 (centralized). Logistics: Monika Dommann, "Material manövrieren: Eine Be-

griffsgeschichte der Logistik," *Wege und Geschichte* 2/2009 (February 2010), 13–17. On the New Deal and liberalism, see, for instance, H.W. Brands, *The Strange Death of American Liberalism* (New Haven and London: Yale University Press, 2001), 19–26; on the comparison between New Deal and Fascism/Stalinism, see Wolfgang Schivelbusch, *Entfernte Verwandtschaft: Faschismus, Nationalsozialismus, New Deal 1933–1939* (Munich and Vienna: Carl Hanser Verlag, 2005).

161. Cecilia Tichi, *Shifting Gears: Technology, Literature, Culture in Modernist America* (Chapel Hill and London: University of North Carolina Press, 1987), 123; Sutter, *Driven Wild*, 151–162; taming: Wilbur Zelinsky, "The imprint of central authority," in: Michael P. Conzen (ed.), *The Making of the American Landscape* (Boston, London, Sydney and Wellington: Unwin Hyman, 1990) 311-334, here: 325.
162. According to the French socialist regional historian Maurice Wolkowitz, "La bourgeoisie française devant l'évolution des techniques de transport," *La pensée: revue du rationalisme moderne* 39 (November 1951), 76–86, here 85, coordination is "first and foremost a problem of profit redistribution between transporters."
163. This has recently been done for the United Kingdom and Germany by Schlimm, *Ordnungen des Verkehrs*.
164. Mom, "Clashes of Cultures"; Mom, "Struggle of the Systems."
165. John Pucher, "Urban Travel Behavior as the Outcome of Public Policy: The Example of Modal-Split in Western Europe and North America," *Journal of the American Planning Association* 54 (Autumn 1988), 509–520, here 517.
166. Leonard Rapport, "Trucker and Builder," in *These Are Our Lives: As Told by the People and Written by Members of the Federal Writers' Project of the Works Progress Administration in North Carolina, Tennessee, and Georgia* (Chapel Hill: University of North Carolina Press, 1939), 317–323, here 319.

CONCLUSION

Transcendence and the Automotive Production of Mobility

> "And when the automobile came we found
> Our incorrupt opinion safe and sound
> Inoculated only by the schism
> For ever proof against all Socialism."[1]

In April 1940, more than half a year after the outbreak of the Second World War, the trains in Germany were so overfilled with "pleasure travelers" that drafted workers and soldiers coming back from the front for some days off had to stay back at the platforms. A year later the situation had not changed. Similar scenes are known from the Netherlands: the touring club ANWB reported in 1941 that it had sold double the number of camping permits as the year before and the number of travel inquiries from members had increased from 50,000 to 80,000, testifying to the "unimpaired travel desire" of the Dutch. Because of a lack of cars and fuel, 400 km of steam tramway lines were reopened before the entire mobility system collapsed two years later, and motorized mobility became a military monopoly.[2] Meanwhile, in the United States, according to a jubilant brochure of car and touring club AAA in 1941, "practically everyone who had a car, some spare time, and a bit of extra cash went for a vacation trip along the nation's highways."[3]

In this closing chapter we will summarize half a century of Atlantic automobilism, which had led to and enabled this remarkable automotive bacchanal in the midst of a world war. We will do so by first reconstructing our story as based upon the more traditional sources, such as archives, trade journals, and secondary literature (as told in the chapters 1, 3, 4, 5, and 7). In order to further qualify and detail our insights into the 'automotive adventure' as it evolved during the first four and a half decades of motorized road mobility, we have used belletristic literature, travelogues, songs, and movies (presented in chapters 2 and 6, and partially in chapters 3 and 5). We will then investigate how much we owe to these more unconventional sources. At the end of this chapter we will reflect on some methodological issues (introduced in the introduction and chapter 2) and close by formulating some desiderata regarding future research.

Crossing Borders: Half a Century of North-Atlantic Automobilism

During the first phase (1895 until WWI) the elite, consisting of the bourgeoisie and the aristocracy, invented automotive adventure. Young citizens, "with modern opinions (and) a new lifestyle," created a societal niche for themselves, not controlled by the state, in which they developed a tripartite, motorized adventure of racing, touring, and tinkering, heavily borrowing from technical and cultural predecessors such as the horse and the carriage, industrial propulsion machines and especially the bicycle movement.[4] Despite their declared 'natural sublime,' the reliability of their contraptions was so problematic that a large part of their adventure was molded by 'technological sublime': speedy, extra-urban touring often did not allow much gazing into the environment, because the car's technology absorbed most of the attention. "Machines work," Wiebe Bijker concluded lately, "only because humans make up for the social failures of these machines."[5] Bijker would have been even more correct if he had included the *technical* failures as well, but given their 'social construction' it is understandable he did not. Based on our concept of the 'dual nature of technology,' we do: such failures are compensated by organizational and other social innovations, as we saw, such as the hiring of chauffeurs, the study of technical handbooks, and, somewhat later, after the example of the electric vehicle, the garage system.

Although technologically speaking this elite niche was a dead end (their cars evolved from light voiturettes into very heavy vehicles for the extended family of about seven passengers), and although most of the driving and maintenance was done by chauffeurs rather than the owners themselves, they managed to overcome a fierce societal resistance that we identified as 'hooliganism' and that led to a surprisingly easy acceptance by the national authorities, as the many early examples of national and even transnational legislation and regulation testify. This eagerness to accommodate a controversial contraption may well have been fueled by the middle and lower layers of the middle classes that right at the end of the first phase were becoming enthralled by the car's possibilities. An important element in this fascination must have been the movement's overwhelmingly masculine, if not macho, character, expressed, among others, by a condescending attitude to nonusers and enabling (according to the historical actors themselves: inviting, if not enforcing) aggressive and violent behavior, the mirror image of the opponents' hooliganism.

If the First World War had not intervened, the first signs of a shift toward lighter cars driven by 'lower' echelons of the middle class might have allowed a first domestication of the aggressiveness and violence of the first phase. In a way, this was especially visible in the United States, where automobilism got appropriated by the midwestern farmer and his family, and where the movement was much less interrupted by the war than in Europe, even after the United States plunged into the European war scene itself. Whether in the United States or Europe, however, the motorized hostilities functioned as a catalyst of tendencies that were already present during the prewar years, such as the conviction that 'adventure' was (near-)deadly, the insight that maintenance and central control of that maintenance had to rescue the seemingly inherent unreliability of the technology and the fact that automotive tourism did not stop at the borders of the killing fields. It also led to a linguistic turn, from 'pleasure' car to 'passenger' car, despite the underlying continuity of the pleasurable sides of automobilism. Car drivers, whether as illegal tourists or as sanctioned war correspondents or ambulance drivers, ventured near the frontlines, in the midst of a multimodal spectrum of airplanes, trains, horses, bicycles, trucks, motor buses, and tanks—and pedestrians, of course.[6] Whereas the machine had entered the garden in the prewar period in the form of the car, the war made the garden, the landscape itself, into a killing machine of murderous mobility, despite the immobility of trench warfare. Airplanes now became the true adventure machines, setting the car free for a shift from grenade to cocoon, weapon to womb. Thus, for many the car now became an ersatz contraption, the opposite of the individual's fantasy, a pathetic object for a massive projection of insatiable desires.

This happened during the second phase (WWI–WWII) in which not only the technology of the car, but also the culture around it underwent some major shifts. The first shift was geographical: the United States became the new gravity center of Atlantic automobilism, laying the foundation for a second model of motorization driven by consumption, seen as a necessity of (an increasingly urbanized) life. 'Comfort' was the key term in this process. While it was presented as the realization of the European promise, it was a different project, partially based on a new type of user. But the emergence (and spectacular character) of American automobilism should not make us forget that all other shifts happened on both sides of the Atlantic ocean in more or less the same relative scale (except for the highly overrated car densities) and in any case at the same time, which made us question the 'master narrative' of a sole American 'model' of motorization mimicked elsewhere a bit later. Indeed, the shift from the elite extended family to the middle-class nuclear family and from a focus on the artifact to the acknowledgement of the system character of the movement happened in both 'models' at more or less the same time. The same is true for a shift in automotive adventure from a pure celebration of speed toward a more spatial celebration of car use, whereas the technological shift from an open (clam) 'shell' to a (closed) 'capsule' also took place on both continents at more or less the same moment. The car now became "a site of familial contention, more so than the house would be, because in a house people can always retreat to different spheres as they so desire."[7] It was a truly transnational discourse, where Europe looked to the United States as the promised land of democratized motorized mobility and the United States looked to Europe for inspiration in technology and styling, albeit to a decreasing extent. For Europe this shift is most eloquently expressed in the work of Heinrich Hauser, whom we identified as the 'inventor' of the new adventure, consisting of a transcendental, multisensorial experience of 'fast slowness,' a relaxing and dreamlike state of automobilism on the *Autobahnen* and the secondary roads shaped after them, a state of unity with the car as substitute for the woman. In this shift of technology and use, the 'old' car was not abandoned (and would reappear at the end of the century as the SUV), but alongside this old car a new, light, but closed-car type became mainstream: the 'affordable family sedan.' Technologically speaking, this vehicle underwent two major shifts resulting in a virtual reinvention of the car: a shift toward a layman's car (undertaken by the chassis engineers in the heydays of their influence), and a shift toward a high-speed family car (with the additional help of new types of engineers competent in the car's dynamic behavior, and comfort). 'Flight' kept being seen as quintessential, also in this phase: the more the car and its experience could be likened to the airplane, the new adventure machine, the better. In this context, Sherwood Anderson's (this other visionary of a 'feminine future') remark about the woman being less affected by the 'machine' because "every woman has a life within herself" (see chapter 5) opens a whole trajectory of Freudian interpretation we will not pursue. Suffice it to say that the transcendence the woman is capable of literally (the experience beyond her own body), the man reshaped as a fleeting—ever to be repeated—experience while driving.

The car 'explosion' during the 1920s was an urban phenomenon; we called it 'urbanizing,' as it happened in the small towns and in the suburbs, Leo Marx's 'middle landscapes' in which technology was as much at home as 'nature.' And again, it was highly controversial: shortly after the war resistance against the car was revived, even in the United States. It was the 'middling sorts,' the petty-bourgeois small family that pushed automobilism into an increasingly long-range practice that consisted at least half (and very probably, most) of pleasure driving, touring, pleasure shopping, and increasingly long-range tourism. Our analysis showed how the 'toy-to-tool myth,' which explains the success of the car out of its (seemingly new) utilitarian properties, was in fact a marketing tool of the car proponents, a way to sell the car as a 'necessity' at the societal level and as an alibi for indulging in the fun of driving at the personal level. In the process automotive adventure became more variegated, realized by youth who discovered the courting potentials of the car, and women and blacks who used the car as a vehicle of liberation, escape, and independence.

The speed and geographical extension of the spread of the car in the Atlantic world during the interwar years is an indicator of the success or failure of the middle classes to establish themselves as the crucial societal power. No doubt this was proven most convincingly in the United States, on the basis of the emancipation of the wealthy farmer—and the small town. It was there that the myth of the car as a societal necessity emerged, but it was formulated by the producers' and consumers' organizations in the urban centers. The car's hedonistic culture drove consumer culture and was shaped by it. The dominance of 'nonserious mobility' even can be read as a feminization of car society, if understood as Virginia Scharff suggested: as an alibi for men to allow other practices than the purely violent and aggressive ones from the prewar elite niche.[8] The car became a "backbone of a consumption-driven society," a multiplier of consumptive practices, of space, as a tool for shopping—in short, as a vehicle of modernity.[9]

From a driver-mechanic, the automotive subject evolved into an 'automobilist' who typically drove his family around, or better: as a collective subject called nuclear family, consisting of a male driver and a female adult and two or three children as 'passengers.' This driving around happened in ever increasing flows, and made the collective subject of the traveling family into a monad within a swarm, a multitude without a leader, ready, although reluctantly, to become regulated. They were "collectively orchestrated without being the product of the orchestrating action of a conductor," in the words of Pierre Bourdieu.[10] Their movement stood for the general "organization of consent" by delegating powers, and taming it, as Victoria de Grazia has convincingly shown for fascist Italy.[11] This semi-collective character of interwar motoring, together with its urban(ized) base (the fact that the movement was one of sedentary people in the first place), enhanced its distinctive characteristics and necessitated the creation of a hierarchy of models in parallel to the social hierarchy. "The Fiats and Renaults of the workers may now push to the campgrounds of the Riviera," a history of post-WWI European bourgeois culture opined, "but the Mercedes and Jaguars still convey their masters to Cap d'Antibes or Santa Margherita."[12] And yet, the middle-class character of the interwar car culture was a cultural phenomenon rather than an issue of income, as the workers on both continents refrained, generally speaking, from acquiring cars, despite an extensive second-hand market and the possibility of purchase on credit, or from acquiring a car for common use by several families. The emphasis on the income differences on both sides of the Atlantic Ocean (and the price differences for energy and fuel), which one can find in many histories of the car, should therefore be relativized.

Workers, and a larger proportion of the middle classes in Europe than in the United States, were traveling in trains, tramways, buses, bicycles, mopeds, and, mostly, on foot. Several of these modalities (such as the bus, the bicycle, and the moped) also benefitted from the emergence of the car as a mass phenomenon, especially when (for bicycles and mopeds) urban streets were improved and (for buses) an intercity network of well-paved roads was constructed, followed by a special network of freeways (in Germany) and state roads (in the United States).[13] Although the differences were great between European countries, and especially between the European countries together and the United States, seen from a 'mobility perspective' (i.e., if we abandon for a moment the exclusive focus on the automobile), the fledgling 'car societies' were much more alike. While the Dutch compensated their modest motorization (and ditto train use) by riding their bikes, Germans discovered adventure on their motorcycles and mopeds. The car's ubiquity, as we have called it, in the 1930s generated a curious mix of experiences and expectations around the car, by users and nonusers: within a generation most of us would be motorists.

So, despite the differences, the history of the persistence phase of automobilism, which at the same time is the emergent phase of the 'automobile system,' reveals a remarkable congruence of national stories, which can perhaps best be summarized as decentralization through centralization. When the car appeared in Europe it found an infrastructure of a road network that was shaped during centuries of ad-hoc adjustment such that almost every point in the landscape was accessible. It was also considered to be a public good, to be paid for by the community. Once accepted as new road user, automobilism adjusted this system to its speed (by straightening, widening, paving the old roads) and simply could pick from the existing infrastructure those roads that it needed, changing their place in the system's hierarchy, and add new stretches against relatively low costs. By constructing a centralized command center, often outside parliamentary control, the gravity centers of road building and coordination could be pushed one gubernational level up: from the local to the regional. This history (told in chapter 7) also reveals that despite the centralized structure and despite the rather straightforward execution of a series of Road Plans from the middle of the 1920s, the resulting 'automobile system' was not the end point of a linear sequence of events, or of a simple passing of road authority one level up. Rather, the eventual configuration could not have been predicted, because car diffusion happened largely behind the backs of the planners and policy makers. In the 'struggle of the systems' they were focused upon the crisis of the railways and waterways, while the car, as a 'pleasure vehicle' with not much economic value in the serious business of freight and passenger transport, was largely neglected, a fateful fact in view of the post-WWII urban mobility crises all over the world.

In trying to *explain* the eventual shape of this converging structure, we have emphasized three crucial factors. The first and foremost factor is no doubt the central role of the experts who, through their international associations and through the dozens of periodicals and publications, knew a great deal about their foreign colleagues' practices and thus could form a powerful element in a general movement toward technocratic management of large national projects. Second, this technocracy movement was carried by civil engineers who could develop their planning at a safe distance of democratic control, often in a semi-autonomous executive body. And third, in doing so they could reshape planning into a sectoral issue, effectively bypassing urban and rural planners who often did not agree with their approach.

When looking at early twentieth-century road motorization, the analysis in the previous chapters has tried to make it plausible for one to speak of an Atlantic mobility culture as part of Western modernization of society. Such an analysis from a transnational perspective reveals common patterns among 'mobility communities' of car, motorcycle, and bus users and emphasizes continuity rather than revolutionary change, which is all the more true if one takes the behavior of users as the point of departure rather than their vehicles. All relevant elements of Western modernity (industrialization, individualization, mass consumption, a fascination for speed and efficiency, a belief in 'progress' and a reshaping of the past, centralization and the nation state, to name only a few), including its contradictions, were not only present in motorized mobility, but this motorized mobility on its turn decisively shaped modernity itself. In most countries this was largely a liberal, republican project, although the many ambivalences of automobilism may have caused a remarkable modesty in (neo)liberalism to claim the 'success of the car' in Western societies, as if it was and still is embarrassed about its 'dark sides.'[14] In any case, Atlantic mobile modernity clearly sets itself apart from, say, the Latin-American modernization process, where a strong centralized state was much less apparent, or from Portugal, where the lack of a middle class led to a different trajectory of motorization.

This aggregated, top-down view, however, does not account for the many facets and differences among the multiple modernities on a subnational level. To take only one example: what for bicycle and car users took the shape of individual freedom of mobility, was seen by those in (central) power and their fellows such as planners and traffic engineers as fragmentation, as anti-systemic behavior very difficult to control and discipline. The atomized swarm of car users may have resonated with the general individualizing trend in consumption society (the same forces that also governed the fragmentation of the family as discussed in chapter 4), but the planning paradigm of Western societies assumed at the same time that these swarms had to be disciplined and lured (for instance through the construction of freeways) into much more easily controllable flows. The same happened to the cycling and pedestrian swarms in the cities, which were forced into flows on sidewalks and bicycle paths, on zebra crossings, waiting before traffic lights, following stripings on the road and barriers to prevent them from leaving their 'ordered' territory.[15]

Against this background, at the present stage of research, European early twentieth-century mobility can perhaps best be characterized by putting it in opposition to the American case, rather than reinforcing an American exceptionalism, because there is not any. This is not to say that there are no large differences, but the ubiquity of the automobile in Western society seems to grant every country its own specificities: America its wealthy farmers, Germany its violent car practices and its motorcycles in the 1930s, the Netherlands its bicycles and buses and its high level of urbanization, Switzerland its decentralized auto-hostility, France its opposite in the radiance of Paris. Before we know it we enter the shadowy realm of national stereotypes, as we have seen in chapter 5.

Despite these differences all countries investigated in this study showed similar trajectories: the construction of a garage repair and maintenance infrastructure with preventive 'service' to counter the still rampant unreliability of the car's technology, the planning and building of high-speed national and regional roads, the taming of the truck and bus to protect the railway interests and those of the large enterprises, and the concerns about the exploding traffic fatalities. The latter revealed that the petty-bourgeois family car project was built upon a shaky foundation, still largely characterized by (covert) violence and (hidden) aggression, just like the expansion of the system toward the 'periphery' had clearly violent and aggressive traits.

From the moment that the car became one of the elements of the 'automobile system' it should be studied as a nonseparable part of this system. From then on it does not make much sense any more to investigate the vehicle's 'impact' on society. That would, to borrow a phrase from Leo Marx, be the same as "to speak of the impact of the bone structure on the human body."[16] Meanwhile the construction of the car as an individualistic vehicle continued. An early example is the post-WWI General Strike in the United Kingdom, when trucks drove around London with banners "By permission of the T.U.C." It enraged an English citizen, owner of a car, who started to drive around with the banner: "By permission of my own bl—dy self."[17]

Crossing Boundaries: Adventure, Fiction, and the Explanation of the Car's Persistence

Although the embedding of the car in its system may be a powerful argument in the controversy between the psychological and the systemic explanations as introduced in the introduction, it does not mean that we can stop here when it comes to explaining the persistence of the car. Showing how the car became part of a system boils down to a mere tautology if we do not account for the reasons motorists were willing to become a part of this system, and to answer questions about how they were willing to do so, defined as what type of subjects, doing what kind of practices. For this we need to understand the motives of would-be car buyers, users, and nonusers, for the simple reason that the automobile system could not have been enforced on a reluctant society without at least the partial consent of its citizens. To investigate this further we took more than five hundred novels, poems, paintings, songs, and films into consideration, highbrow and lowbrow, popular and elite, to dig deeper into the automotive adventure that so much pervaded automotive desires. What, then, did this body of unconventional sources bring, more than the conventional sources could? To answer these questions we distinguished between three levels of analysis: content, symbols, and affinities.

As far as their content is concerned, the utterances from the first phase first of all revealed a strong interest in the past, as if the authors were afraid to lose it in the face of modernization. Second, the car appeared to be a powerful vehicle of fantasy, based upon a bodily, multisensorial experience of movement. This not only generated special visual effects (such as inversion, the apparent movement of the environment around the car as a shell for narcissistic travelers), but also made the sensation of 'flight' into a central, crucial element of automotive driving practice. Expressed in dozens of artistic and bread-and-butter texts, the description of the floating experience in a situation that the car cannot have been free of shocks and vibrations at all, indicates that historiography's emphasis on the visual aspects (and the relationship with early film) should be relativized and rethought: the visual metaphor seems to have functioned as a smokescreen behind which a multisensorial practice was hidden based upon a haptic relationship with the car and its road infrastructure. It was this angst-lust of the speedy trip, the mixture of fear and rush that led to a multisensorial transcendent experience, as an escape from earth in all possible meanings of the term, literal and figurative. Third, the utterances revealed new, aggressive elements of the automotive adventure, erotic and sexual on the one hand (Proust's rape of villages), colonial on the other (the conquest of the southern and exotic 'periphery'). Popular culture (especially ultra-short silent movies) exposed this violent trait in an extraordinary form, for nonusers too: while the cruelty of limbs flying around may have been buffered in soothing slapstick, it nonetheless exposed some basic motivational drives of motorists and onlookers alike.

During the second phase open violence and aggression went underground, while popular culture continued to expose it (in the form of the car chase, or the race ending in a violent accident) as one of the quintessential characteristics of automotive practice. The traffic accident now became the central, narrative-interrupting deus ex machina, its lethal threat always in the background, jeopardizing the carefully undertaken process of domestication of the automotive adventure. This domestication was especially actively pursued in middlebrow culture, where a 'balance' between the breakneck speed of the elite (Arlen's *Green Hat*), the explicit cruelty of the workers and the marginals (Caldwell's *Tobacco Road*), and the family-denying nomadism of the hobos (*Boxcar Bertha*) was sought. In other words, nineteenth-century values had to be "safeguarded ... against modernism," a process fully parallel to the domestication of the car itself. Thus, domestication boiled down to taking a nineteenth-century technology, proposed by the aristocracy, into the home of the nuclear family.[18]

The domestication of the car adventure was accompanied by a parallel process of encapsulation on the technical side, the car as shell evolving into the car as capsule, enabled by the closing of the car body as part of a general development toward the affordable family car, which in the process became a contraption for the combination of 'pleasurable' high-speed touring and 'comfortable' city use. Domestication must also be understood literally: just like "the refrigerator brought the market into the kitchen" (and later, "the television brought the movie theater into the living room"), "the automobile brought the trolley into the garage."[19] The car, as commodity, literally transformed public into private.[20] It could do so because its "ontology" was similar to that of consumption: satisfying ever fleeting desires, combining modernity and nostalgia, present and past, with the consumer as king. The Marxian (Karl) circulation of the commodity had its parallel Marxian (Leo) circulation on wheels, in the garden.[21]

Although some interwar utterances dealt with the nuclear family as the new automotive subject, most travelogues and autopoetic novels denied its existence and kept treating the car as a vehicle for the individual, if not individualism. This is not to deny that several interwar tendencies in artistic utterances ran counter to this dominant trend. The 'traffic novel' (Helmuth Lethen), for instance, introduced the systemic aspect of the automotive experience by entering, for the first time, into the city and its flows. Motorcycle culture, especially in Germany, tried to rescue the violent, narcissistic aspects of the automotive experience, but there too, the background was collective (the club, the gang) rather than individualized. The extended family of deracinated farmers (Steinbeck's *The Grapes of Wrath*) personified a kind of social or socialist future, whereas in the new Soviet Union the car was instrumentalized as a revolutionary vehicle (Shklovsky). Nonetheless, the fundamental experience was highly individual, and consisted of the cyborg feeling, the magic unity with the technology. Interwar artistic utterances are especially prolific in exposing this magical new feeling of transcendence, of being a part of a high-speed swarm, adorned with power and characterized by an increased narcissism, where the 'trip' (the celebration of movement on itself, whether caused by the car, or by drugs) mattered, rather than 'nature.' This new feeling morphed, in some pockets of highbrow autopoetics, into a longing for slowness, which we analyzed as being related to the feeling of monotony while driving at high cruising speed on the new, smooth highways and freeways. Thus, mainstream adventure was reinvented along with the reinvention of the car's technical structure, represented by writers such as Sinclair Lewis (United States), Henry Morton (United Kingdom), Paul Morand (France), and especially Heinrich Hauser (Germany).

In this form the car also became a vehicle for personal emancipation of women, black, and other ethnic groups. Although the road experiences of black and other ethnic groups still wait to be re- and deconstructed, we have spent quite some space in trying to give the car experience by women, and the gendering of these experiences, its role in the emergence of Atlantic automotive adventure. We have seen on several occasions that the female car adventure was often no less aggressive and violent as men's, for instance in 'colonial' travelogues. But not only the It-Girls in the German novels by Keun and Fleisser were excluded from a still very masculine mainstream practice; Hans Falada's Johannes Pinneberg also became 'streetwise' through the use of the taxi, the bus, and most of all through daydreaming. By the arrival of the 1930s one did not need to own a car to feel part of an automotive society in the making. The production of mobility was not only realized through the consumption of the car.

And yet, the proliferation of automotive adventures should not divert our attention from the fundamental fact that the persistence phase of Atlantic automobilism was dominated by the white middle-class male and his family, perhaps best represented by German travel-writer Heinrich Hauser, whose remarkable amalgam of modernity and nostalgia, fear for women and desire for community, resulted in the (literal and figurative) construction of a 'mobile home,' as analyzed in chapter 6.

Perhaps because of its close connection to modern consumption theories, mobility studies failed to interpret automotive experience as transcendental.[22] In this respect we are not very much helped by belletristic literature either, as its emphasis on the Quest of the Self and its preference for the individualistic aspects of the car experience is rather misleading. We have found, on the contrary, by combining highbrow, middlebrow, and lowbrow sources with our analysis of the more traditional sources, that the protection the car provided helped to reconcile self and system by enabling, through self-transcendence, the experience of the 'magic' and 'myths' of the car driving experience. The kick of transgressing 'normalcy,' the domesticated play with modest adventure, appeased and reconciled the car driver with the system, helped to forget that the stability of the system consisted (and consists) in its routines and its adherence to the rules. Borrowing from Zygmunt Bauman, a recent study of the risk society formulates it like this: "Lack of security and uncertainty [functions as] disciplining measures of post-modern societies. . . . Risk is danger made calculable through partitioning, and the culture of risk is based upon the conviction that calculability represents a sufficient form of controllability."[23] Just like motorists from an early date started to delegate skills to the machine through the cyborg effect, motorists of the second phase transcended their experiences toward the collective—the swarm, the system—in an effort to rescue adventure in a controlled society, to reconcile anarchism and order. The price we pay for this pervasive convergence of individual and societal interests is the annual death toll, and a couple of other nuisances at that. Seen from another angle, adventure, in its new petty-bourgeois shape, is not a rejection of certainty, as a German conference announcement on 'Adventures' claimed in 2012; on the contrary, adventure is a test of safety, it is an effort to regain certainty through transgressing its boundaries.[24] The transcendence, needed to enact that practice, represents a "redemption from the chaotic and the paradoxical," or in Nietzsche's terms, "the restless heart seeks rest."[25] The "kick of border crossing" (border used in all possible nuances of meanings, from physical frontier and the boundaries of the body to the limits of one's individuality) is a violent act afforded by the car.[26]

This is one (the 'soft') side of the explanation of the car's persistence. The 'hard' side becomes visible when we realize how the many shifts from the first to the second phase illustrate the incredible versatility of the car as technology, perhaps only paralleled by the personal computer and, lately, the cell phone. Calling the car postmodern[27] is denying its tripartite character of being premodern (restoring the foreground, visiting the past), modern (individualized adventure), *and* postmodern (the malleable body, the visit to nonplaces such as freeways and airports) at the same time. The car, indeed, is also metaphorically an empty cocoon to be filled with all kinds of practices called mobilities, just like capitalism itself "can accommodate itself to all civilisations." Because of this structural homology between (capitalist?) society and the car, system and artifact, the car could play a catalytic and at the same time soothing role in the transfer to modernity and postmodernity.[28]

This versatility is also illustrated by the increasing variation into a multitude of adventures, rather than the single tripartite adventure of the first phase. Quite different from other commodities such as the refrigerator, or the vacuum cleaner, the car appeared to enable a very wide array of adventurous practices. When that threatened to be constrained by the technology, its properties were adjusted to enable the new functions, as we have seen in the development of the closed car (chapter 5). The car as commodity enabled the desire, as soon as it was felt, to be satisfied, but only momentarily, as the desire returned soon. It was like smoking a cigarette, and as such it was the very core of the practice of consumption. It was a minority project generating tremendous promises and expectations for a majority. At the same time the road system around it was adjusted in this phase as well, ushering into a 'car system' as a prefiguration of the post-WWII 'car society': by the outbreak of the war the entire system was ready to be exported, so to speak, beyond the Atlantic realm, while the organization of society itself started to get adjusted to this system as well, in the form of suburbs without public transport, and a network of railway-like super-roads that directly attacked a society based on heavy industry, railways, and a strong, if not authoritarian central government.

This was expressed at the second level of our analysis, the symbolic level. The car and its components became a reservoir for societal symbolism, as an icon of a particular kind of domesticated, automotive culture. Around it, a petit-bourgeois, lower-middle-class culture coagulated, celebrating the nuclear family, experimenting with new values of civility (especially through the elimination of extremes, such as the 'roadhog') and creating a narcissistic, individualist fantasy that increasingly, as the traffic became denser, evolved into an illusion. The car stood for "a new form of subjectivity: privileged, individualistic and phenomenological."[29] It was the culture of Lewis's Babbitts, and the expectation, the hope of Fallada's Pinnebergs, who meanwhile became streetwise in the motor bus and the taxi. The ubiquity of the car was already celebrated at a moment that only (a part of) the middle class had actually motorized: a better proof for the emergence of lower-middle-class society is hardly possible.

It was at the third level, of the affinities between writing and driving, however, that our analysis appeared to be the most fruitful, and generated results that make this book different from earlier ones. During the first phase driving and writing were so closely intertwined that we can confidently claim that it was the literary avant-garde that shaped automotive culture decisively, in its masculine lust for aggression and violence, its angst-lust for high speed and the threatening crash, and the open celebration of danger and near-death. The driver was a poet from the very start of the 'movement.' This is not to say that there were no alternatives: some tried to overcome their ambivalent modernism (a combination of indulging in admiration for the technology and a desire to restore older, nineteenth-century experiences) by the sublime, the myth, the miraculous act, as secularized religion. For the Futurists, this dichotomy was unbearable, and they fled in ecstasy, to play with death, to experience the thrill of danger, setting an example for all avant-gardes to come.

During the second phase the affinities became more subtle, mostly through the distancing effect of the automotive capsule, which, at the behavioral side, enabled and necessitated a cool, cold and harsh attitude (Lethen). Our analysis was prolific here, not so much because literati were better observers, but because post-expressionist art went through a similar process of *sachlichkeit*. A filmic, impressionist style (focusing upon the surface rather than the inner structure) permeated both writing and driving.

The capsule not only protected the occupants from the hostile Other (the people, the environment, the colonial periphery), it also provided the basis for the new form of automotive aggression that made the car into a weapon in the 'war on the road' (Remarque). Whereas literary avant-gardes used irony to distance themselves from 'the mob,' motoring avant-garde used the closed car itself as a distancing device to the Other.[30] This 'ironic car' became a home away from home for the narcissistic driver, who nurtured the illusion that he was alone and that the world moved around him, but who knew, in the back of his head, that he was part of a flow, a swarm, a multitude ready to be controlled. If we are wondering how and why swarms of car drivers agreed to be controlled so easily, the secret of it lies in the relationship between 'rush' and 'control,' as revealed in a recent study into the tension between 'rush' and 'dictatorship.' "Ecstatic and excessive transcendence of the daily routine" seemed to contradict with "claims of all-permeating control," but "nonetheless dictatorships of authoritarian character allowed themselves forms of rush, albeit in strictly institutionalized forms." Similarly, interwar motorizing societies afforded their citizens "times and places of limited delimitation," spaces of "controlled loss of control" such as the speed bacchanals of the freeways. Rhythmic flow, according to cultural historian Joel Dinerstein, is the "unifying aesthetic principle of Machine Age modernism."[31]

In general, subjectivism and control have consumption as intermediary, and hence the car.[32] The car, since then, represents "the pendulum ... between freedom and safety" that constitutes, in the words of Zygmunt Bauman, history.[33] That is also why forms of mobility deviating from the sedentarist model were experienced as threatening: "nomadic violence is ... opposed to state apparatus violence: the tribe is the counterarmy, that is to say the space where the warriors rule. Is this why nomads have always been persecuted as dangerous criminals by the state?" The nomad is subversive not so much because she travels: "it is the subversion of set conventions that defines the nomadic state."[34]

It was this transcendental experience of the car-driver cyborg, this 'hyper-experience' (chapter 6), that in our analysis forms the basic explanation for the persistence of Atlantic automobilism: we have all become super-humans, we all can fly, albeit briefly, during the commute. For the first phase flight (and in a broader context, adventure) formed the central characteristic for emergent automobilism. For the second phase of automobilism's persistence, the car swarm blew up nineteenth-century society, including its mobility network, just like modern literature blew up its own language (William Carlos Williams, the surrealists). If the car appeared able to help transcend the motorist into the technological sublime, it did so by sublimating (in a Freudian sense) violence and aggression into the very act of driving. On several occasions (in chapters 3 and 6) we have pointed at a third form of sublime (next to the natural and the technological): the 'sublime of violence.' It is this feeling, the violent drive, that triggers the transcendental experience and, for instance, made Eddie Rickenbacker into a semi-God (see chapter 7). The car, in Leo Marx's and David Nye's terms, was a special form of technological sublime, while it at the same time transported the users toward the sublime in a double sense: into a sublime state as car driver and into the sublime of technology in nature. "Each of us is born with tendencies towards violence and sexual gratification," a recent biography of the writer J. G. Ballard claims, referring to Freud, "Society seeks to limit these, since they threaten order. In doing so, it stifles the 'pleasure principle,' our most potent motivation." The car, in other words, represented the ironic mobile sublime, enabling transcendence, but tongue-in-cheek, in a distanced way, just like the car itself distanced the automobile subjects from the Other.[35]

In both phases analyzed here it was the car's transcendent, sublimating affordance, the experience the car enabled that one was more than just a Babbitt, that constituted the very core of the Western automotive experience. In other words, car system's persistence can be explained by a rationalization of a deeply corporeal interdependence of driver and vehicle as adventure machine: the cyborg condition is the core, but no less important is the hiding of this core under a discourse of rationality and societal 'necessity.'[36] The open culture of seriousness and 'transport' rests and feeds upon a covert culture (Leo Marx) of tamed hedonism and fun. In this context the accident and the defect acquired a special meaning: as a prefiguration of death, they made violence 'structural.' Since then, any policy to change this culture will have to be aware of this hidden dimension of Atlantic automobilism.

Some Closing Remarks on Methodology and Future Research

Leo Marx, Wolfgang Schivelbusch, Marc Desportes, and a touch of Urry: their work helped us develop the methodology to analyze the utterances produced by and on their turn shaping the high- and middlebrow cultures of the first fifty years of Atlantic automobilism. Likewise, Theodore Schatzki and several others representing 'social practice theory' helped us focus on the 'doings and sayings' (the sayings being doings as well) of several generations of motorists in several countries in Europe and North America, who carved out for themselves a special niche, quickly developing into a flow of swarms, who produced, in the words of Gareth Stedman Jones, "authorless meanings" uncontrolled (and later very difficult to control) by the authorities.[37] These meanings, as developed in the previous chapters and wrapped up in this chapter, form "at least some of the total of sufficient conditions" necessary for an explanation.[38] They are mostly derived from highbrow and some middlebrow literary sources, because literature does more than what many other utterances of popular culture are able to: while movies, for instance, stress the *technics* of viewing, literature is forced to *translate* between senses and *verbalize* these experiences, just as the historian is forced to do. We read what has been felt, seen, and heard.

For this explanation, and based on our empirics, we needed to go beyond the wisdom of the scholars mentioned: we realized that our machine in the garden was not a train, but a car, in a garden where not so much a touristic gaze as a glance was produced, upon a multisensorial, although largely hidden, substrate. Nonetheless, there is a remarkable parallel between some psychological theories of perceptions and literary-inspired studies of travel such as Schivelbusch's. James Gibson, for instance, the founder of "ecological psychology," started his research by studying "the visual techniques of the cinematographers." As a young boy he accompanied his father and learned "what the world looked like from a railroad train and how it seemed to flow inward when seen from the rear of the platform and expand outward when seen from the locomotive," thus inventing the psychological, bodily base for what we have coined the 'inversion' effect (see chapter 2).[39] Later, when the car became ubiquitous, he would develop a "field theory" of car driving, where he coined concepts such as destination, collision, and path. His reasoning appeared remarkably familiar with what we have analyzed in literary utterances, backed by the theories of Leo Marx and Desportes: having bought his first car (a Model T, in 1929) at twenty-five, he observed how the car driver not only "sees the 'flowing' and 'streaming' of optical texture," as the train traveler does, "but controls it as well." "Locomotion," as he called it (in a typical reflex, seeing the car as mobility's alpha and omega), "is a special kind of space perception [as it] imposes a gradient of velocities over the static gradient of textures."[40]

Remarkably, Gibson's contribution to psychology is, among others, that the eye is *not* like a camera (the camera metaphor from cognitive psychology, and several mobility historians in their wake, is related to the panorama, and the train), but part of an active sensorial system to act in the environment.[41] During the Second World War, fully funded by the American army, Gibson conducted experiments of the perception by airplane pilots and developed his theories about the 'eye' that have remarkable parallels with what we have seen to be typical of the travelogue's and the autopoetic novel's 'I': "perception of the environment is ... accompanied by proprioception of the active self." This perception of the position and movements of one's own body is a skill that can be learned.[42]

It is this opposition between a 'panoramic' and a 'locomotive' view that allows the psychologist to join the trio, which is further constituted by the driver and the novelist, through a "theory of observational activity." In order to be able to develop this theory Gibson had to "transcend the dualism of mind and body that has," he found, "so disrupted the growth of all forms of psychology." As explained in the introduction, it was the 'affordance' concept that enabled this transcendence, just like the historical car itself afforded the motorist to self-transcend toward the controlled swarm. Here is the affinity: in a metaphorical sense the motorist did what Gibson saw as fundamental to knowing the world: "obtaining stimulation" rather than having it "imposed." Gibson wanted "to explain such basic facts of daily life as why 'ice looks cold; fur looks soft; a pair of pliers look graspable; bread looks edible; a friend looks friendly.'" A car, from this perspective, not only looks driveable, but also lethal, and protective, dependent on whether one's eye is that of a driver, a pedestrian, or a passenger. The study of psychology, Gibson concluded in his ultimate monograph explaining his theory, "must begin with ecology, not with physics and physiology." Perception, then, encompass "exploratory activities [that] are intrinsically motivating."[43]

Gibson's insights, backed by philosophers such as Schatzki and others (see the introduction), have consequences for the interpretation of our (literary) sources, as they necessitate, perhaps, the formulation of a new theory of 'writing in movement' in all possible meanings of this phrase. The act of writing has been called a form of Gibsonian 'activity,' but the foundation of this writing activity is in itself a practice of perception, an active exploration of the world, a performance in and of itself. In other words, if we analyze the perception described in novels, we acquire knowledge about the protagonists' motivations. This is not to say, of course, that these motivations are 'correct,' as we have seen in the case of the Quest of the Self versus being a part of the swarm on the road. Literature, from this perspective, is useful as an equivalent of the tests in Gibson's theory development: it reveals mechanisms not to be extracted from traditional sources such as the 'flight' experience, or 'covert aggression.'

And yet, in several respects the analysis in this book needs further support. One actor, for instance, that is painfully missing from this narrative is the passenger, whose social practice, "passengering," not only deviates from that of the driver, but also would more easily enable a connection with other mobility forms in which the passenger (as an actor being driven) dominates, such as the taxi, the bus, and the train.[44] It would also include the experience of women and ethnic groups more easily into the road-mobility equation, at least for the first two phases.

Perhaps the embarrassment of the exclusive character of automobilism (by some historians reinterpreted as inclusive, because automobilism always expanded to new groups of users[45]), as well as the covert aspects of its lethal character, is the reason for "contemporary liberal theory's particular silence on automobility," where one would expect a continuous cry of triumph once automobilism became ubiquitous in the liberal West. For Sudhir Chella Rajan, "it is a potential source of embarrassment to liberals that the costs, particularly the risks of causing death to others through accidents and pollution, may well undermine the Kantian principle of treating humans as ends in themselves," a conclusion that tends to jeopardize the iconic character of the car as a product and producer of freedom and equality.[46] Such embarrassment may also haunt mobility studies, when it comes to giving the car its proper due.[47]

Another source of embarrassment may reside in the fact that, up to now, automobilism's story has been nearly exclusively told as a Western story, reduced to a limited number of well-analyzed North-Atlantic countries. Indeed, the history presented here could have been enriched by pointing at the late motorization in Portugal, and no doubt several other countries in the southern European 'periphery.' We could have pondered on the absence of freeways in Norway until far into the post-WWII years to support the thesis, presented in chapter 7, of the low infrastructural need of the car (literally, but also as explanation for its success). Or I could have referred to Australia's ambiguous relationship to the 'landscape' to emphasize the relative scarcity of the European and American 'models,' including camels or a frontier that never seems to get closed.[48] But in the latter case we enter complexes of mobilities that, indeed, start to differ so much from the cases dealt with in the previous chapters that a novel approach seems called for. If American automobility historian James Flink claims that "the appeals of the car were universal, not culturally determined," he is only right (meaning, he is wrong) if we limit universality to the West. Selecting only the obvious candidates out of the plethora of "multiple mobilities" reinforces the dichotomy between the West and the Rest.[49]

After 1920 it was the system rather than the artifact that had to be transferred. This appeared to be a difficult undertaking, as is testified by several cases of outright rejection, for instance when the railway appeared in China.[50] Not only that, different cultural practices (related to extreme poverty causing large forms of internal migration) and different vehicles (such as the collectively owned bus, or the rickshaw, or the one-wheeled barrow for passenger transportation) necessitate a different analytical framework in which, for instance, the automotive adventure may well play a much less pronounced role as in the West. A merger of the two narratives, leading to a reformulation of the Western story on the basis of the insights taken from the Eastern and Southern stories, may form the basis of what I elsewhere called New Mobility Studies, in which also, as explained in the introduction, communication has regained its proper place.[51] Such a new paradigm should address the two "central deficits," according to Valeska Huber, of current mobility studies: their uncritical acceptance of the boundaries of each and every modality, and their refusal to connect them in an explicit way. Others accuse transport historians of "giv(ing) credit to the automobile for elements of modernity that were actually pioneered by the railroad industry," an accusation that holds true, indeed, in those histories and analyses of the car obsessed by the concept of speed.[52] It is to be expected that a theory of affinity between writing (and in general, cultural production) and being mobile (the production of experiences of mobility) may well provide, also in these cases, a privileged entry into the motifs and motives of the historical actors, thus providing a (more) solid basis for the explanation of global mobility. Approaching the car as a medium instead of as a transportation device may better expose the manners motorists could experience *Gemeinschaft* (the 'integrating power of mobility,' in Tim Cresswell's words) while living out their individuality in the *Zerstreuung* (Kracauer) of their automotive dance.[53]

And yet, it seems that not all actors are well served by such a 'mediated mobility theory': the subversive (against the grain) and the subaltern (without its proper voice) forms of mobility need their own methodologies. The 'periphery,' whether in the European or American South, formed the scene of a hybrid culture, where the fledgling car system collided with different mobility cultures, such as the jitneys in the southern American states exploited by blacks as a semi-collective form of transport. Likewise, eastern Europeans also have a car culture clearly deviating from the northwestern European and North American 'norms.'[54] Such peripheral mobility cultures may well form a third 'model' next to the western European and the North-American ones, as a transitional model between the North-Atlantic and a fourth (Australian) and even a fifth model, the latter encompassing many of the developing countries. Despite the rickshaws in Asian towns and the hammocks carried by indigenes in the African colonies, however, we should not underestimate the power of the Western 'model': when Clärenore Stinnes went on her world trip by car in 1927–1929 she found Beirut "bustling with motorized traffic" ("are you still in Asia or are you back in Europe," she asked herself), while Teheran was a "disappointingly modern city" with traffic police and Paris kitsch in its shops.[55] Such early indications of a much messier spread of Western elements, coupled with the reluctance of 'peripheries' to simply follow the first model's motorization, challenge the dichotomy between a motorized model (or set of models) and the motorizing 'rest.'[56] They deserve to be analyzed in their own right instead of as diversions from or aberrations of the hegemonic mainstream, as a possible future that did not (yet) materialize.

However this may be, one thing is certain: when the British philosopher John Gray depicted, in a recent interview, Christian afterlife as a place where "souls err around eternally, not haunted by jealousy, aggression and all other nasty traits that died with our earthly body," he forgot to mention that in such a place the car would never have been invented.[57] No doubt, heaven is a boring place. Meanwhile, in the early 1980s, when 'mass motorization' was in the full process of spreading over the entire globe, French psychologist Jean Piaget was interviewing children about their "definition of life." Referring to their computerized toys they saw 'life' as something psychological, as having 'emotions'—in sharp contrast, Piaget observed, with children during the interwar years who saw life as the ability to "moving of one's own accord."[58] Perhaps some of Piaget's children have developed another type of adventure while playing with their toys, and don't need the car for that purpose anymore.

Notes

1. John Davidson, "The testament of Sir Simon Simplex concerning automobilism," in Davidson, *Fleet Street and Other Poems* (London: Grant Richards, 1909), 100–110, here 107.
2. Karolina Dorothea Fell, *Kalkuliertes Abenteuer: Reiseberichte deutschsprachiger Frauen (1920–1945)* (Stuttgart and Weimar: J. B. Metzler, 1998), 39; Gijs Mom and Ruud Filarski, *Van transport naar mobiliteit: De mobiliteitsexplosie (1895–2005)* (Zutphen: Walburg Pers, 2008), 242.
3. *Americans on the Highway: A Survey of Recent Trends of Tourist Travel*, 1941 edition (Washington, DC: American Automobile Association, 1941), 3.
4. Donald Weber, "Automobilisering en de overheid in België vóór 1940: Besluitvormingsprocessen bij de ontwikkeling van een conflictbeheersingssysteem" (dissertation, University of Ghent, 2008), 199.

5. Wiebe Bijker, book review of Harry Collins, *Tacit and Explicit Knowledge* (Chicago, 1910), in *Technology and Culture* 52, no. 4 (October 2011), 809–810, here 809.
6. As a matter of fact, military historian Martin van Creveld shows how the speed of the armies in WWII did not differ very much from that during WWI or even the Franco-Prussian war of 1870, because of the "footslogging infantry." Martin van Creveld, *Supplying War: Logistics from Wallenstein to Patton* (Cambridge and New York: Cambridge University Press, 1986), 147.
7. Catherine Simpson, "Volatile Vehicles: When Women Take the Wheel; Domestic Journeying & Vehicular Moments in Contemporary Australian Cinema," in Lisa French (ed.), *Womenvision: Women and the Moving Image in Australia* (Melbourne: Damned Publishing, 2003), 197–210, here 200.
8. Geoff Eley, "Is all the world a text? From social history to the history of society two decades later," in Gabrielle M. Spiegel (ed.), *Practicing History: New Directions in Historical Writing after the Linguistic Turn* (New York and London: Routledge, 2005), 35–61, here 51; Virginia Scharff, *Taking the Wheel: Women and the Coming of the Motor Age* (Albuquerque, NM: University of New Mexico Press, 1991).
9. Wolfgang Munzinger, *Das Automobil als heimliche Romanfigur: Das Bild des Autos und der Technik in der nordamerikanischen Literatur von der Jahrhundertwende bis nach dem 2. Weltkrieg* (Hamburg: LIT Verlag, 1997), 117.
10. Quoted in Sudhir Chella Rajan, "Automobility and the liberal disposition," *The Sociological Review* 54, no. s1 (October 2006), 113–129, here 122.
11. Victoria de Grazia, *The Culture of Consent: Mass Organization of Leisure in Fascist Italy* (Cambridge and New York: Cambridge University Press, 1981), 1.
12. Charles S. Maier, *Recasting Bourgeois Europe: Stabilization in France, Germany and Italy in the Decades after World War I* (Princeton, NJ: Princeton University Press, 1988), 3.
13. Robert C. Post, *Urban Mass Transit: The Life Story of a Technology* (Westport, CT, and London: Greenwood Press, 2007), 150.
14. Rajan, "Automobility and the liberal disposition."
15. Barbara Schmucki, *Der Traum vom Verkehrsfluss: Städtische Verkehrsplanung seit 1945 im deutsch-deutschen Vergleich* (Frankfurt and New York: Campus Verlag, 2001).
16. Leo Marx, "*Technology*: The Emergence of a Hazardous Concept," *Social Research* 64, no. 3 (Fall 1997), 965–988, here 981.
17. Rosa Maria Bracco, *Merchants of Hope; British Middlebrow Writers and the First World War, 1919 - 1939* (Providence and Oxford: Berg, 1993), 53.
18. *Merchants of Hope*, 12.
19. John Nerone, "The Public and the Party Period," in Jeremy Packer and Craig Robertson (eds.), *Thinking with James Carey: Essays on Communications, Transportation, History* (New York: Peter Lang, 2006), 157–176, here 157.
20. Fabian Kröger, "Automobile DNS in der Kontrollgesellschaft," in Gerburg Treusch-Dieter, Claudia Gehrke, and Ronald Düker (eds.), *Auto: Konkursbuch* (Tübingen: Konkursbuch Verlag Claudia Gehrke, 2004), 161–170, here 161.
21. Valeska Huber, "Multiple Mobilities: Über den Umgang mit verschiedenen Mobilitätsformen um 1900," *Geschichte und Gesellschaft* 36 (2010), 317–341, here 324.
22. Stefan Schwarzkopf, "The Political Theology of Consumer Sovereignty: Towards an Ontology of Consumer Society," *Theory, Culture & Society* 28, no. 3 (2011), 106–129.

23. Herfried Münkler, "Strategien der Sicherung: Welten der Sicherheit und Kulturen des Risikos; Theoretische Perspektiven," in Herfried Münkler, Matthias Bohlender, and Sabine Meurer (eds.), *Sicherheit und Risiko: Über den Umgang mit Gefahr im 21. Jahrhundert* (Bielefeld: transcript Verlag, 2010), 11–34, here 25, 27.
24. Announceent of conference "Abenteuer: Paradoxien zwischen Sicherheit und Ausbruch—München 09/12" on H-SOZ-U-KULT@H-NET.MSU.EDU (last accessed 26 July 2012).
25. Sybe Schaap, *Het rancuneuze gif: De opmars van het onbehagen* (Budel: Damon, 2012; first ed. 2012), 16.
26. Peter Giesen, "Vroeger knokken op de kermis, nu losgaan in Haren," *De Volkskrant* (24 September 2012), 1.
27. Marita Sturken and Douglas Thomas, "Introduction: Technological Visions and the Rhetoric of the New," in Marita Sturken, Douglas Thomas, and Sandra J. Ball-Rokeach (eds.), *Technological Visions: The Hopes and Fears that Shape New Technologies* (Philadelphia, PA: Temple University Press, 2004), 1–18, here 9–10; for an exposure of the tension between the modern and the post-modern aspects of the car, see Marita Sturken, "Mobilities of Time and Space; Technologies of the Modern and the Postmodern," in Sturken, Thomas, and Ball-Rokeach, *Technological Visions*, 71–91, here 72–73.
28. Slavoj Žižek, *Violence: Six Sideways Reflections* (New York: Picador, 2008), 79 (capitalism), 82 (modernism).
29. Lynda Nead, "Paintings, Films and Fast Cars: A Case Study of Hubert von Herkomer," *Art History* 25, no. 2 (April 2002), 240–255, here 250.
30. Bracco, *Merchants of Hope*, 6.
31. Árpád von Klimó and Malte Rolf, "Rausch und Diktatur: Emotionen, Erfahrungen und Inszenierungen totalitärer Herrschaft," in von Klimó and Rolf (eds.), *Rausch und Diktatur: Inszenierung, Mobilisierung und Kontrolle in totalitären Systemen* (Frankfurt and New York: Campus Verlag, 2006), 11–43, here 12; Joel Dinerstein, *Swinging the Machine: Modernity, Technology, and African American Culture between the World Wars* (Amherst and Boston: University of Massachusetts Press, 2003), 8.
32. Stefan Schwarzkopf, "Kontrolle statt Rausch? Marktforschung, Produktwerbung und Verbraucherlenkung im Nationalsozialismus zwischen Phantasien von Masse, Angst und Macht," in von Klimó and Rolf, *Rausch und Diktatur*, 193–216, here 196.
33. Peter Giesen, "'Wat er ook gebeurt, het ligt aan jezelf,'" *De Volkskrant* (4 February 2012), "Het Vervolg," 6–7 (interview with Zygmunt Bauman).
34. Rosi Braidotti, *Nomadic Subjects: Embodiment and Sexual Difference in Contemporary Feminist Theory* (New York: Columbia University Press, 1994), 27 (state), 5 (conventions).
35. On the relationship between sublimation, civilization, and violence, see Étienne Balibar, *Der Schauplatz des Anderen: Formen der Gewalt und Grenzen der Zivilität (Aus dem Französischen von Thomas Laugstien)* (Hamburg: Hamburger Edition HIS Verlagsgesellschaft, 2006; transl. of *La crainte des masses*, 1997), 259; John Baxter, *The Inner Man: The Life of J.G. Ballard* (London: Weidenfeld & Nicolson, 2011), 34.
36. I thank Kurt Möser, during a workshop in Berlin on 13 and 14 June 2008, for helping me formulate this conclusion.
37. Gareth Stedman Jones, "The determinist fix: Some obstacles to the further development of the linguistic approach to history in the 1990s," in Spiegel (ed.), *Practicing History*, 62–75, here 71.

38. Robert Schnepf, *Geschichte erklären: Grundprobleme und Grundbegriffe* (Göttingen: Vandenhoeck & Ruprecht, 2011), 67.
39. Edward S. Reed, *James J. Gibson and the Psychology of Perception* (New Haven, CT, and London: Yale University Press, 1988), 16. The first quote is Reed's paraphrase.
40. Ibid., 82–83.
41. James J. Gibson and Laurence E. Crooks, "A Theoretical Field-Analysis of Automobile-Driving," *American Journal of Psychology* 51, no. 3 (July 1938), 453–471; Reed, *James J. Gibson*, 22 (controls it), 26 (not a camera), 75 (field theory), 135 (panorama).
42. This affinity between the novelist and the psychologist is realized through a theory of what Gibson called the "retinal motion gradient," the physical basis for the "perception ... conceived ... as the awareness of an active self in the environment." Reed, *James J. Gibson*, 84–85. This theory does not make a "strong dichotomy between the biological and the cognitive self," it is, in other words, "extero- and proprio-ception ... together" (ibid., 87).
43. Ibid., 182 (transcend), 186 (fur), 198 (skills), 198 (obtaining), 230 (ecology), 235 (motivating); all quotes are paraphrases by Reed, except the inner part of the nested quote. See also James J. Gibson, *The Ecological Approach to Visual Perception* (Boston and London: Houghton Mifflin Company, 1979).
44. Eric Laurier, "Driving: Pre-Cognition and Driving," in Tim Cresswell and Peter Merriman (eds.), *Geographies of Mobilities: Practices, Spaces, Subjects* (Farnham and Burlington: Ashgate, 2011), 69–81, here 70.
45. Mathieu Flonneau, *L'Autorefoulement et ses limites: Raisonner l'impensable mort de l'automobile* (Paris: Descartes & Cie, 2010).
46. Rajan, "Automobility and the liberal disposition," 117, 119.
47. For instance, when Colin Pooley, after an excellent analysis of British commuting during the last century, concludes that "lack or access to a car would not have made a fundamental difference to the ways in which people lived their lives in Britian." Colin G. Pooley, "Landscapes without the car: A counterfactual historical geography of twentieth-century Britain," *Journal of Historical Geography* 36 (2010), 266–275, here 274. Or John Modell, who observed how in California at the end of the 1920s "nearly one in three high school junior boys never drove," whereas the opposite, that two-thirds did, is the true spectacular result.
48. I thank Luísa Sousa, Bård Toldness, and Georgine Clarsen, respectively, for discussions on the specificities of the countries they are currently studying, and are in the process of writing about.
49. Greg Thompson, "'My Sewer'; James J. Flink on His Career Interpreting the Role of the Automobile in Twentieth-century Culture," in Peter Norton et al. (eds.), *Mobility in History: The Yearbook of the International Association for the History of Transport, Traffic and Mobility*, vol. 4 (New York and Oxford: Berghahn Journals, 2013), 3–17, here 3; Gijs Mom, "Subversive Mobility: An Uneasy Overview," in Mathieu Flonneau (ed.), *Automobile: Les cartes du désamour; Généalogies de l'anti-automobilisme* (Paris: Descartes & Cie, 2009), 69–78, here 70.
50. Albert Feuerwerker, *China's Early Industrialization: Sheng Hsuan-huai (1844–1916) and Mandarin Enterprise* (Cambridge, MA: Harvard University Press, 1958), 269. I thank Edward Rhoads for suggesting this source.
51. Gijs Mom (together with Georgine Clarsen, Nanny Kim, Cotten Seiler, Kurt Möser, Dorit Müller, Charissa Terranova and Rudi Volti), "Editorial," *Transfers* 1,

no. 1 (Spring 2011), 1–13. For an example of a parallel technology with comparable characteristics, including transcendence, see Robert MacDougall, "The Wire Devils: Pulp Thrillers, the Telephone, and Action at a Distance in the Wiring of a Nation," *American Quarterly* 58, no. 3 (September 2006), 715–741.
52. Huber, "Multiple Mobilities," 324–325; Dinerstein, *Swinging the Machine,* 66.
53. Renate Berger, "'Rotkäppchen, Grossmutter und Wolf in einer Person'; Valeska Gert—*bad girl* des neuen Tanzes," in Julia Freytag and Alexandra Tacke (eds.), *City Girls: Bubiköpfe & Blaustrümpfe in den 1920er Jahren* (Cologne, Weimar, and Vienna: Böhlau Verlag, 2011; Literatur—Kultur—Geschlecht; Studien zur Literatur- und Kulturgeschichte, eds. Anne-Kathrin Reulecke and Ulrike Vedder, in connection with Inge Stephan und Sigrid Weigel; Kleine Reihe, Band 29), 177–188, here 181. *Zerstreuung* is difficult to translate in English; it means dispersion and distraction, and at the same time amusement.
54. Lewis H. Siegelbaum, "Introduction," in Siegelbaum (ed.), *The Socialist Car: Automobility in the Eastern Bloc* (Ithaca, NY, and London: Cornell University Press, 2011), 1–13, here 12–13.
55. Sasha Disko, "'The World is My Domain'—Technology, Gender and Orientalism in German Interwar Motorized Adventure Literature," *Transfers* 1, no. 3 (Fall 2011), 44–63, here 50–51 (the first quote is a paraphrase by Disko).
56. Gijs Mom, "Droit à la mobilité, liberté ou contrainte? Plaidoyer pour des recherches nouvelles sur la mobilité/Freedom of Mobility; A Plea for new mobility studies," in Christophe Gay et al., *Mobile Immobile: Quels choix, quels droits pour 2030; Choices and rights for 2030,* vol. 2 (n.p. [Paris]: Éditions de l'Aube/Forum vie mobiles, 2011), 143–151; Mom, "Subversive Mobility: An Uneasy Overview."
57. Peter Giesen, "'Het is irreëel te geloven in eeuwige vrede,'" *De Volkskrant* (16 April 2011), "Het Vervolg," 8. The quote is a paraphrase by Giesen.
58. Sherry Turkle, "'Cyborg Babies and Cy-Dough-Plasm': Ideas about Self and Life in the Culture of Simulation," http://web.mit.edu/sturkle/www.cyborg_babies.html (last accessed 8 December 2011), 2–3.

Bibliography

Periodicals

ADAC Motorwelt (1927–1933)
Allgemeine Automobil-Zeitung (1908–1910; 1927–1930)
American Motorist (1914–1941)
De Auto (1905–1906; 1908–1918)
Het Automobielbedrijf (1931–1936)
The Autocar (1920–1934)
Autobusdienst (1923–1942)
Autokampioen (1930–1942)
Auto-Leven (1914–1931)
Automotive Industries (1919–1934)
Bedrijfsauto (1926–1928; 1930–1942)
Bovag (1937–1940)
Bulletin de l'AIPCR (1951–1952; 1956–1961)
The Car Illustrated (1909–1910; 1913)
The Club Journal (1910–1920)
The Horseless Age (1895–1915)
Journal of the Society of Automotive Engineers (1917–1940)
Locomotion Automobile (1894–1906)
Motor Travel (successor of *The Club Journal*) (1920–1931)
Der Motorfahrer (1908–1909; 1913–1920)
Motorkampioen (1930)
Het Motorrijwiel en de Kleine Auto (later: *Het Motorrijwiel en de Lichte Auto*) (1922–1926)
NELA Bulletin (1908–1926)
Outing (1905–1911; 1917–1923)
La Revue du Touring-Club de France (1920–1941)
La vie automobile (1903–1914)

Books and Secondary Articles*

50 Jahre ADAC im Dienste der Kraftfahrt (Munich: ADAC, n.d. [1953])
100 Jahre AvD: 100 Jahre Mobilität (Königswinter: Heel Verlag, 1999)
"La 628–E8, 'le nouveau jouet de Mirbeau,'" *Cahiers Octave Mirbeau* 15 (March 2008), 54–67
"10 Jahre NAG," *Stahlrad und Automobil* (7 January 1912), 12–17

---

*In order to keep this bibliography within reasonable limits, primary articles, mostly from the journals mentioned above, have not been included here (except for some short autopoetic texts).

John H. Abner and George L. Andrews, "Grease Monkey to Knitter," in *These Are Our Lives: As Told by the People and Written by Members of the Federal Writers' Project of the Works Progress Administration in North Carolina, Tennessee, and Georgia* (Chapel Hill: University of North Carolina Press, 1939), 165–179

Hans Achterhuis, *Met alle geweld: Een filosofische zoektocht* (Rotterdam: Lemniscaat, 2008)

Paul Adam, *Le troupeau de Clarisse* (Paris: Ernest Flammarion, n.d.)

Henry Adams, *The Education of Henry Adams*, edited with an introduction by Jean Gooder (New York and London: Penguin Books, 1995; first ed. 1906)

Jon Adams, "Real Problems with Fictional Cases," in Peter Howlett and Mary S. Morgan (eds.), *How Well Do Facts Travel? The Dissemination of Reliable Knowledge* (Cambridge and New York: Cambridge University Press, 2011), 167–191

Mark Adams, "The Automobile—A Luxury Becomes a Necessity," in Walton Hamilton et al., *Price and Price Policies* (New York and London: McGraw-Hill, 1938), 27–81

Michael C. C. Adams, *The Great Adventure: Male Desire and the Coming of World War I* (Bloomington and Indianapolis: Indiana University Press, 1990)

Peter Adey, *Mobility*, Key Ideas in Geography, eds. Sarah Holloway and Gill Valentine (London and New York: Routledge, 2010)

Theodor W. Adorno, "Über den Fetischcharakter in der Musik und die Regression des Hörens," in *Dissonanzen: Einleitung in die Musiksoziologie*, Gesammelte Schriften Band 14 (Frankfurt: Suhrkamp Verlag, 1973), 14–50

Judith Adler, "Origins of Sightseeing," *Annals of Tourism Research* 16 (1989), 7–29

——, "Travel as Performed Art," *The American Journal of Sociology* 94, no. 6 (May 1989), 1366–1391

AIPCR—PIARC 1909–1969 (Paris: Association Internationale Permanente des Congrès de la Route/Permanent International Association of Road Congresses, 1970)

Gerd Albers, *Zur Entwicklung der Stadtplanung in Europa: Begegnungen, Einflüsse, Verflechtungen* (Braunschweig and Wiesbaden: Vieweg, 1997)

Daniel M. Albert, "Order out of Chaos: Automobile Safety, Technology and Society 1925 to 1965" (dissertation, University of Michigan, 1997)

——, "Primitive Drivers: Racial Science and Citizenship in the Motor Age," *Science as Culture* 10, no. 3 (September 2001), 327–351

——, "Psychotechnology and Insanity at the Wheel," *Journal of the History of the Behavioral Sciences* 35, no. 3 (Summer 1999), 291–305

Derek H. Aldcroft (ed.), *Europe's Third World: The European Periphery in the Interwar Years*, Modern Economic and Social History Series, ed. Derek H. Aldcroft (Aldershot and Burlington, VT: Ashgate, 2006)

Karen Alkalay-Gut, "Sex and the Single Engine: E.E. Cummings' Experiment in Metaphoric Equation," *Journal of Modern Literature* 20, no. 2 (Winter 1996), 254–258

Fons Alkemade, "De auto in de Nederlandse literatuur en speelfilm," report for Foundation for the History of Technology, Eindhoven, The Netherlands (Eindhoven: Nederlands Centrum voor Autohistorische Documentatie NCAD, August/September 2000)

Thomas Alkemeyer, "Mensch-Maschinen mit zwei Rädern: Überlegungen zur riskanten Aussöhnung von Körper, Technik und Umgebung," in Gunter Gebauer et al. (eds.), *Kalkuliertes Risiko: Technik, Spiel und Sport an der Grenze* (Frankfurt and New York: Campus Verlag, 2006), 225–246

Ann Taylor Allen, "Forum: 'The Future is Ours': Feminists Imagine Europe in 1911," http://hsozkult.geschichte.hu-berlin.de/forum/type=diskussion&id=1214 (last accessed 24 May 2010)

Donna Allen, *Fringe Benefits: Wages or Social Obligation? An Analysis with Historical Perspectives from Paid Vacations* (Ithaca, NY: Cornell University Press, 1964)

Katherine Allen, *Bleeding Hearts: A Solution to the Automobile Tragedy* (San Antonio, TX: The Naylor Company, 1941)

Karsten Alnæs, *De geschiedenis van Europa, 1900–heden: Onbehagen* (Amsterdam and Antwerpen: Ambo/Standaard Uitgeverij, 2007 (transl. *Historien om Europa 4: Mørkets Tid* (2006), by Lucy Pijttersen, Kim Snoeijing, and Carla Joustra)

Ala Alryyes, "Description, the Novel and the Senses," *Senses & Society* 1, no. 1 (March 2006), 53–70

Andrew Gavin Altman, "Motoring for the Million? Cars and Class in Pre-1952 Britain" (unpublished PhD thesis, Boston College, Graduate School of Arts & Sciences, 3 April 1997)

Joseph A. Amato, *On Foot: A History of Walking* (New York and London: New York University Press, 2004)

America's Highways 1776–1976: A History of the Federal-Aid Program (Washington: US Department of Transportation, Federal Highway Administration, 1976)

Mehdi Parvizi Amineh, "Er bestaat helemaal geen islamitische beschaving," *Volkskrant* (11 October 2001)

Edward Anderson, *Thieves Like Us* (New York: The American Mercury, 1937)

Gregory Anderson, "Angestellte in England 1850–1914," in Jürgen Kocka (ed.), *Angestellte im europäischen Vergleich: Die Herausbildung angestellter Mittelschichten seit dem späten 19. Jahrhundert, Geschichte und Gesellschaft,* Sonderheft 7 (Göttingen: Vandenhoeck & Ruprecht, 1981), 59–73

Sherwood Anderson, *Perhaps Women* (Mamaroneck, NY: Paul P. Appel, 1970; first ed. 1931)

——, *Winesburg, Ohio: With an Introduction by Malcolm Cowley* (New York, Toronto, and London: Penguin, 1960; first ed. 1919)

Annual Report of the Secretary of the Interior for the Fiscal Year Ended June 30, 1918 (Washington, DC: Government Printing Office, 1918)

Guillaume Apolinnaire, "Zone, recueil Alcools," http://www.bac-facile.fr/commentaires/1064-guillaume-apollinaire-alcools-zone.html (last accessed 20 November 2009)

——, "La Petite Auto / The Little Car," in Donald Revell, *The Self-Dismembered Man: Selected Later Poems by Guillaume Apolinnaire,* transl. Donald Revell (Middletown, CT: Wesleyen University Press, 2004), 68–75

Victor Appleton, *Tom Swift and his Electric Runabout OR The Speediest Car on the Road* (New York: Grosset & Dunlap, 1910)

Michael Arlen, *The Green Hat: A Romance for a Few People* (Woodbridge, Suffolk: The Boydell Press, 1983; first ed. 1924)

Tim Armstrong, *Modernism, Technology, and the Body: A Cultural Study* (Cambridge: Cambridge University Press, 1998)

Paul Aron, "De *La 628–E8* d'Octave Mirbeau à *La 629–E9* de Didier de Roulx," in Éléonore Reverzy and Guy Ducrey (eds.), *L'Europe en automobile: Octave Mirbeau écrivain voyageur,* Configurations littéraires, collection de l'Université Marc Bloch de Strasbourg (Strasbourg: Presses Universitaires de Strasbourg, 2009), 231–238

John Arquilla and David Ronfeldt, *Swarming & the Future of Conflict* (Rand National Defence Research Institute, n.d.)

Elodie Arriola, "Des automobiles et des Hommes: Les débuts de l'Automobile Club Dauphinois (1899-1904), Volume I" (Master's thesis 1, "Sciences humaines et sociales"; Grenoble, Université Mendès-France, academic year 2007-2008)

Association Internationale Permanente des Congrès de la Route, *Ier Congrès International de la Route* (Paris, 1908), *Compte rendu des travaux du congrès* (Paris: Imprimerie Générale Lahure, 1909)

———, *IIe Congrès International de la Route, Bruxelles, 1910: Compte rendu des travaux du congrès* (Paris: Imprimerie Générale Lahure, 1911)

———, *IIIe Congrès International de la Route, Londres 1913: Compte rendu des travaux du congrès* (Paris: Imprimerie Oberthür, 1913)

———, *IVe Congrès International de la Route, Seville 1923; Compte rendu des travaux du congrès* (Paris, 1923)

———, *IIIe Congrès International de la Route, Washington 1930; Compte rendu des travaux du congrès* (Paris, 1930)

———, *Ve Congrès International de la Route, La Haye 1938; Compte rendu des travaux du congrès* (Paris, 1938)

Diane Atkinson, *Elsie and Mairi Go To War: Two Extraordinary Women on the Western Front* (London: Preface Publishing, 2009)

"L''Auto' et les artistes," *Les Annales politiques et littéraires*, no. 1274 (24 November 1907), 491-493

The Automobile and the Village Merchant: The Influence of Automobiles and Paved Roads on the Business of Illinois Village Merchants (Urbana: University of Illinois, 1928), *University of Illinois Bulletin* 25, no. 41 (12 June 1928); Bureau of Business Research, College of Commerce and Business Administration, Bulletin no. 19 (April 1928)

Enid Bagnold, *A Diary Without Dates*, with a new introduction by Monica Dickens (London: Virago / William Heinemann 1978; first ed. 1918)

———, *The Happy Foreigner* (Middlesex: The Echo Library, 2007; reprint of 1920)

Philip Bagwell, *The Transport Revolution from 1770* (London: B. T. Batsford, 1974)

Bernard Bailyn, *Atlantic History: Concept and Contours* (Cambridge, MA, and London: Harvard University Press, 2005)

David Bakhurst, "Ilyenkov on Aesthetics: Realism, Imagination, and the End of Art," *Mind, Culture, and Activity* 8, no. 2 (2001), 187-199

Gianfranca Balestra, "Edith Wharton, Henry James, and 'the Proper Vehicle of Passion,'" in Francesca Bisutti De Riz and Rosella Mamoli Zorzi, in cooperation with Marina Coslovi (eds.), *Technology and the American Imagination: An Ungoing Challenge; A.I.S.N.A. Associazione Italiana di Studi Nord-Americani, Atti del Dodicesimo Convegno Biennale, Università di Venezia, 28-30 ottobre 1993*, vol. VIII, n. 10-1994, Rivista di Studi Anglo-Americani, Annuario dell' A.I.S.N.A. (Venezia: Edizioni Supernova, 1994), 595-604

Étienne Balibar, *Der Schauplatz des Anderen: Formen der Gewalt und Grenzen der Zivilität (Aus dem Französischen von Thomas Laugstien)*, transl. of *La crainte des masses*, 1997 (Hamburg: Hamburger Edition HIS Verlagsgesellschaft, 2006)

Dag Balkmar, "Men, Cars and dangerous driving—Affordances and driver-car interaction from a gender perspective" (paper presented at Past Present Future, 14-17 June 2007, Umeå, Sweden), citeseerx.ist.psu.edu/viewdoc/download?doi=10.1.1.124.2591... (last accessed 10 November 2010)

Shelley Baranowski, "Radical Nationalism in an International Context: Strength through Joy and the Paradoxes of Nazi Tourism," in John K. Walton (ed.), *Histories of Tourism: Representation, Identity and Conflict*, vol. 6, Tourism and Cultural

Change, eds. Mike Robinson and Alison Phipps (Clevedon, Buffalo, and Toronto: Channel View Publications, 2005), 125–143

Harold Barger, *The Transportation Industries 1889–1946: A Study of Output, Employment, and Productivity* (New York: National Bureau of Economic Research, 1951)

T. C. Barker, "The spread of motor vehicles before 1914," in Charles P. Kindleberger and Guido di Tella (eds.), *Economics in the Long View: Essays in Honor of W.W. Rostow*, vol. 2: Applications and cases, part I (London and Basingstoke, 1982), 149–167

Theo Barker (ed.), *The Economic and Social Effects of the Spread of Motor Vehicles: An International Centenary Tribute* (Houndsmill and London, 1987)

Julian Barnes, "Introduction by Julian Barnes," in Edith Wharton, *A Motor-Flight Through France*, Picador Travel Classics (London and Basingstoke: Picador/MacmIllan, 1995), 1–14

Cécile Barraud, "Octave Mirbeau, un 'batteur d'âmes' à l'horizon de la Revue Blanche," *Cahiers Octave Mirbeau* 15 (March 2008), 92–101

Hal S. Barron, *Mixed Harvest: The Second Great Transformation in the Rural South, 1870–1930* (Chapel Hill and London: University of North Carolina Press, 1997)

Heather B. Barrow, "The Automobile in the Garden: Henry Ford, Suburbanization, and the Detroit Metropolis, 1919–1940" (unpublished dissertation, University of Chicago, Department of History, June 2005)

Michael Bartholomew, "H.V. Morton's English Utopia," in Christopher Lawrence and Anna-K. Mayer (eds.), *Regenerating England: Science, Medicine and Culture in Inter-War Britain*, the Welcome Institute Series in the History of Medicine (Amsterdam and Atlanta, GA: Rodopi, 2000), 25–44

Susan Bassnett, "Travel writing and gender," in Peter Hulme and Tim Youngs (eds.), *The Cambridge Companion to Travel Writing* (Cambridge: Cambridge University Press, 2002), 225–241

Zygmunt Bauman, *The Individualized Society* (Cambridge: Polity Press, 2004; reprint of first ed. 2001)

Hans Baumann, *Deutsches Verkehrsbuch* (Berlin: Deutsche Verlagsgesellschaft, 1931)

Winfried Baumgart, "Eisenbahnen und Kriegsführung in der Geschichte," *Technikgeschichte* 38, no. 3 (1971), 191–219

Ulrike Baureithel, "Motorisierung der Seelen: Anmerkungen zu Arnolt Bronnens Konzeption der Mensch-Maschine-Symbiose," in Wolfgang Bialas and Burkhard Stenzel (eds.), *Die Weimarer Republik zwischen Metropole und Provinz: Intellektuellendiskurse zur politischen Kultur* (Weimar, Cologne, and Vienna: Böhlau Verlag, 1996), 131–142

John Baxter, *The Inner Man: The Life of J.G. Ballard* (London: Weidenfeld & Nicolson, 2011)

Angelo Bazzanella, "Die Stimme der Illiteraten: Volk und Krieg in Italien 1915–1918," in Klaus Vondung (ed.), *Kriegserlebnis: Der Erste Weltkrieg in der literarischen Gestaltung und symbolischen Deutung der Nationen* (Göttingen: Vandenhoeck & Ruprecht, 1980), 334–351

David Beasley, *Who Really Invented the Automobile: Skullduggery at the Crossroads*; rev. ed. (Simcoe, ON: Davus Publishing, 1997); first published as *The Suppression of the Automobile* (Westport, CT: Greenwood Press, 1988)

Robert A. Beauregard, *When America Became Suburban* (Minneapolis and London: University of Minnesota Press, 2006)

Daniel R. Beaver, "Politics and Policy: The War Department Motorization and Standardization Program for Wheeled Transport Vehicles, 1920–1940," *Military Affairs* 47, no. 3 (October 1983), 101–108
———, "The Problem of American Military Supply, 1890–1920," in Benjamin Franklin Cooling (ed.), *War, Business, and American Society: Historical Perspectives on the Military-Industrial Complex* (Port Washington, NY, and London: Kennikat Press, 1977), 73–92
Stefan Beck, *Umgang mit Technik: Kulturelle Praxen und kulturwissenschaftliche Forschungskonzepte* (Berlin: Akademie Verlag, 1997)
Daniel Beck, "Unter Zugzwang; Die Schweizerischen Bundesbahnen und das Automobil 1945–1975" (Master's thesis, Historisches Institut der Universität Bern, January 1999)
Ulrich Beck, *Risikogesellschaft: Auf dem Weg in eine andere Moderne* (Frankfurt: Suhrkamp Verlag, 1986)
Jens Peter Becker, *Das Automobil und die amerikanische Kultur* (Trier: Wissenschaftliche Verlag Trier, 1989)
Sabina Becker, *Urbanität und Moderne: Studien zur Grossstadtwahrnehmung in der deutschen Literatur 1900–1930*, Saarbrücker Beiträge zur Literaturwissenschaft, eds. Karl Richter, Gerhard Sauder, and Gerhard Schmidt-Henkel, Band 39 (St. Ingbert: Werner J. Röhrig Verlag, 1993)
Sven Beckert, "Propertied of a Different Kind: Bourgeoisie and Lower Middle Class in the Nineteenth-Century United States," in Burton J. Bledstein and Robert D. Johnston (eds.), *The Middling Sorts: Explorations in the History of the American Middle Class* (New York and London: Routledge, 2001), 285–295
Karen Beckman, *Crash: Cinema and the Politics of Speed and Stasis* (Durham, NC, and London: Duke University Press, 2010)
Jörg Beckmann, "Mobility and Safety," *Theory, Culture & Society* 21, no. 4/5 (2004), 81–100
———, *Risky Mobility: The Filtering of Automobility's Unintended Consequences* (Copenhagen: Copenhagen University, Sociological Institute: 2001)
Bedaux et al., "Die Unfälle auf den Strassen" (Report 46, PIARC Conference, The Hague, 1938) [see n. 26 on p. 621]
Gail Bederman, *Manliness & Civilization: A Cultural History of Gender and Race in the United States, 1880–1917* (Chicago and London: University of Chicago Press, 1995)
John A. Beeler, *Report to the City of Atlanta on a Plan for Local Transportation, December, 1924* (New York: The Beeler Organization, 1924)
Gillian Beer, "The island and the aeroplane: The case of Virginia Woolf," in Homi K. Bhabha (ed.), *Nation and Narration* (London and New York: Routledge, 1990), 265–290
Howard W. Beers, "A Portrait of the Farm Family in Central New York State," *American Sociological Review* 2, no. 5 (October 1937), 591–600
Anne Beezer, "Women and 'Adventure Travel' Tourism," *New Formations* 21 (1993), 119–130
Laura L. Behling, "'The Woman at the Wheel': Marketing Ideal Womanhood, 1915–1934," *Journal of American Culture* 20, no. 3 (Fall 1997), 13–30
Warren James Belasco, *Americans on the Road: From Autocamp to Motel, 1910–1945* (Cambridge, MA, and London: MIT Press, 1979)
Russell Belk, "Possessions and the Extended Self," *Journal of Consumer Research* 15, no. 2 (September 1988), 139–168

René Bellu, *Toutes les Renault* (Paris, 1979)
Marlou Belyea, "The Joy Ride and the Silver Screen: Commercial Leisure, Delinquency and Play Reform in Los Angeles, 1900–1980" (unpublished dissertation, Boston University Graduate School, 1983)
Reinhard Bendix, "Tradition and Modernity Reconsidered," *Comparative Studies in Society and History* 9, no. 3 (1967), 292–346
Silas Bent, "Speed on the Highways: Study of the Auto and Its Safety Problems," *Current History* 45, no 3 (December 1936), 95–99
Wilfried Berg et al., *Das Reich als Republik und in der Zeit des Nationalsozialismus, Deutsche Verwaltungsgeschichte, Band 4*, eds. Kurt G.A. Jeserich, Hans Pohl, and Georg-Christoph von Unruh (Stuttgart: Deutsche Verlags-Anstalt, 1985)
Leo van Bergen, *Before My Helpless Sight: Suffering, Dying and Military Medicine on the Western Front, 1914–1918* (Franham and Burlington, VT: Ashgate, 2009)
Michael L. Berger, "The Car's Impact on the American Family," in Martin Wachs and Margaret Crawford (with the assistance of Susan Marie Wirka and Taina Marjatta Rikala), *The Car and the City: The Automobile, The Built Environment, and Daily Urban Life* (Ann Arbor: University of Michigan Press, 1992), 57–74
——, *The Devil Wagon in God's Country: The Automobile and Social Change in Rural America, 1893–1929* (Hamden, CT: Archon Books, 1979)
——, "Women drivers! The emergence of folklore and stereotypic opinions concerning feminine automotive behavior," *Women's Studies International Forum* 9, no. 3 (1986), 257–263
Molly W. Berger, "Popular Culture and Technology in the Twentieth Century," in Carroll Pursell (ed.), *A Companion to American Technology*, Blackwell Companions to American History (Malden, Oxford, and Carlton: Blackwell Publishing, 2005), 385–405
Renate Berger, "'Rotkäppchen, Grossmutter und Wolf in einer Person'; Valeska Gert—bad girl des neuen Tanzes," in Julia Freytag and Alexandra Tacke (eds.), *City Girls: Bubiköpfe & Blaustrümpfe in den 1920er Jahren*, Literatur—Kultur—Geschlecht; Studien zur Literatur- und Kulturgeschichte, eds. Anne-Kathrin Reulecke und Ulrike Vedder, in connection with Inge Stephan und Sigrid Weigel; Kleine Reihe, Band 29 (Cologne, Weimar, and Vienna: Böhlau Verlag, 2011), 177–188
Volker Berghahn, *Europa im Zeitalter der Weltkriege: Die Entfesselung und Entgrenzung der Gewalt*, Europäische Geschichte, ed. Wolfgang Benz (Frankfurt: Fischer Taschenbuch Verlag, 2002)
Hartmut Berghoff, "Träume und Alpträume: Konsumpolitik im Nationalsozialistischen Deutschland," in Heinz-Gerhard Haupt and Claudius Torp (eds.), *Die Konsumgesellschaft in Deutschland 1890 – 1990; Ein Handbuch* (Frankfurt and New York: Campus Verlag, 2009), 268–288
Edo J. Bergsma and L. C. Steffelaar, *Het derde Wegen-Congres te Londen, 23–28 Juni 1913* (The Hague: ANWB, Wegen-Commissie, 1913)
Helmuth Berking, "Raumtheoretische Paradoxien im Globalisierungsdiskurs," in Berking (ed.), *Die Macht des Lokalen in einer Welt ohne Grenzen* (Frankfurt and New York: Campus Verlag, 2006), 7–22
Michael Berkowitz, "A 'New Deal' for Leisure: Making Mass Tourism during the Great Depression," in Shelley Baranowski and Ellen Furlough (eds.), *Being Elsewhere: Tourism, Consumer Culture, and Identity in Modern Europe and North America* (Ann Arbor: University of Michigan Press, 2004), 184–212
Tristan Bernard, *Sur les grands chemins* (Paris: Librairie Paul Ollendorff, n.d.; first ed. 1911)

——, *Les veillées du chauffeur* (Paris: Société d'éditions littéraires et artistiques, Librairie Paul Ollendorf, 1909), http://www.answers.com/topic/tristan-bernard (last accessed 2 July 2009)

Catherine Bertho Lavenir, "Autos contre piétons: La guerre est déclarée," *L'Histoire* no. 230 (March 1999), 80–85

——, "Manières de circuler en France depuis 1880," *Le Mouvement Social* no. 192 (July–September 2000), 3–8

——, "Normes de comportement et contrôle de l'espace: Le Touring Club de Belgique avant 1914," *Le Mouvement Social* no. 178 (January/March 1997), 69–87

——, *La roue et le stylo: Comment nous sommes devenus touristes* (Paris: Editions Odile Jacob, April 1999)

John Betjeman, "Slough," http://www.cdr.stanford.edu/intuition/Slough.html (last accessed 5 December 2011)

Karl-Heinz Bette, *X-treme: Zur Soziologie des Abenteur- und Risikosports* (Bielefeld: transcript Verlag, 2004)

Matthias Bickenbach, "Robert Musil und die neue Gesetze des Autounfalls," in Christian Kassung (ed.), *Die Unordnung der Dinge: Eine Wissens- und Mediengeschichte des Unfalls* (Bielefeld: transcript Verlag, 2009), 89–116

Matthias Bickenbach and Michael Stolzke, "Schrott: Bilder aus der Geschwindigkeitsfabrik; Eine fragmentarische Kulturgeschichte des Autounfalls," http://www.textur.com/schrott/ (last accessed 26 April 2011)

Otto Julius Bierbaum, *Eine empfindsame Reise im Automobil von Berlin nach Sorrent und zurück an den Rhein: In Briefen an Freunde beschrieben* (Munich: Albert Langen—Georg Müller Verlag, 1979; reprint of 1903 edition)

——, "Das höllische Automobil; Ein Märchen für sämtliche Alters- und Rangklassen nach einer Idee Alf Bachmanns," in Bierbaum, *Mit der Kraft; Automobilia von Otto Julius Bierbaum* (Berlin: Bard Marquart, 1906), 283-311

——, *Das höllische Automobil: Novellen* (Vienna and Leipzig: Wiener Verlag, 1905)

——, "Eine kleine Herbstreise im Automobil" and "Kleine Reise," *Die Yankeedoodle-Fahrt und andere Reisegeschichten: Neue Beiträge zur Kunst des Reisens* (Munich: Georg Müller, 1910)

——, "Philister contra Automobil," in Bierbaum, *Mit der Kraft,* 325

Wiebe Bijker, [Book review of Harry Collins, *Tacit and Explicit Knowledge* (Chicago, 1910)]," *Technology and Culture* 52, no. 4 (October 2011), 809–810

——, "The Diabolical Symphony of the Mechanical Age: Technology and Symbolism of Sound in European and North American Noise Abatement Campaigns, 1900–40," *Social Studies of Science* 31, no. 1 (February 2001), 37–70

Karin Bijsterveld, *Mechanical Sound: Technology, Culture, and Public Problems of Noise in the Twentieth Century* (Cambridge, MA, and London: MIT Press, 2008)

——, Eefje Cleophas, Stefan Krebs and Gijs Mom, *Sound and Safe: A History of Listening Behind the Wheel* (Oxford: Oxford University Press, 2014)

Richard Birkefeld and Martina Jung, *Die Stadt, der Lärm und das Licht: Die Veränderung des öffentlichen Raumes durch Motorisierung und Elektrifizierung* (Seelze [Velber]: Kallmeyer'sche Verlagsbuchhandlung GmbH, 1994)

David Bissell, "Vibrating materialities: Mobility—body—technology relations," *Area* 42, no. 4 (2010), 479–486

Kaj Björkqvist, "Sex Differences in Physical, Verbal, and Indirect Aggression: A Review of Recent Research," *Sex Roles* 30, no. 3/4 (February 1994), 177–188

Marc K. Blackburn, "A New Form of Transportation: The Quartermaster Corps and Standardization of the United States Army's Motor Trucks 1907–1939," (dissertation, Temple University, August 1992)

―――, *The United States Army and the Motor Truck: A Case Study in Standardization*, Contributions in Military Studies, Number 163 (Westport, CT, and London: Greenwood Press, 1996)

Susan Blake, "A Woman's Trek: What Difference Does Gender Make?" in Nupur Chaudhuri and Margaret Strobel (eds.), *Western Women and Imperialism: Complicity and Resistance* (Bloomington and Indianapolis: Indiana University Press, 1992), 19–34

David Blanke, *Hell on Wheels: The Promise and Peril of America's Car Culture, 1900–1940* (Lawrence: University Press of Kansas, 2007)

Maili Blauw, "Met of zonder vergunning? Wetgeving, regulering en toezicht op het vervoerbedrijf in Nederland in de negentiende en twintigste eeuw," in *150 jaar toezicht Verkeer en Waterstaat: Rampen, wetten en inspectiediensten* (Den Haag, 2004), 181–242

Carey S. Bliss, *Autos Across America: A Bibliography of Transcontinental Travel: 1903–1940* (Los Angeles: Dawson's Book Shop, 1972)

Greet de Block, "Engineering the territory: Technology, space and society in 19th and 20th century Belgium" (unpublished dissertation, Katholieke Universiteit Leuven; manuscript for preliminary defence, October 2011)

W. P. Blockmans, C. A. Davids, and E. K. Grootes, "Inleiding," *Tijdschrift voor Sociale Geschiedenis* 10, no. 35 (August 1984), 223–227

Philipp Blom, *The Vertigo Years: Europe, 1900–1914* (New York: Basic Books, 2008)

Brooke L. Blower, *Becoming Americans in Paris: Transatlantic Politics and Culture between the World Wars* (Oxford and New York: Oxford University Press, 2011)

Léon Bloy, *Journal I, 1892–1907 (Édition établie, présentée et annotée par Pierre Glaudes)* (Paris: Robert Laffont, 1999)

Manuela Boatcă, "Semiperipheries in the World-System: Reflecting Eastern European and Latin American Experiences," *Journal of World-Systems Research* 12, no. 11 (December 2006), 321–346

Steffen Böhm et al., "Introduction: Impossibilities of automobility," *The Sociological Review* 54, no. s1 (October 2006), 3–16

Michele Bonino and Massimo Moraglio, *Inventare gli spostamenti: Storia e immagini dell'autostrada Torino—Savona/Inventing Movement: History and Images of the A6 Motorway* (Torino, London, Venice, and New York: Umberto Allemandi, 2006)

Henri Bonnet, *Marcel Proust de 1907 à 1914: Bibliographie complémentaire (II): Index général des bibliographies (I et II) et une étude: Du Côté de chez Swann dans A la Recherche du Temps Perdu* (Paris: A. G. Nizet, 1976)

Daniel J. Boorstin, *The Image: A Guide to Pseudo-Events in America* (New York and Evanston, IL: Harper & Row, 1961)

Iain Borden, "Driving," in Matthew Beaumont and Gregory Dart (eds.), *Restless Cities* (London: Verso, 2010), 98–121

Mary Borden, *The Forbidden Zone* (London: William Heineman, 1929)

Kevin L. Borg, *Auto Mechanics: Technology and Expertise in Twentieth-Century America* (Baltimore, MD: Johns Hopkins University Press, 2007)

Peter Borscheid, "Lkw kontra Bahn; Die Modernisierung des Transports durch den Lastkraftwagen in Deutschland bis 1939," in Harry Niemann and Armin Hermann (eds.), *Die Entwicklung der Motorisierung im Deutschen Reich und den Nachfolgestaaten: Stuttgarter Tage zur Automobil- und Unternehmensgeschichte: Eine Veranstaltung von Mercedes-Benz Archiv* (Stuttgart: Franz Steiner Verlag, 1995), 23–38

———, *Das Tempo-Virus: Eine Kulturgeschichte der Beschleunigung* (Frankfurt and New York: Campus Verlag, 2004)
Mineke Bosch, *Een onwrikbaar geloof in rechtvaardigheid: Aletta Jacobs 1854-1929* (n.p.: Balans, 2005), 279-283
Scott L. Bottles, *Los Angeles and the Automobile: The Making of the Modern City* (Berkeley, Los Angeles, and London: University of California Press, n.d. [1987])
[Général] Boullaire, *La division légère automobile*, Extrait de la *Revue Militaire Française*, April, May, and June 1924 (Paris: Berger-Levrault Éditeurs, 1924)
S. M. Bowden, "Demand and Supply Constraints in the Inter-War UK Car Industry: Did the Manufacturers Get it Right?" *Business History* 33, no. 2 (April 1991), 241-267
Sue Bowden and Paul Turner, "The Demand for Consumer Durables in the United Kingdom in the Interwar Period," *The Journal of Economic History* 53, no. 2 (June 1993), 244-258
———, "Some Cross-Section Evidence on the Determinants of the Diffusion of Car Ownership in the Inter-War UK Economy," *Business History* 35, no. 1 (1993), 55-69
A. A. van den Braembussche, "Historical Explanation and Comparative Method: Towards a Theory of the History of Society," *History and Theory* 28, no. 1 (February 1989), 1-24
Rosi Braidotti, *Nomadic Subjects: Embodiment and Sexual Difference in Contemporary Feminist Theory* (New York: Columbia University Press, 1994)
H. W. Brands, *The Strange Death of American Liberalism* (New Haven, CT, and London: Yale University Press, 2001)
Susanne Brandt, "Le voyage aux champs de bataille," *Vingtième Siècle: Revue d'histoire* 41, no. 1 (1994), 18-22
Herbert M. Bratter, *The Promotion of Tourist Travel by Foreign Countries*, US Department of Commerce, Bureau of Foreign and Domestic Commerce; Trade Promotion Series—No. 113 (Washington, DC: Government Printing Office, 1931)
Andreas Braun, *Tempo, Tempo! Eine Kunst- und Kulturgeschichte der Geschwindigkeit im 19. Jahrhundert* (Frankfurt: Anabas-Verlag, 2001)
Hans-Joachim Braun, "Lärmbelastung und Lärmbekämpfung in der Zwischenkriegszeit," in Günther Bayerl and Wolfhard Weber (eds.), *Sozialgeschichte der Technik: Ulrich Troitzsch zum 60. Geburtstag* (Münster, New York, Munich, and Berlin: Waxmann Verlag, 1998), 251-258
Hans-Joachim Braun and Elena Ungeheuer, "Formt die Technik die Musik? Montageästhetik zu Beginn des 20. Jahrhunderts," *Berichte zur Wissenschaftsgeschichte* 31, no. 3 (2008), 211-225
Helmut Braun and Christian Panzer, "The Expansion of the Motor-Cycle Industry in Germany and in Great Britain (1918 until 1932)," *Journal of European Economic History* 32, no. 1 (2003), 25-59
Peter J. Brenner, "Einleitung," in Brenner (ed.), *Der Reisebericht: Die Entwicklung einer Gattung in der deutschen Literatur* (Frankfurt: Suhrkamp Verlag, 1989), 7-14
———, "Schwierige Reisen; Wandlungen des Reiseberichts in Deutschland 1918-1945," in Brenner, *Reisekultur in Deutschland: Von der Weimarer Republik zum "Dritten Reich"* (Tübingen: Max Niemeyer Verlag, 1997), 127-176
Bressot, Giguet, and Delaigue, "Massnahmen zur Trennung der Verkehrsarten auf der Strasse" (report 60, PIARC Conference, The Hague, 1938) [see n. 26 on p. 621]
Ashleigh Brilliant, "Some Aspects of Mass Motorization in Southern California, 1919-1929," *Southern California Quarterly* 47, no. 2 (July 1965), 191-208

Adri van den Brink and Marijn Molema, "The origins of Dutch rural planning: A study of the early history of land consolidation in the Netherlands," *Planning Perspectives* 23 (October 2008), 427–453

Douglas Brinkley, *The Wilderness Warrior: Theodore Roosevelt and the Crusade for America* (New York: Harper, 2009)

Michael L. Bromley, "Scorching Through 1902: 'The Automobile Terror'; The Year in Automobiles and Deaths in *The New York Times*," *Automotive History Review* (Summer 2003), 20–25

——, *William Howard Taft and the First Motoring Presidency, 1909–1913* (Jefferson, NC, and London: McFarland, 2003)

Peter Brooker and Peter Widdowson, "A Literature for England," in Robert Colls and Philip Dodd (eds.), *Englishness: Politics and Culture 1880–1920* (London, Sydney, Dover, NH: Croom Helm, 1986), 116–163

Edna Brooks, *The Khaki Girls Behind the Lines: Or, Driving With the Ambulance Corps* (New York: Cupples & Leon Company, 1918)

——, *The Khaki Girls of the Motor Corps: Or, Finding Their Place in the Big War* (New York: Cupples & Leon Company, 1918)

Sammy Kent Brooks, "The Motorcycle in American Culture: From Conception to 1935" (unpublished dissertation, George Washington University, Washington, DC, 1975)

Cecil Kenneth Brown, *The State Highway System of North Carolina: Its Evolution and Present Status*, the University of North Carolina Social Studies Series, ed. Howard W. Odum (Chapel Hill: University of North Carolina Press, 1931)

Clair Brown, *American Standards of Living 1918–1988* (Oxford and Cambridge, MA: Blackwell, 1994)

E. K. Brown, *Willa Cather: A critical Biography, Completed by Leon Edel* (New York: Alfred A. Knopf, 1953)

Jeffrey Brown, "From Traffic Regulation to Limited Ways: The Effort to Build a Science of Transportation Planning," *Journal of Planning History* 5, no. 1 (February 2006), 3–34

——, "A Tale of Two Visions: Harland Bartholomew, Robert Moses, and the Development of the American Freeway," *Journal of Planning History* 4, no. 1 (February 2005), 3–32

Blaine A. Brownell, "The Notorious Jitney and the Urban Transportation Crisis in Birmingham in the 1920's," *Alabama Review* 25 (April 1972), 105–118

——, "A Symbol of Modernity: Attitudes Toward the Automobile in Southern Cities in the 1920s", *American Quarterly* 24 (March 1972), 20–44

Edmund de S. Brunner and J. H. Kolb, *Rural Social Trends* (New York and London: McGraw-Hill Book Company, 1933)

Ulrike Brunotte and Rainer Herrn, "Statt einer Einleitung: Männlichkeit und Moderne—Pathosformeln, Wissenskulturen, Diskurse," in Ulrike Brunotte and Rainer Herrn (eds.), *Männlichkeiten und Moderne: Geschlecht in den Wissenskulturen um 1900* (Bielefeld: transcript Verlag, 2008), 9–23

Roger A. Bruns, *The Damndest Radical: The Life and Work of Ben Reitman, Chicago's Celebrated Social Reformer, Hobo King, and Whorehouse Physician* (Urbana and Chicago: University of Illinois Press, 1987)

C. D. Buchanan, *Mixed Blessing: The Motor in Britain* (London: Leonard Hill, 1958)

Susan Buck-Morss, "The Flaneur, the Sandwhichman and the Whore: The Politics of Loitering," *New German critique* 39 (1986), 99–141

[Bureau of Agricultural Economics], "Horses, Mules, and Motor Vehicles: Year Ended March 31, 1924; With Comparable Data for Earlier Years," Statistical Bulletin no. 5 (Washington, DC: United States Department of Agriculture, January 1925)

[The Bureau of Public Roads, US Department of Agriculture/The Connecticut State Highway Department], *Report of a Survey of Transportation on the State Highway System of Connecticut* (Washington, DC: Government Printing Office, 1926)

David Burgess-Wise, "Fiction of the Motor Car," *Automobile Quarterly* 31, no. 1 (1993), 89–101

John C. Burnham, *Accident Prone: A History of Technology, Psychology, and the Misfits of the Machine Age* (Chicago and London: University of Chicago Press, 2009)

Regine Buschauer, *Mobile Räume: Medien- und diskursgeschichtliche Studien zur Tele-Kommunikation* (Bielefeld: transcript Verlag, 2010)

Kenneth Button, "Freight transport," in David Banister and Joseph Berechman (eds.), *Transport in a Unified Europe: Policies and Challenges*, vol. 24, Studies In Regional Science and Urban Economics, eds. L. Anselin et al. (Amsterdam, London, New York, and Tokyo: North-Holland, 1993), 143–170

Cyriel Buysse, "De laatste ronde," in Buysse, *Reizen van toen: Met de automobiel door Frankrijk*, edited and introduced by Luc van Doorslaer (Antwerpen and Amsterdam: Manteau, 1992), 203–260

——, "Per auto," in Buysse, *Reizen van toen*, 111–199

James Leo Cahill, "How It Feels to Be Run Over: Early Film Accidents," *Discourse* 30, no. 3 (Fall 2008), 289–316

Lendol Calder, *Financing the American Dream: A Cultural History of Consumer Credit* (Princeton, NJ: Princeton University Press, 1999)

Craig Calhoun, "Explanation in Historical Sociology: Narrative, General Theory, and Historically Specific Theory," *American Journal of Sociology* 3 (November 1998), 846–871

J. Caloni, *Comment Verdun fut sauvé* (Paris: Étienne Chiron, 1924)

Marc Camiolo, "Anti-convivialité automobile," *Revue d'Allemagne et des Pays de langue allemande* 43, no. 1 (2011), 119–129

Hans van der Cammen and Len de Klerk, *Ruimtelijke ordening: Van grachtengordel tot Vinex-wijk* (Utrecht: Het Spectrum, 2003)

Angus Campbell and Philip E. Converse (eds.), *The Human Meaning of Social Change* (New York: Russell Sage Foundation, 1972)

Mary Baine Campbell, "Travel writing and its theory," in Peter Hulme and Tim Youngs (eds.), *The Cambridge Companion to Travel Writing* (Cambridge: Cambridge University Press, 2002), 261–278

Theodore Caplow, Howard M. Bahr (ed.), et al., *Recent Social Trends in the United States 1960–1990*, Series Comparative Charting of Social Change, ed. Simon Langlois (Frankfurt, Montreal and Kingston, London, and Buffalo: Campus Verlag / McGill-Queen's University Press, 1991)

Paul Carmignani, "L'Automobile dans la fiction américaine: Pièces détachées et morceaux choisis," in Frédéric Monneyron and Joël Thomas (eds.), *Automobile et Littérature*, Collection Études (n.p.: Presses Universitaires de Perpignan, 2005), 121–158

David Carr, "Narrative Explanation and its Malcontents," *History and Theory* 47 (February 2008), 19–30

——, "Narrative and the Real World: An Argument for Continuity," in Brian Fay, Philip Pomper, and Richard T. Vann (eds.), *History and Theory: Contemporary Readings* (Malden, MA, and Oxford: Blackwell, 1998), 137–152

Reginald P. Carr, *Anarchism in France: The case of Octave Mirbeau* (Manchester: Manchester University Press, 1977)

Henri Carré, *La véritable histoire des taxis de la Marne (6, 7 et 8 Septembre 1914)* (Paris: Librairie Chapelot, 1921)

Noël Carroll, "Interpretation, History, and Narrative," in Brian Fay, Philip Pomper and Richard T. Vann (eds.), *History and Theory; Contemporary Readings* (Malden, MA and Oxford: Blackwell, 1998), 34–56

William C. Carter, *The Proustian Quest* (New York and London: New York University Press, 1992), 9.

Norman Miller Cary Jr., "The Use of the Motor Vehicle in the United States Army, 1899–1939" (unpublished dissertation, University of Georgia, 1980)

Roger N. Casey, *Textual Vehicles: The Automobile in American Literature* (New York and London: Garland Publishing, 1997)

Willa Cather, "The Bohemian Girl" (1912), in *Willa Cather's Collected Short Fiction, 1892–1912: Introduction by Mildred R. Bennett* (Lincoln: University of Nebraska Press, 1965; first ed. 1965), 3–41

——, "Tommy, the Unsentimental," in *Willa Cather's Collected Short Fiction*, 473–480

Archie Chadbourne, "Debt is the Only Adventure a Poor Man Can Count On," *The American Magazine* 104 (December 1927), 44–45, 108–112

Dipesh Chakrabarty, *Provincializing Europe: Postcolonial Thought and Historical Difference* (Princeton, NJ, and Oxford: Princeton University Press, 2000)

Roy D. Chapin, "The Motor's Part in Transportation," *Annals of the American Academy of Political and Social Sciences* 116 (The Automobile: Its Province and Its Problems; November 1924), 1–8

Jean-Claude Chesnais, "La mortalité par accidents en France depuis 1826," *Population* (French edition) 29, no. 6 (November–December 1974), 1097–1136

Kristen Stromberg Childers, *Fathers, Families, and the State in France 1914–1945* (Ithaca, NY, and London: Cornell University Press, 2003)

William R. Childs, *Trucking and the Public Interest: The Emergence of Federal Regulation 1914–1940* (Knoxville: The University of Tennessee Press, 1985)

David Chirico, "The Travel Narrative as a (Literary) Genre," in Wendy Bracewell and Alex Drace-Francis (eds.), *Under Eastern Eyes: A Comparative Introduction to East European Travel Writing on Europe* (Budapest and New York: Central European University Press, 2008), 27–59

Kimberly Chuppa-Cornell, "The U.S. Women's Motor Corps in France, 1914–1921," *The Historian* 56, no. 3 (Spring 1994), 465–476

Robert Clair, "Again the Woman Driver!" *Journal of American Insurance* 8, no. 8 (1931), 13–15

Carroll D. Clark and Cleo E. Wilcox, "The House Trailer Movement," *Sociology and Social Research* 22, no. 6 (July–August 1938), 503–519

Deborah Clarke, "Anxiously Popular: Women and the Automobile Culture of the Early 20th Century" (Paper delivered at the Society of Automotive History conference, Los Angeles, March 2000)

Sally Clarke, "Consumers, Information, and Marketing Efficiency at GM, 1921–1940," *Business and Economic History* 25, no. 1 (Fall 1996), 186–195

Constance Classen, "Engendering Perception: Gender Ideologies and Sensory Hierarchies in Western History," *Body & Society* 3, no. 2 (1997), 1–19

——, "Foundations for an anthropology of the senses," *International Social Science Journal* 49, no. 153 (1997), 401–412

——, *The Color of Angels: Cosmology, gender and the aesthetic imagination* (London and New York: Routledge, 1998)

John A. Clausen, *American Lives: Looking Back at the Children of the Great Depression* (Berkeley, Los Angeles, and London: University of California Press, 1993)

W. C. Clemens et al., "Massnahmen zur Trennung der Verkehrsarten auf der Stasse" (report 61, PIARC Conference, The Hague, 1938)

H. Cleyndert Azn, "De beweging voor roadside improvement in Amerika," *Tijdschrift voor volkshuisvesting en stedebouw* 13 (1932), 30–32

Evelyn Cobley, *Modernism and the Culture of Efficiency: Ideology and Fiction* (Toronto, Buffalo, and London: University of Toronto Press, 2009)

Christina Cogdell, "The Futurama Recontextualized: Normal Bel Geddes's Eugenic 'World of Tomorrow,'" *American Quarterly* 52, no. 2 (June 2000), 193–245

Steven Cohan and Ina Rae Hark (eds.), *The Road Movie Book* (London and New York: Routledge, 1997)

Debra Rae Cohen, "Culture and the 'Cathedral': Tourism as Potlatch in One of Ours," *Cather Studies* 6 (2006), 1–18, http://cather.unl.edu/cs006_cohen.html (last accessed 23 September 2014)

Lizabeth Cohen, *Making a New Deal: Industrial Workers in Chicago, 1919–1939* (Cambridge and New York: Cambridge University Press, 1990)

Michael Cole, "Cross-Cultural Research in the Sociohistorical Tradition," *Human Development* 31 (1988), 137–157

Martin Collins, "Introduction," *History and Technology* 26, no. 4 (December 2010), 359–360 ("*History and Technology* Forum")

Randall Collins, *Violence: A Micro-sociological Theory* (Princeton, NJ, and Oxford: Princeton University Press, 2008)

Michel Collomb, *La littérature Art Déco; Sur le style d'époque* (Paris: Méridiens Klincksieck, 1987)

"Confessions of a Ford Dealer," published in *Harper's Monthly Magazine* in 1927 and reproduced in George E. Mowry (ed.), *The Twenties: Fords, Flappers & Fanatics* (Englewood Cliffs, NJ: Prentice-Hall, 1963), 26–31

Peter Conrad, *Modern Times, Modern Places* (New York: Alfred A. Knopf, 1999)

R. W. Connell, *Masculinities* (Cambridge and Oxford: Polity Press, 1995)

Edward Constant, "Recursive practice and the evolution of technological knowledge," in John Ziman (ed.), *Technological Innovation as an Evolutionary Process* (Cambridge: Cambridge University Press, 2000), 219–233

Jane Conway, *Mary Borden: A Woman of Two Wars* (n.p.: Munday Books, 2010)

Michael P. Conzen (ed.), *The Making of the American Landscape* (Boston, London, Sydney, and Wellington: Unwin Hyman, 1990)

Lorraine Coons and Alexander Varias, *Tourist Third Cabin: Steamship Travel in the Interwar Years* (New York and Houndsmill, Basingstoke: Palgrave MacMillan, 2003)

Stanley Cooperman, *World War I and the American Novel* (Baltimore, MD: Johns Hopkins Press, 1967)

Marie Corelli, *The Devil's Motor: A Fantasy* (London: Hodder & Stoughton, n.d. [1910])

Joseph J. Corn, *user unfriendly: Consumer Struggles with Personal Technologies, from Clocks and Sewing Machines to Cars and Computers* (Baltimore, MD: Johns Hopkins University Press, 2011)

Joseph J. Corn, "Work and Vehicles: A Comment and Note," in Martin Wachs and Margaret Crawford (with the assistance of Susan Marie Wirka and Taina Marjatta Rikala), *The Car and the City: The Automobile, The Built Environment, and Daily Urban Life* (Ann Arbor: University of Michigan Press, 1992), 25–38

Lewis A. Coser (ed.), *Sociology Through Literature: An Introductory Reader* (Englewood Cliffs, NJ: Prentice-Hall, 1963)

Alan Costall and Arthur Still (eds.), *Cognitive Psychology in Question* (Sussex: The Harvester Press, 1987)
Louis Couperus, *De boeken der kleine zielen: Het late leven, eerste deel* [The books of the small souls; first part: the late life] (Amsterdam, n.d. [1901])
Donald R. G. Cowan, "Regional Consumption and Sales Analysis," *The Journal of Business of the University of Chicago* 8, no. 4 (1935), 345–381
Malcolm Cowley, *Exile's Return: A Literary Odyssey of the 1920s*, edited and with an introduction by Donald W. Faulkner (New York and London: Penguin Books, 1994; first ed. 1934)
Sandra J. Coyner, "Class consciousness and consumption: The new middle class during the Weimar Republic," *Journal of Social History* 10, no. 3 (Spring 1977), 310–331
Sandra Jean Coyner, "Class Patterns of Family Income and Expenditure during the Weimar Republic: German White-Collar Employees as Harbingers of Modern Society" (PhD dissertation, Rutgers University, 1975)
Finla Goff Crawford, *Motor Fuel Taxation in the United States* (Syracuse, NY: published by author, 1939)
Pierre de Crawhez, *Les Grands Itinéraires en Automobile à travers l'Europe, l'Algérie et la Tunisie: Annuaire de l'A.C.N.L.* (Namen: Bertrand, 1906)
Tim Cresswell, "Embodiment, power and the politics of mobility: The case of female tramps and hobos," *Transactions of the Institute of British Geographers* NS 24 (1999), 175–192
——, "Mobility as resistance: A geographical reading of Kerouac's 'On the road,'" *Transactions of the Institute of British Geographers* NS 18 (1993), 249–263
——, *On the Move: Mobility in the Modern Western World* (New York and London: Routledge, 2006)
——, *The Tramp in America* (London: Reaktion Books, 2001)
Tim Cresswell and Deborah Dixon, "Introduction: Engaging Film," in Cresswell and Dixon (eds.), *Engaging Film: Geographies of Mobility and Identity* (Lanham, Boulder, CO, New York, and Oxford: Rowman & Littlefield, 2002), 1–10
Martin van Creveld, *Supplying War: Logistics from Wallenstein to Patton* (London and New York: Cambridge University Press, 1986; reprint of first ed., 1977)
Fabio Crivellari et al., "Einleitung: Die Medialität der Geschichte und die Historizität der Medien," in Crivellari et al., in cooperation with Sven Grampp (eds.), *Die Medien der Geschichte: Historizität und Medialität in interdisziplinärer Perspektive* (Konstanz: UVK Verlagsgesellschaft, 2004), 9–45
Fabio Crivellari and Marcus Sandl, "Die Medialität der Geschichte: Forschungsstand und Perspektiven einer interdisziplinären Zusammenarbeit von Geschichts- und Medienwissenschaften," *Historische Zeitschrift* 277 (2003), 619–654
Frederic W. Cron, "Highway Design for Motor Vehicles—A Historical Review," *Public Roads* 38, no. 3 (December 1974), 85–95
Geoffrey Crossick, "The Emergence of the Lower Middle Class in Britain: A Discussion," in Crossick (ed.), *The Lower Middle Class in Britain 1870–1914* (London: Croom Helm, 1977), 11–60
Geoffrey Crossick and Heinz-Gerhard Haupt, "Shopkeepers, master artisans and the historian: The petite bourgeoisie in comparative focus," in Crossick and Haupt (eds.), *Shopkeepers and Master Artisans in Nineteenth-Century Europe* (London and New York: Methuen, 1984), 3–31
John E. Crowley, "The Sensibility of Comfort," *American Historical Review* 104, no. 3 (June 1999), 749–782

Melba Cuddy-Keane et al., "The Heteroglossia of History, Part One: The Car," in Beth Rigel Daugherty and Eileen Barrett (eds.), *Virgina Woolf: Texts and Contexts; Selected Papers from the Fifth Annual Conference on Virginia Woolf: Otterbein College, Westerville, Ohio, June 15–18, 1995* (New York: Pace University Press, 1996), 71–80

Ernst Robert Curtius, *Marcel Proust*, Bibliothek Suhrkamp, Band 28 (Berlin and Frankfurt: Suhrkamp Verlag, 1952; first ed. 1925)

Richard O. Curry and Lawrence B. Goodheart, "Individualism in Trans-National Context," in Curry and Goodheart (eds.), *American Chameleon: Individualism in Trans-National Context* (Kent, OH, and London: Kent State University Press, 1991), 1–19

Peter Dalsgaard and Karen Johanne Kortbek, "Staging Urban Atmospheres in Interaction Design" (Nordic Design Research Conferences, Engaging Artefacts 2009, Oslo), http://www.nordes.org (last accessed 13 April 2011)

Sara Danius, "The Aesthetics of the Windshield: Proust and the Modernist Rhetoric of Speed," *Modernism/Modernity* 8, no. 1 (2001), 99–126

——, "Orpheus and the Machine: Proust as Theorist of Technological Change, and the Case of Joyce," *Forum for Modern Language Studies* 37, no. 2 (2001), 127–140

W. Dale Dannefer, "Driving and Symbolic Interaction," *Sociological Inquiry* 47, no. 1 (1977), 33–38

Gabriele D'Annunzio, *Forse che si forse che no* (Milan: Presso I Fratelli Treves, 1910)

——, *Vielleicht—vielleicht auch nicht: Roman: Mit einem Nachwort von László F. Földényi* (Munich: Matthes & Seitz Verlag, 1989; transl. Karl Vollmöller, 1910)

John Davidson, *Fleet Street and Other Poems* (London: Grant Richards, 1909)

Donald Finley Davis, *Conspicuous Production: Automobiles and Elites in Detroit, 1899–1933* (Philadelphia, PA: Temple University Press, 1988)

Donald F. Davis, "Dependent Motorization: Canada and the Automobile to the 1930s," *Revue d'études canadiennes* 21, no. 3 (Fall 1986), 106–132

——, "Competition's Moment: The Jitney-Bus and Corporate Capitalism in the Canadian City, 1914–29," *Urban History Review/Revue d'histoire urbaine* 18, no. 2 (October 1989), 101–122

Timothy Davis, "Mount Vernon Memorial Highway and the Evolution of the American Parkway" (unpublished dissertation, University of Texas at Austin, December 1997)

——, "The Rise and Decline of the American Parkway," in Christof Mauch and Thomas Zeller (eds.), *The World Beyond the Windshield: Roads and Landscapes in the United States and Europe* (Athens, OH, and Stuttgart: Ohio University Press / Franz Steiner Verlag, 2008), 35–58

Graeme Davison (with Sheryl Yelland), *Car Wars: How the Car Won Our Hearts and Conquered Our Cities* (Crows Nest: Allen & Unwin, 2004)

Uwe Day, "Mythos ex machina; Medienkonstrukt 'Silberpfeil' als massenkulturelle Ikone der NS-Modernisierung" (dissertation, Universität Bremen, November 2004)

J. S. Dean, *Murder Most Foul: A Study of the Road Deaths Problem* (London: George Allen & Unwin, 1947)

Charles L. Dearing, *American Highway Policy* (Washington, DC: The Brookings Institution, 1941)

Hervé Debacker, "L'automobile à Orléans de 1897 à 1913: Les débuts d'un nouvel espace identitaire" (Mémoire de maîtrise dirigé par Marie-Claude Blanc-Chaleard, Université d'Orléans, Département d'histoire, Année universitaire 1998–1999)

Wilhelm Deist, "La mort et le deuil: La 'moral' des troupes allemandes sur le front occidental à la fin de l'année 1916," in Jean-Jacques Becker, Jay M. Winter, Gerd

Krumeich, Annette Becker and Stéphane Audoin-Rouzeau, *Guerre et cultures 1914 – 1918* (Paris : Armand Colin, 1994), 91–102

Walter Delabar, "Linke Melancholie? Erich Kästners *Fabian*," in Jörg Döring, Christian Jäger, and Thomas Wegmann (eds.), *Verkehrsformen und Schreibverhältnisse: Mediale Wandel als Gegenstand und Bedingung von Literatur im 20. Jahrhundert* (Opladen: Westdeutscher Verlag, 1996), 15–34

Louis Delanney, *Strassenverwaltung und –finanzierung in fünfzehn Ländern* (Paris: Internationale Handelskammer, June 1937)

Delemer et al., "Les moyens propres à assurer la sécurité de la circulation" (report 60, PIARC Conference, Munich, 1934)

Peter Demetz, *Worte in Freiheit: Der italienische Futurismus und die deutsche literarische Avantarde (1912–1934): Mit einer ausführlichen Dokumentation* (Munich: R. Piper, 1990)

Eugène Demolder, *L'Espagne en Auto: Impressions de voyage* (Paris: Société du Mercure de France, 1906; second ed.)

Priscilla Denby, "The Self Discovered: The Car in American Folklore and Literature" (unpublished dissertation, Indiana University, May 1981)

Kingsley Dennis and John Urry, *After the Car* (Cambridge and Malden, MA: Polity Press, 2009)

Jacques Derouard, *Maurice Leblanc: Arsène Lupin malgré lui* (n.p.: Librairie Séguier, 1989)

Marc Desportes, "L'Ère technique de la spatialité urbaine: Genèse et expérience des aménagements techniques et urbains. Le cas des infrastructures routières, 1900–1940" (dissertation, Université de Paris VIII, Institut Français d'Urbanisme, Doctorat "Urbanisme et Aménagement," 1995)

———, *Paysages en mouvement: Transports et perception de l'espace XVIIIe–XXe siècle* (n.p. [Paris]: Gallimard, 2005)

Marc Desportes and Antoine Picon, *De l'espace au territoire: L'aménagement en France, XVIe–XXe siècles* (Paris: Presses de l'École Nationale des Ponts et Chaussées, 1997)

Cynthia Colomb Dettelbach, *In the Driver's Seat: The Automobil in American Literature and Popular Culture* (Westport, CT, 1976)

Lothar Diehl, "Tyrannen der Landstrassen: Die Automobilkritik um 1900," *Kultur & Technik* no. 3 (1998), 51–57

Eugen Diesel, "Autoreise 1905," in Rüdiger Kremer and Wolfgang Rumpf, *Baby, won't you drive my car? Dichter am Steuer* (Frankfurt: Fischer Taschenbuch Verlag, 1999), 31–44

Joel Dinerstein, *Swinging the Machine: Modernity, Technology, and African American Culture between the World Wars* (Amherst and Boston: University of Massachusetts Press, 2003)

Robert Dinse, *Das Freizeitleben der Grossstadtjugend: 5000 Jungen und Mädchen berichten: Zusammengestellt und bearbeitet in Verbindung mit dem Deutschen Archiv für Jugendwohlfahrt von ––, Stadtjugendpfleger, Berlin,* Schriftenreihe des Deutschen Archivs für Jugendwohlfahrt, Heft 10 (Eberswalde-Berlin: Verlagsgesellschaft R. Müller, n.d. [1932?])

Sasha Disko, "Men, Motorcycles and Modernity: Motorization in the Weimar Republic" (unpublished dissertation, New York University, May 2008)

Colin Divall, "Mobilizing the History of Technology," *Technology and Culture* 51, no. 4 (October 2010), 938–960

Colin Divall and George Revill, "Cultures of transport: Representation, practice and technology," *Journal of Transport History* (Third Series) 26, no. 1 (March 2005), 99–111

Bryony Dixon, *100 Silent Films*, BFI Screen Guides (Houndsmill and New York: Palgrave Macmillan, 2011)

Robert Dixon, *Prosthetic Gods: Travel, Representation and Colonial Governance* (St Lucia: University of Queensland Press in association with the API Network, 2001)

Anselm Doering-Manteuffel, "Konturen von 'Ordnung' in den Zeitschichten des 20. Jahrhunderts," in Thomas Etzemüller (ed.), *Die Ordnung der Moderne. Social Engineering im 20. Jahrhundert* (Bliefeld: transcript Verlag, 2009) 41–64

Lubomír Doležel, *Possible Worlds of Fiction and History* (Baltimore, MD: Johns Hopkins University Press, 2010)

Rob Doll, "An Interpretation [of] E.M. Forster's Howards End" (2000), http://www.emforster.info/pages/howardsend.html (last accessed 27 July 2009)

Monika Dommann, "Material manövrieren: Eine Begriffsgeschichte der Logistik," *Wege und Geschichte* 2/2009 (February 2010), 13–17

James Rood Doolittle, *The Romance of the Automobile Industry: Being the Story . . .* (etc.) (New York, 1916)

Luc van Doorslaer, "Cyriel Buysse en de automobiel," in Cyriel Buysse, *Reizen van toen: Met de automobiel door Frankrijk*, edited and introduced by Luc van Doorslaer (Antwerpen and Amsterdam: Manteau, 1992), 7–22

[Aimé] Doumenc, *Les transports automobiles sur le front français 1914 – 1918*[4] (Paris: Plon-Nourrit et Cie., 1920)

André Doumenc, "Les transports automobiles pendant la guerre de 1914–1918," in Gérard Canini (ed.), *Les fronts invisibles: Nourrir—fournir—soigner; Actes du colloque international sur La logique des armées au combat pendant la Première Guerre Mondiale organise à Verdun les 6, 7, 8 juin 1980 sous la présidence du general d'Armée F. Gambiez, membre de l'Institut* (Nancy: Presses Universitaires de Nancy, 1984), 371–380

James A. Doyle, *The Male Experience* (Madison, WI, and Dubuque, IA: Wm. C. Brown Communications, 1995; first ed. 1983)

Theodore Dreiser, *A Hoosier Holiday*, with illustrations by Franklin Booth and an introduction by Douglas Brinkley (Bloomington and Indianapolis: Indiana University Press, 1997; first ed. 1916)

——, *A Traveler at Forty*, edited by Renate von Bardeleben; the Dreiser Edition, eds. Thomas P. Riggio et al. (Urbana and Chicago: University of Illinois Press, 2004; first ed. 1913)

W. Drukker and Timo de Rijk, "American Influences on Dutch Material Culture and Product Design," in Hans Krabbendam, Cornelis A. van Minnen, and Giles Scott-Smith (eds.), *Four Centuries of Dutch-American Relations 1609–2009* (Amsterdam: Boom, 2009)

Diane Drummond, "The impact of the railway on the lives of women in the nineteenth-century city," in Ralf Roth and Marie-Noëlle Polino (eds.), *The City and the Railway in Europe*, Historical Urban Studies, eds. Richard Rodger and Jean-Luc Pinol (Aldershot: Ashgate, 2003), 237–255

Antoine Dubois, "La motocyclette, modernité et véhicule social: Émergence et resistance, l'exemple du département du Cher, 1899–1914" (Thèse de Maîtrise, Université François Rabelais U.E.R., Tours 1989)

Jean-Pierre Dubost, "Ubu automobile," *Revue des sciences humaines* 3 (1986), 161–180

Hyacinthe Dubreuil, *Les codes de Roosevelt et les perspectives de la vie sociale* (Paris: Éditions Bernard Grasset, 1934)

Baron F. Duckham, "Early motor vehicle licence records and the local historian," *The local historian* 17 (1987), 351–357

Stefan Dudink, Karen Hagemann, and John Tosh (eds.), *Masculinities in Politics and War: Gendering Modern History* (Manchester and New York: Manchester University Press, 2004)

Enda Duffy, *The Speed Handbook: Velocity, Pleasure, Modernism*, Post-Contemporary Interventions, eds. Stanley Fish and Fredric Jameson (Durham, NC, and London: Duke University Press, 2009)

Larry Duffy, *Le Grand Transit Moderne: Mobility, Modernity and French Naturalist Fiction* (Amsterdam and New York: Rodopi, 2005)

Jost Dülffer, "Attentes allemandes, attentes italiennes: Préfigurations de la guerre en Allemagne avant 1914," in Jean-Jacques Becker, Jay M. Winter, Gerd Krumeich, Annette Becker and Stéphane Audoin-Rouzeau, *Guerre et cultures 1914 – 1918* (Paris : Armand Colin, 1994) 65–78

James A. Dunn, Jr., "The Importance of Being Earmarked: Transport Policy and Highway Finance in Great Beritain and the United States," *Comparative Studies in Society and History* 20, no. 1 (January 1978), 29–53

Alastair Durie, "No holiday this year? The depression of the 1930s and tourism in Scotland," *Journal of Tourism History* 2, no. 2 (August 2010), 67–82

André Duvignac, *Histoire de l'armée motorisée* (Paris: Imprimerie Nationale, 1947)

Carol Dyhouse, "Mothers and Daughters in the Middle-Class Home, c. 1870–1914," in Jane Lewis (ed.), *Labour and Love: Women's Experience of Home and Family, 1850–1940* (Oxford: Basil Blackwell, 1986), 31–47

Anne-Katrin Ebert, "Cycling towards the Nation: The Use of the Bicycle in Germany and the Netherlands, 1880–1940," *European Review of History* 11, no. 3 (2004), 347–364

———, "Ein Ding der Nation? Das Fahrrad in Deutschland und den Niederlanden, 1880–1940. Eine vergleichende Konsumgeschichte" (doctoral dissertation, Universität Bielefeld, Fakultät für Geschichtswissenschaft, Philosophie und Theologie, January 2009)

Wilfried Echterhoff, "Geschichte der Verkehrspsychologie, Teil 1," *Zeitschrift für Verkehrssicherheit* 36, no. 2 (1990), 50–70

———, "Geschichte der Verkehrspsychologie, Teil 2," *Zeitschrift für Verkehrssicherheit* 36, no. 3 (1990), 98–112

Ross D. Eckert and George W. Hilton, "The Jitneys," *Journal of Law and Economics* 25 (October 1972), 293–325

Heidrun Edelmann, "Der Umgang mit dem Rückstand. Deutschlands Automobilindustrie in der Zwischenkriegszeit," in Rudolf Boch (ed.), *Geschichte und Zukunft der deutschen Automobilindustrie: Tagung im Rahmen der "Chemnitzer Begegnungen" 2000* (Stuttgart: Franz Steiner Verlag, 2001), 41–48.

———, *Vom Luxusgut zum Gebrauchsgegenstand: Die Geschichte der Verbreitung von Personenkraftwagen in Deutschland* (Frankfurt: Verband der Automobilindustrie, 1989)

Tim Edensor, "Automobility and National Identity: Representation, Geography and Driving Practice," *Theory, Culture & Society* 21, no. 4/5 (2004), 101–120

Ronald Edsforth, *Class Conflict and Cultural Consensus: The Making of a Mass Consumer Society in Flint, Michigan*, Class and Culture Series, ed. Milton Cantor and Bruce Laurie (New Brunswick, NJ, and London: Rutgers University Press, 1987)

Hanno Ehrlicher, *Die Kunst der Zerstörung: Gewaltphantasien und Manifestationspraktiken europäischer Avantgarden,* Studien aus dem Warburg-Haus, eds. Wolfgang Kemp et al., Band 4 (Berlin: Akademie Verlag, 2001)

Henning Eichberg, "'Join the army and see the world': Krieg als Touristik—Tourismus als Krieg," in Dieter Kramer and Ronald Lutz (eds.), *Reisen und Alltag: Beiträge zur kulturwissenschaftlichen Tourismusforschung* (Frankfurt: Institut für Kulturanthropologie und Europäische Ethnologie der Universität Frankfurt am Main, 1992), 207–228

Carl Einstein, *Bebuquin,* ed. Erich Kleinschmidt, Reclams Universal-Bibliothek no. 8057 (Stuttgart: Philipp Reclam jun., 2008)

Christiane Eisenberg, *"English sports" und Deutsche Bürger: Ein Gesellschaftsgeschichte 1800–1939* (Paderborn, Munich, Vienna, and Zürich: Ferdinand Schöningh, 1999)

———, "Massensport in der Weimarer Republik; Ein statistischer Überblick," *Archiv für Sozialgeschichte* 33 (1993), 137–177

Modris Eksteins, *Tanz über Graben: Die Geburt der Moderne und der Erste Weltkrieg,* transl. from *Rites of Spring: The Great War and the Birth of the Modern Age,* 1989, by Bernhard Schmid (Reinbek bei Hamburg: Rowohlt Verlag, 1990)

Geoff Eley, "Is all the world a text? From social history to the history of society two decades later," in Gabrielle M. Spiegel (ed.), *Practicing History: New Directions in Historical Writing after the Linguistic Turn,* Rewriting Histories, ed. Jack R. Censer (New York and London: Routledge, 2005), 35–61

Norbert Elias, "Technization and Civilization," *Theory, Culture & Society* 12 (1995), 7–42

Norbert Elias and Eric Dunning, *Quest for Excitement: Sport and Leisure in the Civilizing Process* (Oxford and New York: Basil Blackwell, 1986)

Jane Elliott, *Using Narrative in Social Research: Qualitative and Quantitative Approaches* (Thousand Oaks, CA: Sage, 2009; reprint of 2005 edition)

Cliff Ellis, "Lewis Mumford and Norman Bel Geddes: The highway, the city and the future," *Planning Perspectives* 20 (January 2005), 51–68

Christopher Endy, *Cold War Holidays: American Tourism in France,* the New Cold War History, ed. John Lewis Gladdis (Chapel Hill and London: University of North Carolina Press, 2004)

"Épilogue; La Mort tient le volant ... ," in F. T. Marinetti, *La Ville charnelle* (Paris: E. Sansot, 1908), 221–229

Ralph C. Epstein, *The Automobile Industry: Its Economic and Commercial Development* (Chicago and New York, 1928)

Thomas Etzemüller (ed.), *Die Ordnung der Moderne: Social Engineering im 20. Jahrhundert* (Bielefeld: transcript Verlag, 2009), 153–169

Rhys Evans and Alexandra Franklin, "Equine Beats: Unique Rhythms (and Floating Harmony) of Horses and Riders," in Tim Edensor (ed.), *Geographies of Rhythm: Nature, Place, Mobilities and Bodies* (Farnham and Burlington: Ashgate, 2010), 173–185

Bernard Fabienne, "Les femmes et l'automobile de 1900 à 1930" (Thèse de Maîtrise d'Histoire Contemporaine, Université Blaise Pascal, 1977)

Dietmar Fack, *Automobil, Verkehr und Erziehung: Motorisierung und Sozialisation zwischen Beschleunigung und Anpassung 1885–1945* (Opladen: Leske + Budrich, 2000)

———, "Das deutsche Kraftfahrschulwesen und die technisch-rechtliche Konstitution der Fahrausbildung 1899–1943," *Technikgeschichte* 67, no. 2 (2000), 111–138

Delia Falconer, "'The Poetry of the Earth is Never Dead': Australia's Road Writing," *Journal of the Association for the Study of Australian Literature*, Special Issue: Australian Literature in a Global World (2009), 1–16

Hans Fallada, *Kleiner Mann—was nun?* (Hamburg: Rowohlt Taschenbuch Verlag, 2009; first ed. 1932)

Family Expenditures in Selected Cities, 1935–36, vol. VI: Travel and Transportation (Washington, DC: United States Government Printing Office, 1940)

John T. Faris, *Roaming American Playgrounds* (New York: Farrar & Rinehart, 1934)

Paula S. Fass, *The Damned and the Beautiful: American Youth in the 1920's* (New York: Oxford University Press, 1977)

Étienne Faugier, "L'introduction du système automobile et ses impacts sur les campagnes du département du Rhône entre 1900 et 1939" (unpublished Master's thesis in Histoire Contemporaine [Études Rurales], Université Lumière Lyon 2, Département d'histoire, September 2007)

Leo Faust, "Het ontwaken," *De Auto* (1909)

Mike Featherstone (ed.), *Global Culture: Nationalism, Globalization and Modernity* (Thousand Oaks, CA: Sage, 1990)

Michael R. Fein, *Paving the Way: New York Road Building and the American State, 1880–1956* (Lawrence: University Press of Kansas, 2008)

Karolina Dorothea Fell, *Kalkuliertes Abenteuer: Reiseberichte deutschsprachiger Frauen (1920–1945)*, Ergebnisse der Frauenforschung, eds. Anke Bennholdt-Thomsen et.al., Band 49 (Stuttgart and Weimar: J. B. Metzler, 1998)

Anton Fendrich, *Mit dem Auto an der Front: Kriegserlebnisse* (Stuttgart: Franckh'sche Verlagshandlung, n.d.)

Theodore N. Ferdinand, "The Criminal Patterns of Boston since 1849," *American Journal of Sociology* 73, no. 1 (July 1967), 84–99

Olaf von Fersen (ed.), *Ein Jahrhundert Automobiltechnik: Nutzfahrzeuge* (Düsseldorf, 1987)

Albert Feuerwerker, *China's Early Industrialization: Sheng Hsuan-huai (1844–1916) and Mandarin Enterprise* (Cambridge, MA: Harvard University Press, 1958)

Mark Fiege, "The Weedy West: Mobile Nature, Boundaries, and Common Space in the Montana Landscape," *Western Historical Quarterly* 36, no. 1 (Spring 2005), 22–47

Ruud Filarski (in cooperation with Gijs Mom), *Shaping Transport Policy: Two centuries of struggle between the public and private sector—A comparative perspective* (The Hague: Sdu Uitgevers, 2011)

Ruud Filarski and Gijs Mom, *Van transport naar mobiliteit: De transportrevolutie (1800–1900)* (Zutphen: Walburg Pers, 2008)

Claude S. Fischer, *Made in America: A Social History of American Culture and Character* (Chicago and London: University of Chicago Press, 2010)

Claude S. Fischer and Glenn R. Carroll, "Telephone and Automobile Diffusion in the United States, 1902–1937," *American Journal of Sociology* 93, no. 5 (March 1988), 1153–1178

Wolfram Fischer, "Wirtschaft, Gesellschaft und Staat in Europa 1914–1980," in Wolfram Fischer (ed.), *Europäische Wirtschafts- und Sozialgeschichte vom Ersten Weltkrieg bis zur Gegenwart* (Stuttgart: Klett-Cotta, 1987), 1–221

Marion Fish, *Buying for the Household as Practiced by 368 Farm Families in New York, 1928–29*, Bulletin 561; Contribution from Studies in Home Economics (Ithaca, NY: Cornell University Agricultural Experiment Station, June 1933)

Thomas Fitzel, "Schmutz und Geschwindigkeit oder Warum das *Tempo* Tempo heisst," in Jörg Döring, Christian Jäger, and Thomas Wegmann (eds.), *Verkehrsformen und Schreibverhältnisse: Mediale Wandel als Gegenstand und Bedingung von Literatur im 20. Jahrhundert* (Opladen: Westdeutscher Verlag, 1996), 74–98

Bradley Flamm, "Putting the brakes on 'non-essential' travel: 1940s wartime mobility, prosperity, and the US Office of Defense Transportation," *Journal of Transport History*, Third Series, 27, no. 1 (March 2006), 71–92

Marieluise Fleisser, *Gesammelte Werke, Erster Band: Dramen*, edited by Günther Rühle (Frankfurt: Suhrkamp Verlag, 1983; first ed. 1972)

Hans Flesch von Bruningen, "Der Satan," *Die Aktion* 4, no. 30 (25 July 1914), columns 658–670

Reiner Flik, *Von Ford lernen? Automobilbau und Motorisierung in Deutschland bis 1933* (Cologne, Weimar, and Vienna: Böhlau Verlag, 2001)

James J. Flink, *The Car Culture* (Cambridge, MA, and London: MIT Press, 1975)

——, "*The Car Culture* Revisited: Some Comments on the Recent Historiography of Automotive History" in David L. Lewis and Laurence Goldstein (eds.), *The Automobile and American Culture* (Ann Arbor: University of Michigan Press, 1983), 89–104

——, "Three Stages of American Automobile Consciousness," *American Quarterly* 24 (October 1972), 451–473

——, "The Ultimate Status Symbol: The Custom Coachbuilt Car in the Interwar Period," in Martin Wachs and Margaret Crawford (with the assistance of Susan Marie Wirka and Taina Marjatta Rikala), *The Car and the City: The Automobile, The Built Environment, and Daily Urban Life* (Ann Arbor: University of Michigan Press, 1992), 154–166

Mathieu Flonneau (ed.), Automobile: Les cartes des désamour; Généalogies de l'antiautomobilisme (Paris: Descartes & Cie, 2009) (Collection "Cultures Mobiles," eds.: Mathieu Flonneau and Arnaud Passalacqua)

Mathieu Flonneau, L'Autorefoulement et ses limites; Raisonner l'impensable mort de l'automobile (Paris: Descartes & Cie, 2010) (Collection "Cultures mobiles," eds.: Mathieu Flonneau and Arnaud Passalacqua)

Mathieu Flonneau, *Les cultures du volant: Essai sur les mondes de l'automobilisme XXe–XXIe siècles*, Collection Mémoires / Culture no. 141 (Paris: Éditions Autrement, 2008), 65–66

——, "The History of Mobility in France: A Recent, but Now Accepted, Turn," in Peter Norton et al. (eds.), *Mobility in History: Reviews and Reflections (T²M Yearbook 2012)* (Neuchâtel: Alphil, 2011), 51–62

——, *Paris et l'automobile: Un siècle de passions* (Paris: Hachette Littératures, 2005)

——, "Read Tocqueville, or drive? A European perspective on US 'automobilization,'" *History and Technology* 26, no. 4 (December 2010), 379–388

——, "Une histoire des taxis parisiens au cours de la première moitié du XXe siècle: Entre innovation, occasions manquées et blocages," in *Paris et Ile-de-France: Mémoires publiés par la Fédération des Sociétés Historiques et Archéologiques de Paris et de L'Ile-de-France, Tome 62* (Paris, 2011), 37–53

Douwe Fokkema and Elrud Ibsch, *Modernist Conjectures: A Mainstream in European Literature 1910–1940* (London: C. Hurst & Company, 1987)

James W. Follin, "Taxation of Motor Vehicles in the United States," *Annals of the American Academy of Political and Social Sciences* 116 (The Automobile: Its Province and Its Problems) (November 1924), 141–159

T. W. Forbes, "The Normal Automobile Driver as a Traffic Problem," *Journal of General Psychology* 20 (1939), 471–474

Mark Ford, *Raymond Roussel and the Republic of Dreams* (Ithaca, NY: Cornell University Press, 2000)

Ellen Forest, *Een vacantiereis van 23.000 k.m. per Hudson door Europa* (Amersfoort: S.W. Melchior, n.d.)

E. M. Forster, *Howards End: Introduction and notes by David Lodge* (London and New York: Penguin Books, 2000) (first ed.: 1910)

Mark Foster, "City Planners and Urban Transportation; The American Response, 1900–1940," *Journal of Urban History* 5, no. 3 (May 1979), 365–396

———, *From Streetcar to Superhighway: American City Planners and Urban Transportation, 1900–1940*, Technology and Urban Growth Series, eds. Blaine A. Brownell et al. (Philadelphia: Temple University Press, 1981)

———, "In the Face of 'Jim Crow': Prosperous Blacks and Vacations, Travel and Outdoor Leisure, 1890–1945," *Journal of Negro History* 84, no. 2 (Spring 1999), 130–149

———, "The Role of the Automobile in Shaping a Unique City: Another Look," in Martin Wachs and Margaret Crawford (with the assistance of Susan Marie Wirka and Taina Marjatta Rikala), *The Car and the City: The Automobile, The Built Environment, and Daily Urban Life* (Ann Arbor: University of Michigan Press, 1992), 186–193

Stephen Fox, *John Muir and his Legacy: The American Conservation Movement* (Boston and Toronto: Little, Brown and Company, 1981)

Allen J. Frantzen, *Bloody Good: Chivalry, Sacrifice, and the Great War* (Chicago and London: University of Chicago Press, 2004)

Kathleen Franz, "'The Open Road': Automobility and Racial Uplift in the Interwar Years," in Bruce Sinclair (ed.), *Technology and the African-American Experience: Needs and Opportunities for Study* (Cambridge, MA, and London: MIT Press, 2004), 131–153

———, *Tinkering: Consumers Reinvent the Early Automobile* (Philadelphia, PA: University of Pennsylvania Press, 2005)

Uwe Fraunholz, *Motorphobia: Anti-automobiler Protest in Kaiserreich und Weimarer Republik*, Kritische Studien zur Geschichtswissenschaft, eds. Helmut Berding et al., Band 156 (Göttingen: Vandenhoeck & Ruprecht, 2002)

Christine Frederick, "New Wealth, New Standards of Living, and Changed Family Budgets," *Annals* (American Academy of Political and Social Science) 115 (September 1924), 74–82

Christine McGaffey Frederick, "The Commuter's Wife and the Motor Car," *Suburban Life* no. 13 (July 1912), 13–14, 46

Libbie Freed, "Networks of (colonial) power: Roads in French central Africa after World War I," *History and Technology* 26, no. 3 (September 2010), 203–233

Julia Freytag, "'Lebenmüssen ist eine einzige Blamage': Marieluise Fleissers Blick auf stumme Provinzheldinnen und Buster Keaton," in Julia Freytag and Alexandra Tacke (eds.), *City Girls: Bubiköpfe & Blaustrümpfe in den 1920er Jahren*, Literatur—Kultur—Geschlecht; Studien zur Literatur- und Kulturgeschichte, eds. Anne-Kathrin Reulecke and Ulrike Vedder, in connection with Inge Stephan und Sigrid Weigel, Kleine Reihe, Band 29 (Cologne, Weimar, and Vienna: Böhlau Verlag, 2011), 91–109

"Friedrich Schnack," http://de.wikipedia.org/wiki/Friedrich_Schnack (last accessed 29 October 2011)

Ann Frodi, Jacqueline Macaulay, and Pauline Ropert Thome, "Are Women Always Less Aggressive Than Men? A Review of the Experimental Literature," *Psychological Bulletin* 84, no. 4 (1977), 634–660

Wayne E. Fuller, "Good Roads and Rural Free Delivery of Mail," *Mississippi Valley Historical Review* 42 (June 1955), 67–83

Ellen Furlough, "Making Mass Vacations: Tourism and Consumer Culture in France, 1930s to 1970s," *Comparative studies in society and history* 40 (1998), 247–286

Paul Fussell, *Abroad: British Literary Traveling Between the Wars* (Oxford, New York, Toronto, and Melbourne: Oxford University Press, 1980)

——, *The Great War and Modern Memory* (Oxford: Oxford University Press, 2000; first ed. 1975)

Lothar Gall, "Eisenbahn in Deutschland: Von den Anfängen bis zum Ersten Weltkrieg," in Lothar Gall and Manfred Pohl (eds.), *Die Eisenbahn in Deutschland: Von den Anfängen bis zur Gegenwart* (Munich: Verlag C. H. Beck, 1999), 143–163

George H. Gallup, *The Gallup Poll: Public Opinion 1935–1971*, vol. 1, 1935–1948 (New York: Random House, 1972)

Garet Garrett, "Business," in Harold E. Stearns, *Civilization in the United States: An Inquiry by Thirty Americans* (New York: Harcourt, Brace and Company, 1922), 397–415

Jane Garrity, "Virginia Woolf, Intellectual Harlotry, and 1920s British *Vogue*," in Pamela L. Caughie (ed.), *Virginia Woolf in the Age of Mechanical Reproduction*, Border Crossings, ed. Daniel Albright (New York/London: Garland, 2000), 185–218

David Gartman, *Auto Opium: A Social History of American Automobile Design* (New York and London: Routledge, 1994)

Ellen Gruber Garvey, "Reframing the Bicycle: Advertising-Supported Magazines and Scorching Women," *American Quarterly* 47, no. 1 (March 1995), 66–101

Oscar Gaspari, "Cities against States? Hopes, Dreams and Shortcomings of the European Municipal Movement, 1900–1960," *Contemporary European History* 11, no. 4 (2002), 597–621

Lettie Gavin, *American Women in World War 1: They Also Served* (Niwot: University Press of Colorado, 1997)

"Théophile Gautier," http://en.wikipedia.org/wiki/ThpercentC3percentA9ophile_Gautier (last accessed 13 March 2010)

Kostas Gavroglu et al., "Science and Technology in the European Periphery: Some Historiographical Reflections," *History of Science* 46 (2008), 153–175

Peter Gay, *The Bourgeois Experience, Victoria to Freud*, vol. II: The Tender Passion (New York and Oxford: Oxford University Press, 1986)

——, *The Bourgeois Experience, Victoria to Freud*, vol. III: The Cultivation of Hatred (New York and London: W.W. Norton Company, 1993)

——, *Weimar Culture: The Outsider as Insider* (New York and London: W. W. Norton & Company, 2001)

Gunter Gebauer et al. (eds.), *Kalkuliertes Risiko: Technik, Spiel und Sport an der Grenze* (Frankfurt and New York: Campus Verlag, 2006)

Gedenkboek der Werkstakingen van 1903 (Actestukken der Samenzwering), Deel I: Voorspel. Eerste en Tweede Periode (Wageningen: N. V. Drukkerij "Vada", 1903)

Gedenkboek der Werkstakingen van 1903 (Actestukken der Samenzwering), Deel II: Derde en Vierde Periode, Naspel, Bijlage en Register (Wageningen: N. V. Drukkerij "Vada", 1904)

Gedenkboek van het 25-jarig bestaan der Koninklijke Nederlandsche Automobiel Club, 1898-3 juli 1923 (Haarlem, 1923)

Generale rapporten: VIIIe Congres 's-Gravenhage 1938 (n.p., n.d.) [see n. 26 on p. 621]

Generalinspektor für das deutsche Strassenwesen, Der Kraftverkehr auf Reichsautobahnen, Reichs- und Landstrassen im Dritten Reich I: Die dritte allgemeine Verkehrszählung 1936/37 auf Reichsautobahnen, Reichsstrassen und Landstrassen I. Ordnung: Die Verkehrsentwicklung auf den Reichsautobahnen bis Dezember 1938: Die dritte österreichische Verkehrszählung (Winterhalbjahr 1937/38): Die Zusatzverkehrszählung 1936/37 auf Reichsautobahnen, Reichs- und Landstrassen (Berlin: Volk und Reich Verlag, 1939)

James J. Gibson, The Ecological Approach to Visual Perception (Boston and London: Houghton Mifflin Company, 1979)

James J. Gibson and Laurence E. Crooks, "A Theoretical Field-Analysis of Automobile-Driving," American Journal of Psychology 51, no. 3 (July 1938), 453-471

Thomas Gibson, Road Haulage by Motor in Britain: The First Forty Years (Aldershot, Burlington, Singapore, and Sydney: Ashgate, 2001)

André Gide, Journal sans dates, La Nouvelle Revue française no. 21 (1 September 1910), 336-341

——, Voyage au Congo, suivi de Le retour du Tchad: Carnets de Routes (n.p., n.y [Paris: Gallimard, 1985]; first ed. 1927 and 1928)

Peter Giesen, "'Het is irreëel te geloven in eeuwige vrede,'" De Volkskrant (16 April 2011), "Het Vervolg," 8 (interview with John Gray)

——, "Vroeger knokken op de kermis, nu losgaan in Haren," De Volkskrant (24 September 2012), 1

——, "'Wat er ook gebeurt, het ligt aan jezelf,'" De Volkskrant (4 February 2012) "Het Vervolg," 6-7 (interview with Zygmunt Bauman)

John S. Gilkeson, Anthropologists and the Rediscovery of America, 1886-1965 (Cambridge and New York: Cambridge University Press, 2010)

John S. Gilkeson, Jr., Middle-Class Providence, 1820-1940 (Princeton, NJ: Princeton University Press, 1986)

Maurice Girard, L'automobile fait son cinema (n.p. [Paris]: Du May, n.d.)

Guillermo Giucci, The Cultural Life of the Automobile: Roads to Modernity, Llilas Translations from Latin America Series (Austin: University of Texas Press / Teresa Lozano Long Institute of Latin American Studies, 2012; transl. from A vida cultural del automóvil: Rutas de la modernidad cinética (2007) by Anne Mayagoitia and Debra Nagao)

Hermann Glaser, Das Automobil: Eine Kulturgeschichte in Bildern (Munich: Verlag C. H. Beck, 1986)

Pierre Glaudes, "Introduction générale," in Léon Bloy, Journal I, 1892-1907 (Édition établie, présentée et annotée par Pierre Glaudes) (Paris: Robert Laffont, 1999), VII-LXXV

Anke Gleber, "Die Erfahrung der Moderne in der Stadt; Reiseliteratur der Weimarer Republik," in Peter J. Brenner (ed.), Der Reisebericht; Die Entwicklung einer Gattung in der deutschen Literatur (Frankfurt am Main: Suhrkamp Verlag, 1989; suhrkamp taschenbuch 2097; suhrkamp taschenbuch materialien), 463-489

Ellen Walker Glenn, "The Androgynous Woman Character in the American Novel" (unpublished dissertation, University of Colorado, 1980)

Hiltrud Gnüg, Kult der Kälte: Der klassische Dandy im Spegel der Weltliteratur (Stuttgart: J. B. Metzlersche Verlagsbuchhandlung, 1988)

Stephen B. Goddard, *Getting There: The Epic Struggle between Road and Rail in the American Century* (New York: Harper Collins, 1994)

Lawrence B. Goodheart and Richard O. Curry, "A Confusion of Voices: The Crisis of Individualism in Twentieth-Century America," in Richard O. Curry and Lawrence B. Goodheart (eds.), *American Chameleon: Individualism in Trans-National Context* (Kent, OH, and London: Kent State University Press, 1991), 188–212

Paul Grabein, *Im Auto durch Feindesland: Sechs Monate im Autopark der Obersten Heeresleitung* (Berlin and Vienna: Verlag Ullstein & CO, 1916)

Grith Graebner, "Dem Leben unter die Haut kriechen..." *Heinrich Hauser—Leben und Werk: Eine kritisch-biographische Werk-Bibliographie* (Aachen: Shaker Verlag, 2001; D38 [Dissertation Universität Cologne])

Nanette Hope Graf, "The Evolution of Willa Cather's Judgment of the Machine and the Machine Age in her Fiction" (unpublished doctoral dissertation, University of Nebraska, Lincoln, 1991)

Kenneth Grahame, *The Wind in the Willows* (New York: Dell, 1969)

Elke Gramespacher, "Sport—Bewegungen—Geschlechter," *Geschlechter—Bewegungen—Sport* (Freiburger GeschlechterStudien 23 [2009]), 13–30

Greg Grandin, *Fordlandia: The Rise and Fall of Henry Ford's Forgotten Jungle City* (New York: Metropolitan Books / Henry Holt and Company, 2009)

Robert Graves and Alan Hodge, *The Long Week-end: A Social History of Great Britain 1918–1939* (London: Faber and Faber, 1940)

Victoria de Grazia, *The Culture of Consent: Mass Organization of Leisure in Fascist Italy* (Cambridge and London: Cambridge University Press, 1981)

Martin Green, *Seven Types of Adventure Tale: An Etiology of a Major Genre* (University Park: Pennsylvania State University Press, 1991)

James N. Gregory, *The Southern Diaspora: How the Great Migrations of Black and White Southerners Transformed America* (Chapel Hill: University of North Carolina Press, 2005)

Roger Griffin, *Modernism and Fascism: The Sense of a Beginning under Mussolini and Hitler* (Houndsmill and New York: Palgrave, 2007)

Charles Grivel, "Automobile et cinéma I," in Michel Bouvier, Michel Lartouce, and Lucie Roy (eds.), *Cinéma: acte et présence* (Quebec: Éditions Nota Bene, 1999), 63–85

——, "D'un écran automobile," in Jochen Mecke and Volker Roloff (eds.), *Kino-/(Ro)Mania: Intermedialität zwischen Film und Literatur,* Siegener Forschungen zur romanischen Literatur- und Medienwissenschaft, ed. Volker Roloff, Band 1 (Tübingen: Stauffenburg Verlag, 1999), 47–77

——, "Photographie, littérature, peinture, cinéma: la guerre des arts," in Beate Ochsner and Charles Grivel (eds.), *Intermediale: Kommunikative Konstellationen zwischen Medien,* Siegener Forschungen zur romanischen Literatur- und Medienwissenschaft, ed. Volker Roloff, Band 10 (Tübingen: Stauffenburg Verlag, 2001), 11–44

——, "Travel Writing," in Hans Ulrich Gumbrecht and K. Ludwig Pfeiffer (eds.), *Materialities of Communication* (Stanford, CA: Stanford University Press, 1994; partially transl. by William Whobrey from *Materialität der Kommunikation,* 1988), 242–257

John A. Groeger, *Understanding Driving: Applying Cognitive Psychology to a Complex Everyday Task* (Hove: Psychology Press, 2000; reprint 2001)

Grossjohan, Mallison, and Temme, "Construction and maintenance" (report 22, PIARC Conference, The Hague, 1938) [see n. 26 on p. 621]

Götz Grossklaus, "Der Naturraum des Kulturbürgers," in Götz Grossklaus and Ernst Oldemeyer, *Natur als Gegenwelt: Beiträge zur Kulturgeschichte der Natur* (Karlsruhe: von Loeper Verlag, 1983)

Dave Grossman, *On Killing: The Psychological Cost of Learning to Kill in War and Society* (New York, Noston, and London: Back Bay Books / Little, Brown and Company, 2009)

Jan Erik Grunveld, *Per ATO en per spoor: 20 jaar omstreden autobushistorie* (Utrecht, 1987)

Catherine Gudis, *Buyways: Billboards, Automobiles, and the American Landscape*, Cultural Spaces series, ed. Sharon Zukin (New York and London: Routledge, 2004)

———, "Driving consumption," *History and Technology* 26, no. 4 (December 2010), 369–378

Anette Gudjons, "Die Entwicklung des 'Volksautomobils' von 1904 bis 1945 unter besonderer Berücksichtigung des 'Volkswagens': Ein Beitrag zu Problemen der Sozial-, Wirtschafts- und Technikgeschichte des Automobils" (unpublished dissertation, Universität Hannover, 1988)

Jürgen Gunia, "Extreme Diskurse: Anmerkungen zur Kritik medialer Beschleunigung bei Günther Anders und Paul Virilio," in Leonhard Fuest and Jörg Löffler (eds.), *Diskurse des Extremen: Über Extremismus und Radikalität in Theorie, Literatur und Medien*, Film—Medium—Diskurs, eds. Oliver Jahraus and Stefan Neuhaus, Band 6 (Würzburg: Königshausen & Neumann, 2005), 173–188

Ted Robert Gurr (ed.), *Violence in America*, vol. 2: Protest, Rebellion, Reform; Violence, Cooperation, Peace: An International Series, eds. Francis A. Beer and Ted Robert Gurr (Thousand Oaks, CA: Sage, 1989; first ed. 1969, in one volume)

Hermann Gurtner, *Automobil, Tourismus, Hotellerie: Eine Untersuchung über die Bedeutung des Automobils* (Bern: Touring-Club der Schweiz, 1946)

Owen D. Gutfreund, *Twentieth-Century Sprawl: Highways and the Reshaping of the American Landscape* (Oxford: Oxford University Press, 2004)

Rüdiger Hachtmann, *Tourismus-Geschichte*, Grundkurs neue Geschichte, eds. Manfred Hettling, Martin Sabrow and Hans-Ulrich Thamer (Göttingen: Vandenhoeck & Ruprecht, 2007)

———, "Tourismusgeschichte—ein Mauerblümchen mit Zukunft! Ein Forschungsüberblick," in H-Soz-u-Kult 06.10.2011, http://hsozkult.geschichte.hu-berlin.de/forum/2011–10–001 (last accessed 6 October 2011)

F. Hageman, "De Truc," *De Auto* (1909), 119–121, 146–148

Cindy Hahamovitch, *The Fruits of Their Labor: Atlantic Coast Farmworkers and the Making of Migrant Poverty, 1870–1945* (Chapel Hill and London: University of North Caroline Press, 1997)

Hans Peter Hahn, "The Appropriation of Bicycles in Africa: Pragmatic Approaches to Sustainability" (paper presented at the workshop "Re/Cycling Histories: Users and the Paths to Sustainability in Everyday Life," Rachel Carson Center, Munich, 27–29 May 2011)

———, "Global Goods and the Process of Appropriation," in Peter Probst and Gerd Spittler (eds.), *Between Resistance and Expansion: Dimensions of Local Vitality in Africa* (Münster: Lit, 2003), 213–231

———, *Materielle Kultur: Eine Einführung* (Berlin: Dietrich Reimer Verlag, 2005)

Frank A. Haight, "Accident Proneness: The History of an Idea" (Irvine: Institute of Transportation Studies, August 2001; UCI-ITS-WP-01–2)

Wolfgang Hallet and Birgit Neumann, "Raum und Bewegung in der Literatur: Zur Einführung," in Hallet and Neumann (eds.), *Raum und Bewegung in der Litera-*

tur: Die Literaturwissenschaften und der Spatial Turn (Bielefeld: transcript Verlag, 2009), 11–32

Walton Hamilton et al., *Price and Price Policies* (New York and London: McGraw-Hill, 1938)

Richard Hamilton, "Die soziale Basis des Nationalsozialismus: Eine kritische Betrachtung," in Heinz-Gerhard Haupt and Claudius Torp (eds.), *Die Konsumgesellschaft in Deutschland 1890 – 1990; Ein Handbuch* (Frankfurt and New York: Campus Verlag, 2009), 354–375

———, *Who Voted for Hitler?* (Princeton, NJ: Princeton University Press, 1982)

Shane Hamilton, *Trucking Country: The Road to America's Wal-Mart Economy* (Princeton, NJ, and Oxford: Princeton University Press, 2008)

Christa Hämmerle, "Zur Relevanz des Connell'schen Konzept hegemonialer Männlichkeit für 'Militär und Männlichkeit/en in der Habsburgermonarchie (1868–1914/18),'" in Martin Dinges (ed.), *Männer—Macht—Körper: Hegemoniale Männlichkeiten vom Mittelalter bis heute* (Frankfurt and New York: Campus Verlag, 2005), 103–121

Keith Hanley and John K. Walton, *Constructing Cultural Tourism: John Ruskin and the Tourist Gaze*, Tourism and Cultural Change, eds. Mike Robinson and Alison Phipps (Bristol, Buffalo, and Toronto: Channel View Publications, 2010)

Susan Handy, Lisa Weston, and Patricia Mokhtarian, "Driving by choice or necessity?" *Transportation Research Part A* 39 (2005), 183–203

Kevin Hannam, Mimi Sheller, and John Urry, "Editorial: Mobilities, Immobilities and Moorings," *Mobilities* 1, no. 1 (March 2006), 1–22

Han Harmsze, *75 jaar motorleven in Nederland* (Den Haag: Nortier & Harmsze, n.d. [1979])

David M. Hart, "Herbert Hoover's Last Laugh: The Enduring Significance of the 'Associative State' in the U.S.A.," *Journal of Policy History* 10, no. 3 (1998), 419–444

Friedrich Hartmannsgruber, "'. . . ungeachtet der noch ungeklärten Finanzierung': Finanzplanung und Kapitalbeschaffung bei den Bau der Reichsautobahnen 1933–1945," *Historische Zeitschrift* 278 (2004), 625–681

Michael Hascher, *Politikberatung durch Experten: Das Beispiel der deutschen Verkehrspolitik im 19. und 20. Jahrhundert* (Frankfurt and New York: Campus Verlag, 2006)

Gilbert Hatry, "Les rapports gouvernement, armée, industrie privée pendant la Première Guerre mondiale: Le cas des usines Renault," in Gérard Canini (ed.), *Les fronts invisibles: Nourrir—fournir—soigner; Actes du colloque international sur La logique des armées au combat pendant la Première Guerre Mondiale organise à Verdun les 6, 7, 8 juin 1980 sous la présidence du general d'Armée F. Gambiez, membre de l'Institut* (Nancy: Presses Universitaires de Nancy, 1984), 171–188

Barbara Haubner, *Nervenkitzel und Freizeitvergnügen: Automobilismus in Deutschland 1886–1914* (Göttingen: Vandenhoeck & Ruprecht, 1998)

Cornelius W. Hauck, "1907–1911: The Highwheeler Era," *Antique Automobile* 27, no. 2 (March/April 1963), 72–79

Heinz-Gerhard Haupt, "Der Adel in einer entadelten Gesellschaft: Frankreich seit 1830," in Hans-Ulrich Wehler (ed.), *Europäischer Adel 1750–1950,* Geschichte und Gesellschaft: Zeitschrift für Historische Sozialwissenschaft, Sonderheft 13 (Göttingen: Vandenhoeck & Ruprecht, 1990)

Heinrich Hauser, "Autowandern, eine wachsende Bewegung," *Die Strasse* (1936), Heft 14, 455–457

——, *Fahrten und Abenteuer im Wohnwagen: Herausgegeben, kommentiert und mit einem Nachwort versehen von Robert Hilgers* (Stuttgart: DoldeMedien Verlag, 2004; reprint of original edition of 1935)
——, "Herbstfahren im Auto," *Frankfurter Zeitung, Reiseblatt* 79 (21 October 1934)
——, *Die letzten Segelschiffe: Schiff, Mannschaft, Meer und Horizont* (Berlin: S. Fischer Verlag, 1932, Elfte bis vierzehnte Auflage; first ed. 1930)
——, *Schwarzes Revier, Herausgegeben von Barbara Weidle: Mit einem Vorwort von Andreas Rossmann* (Bonn: Weidle Verlag, 2010; first ed. 1930)
——, *Süd-Ost-Europa ist erwacht: Im Auto durch acht Balkanländer* (Berlin: Rowohlt, 1938)
Ellis W. Hawley, *The Great War and the Search for a Modern Order: A History of the American People and Their Institutions, 1917–1933* (New York: St. Martin's Press, 1979)
Kimberley J. Healey, *The Modernist Traveler: French Detours, 1900–1930* (Lincoln and London: University of Nebraska Press, 2003)
Thomas M. Heaney, "'The Call of the Open Road': Automobile Travel and Vacations in American Popular Culture, 1935–1960" (unpublished dissertation, University of California, Irvine, 2000)
John A. Heitmann, *The Automobile and American Life* (Jefferson, NC, and London: McFarland, 2009)
Paul Hekkert, "Design aesthetics: Principles of pleasure in design," *Psychology Science* 48, no. 2 (2006), 157–162
Hellich et al., "Budget des Routes" (report 47, PIARC Conference, Washington, 1930) [see n. 26 on p. 621]
Félicien Hennequin, *L'évolution automobiliste en France de 1899 à 1905* (Rapport de la Commission extraparlementaire de la circulation des automobiles) (Paris: Imprimerie Nationale, 1905)
A. M. Henniker, *Transportation on the Western Front 1914–1918* (History of the Great War, Based on official documents by Direction of the Historical Section of the Committee of Imperial Defence), the Battery Press Great War Series, 22 (London and Nashville, TN: The Imperial War Museum / The Battery Press, 1992; first ed. 1937)
Manfred Henningsen, "Das amerikanische Selbstverständnis und die Erfahrung des Grossen Kriegs," in Klaus Vondung (ed.), *Kriegserlebnis; Der Erste Weltkrieg in der literarischen Gestaltung und symbolischen*, 368–386
Joseph Henrich, Robert Boyd, and Peter J. Richerson, "The puzzle of monogamous marriage," *Philosophical Transactions of the Royal Society B*, 367 (2012), 657–669
John H. Hepp, IV, *The Middle-Class City: Transforming Space and Time in Philadelphia, 1876–1926* (Philadelphia: University of Pennsylvania Press, 2003)
Cecil Hepworth, "How It Feels To Be Run Over," in *The Movies Begin: A Treasure of Early Cinema 1894–1913* (New York: Kino Video, 2002), vol. 3
Anke Hertling, "Angriff auf eine Männerdomäne: Autosportlerinnen in den zwanziger und dreissiger Jahren," *Feministische Studien* 30, no. 1 (2012), 12–29
Hermann Hesse, *Gesammelte Schriften: Vierter Band* (n.p. [Frankfurt]: Suhrkamp Verlag, 1968)
Aaltje Hessels, *Vakantie en vakantiebesteding sinds de eeuwwisseling: Een sociologische verkenning ten behoeve van de sociale en ruimtelijke planning in Nederland* (Assen: Van Gorcum & Comp., 1973)
Paul Heuzé, *La Voie Sacrée: Le Service automobile à Verdun (Février–Août 1916)* (Paris: La Renaissance du Livre, 1919)

Charlotte Heymel, *Touristen an der Front: Das Kriegserlebnis 1914–1918 als Reiseerfahrung in zeitgenössischen Reiseberichten* (Berlin: Lit Verlag Dr. W. Hopf, 2007)
John Hibbs, *The History of British Bus Services* (Newton Abbot: David & Charles, 1968)
John Higham, "The Reorientation of American Culture in the 1890's," in John Weiss (ed.), *The Origins of Modern Consciousness* (Detroit, MI: Wayne State University Press, 1965), 25–48
Jordan Jay Hillman, *The Parliamentary Structuring of British Road-Rail Freight Coordination* (Evanston, IL: The Transportation Center, Northwestern University, 1973)
Michael Hochgeschwender, "The Nobles Philosophy and Its Most Efficient Use: Zur Geschichte des *social engineering* in den USA, 1910–1965," in Thomas Etzemüller (ed.), *Die Ordnung der Moderne. Social Engineering im 20. Jahrhundert* (Bielefeld: transcript Verlag, 2009), 171–197
Dorothee Hochstetter, *Motorisierung und "Volksgemeinschaft": Das Nationalsozialistische Kraftfahrkorps (NSKK), 1931–1945* (Munich: R. Oldenbourg Verlag, 2005)
Rudolf Hoenicke, *Die amerikanische Automobilindustrie in Europa* (Bottrop: Buchdruckerei Wilhelm Postberg, 1933)
Rudolf Hoffmann, "Der Kraftverkehr auf deutschen Strassen Pfingsten 1938," *Die Strasse* 5, no. 12 (1938), 388–392
Walther G. Hoffmann (in cooperation with Franz Grumbach and Helmut Hesse), *Das Wachstum der deutschen Wirtschaft seit der Mitte des 19. Jahrhunderts*, Enzyklopädie der Rechts- und Staatswissenschaft, eds. W. Kunkel, H. Peters, and E. Preiser; Abteilung Staatswissenschaft (Berlin, Heidelberg, and New York: Springer-Verlag, 1965)
Wolfgang Hoffmann-Harnisch, *Wunderland Brasilien: Eine Fahrt mit Auto, Bahn und Flugzeug*, Band 573 (Hamburg: Deutsche Hausbücherei, 1938)
Harold C. Hoffsommer, *Relations of Cities and Larger Villages to Changes in Rural Trade and Social Areas in Wayne County, New York* (Ithaca, NY: Cornell University Agricultural Experiment Station, Bulletin 582, February 1934)
Reinhold Hohl, "Marcel Proust in neuer Sicht; Kubismus und Futurismus in seinem Romanwerk," *Die neue Rundschau* 88 (1977), 54–72
Rosan Hollak, "De logica van de lijn," *NRC Handelsblad* (4 February 2011), Cultureel Supplement, 11
I. B. Holley, Jr., *The Highway Revolution, 1895–1925: How the United States Got Out of the Mud* (Durham, NC: Carolina Academic Press, 2008)
Arnold Holtz, *Im Auto zu Kaiser Menelik* (Berlin-Charlottenburg: VITA, Deutsches Verlagshaus, n.d. [1908])
William Holtz, "Sinclair Lewis, Rose Wilder Lane, and the Midwestern Short Novel," *Studies in Short Fiction* 24, no. 1 (1987), 41–48
Marie Holzer, "Das Automobbil," *Die Aktion* 2, no. 34 (21 August 1912), columns 1072–1073
Ari and Olive Hoogenboom, *A History of the ICC: From Panacea to Palliative* (New York: W. W. Norton & Company, 1976)
Eva Horn and Lucas Marco Gisi (eds.), *Schwärme—Kollektive ohne Zentrum: Eine Wissensgeschichte zwischen Leben und Information*, Masse und Medium 7 (Bielefeld: transcript Verlag, 2009)
Craig Horner, "'Modest Motoring' and the Emergence of Automobility in the United Kingdom," *Transfers* 2, no. 3 ((Winter 2012), 56–75
Craig Horner and Julian Greaves, "Mobility Spotting: Running Off the Rails in the Transport Historiography of the United Kingdom," in Gijs Mom et al. (eds.), *Mobil-*

ity in History: Themes in Transport (T²M Yearbook 2011) (Neuchâtel: Alphil, 2010), 151-158

Richard Hornsey, "'He Who Thinks, in Modern Traffic, is Lost': Automation and the Pedestrian Rhythms of Interwar London," in: Tim Edensor (ed.), *Geographies of Rhythm; Nature, Place, Mobilities and Bodies* (Farnham and Burlington: Ashgate, 2010), 99-112

Jeffrey M. Hornstein, "The Rise of the Realtor®: Professionalism, Gender, and Middle-Class Identity, 1908-1950," in Burton J. Bledstein and Robert D. Johnston (eds.), *The Middling Sorts; Explorations in the History of the American Middle Class* (New York and London: Routledge, 2001, 217-233

Daniel Horowitz, *The Morality of Spending: Attitudes Toward the Consumer Society in America, 1875-1940* (Baltimore, MD, and London: Johns Hopkins University Press, 1985)

Roger Horowitz, "Introduction," in Horowitz (ed.), *Boys and Their Toys? Masculinity, Technology, and Class in America* (New York and London: Routledge, 2001), 1-10

Gerhard Horras, *Die Entwicklung des deutschen Automobilmarktes bis 1914* (Munich, 1982)

Hours, Wages, and Working Conditions in the Intercity Motor Transport Industries, Part I: Motor-Bus Transportation (Washington, DC: Federal Coordinator of Transportation, Section of Research, Section of Labor Relations, 1936)

Hours, Wages, and Working Conditions in the Intercity Motor Transport Industries, Part III: Considerations Common to Motor-Bus and Motor-Truck Transportation (Washington, DC: Federal Coordinator of Transportation, Section of Research, Section of Labor Relations, 1936)

How American Buying Habits Change (Washington, DC: US Department of Labor, n.d. [1959?])

"How Many Dogs Have You Killed?" *The Club Journal* 2, no. 25 (18 March 1911), 995-999; no. 26 (1 April 1911), 1069-1072

"How To See Europe: Members Who Have Toured the Continent Give Those Going Abroad the Benefit of Their Experiences," *The Club Journal* 3, no. 2 (29 April 1911), 108-109

David Howes, "Introduction: Commodities and cultural borders," in Howes (ed.), *Cross-Cultural Consumption: Global Markets, Local Realities* (London and New York: Routledge, 1996), 1-16

——, "Introduction: 'To Summon All the Senses,'" in Howes (ed.), *The Varieties of Sensory Experience: A Sourcebook in the Anthropology of the Senses* (Toronto, Buffalo, and London: University of Toronto Press, 1991), 3-21

Valeska Huber, "Multiple Mobilities: Über den Umgang mit verschiedenen Mobilitätsformen um 1900," *Geschichte und Gesellschaft* 36 (2010), 317-341

Gerhard Huck, "Der Reisebericht als historische Quelle," in Gerhard Huck and Jürgen Reulecke (eds.), *... und reges Leben ist überall sichtbar! Reisen im Bergischen Land um 1800*, Bergische Forschungen; Quellen und Forschungen zur bergischen Geschichte, Kunst und Literatur, eds. Wolfgang Köllmann and Jürgen Reulecke (Neustadt an der Aisch: Druck und Verlag Ph.C.W. Schmidt, 1978), 27-44

Christian Huck and Stefan Bauernschmidt (eds.), *Travelling Goods, Travelling Moods; Varieties of Cultural Appropriation (1850 - 1950)* (Frankfurt and New York: Campus Verlag, 2012)

Remo Hug, *Gedichte zum Gebrauch: Die Lyrik Erich Kästners; Besichtigung, Beschreibung, Bewertung* (Würzburg: Königshausen & Neumann, 2006)

Thomas P. Hughes, *Networks of Power: Electrification in Western Society, 1880–1930* (Baltimore, MD, and London, 1983)

Peter J. Hugill, "Good roads and the automobile in the United States 1880–1929," *Geographical Review* 72 (July 1982), 327–349

——, "The Rediscovery of America: Elite Automobile Touring," *Annals of Tourism Research* 12, no. 3 (1985), 435–447

——, "Technology and Geography in the Emergence of the American Automobile Industry, 1895–1915," in Jan Jennings (ed.), *Roadside America: The Automobile in Design and Culture* (Ames, IA: Iowa State University Press, for the Society for Commercial Archaeology, 1990), 29–39

Auke Hulst, "Van loopgraaf naar bergtop," *NRC Handelsblad* (16 December 2011), "Boeken," 1–2

Seth K. Humphrey, "Our Delightful Man-Killer," *The Atlantic Monthly* 148 (1931), 724–730

Aldous Huxley, *Along the Road: Notes and Essays of a Tourist* (London: Chatto & Windus, 1948; first ed. 1925)

Sarah E. Igo, *The Averaged American: Surveys, Citizens, and the Making of a Mass Public* (Cambridge, MA, and London: Harvard University Press, 2007)

Tim Ingold, *The Appropriation of Nature: Essays on Human Ecology and Social Relations*, Themes in Social Anthropology (Manchester: Manchester University Press, 1986)

——, *The Perception of the Environment: Essays on Livelihood, Dwelling and Skill* (London and New York: Routledge, 2000)

International Labour Office, *A Contribution to the Study of International Comparisons of Costs of Living: An Enquiry into the Cost of Living of Certain Groups of Workers in Detroit (U.S.A.) and Fourteen European Towns*, the World Economy: A Garland Series, ed. Mira Wilkins (New York and London: Garland, 1983; reprint of 1932 edition)

Joseph Anthony Interrante, "A Movable Feast: The Automobile and the Spatial Transformation of American Culture 1890–1940" (PhD thesis, Department of History, Harvard University, Cambridge, MA, June 1983)

Robert E. Ireland, *Entering the Auto Age: The Early Automobile in North Carolina, 1900–1930* (Raleigh: Division of Archives and History, North Carolina Department of Cultural Resources, 1990)

M.C. Isacco, E. Mellini, and A. Mercanti, *Les moyens propres à assurer la sécurité de la circulation . . . etc.*," report, Association Internationale Permanente des Congrès de la Route, *VIIe Congrès Munich—1934* (Paris: Imprimerie Oberthür, n.d. [1934])

Wolfgang Iser, "Texts and Readers," *Discourse Processes* 3, no. 4 (1980), 327–343

Alan A. Jackson, *The Middle Classes 1900–1950* (Nairn: David St. John Thomas Publisher, 1991)

Julian Jackson, "'Le temps des loisirs': Popular tourism and mass leisure in the vision of the Front Populaire," in Martin S. Alexander and Helen Graham (eds.), *The French and Spanish Popular Fronts: Comparative Perspectives* (Cambridge and New York: Cambridge University Press, 1989), 226–239

Margaret C. Jacob, "Science Studies after Social Construction: The Turn toward the Comparative and the Global," Victoria E. Bonnell and Lynn Hunt (eds.), *Beyond the Cultural Turn: New Directions in the Study of Society and Culture* (Berkeley, Los Angeles, and London: University of California Press, 1999), 95–120

Aletta Jacobs, *Herinneringen: Autobiografie* (n.p. [Amsterdam]: Contact, 1995; first ed. 1924)

Frederic Cople Jaher, *The Urban Establishment: Upper Strata in Boston, New York, Charleston, Chicago, and Los Angeles* (Urbana, Chicago, and London: University of Illinois Press, 1982)

John A. Jakle, "Landscapes redesigned for the automobile," in Michael P. Conzen (ed.), *The Making of the American Landscape* (Boston, London, Sydney, and Wellington: Unwin Hyman, 1990), 293–310

John A. Jakle and Keith A. Sculle, *Lots of Parking: Land Use in a Car Culture* (Charlottesville and London: University of Virginia Press, 2004)

——, *Motoring: The Highway Experience in America* (Athens and London: University of Georgia Press, 2008)

Lawrence James, *The Middle Class: A History* (London: Little and Brown, 2006)

Pearl James, "The 'Enid Problem': Dangerous Modernity in One of Ours," Cather Studies 6 (2006), 1–33, http://cather.unl.edu/cs006_James.html (last accessed 18 September 2011)

George Kirkham Jarvis, "The diffusion of the automobile in the United States: 1895–1969" (unpublished dissertation, University of Michigan, Ann Arbor, 1972)

Robin Jarvis, *Romantic Writing and Pedestrian Travel* (Houndsmill, London, and New York: MacMillan Press / St. Martin's Press, 1997)

Glen Jeansonne, "The Automobile and American Morality," *Journal of Popular Culture* 8, no. 1 (Summer 1974), 125–131

Rainer Jeglin and Irmgard Pickerodt, "Weiche Kerle in harter Schale; Zu *Drei Kameraden*," in Thomas F. Schneider (ed.), *Erich Maria Remarque: Leben, Werk und weltweite Wirkung*, Schiften des Erich Maria Remarque-Archivs, Band 12, eds. Thomas F. Schneider and Tilman Westphalen (Osnabrück: Universitätsverlag Rasch, 1998), 217–234

Richard Jensen, "The Lynds Revisited," *Indiana Magazine of History* 75 (1979), 303–319

Jepson, "How to drive on a superhighway—and how to drive away from it alive," *Changing Times* 10 (June 1956), 13–15

Jerome K. Jerome, *Three Men on the Bummel* (n.p.: Penguin Popular Classics, n.d.; first ed. 1900)

Leif Jerram, *Streetlife: The Untold History of Europe's Twentieth Century* (Oxford and New York: Oxford University Press, 2011)

Sir Francis Jeune, "The charms of driving in motors," in Alfred C. Harmsworth, *Motors and motor-driving* (London and Bombay: Longmans, Green, and Co., 1902), 341–345

Derek Jewell, *Man & Motor: The 20th Century Love Affair* (London: Hodder and Stoughton, n.d. [1966])

Ann Johnson, *Hitting the Brakes: Engineering Design and the Production of Knowledge* (Durham, NC, and London: Duke University Press, 2009)

Robert D. Johnston, "Conclusion: Historians and the American Middle Class," in Burton J. Bledstein and Robert D. Johnston (eds.), *The Middling Sorts: Explorations in the History of the American Middle Class* (New York and London: Routledge, 2001), 296–306

——, *The Radical Middle Class: Populist Democracy and the Question of Capitalism in Progressive Era Portland, Oregon* (Princeton, NJ, and Oxford: Princeton University Press, 2003)

Gareth Stedman Jones, "The determinist fix: Some obstacles to the further development of the linguistic approach to history in the 1990s," in Gabrielle M. Spiegel (ed.), *Practicing History: New Directions in Historical Writing after the Linguistic*

Turn, Rewriting Histories, ed. Jack R. Censer (New York and London: Routledge, 2005), 62–75

Joseph Jones, *The Politics of Transport in Twentieth-Century France* (Kingston and Montreal: McGill-Queen's University Press, 1984)

Sicco de Jong, *Geschiedenis eener Nederlandsche Vereeniging: RAI 1893–1968* (Bussum: C. A. J. van Dishoeck, 1968)

Johan W. D. Jongma, *Geschiedenis van het Nederlandse wegvervoer* (Drachten and Leeuwarden: FPB Uitgevers, 1992)

Herbert Jost, "Selbst-Verwirklichung und Seelensuche; Zur Bedeutung des Reiseberichts im Zeitalter des Massentourismus," in Peter J. Brenner (ed.), *Der Reisebericht: Die Entwicklung einer Gattung in der deutschen Literatur*, suhrkamp taschenbuch 2097; suhrkamp taschenbuch materialien (Frankfurt: Suhrkamp Verlag, 1989), 490–507

Ernst Jünger, *Das Abenteuerliche Herz (Sämtliche Werke, Zweite Abteilung: Essays, Band 9: Essays III)* (Stuttgart: Klett-Cotta, 1979)

Harald Jürgensen and Dieter Aldrup, *Verkehrspolitik im europäischen Integrationsraum* (Baden-Baden: Nomos Verlagsgesellschaft, 1968)

Hartmut Kaelble, *Nachbarn am Rhein: Entfremdung und Annäherung der französischen und deutschen Gesellschaft seit 1880* (Munich: Verlag C. H. Beck, 1991)

Alexander Kaempfe, "Viktor Schklowskij und sein Zoo," in: Viktor Shklovsky (transliterated as "Schklowskij"), *Zoo; oder Briefe nicht über die Liebe* (Frankfurt am Main: Suhrkamp Verlag, 1965; edition suhrkamp 130) (transl. from Russian by Alexander Kaempfe from original ed. of 1923) 121–144

Michael Kammen, "The Problem of American Exceptionalism: A Reconsideration," *American Quarterly* 45, no. 1 (March 1993), 1–43

Oskar Kanehl, "Schrei eines Überfahrenen," *Die Aktion* 4, no. 6 (7 February 1914), columns 119–120

Caren Kaplan, *Questions of Travel: Postmodern Discourses of Displacement* (Durham, NC, and London: Duke University Press, 1996)

Walter Kappacher, "Vor dem Rennen in den Wicklow Mountains," *Frankfurter Allgemeine Zeitung* (9 July 2011), "Bild ind Zeiten," Z1–Z2

Alan L. Karras, "The Atlantic world as a unit of study," in Alan L. Karras and J. R. McNeill (eds.), *Atlantic American Societies: From Columbus through abolition 1492–1888* (London and New York: Routledge, 1992), 1–15

Christiane Katz, "Städtische Pfade—Die Automobilisierung der Stadtverwaltung und Berufsfeuerwehr in Aachen vor dem Ersten Weltkrieg," *Zeitschrift des Aachener Geschichtsvereins* 113/114 (2012), 207–230

Harold Katz, *The Decline of Competition in the Automobile Industry, 1920–1940*, Dissertations in American Economic History, eds. Stuart Bruchey and Eleanor Bruchey (New York: Arno Press, 1970)

Stefan Kaufmann, "Technik am Berg: Zur technischen Strukturierung von Risiko- und Naturerlebnis," in Gunter Gebauer et al. (eds.), *Kalkuliertes Risiko: Technik, Spiel und Sport an der Grenze* (Frankfurt and New York: Campus Verlag, 2006), 99–124

Christian Kehrt, "'Das Fliegen ist immer noch ein gefährliches Spiel'—Risiko und Kontrolle der Flugzeugtechnik von 1908 bis 1914," in Gebauer et al. (eds.), *Kalkuliertes Risiko*, 199–224

Christian Kehrt, *Moderne Krieger; Die Technikerfahrungen deutscher Militärpiloten 1910 – 1945* (Paderborn, München, Wien and Zürich: Ferdinand Schöningh, 2010; Krieg in der Geschichte, eds.: Stig Förster, Bernhard R. Kroener, Bernd Wegner and Michael Werner; Band 58)

David Keir and Bryan Morgan, *Golden Milestone: 50 Years of the AA* (London: The Automobile Association, 1955)
Christine Keitz, "Die Anfänge des modernen Massentourismus in der Weimarer Republik," *Archiv für Sozialgeschichte* 33 (1993), 179–209
Andreas Kelletat, "Philologie, Textwissenschaft und Kulturwissenschaft: Konkurrenz oder friedliche Koexistenz?" (lecture in the series "Wege der Kulturwissenschaft," Johannes Gutenberg Universität, Mainz) http://www.fask.uni-mainz.de/inst/ro manistik/ringvorlesung/vortrag_AFK/vortrag-A... (last accessed 10 Fenruary 2008)
Roger Kempf, "Sur quelques véhicules," *L'Arc* 47 (1971), 47–57
[Mary] Kennard, *The Motor Maniac: A Novel* (London: Hutchinson & Co., 1902)
Stephen Kern, *The Culture of Time and Space 1880–1918* (London, 1983)
——, *A Cultural History of Causality: Science, Murder Novels, and Systems of Thought* (Princeton, NJ, and Oxford: Princeton University Press, 2004)
Irmgard Keun, *Gilgi – eine von uns; Roman⁴* (Berlin: List Verlag, 2008; first ed.: 1931)
——, *Das kunstseidene Mädchen: Roman, mit Materialien, ausgewählt von Jörg Ulrich Meyer-Bothling*, Editionen für den Literaturunterricht, ed. Thomas Kopfermann (Leipzig, Stuttgart, and Düsseldorf: Erbst Klett Schulbuchverlag Leipzig)
Charles P. Kindleberger, *Manias, Panics, and Crashes: A History of Financial Crises* (London and Basingstoke, 1978)
Celia M. Kingsbury, "'Squeezed into an Unnatural Shape': Bayliss Wheeler and the Element of Control in One of Ours," *Cather Studies* 6 (2006), 1–14, http://cather.unl.edu/cs006_kingsbury.html (last accessed 18 September 2011)
Rudyard Kipling, "Carmen Circulare," in "The Muse Among The Motors 1900–1930," in M. M. Kaye (ed.), *Rudyard Kipling: The Complete Verse* (London: Kyle Cathie Ltd., 1996), 555–581
——, "The Muse Among The Motors 1900–1930," in Kaye (ed.), *Rudyard Kipling: The Complete Verse* (London: Kyle Cathie Ltd., 1996), 555–581
——, "Steam Tactics. The Necessitarian," in Kipling, *Traffics and Discoveries* (n.p.: Kessinger Publishing's Rare Prints, n.d. [1904]; reprint), 112–161
——, "'They' The Return of the Children," in Kipling, *Traffics and Discoveries* (n.p.: Kessinger Publishing's Rare Prints, n.d. [1904]; reprint), 205–249
——, "The Village that Voted the Earth was Flat" (1913), in Kipling, *A Diversity of Creatures* (1917), http://whitewolf.newcastle.edu.au/words/authors/K/KiplingRudyard/prose/DiversityO... (last accessed 14 May 2008)
Lynne Kirby, *Parallel Tracks: The Railroad and Silent Cinema* (Exeter: University of Exeter Press, 1997)
David A. Kirsch and Gijs P. A. Mom, "From Service tot Product Based Mobility Concepts: Technical Choice and the History of the Electric Vehicle Company," *Business History Review* 76 (Spring 2002), 75–110
Henry Kistemaeckers, *Monsieur Dupont chauffeur; Nouveau roman comique de l'automobilisme* (Paris: Bibliothèque-Charpentier, 1908) (troisième mille)
Dietmar Klenke, *"Freie Stau für freie Bürger": Die Geschichte der bundesdeutschen Verkehrspolitik 1949–1994* (Darmstadt: Wissenschaftliche Buchgesellschaft, 1995)
Betsy Klimasmith, *At Home in the City: Urban Domesticity in American Literature and Culture, 1850–1930*, Becoming Modern: New Nineteenth-Century Studies, eds. Sarah Way Sherman et al. (Durham, NH: University of New Hampshire Press, Published by the University Press of New England, 2005)
Árpád von Klimó and Malte Rolf, "Rausch und Diktatur: Emotionen, Erfahrungen und Inszenierungen totalitärer Herrschaft," in von Klimó and Rolf (eds.), *Rausch und*

Diktatur: Inszenierung, Mobilisierung und Kontrolle in totalitären Systemen (Frankfurt and New York: Campus Verlag, 2006), 11–43

Ronald Kline, *Consumers in the Country: Technology and Social Change in Rural America* (Baltimore, MD, and London: Johns Hopkins University Press, 2000)

Ronald Kline and Trevor Pinch, "Users as Agents of Technological Change: The Social Construction of the Automobile in the Rural United States," *Technology and Culture* 37, no. 4 (October 1996), 763–795

F. D. Klingender, *The Condition of Clerical Labour in Britain* (London: Martin Lawrence, 1935)

Nadine Klopfer, "'Clean Up': Stadtplanung und Stadtvisionen in New Orleans, 1880er–1920er Jahre," in in: Thomas Etzemüller (ed.), *Die Ordnung der Moderne. Social Engineering im 20. Jahrhundert* (Bielefeld: transcript Verlag, 2009) 153–169

Bettina Knapp, *Maurice Maeterlinck* (Boston: Twayne Publishers, 1975)

Koen Koch, "'De laatste dagen der mensheid': De catastrofale Eerste Wereldoorlog als cesuur," *Geschiedenis Magazine* (November 2007), 33–36

Jürgen Kocka, "The Middle Classes in Europe," in *Journal of Modern History* 67, no. 4 (December 1995), 783–806

——, *White Collar Workers in America 1890–1940: A Social-Political History in International Perspective* (Thousand Oaks, CA: Sage Publications, 1980)

Paul A. C. Koistinen, *Mobilizing for Modern War: The Political Economy of American Warfare, 1865–1919* (Lawrence: University Press of Kansas, 1997)

Elke Kölcke, "Die Sommerfrische—vom 'reisenden Mann' zum 'Familienurlaub,'" in Wiebke Kolbe, Christian Noack, and Hasso Spode (eds.), *Voyage: Jahrbuch für Reise- & Tourismusforschung 2009, Studies on Travel & Tourism*, Band 8 (Sonderband), eds. Hasso Spode et al. (Munich and Vienna: Profil Verlag, 2009), 35–45

Gottfried Kölwel, "So stand ich vor dem Sterben . . . ," *Die Aktion* 4, no. 12 (21 March 1914), columns 258–259

René König, "Zur Soziologie der Zwanziger Jahre, oder Ein Epilog zu zwei Revolutionen, die niemals stattgefunden haben, und was daraus für unsere Gegenwart resultiert," in Leonhard Reinisch (ed.), *Die Zeit ohne Eigenschaften: Ein Bilanz der zwanziger Jahre* (Stuttgart: W. Kohlhammer Verlag, 1961), 82–118

Wolfgang König, "Autocrash und Kernkraft-GAU: Zum Umgang mit technischen Risiken," in Herfried Münkler, Matthias Bohlender, and Sabine Meurer (eds.), *Sicherheit und Risiko: Über den Umgang mit Gefahr im 21. Jahrhundert* (Bielefeld: transcript Verlag, 2010), 207–222

——, *Geschichte der Konsumgesellschaft*, Vierteljahresschrift für Sozial- und Wirtschaftsgeschichte, Beihefte, eds. Hans Pohl et al. (Stuttgart: Franz Steiner Verlag, 2000)

——, *Volkswagen, Volksempfänger, Volksgemeinschaft: "Volksprodukte" im Dritten Reich: Vom Scheitern einer nationalsozialistischen Konsumgesellschaft* (Paderborn, Munich, Vienna, and Zürich: Ferdinand Schöningh, 2004)

——, "Wilhelm II. und das Automobil: Eine Technik zwischen Transport, Freizeitvergnügen und Risiko," in Gunter Gebauer et al. (eds.), *Kalkuliertes Risiko: Technik, Spiel und Sport an der Grenze* (Frankfurt and New York: Campus Verlag, 2006), 179–198

——, *Wilhelm II. und die Moderne: Der Kaiser und die technisch-industrielle Welt* (Paderborn, Munich, Vienna, and Zürich: Ferdinand Schöningh, 2007)

P. G. Konody, *Through the Alps to the Apennines* (London: Kegan Paul, Trench, Trübner & Co, 1911)

Nancy Koppelman, "One for the Road: Mobility in American Life, 1787–1905" (unpublished doctoral dissertation, Emory University, 1999)

Christopher Kopper, *Handel und Verkehr im 20. Jahrhundert* (Munich: R. Oldenbourg Verlag, 2002)

———, "Eine komparative Geschichte des Massentourismus im Europa der 1930er bis 1980er Jahre: Deutschland, Frankreich und Grossbritannien im Vergleich," *Archiv für Sozialgeschichte* 49 (2009), 129–148

———, "Mobile Exceptionalism? Passenger Transport in Interwar Germany," *Transfers* 3 no. 2 (Summer 2013) 89–107

Christopher Kopper and Heike Wolter, "Mobility History in Germany: One Field, Many Perspectives," in Gijs Mom et al. (eds.), *Mobility in History: Themes in Transport (T²M Yearbook 2011)* (Neuchâtel: Alphil, 2010), 159–173

Barbara Korte, "Erfahrungsgeschichte und die 'Quelle' Literatur: Zur Relevanz genretheoretischer Reflexion am Beispiel der britischen Literatur des Ersten Weltkriegs," in Jan Kusber et al. (eds.), *Historische Kulturwissenschaften: Positionen, Praktiken und Perspektiven*, Mainzer Historische Kulturwissenschaften, Band 1 (Bielefeld: transcript Verlag, 2010), 143–159

Barbara Korte, Ralf Schneider, and Claudia Sternberg, "Einleitung und Grundlegung," in Korte, Schneider, and Sternberg (eds.), *Der Erste Weltkrieg und die Mediendiskurse der Erinnerung in Grossbritannien: Autobiographie—Roman—Film (1919–1999)*, Film—Medium—Diskurs, eds. Oliver Jahraus and Stefan Neuhaus, Band 15 (Würzburg: Königshausen & Neumann, 2005), 1–32

Reinhart Koselleck, "'Neuzeit': Zur Semantik moderner Bewegungsbegriffe," in Koselleck, *Vergangene Zukunft: Zur Semantik geschichtlicher Zeiten* (Frankfurt: Suhrkamp Verlag, 1979), 300–348

Rudy Koshar, "Cars and Nations: Anglo-German Perspectives on Automobility between the World Wars," *Theory, Culture & Society* 21, no. 4/5 (2004), 121–144

———, *German Travel Cultures* (Oxford and New York: Berg, 2000)

———, "Germans at the Wheel: Cars and Leisure Travel in Interwar Germany," in Koshar (ed.), *Histories of Leisure* (Oxford and New York: Berg, 2002), 215–230

———, "On the History of the Automobile in Everyday Life," *Contemporary European History* 10, no. 1 (2001), 143–154

———, "Organic Machines: Cars, Drivers, and Nature from Imperial to Nazi Germany," in Thomas Lekan and Thomas Zeller (eds.), *Germany's Nature: Cultural Landscapes and Environmental History* (New Brunswick, NJ, and London: Rutgers University Press, 2005), 111–139

———, "'What ought to be seen': Tourists' Guidebooks and National Identities in Modern Germany and Europe," *Journal of Contemporary History* 33, no. 3 (1998), 323–340

Barbara Kosta, "Die Kunst des Rauchens: Die Zigarette und die neue Frau," in Julia Freytag and Alexandra Tacke (eds.), *City Girls: Bubiköpfe & Blaustrümpfe in den 1920er Jahren*, Literatur—Kultur—Geschlecht; Studien zur Literatur- und Kulturgeschichte, eds. Anne-Kathrin Reulecke and Ulrike Vedder, in connection with Inge Stephan und Sigrid Weigel; Kleine Reihe, Band 29 (Cologne, Weimar, and Vienna: Böhlau Verlag, 2011), 143–158

Siegfried Kracauer, "Die Reise und der Tanz," in Siegfried Kracauer, *Das ornament der Masse: Essays* (Frankfurt: Suhrkamp Verlag, 1977)

Alan Kramer, "Les 'atrocités allemandes': Mythologie populaire, propagande et manipulations dans l'armée allemande," in Jean-Jacques Becker et al., *Guerre et cultures 1914–1918* (Paris: Armand Colin, 1994), 147–164

Beth Kraig, "Woman At The Wheel: A History Of Women And the Automobile In America" (unpublished doctoral dissertation, University of Washington, 1987)

Alan Kramer, *Dynamic of Destruction: Culture and Mass Killing in the First World War, the Making of the Modern World*, eds. Chris Bayly, Richard J. Evans, and David Reynolds (Oxford and New York: Oxford University Press, 2007)

Thomas Krämer-Bodoni, Herbert Grymer, and Marianne Rodenstein, *Zur sozio-ökonomischen Bedeutung des Automobils* (Frankfurt: Suhrkamp Verlag, 1971)

Stefan Krebs, "Closing the Body: Car Technology, Acoustic and Driving Experience" (paper presented at the T^{2}M Conference, Lucerne, Switzerland, 5–7 November 2009)

——, "'Notschrei eines Automobilisten' oder die Herausbildung des deutschen Kfz-Handwerks in der Zwischenkriegszeit," *Technikgeschichte* 79, no. 3 (2012), 185–206

Arthur Krim, "Route 66: Auto River of the American West," in Donald G. Janelle (ed.), *Geographical Snapshots of North America: Commemorating the 27th Congress of the International Geographical Union and Assembly* (New York and London: Guilford Press, 1992), 30–33

Peter Kroes and Anthonie Meijers, "Introduction: The dual nature of technical artefacts," *Studies in History and Philosophy of Science* 37 (2006), 1–4

Fabian Kröger, "Automobile DNS in der Kontrollgesellschaft," in Gerburg Treusch-Dieter, Claudia Gehrke, and Ronald Düker (eds.), *Auto: Konkursbuch*, Konkursbuch 42 (Tübingen: Konkursbuch Verlag Claudia Gehrke, 2004), 161–170

Ulrich Kubisch, "Häuser am Haken: Der Wohnwagen: Geschichte, Technik, Urlaubsfreuden," *Kultur & Technik* no. 3 (1998), 11–18

Oliver Kühlschelm, "Konsumgüter und Nation: Theoretische und methodische Überlegungen," *Österreichische Zeitschrift für Geschichtswissenschaften* 21, no. 2 (2010), 19–49

Klaus Kuhm, *Das eilige Jahrhundert: Einblicke in die automobile Gesellschaft* (Hamburg: Junius Verlag, 1995)

——, *Moderne und Asphalt: Die Automobilisierung als Prozess technologischer Integration und sozialer Vernetzung* (Pfaffenweiler: Centaurus-Verlagsgesellschaft, 1997)

K. Kühner, *Geschichtliches zum Fahrzeugantrieb* (Friedrichshafen, 1965)

P. Kuin and H. J. Keuning, *Het vervoerswezen*, De Nederlandse volkshuishouding tussen twee wereldoorlogen, III (Utrecht and Brussels: Uitgeverij Het Spectrum, 1952)

Richard F. Kuisel, "Vichy et les origines de la planification économique (1940–1946)," *Le mouvement social* 98 (1977), 77–101

Patrick Kupper, "Translating Yellowstone: Early European National Parks, Weltnaturschutz and the Swiss Model," in Bernhard Gissibl, Sabine Höhler, and Patrick Kupper (eds.), *Civilizing Nature: National Parks in a global historical perspective* (New York and Oxford: Berghahn Books, 2012), 123–139

Samuel B. Kutash, "Psychoanalytic Theories of Aggression," in Irwin L. Kutash et al., *Violence: Perspectives on Murder and Aggression* (San Francisco, CA, Washington, DC, and London: Josey-Bass Publishers, 1978), 7–28

Hazel Kyrk, *Economic Problems of the Family* (New York and London: Harper & Brothers, 1933)

Hazel Kyrk, Day Monroe, Kathryn Cronister, and Margaret Perry, *Family Expenditures for Household Operation; Five Regions* (Washington, D.C.: United States Department of Agriculture, The Bureau of Home Economics in cooperation with the

Work Projects Administration, 1941) (Consumer Purchase Study, Urban and Village Series, Miscellaneous Publication No. 432)

Dirk van Laak, "Zwischen 'organisch' und 'organisatorisch': 'Planung' als politische Leitkategorie zwischen Weimar und Bonn," in Burkard Dietz, Helmut Gabel, and Ulrich Tiedau (eds.), *Griff nach dem Westen: Die "Westforschung" der völkisch-nationalen Wissenschaften zum nordeuropäischen Raum (1919–1960), Teilband I*, Studien zur Geschichte und Kultur Nordwesteuropas, Band 6, ed. Horst Lademacher (Münster, New York, Munich, and Berlin: Waxmann 2003), 67–90

Paolo Laconi, "De *Summer of Love* in vrijstaat Fiume," *Historisch Nieuwsblad* 18, no. 9 (November 2009), 42–50

Brian Ladd, *Autophobia: Love and Hate in the Automotive Age* (Chicago and London: University of Chicago Press, 2008)

Samuel Lair, "La 628–E8, 'le nouveau jouet de Mirbeau,'" *Cahiers Octave Mirbeau* 15 (March 2008), 54–67

———, *Mirbeau, l'iconoclaste*, Critiques Littéraires, ed. Maguy Albet (Paris: L'Harmattan, 2008)

Catharina Landström, "A Gendered Economy of Pleasure: Representations of Cars and Humans in Motoring Magazines," *Science Studies* 19, no. 2 (2006), 31–53

Jean-Paul Laplagne, "La femme et la bicyclette à l'affiche," in Pierre Arnaud and Thierry Terret (eds.), *Histoire du sport féminin: Tome I; Le sport au féminin: histoire et identité*, Collection "Espaces et Temps du Sport," ed. Pierre Arnaud (Paris and Montreal: L'Harmattan, 1996), 83–94

Valery Larbaud, *A.O. Barnabooth: Son journal intime* (Paris: Éditions de la Nouvelle Revue Française, 1932^{26})

———, *Aux couleurs de Rome* (Paris: Gallimard, 1938^2)

Jonas Larsen, "Tourism Mobilities and the Travel Glance: Experiences of Being on the Move," *Scandinavian Journal of Hospitality and Tourism* 1, no. 2 (2001), 80–98

Torben Huus Larsen, *Enduring Pastoral: Recycling the Middle Landscape Ideal in the Tennessee Valley* (Amsterdam and New York: Rodopi, 2010)

"Lastkraftwagen in der Volkswirtschaft," *Wochenbericht des Instituts für Konjunkturforschung* 3, no. 6 (7 May 1930), 23–24

E. Lauber, "Anforderungen des Automobilisten an Strassenbau und Strassenverkehr," *Der Strassenbau* 18, no. 26 (10 September 1927), 445–449

Eric Laurier, "Driving: Pre-Cognition and Driving," in Tim Cresswell and Peter Merriman (eds.), *Geographies of Mobilities: Practices, Spaces, Subjects* (Farnham and Burlington: Ashgate, 2011), 69–81

James M. Laux, *The European Automobile Industry* (New York, 1992)

———, *In First Gear: The French Automobile Industry to 1914* (Liverpool, 1976)

———, "Trucks in the west during the first world war," *Journal of Transport History* 6, no. 2 (September 1985), 64–70

Beatrice Lavarini, "Bonnard und seine Liebe zum Automobil," in Gert Schmidt (ed.), "Automobil und Kultur; Nürnberger SFZ-Kolloquien 1999 und 2000," Schriftenreihe des Sozialwissenschaftlichen Forschungszentrums der Friedrich-Alexander-Universität Erlangen-Nürnberg, vol. 8 (Nürnberg, 2001), 186–202

Maud Lavin, *Push Comes To Shove: New Images of Aggressive Women* (Cambridge, MA, and London: MIT Press, 2010)

Patrick Laviolette, *Extreme Landscapes of Leisure: Not a Hap-Hazardous Sport* (Farnham and Burlington: Ashgate, 2011)

C. H. Lawshe, Jr., *A Review of the Literature Related to the Various Psychological Aspects of Highway Safety: A Progress Report of an Investigation Conducted by The*

Engineering Experiment Station of Purdue University, In cooperation with State Highway Commision of Indiana (Lafayette, IN: Purdue University, April 1939), Research Series no. 66, Highway Research Bulletin, no. 2, Engineering Bulletin 23 (1939), pt. 2A

T. J. Jackson Lears, "From Salvation to Self-Realization; Advertising and the Therapeutic Roots of Consumer Culture, 1880 –1930," in Richard Wightman Fox and T. J. Jackson Lears (eds.), *The Culture of Consumption: Critical Essays in American History, 1880–1980* (New York: Pantheon Books, 1983), 1–38

———, *No Place of Grace: Antimodernism and the Transformation of American Culture 1880–1920* (Chicago and London: University of Chicago Press, 1984)

———, *Rebirth of a Nation: The Making of Modern America, 1877–1920* (New York: HarperCollins, 2009)

Die Lebenshaltung von 2000 Arbeiter-, Angestellten- und Beamtenhaushaltungen: Erhebungen von Wirtschaftsrechnungen im Deutschen Reich vom Jahre 1927/28, Teil I, Gesamtergebnisse (Bearbeitet im Statistischen Reichsamt), Einzelschriften zur Statistik des Deutschen Reichs, nr. 221 (Berlin: Verlag von Reimar Hobbing, 1932)

Maurice Leblanc, *Gueule-Rouge, 80–Chevaux* (Paris: Société d'éditions littéraires et artistiques, 1904₃; first ed. 1904)

———, "The Most Amazing True Story of Crime ever told: 'The Auto-Bandits of Paris,' by the Author of 'Arsene Lupin'; Sent by Cable to the New York Times," *New York Times* (5 May 1912), http://query.nytimes.com/mem/archive-free/pdf?res=9F0DE4DC153CE633A25756C0A9639C946396D6CF (last accessed 15 March 2010)

———, *Voici des ailes: Roman* (Paris: Phébus, 1999; first ed. 1898)

Hermione Lee, *Edith Wharton* (New York: Alfred A. Knopf, 2007)

———, *Virginia Woolf* (New York: Vintage Books, 1999; first ed. 1997)

Vernon Lee, "Gospels of Anarchy," in Andrea Broomfield and Sally Mitchell (eds.), *Prose by Victorian Women: An Anthology* (New York and London: Garland Publishing, 1996), 711–729

———, "The Motor-car and the Genius of Places," in Lee, *The Enchanted Woods and Other Essays on the Genius of Places* (London and New York: The Bodley Head, 1905), 95–113

Ariane Leendertz, "Ordnung, Ausgleich. Harmonie: Koordinaten raumplanerischen Denkens in Deutschland, 1920 bis 1970," in Thomas Etzemüller (ed.), *Die Ordnung der Moderne. Social Engineering im 20. Jahrhundert* (Bielefeld: transcript Verlag, 2009), 129–150

———, *Ordnung schaffen: Deutsche Raumplanung im 20. Jahrhundert*, Beiträge zur Geschichte des 20. Jahrhunderts, ed. Norbert Frei, Band 7 (Göttingen: Wallstein Verlag, 2008)

Henri Lefebvre, *Everyday Life in the Modern World* (New Brunswick, NJ, and London: Transaction Books, 1984; transl. from *La vie quotidienne dans le monde moderne*, 1971, by Sacha Rabinovitch)

Herbert Leibowitz, *"Something Urgent I Have to Say to You": The Life and Works of William Carlos Williams* (New York: Farrar, Straus and Giroux, 2011)

Paul Lerner, *Hysterical Men: War, Psychiatry, and the Politics of Trauma in Germany, 1890–1930* (Ithaca, NY, and London: Cornell University Press, 2003)

Friedrich Lenger, "Der Stadt-Land-Gegensatz in der europäischen Geschichte des 19. und 20. Jahrhunderts—ein Abriss," *comparativ: Zeitschrift für Globalgeschichte und vergleichende Gesellschaftsforschung* 18, no. 2 (2008), 57–70

Harro van Lente, *Promising Technology: The Dynamics of Expectations in Technological Developments* (Delft: Eburon, 1993)

T. Arthur Leonard, *Adventures in Holiday Making: Being the story of the Rise and Development of a People's Holiday Movement* (London: Holiday Fellowship, n.d. [1934])

Paul M. Leonardi, "From Road to Lab to Math: The Co-evolution of Technological, Regulatory, and Organizational Innovations for Automotive Crash Testing," *Social Studies of Science* 40, no. 2 (April 2010), 243–274

A. N. Leont'ev, *Activity, Consciousness, and Personality*, transl. Marie J. Hall (Englewood Cliff, NJ: Prentice Hall, 1978)

Esther Leslie, "Flâneurs in Paris and Berlin," in Rudy Koshar (ed.), *Histories of Leisure* (Oxford and New York: Berg, 2002), 61–77

Stuart W. Leslie, *Boss Kettering* (New York: Columbia University Press, 1983)

Corey Todd Lesseig, "Automobility and Social Change: Mississippi, 1909–1939" (dissertation, University of Mississippi, May 1997)

——, *Automobility: Social Changes in the American South 1909–1939* (New York and London: Routledge, 2001)

Hemut Lethen, *Cool Conduct: The Culture of Distance in Weimar Germany* (Berkeley, Los Angeles, and London: University of California Press, 2002; transl. *Verhaltenslehre der Kälte*, by Don Reneau)

——, "Von Geheimagenten und Virtuosen: Peter Sloterdijks Schulbeispiele des Zynismus aus der Literatur der Weimarer Republik," in *Peter Sloterdijks "Kritik der zynischen Vernunft"* (Frankfurt: Suhrkamp, 1987), 324–355

Harvey Levenstein, *Seductive Journey: American Tourists in France from Jefferson to the Jazz Age* (Chicago and London: University of Chicago Press, 1998)

P. H. Levin, "Highway Inquiries: A Study in Governmental Responsiveness," *Public Administration* 57, no. 1 (1979), 21–49

Melvin R. Levin and Norman A. Abend, *Bureaucrats in Collision: Case Studies in Area Transportation Planning* (Cambridge, MA, and London: MIT Press, 1971)

David Lewis, Dennis Rodgers, and Michael Woolcock, "The Fiction of Development: Literary Representation as a Source of Authoritative Knowledge," *Journal of Development Studies* 44, no. 2 (February 2008), 198–216

Claudia Lieb, *Crash: Der Unfall der Moderne* (Bielefeld: Aisthesis Verlag, 2009)

Alphonse van Lier, "Een rit voor Geluk," *De Auto* (1909), 64–67, 90–93

Michael Lind, *Land of Promise: An Economic History of the United States* (New York: Harper-Collins, 2012)

Mathea Francisca Antonia Linders-Rooijendijk, *Gebaande wegen voor mobiliteit en vrijetijdsbesteding: De ANWB als vrijwillige associatie 1883–1937* (Heeswijk, 1989)

Richard Lingeman, *Theodore Dreiser: An American Journey 1908–1945* (New York: G. P. Putnam's Sons, 1990)

Arthur S. Link, "What Happened to the Progressive Movement in the 1920's?" *American Historical Review* 64, no. 4 (July 1959), 833–851

Jürgen Link, "'Einfluss des Fliegens!—Auf den Stil selbst!' Diskursanalyse des Ballonsymbols," in Jürgen Link and Wulf Wülfing (eds.), *Bewegung und Stillstand in Metaphern und Mythen: Fallstudien zum Verhältnis von elementarem Wissen und Literatur im 19. Jahrhundert*, Sprache und Geschichte, eds. Reinhart Koselleck and Karlheinz Stierle, Band 9 (Stuttgart: Klett-Cotta, 1984), 149–163

Jürgen Link and Siegfried Reinecke, "'Autofahren ist wie das Leben': Metamorphosen des Autosymbols in der deutschen Literatur," in Harro Segeberg (ed.), *Technik in der Literatur: Ein Forschungsüberblick und zwölf Aufsätze*, suhrkamp taschenbuch wissenschaft 655 (Frankfurt: Suhrkamp Verlag, 1987), 436–482

Jeff Lipkes, *Rehearsals: The German Army in Belgium, August 1914* (Leuven: Leuven University Press, 2007)

Frank Lippert, *Lastkraftwagenverkehr und Rationalisierung in der Weimarer Republik: Technische und ökonomische Aspekte fertigungsstruktureller und logistischer Wandlungen in den 1920er Jahren* (Frankfurt, Berlin, Bern, New York, Paris, and Vienna: Peter Lang, 1999)

——, "Oekonomische Dimensionen des Lkw-Verkehrs in der Weimarer Republik: Zur Interdependenz von industrieller Rationalisierung und logistischer Flexibilisierung in den 1920er Jahren," *Zeitschrift für Unternehmensgeschichte* 42, no. 2 (1997), 185–216

Giovanni Lista, *F.T. Marinetti; L'Anarchiste du futurisme* (Paris: Séguier, 1995; Les Biographies de Séguier)

Kenneth Little, "On Safari: The Visual Politics of a Tourist Representation," in David Howes (ed.), *The Varieties of Sensory Experience: A Sourcebook in the Anthropology of the Senses* (Toronto, Buffalo, and London: University of Toronto Press, 1991), 148–163

David W. Lloyd, *Battlefield Tourism: Pilgrimage and the Commemoration of the Great War in Britain, Australia and Canada, 1919–1939* (Oxford and New York: Berg, 1998)

Orvar Löfgren, "Know Your Country: A Comparative Perspectrive on Tourism and Nation Building in Sweden," in Shelley Baranowski and Ellen Furlough (eds.), *Being Elsewhere: Tourism, Consumer Culture, and Identity in Modern Europe and North America* (Ann Arbor: University of Michigan Press, 2004), 137–154

Sylvère Lotringer and Paul Virilio, *The Accident of Art*, transl. Michael Taormina (New York: Semiotext(e), 2005)

Jean-Louis Loubet, "La naissance du modèle automobile français (1934–1973)," *Culture technique* 25 (1992), 73–82

David Louter, *Windshield Wilderness: Cars, Roads, and Nature in Washington's National Parks* (Seattle and London: University of Washington Press, 2006)

Bill Luckin, "War on the Roads: Traffic Accidents and Social Tension in Britain, 1939–45," in Roger Cooter and Bill Luckin (eds.), *Accidents in History: Injuries, Fatalities and Social Relations* (Amsterdam and Atlanta: Rodopi, 1997), 234–254

David N. Lucsko, "John Bell Rae and the Automobile; 1959, 1965, 1971, 1984 (Classics Revisited)," *Technology and Culture* 50, no. 4 (October 2009), 894–914

——, *The Business of Speed: The Hot Rod Industry in America, 1915–1990* (Baltimore, MD: Johns Hopkins University Press, 2008)

Irmela van der Lühe, *Erika Mann: Eine Biographie* (Frankfurt: Fischer Taschenbuch Verlag, 2001; first ed. 1996)

Carlo Luna, "One hundred years of Italy on the move," in *Motu proprio: ACI's first hundred years, heading onward* (Rome: Automobile Club d'Italia, October 2005), 21–76

George A. Lundberg, Mirra Komarosky, and Mary Alice McInerney, *Leisure: A Suburban Study* (New York: Columbia University Press, 1934)

Matthias Luserke-Jaqui, "Carl Einstein: Bebuquin (1912) als Anti-Prometheus oder Plädoyer für das 'zerschlagene Wort,'" in Matthias Luserke-Jaqui (in cooperation with Monika Lippke) (ed.), *Deutschsprachige Romane der klassischen Moderne* (Berlin and New York: Walter de Gruyter, 2008), 110–127

Wolf Dieter Lützen, "Radfahren, Motorsport, Autobesitz: Motorisierung zwischen Gebrauchswerten und Statuserwerb," in Wolfgang Ruppert (ed.), *Die Arbeiter: Lebensformen, Alltag und Kultur von der Frühindustrialisierung bis zum "Wirtschaftswunder"* (Munich: Verlag C. H. Beck, 1986), 369–377

Robert S. Lynd and Helen Merrell Lynd, *Middletown in Transition: A Study in Cultural Conflicts* (New York: Harcourt, Brace and Company, 1937)

Robert S. Lynd and Helen Merrell Lynd, *Middletown; A Study in American Culture* (New York: Harcourt, Brace and Company, 1929)

Stephen Lyng, "Edgework: A Social Psychological Analysis of Voluntary Risk Taking," *The American Journal of Sociology* 95, no. 4 (January 1990), 851–886

Robert MacDougall, "The Wire Devils: Pulp Thrillers, the Telephone, and Action at a Distance in the Wiring of a Nation," *American Quarterly* 58, no. 3 (September 2006), 715–741

John M. MacKenzie, "T.E. Lawrence: The Myth and the Message," in Robert Giddings (ed), *Literature and Imperialism*, Insights, ed. Clive Bloom (Houndsmill and London: MacMillan, 1991), 150–181

Phil Macnaghten and John Urry, *Contested Natures* (Thousand Oaks, CA: Sage, 1999)

Maurice Maeterlinck, "L'Accident," in Maeterlinck, *L'Intelligence des fleurs*, Troisième mille (Paris: Fasquelle, 1907), 237–253

——, "Éloge de la boxe," in Maeterlinck, *L'Intelligence des fleurs*, Troisième mille (Paris: Fasquelle, 1907), 183–187

——, "En automobile," in Maeterlinck, *Le Double jardin* (Paris: Bibliothèque-Charpentier, 1904), 51–65

——, "On a Motor-Car," in Maeterlinck, *The Double Garden*, transl. by Alexander Teixeira de Mattos (London: George Allen & Company, 1915; first ed. May 1904), 139–152

Charles S. Maier, *Recasting Bourgeois Europe: Stabilization in France, Germany and Italy in the Decades after World War I* (Princeton, NJ: Princeton University Press, 1988; first ed. 1975)

Ben Malbon, *Clubbing: Dancing, Ecstasy and Vitality*, Critical Geographies, eds. Tracey Skelton and Gill Valentine (London and New York: Routledge, 1999)

Thomas Peter Maloney, "Essex motor car registrations, 1904" (unpublished MSc thesis, University of London, 1986)

Dave Mann and Ron Main, *Races, Chases & Crashes: A Complete Guide to Car Movies and Biker Flicks* (Osceola, WI: Motorbooks International, 1994)

Erika and Klaus Mann, *Rundherum: Abenteuer einer Weltreise: Mit Originalfotos: Nachwort von Uwe Naumann* (Reinbek bei Hamburg: Rowohlt Taschenbuch Verlag, 2001; first ed. 1929)

Klaus Mann, "Speed," in Mann, *Speed: Die Erzählungen aus dem Exil* (Reinbek bei Hamburg: Rowohlt, 2003; first ed. 1990), 105–150

Thomas Mann, *Tristan, Tonio Kröger, Der Tod in Venedig, Mario und der Zauberer, Meistererzählungen: Königliche Hoheit, Roman*, Nachwort von Hans Wysling (Zürich: Manesse Verlag, 2004)

Paul Mariani, *William Carlos Williams: A New World Naked* (New York: McGraw-Hill Book Company, 1981)

Marco Mariano, "Introduction," in Mariano (ed.), *Defining the Atlantic Community: Culture, Intellectuals, and Policies in the Mid-Twentieth Century* (New York and London: Routledge, 2010), 1–10

Jane Marcus, "Afterword: Corpus/Corps/Corpse; Writing the Body in/at War," in Helen Zenna Smith, *Not so Quiet . . . Stepdaughters of War*, afterword by Jane Marcus

(New York: The Feminist Press at the City University of New York, 1989; first ed. 1930), 241–300

"Marie Corelli 1855–1924," Literary Heritage West Midlands, http://www3.shropshire-cc.gov.uk/people/corelli.htm (last accessed 9 March 2010)

Jean Marigny, "Voitures fantastiques," in Frédéric Monneyron and Joël Thomas (eds.), *Automobile et Littérature,* Collection Études (n.p.: Presses Universitaires de Perpignan, 2005), 173–187

F. T. Marinetti, "À Mon Pégase," in Marinetti, *La Ville charnelle* (Paris: E. Sansot, 1908), 169–172

——, "The Founding and Manifesto of Futurism 1909," in Umbro Apollonio (ed.), *Futurist Manifestos* (New York: The Viking Press, 1973)

E. Marion, "Autorités chargées de la construction et de l'entretien des Routes, etc." (report 55, PIARC Conference, London, 1913)

Margaret Marsh, *Suburban Lives* (New Brunswick, NJ, and London: Rutgers University Press, 1990)

Jürgen Martschukat and Olaf Stieglitz, *Geschichte der Männlichkeiten* (Frankfurt and New York: Campus Verlag, 2008)

Leo Marx, *The Machine in the Garden: Technology and the Pastoral Idea in America* (Oxford and New York: Oxford University Press, 2000; first ed. 1964)

——, *The Pilot and the Passenger: Essays on Literature, Technology, and the Culture in the United States* (New York and Oxford: Oxford University Press, 1988)

——, "Technology: The Emergence of a Hazardous Concept," *Social Research* 64, no. 3 (Fall 1997), 965–988

Philip Massey, "The Expenditure of 1,360 British Middle-Class Households in 1938–39," *Journal of the Royal Statistical Society* 105, no. 3 (1942), 159–196

Kate Masterson, "The Monster in the Car: A Study of the Twentieh-Century Woman's Passion for the Motor Speed-Mania and Its Attendant Evils and Vagaries," *Lippincott's Monthly Magazine* 86 (10 August 1910), 204–211

Hervé Mathe, "Problème de la conduit de la démarche logistique: le concept de logistique dans une histoire contemporaine de la pensée sur la guerre," in Gérard Canini (ed.), *Les fronts invisibles: Nourrir—fournir—soigner; Actes du colloque international sur La logique des armées au combat pendant la Première Guerre mondiale organisé à Verdun les 6, 7, 8 juin 1980 sous la présidence du général d'Armée F. Gambiez, membre de l'Institut* (Nancy: Presses Universitaires de Nancy, 1984), 21–39

David Matless, *Landscapes and Englishness* (London: Reaktion Books, 1998)

Jean-Philippe Matthey, "Un aspect de la motorisation dans le canton de Vaud: La question de l'interdiction dominicale de la circulation automobile (1906–1923)" (unpublished Master's thesis, Université de Lausanne, Faculté des Lettres, 1992)

Christof Mauch and Thomas Zeller, "Introduction," in Mauch and Zeller (eds.), *The World Beyond the Windshield: Roads and Landscapes in the United States and Europe* (Athens and Stuttgart: Ohio University Press / Franz Steiner Verlag, 2008), 1–13

—— (eds.), *The World Beyond the Windshield: Roads and Landscapes in the United States and Europe* (Athens and Stuttgart: Ohio University Press / Franz Steiner Verlag, 2008)

Claude Mauriac, *Marcel Proust in Selbstzeugnissen und Bilddokumenten,* transl. from the French by Eva Rechel-Mertens, rowohlts monographien, ed. Kurt Kuenberg (Hamburg: Rowohlt Taschenbuch Verlag, 1976; first ed. 1958)

Arno J. Mayer, *The Persistence of the Old Regime: Europe to the Great War* (London: Croom Helm, 1981)

Enda McCaffrey, "*La 628–E8*: La voiture, le progrès et la postmodernité," *Cahiers Octave Mirbeau* 6 (1999), 122–141

Tom McCarthy, *Auto Mania: Cars, Consumers, and the Environment* (New Haven, CT, and London: Yale University Press, 2007)

Curt McConnell, *Coast to Coast by Automobile: The Pioneering Trips, 1899–1908* (Stanford, CA: Stanford University Press, 2000)

———, *"A Reliable Car and a Woman Who Knows It": The First Coast-to-Coast Auto Trips by Women, 1899–1916* (Jefferson, NC, and London: McFarland, 2000)

Alexis McCrossen, *Holy Day, Holiday: The American Sunday* (Ithaca, NY, and London: Cornell University Press, 2000)

Linda McDowell, "Off the road: Alternative views of rebellion, resistance and 'the beats,'" *Transactions of the Institute of British Geographers* NS 21 (1996), 412–419

Marcia Phillips McGowan, "A Nearer Approach to the Truth': Mary Borden's *Journey Down a Blind Alley*," *War, Literature and the Arts*, http://wlajournal.com/23_1/images/mcgowan.pdf (last accessed 30 March 2013)

Michael McGerr, *A Fierce Discontent: The Rise and Fall of the Progressive Movement in America, 1870–1920* (Oxford and New York: Oxford University Press, 2003)

Stephen L. McIntyre, "The Failure of Fordism: Reform of the Automobile Repair Industry, 1913–1940," *Technology and Culture* 41, no. 2 (April 2000), 269–299

———, "'The Repair Man Will Gyp You': Mechanics, Managers, and Customers in the Automobile Repair Industry, 1896–1940" (unpublished dissertation, University of Missouri-Columbia, May 1995)

Kate McLoughlin, "Edith Wharton, War Correspondent," *Edith Wharton Review* 21, no. 2 (Fall 2005), 1–10

Clay McShane, "The Carriage and the Origin of Urban Highways" (paper presented at the T²M Conference, Dearborn, MI, 5 November 2004)

———, *Down the Asphalt Path: The Automobile and the American City*, Columbia History of Urban Life, ed. Kenneth T. Jackson (New York: Columbia University Press, 1994)

———, "The Origins and Globalization of Traffic Control Signals," *Journal of Urban History* 25, no. 3 (March 1999), 379–404

Clay McShane and Gijs Mom, "Death and the Automobile: A Comparison of Automobile Ownership and Fatal Accidents in the United States and the Netherlands, 1910 – 1980" (unpublished paper presented at the ICOHTEC conference, Prague, 22–26 August 2000)

Clay McShane and Joel A. Tarr, *The Horse in the City: Living Machines in the Nineteenth Century*, Animals, History, Culture, ed. Harriet Ritvo (Baltimore, MD: Johns Hopkins University Press, 2007)

Katharine Mechler, "General Motors: Innovations in American Social Class Structure," *Automotive History Review* (Spring 2007), 4–25

Emil Merkert, *Personenkraftwagen, Kraftomnibus und Lastkraftwagen in den Vereinigten Staaten von Amerika: Mit besonderer Berücksichtigung ihrer Beziehungen zu Eisenbahn und Landstrasse* (Berlin: Springer, 1930)

Christoph Maria Merki, "Die 'Auto-Wildlinge' und das Recht: Verkehrs(un)sicherheit in der Frühzeit des Automobilismus," in Harry Niemann and Armin Hermann (eds.), *Geschichte der Strassenverkehrssicherheit im Wechselspiel zwischen Fahrzeug, Fahrbahn und Mensch* (Bielefeld: Delius & Klasing, 1999), 51–73

———, *Der holprige Siegeszug des Automobils 1895–1930: Zur Motorisierung des Strassenverkehrs in Frankreich, Deutschland und der Schweiz* (Vienna, Cologne, and Weimar: Böhlau, 2002)

Peter Merriman, *Driving Spaces: A Cultural-Historical Geography of England's M1 Motorway* (Malden, MA, Oxford, and Carlton, Australia: Blackwell, 2007)

———, *Mobility, Space and Culture*, International Library of Sociology, ed. John Urry (London and New York: Routledge, 2012)

Mike Michael, "Co(a)gency and the Car: Attributing Agency in the Case of 'Road Rage,'" in Brita Brenna, John Law, and Ingunn Moser (eds.), *Machines, Agency and Desire* (n.p.: Centre for Technology and Culture, 1998), 125–141

Pierre Michel, "Introduction," in Octave Mirbeau, *Oeuvre romanesque*, vol. 3 (Paris, Buchet, and Chastel: Pierre Zech / Société Octave Mirbeau, 2001), 260–277

———, "Maeterlinck et *La 628-E8*," *Cahiers Octave Mirbeau* 9 (2002), 239–247

———, *Un moderne: Octave Mirbeau* (Cazaubon: Eurédit, 2004)

———, "Octave Mirbeau et le concept de modernité," *Cahiers Octave Mirbeau* 4 (1997), 11–32

Pierre Michel and Jean-Claude Delauney, "Les épreuves corrigées de *La 628-E8*," *Cahiers Octave Mirbeau* 15 (March 2008), 209–217.

Pierre Michel and Jean-François Nivet, *Octave Mirbeau, l'imprécateur au cœur fidèle: Biographie* (Paris: Librairie Séguier, 1990)

Alfred C. Mierzejewski, *Hitler's Trains: The German National Railways & The Third Reich* (Stroud: Tempus, 2000)

Reijo Miettinen, "The Riddle of Things: Activity Theory and Actor-Network Theory as Approaches to Studying Innovations," *Mind, Culture, and Activity* 6, no. 3 (1999), 170–195

P. J. Mijksenaar, "Wie, waar en hoe men in Holland over de bruggen komt," *Het Motorrijwiel en de Populaire Auto* 23 (1935), 898–900

"Military Automobiles," *Horseless Age* 8, no. 15 (10 July 1901), 335

E. Willard Miller, "The Pennsylvania Turnpike," in Donald G. Janelle (ed.), *Geographical Snapshots of North America: Commemorating the 27th Congress of the International Geographical Union and Assembly* (New York and London: Guilford Press, 1992), 428–431

C. Wright Mills, *White Collar: The American Middle Classes* (New York: Oxford University Press, 1953)

Charles M. Mills, *Vacations for Industrial Workers* (New York: The Ronald Press Company, 1927)

Francis Miltoun, *The Automobilist Abroad* (London: Brown, Langham & Company, 1907)

———, *Italian Highways and Byways from a Motor Car* (London: Hodder & Stoughton, 1909)

[Ministry of Transport], *Report of the Conference on Rail and Road Transport, 29th July, 1932* (London: HMSO, 1932)

Makiko Minow-Pinkney, "Virginia Woolf and the Age of Motor Cars," in Pamela L. Caughie (ed.), *Virginia Woolf in the Age of Mechanical Reproduction*, Border Crossings, ed. Daniel Albright (New York and London: Garland, 2000), 159–182

Steven Mintz and Susan Kellogg, *Domestic Revolutions: A Social History of American Family Life* (New York: The Free Press, 1988)

[Octave Mirbeau], *Sketches of a Journey: Travels in an Early Motorcar. From Octave Mirbeau's journal 'La 628-E8'. With illustrations by Pierre Bonnard* (London: Philip Wilson Publishers, in association with Richard Nathanson, 1989)

David Mitchell, *Women on the Warpath: The Story of the Women of the First World War* (London: Jonathan Cape, 1966)
Richard G. Mitchell, Jr., *Mountain Experience: The Psychology and Sociology of Adventure* (Chicago and London: University of Chicago Press, 1983)
William J. Mitchell, Christopher E. Borroni-Bird, and Lawrence D. Burns, *Reinventing the Automobile: Personal Urban Mobility for the 21st Century* (Cambridge, MA, and London: MIT Press, 2010)
John Modell, *Into One's Own: From Youth to Adulthood In The United States 1920–1975* (Berkeley, Los Angeles, and London: University of California Press, 1989)
Marco Modenesi, "Locomotions nouvelles: Automobiles et écrivains à la fin du XIXe siècle," in Frédéric Monneyron and Joël Thomas (eds.), *Automobile et Littérature*, Collection Études (n.p.: Presses Universitaires de Perpignan, 2005), 25–36
L. Moholy-Nagy, *vision in motion* (Chicago: Paul Theobald and Company, 1956; first ed. 1947)
Arwen P. Mohun, "Designed for Thrills and Safety: Amusement Parks and the Commodification of Risk, 1880–1929," *Journal of Design History* 14, no. 4 (2001), 291–306
Norman T. Moline, *Mobility and the Small Town, 1900–1930: Transportation Change in Oregon, Illinois* (Chicago: University of Chicago Press, 1971)
Gijs Mom, "Building an Automobile System: PIARC and Road Safety (1908–1938)" (Proceedings of the PIARC Conference, Paris, 17–23 September 2007; on CD-ROM)
——, "Civilized adventure as a remedy for nervous times: Early automobilism and fin the siecle culture," *History of Technology* 23 (2001), 157–190
——, "Clashes of Cultures: Road vs. Rail in the North-Atlantic World during the Inter-War Coordination Crisis," in Christopher Kopper and Massimo Moraglio (eds.), *The Organization of Transport: A History of Users, Industry, and Public Policy* (London and New York: Routledge, forthcoming 2015)
——, "Constructing Multifunctional Networks: Road Building in the Netherlands, 1810–1980," in Mom and Laurent Tissot (eds.), *Road History: Planning, Building and Use* (Lausanne: Alphil, 2007), 33–62
——, "Constructing the State of the Art: Innovation and the Evolution of Automotive Technology (1898–1940)," in Rolf-Jürgen Gleitsmann and Jürgen E. Wittmann (eds.), *Innovationskulturen um das Automobil: Von gestern bis morgen: Stuttgarter Tage zur Automobil- und Unternehmensgeschichte 2011*, Wissenschaftliche Schriftenreihe der Mercedes-Benz Classic Archive, Band 16 (Stuttgart: Mercedes-Benz Classic Archive, 2012), 51–75
——, "Decentering highways: European national road network planning from a transnational perspective," in Hans-Liudger Dienel and Hans-Ulrich Schiedt (eds.), *Die moderne Strasse: Planung, Bau und Verkehr vom 18. bis zum 20. Jahrhundert*, Deutsches Museum, Beiträge zur historischen Verkehrsforschung, Band 11; eds. Helmuth Trischler, Christopher Kopper, and Hans-Liudger Dienel (Frankfurt: Campus Verlag, 2010), 77–100
——, "Diffusion and technological change: Culture, technology and the emergence of a 'European car,'" *Jahrbuch für Wirtschaftsgeschichte* no. 1 (2007), 67–82
——, "Droit à la mobilité, liberté ou contrainte? Plaidoyer pour des recherches nouvelles sur la mobilité/Freedom of Mobility; A Plea for new mobility studies," in Christophe Gay et al., *Mobile Immobile: Quels choix, quels droits pour 2030: Choices and rights for 2030*, vol. 2 (n.p. [Paris]: Éditions de l'Aube / Forum vie mobiles, 2011), 143–151

―――, "The Dual Nature of Technology: Automotive Ignition Systems and the Evolution of the Car," in Christian Huck and Stefan Bauernschmidt (eds.), *Travelling Goods, Travelling Moods: Varieties of Cultural Appropriation (1850–1950)* (Frankfurt and New York: Campus Verlag, 2012), 189–207

―――, *The Electric Vehicle: Technology and Expectations in the Automobile Age* (Baltimore, MD: Johns Hopkins University Press, 2004)

―――, *The Evolution of Automotive Technology: A Handbook* (Warrendale: SAE International, forthcoming 2015)

―――, "Fox Hunt: Materials Selection and Production Problems of the Edison Battery (1900–1910)," in Hans-Joachim Braun and Alexandre Herlea (eds.), *Materials: Research, Development and Applications (Proceedings of the XXth International Congress of History of Science [Liège, 20–26 July 1997])* (Turnhout: Brepols, 2002), 147–154

―――, "Frozen History: Limitations and Possibilities of Quantitative Diffusion Studies," in Ruth Oldenziel and Adri de la Bruhèze (eds.), *Manufacturing Technology: Manufacturing Consumers: The Making of Dutch Consumer Society* (Amsterdam; aksant, 2008), 73–94

―――, "'The future is a shifting panorama': The role of expectations in the history of mobility," in Weert Canzler and Gert Schmidt (eds.), *Zukünfte des Automobils: Aussichten und Grenzen der autotechnischen Globalisierung* (Berlin: edition sigma, 2008), 31–58

―――, "'Historians Bleed Too Much': Recent Trends in the State of the Art in Mobility History," in Peter Norton et al. (eds.), *Mobility in History: Reviews and Reflections (T²M Yearbook 2012)* (Neuchâtel: Alphil, 2011), 15–30

―――, "Inter-artefactual Technology Transfer: Road Building Technology in the Netherlands and the Competition between Bricks, Macadam, Asphalt and Concrete," *History and Technology* 20, no. 1 (April, 2004), 3–23

―――, "Mobility for Pleasure: A Look at the Underside of Dutch Diffusion Curves (1920–1940)," *TST Revista de Historia: Transportes, Servicios y Telecomunicaciones* no. 12 (June 2007), 30–68

―――, "Orchestrating Automobile Technology: Comfort, Mobility Culture, and the Construction of the 'Family Touring Car', 1917 – 1940," *Technology and Culture* 55 No. 2 (April 2014) 299-325

―――, "Die Prothetisierung des Autos; Kultur und Technik bei der Bedienung des Automobils" (unpublished presentation, Workshop at the Museum für Technik und Arbeit, Mannheim, 26–27 January 1995)

―――, "Roads without Rails; European Highway-Network Building and the Desire for Long-Range Motorized Mobility," *Technology and Culture* 46, no. 4 (October 2005), 745–772

―――, "Struggle of the Systems: Freight mobility from a Transatlantic Perspective, 1920 - 2000," in: Gérard Duc, Olivier Perroux, Hans-Ulrich Schiedt, François Walter (eds.), *Histoire des transports et de la mobilité. Entre concurrence modale et coordination (de 1918 à nos jours)/ Transport and mobility history. Between modal competition and coordination (from 1918 to the present)* (Neuchâtel: Alphil, 2014)

―――, "Subversive Mobility: An Uneasy Overview," in Mathieu Flonneau (ed.), *Automobile: Les cartes du désamour: Généalogies de l'anti-automobilisme*, Collection "Cultures Mobiles," eds. Mathieu Flonneau and Arnaud Passalacqua (Paris: Descartes & Cie, 2009), 68–78

———, "Translating Properties into Functions (and Vice Versa): Design, User Culture and the Creation of an American and a European car (1930–1970)," *Journal of Design History* 20, no. 2 (2007), 171–181

———, "Transporting mobility: The emergence of two different car societies in a transatlantic perspective," in Hans Krabbendam, Cornelis A. van Minnen, and Giles Scott-Smith (eds.), *Four Centuries of Dutch-American Relations 1609–2009* (Amsterdam: Boom, 2009), 819–830

———, "What kind of transport history did we get? Half a century of JTH and the future of the field," *Journal of Transport History* 24, no. 2 (September 2003), 121–138

——— (together with Georgine Clarsen et al.), "Editorial," *Transfers* 1, no. 1 (Spring 2011), 1–13

Gijs Mom and Ruud Filarski, *Van transport naar mobiliteit: De mobiliteitsexplosie (1895–2005)* (Zutphen: Walburg Pers, 2008)

Gijs Mom and David Kirsch, "The Holy Road to the Automobile System: Interactions between the Military and Early Automobilism, 1898–1920" (unpublished paper presented at the annual SHOT Conference in San Jose, CA, 4–7 October 2001)

Gijs Mom, Gordon Pirie, and Laurent Tissot (eds.), *Mobility in History: The State of the Art in the History of Transport, Traffic and Mobility* (Neuchâtel: Alphil, 2009)

Gijs Mom, Peter Staal, and Johan Schot, "Civilizing motorized adventure: Automotive technology, user culture and the Dutch Touring Club as mediator," in Ruth Oldenziel and Adri de la Bruhèze (eds.), *Manufacturing Technology: Manufacturing Consumers: The Making of Dutch Consumer Society* (Amsterdam: aksant, 2008), 141–160

Gijs Mom and Laurent Tissot (eds.), *Road History: Planning, Building and Use* (Lausanne: Alphil, 2007)

Wolfgang J. Mommsen, "Conclusion," in: Jean-Jacques Becker et al., *Guerre et cultures 1914–1918* (Paris: Armand Colin, 1994), 429–442

Sidney Monas, "Driving Nails with a Samovar: A Historical Introduction," in Viktor Shklovsky, *A Sentimental Journey: Memoirs, 1917–1922*, translation and literary introduction by Richard Sheldon; historical introduction by Sidney Monas (n.p.: Dalkey Archive Press, 2004; first ed. 1923; first English ed. 1970), xxvii–xlvii

Eric H. Monkkonen, "Regional Dimensions of Tramping, North and South, 1880–1910," in Monkkonen (ed.), *Walking to Work: Tramps in America, 1790–1935* (Lincoln and London: University of Nebraska Press, 1984), 189–211

Frédéric Monneyron and Joël Thomas, *L'Automobile: Un imaginaire contemporain* (Paris: Éditions Imago, 2006)

———, "Introduction," in Monneyron and Thomas (eds.), *Automobile et Littérature*, Collection Études (n.p.: Presses Universitaires de Perpignan, 2005), 7–11

Day Monroe et al., *Family Expenditures for Automobile and Other Transportation: Five Regions* (Washington, DC, 1941)

Ileen Montijn, *Leven op stand 1890–1940* (Amsterdam: Thomas Rap, 1998)

James F. Moore, "Predators and Prey: A New Ecology of Competition," *Harvard Business Review* (May–June 1993), 75–86

Shaun Moores, "Television, Geography and 'Mobile Privatization,'" *European Journal of Communication* 8 (1993), 365–379

Massimo Moraglio, "Real ambition or just coincidence? The Italian fascist motorway projects in inter-war Europe," *The Journal of Transport History*, Third Series, 30, no. 2 (December 2009), 168–182

———, "The Renewal of the Italian Road Network in the 1920s," *Transfers* 2, no. 1 (Spring 2012), 87–105

———, "Transferring Technology, Shaping Society: Traffic Engineering in PIARC Agenda, in the early 1930s," *Technikgeschichte* 80, no. 1 (2013), 13–32

Joe Moran, "Crossing the Road in Britain, 1931–1976," *The Historical Journal* 49, no. 2 (2006), 477–496

Jan Morris (ed.), *Travels with Virginia Woolf* (London: Pimlico, 1993)

Kurt Möser, "The dark side of 'automobilism', 1900–30: Violence, war and the motor car," *Journal of Transport History*, Third Series, 24, no. 2 (September 2003), 238–258

———, "Einparken zwischen Kompetenzlust und Automatisierungsdruck," *Wege und Geschichte/Les chemins et l'histoire/Strade e storia* no. 1 (2009), 34–38.

———, *Fahren und Fliegen in Frieden und Krieg: Kulturen individueller Mobilitätsmaschinen 1880–1930* (Heidelberg, Ubstadt-Weiher, Neustadt a.d.W., and Basel: Verlag Regionalkultur, 2009)

———, *Geschichte des Autos* (Frankfurt and New York: Campus Verlag, 2002)

———, "'Der Kampf des Automobilisten mit der Maschine'- Eine Skizze der Vermittlung der Autotechnik und des Fahrenlernens im 20. Jahrhundert," in Lars Bluma, Karl Pichol, and Wolfhard Weber (eds.), *Technikvermittlung und Technikpopularisierung—Historische und didaktische Perspektiven* (Münster, New York, Munich, and Berlin: Waxmann, 2004), 89–102

———, "World War One and the Creation of Desire for Cars in Germany," in Susan Strasser, Charles McGovern, and Matthias Judt (eds.), *Getting and Spending: European and American Consumer Societies in the Twentieth Century* (Washington, DC, 1998), 195–222

———, "Zwischen Systemopposition und Systemteilnahme: Sicherheit und Risiko im motorisierten Strassenverkehr 1890–1930," in Harry Niemann and Armin Hermann (eds.), *Geschichte der Strassenverkehrssicherheit im Wechselspiel zwischen Fahrzeug, Fahrbahn und Mensch* (Bielefeld: Delius & Klasing, 1999; (DaimlerChrysler Wissenschaftliche Schriftenreihe, Band 1)

Marina Moskowitz, *Standard of Living: The Measure of the Middle Class in Modern America* (Baltimore, MD: Johns Hopkins University Press, 2004)

Motor-Vehicle Traffic Conditions in the United States, Part 6: The Accident-Prone Driver (House Documents, 75th Congress, 3d Session [3 January –16 June 1938]) (Washington, DC: United States Government Printing Office, 1938)

Harold G. Moulton, *The American Transportation Problem (Prepared for the National Transportation Committee)* (Washington, DC: The Brookings Institution, 1933)

George E. Mowry (ed.), *The Twenties: Fords, Flappers & Fanatics* (Englewood Cliffs, NJ: Prentice-Hall, 1963)

José C. Moya, "Modernization, Modernity, and the Trans/formation of the Atlantic World in the Nineteenth Century," in Jorge Cañizares-Esguerra and Erik R. Seeman (eds.), *The Atlantic in Global History 1500–2000* (Upper Saddle River, NJ: Pearson Education / Prentice Hall, 2007), 179–197

John Henry Mueller, "The Automobile: A Sociological Review" (unpublished doctoral dissertation, University of Chicago, August 1928)

Richard Mullen and James Munson, *"The Smell of the Continent': The British Discover Europe 1814–1914* (London, Basingstoke, and Oxford: Macmillan, 2009)

Dorit Müller, "Transfers between Media and Mobility: Automobilism, Early Cinema and Literature, 1900–1920," *Transfers* 1, no. 1 (Spring 2011), 53–75

———, *Gefährliche Fahrten: Das Automobil in Literatur und Film um 1900*, Epistemata, Würzburger wissenschaftliche Schriften, Reihe Literaturwissenschaft, Band 486 (Würzburg: Königshausen & Neumann, 2004)

Wolfgang Müller-Funk, *Literatur als geschichtliches Argument: Zur ästhetischen Konzeption und Geschichtsverarbeitung in Lion Feutchwangers Romantrilogie Der Wartesaal,* Europäische Hochschulschriften, Reihe I, Deutsche Sprache und Literatur, Band 415 (Frankfurt am Main/Bern: Peter D. Lang, 1981)

Corinne Mulley, "The background to bus regulation in the 1930 Road Traffic Act: Economic, political and personal influences in the 1920s," *Journal of Transport History,* Third Series, 4, no. 2 (September 1983), 1–19

Herfried Münkler, "Strategien der Sicherung: Welten der Sicherheit und Kulturen des Risikos; Theoretische Perspektiven," in Herfried Münkler, Matthias Bohlender, and Sabine Meurer (eds.), *Sicherheit und Risiko: Über den Umgang mit Gefahr im 21. Jahrhundert* (Bielefeld: transcript Verlag, 2010), 11–34

Marianne Muse, *The Standard of Living on Specific Owner-Operated Vermont Farms,* University of Vermont and State Agricultural College, Bulletin 340 (Burlington, VT: Free Press Printing Co., June 1932)

Gabe Mythen, *Ulrich Beck: A Critical Introduction to the Risk Society* (London and Sterling, VA: Pluto Press, 2004)

Jelena Novaković, "La vitesse dand *La 628-E8* de Mirbeau et *L'homme pressé* de Paul Morand," in Éléonore Reverzy and Guy Ducrey (eds.), *L'Europe en automobile: Octave Mirbeau écrivain voyageur,* Configurations littéraires, collection de l'Université Marc Bloch de Strasbourg (Strasbourg: Presses Universitaires de Strasbourg, 2009), 255–267

Nebojša Nakicenovic, "The Automobile Road to Technological Change; Diffusion of the Automobile as a Process of Technological Substitution," *Technological Forecasting and Social Change* 29 (1986), 309–340

Jacques Nantet, "Marcel Proust et la vision cinématographique," *La revue des lettres modernes: histoire des idées et des littératures* 5, no. 36–38 (1958), 307–313

A. F. Napp-Zinn, "Eigenart, Stand und Probleme der Verkehrswirtschaft der Niederlande," *Zeitschrift für Verkehrswissenschaft* 14, no. 3 (1937), 157–172

Anedith Jo Bond Nash, "Death on the Highway: The Automobile Wreck in American Culture, 1920–40" (unpublished dissertation, University of Minnesota, June 1983)

Roderick Nash, *The Nervous Generation: American Thought, 1917–1930* (Chicago: Rand McNally College Publishing Company, 1970)

Richard Nate, *Amerikanische Träume: Die Kultur der Vereinigten Staaten in der Zeit des New Deal* (Würzburg: Königshausen & Neumann, 2003)

Elinor Nauen (ed.), *Ladies, Start Your Engines: Women Writers on Cars and the Road* (Boston and London: Faber and Faber, 1996)

Petra Naumann-Winter, "'Das Radfahren der Damen': Bildbetrachtungen zum Diskurs über Modernisierung und Technisierung um 1900," in Christel Köhle-Hezinger, Martin Scharfe, and Rolf Wilhelm Brednich (eds.), *Männlich. Weiblich. Zur Bedeutung der Kategorie Geschlecht in der Kultur: 31. Kongress der Deutschen Gesellschaft für Volkskunde, Marburg 1997* (New York, Munich, and Berlin: Waxmann, 1999), 430–443

Lynda Nead, "Paintings, Films and Fast Cars: A Case Study of Hubert von Herkomer," *Art History* 25, no. 2 (April 2002), 240–255

Nicolas Neiertz, *La coordination des transports en France: De 1918 à nos jours* (Paris: Ministère de l'Économie, des Finances et de l'Industrie, Comité pour l'histoire économique et financière, 1999)

Michael Nerlich, *Abenteuer oder das verlorene Selbstverständnis der Moderne: Von der Unaufhebbarkeit experimentalen Handelns* (Munich: Gerling Akademie Verlag, 1997)

John Nerone, "The Public and the Party Period," in Jeremy Packer and Craig Robertson (eds.), *Thinking with James Carey: Essays on Communications, Transportation, History*, Intersections in Communications and Culture; Global Approaches and Transdisciplinary Perspective, vol. 15; eds. Cameron McCarthy and Angharad N. Valdivia (New York: Peter Lang, 2006), 157–176

Patricia D. Netzley, *The Encyclopedia of Women's Travel and Exploration* (Westport, CT: Oryx Press, 2001)

Joan Neuberger, *Hooliganism: Crime, Culture, and Power in St. Petersburg, 1900–1914*, Studies on the History of Society and Culture, eds. Victoria E. Bonnell and Lynn Hunt (Berkeley, Los Angeles, and London: University of California Press, 1993)

Patricia D. Netzley, *The Encyclopedia of Women's Travel and Exploration* (Westport, CT: Oryx Press, 2001)

Shelley Kaplan Nickles, "Object Lessons: Household Appliance Design and the American Middle Class, 1920–1960" (unpublished doctoral dissertation, University of Virginia, January 1999), 70

J. Nicod, "Les autoroutes de l'Europe Occidentale et la formation d'un réseau de grandes routes européennes," *L'information géographique* 19, no. 1 (1955), 3–19

Becky M. Nicolaides, *My Blue Heaven: Life and Politics in the Working-Class Suburbs of Los Angeles, 1920–1965* (Chicago and London: University of Chicago Press, 2002)

Paul Nieuwenhuis and Peter Wells, "The all-steel body as a cornerstone to the foundations of the mass production car industry," *Industrial and Corporate Change* 16, no. 2 (2007), 183–211

Wolfram Nitsch, "Fantasmes d'essence: Les automobiles de Proust à travers l'histoire du texte," in Rainer Warning and Jean Milly (eds.), *Marcel Proust: Écrire sans fin*, Textes et Manuscrits, ed. Louis Hay (Paris: CNRS Éditions, 1996), 125–141

John Nolen and Henry V. Hubbard, *Parkways and Land Values* (Cambridge, MA: Harvard University Press, 1937)

Peter Norton, "Americans' Affair of Hate with the Automobile: What the 'Love Affair' Fiction Concealed," in Mathieu Flonneau (ed.), *Automobile: Les cartes des désamour: Généalogies de l'anti-automobilisme,* Collection "Cultures Mobiles," eds. Mathieu Flonneau and Arnaud Passalacqua (Paris: Descartes & Cie, 2009), 93–104

Peter David Norton, *Fighting Traffic: The Dawn of the Motor Age in the American City* (Cambridge, MA, and London: MIT Press, 2008)

———, "Street Rivals: Jaywalking and the Invention of the Motor Age Street," *Technology and Culture* 48, no. 2 (April 2007), 331–359

David E. Nye, *American Technological Sublime* (Cambridge, MA, and London: MIT Press, 1994)

Robert A. Nye, "Degeneration, Neurasthenia and the Culture of Sport in Belle Epoque France," *Journal of Contemporary History* 17, no. 1 (special issue "Decadence") (January 1982), 51–68

Sean O'Connell, *The Car and British Society: Class, Gender and Motoring, 1896–1939*, Studies in Popular Culture, ed. Jeffrey Richards (Manchester and New York: Manchester University Press, 1998)

———, "From Toad of Toad Hall to the 'Death Drivers' of Belfast: An Exploratory History of 'Joyriders,'" *British Journal of Criminology* 46 (2006), 455–469

———, "Gender and the Car in Inter-war Britain," in Moira Donald and Linda Hurcombe (eds.), *Gender and Material Culture in Historical Perspective* (London and New York: MacMillan Press / St. Martin's Press, 2000), 175–191

Klaus-Dieter Oelze, *Das Feuilleton der Kölnischen Zeitung im Dritten Reich*, Regensburger Beiträge zur deutschen Sprach- und Literaturwissenschaft, ed. Bernhard Gajek; Reihe B/Untersuchungen, Bd. 45 (Frankfurt, Bern, New York, and Paris: Peter Lang, 1990)

F. W. Ogilvie, *The Tourist Movement: An Economic Study* (London: P. S. King & Son, 1933)

Ernst Oldemeyer, "Entwurf einer Typologie des menschlichen Verhältnisses zur Natur," in Götz Grossklaus and Ernst Oldemeyer, *Natur als Gegenwelt: Beiträge zur Kulturgeschichte der Natur* (Karlsruhe: von Loeper Verlag, 1983), 15–42

Martha L. Olney, *Buy Now Pay Later: Advertising, Credit, and Consumer Durables in the 1920s* (Chapel Hill and London: University of North Carolina Press, 1991)

Janet Oppenheim, *"Shattered Nerves": Doctors, Patients, and Depression in Victorian England* (New York and Oxford: Oxford University Press, 1991)

Kevin H. O'Rourke and Jeffrey G. Williamson, *Globalization and History: The Evolution of a Nineteenth-Century Atlantic Economy* (Cambridge, MA, and London: MIT Press, 2000; first ed. 1999)

Jean Orselli, "Usages et usagers de la route: Pour une histoire de moyenne durée, 1860–2008" (doctoral dissertation, Université de Paris I Panthéon Sorbonne, 2009)

Eli Osman and Ching-tse Lee, "Sociological Theories of Aggression," in Irwin L. Kutash, Samuel B. Kutash, Louis B. Schlesinger and Associates, *Violence; Perspectives on Murder and Aggression* (San Francisco, Washington and London: Josey-Basstiessen Publishers, 1978), 58–73

Wa. Ostwald, "Zwecksport," *Public Roads* 12, no. 29 (18 July 1914), 21–24

Didier Ottinger, "Cubism + Futurism = Cubofuturism," in Didier Ottinger (ed.), *Futurism* (Paris and Milan: Éditions du Centre Pompidou/5 Continents Editions, 2009; orig. title: *Le Futurisme à Paris: Une avant-garde explosive*, 2008), 20–41

Helmut Otto, "Die Herausbildung des Kraftfahrwesens im deutschen Heer bis 1914," *Militärgeschichte* 3 (1989), 227–236

Marthe Oulié, *Bidon 5: En rallye à travers le Sahara* (n.p. [Paris]: Ernest Flammarion, n.d. [1931; cinquième mille])

Marthe Oulié and Hermine de Saussure, *La croisière de "Perlette": 1700 milles dans la Mer Égée* (n.p. [Paris]: Librairie Hachette, 1926)

Ian Ousby, *The Englishman's England: Taste, Travel and the Rise of Tourism* (Cambridge and New York: Cambridge University Press, 1990)

Richard Overy, "Heralds of Modernity: Cars and Planes from Invention to Necessity," in Mikuláš Teich and Roy Porter (eds.), *Fin de siècle and its legacy* (Cambridge and New York: Cambridge University Press, 1990), 54–79

Wilfred Owen, *Transportation and World Development* (Baltimore, MD: Johns Hopkins University Press, 1987)

Ted Ownby, "The Snopes Trilogy and the Emergence of Consumer Culture," in Donald M. Kartiganer and Ann J. Abadie (eds.), *Faulkner and Ideology: Faulkner and Yoknapatawpha, 1992* (Jackson: University Press of Mississippi, 1995), 95–128

Cord Pagenstecher, *Der bundesdeutsche Tourismus: Ansätze zu einer Visual History: Urlaubsprospekte, Reiseführer, Fotoalben 1950–1990*, Schriftenreihe *Studien zur Zeitgeschichte*, Band 34 (Hamburg: Verlag Dr. Kovač, 2003)

Jeremy Packer, "Automobility and apparatuses: Commentary on Cotten Seiler's *Republic of Drivers*," *History and Technology* 26, no. 4 (December 2010), 361–368

——, "Becoming Bombs: Mobilizing Mobility in the War of Terror," *Cultural Studies* 20, no. 4–5 (July–September 2006), 378–399

––––, *Mobility Without Mayhem: Safety, Cars, and Citizenship* (Durham, NC, and London: Duke University Press, 2008)

Jan Palmowski, "Travels with Baedeker—The Guidebook and the Middle Classes in Victorian and Edwardian Britain," in Rudy Koshar (ed.), *Histories of Leisure*, Leisure, Consumption and Culture, ed. Koshar (Oxford and New York: Berg, 2002), 105–130

Volker Pantenburg, "Weltstadt in Flegeljahren: Ein Bericht über Chicago" (film review, 30 November 2003) (http://newfilmkritik.de/archiv/2003-11/weltstadt-in-flegel jahren-ein-bericht-uber-chicago/; last accessed 26 September 2014)

Mika Pantzar, "Do Commodities Reproduce Themselves Through Human Beings? Toward an Ecology of Goods," *World Futures* 38 (1993) 201–224

Franz Urban Pappi, "The Petite Bourgeoisie and the New Middle Class: Differentiation or Homogenisation of the Middle Strata in Germany," in Frank Bechhofer and Brian Elliott (eds.), *The Petite Bourgeoisie: Comparative Studies of the Uneasy Stratum* (New York: St. Martin's Press, 1981), 105–120

Alfons Paquet, *Weltreise eines Deutschen: Landschaften, Inseln, Menchen, Städte* (Berlin: Büchergilde Gutenberg, 1934)

Jennifer Parchesky, "Women in the driver's seat: The auto-erotics of early women's films," *Film History* 18, no. 2 (2006), 174–184

Julius H. Parmelee and Earl R. Feldman, "The Relation of the Highway to Rail Transportation," in Jean Labatut and Whearon J. Lane (eds.), *Highways in our National Life: A Symposium* (New York: Arno Press, A New York Times Company, 1972), 227–239

Richard Parry, *The Bonnot Gang* (n.p. [London]: Rebel Press, 1987)

Joris van Parys, *Het leven, niets dan het leven: Cyriel Buysse en zijn tijd* (Antwerp and Amsterdam: Houtekiet/Atlas, 2007)

G. Sydney Paternoster, *The Lady of the Blue Motor* (New York: Grosset & Dunlap, 1907)

Phil Patton, *Open Road: A Celebration of the American Highway* (New York: Simon and Schuster, 1986)

E. George Payne, *A Program of Education in Accident Prevention, with Methods and Results*, US Department of the Interior, Bureau of Education, Bulletin, no. 32 (Washington, DC: Government Printing Office, 1922)

Susan Pedersen, *Family, Dependence, and the Origins of the Welfare State: Britain and France, 1914–1945* (Cambridge: Cambridge University Press, 2006; reprint; first ed. 1993)

Annegret Pelz, "City Girls im Büro: Schreibkräfte mit Bubikopf," in Julia Freytag and Alexandra Tacke (eds.), *City Girls: Bubiköpfe & Blaustrümpfe in den 1920er Jahren, Literatur—Kultur—Geschlecht; Studien zur Literatur- und Kulturgeschichte*, eds. Anne-Kathrin Reulecke and Ulrike Vedder, in connection with Inge Stephan und Sigrid Weigel; Kleine Reihe, Band 29 (Cologne, Weimar, and Vienna: Böhlau Verlag, 2011), 35–53

J. A. G. Pennington, *Advertising Automotive Products in Europe*, US Department of Commerce, Bureau of Foreign and Domestic Commerce, Trade Information Bulletin, no. 462 (February 1927)

Marjorie Perloff, *The Futurist Moment: Avant-Garde, Avant Guerre, and the Language of Rupture* (Chicago and London: University of Chicago Press, 1986)

Charles Perrow, *Normal Accidents: Living with High-Risk Technologies* (New York: Basis Books, Inc., Publishers, 1984)

Michael Persson, "Techniek raakt basale babysnaar," *Technisch Weekblad* (2 February 2008), 13

John Durham Peters, "Technology and Ideology: The Case of the Telegraph Revisited," in Jeremy Packer and Craig Robertson (eds.), *Thinking with James Carey: Essays on Communications, Transportation, History*, Intersections in Communications and Culture; Global Approaches and Transdisciplinary Perspective, vol. 15; eds. Cameron McCarthy and Angharad N. Valdivia (New York: Peter Lang, 2006), 137-155

Jennifer Lynn Peterson, "'The Nation's First Playground': Travel Films and the American West, 1895-1920," in Jeffrey Ruoff (ed.), *Virtual Voyages; Cinema and Travel* (Durham and London: Duke University Press, 2006),, 79-98

J. F. R. Philips, "Recreatie en toerisme, spiegelbeeld of reactie?" in K. P. C. de Leeuw, M. F. A. Linders-Rooijendijk, and P. J. M. Martens (eds.), *Van ontspanning en inspanning: Aspecten van de geschiedenis van de vrije tijd* (Tilburg, 1995), 35-62

Meinrad Pichler, "Von Aufsteigern und Deklassierten: Ödön von Horváths literarische Analyse des Kleinbürgertums und ihr Verständnis zu den Aussagen der historischen Sozialwissenschaften," in Traugott Krischke (ed.), *Ödön von Horváth*, suhrkamp taschenbuch materialien, vol. 2005 (Frankfurt: Suhrkamp, 1981), 67-86

George W. Pierson, "The M-Factor in American History," *American Quarterly* 14, no. 2, part 2: Supplement (Summer 1962), 275-289

——, *The Moving American* (New York: Alfred A. Knopf, 1973)

Trevor Pinch, "The Social Construction of Technology: A Review," in Robert Fox (ed.), *Technological Change: Methods and Themes in the History of Technology* (Amsterdam: Harwood Academic Publishers, 1996)

R. G. Pinney, *Britain—destination for tourists?* (London: The Travel and Industrial Development Association of Great Britain and Ireland, November 1944; second revised edition; first ed. April 1944)

Gordon Pirie, "Non-urban Motoring in Colonial Africa in the 1920s and 1930s," *South African Historical Journal* 63, no. 1 (March 2011), 38-60

Gaëtan Pirou, *Les cadres de la vie économique: Les transports* (Bordeaux: Recueil Sirey, 1942)

Jean-Luc Piveteau, "La voiture, signe et agent d'une nouvelle relation de l'homme à l'espace," *Cahiers de l'Institut de Géographie* 7 (1990), 45-55

Dan Fellows Platt, *Through Italy with Car and Camera* (New York and London: G. P. Putnam's Sons, 1908)

F. R. Pleasonton, "The Automotive Industry: A Study of the Facts of Automobile Production and Consumption in the United States," *Annals of the American Academy of Political and Social Science* 97 (September 1921), 107-127

William Plowden, *The Motor Car and Politics in Britain* (Harmondsworth and Ringwood, Australia: Penguin, 1971)

Ieme van der Poel, "Inleiding," in Marcel Proust, *Op zoek naar de verloren tijd: De kant van Swann*, transl. of *À la recherche du temps perdu: De côté de Swann*, by Thérèse Cornips; introduced and annotated by Ieme van der Poel and Tom Hoenselaars (Amsterdam: De Bezige Bij, 2009), 7-38

Mark Polizotti, "Patabiographical: The portrait of a louche French author and his kingly alter ego," *Bookforum* (December-Januray 2012), 36-37

Marcus Popplow, "Contextualization and Public Impact: Converging or Conflicting Trends in Recent German Mobility History?" in Peter Norton et al. (eds.), *Mobility in History: Reviews and Reflections (T²M Yearbook 2012)* (Neuchâtel: Alphil, 2011), 63-72

Colin G. Pooley, "Landscapes without the car: a counterfactual historical geography of twentieth-century Britain," *Journal of Historical Geography* 36 (2010), 266–275

Colin G. Pooley and Jean Turnbull, "Commuting, transport and urban form: Manchester and Glasgow in the mid-twentieth century," *Urban History* 27, no. 3 (2000), 360–383

Colin Pooley, Jean Turnbull, and Mags Adams, "The impact of new transport technologies on intraurban mobility: A view from the past," *Environment and Planning A* 38 (2006), 253–267

——, *A Mobile Century? Changes in Everyday Mobility in Britain in the Twentieth Century* (Aldershot: Ashgate, 2005)

Rémy Porte, *La Direction des Services Automobiles des Armées et la Motorisation des Armées Françaises (1914–1919)* (Panazol: Charles Lavauzelle, 2004)

Robert C. Post, *Urban Mass Transit: The Life Story of a Technology*, Greenwood Technographies (Westport, CT, and London: Greenwood Press, 2007)

Christopher Thomas Potter, "Motorcycle Clubs in Britain During the Interwar Period, 1919–1939: Their Social and Cultural Importance," *International Journal of Motorcycle Studies* 1 (March 2005), http://ijms.nova.edu/March 2005/IJMS_ArtclPotter0305.html (last accessed 16 July 2010)

Anne-Cécile Pottier-Thoby, "*La 628–E8*, opus futuriste?" *Cahiers Octave Mirbeau* 8 (2001), 106–120

Howard L. Preston, *Automobile Age Atlanta: The Making of a Southern Metropolis 1900–1935* (Athens: University of Georgia Press, 1979)

Howard Lawrence Preston, *Dirt Roads to Dixie: Accessibility and Modernization in the South, 1885–1935* (Knoxville: University of Tennessee Press, 1991)

Susan Priestley, *The Crown of the Road: The Story of the RACV* (Melbourne: The MacMillan Company of Australia, 1983)

Ronald Primeau, *Romance of the Road: The Literature of the American Highway* (Bowling Green, OH: Bowling Green State University Popular Press, 1996)

Michael Prinz, "Das Ende der Standespolitik: Voraussetzungen, Formen und Konsequenzen mittelständischer Interessenpolitik in der Weimarer Republik am Beispiel des Deutschnationalen Handlungsgehilfenverband," in Jürgen Kocka (ed.), *Angestellte im europäischen Vergleich: Die Herausbildung angestellter Mittelschichten seit dem späten 19. Jahrhundert*, Geschichte und Gesellschaft, Sonderheft 7 (Göttingen: Vandenhoeck & Ruprecht, 1981), 331–353

Marcel Proust, *Op zoek naar de verloren tijd: De kant van Swann*, transl. of *À la recherche du temps perdu: De côté de Swann*, by Thérèse Cornips; introduced and annotated by Ieme van der Poel and Tom Hoenselaars (Amsterdam: De Bezige Bij, 2009)

John Pucher, "Urban Travel Behavior as the Outcome of Public Policy: The Example of Modal-Split in Western Europe and North America," *Journal of the American Planning Association* 54 (Autumn 1988), 509–520

Hans-Jürgen Puhle (ed.), *Bürger in der Gesellschaft der Neuzeit: Wirtschaft—Politik—Kultur*, Bürgertum; Beiträge zur europäischen Gesellschaftsgeschichte, eds. Wolfgang Mager et al., Band 1 (Göttingen: Vandenhoeck & Ruprecht, 1991)

Carroll Purcell, "American Boys, their Books, and the Romance of Modern Transport" (paper presented at a workshop on "gendered mobility," University of Wollongong, 15–16 December 2010)

Paul Raabe, "Einleitung," in Raabe (ed.), *Ich schneide die Zeit aus; Expressionismus und Politik in Franz Pfemferts "Aktion"* (München: Taschenbuch-Verlag, 1964; dtv-dokumente), 7-15

Rüdiger Rabenstein, *Radsport und Gesellschaft: Ihre sozialgeschichtlichen Zusammenhänge in der Zeit von 1867 bis 1914* (Hildesheim, Munich, and Zurich, 1991)
Joachim Radkau, "Auto-Lust: Zur Geschichte der Geschwindigkeit," in Tom Koenigs and Roland Schaeffer (eds.), *Fortschritt vom Auto? Umwelt und Verkehr in den 90er Jahren: Kongress des Umwelt Forum Frankfurt a.M. am 5./6. Oktober 1990* (Munich: Raben Verlag von Wittern, 1991), 113–130
——, "'Die Nervosität des Zeitalters,' Die Erfindung von Technikbedürfnissen um die Jahrhundertwende," *Kultur & Technik* no. 3 (1994), 51–57
——, *Das Zeitalter der Nervosität: Deutschland zwischen Bismarck und Hitler* (Munich and Vienna, 1998)
John B. Rae, *American Automobile Manufacturers: The First Forty Years* (Philadelphia, PA, and New York, 1959)
——, *The American Automobile: A Brief History* (Chicago and London, 1965)
——, *The American Automobile Industry* (Boston, 1984)
——, "The Electric Vehicle Company: A Monopoly that Missed," *Business History Review* 29, no. 4 (December 1955), 298–311
——, *The Road and the Car in American Life* (Cambridge, MA, and London: MIT Press, 1971)
——, "Why Michigan?" in David L. Lewis and Laurence Goldstein (eds.), *The Automobile and American Culture* (Ann Arbor: University of Michigan Press, 1983), 1–9
Suzanne Raitt, "May Sinclair and the First World War," *Ideas* 6, no. 2 (1999), http://nationalhumanitiescenter.org/ideasv62/raitt.htm (last accessed 5 March 2012)
Sudhir Chella Rajan, "Automobility and the liberal disposition," *Sociological Review* 54, no. s1 (October 2006), 113–129
Stephan Rammler, *Mobilität in der Moderne: Geschichte und Theorie der Verkehrssoziologie* (Berlin: edition sigma, 2001)
Alice Huyler Ramsey, *Veil, Duster, and Tire Iron* (Covina, CA: published by author, 1962; first ed. 1961)
Michele Ramsey, "Selling Social Status: Woman and Automobile Advertisements from 1910–1920," *Women and Language* 28, no. 1 (Spring 2005), 26–38
Teresa Ransom, *The Mysterious Miss Marie Corelli: Queen of Victorian Bestsellers* (Stroud: Sutton Publishing, 1999)
Leonard Rapport, "Trucker and Builder," in *These Are Our Lives: As Told by the People and Written by Members of the Federal Writers' Project of the Works Progress Administration in North Carolina, Tennessee, and Georgia* (Chapel Hill: University of North Carolina Press, 1939), 317–323
Hayagreeva Rao, "Institutional activism in the early American automobile industry," *Journal of Business Venturing* 19 (2004), 359–384
André Rauch, *Vacances en France de 1830 à nos jours* (Paris: Hachette Littératures, 2001; extended edition; first ed. 1996)
Recent Social Trends: Report of the President's Research Committee on Social Trends, vol. I (New York and London: McGraw-Hill Book Company, Inc., 1933)
Andreas Reckwitz, *Das hybride Subjekt: Eine Theorie der Subjektkulturen von der bürgerlichen Moderne zur Postmoderne* (Weilerswist: Velbrück Wissenschaft, 2006)
——, "Toward a Theory of Social Practices: A Development in Culturalist Theorizing," *European Journal of Social Theory* 5, no. 2 (2002), 243–263
Arthur Redding, *Raids on Human Consciousness: Writing, Anarchism, and Violence*, Cultural Frames, Framing Culture Series, ed. Robert Newman (Columbia: University of South Carolina Press, 1998)

Edward S. Reed, *James J. Gibson and the Psychology of Perception* (New Haven, CT, and London: Yale University Press, 1988)
Edward Reed and Rebecca Jones (eds.), *Reasons for Realism: Selected Essays of James J. Gibson* (Hillsdale, NJ, and London: Lawrence Erlbaum Ass., 1982)
Louis Raymond Reid, "The Small Town," in Harold E. Stearns, *Civilization in the United States; An Inquiry by Thirty Americans* (New York: Harcourt, Brace and Company, 1922), 285–296
Wolfgang Reif, "Exotismus im Reisebericht des frühen 20. Jahrhunderts," in Peter J. Brenner (ed.), *Der Reisebericht: Die Entwicklung einer Gattung in der deutschen Literatur* (Frankfurt: Suhrkamp Verlag, 1989), 434–462
Siegfried Reinecke, *Autosymbolik in Journalismus, Literatur und Film: Strukturalfunktionale Analysen vom Beginn der Motorisierung bis zur Gegenwart* (Bochum: Universitätsverlag Dr. N. Brockmeyer, 1992)
——, *Mobile Zeiten: Eine Geschichte der Auto-Dichtung* (Bochum, 1986)
Dominique Renouard, *Les transports de marchandises par fer, route et eau depuis 1850* (Paris: Librairie Armand Colin, 1960)
Donald Revell, http://www.poets.org/viewmedia.php/prmMID/19454 (last accessed 19 November 2009)
Georges Reverdy, *Les routes de France du XXe siècle 1900–1951* (Paris: Presses de l'école nationale des ponts et chaussées, 2007)
"A Review of Findings by the President's Research Committee on Social Trends," in *Recent Social Trends: Report of the President's Research Committee on Social Trends*, vol. I (New York and London: McGraw-Hill Book Company, Inc., 1933), xi–lxxv
George Revill, "Perception, Reception and Representation: Wolfgang Schivelbusch and the Cultural History of Travel and Transport," in Peter Norton et al. (eds.), *Mobility in History: Reviews and Reflections (T²M Yearbook 2012)* (Neuchâtel: Alphil, 2011), 31–48
David A. Revzan, "Trends in Economic Activity and Transportation in the San Francisco Bay Area," in *Parking as a Factor in Business, containing five papers presented at the thirty-second annual meeting January 13–16, 1953*, Highway Research Board, Special Report 11 (Washington, DC, 1953), 161–308
Lillian Rhoades, "One of the Groups Middletown Left Out," *Opportunity, Journal of Negro Life* (March 1933), 75–77, 93
Jean-Claude Richez and Léon Strauss, "Un temps nouveau pour les ouvriers: Les congés payés (1930–1960)," in Alain Corbin, *L'avènement des loisirs 1850–1960* (Paris and Rome: Aubier/Laterza, 1995), 376–412
Edward V. Rickenbacker, *Rickenbacker* (Englewood Cliffs, NJ: Prentice-Hall, 1967; fifth printing, December 1967)
David Riesman (in collaboration with Reuel Denney and Nathan Glazer), *The Lonely Crowd: A Study of the Changing American Character*, Studies in National Poetry (London: Geoffrey Cumberlege/Oxford University Press, 1953; first ed. 1950)
Ann Rigney, "De lokroep van het verleden: Literatuur als historische bron," [The Seductive Call of the Past; Literature as Historical Source] *Feit & fictie* 4, no. 3 (Summer 1999), 82–98
Karl Riha, "Die Dichtung des deutschen Frühexpressionismus," in York-Gothart Mix (ed.), *Naturalismus, Fin de siècle, Expressionismus 1890–1918* (Munich: Deutsche Taschenbuch Verlag, May 2000), 454–469
Gregor M. Rinn, "Das Automobil als nationales Identifikationssymbol: Zur politischen Bedeutungsprägung des Kraftfahrzeugs in Modernitätskonzeptionen des 'Dritten

Reichs' und der Bundesrepublik" (unpublished dissertation, Humboldt-Universität, Berlin, 2008)
Joseph C. Robert, *Ethyl: A History of the Corporation and the People Who Made It* (Charlottesville: University Press of Virginia, 1983)
Eric Robertson, *Writing Between The Lines: René Schickele, 'Citoyen français, deutscher Dichter' (1883–1940)*, Internationale Forschungen zur Allgemeinen und Vergleichenden Literaturwissenschaft, 11, ed. Alberto Martino (Amsterdam and Atlanta, GA: Rodopi, 1995)
Mark A. Robison, "Recreation in World War I and the Practice of Play in One of Ours," *Cather Studies* 6 (2006), 1–22, http://cather.unl.edu/cs006_robison.html (last accessed 18 September 2011)
Peter Rocholl, *Vergleichende Analyse der Entwicklung des Personenkraftverkehrs im westeuropäischen Wirtschaftsraum* (Düsseldorf: Droste Verlag, 1962)
Daniel T. Rodgers, *Atlantic Crossings: Social Politics in a Progressive Age* (Cambridge, MA, and London: The Belknap Press of Harvard University Press, 1998)
Heidi Rohde, *Transportmodernisierung contra Verkehrsbewirtschaftung: Zur staatlichen Verkehrspolitik gegenüber dem Lkw in den 30er Jahren* (Frankfurt: Peter Lang, 1999)
Horst Rohde, "Facktoren [sic] der deutschen Logistik im Ersten Weltkrieg," in Gérard Canini (ed.), *Les fronts invisibles: Nourrir—fournir—soigner; Actes du colloque international sur La logique des armées au combat pendant la Première Guerre Mondiale organisé à Verdun les 6, 7, 8 juin 1980 sous la présidence du général d'Armée F. Gambiez, membre de l'Institut* (Nancy: Presses Universitaires de Nancy, 1984), 103–122
Chris Rojek and John Urry, "Transformations of Travel and Theory," in Rojek and Urry (eds.), *Touring Cultures: Transformations of Travel and Theory* (London and New York: Routledge, 1997)
William H. Rollins, "Whose Landscape? Technology, Fascism, and Environmentalism on the National Socialist *Autobahn*," *Annals of the Association of American Geographers* 85, no. 3 (1995), 494–520
Nancy Tillman Romalov, "Mobile Heroines: Early Twentieth-Century Girls' Automobile Series," *Journal of Popular Culture* 28, no. 4 (Spring 2005), 231–243
James Rorty, *Where Life is Better: An Unsentimental American Journey* (New York: Reynal & Hitchcock / John Day, 1936)
Hartmut Rosa, *Beschleunigung: Die Veränderung der Zeitstrukturen in der Moderne*, suhrkamp taschenbuch wissenschaft 1760 (Frankfurt: Suhrkamp Verlag, 2005)
———, "Social Acceleration: Ethical and Political Consequences of a Desynchronized High-Speed Society," in Rosa and William E. Scheuerman (eds.), *High-Speed Society: Social Acceleration, Power, and Modernity* (University Park: Pennsylvania State University Press, 2009), 77–111
Mark H. Rose, Bruce Seely, and Paul F. Barrett, *The Best Transportation System in the World: Railroads, Trucks, Airlines, and American Public Policy in the Twentieth Century* (Columbus: Ohio State University Press, 2006)
Mechtild Rössler, "Applied geography and area research in Nazi society: Central place theory and planning, 1933 to 1945," *Environment and Planning D: Society and Space* 7 (1989), 419–431.
———, "Geography and Area Planning under National Socialism," in Margit Szöllösi-Janze, *Science in the Third Reich* (Oxford and New York: Berg, 2001), 59–78

Nadine Rossol, "Tagber: Gemeinschaftsdenken in Europa 1900–1938. Urspruenge des schwedischen 'Volksheim' im Verglecih,"H-SOZ-U-KULT@H-NET.MSU.EDU, 16 January 2012

Paul A. Roth, "Hearts of darkness: 'Perpetrator history' and why there is no why," *History of the Human Sciences* 17, no. 2/3 (2004), 211–251

———, "How Narratives Explain," *Social Research* 56, no. 2 (Summer 1989), 449–478

Ellen K. Rothman, *Hands and Hearts: A History of Courtship in America* (New York: Basic Books, Inc., Publishers, 1984)

Hermann Röttger, *Die Struktur der deutschen Automobil-Einzelhandels mit Personenkraftwagen* (n.p., n.d.) (dissertation, Universität Köln, 1940)

Raymond Roussel, *Comment j'ai écrit certains de mes livres*, Série "Fins de siècles," ed. Hubert Juin (n.p.: Jean-Jacques Pauvert, 1977)

Claude Rouxel, "Frankreich auf der Suche nach dem idealen Taxi: Mit dem Taxi zur Front," in Ulrich Kubisch et al., *Taxi: Das mobilste Gewerbe der Welt*, Schriftenreihe des Museums für Verkehr und Technik Berlin, Band 12 (Berlin, n.d. [1993]), 263–284

———, *La grande histoire des taxis français 1898–1988* (Pontoise, 1989)

Sheila Rowbotham, *Dreamers of a New Day: Women Who Invented the Twentieth Century* (London and New York: Verso, 2010)

[Royal Commission on Transport], *The Co-ordination and Development of Transport: Final Report* (London: His Majesty's Stationery Office, 1931)

Éléonore Roy-Reverzy, "*La 628–E 8* ou la mort du roman," *Cahiers Octave Mirbeau* 4 (1997), 257–266

Susan Sessions Rugh, *Are We There Yet? The Golden Age of American Family Vacations* (Lawrence: University Press of Kansas, 2008)

Günther Rühle, "Leben und Schreiben der Marieluise Fleisser aus Ingolstadt," in Marieluise Fleisser, *Gesammelte Werke, Erster Band: Dramen*, edited by Günther Rühle (Frankfurt: Suhrkamp Verlag, 1983; first ed. 1972), 5–60

Edmund Russell, "Introduction: The Garden in the Machine: Toward an Evolutionary History of Technology," in Susan R. Schrepner and Philip Scranton (eds.), *Industrializing Organisms: Introducing Evolutionary History*, Hagley Perspectives on Business and Culture, vol. 5; eds. Philip Scranton and Roger Horowitz (New York and London: Routledge, 2004), 1–16

Beatriz Russo, *Shoes, Cars and Other Love Stories: Investigating the Experience of Love for Products* (Delft: VSSD, 2010)

Beatriz Russo, Stella Boess, and Paul Hekkert, "'What's Love Got to Do With It?' The Experience of Love in Person–Product Relationships," *Design Journal* 14, no. 1 (2011), 8–27

Beatriz Russo and Paul Hekkert, "On the Conceptualization of the Experience of Love: The Underlying Principles" (paper submitted to Designing Pleasurable Products Conference—DPPI'07), http://studiolab.io.tudelft.nl/russo/stories/storyReader$20 (last accessed 18 March 2011)

A. J. C. Rüter, *De spoorwegstakingen van 1903: Een spiegel der arbeidersbeweging in Nederland* (Leiden: E. J. Brill, 1935)

Wolfgang Sachs, *For the Love of the Automobile: Looking back into the History of Our Desires*, transl. Don Reneau (Berkeley, Los Angeles, and London: University of California Press, 1992)

———, *Die Liebe zum Automobil: Ein Rückblick in die Geschichte unserer Wünsche* (Reinbeck bei Hamburg: Rowohlt Verlag, 1984)

Fritz Sager, "Spannungsfelder und Leitbilder in der schweizerischen Schwerverkehrspolitik 1932 bis 1998," *Schweizerische Zeitschrift für Geschichte* 49, no. 4 (1999), 307–332

Monique de Saint Martin, *L'Espace de la noblesse*, Collection Leçons de choses, ed. Luc Boltanski (Paris: Éditions Métailié, 1993)

Hiroya Sakamoto, "La genèse des 'littératures automobiles': Histoire d'une polémique en 1907 et au-delà," *La voix du regard* no. 19 (2006–2007; Dossier: en voiture!), 31–42

Adelheid von Saldern, "Massenfreizeitkultur im Visier: Ein Beitrag zu den Deutungs- und Einwirkungsversuchen während der Weimarer Republik," *Archiv für Sozialgeschichte* 33 (1993), 21–58

Jessica Amanda Salmonson, "Marie Corelli and her Occult Tales," http://www.victorianweb.org/authors/corelli/salmonson1.html (last accessed 9 March 2010)

Nicole Samuel, "Histoire et sociologie du temps libre en France," *Revue internationale des sciences sociales* 107 (1986), 53–68

Marion Sanders, *Dorothy Thompson: A Legend in Her Time* (Boston: Houghton Mifflin, 1973)

Stefan Sandmeier, "Vom Eisenbahn zur Verkehrsplanung; Die Institutionalisierungsgeschichte des Verkehrswesens an der ETH Zürich," in Stefan Sandmeier, Andreas Frei, and Kay W. Axhausen, *125 Jahre Verkehrswesen an der ETH Zürich* (Zürich: Institut für Verkehrsplanung und Transportsysteme, ETH Zürich, 2008), 5–33

Dietmar Sauermann, "Das Bürgertum im Spiegel von Gästebüchern des Sauerlandes," in Dieter Kramer and Ronald Lutz (eds.), *Reisen und Alltag: Beiträge zur kulturwissenschaftlichen Tourismusforschung* (Frankfurt: Institut für Kulturanthropologie und Europäische Ethnologie der Universität Frankfurt am Main, 1992), 81–99

W. O. Saunders, "Business Is a Pleasure," in *These Are Our Lives: As Told by the People and Written by Members of the Federal Writers' Project of the Works Progress Administration in North Carolina, Tennessee, and Georgia* (Chapel Hill: University of North Carolina Press, 1939), 277–289

Alfred Sauvy, "L'automobile en France depuis la guerre," *Bulletin de la Statistique générale de la France* 22 (October 1932–September 1933), 565–615

———, "Production et vente d'automobiles en France," *Bulletin de la Statistique générale de la France* 22 (October 1932–September 1933), 57–66

Christopher I. Savage, *An Economic History of Transport* (London: Hutchinson University Library, 1966)

Jon Savage, *Teenage: The Creation of Youth Culture* (London: Pimlico, 2007)

Sybe Schaap, *Het rancuneuze gif: De opmars van het onbehagen*[4] (Budel: Damon, 2012; first ed. 2012)

Rolf Schäfer, *Die tödliche Verkehrsunfälle in Nordbaden in den Jahren 1929–1934* (Heidelberg: Brausdruck, 1935)

Richard P. Scharchburg, *Carriages Without Horses* (Warrendale, PA, 1993)

Virginia Scharff, "Gender, Electricity, and Automobility," in Martin Wachs and Margaret Crawford (with the assistance of Susan Marie Wirka and Taina Marjatta Rikala), *The Car and the City: The Automobile, The Built Environment, and Daily Urban Life* (Ann Arbor: University of Michigan Press, 1992), 75–85

———, "Putting wheels on women's sphere," in Cheris Kramarae (ed.), *Technology and Women's Voices: Keeping in Touch* (New York and London, 1988), 135–146

———, *Taking the Wheel: Women and the Coming of the Motor Age* (New York, Toronto, Oxford, Singapore, and Sydney, 1991)

―――, *Twenty Thousand Roads: Women, Movement, and the West* (Berkeley and London: University of California Press, 2003)
Theodore R. Schatzki, "Nature and Technology in History," *History and Theory*, Theme Issue 42 (December 2003), 82–93
―――, *Social Practices: A Wittgensteinian Approach to Human Activity and the Social* (Cambridge: Cambridge University Press, 1996)
Karl E. Scheibe, "Self-Narratives and Adventure," in Theodore R. Sarbin (ed.), *Narrative Psychology: The Storied Nature of Human Conduct*, Praeger Special Studies / Praeger Scientific (New York, Westport, CT, and London: Praeger, 1986), 129–151
René Schickele, *Benkal der Frauentröster* (Berlin: Paul Cassirer, n.d.; "Zweites bis viertes Tausend," first ed. 1913)
Friedrich Schildberger, "Die Entstehung der Automobilindustrie in den Vereinigten Staaten" (manuscript Daimler Benz archives, Stuttgart), 87–88
Willem Schinkel, *Aspects of Violence: A Critical Theory*, Cultural Criminology, ed. Mike Presdee (Houndsmill and New York: Palgrave Macmillan, 2010)
―――, *De nieuwe democratie: Naar andere vormen van politiek* (Amsterdam: De Bezige Bij, 2012)
[Albert von Schirnding], "Nachwort von Albert von Schirnding," in Thomas Mann, *Königliche Hoheit: Roman* (Nachwort von Albert von Schirnding); *Gesammelte Werke in Einzelbänden, Frankfurter Ausgabe*, ed. Peter de Mendelssohn (Frankfurt: S. Fischer Verlag, 1984), 369–391
Frank Schipper, *Driving Europe: Building Europe on Roads in the Twentieth Century* (Amsterdam: aksant, 2008)
Frank Schipper, Vincent Lagendijk, and Irene Anastasiadou, "New Connections for an Old Continent: Rail, Road and Electricity in the League of Nations' Organisation for Communications and Transit," in Alexander Badenoch and Andreas Fickers (eds.), *Materialising Europe: Transnational Infrastructures and the Project of Europe* (Basingstoke: Palgrave, 2010), 113–143
Wolfgang Schivelbusch, *Entfernte Verwandtschaft: Faschismus, Nationalsozialismus, New Deal 1933–1939* (Munich and Vienna: Carl Hanser Verlag, 2005)
―――, *Geschichte der Eisenbahnreise: zur Industrialisierung von Raum und Zeit im 19. Jahrhundert* (Munich and Vienna, 1977)
Anette Schlimm, *Ordnungen des Verkehrs: Arbeit an der Moderne—deutsche und britische Verkehrsexpertise im 20. Jahrhundert* (Bielefeld: transcript, 2011)
Wolfgang Schmale, *Geschichte der Männlichkeit in Europa (1450–2000)* (Vienna, Cologne, and Weimar: Böhlau Verlag, 2003)
Wieland Schmied, "Der kühle Blick: Der Realismus der Zwanzigerjahre," in Schmied (ed.), *Der kühle Blick: Realismus der zwanziger Jahre, 1. Juni–2. September 2001* (Munich, London, and New York: Prestel Verlag / Kunsthalle der Hypo-Kulturstiftung, 2001), 9–36
Peter J. Schmitt, *Back to Nature: The Arcadian Myth in Urban America* (New York: Oxford University Press, 1969)
Uwe Schmitt, "Wir konnten uns nirgends beschwerden," *Die Welt* (17 December 2011), 8
Barbara Schmucki, *Der Traum vom Verkehrsfluss: Städtische Verkehrsplanung seit 1945 im deutsch-deutschen Vergleich* (Frankfurt and New York: Campus Verlag, 2001)
Friedrich Schnack, *Das Zauberauto: Ein Roman* (Leipzig: Insel-Verlag, 1928)
Jeffrey T. Schnapp, "Crash (Speed as Engine of Individuation)," *Modernism/Modernity* 6, no. 1 (1999), 1–49

Florian Schneider, "Der Urwald der Moderne: Über Robert Musils Glosse *Wer hat dich, du schöner Wald . . . ?*" transcript: *Zeitschrift für Kulturwissenschaften* no. 2, special issue "Politische Ökologie," eds. Sebastian Giessmann et al. (2009), 79–90

John C. Schneider, "Tramping Workers, 1890–1920: A Subcultural View," in Eric H. Monkkonen (ed.), *Walking to Work: Tramps in America, 1790–1935* (Lincoln and London: University of Nebraska Press, 1984), 212–234

Robert Schnepf, *Geschichte erklären: Grundprobleme und Grundbegriffe* (Göttingen: Vandenhoeck & Ruprecht, 2011)

Thomas F. Schneider, "Die Wiederkehr der Kriege in der Literatur: Voraussetzungen und Funktionen 'pazifistischer' und 'bellizistischer' Kriegsliteratur vom Ersten Weltkrieg bis zum Dritten Golfkrieg," *Osnabrücker Jahrbuch Frieden und Wissenschaft* 12 (2005), 201–221

Rainer Schönhammer, *In Bewegung: Zur Psychologie der Fortbewegung* (Munich, 1991)

Matthias Schöning, *Versprengte Gemeinschaft: Kriegsroman und intellektuelle Mobilmachung in Deutschland 1914–33* (Göttingen: Vandenhoeck & Ruprecht, 2009)

Mark Schorer, *Sinclair Lewis: An American Life* (New York, Toronto, and London: McGraw-Hill, 1961)

Peter Schöttler, "Eine Art 'Generalplan West': Die Stuckart-Denkschrift vom 14. Juni 1940 und die Planungen für eine deutsch-französische Grenze im Zweiten Weltkrieg," *Social. Geschichte* 18, no. 3 (2003), 83–131

Dominik Schrage, "Der Konsum in der deutschen Soziologie," in Heinz-Gerhard Haupt and Claudius Torp (eds.), *Die Konsumgesellschaft in Deutschland 1890 – 1990; Ein Handbuch* (Frankfurt and New York: Campus Verlag, 2009), 319–334

Mary Suzanne Schriber, *Writing Home: American Women Abroad 1830–1920* (Charlottesville and London: University Press of Virginia, 1997)

Jürgen Schröder, "Das Spätwerk Ödön von Horváths," in Traugott Krischke (ed.), *Ödön von Horváth,* suhrkamp taschenbuch materialien, vol. 2005 (Frankfurt: Suhrkamp, 1981), 125–155

Morris S. Schulzinger, *The Accident Syndrome: The Genesis of Accidental Injury* (Springfield, IL: Charles C. Thomas, 1956)

Dirk Schümer, *Zu Fuss: Eine kurze Geschichte des Wanderns* (Munich: Malik / Piper Verlag, 2010)

George S. Schuyler, "Traveling Jim Crow," *The American Mercury* 20, no. 80 (August 1930), 423–432

Michael W. Schuyler, *The Dread of Plenty: Agricultural Relief Activities of the Federal Government in the Middle West, 1933–1939* (Manhattan, KS: Sunflower University Press, 1989)

Kees Schuyt and Ed Taverne, *1950 Welvaart in zwart-wit* (Den Haag: Sdu Uitgevers, 2000)

Carlos A. Schwantes, "The West Adapts the Automobile: Technology, Unemployment, and the Jitney Phenomenon of 1914–1917," *Western Historical Quarterly* 16, no. 3 (July 1985), 307–326

Stefan Schwarzkopf, "Kontrolle statt Rausch? Marktforschung, Produktwerbung und Verbraucherlenkung im Nationalsozialismus zwischen Phantasien von Masse, Angst und Macht," in Árpád von Klimó and Malte Rold (eds.), *Rausch und Diktatur: Inszenierung, Mobilisierung und Kontrolle in totalitären Systemen* (Frankfurt and New York: Campus Verlag, 2006), 193–216

Joan Scott, "The evidence of experience," in Gabrielle M. Spiegel (ed.), *Practicing History: New Directions in Historical Writing after the Linguistic Turn* (New York and London: Routledge, 2005), 199–216

Peter Scott, "British Railways and the Challenge from Road Haulage: 1919–39," *Twentieth Century British History* 13, no. 2 (2002), 101–120

——, "Public-Sector Investment and Britain's Post-War Economic Performance: A Case Study of Roads Policy," *Journal of European Economic History* 34, no. 2 (2005), 391–418

W. S. [Webster Schott], "Introduction: Beautiful Blood, Beautiful Brain," in William Carlos Williams, *Imaginations: Kora in Hell, Spring and All, The Great American Novel, The Descent of Winter, A novelette and Other Prose*, edited with introductions by Webster Schott (New York: New Directions Books, 1970), ix–xviii

——, "Introduction: The Great American Novel," in Williams, *Imaginations*, 155–157

Hans Ulrich Seeber, *Mobilität und Moderne: Studien zur englischen Literatur des 19. und 20. Jahrhunderts*, Anglistische Forschungen, Band 371, eds. Rüdiger Ahrens, Heinz Antor, and Klaus Stierstorfer (Heidelberg: Universitätsverlag Winter, 2007)

Bruce Seely, "The Automotive Transportation System: Cars and Highways in Twentieth-Century America," in Carroll Pursell (ed.), *A Companion to American Technology*, Blackwell Companions to American History (Malden, Oxford, and Carlton: Blackwell Publishing, 2005), 233–254

Spencer D. Segalla, "Re-Inventing Colonialism: Race and Gender in Edith Wharton's *In Morocco*," *Edith Wharton Review* 17, no. 2 (Fall 2001), 22–30

Harro Segeberg, "Literarische Kino-Ästhetik; Ansichten der Kino-Debatte," in Corinna Müller and Harro Segeberg (eds.), *Die Modellierung des Kinofilms: Zur Geschichte des Kinoprogramms zwischen Kurzfilm und Langfilm 1905/06–1918*, Mediengeschichte des Films, ed. Segeberg, in cooperation with Knut Hickethier and Corinna Müller, Band 2 (Munich: Fink, 1998), 193–219

——, "Literatur im Film: Modelle der Adoption," in Segeberg (ed.), *Mediale Mobilmachung II: Das Kino der Bundesrepublik Deutschland als Kulturindustrie (1950–1962)*, Mediengeschichte des Films, ed. Segeberg, in cooperation with Knut Hickethier, Corinna Müller, and the Metropolis-Kino Hamburg, Band 6 (Munich: Wilhelm Fink Verlag, 2009), 411–431

——, *Literatur im Medienzeitalter: Literatur, Technik und Medien seit 1914* (Darmstadt: Wissenschaftliche Buchgesellschaft, 2003)

—— (ed.), *Die Mobilisierung des Sehens: Zur Vor- und Frühgeschichte des Films in Literatur und Kunst*, Mediengeschichte des Films, ed. Segeberg, in cooperation with Knut Hickethier and Carinna Müller, Band 1 (Munich: Wilhelm Fink, 1996)

Joop Segers, "Benzinestations: Een geschiedenis van de benzinedistributie in Nederland," *Industriële Archeologie* no. 13 (1984), 164–180

Hans-Christoph Graf von Seherr-Thoss, *75 Jahre ADAC 1903–1978: Tagebuch eines Automobilclubs* (Munich: ADAC Verlag, 1978)

Cotten Seiler, "Anxiety and Automobility: Cold War Individualism and the Interstate Highway System" (dissertation, University of Kansas, 2002)

——, "Author response: The end of automobility," *History and Technology* 26, no. 4 (December 2010), 389–397

——, *Republic of Drivers: A Cultural History of Automobility in America* (Chicago and London: University of Chicago Press, 2008)

Kristin Semmens, "'Tourism and Autarky are Conceptually Incompatible': International Tourism Conferences in the Third Reich," in Eric G. E. Zuelow (ed.), *Touring*

Beyond the Nation: A Transnational Approach to European Tourism History (Farnham and Burlington: Ashgate, 2011), 195–213

"Sense," http://en.wikipedia.org/wiki/Sense (last accessed 21 June 2010)

William H. Sewell, Jr., "The Concept(s) of Culture," in Victoria E. Bonnell and Lynn Hunt (eds.), *Beyond the Cultural Turn: New Directions in the Study of Society and Culture* (Berkeley, Los Angeles, and London: University of California Press, 1999), 35–61

Ben A. Shackleford, "Masculinity, the Auto Racing Fraternity, and the Technological Sublime; The Pit Stop As a Celebration of Social Roles," in Roger Horowitz (ed.), *Boys and Their Toys? Masculinity, Technology, and Class in America* (New York and London: Routledge, 2001), 229–250

Marguerite S. Shaffer, *See America First: Tourism and National Identity, 1880–1940* (Washington, DC, and London: Smithsonian Institution Press, 2001)

——, "Seeing America First: The Search for Identity in the Tourist Landscape," in David M. Wrobel and Patrick T. Long (eds.), *Seeing and Being Seen: Tourism in the American West* (Lawrence: University Press of Kansas, 2001), 165–193

Richard Sheldon, "Making Armored Cars and Novels: A Literary Introduction," in Viktor Shklovsky, *A Sentimental Journey: Memoirs, 1917–1922*, transl. and literary introduction by Richard Sheldon; historical introduction by Sidney Monas (n.p.: Dalkey Archive Press, 2004; first ed. 1923; first English ed. 1970), ix–xxv

Mimi Sheller and John Urry, "The new mobilities paradigm," *Environment and Planning A* 38 (2006), 207–226

Jennifer Shepherd, "The British Press and Turn-of-the-Century Developments in the Motoring Movement," *Victorian Periodicals Review* 38, no. 4 (Winter 2005), 379–391

Norman G. Shidle, "Lest We Forget—There are Still More Open Cars Than Closed," *Automotove Industries* 55, no. 10 (2 September 1926), 361–362

Viktor Shklovsky, *A Sentimental Journey: Memoirs, 1917–1922*, transl. and literary introduction by Richard Sheldon; historical introduction by Sidney Monas (n.p.: Dalkey Archive Press, 2004; first ed. 1923; first English ed. 1970)

Viktor Shklovsky (transliterated as "Schklowskij"), *Zoo; oder Briefe nicht über die Liebe* (Frankfurt am Main: Suhrkamp Verlag, 1965; edition suhrkamp 130) (transl. from Russian by Alexander Kaempfe from original ed. of 1923)

David Shinar, *Psychology on the Road: The Human Factor in Traffic Safety* (New York, Toronto, Santa Barbara, CA, Chichester, and Brisbane: John Wiley & Sons, 1978)

Edward Shorter, *The Making of the Modern Family* (New York: Basic Books, 1975)

Nicole Shukin, *Animal Capital: Rendering Life in Biopolitical Times*, posthumanities 6, ed. Cary Wolfe (Minneapolis and London: University of Minnesota Press, 2009)

Rolf Peter Sieferle, *Fortschrittsfeinde? Opposition gegen Technik und Industrie von der Romantik bis zur Gegenwart* (Munich: Verlag C. H. Beck, 1984)

Lewis H. Siegelbaum, "Introduction," in Siegelbaum (ed.), *The Socialist Car: Automobility in the Eastern Bloc* (Ithaca, NY, and London: Cornell University Press, 2011), 1–13

——, "Soviet Car Rallies of the 1920s and 1930s and the Road to Socialism," *Slavic Review* 64, no. 2 (Summer 2005), 247–273

Mary Corbin Sies, "North American Suburbs, 1880–1950," *Journal of Urban History* 27, no. 3 (March 2001), 313–346

Immo Sievers, *AutoCars: Die Beziehungen zwischen der englischen und der deutschen Automobilindustrie vor dem Ersten Weltkrieg*, Europäische Hochschulschriften, Reihe III, Bd. 640 (Frankfurt, Berlin, Bern, New York, Paris, Vienna, 1995)

Georg Simmel, "Das Abenteuer," in Simmel, *Philosophische Kultur: Über das Abenteuer; die Geschlechter und die Krise der Moderne; Gesammelte Essays, Mit ein Vorwort von Jürgen Habermas*, Wagenbuchs Taschenbücherei 133 (Berlin: Verlag Klaus Wagenbach, 1986)

——, *Soziologie: Untersuchungen über die Formen der Vergesellschaftung*, Gesammelte Werke: Zweiter Band; Unveränderter Nachdruck der 1923 erschienenen 3. Auflage (Berlin: Duncker & Humblot, 1958)

John K. Simon, "View from the Train: Butor, Gide, Larbaud," *French Review* 36, no. 2 (December 1962), 161–166

May Sinclair, *A Journal of Impressions in Belgium* (New York: The Macmillan Company, 1915), http://www.archive.org/details/journalofimpress00sinciala (last accessed 2 March 2012)

Bayla Singer, "Automobiles and Feminity," *Research in Philosophy and Technology*, vol. 13: Technology and Feminism (Greenwich, CT: JAI Press, 1993), 31–42

Gustav Sjöblom, "The Political Economy of Railway and Road Transport in Britain and Germany, 1918–1933" (unpublished dissertation, Darwin College, University of Cambridge, 16 April 2007)

Raymond W. Smilor, "Cacophony at 34th and 6th: The Noise Problem in America, 1900–1930," *American Studies* 18, no. 1 (1977), 23–38

Helen Zenna Smith, *Not so Quiet . . . Stepdaughters of War*, afterword by Jane Marcus (New York: The Feminist Press at the City University of New York, 1989; first ed. 1930)

Carroll Smith-Rosenberg, "Discourse of Sexuality and Subjectivity: The New Woman, 1870–1936," in Martin Bauml Duberman, Martha Vicinus, and George Chauncey, Jr. (eds.), *Hidden From History: Reclaiming the Gay and Lesbian Past* (New York and Ontario: New American Library, 1989), 264–280

Alfred Edgar Smith, "Through the Windshield," *Opportunity* 11, no. 5 (1933), 142–144

Jane S. Smith, "Plucky Little Ladies and Stout-Hearted Chums: Serial Novels for Girls, 1900–1920," in Jack Salzman (ed.), *Prospects: An Annual of American Cultural Studies, Volume Three* (New York: Burt Franklin & CO., 1977)

Jason Scott Smith, *Building New Deal Liberalism: The Political Economy of Public Works, 1933–1956* (Cambridge: Cambridge University Press, 2005)

Julian Smith, "A Runaway Match: The Automobile in the American Film, 1900–1920," *Michigan Quarterly Review* 19/20 (Fall 1980–Winter 1981), 574–587

Sidonie Smith, *Moving Lives: Twentieth-Century Women's Travel Writing* (Minneapolis and London: University of Minnesota Press, 2001)

"Sorel, Georges," the Columbia Encyclopedia, Sixth Edition, 2001–2007, http://www.bartleby.com/65/so/Sorel-Ge.html (last accessed 20 January 2008)

Beverley Southgate, *History Meets Fiction* (Harlow and London: Pearson Education, 2009)

Jonathan D. Spence, *Chinese Roundabout: Essays in History and Culture* (New York and London: W. W. Norton, 1992)

Nicolas Spinga, "L'introduction de l'automobile dans la société française entre 1900 et 1914: Étude de presse" (unpublished Master's thesis, Histoire contemporaine, Université de Paris X—Nanterre, 1972–1973)

Hasso Spode, "Der Aufstieg des Massentourismus im 20. Jahrhundert," in Heinz-Gerhard Haupt and Claudius Torp (eds.), *Die Konsumgesellschaft in Deutschland 1890–1990: Ein Handbuch* (Frankfurt and New York: Campus Verlag, 2009), 114–128

Reinhard Spree, "Angestellte als Modernisierungsagenten: Indikatoren und Thesen zum reproduktiven Verhalten von Angestellten im späten 19. und frühen 20. Jahrhundert," in Jürgen Kocka (ed.), *Angestellte im europäischen Vergleich: Die Herausbildung angestellter Mittelschichten seit dem späten 19. Jahrhundert*, Geschichte und Gesellschaft, Sonderheft 7 (Göttingen: Vandenhoeck & Ruprecht, 1981), 279–308

David J. St. Clair, *The Motorization of American Cities* (New York, Westport, CT, and London: Praeger, 1986)

G. P. St. Clair, "Trends in Motor-Vehicle Travel, 1936 to 1945," *Public Roads* (October–December 1946), 261–267

Oliver Stallybrass, "Editor's Introduction," in E. M. Forster, *Howards End*, edited by Oliver Stallybrass (London and New York: Penguin Books, 2000; first ed. 1910), 7–17

Dushan Stankovich, *Otto Julius Bierbaum—eine Werkmonographie*, Australisch-Neuseeländische Studien zur deutschen Sprache und Literatur, eds. Gerhard Schulz and John A. Asher with B. L. D. Coghlan, vol. 1 (Bern and Frankfurt: Verlag Herbert Lang, 1971)

Isabelle Stauffer, "Von Hollywood nach Berlin: Die deutsche Rezeption der Flapper-Filmstars Colleen Moore und Clara Bow," in Julia Freytag and Alexandra Tacke (eds.), *City Girls: Bubiköpfe & Blaustrümpfe in den 1920er Jahren*, Literatur—Kultur—Geschlecht; Studien zur Literatur- und Kulturgeschichte, eds. Anne-Kathrin Reulecke and Ulrike Vedder, in connection with Inge Stephan und Sigrid Weigel; Kleine Reihe, Band 29 (Cologne, Weimar, and Vienna: Böhlau Verlag, 2011), 111–126

Peter N. Stearns, *American Cool: Constructing a Twentieth-Century Emotional Style* (New York and London: New York University Press, 1994)

Emmelina Meintje Steg, "Gedragsverandering ter vermindering van het autogebruik; Theoretische analyse en empirische studie over probleembesef, verminderingsbereidheid en beoordeling van beleidsmaatregelen" (n.p., n.d.) (dissertation Rijksuniversiteit Groningen, 1996)

Linda Steg, Charles Vlek, and Goos Slotegraaf, "Intrumental-reasoned and symbolic-affective motives for using a motor car" *Transportation Research Part F: Traffic Psychology and Behaviour* 4, no. 3 (September 2001), 151–169

Bernhard Stegemann, "Autobiographisches aus der Seminar- und Lehrerzeit von Erich Maria Remarque im Roman *Der Weg zurück*," in Thomas F. Schneider (ed.), *Erich Maria Remarque: Leben, Werk und weltweite Wirkung*, Schiften des Erich Maria Remarque-Archivs, Band 12, eds. Thomas F. Schneider and Tilman Westphalen (Osnabrück: Universitätsverlag Rasch, 1998), 57–67

Jesse Frederick Steiner, *Americans at Play: Recent Trends in Recreation and Leisure Time Activities* (New York and London: McGraw-Hill, 1933)

Jesse F. Steiner, *Research Memorandum on Recreation in the Depression Prepared under the Direction of the Committee on Studies in Social Aspects of the Depression* (New York: Social Science Research Council, 1937; reprint: n.p. [New York]: Arno Press, 1972)

Benjamin Steininger, *Raum-Maschine Reichsautobahn*, Kaleidogramme, Bd. 2 (Berlin: Kulturverlag Kadmos, 2005)

B. S. [=Bruno Stephan], "Cyriel Buysse als automobilist," *De Auto* (1911), 1209–1211

Milton R. Stern, *The Golden Moment: The Novels of F. Scott Fitzgerald* (Urbana, Chicago, and London: University of Illinois Press, 1970)

Claudia Sternberg, "Der Erinnerungsdiskurs im Spielfilm und Fernsehspiel," in Barbara Korte, Ralf Schneider, and Claudia Sternberg (eds.), *Der Erste Weltkrieg und* die Mediendiskurse der Erinnerung in Grossbritannien: Autobiographie—Roman—Film (1919–1999), Film—Medium—Diskurs, eds. Oliver Jahraus and Stefan Neuhaus, Band 15 (Würzburg: Königshausen & Neumann, 2005), 243–342

Laurence Sterne, A Sentimental Journey Through France and Italy, transcribed from the 1892 George Bell and Son edition by David Price, http://www.gutenberg.org/cache/epub/804/pg804.txt (last accessed 4 July 2011)

Richard Sterner (in collaboration with Lenore A. Epstein, Ellen Winston, and others), The Negro's Share: A Study of Income, Consumption, Housing and Public Assistance (Westport, CT: Negro University Press, 1971; reprint; first ed. 1943)

David Stevenson, "War by timetable? The railway race before 1914," Past and Present 162 (1999), 161–194

——, With Our Backs to the Wall: Victory and Defeat in 1918 (London and New York: Allen Lane / Penguin, 2011)

Jill Steward and Alexander Cowan, "Introduction," in Alexander Cowan and Jill Steward (eds.), The City and the Senses: Urban Culture Since 1500 (Aldershot: Ashgate, 2007), 1–22

Philip Stewart, "This Is Not a Book Review: On Historical Uses of Literature," Journal of Modern History 66 (September 1994), 521–538

J. J. Stieltjes, "III. Het verkeer te land: C. De exploitatie van het railloos verkeer," in "Het verkeer in Nederland in de XXe eeuw," Tijdschrift van het Koninklijk Nederlandsch Aardrijkskundig Genootschap, Tweede serie, 50 (1933), 331–662

Reinhold Stisser, Der deutsche Automobilexport unter besonderer Berücksichtigung des niederländischen Kraftfahrzeugmarktes (Kiel, 1938)

Manuel Stoffers, "Cycling Cultures: Review Essay," Transfers 1, no. 1 (Spring 2011), 155–162

Bart Stol, "Iedereen wilde oorlog in 1870," Historisch Nieuwsblad (September 2007), 45–49

B. Stolk, 'Hun Auto,' De Auto (1908), 1020–1025, 1053–1055.

Pieter Stokvis and Marita Mathijsen, "Literatuur en maatschappij: Het beeld van de burgerlijke levensstijl in Nederlandse romans 1840–1910," De Negentiende Eeuw 18 (1994), 145–172

Storia Urbana 26, no. 100 (July–September 2002), special issue "La formazione della rete autostradale europea: Italia, Spagna, Francia, Germania"

Dieter Storz, Kriegsbild und Rüstung vor 1914: Europäische Landstreitkräfte vor dem Ersten Weltkrieg, Militärgeschichte und Wehrwissenschaften Band 1, eds. Wolfram Funk et al. (Herford, Berlin, and Bonn: Verlag E. S. Mittler & Sohn, 1992)

Janis P. Stout, The Journey Narrative in American Literature: Patterns and Departures (Westport, CT, and London: Greenwood Press, 1983)

Marilyn Strathern, "Foreword: The mirror of technology," in Roger Silverstone and Eric Hirsch (eds.), Consuming Technologies: Media and Information in Domestic Spaces (London and New York: Routledge, 1992), vii–xiii

Gustav Stratil-Sauer, Fahrt und Fessel: Mit dem Motorrad von Leipzig nach Afghanistan (Münster: Verlagshaus Monsenstein und Vannerdat, n.d.; reprint of 1927 edition by August Scheerl Verlag, Berlin)

Gregor Streim, "Flucht nach vorn zurück; Heinrich Hauser—Portrait eines Schriftstellers zwischen Neuer Sachlichkeit und 'reaktionärem Modernismus,'" Jahrbuch der Deutschen Schillergesellschaft 43 (1999), 377–402

——, "Junge Völker und neue Technik: Zur Reisereportage im 'Dritten Reich', am Beispiel von Friedrich Sieburg, Heinrich Hauser und Margaret Boveri," Zeitschrift für Germanistik 9, no. 2 (1999), 344–359

Frank Stricker, "Affluence for Whom? Another Look at Prosperity and the Working Classes in the 1920s," *Labor History* 24 (1983), 5-33

Ingrid Strohkark, "Die Wahrnemung von 'Landschaft' und der Bau von Autobahnen in Deutschland, Frankreich und Italien vor 1933" (unpublished dissertation, Hochschule der Künste Berlin, 2001)

Mildred Strunk (under the editorial direction of Hadley Cantril), *Public Opinion 1935-1946* (Westport, CT: Greenwood Press, 1978; reprint of 1951 edition, Princeton Universiy Press)

Christophe Studeny, *L'invention de la vitesse: France, XVIIIe-XXe siècle* (n.p. [Paris]: Gallimard, 1995)

Marita Sturken, "Mobilities of Time and Space: Technologies of the Modern and the Postmodern," in Sherry Turkle, "'Spinning' Technology: What We Are Not Thinking about When We Are Thinking about Computers," in Marita Sturken, Douglas Thomas, and Sandra J. Ball-Rokeach (eds.), *Technological Visions: The Hopes and Fears that Shape New Technologies* (Philadelphia, PA: Temple University Press, 2004), 71-91

Marita Sturken and Douglas Thomas, "Introduction: Technological Visions and the Rhetoric of the New," in Marita Sturken, Douglas Thomas, and Sandra J. Ball-Rokeach (eds.), *Technological Visions: The Hopes and Fears that Shape New Technologies* (Philadelphia, PA: Temple University Press, 2004), 1-18

Ruth Suckow, *Country People* (New York: Alfred A. Knopf, 1924), http://www.ruthsuckow.org/home/ (last accessed 10 February 2012)

Otto Suhr, *Die Lebenshaltung der Angestellten: Untersuchungen auf Grund statistischer Erhebungen des Allgemeinen freien Angestelltenbundes*, AFA-Schriften-Sammlung (Berlin: Freier Volksverlag, 1928)

Survey of Tourist Traffic considered as an International Economic Factor (Geneva: League of Nations, Economic Committee, 22 January 1936)

Paul S. Sutter, *Driven Wild: How the Fight against Automobiles Launched the Modern Wilderness Movement* (Seattle and London: University of Washington Press, 2002)

Nina Sylvester, "Das Girl: Crossing Spaces and Spheres. The Function of the Girl in the Weimar Republic" (unpubl. diss., University of California, Los Angeles, 2006)

Jean-Yves Tadié, *Le roman d'aventures* (Paris: Presses Universitiares de France, 1982)

—— (ed), *La littérature française: Dynamique & histoire, II* (Paris: Gallimard, 2007)

Booth Tarkington, *The Magnificent Ambersons* (1918), http://etext.virginia.edu/etcbin/toccer-new2?id=TarMagn.sgm&images=images/mode... (last accessed 22 April 2011)

Joel A. Tarr, *Transportation Innovation and Changing Patterns in Pittsburgh, 1850-1934* (Chicago: Public Works Historical Society, 1978)

Peter Tauber, "Der Krieg als 'welterschütternde Olympiade': Der Sport als Allegorie für den Krieg in Briefen und Gedichten des Ersten Weltkrieges," *Militärgeschichtliche Zeitschrift* 66, no. 2 (2007), 309-330

Harvey Taylor, *A Claim on the Countryside: A History of the British Outdoor Movement* (Edinburgh: Keele University Press, 1997)

Howard Taylor, "Forging the Job: A Crisis of 'Modernization' or Redundancy for the Police in England and Wales, 1900-39," *British Journal of Criminology* 39, no. 1 (Special Issue 1999), 113-135

Susan Tenenbaum, "The Progressive Legacy and the Public Corporation: Entrepreneurship and Public Virtue," *Journal of Policy History* 3, no. 3 (1991), 309-330

Lisa St. Aubin de Terán (ed.), *Indiscreet Journeys: Stories of Women on the Road* (London: Sceptre, 1991; first ed. 1989)

Andrew Thacker, *Moving Through Modernity: Space and Geography in Modernism* (Manchester and New York: Manchester University Press, 2009)
——, "Traffic, gender, modernism," *Sociological Review* 54, no. s1 (October 2006), 175–189
Eugene Thacker, "Networks, Swarms, Multitudes," www.ctheory.net/articvles.aspx ?id=423 (last accessed 6 December 2011)
Klaus Theweleit, *Männerphantasien 1 + 2; Band 1: Frauen, Fluten, Körper, Geschichte; Band 2: Männerkörper—zur Psychoanalyse des weissen Terrors* (Munich and Zürich: Piper, 2009; first ed. 1977 and 1978)
Bruno Thibault, *L'Allure de Morand: Du Modernism au Pétainisme* (Birmingham, AL: Summa Publications, Inc., 1992)
Jan Thiels, "50 Jaar heilige koe: Een analyse van 50 jaar wisselwerking tussen de auto, het toerisme en het Vlaamsnationalisme binnen het kader van VTB-VAB; Van 1922 tot 1971" (Master's thesis, Vrije Universiteit Brussel, 2008–2009)
Jean-Claude Thoenig, *L'Ère des technocrates: Le cas des Ponts et Chaussées* (Paris: Les Éditions d'Organisation, 1973)
Adrienne Thomas, *Die Katrin wird Soldat: Ein Roman aus Elsass-Lothringen* (Amsterdam: Allert de Lange, 1936; 225. Tausend; first ed. 1930)
Robert Paul Thomas, "Style Change and the Automobile Industry During the Roaring Twenties," in Louis P. Cain and Paul J. Uselding (eds.), *Business Enterprise and Economic Change: Essays in Honor of Harold F. Williamson* (n.p.: Kent State University Press, 1973), 118–138
Gregory Thompson, "From Rails to Rubber; from Public to Private: Passenger Transport Modernization in California from 1911 to 1941," http://www.coss.fsu.edu/durp/research/working-papers (last accessed 31 August 2011)
——, "'My Sewer'; James J. Flink on His Career Interpreting the Role of the Automobile in Twentieth-century Culture," in Peter Norton et al. (eds.), *Mobility in History: The Yearbook of the International Association for the History of Transport, Traffic and Mobility*, vol. 4 (New York and Oxford: Berghahn Journals, 2013), 3–17
Mark Thompson, *The White War: Life and Death on the Italian Front 1915–1919* (London: Faber and Faber, 2008)
Nigel Thrift, *Spatial Formations*, Theory, Culture and Society, ed. Mike Featherstone (Thousand Oaks, CA: Sage, 1996)
Cecilia Tichi, *Shifting Gears: Technology, Literature, Culture in Modernist America* (Chapel Hill and London: University of North Carolina Press, 1987)
Laurent Tissot and Béatrice Veyrassat (eds.) (with the collaboration of Michèle Merger & Antoine Glaenzer), *Technological Trajectories, Markets, Institutions: Industrialized Countries, 19th–20th Centuries; From Context Dependency to Path Dependency* (Bern, Berlin, Brussels, Frankfurt, New York, Oxford, and Vienna: Peter Lang, 2002)
Michael Titzmann, "Kulturelles Wissen—Diskurs—Denksystem: Zu einigen Grundbegrifffen der Literaturgeschichtsschreibung," *Zeitschrift für französisch Sprache und Literatur* 99 (1989), 47–61
Steven Tolliday, "Transplanting the American Model? US Automobile Companies and the Transfer of Technology and Management to Britain, France, and Germany, 1928–1962," in Jonathan Zeitlin and Gary Herrigel (eds.), *Americanization and Its Limits: Reworking US Technology and Management in Post-War Europe and Japan* (Oxford and New York: Oxford University Press, 2000), 76–119
Afke van der Toolen, "The making of Aletta Jacobs," *Historisch Nieuwsblad* (September 2009), 34–39

Bård Toldnes, *"Indtil Automobilerne har gaat sin rolige men sikre seiersgang"- Integrering av ny teknologi i perioden 1895–1926* (Trondheim: NTNU, 2007)

"Tom Swift," http://en.wikipedia.org/wiki/Tom_Swift

"Transcendence (philosophy)," http://en.wikipedia.org/wiki/Transcendence_(philosophy) (last accessed 17 June 2011)

B. Traven, *Die Weisse Rose: Roman,* Werkausgabe B. Traven in Einzelbänden, Band 5 (Frankfurt: Diogenes Verlag / Büchergilde Gutenberg, 1983; first ed. 1929)

Helmuth Trischler, "Der epische Konflikt zwischen Schiene und Strasse: Der Güterverkehr der USA seit dem Ende des 19. Jahrhunderts," in Harry Niemann and Armin Hermann (eds.), *100 Jahre LKW: Geschichte und Zukunft des Nutzfahrzeuges,* Stuttgarter Tage zur Automobil- und Unternehmensgeschichte; Eine Veranstaltung des Daimler-Benz Archivs, Stuttgart, Bd. 3 (Stuttgart: Franz Steiner Verlag, 1997), 243–262

Steven Trout, "Willa Cather's One of Ours and the Iconography of Remembrance," *Cather Studies* 4 (2004), 1–15, http://cather.unl.edu/cs004_trout.html (last accessed 18 September 2011)

Henry R. Trumbower et al., "Recensement de la circulation" (report for the Vth PIARC Conference, Milan, 1926).

Sherry Turkle, "Cyborg Babies and Cy-Dough-Plasm: Ideas about Self and Life in the Culture of Simulation," http://web.mit.edu/sturkle/www/cyborg_babies.html (last accessed 8 December 2011)

——, "'Spinning' Technology: What We Are Not Thinking about When We Are Thinking about Computers," in Marita Sturken, Douglas Thomas, and Sandra J. Ball-Rokeach (eds.), *Technological Visions: The Hopes and Fears that Shape New Technologies* (Philadelphia, PA: Temple University Press, 2004), 19–33

Henry S. Turner, "Lessons from Literature for the Historian of Science (and Vice Versa); Reflections on 'Form,'" *Isis* 101 (2010), 578–589

Elisabeth Tworek-Müller, *Kleinbürgertum und Literatur: Zum Bild des Kleinbürgers im bayerischen Roman der Weimarer Republik* (Munich: tuduv-Verlagsgesellschaft, 1985)

Frank Uekötter, "Stark im Ton, schwach in der Organisation: Der Protest gegen den frühen Automobilismus," *Geschichte in Wissenschaft und Unterricht: Zeitschrift des Verbandes der Geschichtslehrer Deutschlands* 54, no. 11 (2003), 658–670

John Urry, "Automobility, Car Culture and Weightless Travel: A discussion paper" (Department of Sociology, Lancaster University), http://www.comp.lancs.ac.uk/sociology/papers/Urry-Automobility.pdf (last accessed 25 August 2011).

——, *Consuming Places* (London and New York: Routledge, 1995)

——, "Inhabiting the car," *Sociological Review* 54, no. s1 (October 2006), 17–31

——, *Sociology Beyond Societies: Mobilities for the Twenty-First Century,* International Library of Sociology, ed. John Urry (London and New York: Routledge, 2000)

——, "The 'System' of Automobility," *Theory, Culture and Society* 21, no. 4/5 (2004), 25–39

——, *The Tourist Gaze: Leisure and Travel in Contemporary Societies* (Thousand Oaks, CA: Sage, 1990)

US Department of Commerce, Bureau of Foreign and Domestic Commerce, *Installment Selling of Motor Vehicles in Europe,* Trade Information Bulletin no. 550, May 1928 (compiled in Automotive Division from reports by representatives of the Department of Commerce)

Bernd Utermöhlen, "Margarete Winter—eine Automobilistin aus Buxtehude," *Technik und Gesellschaft* 10 (1999), 271–279

Richard Vahrenkamp, *Autobahnbau in Hessen bis 1943* (n.p. [Darmstadt]: Hessisches Wirtschaftsarchiv, 2007)

———, "Die Entwicklung der Speditionen in Deutschland 1880 bis 1938," in Hans-Liudger Dienel and Hans-Ulrich Schiedt (eds.), *Die moderne Strasse: Planung, Bau und Verkehr vom 18. bis zum 20. Jahrhundert* (Frankfurt: Campus Verlag, 2010), 309–337

———, "Lastkraftwagen und Logistik in Deutschland 1900 bis 1955: Neue Geschäftsfelder, neue Kooperationsformen und neue Konfliktlinien," *Vierteljahresschrift für Sozial- und Wirtschaftsgeschichte* 95, no. 4 (2008), 430–455

———, *Die logistische Revolution: Der Aufstieg der Logistik in der Massenkonsumgesellschaft*, Deutsches Museum, Beiträge zur Historischen Verkehrsforschung, eds. Helmuth Trischler, Christopher Kopper, and Hans-Liudger Dienel, Band 12 (Frankfurt and New York: Campus Verlag, 2011)

Tom Vanderbilt, *Traffic: Why We Drive the Way We Do (and What It Says About Us)* (London: Allen Lane / Penguin Books, 2008)

Harold G. Vatter, "Has There Been a Twentieth-Century Consumer Durables Revolution?" *Journal of Economic History* 27, no. 1 (March 1967), 1–16

Thorstein Veblen, *Absentee Ownership: Business Enterprise in Recent Times: The Case of America*, with a new introduction by Marion J. Levy (New Brunswick, NJ, and London: Transaction Publishers, 2009; first ed. 1923)

Michael Venhoff, *Die Reichsarbeitsgemeinschaft für Raumforschung (RAG) und die reichsdeutsche Raumplanung seit ihrer Entstehung bis zum Ende des Zweiten Weltkrieges 1945*, Arbeitsmaterial ARL no. 258 (Hannover: Akademie für Raumforschung und Landesplanung, 2000)

Peter-Paul Verbeek, "Cyborg intentionality: Rethinking the phenomenology of human-technology relations," *Phenomenology and the Cognitive Sciences* 7 (2008), 387–395

Jac. Verheij, *Wetten Voor Weg en Water (1923–1998): Het experiment van de Wet Autovervoer Goederen en de Wet Goederenvervoer Binnenscheepvaart en de jaren erna* (Delft: Eburon, 2001)

Verkeerswaarnemingen van den Rijkswaterstaat 1935: Bijlagen (Den Haag, 1936)

Verslag van het vierde internationale wegencongres, gehouden te Sevilla, 1923 (The Hague: Vereeniging het Nederlandsche Wegen-Congres, 1924)

Verslag van het zevende internationale wegencongres, gehouden te München in 1934 (The Hague: Algemeene Landsdrukkerij, 1937)

Laurence Veysey, "The Autonomy of American History Reconsidered," *American Quarterly* 31 No. 4 (Autumn 1979) 455–477

Arthur J. Vidich and Joseph Bensman, *Small Town in Mass Society: Class, Power and Religion in a Rural Community* (Princeton, NJ: Princeton University Press, 1968)

Frédéric Vieban, "L'Image de l'automobile auprès des Français 1930–1950" (unpublished Mémoire de maîtrise, Histoire Contemporaine, Université F. Rabelais, Tours, France, 1987)

Fred W. Viehe, "Black Gold Suburbs: The Influence of the Extractive Indstry on the Suburbanization of Los Angeles, 1890–1930," *Journal of Urban History* 8 (1981), 3–26

Herman Vink et al., "History of innovation" (final student report, Eindhoven University of Technology, Department *Technische Innovatiewetenschappen*, December 2007)

Vincent van der Vinne, *De trage verbreiding van de auto in Nederland 1896–1939: De invloed van ondernemers, gebruikers en overheid* (n.p.: De Bataafsche Leeuw, 2007)

Paul Virilio, *The Art of the Motor* (Minneapolis and London: University of Minnesota Press, 1998; transl. from *L'art du moteur,* 1993, by Julie Rosie)

Morris S. Viteles and Helen M. Gardner, "Women Taxicab Drivers: Sex Differences in Proneness to Motor Vehicle Accidents," *Personnel Journal* 7 (1929), 349–355

Erik van der Vleuten et al., "Europe's system builders: The contested shaping of transnational road, rail, and electricity networks," *Contemporary European History* 16, no. 3 (2007), 321–347

Wolfgang Vogel, "Lilliputaner (Wirkliche Volksautomobile)," *Public Roads* (4 December 1908), 1112–1114

"La Voie Sacrée", *La lettre de la fondation de l'automobile Marius Berliet,* no. 94 (July–August 2001), 9

Heide Volkening, "Working Girl—eine Einleitung," in Sabine Biebl, Verena Mund and Heide Volkening (eds.), *Working Girls: Zur Ökonomie von Liebe und Arbeit,* copyrights, Band 21, eds. Dirk Baecker and Elmar Lampson (Berlin: Kulturverlag Kadmos, 2007), 7–22

Rudi Volti, *Cars and Culture: The Life Story of a Technology,* Greenwood Technographies Series (Westport, CT, and London: Greenwood Press, 2004)

Klaus Vondung, "Einleitung: Propaganda oder Sinndeutung?" in Vondung (ed.), *Kriegserlebnis; Der Erste Weltkrieg in der literarischen Gestaltung und symbolischen Deutung der Nationen* (Göttingen: Vandenhoeck & Ruprecht, 1980), 11–37

Benno Wagner, "Kafkas Poetik des Unfalls," in Christian Kassung (ed.), *Die Unordnung der Dinge: Eine Wissens- und Mediengeschichte des Unfalls* (Bielefeld: transcript Verlag, 2009), 421–454

Louis C. Wagner, "Economic Relationships of Parking to Business in Seattle Metropolitan Area," in *Parking as a Factor in Business, containing five papers presented at the thirty-second annual meeting January 13 – 16, 1953* (Washington, 1953; Highway Research Board, Special Report 11), 53–90

Gilbert Walker, *Road and Rail: An Enquiry into the Economics of Competition and State Control* (London: George Allen and Unwin Limited, 1947)

Daniel J. Walkowitz, *Working with Class: Social Workers and the Politics of Middle-Class Identity* (Chapel Hill and London: University of North Carolina Press, 1999)

Margaret Walsh, *Making Connections: The Long-Distance Bus Industry in the USA* (Aldershot, Burlington, VT, Singapore, and Sydney: Ashgate, 2000)

John K. Walton, "The Demand for Working-Class Seaside Holidays in Victorian England," *Economic History Review* NS 34, no. 2 (May 1981), 249–265

John Kimmons Walton, "The Social Development of Blackpool 1788–1914" (unpublished doctoral dissertation, University of Lancaster, July 1974)

Stephen V. Ward, *Planning the Twentieth-Century City: The Advanced Capitalist World* (Chichester: John Wiley and Sons, 2002)

——, "What did the Germans ever do for us? A century of British learning about and imagining modern town planning," *Planning Perspectives* 25, no. 2 (April 2010), 117–140

A. M. Warnes, "Estimates of Journey-to-Work Distances from Census Statistics," *Regional Studies* 6 (1972), 315–326

Was verbrauchen die Angestellten? Ergebnisse der dreijährigen Haushaltsstatistik des Allgemeinen Freien Angestelltenbundes (Berlin: Freier Volksverlag, 1931)

Bernard Wasserstein, *Barbarij en beschaving: Een geschiedenis van Europa in onze tijd* (n.p. [Amsterdam]: Nieuw Amsterdam, n.d. [2008])

Alvin W. Waters, "The Last of the Glidden Tours: Minneapolis to Glacier Park, 1913," *Minnesota History* (March 1963), 205–215

Evelyn Waugh, *Vile Bodies* (New York, Boston, and London: Back Bay Books, 1999; first ed. 1930)

Melvin M. Webber, "Transportation Planning for the Metropolis," in Leo F. Schnore and Henry Fagin (eds.), *Urban Research and Policy Planning* (Thousand Oaks, CA: Sage, 1967), 389–407

Donald Weber, "Automobilisering en de overheid in België vóór 1940: Besluitvormingsprocessen bij de ontwikkeling van een conflictbeheersingssysteem" (dissertation, University of Ghent, 2008)

Eugen Weber, *France, fin de siècle* (Cambridge, MA, and London, 1986)

——, "Gymnastics and Sports in Fin-de-Siècle France: Opium of the Classes?" *American Historical Review* 76, no. 1 (February 1971), 70–98

Regina Weber, "'Der rechte Mann am rechten Platz': Psychotechnische Eignungsprüfungen und Rationalisierung der Arbeit bei Osram in der 20er-Jahren," *Technikgeschichte* 68, no. 1 (2001), 21–51.

Hans-Ulrich Wehler, "Die Geburtsstunde des deutschen Kleinbürgertums," in Hans-Jürgen Puhle (ed.), *Bürger in der Gesellschaft der Neuzeit: Wirtschaft—Politik—Kultur, Bürgertum; Beiträge zur europäischen Gesellschaftsgeschichte*, eds. Wolfgang Mager et al., Band 1 (Göttingen: Vandenhoeck & Ruprecht, 1991), 199–209

——, "Klasenbildung und Klassenverhältnisse: Bürger und Arbeiter 1800–1914," in Jürgen Kocka (unter Mitarbeit von Elisabeth Müller-Luckner), *Arbeiter und Bürger im 19. Jahrhundert: Varianten ihres Verhältnisses im europäischen Vergleich*, Schriften des Historischen Kollegs, Kolloquien 7 (Munich: R. Oldenbourg Verlag, 1986), 1–27

Julius Weinberger, "Economic Aspects of Recreation," *Harvard Business Review* 15 (1937), 448–463

Lynn Weiner, "Sisters of the Road: Women Transients and Tramps," in Eric H. Monkkonen (ed.), *Walking to Work: Tramps in America, 1790–1935* (Lincoln and London: University of Nebraska Press, 1984), 171–188

Bernd Weisbrod, "The crisis of bourgeois society in interwar Germany," in Richard Bessel (ed.), *Fascist Italy and Nazi Germany: Comparisons and contrasts* (Cambridge: Cambridge University Press, 1997; reprint of 1996 edition), 23–39

Albert P. Weiss and Alvhh R. Lauer, *Psychological Principles in Automotive Driving: Under the Auspices of the National Research Council 1927–1929* (Columbus: Ohio State University, 1930)

Thomas Weiss, "Tourism in America before World War II," *Journal of Economic History* 64, no. 2 (June 2004), 289–327

Christopher Wells, "Car Country: Automobiles, Roads, and the Shaping of the Modern American Landscape, 1890–1929" (dissertation, University of Wisconsin-Madison, 2004)

——, "The Road to the Model T: Culture, Road Conditions, and Innovation at the Dawn of the American Motor Age," *Technology and Culture* 48, no. 3 (July 2007), 497–523

H. G. Wells, *The Wheels of Change* (London: Faber and Faber, 2008; reprint of 1896 edition)

James V. Wertsch (ed.), *Culture, Communication, and Cognition: Vygotskian Perspectives* (Cambridge and New York: Cambridge University Press, 1985)

James West, *Plainville, U.S.A.* (New York: Columbia University Press, 1958; first ed. 1945)
Andrea Wetterauer, *Lust an der Distanz: Die Kunst der Autoreise in der "Frankfurter Zeitung"* (Tübingen: Tübinger Verein für Volkskunde, 2007)
Bernhard Weyergraf (ed.), *Literatur der Weimarer Republik 1918 – 1933* (Munich and Vienna: Carl Hanser Verlag, 1995; Hansers Sozialgeschichte der deutschen Literatur vom 16. Jahrhundert bis zur Gegenwart, Band 8)
Edith Wharton, *A Backward Glance* (New York and Berlin: Globusz Publishing, n.d.; first ed., New York and London: Appleton-Century, 1934) http://www.globusz.com/ebooks/BackardGlace/index.htm (last accessed 22 June 2009)
——, *Fighting France: From Dunkerque to Belfort*, The War on All Fronts, vol. III (New York: Charles Scribner's Sons, 1919; first ed. 1915)
——, *The Marne: A Tale of the War* (London: MacMillan, 1918)
——, *A Motor-Flight Through France*, Picador Travel Classics (London and Basingstoke: Picador/MacmIllan, 1995)
"What Do Folks Use Their Cars For?" *The Literary Digest* (17 November 1923), 66, 68–69
William Morton Wheeler, *Emergent Evolution and the Development of Societies* (New York: W. W. Norton & Company, 1928)
Michael Wildt, *Am Beginn der "Konsumgesellschaft": Mangelerfahrung, Lebenshaltung, Wohlstandshoffnung in Westdeutschland in den fünfziger Jahren*, Forum Zeitgeschichte, Band 3, ed. Forschungsstelle für die Geschichte des Nationalsozialismus in Hamburg (Hamburg: 1995; first ed. 1994)
Hayden White, "Interpretation in History," *New Literary History* 4 (1972), 281–314
——, *Tropics of Discourse: Essays in Cultural Criticism* (Baltimore, MD, and London: Johns Hopkins University Press, 1986; first ed. 1978)
Roger B. White, *Home on the Road: The Motor Home in America* (Washington, DC, and London: Smithsonian Institution Press, 2000)
Peter Whitfield, *Travel: A Literary History* (Oxford: The Bodleian Library, 2011)
Robert H. Wiebe, *The Search for Order 1877–1920* (New York: Hill and Wang, 1967)
Benno von Wiese, "Ödön von Horváth," in Traugott Krischke (ed.), *Ödön von Horváth*, suhrkamp taschenbuch materialien, vol. 2005 (Frankfurt: Suhrkamp, 1981), 7–45
Reynold M. Wik, *Henry Ford and Grass-roots America* (Ann Arbor: University of Michigan Press, 1972)
Willa Cather's Collected Short Fiction, 1892–1912, introduction by Mildred R. Bennett (Lincoln: University of Nebraska Press, 1965; first ed. 1965)
Frances E. Willard, *How I Learned to ride the Bicycle: Reflections of an influential 19th century woman*, introduction by Edith Mayo, ed. Carol O'Hare (Sunnyvale, CA: Fair Oaks Publishing, 1991; rev. ed. of *A Wheel Within a Wheel*, 1895)
Raymond Williams, *Keywords: A Vocabulary of Culture and Society* (New York: Oxford University Press, 1985; revised edition)
William Carlos Williams, *The Autobiography of William Carlos Williams* (New York: New Directions Books, 1967)
——, *A Voyage to Pagany* (New York: The Macaulay Company, 1972; first ed. 1927)
——, "The Great American Novel," in Williams, *Imaginations: Kora in Hell, Spring and All, The Great American Novel, The Descent of Winter, A novelette and Other Prose*, edited with introductions by Webster Schott (New York: New Directions Books, 1970), 158–227
Alice M. Williamson, *The Inky Way* (London: Chapman & Hall, 1931)

———, *The Lightning Conductor Comes Back* (London: Chapman & Hall, 1933).
C. N. Williamson and A. M. Williamson, *The Botor Chaperon* (London: Methuen, 1906; reprint BiblioLife, n.d.)
———, *De kranige chauffeur: Wonderlijke avonturen van een motorwagen*, transl. Ms. Van Heuvelinck (Utrecht: H. Honig, 1913)
———, *De kranige chauffeuse: Vervolg op (De kranige chauffeur)*, transl. mevr. J. P. Wesselink—Van Rossum (Rijswijk: Blankwaardt & Schoonhoven, n.d.)
———, *The Lightning Conductor Discovers America* (Garden City, NY: Doubleday Page & Company, 1916)
———, *The Lightning Conductor: The Strange Adventures of a Motor-Car* (New York: Henry Holt and Company, 1903; eleventh impression, revised and enlarged)
———, *The Lightning Conductress* (London: Methuen, 1916)
———, *The Princess Passes: A Romance of a Motor-Car* (New York: Henry Holt & Company, 1905; first ed. 1903?)
Angus Wilson, *The Strange Ride of Rudyard Kipling: His Life and Works* (London: Pimlico, 1977)
Jay Winter, "The Lost Generation of the First World War; Paul Fussell: *The Great War and Modern Memory* (1975)," in Uffa Jensen et al. (eds.), *Gewalt und Gesellschaft: Klassiker modernen Denkens neu gelesen* (Göttingen: Wallstein Verlag, 2011), 337–350
[Margarete Winter], *5000 Kilometer Autofahrt ohne Chauffeur, Buxtehude 1905* (Stuttgart: Uhlandsche Buchdruckerei, n.d. [1905])
Birgit Winterberg, *Literatur und Technik: Aspekte des technisch-industriellen Fortschritts im Werk Paul Morands*, Europäische Hochschulschriften, Reihe XIII, Französische Sprache und Literatur, Band 168 (Frankfurt, Bern, New York, and Paris: Peter Lang, 1991)
Laurie Winters, "Die Wiederentdeckung des Biedermeier," in Hans Ottomeyer, Klaus Albrecht Schröder, and Laurie Winters (eds.), *Biedermeyer: Die Erfindung der Einfachheit* (Ostfildern: Hatje Cantz Verlag, 2006), 31–55
Sarah Wintle, "Horses, Bikes and Automobiles: New Woman on the Move," in Angelique Richardson and Chris Willis (eds.), *The New Woman in Fiction and Fact: Fin-de-Siècle Feminisms* (Houndsmill and New York: Palgrave MacMillan in association with Institute for English Studies, School of Advanced Study, University of London, 2001), 66–78
Erika Wolf, "The Author as Photographer: Tret'iakov's, Ehrenburg's, and Il'f's Images of the West," *Configurations* 18 (2010), 383–403
Hanna Wolf, *Following America? Dutch Geographical Car Diffusion, 1900 to 1980* (Eindhoven: ECIS, 2010)
James B. Wolf, "Imperial Integration on Wheels: The Car, the British and the Cape-to-Cairo Route," in Robert Giddings (ed), *Literature and Imperialism*, Insights, ed. Clive Bloom (Houndsmill and London: MacMillan, 1991), 112–127
Hellmuth Wolff, *Kampf dem Verkehrsunfall*, Schriften des Seminars für Verkehrswesen an der Martin Luther-Universität Halle-Wittenberg, ed. Hellmuth Wolff, no. 11 (Halle: Akademischer Verlag, 1938)
Leonard Woolf, *Downhill All the Way, An Autobiography of the Years 1919–1939* (New York: Harcourt, Brace & World, Inc.: 1967)
Virginia Woolf, "Evening Over Sussex: Reflections in a Motor Car," in: Woolf, *The Death of the Moth, and other essays* (http://ebooks.adelaide.edu.au/w/woolf/virginia/w91d/chapter2.html; last accessed 23 April 2011)

——, *Mrs. Dalloway* (ebooks@Adelaide, 2010; http://ebooks.adelaide.edu.au/w/woolf/virginia/w91md/complete.html; last accessed 23 April 2011) (first ed. 1925)

Genevieve Wren, "Women in the World of Autos; 5000+ Women Granted Auto Patents between 1912-22," *Wheels* (June 1994), 7-8

Elizabeth Yardley, *A Motor Tour Through France and England: A Record of Twenty-one and a Half Days Automobiling* (London: Stanley Paul & Co, n.d. [1911?])

Clarence Young, *Ned, Bob and Jerry Bound for Home: Or, The Motor Boys on the Wrecked Troopship* (New York: Cupples & Leon Company, 1920)

——, *Ned, Bob and Jerry in the Army: Or, The Motor Boys as Volunteers* (New York: Cupples & Leon Company, 1918)

——, *Ned, Bob and Jerry on the Firing Line: Or, The Motor Boys Fighting for Uncle Sam* (n.p., n.d.; reprint Milton Keynes: Dodo Press, 2009)

Patrick Young, "*La Vieille France* as Object of Bourgeois Desire: The Touring Club de France and the French Regions, 1890-1918," in Rudy Koshar (ed.), *Histories of Leisure* (Oxford and New York: Berg, 2002), 169-189

Paul A. Youngman, *We Are the Machine: The Computer, the Internet, and Information in Contemporary German Literature*, Studies in German Literature, Linguistics, and Culture (Rochester and New York: Camden House, 2009)

Angela Zatsch, *Staatsmacht und Motorisierung am Morgen des Automobilzeitalters* (Konstanz, 1993)

Susan Zeiger, *In Uncle Sam's Service: Women Workers with the American Expeditionary Force, 1917-1919* (Ithaca, NY, and London: Cornell University Press, 1999)

Wilbur Zelinsky, "The imprint of central authority," in: Michael P. Conzen (ed.), *The Making of the American Landscape* (Boston, London, Sydney and Wellington: Unwin Hyman, 1990) 311-334

Thomas Zeller, "Der automobile Blick : Berg- und Alpenstrassen und die Herstellung von Landschaft in Deutschland und den USA im 20. Jahrhundert," in Hans-Liudger Dienel and Hans-Ulrich Schiedt (eds.), *Die moderne Strasse: Planung, Bau und Verkehr vom 18. bis zum 20. Jahrhundert*, Deutsches Museum, Beiträge zur historischen Verkehrsforschung, Band 11, eds. Helmuth Trischler, Christopher Kopper, and Hans-Liudger Dienel (Frankfurt: Campus Verlag, 2010), 265-283

——, "Building and Rebuilding the Landscape of the Autobahn, 1930-70," in: Christof Mauch and Thomas Zeller (eds.), *The World Beyond the Windshield: Roads and Landscapes in the United States and Europe* (Athens, OH and Stuttgart: Ohio University Press/Franz Steiner Verlag, 2008) 125-142

——, *Strasse, Bahn, Panorama; Verkehrswege und Landschaftsveränderung in Deutschland von 1930 bis 1990* (Frankfurt and New York: Campus Verlag, 2002; Deutsches Museum, Beiträge zur Historischen Verkehrsforschung, Band 3; eds.: Helmuth Trischler, Christoph Kopper, Hans-Liudger Dienel)

Robert Ziegler, *The Nothing Machine: The Fiction of Octave Mirbeau*, Faux Titre, 298, Etudes de langue et littérature françaises, eds. Keith Busby et al. (Amsterdam and New York: Rodopi, 2007)

Slavoj Žižek, *Violence: Six Sideways Reflections* (New York: Picador, 2008)

Christa Zorn, *Vernon Lee: Aesthetics, History, and the Victorian Female Intellectual* (Athens: Ohio University Press, 2003)

W. Zweerts de Jong, *De geschiedenis van het Vrijwillig Militair Automobiel Korps, naar verschillende gegevens verzameld en bewerkt* (Haarlem: J. A. Boom, 1918)

Paul Zweig, *The Adventurer* (London: Basic Books, 1974)

Index

Page numbers in italic indicate a definition or central treatment of the term.

AA (Automobile Association), 589
AAA (American Automobile Association), 106, 333, 334, 391, 394, 399, 403
AAAA (Afro-American Automobile Association), 406
ACA (American Automobile Club), 106
acceleration, 85
accident (statistics), 88, 92, 101–102, 128n157, 136, 148, 167, 169, 176, 181, 183–185, 187, 216n112, 220n168, 221n180, 393, 395, 398, 402, 427, 438, 447, 452, 453, 455–456, 458, 466, *471*, 472–473, 475, 477, 483, 493–495, 502, *514–515*, 525, 527, 530–531, 533, 548n175, 575, 579, 592, *595–603*, 606, 644, 650, 653
    accident-prone motorists, 396, 408, 603–605
    female deaths, 597
    traffic deaths, *597–606*
    See also safety
Achterhuis, Hans, 233
ADAC (Allgemeiner Deutsche Automobil-Club), *80–81*, 103, 289, *290–292*, 327–328, 378, 394, 570
Adam, Paul, 170, 179
Adams, Henry, 93, 94, 187, 199
Adams, Jon, 30
Adorno, Theodor, 514, 518
adventure (automotive), 23, 68, *84*, 90, 93, 94, *112–113*, 133, *142–143*, *190*, *204*, 394, *515–516*, 643
    African-American, 406–407
    airplane as, 63, 106, 169, 180, 520, 637
    bicycle as, 64, 166
    car as, 191, 391, 492, 647
    collective, 99–100, 254
    colonialism by car, 155, 156

consumption as, 472
domestication, *392*, 393, 394, 439, 528, 645; of the car, 447, 644; of the travel account, 449
experience in train, 521
extreme, 133, 232
female, 41n4, 95, 96, 134, 144, 153, 157, 158, 175, 269, *407*, 412, 432, 436, 438, 439, 461, 478, 484, *487*, 489, 646
female functional, 438
functional, 108, 157, 490, 567
functional (tinkering), 63, 85, 107–108, 145, 154, 169, 173, 191, 333, 394, 429, 435, *438*, 441–442, 474, 490, 504, 522, 524, *567*
imperialism by car, 160
and literature, 135
maintenance, 567
motorcycle as, 83, 312, 437, 454
multiple (more than tripartite), 29, 112; erotic, 147, 165, 168, 171, 404, 408, 432, 443, 455; imperialist/colonial ('conquest'), 93, 112, 154, 160, 170, 426, 432, 522–523
for nonusers, 135
reformulation of the car, 286
restricted, 98, 106, 133, 160, 175, 303, 333, 374, *392–394*, 425, *439*, 445, 449, 463
semi-collective character of interwar motoring, 639
shift in, 441
spatial (touring), 29, 97, 146, 156, 191, 194–197, 286, *399*, 433–434
spatial automotive, 194–197
speed, 85, *197–200*, *399*, *456*
temporal (racing, speedy touring), 85, 160, 191, 197, 257, 260, *395*, 457, 525
touristic, 87, 92, 250
tripartite, 29, 64, 78, 112, 157, 393

walking as, 86
'adventure machine,' 29, 59, 63, 111, 153, 157, 373, 650
adventure novel, 142–143, 174, 209n46, 431, 515
AEG (*Allgemeine Elektricitäts-Gesellschaft*), 236
affinity, *511, 648–650*, 137, 202–204
affordance, 36, 96, 169, 345, 346, 391, 438, 647, 650, 652
  perception of environment, 652
aggression (aggressive car culture), 60, 66, 75, 89, 93, 144, 147–148, 156–157, 161, 176, 179–182, 188, *190*, 192, 195, 196, *197–200, 228–235*, 337, 395–396, 398, 402, *407–409*, 452–453, 471, 494, 498, 503, 505, 510, 511, 514, 516, 518, 527, 529, 531–532, 605, 637, 739, 643–644, 648–650, 652, 655
  colonial, 156, 191
  female, 156, *408–409*, 433, 435, 490, 646
  See also violence
AIACR (*Association Internationale des Automobile-Clubs Reconnus*), 68, 572, 576
AIT (*Alliance Internationale de Tourisme*, successor of LIAT), 581
Albert, Daniel, 604
alpinism, 86
America-as-model myth, 13, 27, 638, 653
American exceptionalism, 306
Americanization, 306–313, 386
anarchism, anarchist, 76–77, 133, 148, 165, 167, 177, 179, 181, 182, 189, 333, 444, 502, 515, 527
  character of car use, 75–76, 232, *248, 595, 614, 617*
Anderson, Edward, 457
Anderson, Rudolph, 186
Anderson, Sherwood, 409, 450, 476
angst-lust, 93, 167, *190*, 456, 606, 644, 648
annual model change, 385
anti-Americanism, 168, 169
anti-European feelings, 90, 92
anti-urban, 91, 154, 195, 235
ANVF (*Algemeene Nederlandsche Verkeersfederatie*), 594
ANWB (*Algemeene Nederlandsche Wielrijders Bond*), 77, 78, 83, 231, 291, 335, 519, 570, 581, 582, 594, 611
Apollinaire, Guillaume, 141, 177, 227, 239

Appleton, Victor, 175
applied history, 4
Aragon, Louis, 513, 515
Arendt, Hanna, 6
aristocracy as car users, *66–67*, 71
Arlen, Michael, 469, 470, 508
Armstrong, Tim, 204
Arntz, Gerd, 511
associative writing, 446
Astor, John Jacob, 229
automobile. *See* car
autopoetic(s), *141–145*, 151, 156, 160, 168, 171, 177, 186, 191–192, 426, 440, 447, 457, 471, 477, 486, 509, 512, 645, 652
  belletristic literature as a historical source, 28–34
  movie/film, 40, 96, 136–137, 142, 144, 184, 262, 399, 405, 434, 454, 457, 458, 465, 470, 474, 481, 489, 491, 493, 510, 521, 523, 524–528, 611; affinity with driving, 195, 389, 478, 503, 523, 531; cinematic style, 168, 531; silent, 152, *176*, 183–186, 190, 220n169, 521, 524, 563n352, 644; slapstick, 505, 527 (*see also* road: movie)
  symbolic, 648
  symbolic level, 530
  symbols of the car, 200–202
  travel experience, 426
  urban, 456
Autobahn, 291, 326, 328, 388, 390, 391, 451, 457, 589, 590, 592
automatic transmission, 389
automation, 388
automobile sport, 229
automobile system, 41, 286, 426, 565, 566, 613, 618
automobile tests by armies, 229
automobilism
  as collective practice, 134
  as a concept, 37–38
  critique of concept, 37–38
  military influence on, 228
automobility, 18, 24–27, 29, *37–38*, 157, 403, 489, 603, 619, 653
autostrada, 590
Autowandern, 451, 457

Bachelard, Gaston, 201
Baden-Powell, Robert, 93
Bagnold, Enid, 261

Baker, Josephine, 92
Bakker Schut, P., 602
Balfour, Arthur, 75
Balint, Michael, 199
Ball, Hugo, 266
balloon tires, 382
Barthes, Roland, 33
Barton, Bruce, 392
Barzini, Luigi, 108
Basie, Count, 406
battle between the propulsion systems, 229
Battle of the Marne, 241
Battle of the Somme, 239
Baudry de Saunier, Charles-Louis, 67, 81, 82
Bauman, Zygmunt, 606, 646, 649
Becher, Johannes R., 239, 266
Beck, Stefan, 22–23
Beck, Ulrich, 395
Becker, Jens Peter, 474, 494, 509, 519
Beckett, Samuel, 454
Beckman, Karen, 489, 515
Beckmann, Jörg, 396
Behrens, Franz, 266
Beinhorn, Elly, 488
Belasco, Warren, 331, 333
Benjamin, Walter, 400, 482, 500, 510, 521
Benz, Carl, 62
Berber, Anita, 488
Berger, Michael, 94, 95
Bernard, Tristan, 170
Bertho Lavenir, Catherine, 16, 21, 24, 86, 139
Bethmann Hollweg, Theobald von, 240
Betjeman, John, 430
BIAR (*Bureau Internationale des Autoroutes*), 581
bicycle, 65, 87, 156, 158, 165, 166, 193, 194, 228, 309, 609, 614
    craze, 63, 87
    culture, 62, *63–65*
    as sports vehicle, 229
    and walking movement, 447
    women and, 65
biedermeier travel, 60, 133, 148, 193, 196, 394, 397
Bierbaum, Otto Julius, 59, 133, 146–148, 189, 193, 194, 196, 197, 200, 201, 234, 266, 392
Biggles, 431
Blackburn, Marc, 243

blacks, 531
Blanke, David, *14*, 463
Bloem, Walter Julius, 452–453
Bloy, Léon, 148–149, 189
Blue Ridge Parkway, 590
Blum, Léon, 340
Bogart, Humphrey, 527
Bonnot, Jules, 133, 148, 185
Boorstin, Daniel, 392
Borden, Mary, 261
Bordewijk, F., 477, 484
Borg, Kevin, 23
Borrebach, Hans, 520
Bourdieu, Pierre, 22, 35, 639
bourgeoisie as car user, 66–67, 86
Bóveda, Xavier, 502
Boveri, Margret, 488
Boxcar Bertha, 442–444
Boy Scouts, 235, 254
    novels, 176
Boylston, Helen Dore, 432
BPR (Bureau of Public Roads), 215, 587
Braunbeck, Gustav, 81
Brecht, Bertolt, 393, 395, 472, 480, 482, 493–494, 506, 509, 511
Breton, André, 498, 510, 511, 515
BRF (British Road Federation 1932), 594
Bronnen, Arnolt, 390, 497, 504–506, 511
Bryan, Vincent, 186
Buchanan, Colin, 589
Buell, Lawrence, 141
'bumper thumping', 398
Burroughs, John, 90
business use of the car, 72–73, 321, 363n143
Buysse, Cyriel, *167–169*, 193, 195, 255–256

Cagliero, Roberto, 532
Cagney, James, 524, 525
Caldwell, Erskine, 502
Calvert, W. R., 445
camping, 334
Canzler, Weert, 22
Capra, Frank, 527
capsule, 401, 411
car
    as capsule, 286, 373, 401, 411, 503, 644
    as cocoon, 159, 160, 373–374, 455, 503
    as collective symbol, 200
    as commodity, 29, 286, 287–306

as deadly weapon, 267
  as devil's vehicle, 152
  as medium, 21
  as multisensorial room on wheels, 375, 386
  as necessity, 77, 313–328, 322–328, 336, 345, 363n143
  as panzer, 502
  as (part of) corridor, 401
  as shell, 192, *200–202*, 286
  as time machine, 160, 162
car diffusion, 7, *12–13*, 20–21, 27, 48n80, 51n105, 70–72, 103–104, 112, 288–289, 291–292, 298, 310–313, 346, 358n101, 641
Carey, James, 21
"car gypsying," 295, 333
carnets de douane, 572
car occupancy rates, 322
car ownership among blacks, 304
car ownership without use, 312
car prices, 295
Carr, David, 5
Carranza, Venustiano, 243
carriage racing, 64
car (and motorcycle) densities, *68–71*, 289, 292, 293, 299, *301–302*, 312–313, 316, 337, 355n69, 588, 299, 637
car society, 41, 619
car travel as 'non-essential,' 325
car use without ownership, 312
Cather, Willa, 144, *187*, 267, 436, 490–492
Céline, Louis-Ferdinand, 431
Cendrars, Blaise, 156, 426, 499
centralization of road planning, 592–593
Chadbourne, Archie, 285
Chaplin, Charlie, 185, 525, 526
chauffeur, 79, 85, 92, *109–110*, 171, 192
Chesnais, Claude, 596
children and cars, 99, 101, 176, 445–446, 655
Christie, Agatha, 520–521
Chrysler, 383
Churchill, Winston, 586
CIAM (*Congrès Internationaux d'Architecture Moderne*), 602
cityscape, 191, 400. *See also* landscape
Clarke, Deborah, 97, 399
Clarsen, Georgine, 95, 407, 434–435, 571
Clifford, James, 35
closed automobile, 374–386, 464
Colbert, Claudette, 527

cold attitude/persona, 402, 498. *See also* cool gaze
collective
  bicycle culture, 64, 86
  culture, 97, 99
  gaze, 330
  leisure practices, 311
  See also car; family
combined use of single car trips, 322
comfort, *323*, 374, 375, *376*, 381, 383, 389, 411
communication, history of, 21, 28, 33, 83, 137, 203, 225n23, 306, 505, 566, 654
commuting, 108, 305, 308, 309, 315–318, 344. *See also* adventure
congestion, 169, 250, 298, 312, 477, 574, 580, 589, 598, 602, 616
consumerism, consumption, 286, 296, 297, 315, 320
continental crossings by car, 84
convertible/cabriolet, 384
conversion (from a car hater into a car lover), 145, 150, 172, 192, 486, 487, 494
Cook, Thomas, 251
cool gaze, 499
coordination crisis, 606–617. *See also* road: road/railway controversy
Corelli, Marie, 193, 503
Couperus, Louis, 113
Cowley, Malcolm, 428
Crane, Laura Dent, 175
credit, 295, 297, 300, 347
Cresswell, Tim, 489, 529, 654
Csikszentmihalyi, Mihaly, 397, 398
cultural turn, 14
Cummings, Edward Estlin, 267, 431, 509
Cuneo, Joan, 84
cyborg, 146, 160, 164, 166, 172, 192, 194, 204, 390, 391, 397, 439, 447, 456, 461, 474, 476, 478, *497*, 523, 646
    motif/metaphor, 146, 397, *447–456*, 461, 476, 528 (*see also* flight)

DAC (*Deutsche Automobil-Club*), 80
Dada, 266, 458, 512, 515
dancing and car driving, 447–448, 477
D'Annunzio, Gabriele, 180, 239, 287
Dautry, Raoul, 615
Davids, Louis, 521
Davis, Simone Weil, 462
de Certeau, Michel, 30
de Crawhez, Pierre, 156

deep movement, 397
de Gaulle, Charles, 288
de Grazia, Victoria, 639
Delaisi, Francis, 593
De L. Welch, Marie, 440
de Lempicka, Tamara, 499
Deleuze, Gilles, 32, 194
Demolder, Eugène, 155–156, 200
Denby, Priscilla, 201
dèr Mouw, J. A, 509
de Roulx, Didier, 427
Desportes, Marc, 30, *136–139*, 192, 194, 650, 651
Diesel, Eugen, 154
diffusion, 13, 27, 298, 313
    curve, 20, 298, 312, 347, 612
    rate, 298
    studies, classic, 373
    See also car diffusion
diffusionist myth, 373
Dillinger, John, 458
Dinerstein, Joel, 406, 514, 649
Disko, Sasha, 438
distancing, 400, 401
Dixie Highway, 587
Dixon, Winifred, 438–439
Dix, Otto, 488, 498
Döblin, Alfred, 477, 497, 508
Dodge, 383
domesticated hedonism, 425
Donny, Julius, 454
Doolittle, James Rood, 7
Dos Passos, John, 267, 386, 431, 496, 500
Douglas, Mary, 396
Doumenc, Aimé, 245–246
Doyle, Arthur Conan, 133, 258
Draws-Tychsens, Hellmut, 439
Dreiser, Theodore, 60, 187–189, 193, 195, 467, 493, 494–496
dribble spending, 296
drivers license, 319
DSA (*Direction des Services Automobiles*), 248–250
dual nature of technology, 35–36
Duncan, Marise, 520
Durtain, Luc, 455–456
Duryea, Charles, 108
Dutch Ministry of Traffic and Water Management (*Ministerie van Verkeer en Waterstaat*), 2

Earhart, Amelia, 520
Eckstein-Diener, Bertha, 439
economic importance of the automobile, 77
écriture automatique, 512
Edelmann, Heidrun, 11–13
Edison, Thomas, 184
Edward G. Budd Mfg. Co, 383
Edwards, Gus, 186
Ehrenburg, Ilya, 399, 474
Ehrlicher, Hanno, 510
Eiichi, Kiyooka, 522
Einstein, Carl, 182
Eisenstein, Sergei, 528
Eksteins, Modris, 242, 253
electric vehicle, 62, 73, 96, 97, 109, 236
Elias, Norbert, 234, 387, 395, 606
Ellis, Edward, 407
enskilment (Ingold), 387
Enzensberger, Hans-Magnus, 22
Epstein, Ralph, 7
eroticism, 178
Essex, 383
experience of speed, 162
experts, 641
Expressionism, Expressionists, 135, 181, 182, 199, 254, 266, 427, 471
extreme sports, 396, 399

Fallada, Hans, 481, 508
family, as a collective subject, 345, 439, 446
    budget spent on cars, 308
    car, 294, 328–344
    expenditures, 318
    touring car, 411
    See also household
farmers as motorist, 72–73, 85, *103–104*, 188, *299–303*, 311, 314–315, 642
    commuting, 317
Faulkner, William, 492
Federal Aid Highway Act (1916), 106, 332
Fell, Karolina, 435, 488, 489
female travel writers, 156–161
feminism, second wave, 95–96
Fendrich, Anton, 262–265
Ferber, Edna, 201
Field, Edward, 174
Fields, W. C., 527
Fischer, Claude, 86
Fisher Body Company, 383
Fitzgerald, F. Scott, 140, 469, 470, 474, 492–493

flâneur, 400, 458
Fleisser, Marieluise, 439, 448–449, 479–481, 488, 509
Flesch von Bruningen, Hans, 181, 182
flight (metaphor), 79, 144, 151–152, 154, 156, 157, 158, 161, 165–166, 168, 180, 183, 185, *190*, *192–194*, 198, 202, 204, 223n199, 223n203, 259, 329, 334, 386, *398*, 401–402, 411, 437, 441, 447, *451*, 457, 458, 459, 461, 475, 478, 492, 504, 506, 638, 644, 649, 652
Flik, Reiner, 12, 13, 326
Flink, James, 8, 71, 299, 305, 403
Flonneau, Mathieu, 25–26
flow, 397, 460
Flying Ambulance Corps, 269
Foley, Margaret, 96
Fonda, Henry, 527
Ford Model A (1927), 375, 398
Ford Model T, 107, 311, 312, 375, 381, 398, 432, 466, 468, 472, 495, 521, 522, 526, 568, 608, 651
Ford V-8, 457, 458, 484
Ford, Henry, 107, 262, 375, 568, 570
Forest, Ellen, 463
Forster, E. M., 149, 519
Foster, Shirley, 489, 529
Foucault, Michel, 22
Franz, Kathleen, 23
Freeston, Charles L., 154
freeways, 581, 584, 589, 593. *See also* Autobahn, autostrada
French-Belgian group of writer-motorists, 135, 161–170
Freud, Sigmund, 142, 204, 483, 531
Frisch, Rudolf, 156
Frondaie, Pierre, 499
fuel consumption, 569
fuel-supply network, 568
fuel tax, 582
function (relational), 36, 38, 62, 113n2, 158, 167, *190*, 191, 194, 198, 286, 316–317, 322, 326, 334, 362n137, 374, 376, *380–381*, 399, 410–411, *413n10*, 477, 615, 647
    adventurous, 85
    as armor, 476
    as carrier of fantasies, 135
    cocoon, 457, 496, 524
    commuter, 313
    compensatory, 309
    consolating, 25
    door-to-door, 615
    as extension, 182
    as isolated viewpoint, 160
    liberating, 175
    as male vibrator, 92
    as means to seduce, 481
    migration, 328
    as optimistic antidote, 484
    as prosthesis, 192, 461
    shopping, 360n117
    as showfront, 469
    specialized, 411
    speeding, 388
    suburban, 465
    supply, 237
    therapeutic, 85, 199
    as token of extreme wealth, 148
    touring/touristic, 373, 375, 411, 569
    universal, 411
    urban, 388, 411
    user, 38
    utilitarian, 4, *9–12*, 27, 48n82, 73, 80, 83, 95, 100, 107–108, 198, 228, 231, 256, 286, 310, *322–328*, 344–347, 393, 416n57, 464, *504*, 608, 615, 639
    as vehicle for tales of expectation, 175
    See also adventure: functional; car; properties (technical)
'function combiner' (car as), 322
functional complex as commodity, 29
functional embeddedness, 314
functional mix of rural and urban use, 344
functional shift, 27
functional spectrum, 286, 567
functional split, 316, 328
Fussel, Paul, 134, 258, 431
Futurism, Futurists, 59, 176, 180, 199, 287, 427 502

Gable, Clark, 527
Galliene, Joseph, 244–245
garages run by women, 571
Garbo, Greta, 470
Garis, Howard, 175
gasoline tax, 319
Gautier, Théophile, 155–156
Gay, Peter, 31–32, 232, 233, 234, 235
gear-shifting, 388
Geddes, Norman Bel, 590
Geddes, Sir Eric, 230
Geertz, Clifford, 2, 3

Index

Geiger, Theodor, 308
Genet, Jean, 480
Gerron, Kurt, 521
Gershwin, George, 406
Gibson, James, 34–36, 651–652
Giddens, Anthony, 396, 448
Gide, André, *143*, 163–164, 178, 197, 437, 459
Gideon, Siegfried, 16, 400
girl gangs, 405
girls' novels, 175, 519–520
Glidden Tour reliability run, 107
Goebbels, Joseph, 508
Going-to-the-Sun Highway, 332
Goldmann, Lucien, 30
Gooden-Chisholm, Mairi, 269
Good Roads Movement, 332, 587, 588
Göring, Hermann, 501
Grabein, Paul, 262–265
Grahame, Kenneth, 150–151, 193, 197
grammar of the car adventure, 135
Gramsci, Antonio, 505
Grand Tour, 65, 156
Gray, John, 655
Great Migration, 329
Green, Graham, 436
Green, Martin, 489
grenade (car as), 179, 202
Griffith, D.W., 184
Grivel, Charles, 135–136, 204, 398, 513
Guattari, Félix, 32
Gudis, Catherine, 24–25
Guglielminetti, Ernest, 573

HAFRABA (association), 594
Haig, Douglas, 239
Hansa-Lloyd, 236
Harmsworth, Alfred, 136, 170, 229
Hauser, Heinrich, 387, 449–452, 457, 477, 523, 638, 646
Hawks, Howard, 525
Hawthorne, Nathaniel, 139, 432
Heard, Gerald, 531
Hemingway, Ernest, 261, 267, 431
Hepworth, Cecil, 183
Herkomer, Hubert von, 166
Hesse, Hermann, 184
Hessel, Franz, 459
highway 66, 441
Hill, Bertha Chippie, 521
Hinckmans, L. and S., 5–6
hitchhiking, 317, 329, 403, 435, 443, 444, 527, 539n70

women, 442, 443
Hitler, Adolf, 291, 326–327, 328, 339, 345, 457, 500, 501, 503, 605, 339
hobo, 443–444
hoboing de luxe, 333
Hoffmann, Trüdel, 439
Hoffmann-Harnisch, Wolfgang, 434
Holitscher, Arthur, 433–434
Hollaender, Friedrich, 521
Hollweg, Bethmann, 267
Holzer, Marie, 181
homoerotic relationship between men and their cars, 97
Hooker, Katharine, 433
hooliganism, 76, 200, 636–637. *See also* resistance (against the car)
Hoover, Herbert, 296
Horner, Craig, 392
Horras, Gerhard, 11
horse racing, 64
hot rod culture, 398
household expenditures, 296–297, 299, 304, 307–309, 318, 363n145. *See also* family
Huber, Valeska, 654
Hughes, Thomas, 382
Hugill, Peter, 102
Hühnlein, Adolf, 292, 328
Hulme, Kathryn, 439
Huxley, Aldous, 428
hybrid truck, 230

ICC (International Chamber of Commerce), 607, 617
Il'f, Il'ia, 344
illusion of individuality, 396
ILO (International Labour Office), 581, 593
Impressionist style, 168
Ingold, Tim, 34, 401
intermediality, 137
Interrante, Joseph, 317
inversion, 136, 144, 150, 157, 159, 162–163, 167, 179, 183, 204, 264, 457, 469, 474, 490, 492, 644, 651
olfactory, 504
Ireland, Robert, 8
ironic (approach of reality), 140, 166, 432, 461, 468, *497*, 482, 506, 531, 649, 650
'ironic car,' 649
IULA (International Union of Local Authorities), 68
IWW (Industrial Workers of the World), 444

Jacobs, Aletta, 94, 96
James, Fredric, 6
James, Henry, 92, 158
Jarry, Alfred, 145
Jarvis, Adrian, 298, 301
Jeantaud, Charles, 62
Jerome, Jerome K., 144–145, 165
jitney, 472, 612–613
Johnson, Amy, 520
Johst, Hanns, 592
joyriding, 20, 109, 345, *405*
Jünger, Ernst, 397, 398, 403, 449, 500–501, 506, 507, 515

Kafka, Franz, 493, 497
Kaftan, Kurt Gustav, 581
Kanehl, Oskar, 181
Kästner, Erich, 493, 509
Kazin, Alfred, 466
Keaton, Buster, 480, 526, 528
Kehrt, Christian, 35
Kennard, Mary, 145
Kern, Stephen, 4, 34
Keun, Irmgard, 478–479
Keystone Kops, 185, 525. See also autopoetic: movie/film
Kipling, Rudyard, 141, 145–146, 195, 258
Kisch, Egon Erwin, 401, 433, 434, 526
Kistemaeckers, Henry, 166, 169, 201, 202
Kittler, Friedrich, 33
Klemperer, Victor, 456, 508
Knie, Andreas, 22
Knocker, Elsie, 269
Köhler, Hanni, 439
Kölwel, Gottfried, 181
Konody, Paul, 154, 195, 196, 200, 202
Körner, Susanna, 439
Koselleck, Reinhart, 34
Koshar, Rudy, 22
Kracauer, Siegfried, 447–448, 477
Kramer, Lloyd, 6
Krämer-Bodoni, Thomas, 23
Kriéger, Louis Antoine, 229–230
Krim, Arthur, 592
Ku Klux Klan, 444, 467
Kuhm, Klaus, 13
Kyrk, Hazel, 325

Lagrange, Léo, 340
Lainé, Lucien, 581, 594
landscape, 24–25, 27, 36, 39, 59, 85, 88–89, 125n112, 136–138, 145, 147, 153–155, 157–159, 168–169, 181, 191, 194–197, 203, 256, 264, 266, 400, 434, 437, 448–449, 451–452, 454, 455, 494, 501, 568, 590, 637
    Australian, 653
    Impressionist, 136
    movement of, 136 (see also inversion)
    poet, 455
    projection on windscreen, 389
Lane, Rose Wilder, 432
Lang, Fritz, 525
Larbaud, Valery, 178, 179, 454, 460, 499
Larsen, Jonas, 138, 195, 203
Latour, Bruno, 6, 36
Laurel and Hardy, 525, 526
Laux, James, 7–10
Lavin, Maud, 407
Lawrence, D. H., 430
Lawrence, T. E., 431
League of Nations, 593
League against Road Dust (*Ligue contre la poussière sur les routes*), 573
Lear, T. Jackson, 193
Leblanc, Georgette, 178, 202
Leblanc, Maurice, 141, 165–166
Ledwinka, Joseph, 385
Lee, Vernon, 152–153
leisure travel, use of car, 320–321
Leitner, Maria, 439
Lely, Cornelis, 77
Lersch, Heinrich, 502
Lethen, Helmut, 403, 497, 499, 500, 505, 511, 645
Lewis, Sinclair, 460–467, 471, 482, 487, 493, 508
LIAT (*Ligue Internationale des Associations Touriste*), 68, 77, 572, 576
Lieb, Claudia, 493, 515
Light, Alison, 520
Lincoln Highway, 91, 332, 587
Lincoln, Andrew Carey, 176
Lind, Michael, 306
Lipset, Seymour, 483, 532
Liston, Virginia, 521
Lloyd, Harold, 525
logistics, 228, 230, *242–250*, 255, 269, 339, 568, 616
logistic curve, 609 (see also diffusion curve)
long-range family touring car, 386
long-range tourism, 335, 375, 389, 610
Luhman, Niklas, 396
Lumière brothers, 183

Lundberg, Ilse, 439
Lundin, Per, 13, 26
Luscomb, Florence, 96
lynching, 472
Lynd, Robert and Helen, 303, 304, 305–306, 323, 404, 463, 524

Maeterlinck, Maurice, 167, 169, 173, 177
male hysteria, 235
Malevich, 239
Mann, Erika, 431–433, 488
Mann, Klaus, 186, 431–433, 458
Mann, Thomas, 141, 148, 163, 266
Marcus, Jane, 260
marijuana, 458
Marinetti, Filippo Tommaso, 59–60, 179–180, 193, 196, 201, 239
market saturation, 313
marketing, 385
Marsalis, Wynton, 406
Marsman, H., 509
Marx, Groucho, 526
Marx, Karl, 204, 532
Marx, Leo, 23, 32, 89, 139–141, 188, 199, 301, 387, 450, 451, 469, 638, 650, 651
masculinity, masculine (macho) car culture, 20, 29, 62, 65, 78, *84–100*, 106, 112, 134, 144, 150, 166, 175, 179, 180, 196, 199, 210n58, 232, 234–235, 252, 256, 264, 306, 330, 385, 390, 398, 405, *407, 409–410*, 427, 431, 436, 448, 450, 462, 489, 510, 515, 520, 523, 549n184, 636, 646, 648
  car as masculine, 399
mass psychology, 566
master narrative, 7, 26–28, 637
Masterson, Kate, 95
Maxim, Hiram, 108–109
Mayakovsky, Vladimir, 181
McCarthy, Tom, 14
McClintock, Laura, 529
McDowell, Edward, 185
McLuhan, Marshall, 33
McShane, Clay, 10–11, 25, 200
mediated mobility theory, 654
medical doctor as motorist, 71
Méliès, Georges, 184, 185
Melville, Herman, 140
men's studies, 234, 409
Mercedes, 62, 93, 173, 402, 475, 478, 479, 488, 640
Merki, Christoph, 16–18, 232
Messter, Oskar, 184

Métivet, Lucien, 165
middle class, 19, 80, 84, 107, 175, 188, 196, 295, 299, 303, 305, 306, 307, 309, 311, 316, 334, 531, 567, 639, 640
  Angestellte, 290, 296, 306, 308, 309, 326, 340, 448, 497
  Beamte, 308, 309
  black, 405–406
  car culture, 80, 300, 310
  children, 405
  clerks, 307
  family, 330, 331, 344, 403, 460;
  culture, 301
  gendered nature of, 20
  imperialism, 93
  leisure, 20, 306, 337
  Mittelstand, 308, 326, 336, 453, 482, 483, 518, 532
  motorists, 104
  nuclear family, 286
  white collars, 38–39, 68, 239, 296, *305–308*, 341, 356n81, 404, 448, 455, 467–468, 480, 492
  youth, 411
middle landscape, 89, 90, 173, 188, 301
middle state, theory of the, 90
Middletown, 303, 306, 312, 323, 404, 463
Miettinen, Reijo, 36
migration, 328–344
Milburn, George, 472, 473
mileage (incl. modal split), 132n193, 314, 321–322, 325, 329, 361n124, 362n138, 363n150, 610, 613
  of women, 603
Millar, Richard, 4
Mills, C. Wright, 303
Mills, Dorothy, 435, 436
Mills, Kerry, 186
Miltoun, Francis, 154
Mirbeau, Octave, 167, 177–179, 194, 196, 197, 200, 427
Mitton, Geraldine, 269
modal split, 316, 343, 528, 609, 613–614, 618. See also mileage
modest motoring, 78, 160
Moholy-Nagy, László, 511
Moline, Norman, 8–10
Monneyron, Frédéric, 510
Montagu, Lord, 81, 82
moped, motorized bicycle, 640. See also motorcycle (culture)
Morand, Paul, 428, 431, 433, 440, 458–459, 460, 499, 508, 515

Morton, Henry Vollam, 429
Möser, Kurt, 14–16, 23, 139, 232, 252, 270, 387, 389, 390, 501
Moss, Fred A., 383
motives of car use, 4–6, 8, 9, 14, 28, 30–33, 37, 38, 40, 61, 89, 113, 157, 220n169, 303, 322, 325, 372n227, 385, 395, 425–426, *444*, 447, 469, 504, 513, 524, 546n164, 617, *643–644*, 652, 654
    of coldness, 482
    See also cyborg motif
motor bus, 181, 183, 195, 244, 312, 429–430, 434, 476, 493, 523, *608–614*, 631n142, 637, 648
    diffusion of, 608
    wild buses, 609
motorcycle (culture), 82–83, 176, 193, 289, 290, 292, 294, 307, 309, 312, 326, 327, 339, 394, 404, 439, 442, 453, 454, 455, 487, 640, 645
    density, 293
    diffusion of, 292
Motor Field Ambulance Corps, 258–259
Motor-HJ (*Hitler Jugend*), 292
motorist as poet, 204
mountain climbing, 427
Mount Vernon Memorial Highway, 401
Muir, John, 91
Müller, Dorit, 181
(multi)sensorial experience (car driving as), 29, 34, 138, 184, 265, 373, 375, 386, 391, 397, 411, 434, 516, 644. See also senses; transcendence, transcendental experience
Munzinger, Wolfgang, 470
Musil, Robert, 28, 433, 456, 476–477, 511
Mystery Hikes, 394

NACC (National Automobile Chamber of Commerce), 322–323, 376
NAG (*Neue/Nationale Automobil-Gesellschaft*), 236
Nakicenovic, Nebojša, 13
Namag (*Norddeutsche Automobil und Motoren AG*), 236
narcissism, 163, 204, 395, 455, 644
    narcissistic attitude of aristocracy, 66
narrative explanation, 3
NASCAR (National Association for Stock Car Auto Racing), 398
national parks, 91, 106, 331
National Park Service, 106, 331, 332

Nelson, Andrew, 444
neurasthenia, 94, 161, 167, 192
New Deal, 588, 590, 616
New Mobility Studies, 4
New Objectivity, 479, 497, 499, 512
new woman, 145, 152, 411
nickel-iron battery, 176
Nietzsche, Friedrich, 143, 397–398, 449, 500
noise, 39, 85, 93, 152, 169, 187, 232, 253, 264, 374, 383, 386, 388, 406, 411, 455, 457, 481, 496, 573. *See also* (multi)sensorial experience; senses; sound
nonwhite mobility, 405
Normand, Mabel, 96
Norton, Peter, 9, 10
NSKK (*Nationalsozialistische Kraftfahrkorps*), 338–339, 393
nuclear family as automotive subject, 330, 402, 460, 645
Nye, David, 411

O'Connell, Sean, 18–21, 95
OIAR (*Office International des Autoroutes*), 581
Oldfield, Barney, 184
Olmsted, Fredrick Law, 90, 590
Orwell, George, 143, 429–430, 463
the 'Other,' 155, 160, 195, 199, 234, 255, 263, 396, 401, 430, 442, 531, 649
Oulié, Marthe, 523
overdrive, 388, 390

package tours, 336
Packer, Jeremy, 25
Panic of 1907, 100, 112
panoramatic experience, 136
Paquet, Alfons, 433
Paris to Beijing race, 156
park-to-park highway, 332
parkways, 107, 200, *590*, 625n74, 626n75
    Blue Ridge Parkway, 590
    Bronx River Parkway, 320
Parr, Rolf, 510
passengering, 653
pedestrianism, 86
Pennsylvania Turnpike, 391
Penrose, Margaret, 175
performance (in passenger kilometres), 295
periodization of automobile use, 38

periphery, 2, 202, 341, 426, 428, 653
  European, 160
Pershing, John J., 240, 243
Pessoa, Fernando, 427
Pétain, Henri (maréchal), 247
Petrov, Evgenii, 344
petting (by car), 404
petty-bourgeois family car, 643
petty-bourgeois small family, 639
Pfemfert, Franz, 182
phases of car culture, 1
  doom, 1
  emergence, 1, 40, 60–61
  extravagance, 1, 4
  persistence, 1, 4, 27, 40, 346, 643
philistine, 147, 264, 464–465, 467, 480, *482*, 506
Piaget, Jean, 655
PIARC (Permanent International Association of Road Congresses), 68, 572, 595, 596, 599, 600, 601, 602
Picabia, Francis, 458
Picasso, Pablo, 471, 499
Plainville, 325
'playing chicken,' 398
popular culture, 39, 142, 183, 407, 425, *508*, 517–528, 530, 532, 544n127, 644, 651
Pound, Ezra, 239
preventive service (Ford), 568
Priestley, J. B., 428–429, 463
properties (technical), 36, 380, 386, 647. *See also* function (relational)
prosthesis, prosthetic (car), 33, 53n133, 192, 204, *386–391*, 400, 411, 452, 532
  writing as, 532
Proust, Marcel, 141, 161–164, 193, 194, 195, 196, 197, 201, 430
psychotechnics, 603
public transport, 8, 65, 298, 308–309, 316–317, 344, 346, 403, 405, 460, 476, 574, 610, 613–614, 648
Puricelli, Piero, 581, 594

Quest of Self, 31, 39, 141, 144, 201, 209n37, 425, 430, 440, 446, 458, 646, 652

RAC (Royal Automobile Club), 589
race from Paris to Bordeaux, 165
racing, 63, 66, 84, 85, 176, 266, 454. *See also* adventure: temporal
racism, 175, 263, 459
Radkau, Joachim, 92, 200
Rae, John, 7, 11, 71
Raffard, Nicolas-Jules, 62
Rajan, Sudhir Chella, 653
Rammert, Werner, 22
Ramsey, Alice, 84, 95
Rapaport, David, 233
rape, 163, 223n211, 498, 601
Rath, E. J., 489–490
Ray, Man, 458
reality as fiction, 159
reasonable travel, 153
recreation, 88, 307, 309, 322, 333
Red Cross Motor Corps, 269
Reich, Wilhelm, 483
Reitman, Ben, 444
relationship between time and space, 159
reliability of the car, 107, 109, 568, 569
Remarque, Erich Maria, 255, 390, 472–474, 510
Renault cab, 245, 257
Renoir, Jean, 512
repair, 567
  and black employment, 571
  See also adventure: functional (tinkering)
resistance (against the car), 74, 76, 79, 84, 99, 103, 112, 120n68, 148, 152, 154, 182, 202–203, 584, 636, 639
  in the 1920s, 324, 364n153
  against elite touring, 68–83
  in poetry, 520
  See also hooliganism
restlessness, 193
Reuters, Bernd, 473
reverse salient, 382
Rhodes, Cecil, 93
Rickenbacker, Eddie, 427, 606, 650
rickshaw, 654
riding and body engineers, 374
riding quality, 377
Rijkswaterstaat (agency), 584
risk, 199, 397
risk society, 395
risk-taking as sensation, 399
roadability, 375
Road Board (UK), 586
road
  census, 289
  fund, 582, 586
  hogs, 99, 603
  movie, 527, 562n333, 562n339 (*see also* autopoetic movie/film)

network, 321, 572–590
plan, 579, 641
road/railway controversy, 230 (see also coordination crisis)
research laboratories, 579
safety (see safety)
signs, 567
Robbe-Grillet, Alain, 515
Robowtham, Sheila, 411
romantic gaze (tourism), 330
Roosevelt, Alice, 96
Roosevelt, Franklin, 345
Roosevelt, Theodore, 90, 93, 96, 104–105, 106, 234, 257
Rorty, James, 444–445
Rosa, Hartmut, 34
Rosemeyer, Bernd, 488
Ross, Colin, 446
Roth, Paul, 6
rural free delivery, 472
rural ownership of cars (US), 315
Russel, Raymond, 430–431
Ruttmann, Walther, 523

Sachlichkeit, 449, 497, 499, 500, 648
Sachs, Wolfgang, 13, 14, 15, 139, 397
SAE (Society of Automotive Engineers), 375
safety, 100, 594–606
children as victims, 101, 150, 599
cyclists as victims, 599
traffic deaths, 596, 597–606
See also accident
Said, Edward, 438
Saint-Point, Valentine de, 189
saloon, sedan (closed car), 384
Sand, George, 159
saturation crisis of 1923–1924 (US), 300, 310
Scharff, Virginia, 385, 639
Schatzki, Theodore, 35–36, 650
Scheibe, Karl, 196, 398
Schickele, René, 182
Schilperoort, Tom, 81, 82, 255
Schivelbusch, Wolfgang, 30, 136, 162–163, 194, 202, 650, 651
Schlieffen, Alfred von, 243
Schnack, Friedrich, 503–504, 508, 514
Schönhammer, Rainer, 198
Schriber, Mary, 489
Schwitters, Kurt, 452
scientification of car technology, 374

Scorsese, Martin, 444
See America First, 91–92, 104, 106, 111, 331, 340, 342
Seiler, Cotten, 13, 14, 24–26, 95, 514
Sennett, Mack, 185
senses, 33, 164, 194, 374, 383, 401. See also (multi)sensorial experience
Shaw, George Bernard, 97–98
Shepherd, C. K., 453
shimmy, 382
Shklovsky, Viktor, 266–267, 459, 506–507
shopping, 317, 344
SHOT (Society for the History of Technology), 140
Sieferle, Rolf Peter, 89
Simmel, Georg, 143, 204, 394, 400, 408, 515
Sinclair, May, 258–259
Singer, Bayla, 200
skills, 199, 374, 387, 390, 396, 398, 569
bodily, 452
driving, 146, 155, 391
technical, 191–192
Skorphil, Sophia, 439
Sloan Jr., Alfred, 102
Sloterdijk, Peter, 33
slowness, 451, 456, 459–460, 500, 523, 544n127, 638, 645
smell, 164, 169, 195. See also (multi)sensorial experience; senses; sound
Smith, Helen Zenna, 259–261
social practice theory, 35, 650
Sombart, Werner, 143, 449
Sorel, Georges, 189, 234
sound, 164, 195, 374. See also (multi)sensorial experience; noise; senses
Soupault, Philippe, 512
Southern California Timing Association, 398
Southgate, Beverly, 30
Spada, Friedel, 514
speeding on highways, 388
Spinga, Nicolas, 72
sport, 64, 235, 252, 255
Stawell, Rodolph, 432–433
steam car, steam propulsion, 62, 229, 236
Stein, Gertrude, 92, 256–257, 427, 484–485, 512, 513
Steinbeck, John, 329, 440–442, 453, 529
Steiner, Jesse, 325
Steinmetz, George, 6
Sterne, Laurence, 147, 266

Stewart, James, 525
Stiegler, Bernhard, 532
Stinnes, Clärenore, 488, 655
Stokes, Katherine, 175
Stoll, F., 452
Stout, Janis, 440
Stratemeyer, Edward, 175
Stratemeyer Adams, Harriet, 175
Stratil-Sauer, Gustav, 437–438
streamlined car body shape, 385–386
Stricker, Frank, 304
subaltern mobility, 654
sublime, 89–90, 100, 142, 154, 159, 183, 188, *190*, 193, 194, 204, 251, 254, 265, 398, 410, 411, 450, 504, 516, 523, 636, 648, *650*. *See also under* violence
sub-literary and non-literary texts, 31
suburb, suburbanization, 301, 316, 317, 404
Subventionslastwagen, 236
subversive mobility, 50n98, 82, 94, 97, 112, 189, 320, 405, 411, 424n158, 437, 440, 442–443, 457–458, 488, 552n209, 610, 612, 649, 654
suffragette movement, 268
Sunday drive, 88
super balloon tires, 382
Superman (US), 500
Surrealism, Surrealists, 471, 498, 512, 521
Susman, Warren, 323
Swanson, Gloria, 524
swarming, 64, *138–139*, 190, 207n24, *248*, 250, 339, 395–396, *403*, 431, 440, 441, 444, 448, 451, 453, 473, 476–477, 493, 496, 507, 513, 515, 530–531, *565–566*, *591*, 605, *626n76*, 639, 642, 645–646, 649, 651–652
Swope, Ammon, 383
symbols of the car, 200–202
systemic character of the car culture, 530
systems approach, 243, 248

Taft, William Howard, 97, 105, 188
Tank Battle of Cambrai, 242. *See also* battle
Tarde, Gabriel, 566
Tarkington, Booth, 186–187
tax (on cars), 291, 313
    earmarking of car tax for road building, 573
"taxis de la Marne," 243–245, 250

TCF (*Touring Club de France*), 394, 585
technical support services of car clubs, 570
Teilhard de Chardin, Pierre, 265, 522
thanatourism, 250–267
therapeutic characteristics of consumption, 93
Theweleit, Klaus, 457, 501
Thomas, Albert, 581, 593
Thomas, Joël, 510
Thompson, Dorothy, 467
Thrift, Nigel, 34
thrill, 199, 393, 398
Tichi, Cecilia, 139
Tin-can Tourists of the World, 334
tinkering. *See* adventure: functional (tinkering)
Titzmann, Michael, 32
Todt, Fritz, 592
Toklas, Alice, 256
Tomas, Adrienne, 265
Tooker, Helen V., 467–469
touring, 63, 78, 84, 85, 86, 97, 98, 194. *See also* adventure: spatial (touring)
tourism, 64, 85, 318, 328–344
    roots of, 85–94
    in war, 254
    See also thanatourism
tourist gaze (Urry), 29, 138, 411. *See also* travel glance
Tim Hurst Auto Tours, 183, 195
Townsend, Charles, 587
toy to tool myth, 3–7, 5, 12, 13, 639
"trading paint," 398
traffic fund, 582
traffic safety and racism, 604. *See also* safety
trailer movement, 320
train, 19, 78, 87, 90, 104, *136*, 145, 150, 158, 183, 184, 202, 238, 245, 250, 251, 255–256, 258, 265, 316, 321, 329, 333, 338, 339, 342–343, 428, 429, 432–436, 441–443, 446, 448, 453, 455, 459, 462, 464, 479–482, 492–493, 503, 505, 514, 521, 611, 612, 636, 640, 651
    African-Americans and, 406
    attack on, 59, 154, 157, 163, 167–169, 178–179, 194, 613
    commuting by, 317, 460, 467
    as flight, 192, 202
    railroad, 472
    railway network, 230

railways, 238
railway strike in 1903, 231
society, 566
transborder traffic, 567
transcendence, transcendental experience, 142, 144, 392, 393, 397, 401, 411, 440, 450, 504, 512, 515–516, 524, 527, 531, 638, 644–645, 647, 649–650, 659n51
  multisensorial transcendence, 516–517
translation (of consumer wishes), 378–382, 383, 387, 411
transnational approach, 1, 618, 638
travel account, travelogue, 141, 159, 252, 426, 427
travel glance, 138
traveling camera, 138, 185
travel writing, 141
Traven, B., 457, 502
Treatt, Stella Court, 435
Trilling, Lionel, 193
trip length, 329, 332, 612, 618
triptyques, 572
truck, 242, 243, 612, 614
Tucholsky, Kurt, 449, 506, 508, 509–510
Tworek-Müller, Elisabeth, 518
typical European car, 378
typical purchaser of a car (US), 296
Tzara, Tristan, 460, 513

underside of the diffusion curve, 312
universal car, 375
unreliability, 567
Upton, Sinclair, 526
urban car culture, 299
urban middle-class family, 375
urban youth, 404
'urbanizing' car culture, 301
Urry, John, 37–38, 87, 195, 650
US Army motorization, 236
utilitarian functions as alibi, 324
utilitarian side of mobility, 346

vacation (paid), 340, 341
VADs (Voluntary Aid Detachments), 261
Vanderbilt, William, 108
Veblen, Thorsten, 71, 148
Verne, Jules, 152
Vertov, Dzinga, 528
Verwey, Albert, 189
vibrations, 181, 192

violence, 15, 60, 76, 78, 93, 94, 97, 149, 153, 156, *161*, 166, 169, 174, *176*, 177–183, 185, 189, *190*, 191, 197, 199, 200, 202, 204, 219n137, 228, 232, *233–235*, *238*, 239–240, 252, 254–256, 262, 265–266, 280n112, 287–288, 337, 339, 347, 390, 391, 396–397, 398, 401, 403, 405, *407–409*, 411, 425, 430, 447, 453–455, 457, *460*, 471, 475–477, 485, 495–496, 498, 501–502, 505, 507, 510, 511–512, 514, 516, 525–527, 529–531, 598, 605–606, 637, 639, 642–645, 647–650, 657n35
  female, 156, 160, 407–408, 433, 434, 480, 488, 490, 520
  sublime of, 251, 411, 425, 516, 650
Virilio, Paul, 471
visibility, 383, 385
vision, 164
VMAK (Volunteer Military Automobile Corps), 231
Voie Sacrée, 246–247
Volkswagen, 294
Volti, Rudi, 9
Volunteer Motor Ambulance Corps, 256
von Clausewitz, Carl, 230
von Freytag-Loringhoven, Elsa, 487
von Horváth, Ödön, 481–484, 508
von Nathusius, Annemarie, 488
von Richthofen, Wolfram, 402
von Sternberg, Joseph, 525

WAAC (Women's Army Auxiliary Corps), 268
walking, 64, 86, 88, 221n63, 256, 308–309, 332, 340, 346, 393, 405, 447, 468, 513
'walking' (carried through forest), 435
walks of Buster Keaton, 528
Walsh, Margaret, 489
Walsh, Raoul, 527
Waugh, Evelyn, 436, 470–471
Wayne, John, 525
Weber, Max, 143, 498
Weesling, Nellie, 519
Weicker, Alexander, 480
Wells, H. G., 145, 149, 520
Wessem, Constant van, 517
Westchester County, 404
Wetterauer, Andrea, 22–23, 448
Wharton, Edith, 153, 156, 171, 195, 196, 201, 203, 258, 429, 467
Wheeler, William, 566
white collars. See under middle class

White, Hayden, 6, 29
Wilde, Oscar, 498
Wilder, Thornton, 445–446
wilderness, 89
Willard, Frances, 145
Williams, Raymond, 233
Williams, William Carlos, 107, 446, 484, 510, 513, 516
Williamson, Alice and Charlie, 170–174, 195, 197, 199–201, 387
Wilson, Woodrow, 105, 188, 262
windscreen, 389
Winkler, Heinrich, 532
Winter, Margarete, 156–157, 193, 194, 195, 197, 200, 202
Winterberg, Birgit, 459
Wolf, James, 436
Wolff, Janet, 407
women (and the car), 407, 435–436, 442, 531
  as car buyers, 385
  as dangerous drivers, 603
  travelers, 488

Women's Automobile Club (US), 95
Women's Legion Motor Transport Section, 268
Woolf, Leonard, 486–487
Woolf, Virginia, 467, 485–487, 510
working class, 303, 305
  family, 316; as car owner, 297
Wosk, Julie, 407
Wren, Christopher, 515
Wren, Genevieve, 95
Wright, Richard, 323

Yellowstone National Park, 315, 332
Young, Clarence, 174
youth novels, 174, 519

Žižek, Slavoj, 233
Zola, Émile, 144
Zweig, Paul, 516